Lecture Notes in Computer Science 7341

Commenced Publication in 1973
Founding and Former Series Editors:
Gerhard Goos, Juris Hartmanis, and Jan van Leeuwen

Feng Bao Pierangela Samarati
Jianying Zhou (Eds.)

Applied Cryptography and Network Security

10th International Conference, ACNS 2012
Singapore, June 26-29, 2012
Proceedings

Springer

Volume Editors

Feng Bao
Institute for Infocomm Research
1 Fusionopolis Way, #21-01 Connexis, South Tower
Singapore 138632, Singapore
E-mail: baofeng@i2r.a-star.edu.sg

Pierangela Samarati
Università degli Studi di Milano
Dipartimento di Tecnologie dell' Informazione
Via Bramante 65, 26013 Crema (CR), Italy
E-mail: pierangela.samarati@unimi.it

Jianying Zhou
Institute for Infocomm Research
1 Fusionopolis Way, #21-01 Connexis, South Tower
Singapore 138632, Singapore
E-mail: jyzhou@i2r.a-star.edu.sg

ISSN 0302-9743
e-ISSN 1611-3349
ISBN 978-3-642-31283-0
e-ISBN 978-3-642-31284-7
DOI 10.1007/978-3-642-31284-7
Springer Heidelberg Dordrecht London New York

Library of Congress Control Number: 2012939841

CR Subject Classification (1998): K.6.5, E.3, K.4.4, D.4.6, E.4, C.2, J.1, E.1

LNCS Sublibrary: SL 4 – Security and Cryptology

Typesetting: Camera-ready by author, data conversion by Scientific Publishing Services, Chennai, India

Printed on acid-free paper

Springer is part of Springer Science+Business Media (www.springer.com)

Preface

These proceedings contain the papers selected for presentation at the 10th International Conference on Applied Cryptography and Network Security (ACNS 2012), held during June 26–29, 2012, in Singapore. The conference was organized by iTwin, sponsored by AdNovum, and supported by Infocomm Development Authority of Singapore (IDA).

In response to the call for papers, 192 papers from 38 countries were submitted to the conference. These papers were evaluated on the basis of their significance, novelty, technical quality, and practical impact. Reviewing was "double-blind": the identities of reviewers were not revealed to the authors of the papers and author identities were not revealed to the reviewers. The Program Committee meeting was held electronically, yielding intensive discussion over a period of two weeks. Of the papers submitted, 33 were selected for presentation at the conference and inclusion in this Springer volume (LNCS 7341), giving an acceptance rate lower than 18%.

Besides the technical program composed of the papers collated in the proceedings, the conference included a non-archival industrial track. The conference was also featured with 3 keynote speeches, by Moti Yung (co-founder of ACNS) entitled "Applied Cryptography and Network Security - 10 years in the past and 10 years in the future", by Peng Ning entitled "Cloud Computing Infrastructure Security", and by Hongjun Wu entitled "JH in the NIST Hash Function Competition", respectively.

There is a long list of people who volunteered their time and energy to put together the conference and who deserve special thanks. Thanks to the Program Committee members and the external reviewers, for all their hard work in the paper evaluation. Owing to the large number of submissions, the Program Committee members were required to work hard in a short time frame, and we are very thankful to them for the commitment they showed with their active participation in the electronic discussion.

We are also very grateful to all those people whose work ensured a smooth organization process: Xinyi Huang and Giovanni Livraga, Publicity Chairs, for their work in ensuring the wide distribution of the call for papers and participation; Shen-Tat Goh, Organizing Chair, as well as Lux Anantharaman and Kal Takru for taking care of the local organization and Ying Qiu for managing the conference website and EasyChair system.

Last but certainly not least our thanks go to all the authors who submitted papers and all the attendees. We hope you find the program is stimulating and a source of inspiration for your future research and practical development.

April 2012

Feng Bao
Pierangela Samarati
Jianying Zhou

ACNS 2012

10th International Conference on Applied Cryptography and Network Security

Singapore
June 26–29, 2012

Organized by iTwin, Singapore

Sponsored by AdNovum, Singapore

Supported by

Infocomm Development Authority of Singapore (IDA)

General Chair

Jianying Zhou	Institute for Infocomm Research, Singapore

Program Chairs

Feng Bao	Institute for Infocomm Research, Singapore
Pierangela Samarati	Università degli Studi di Milano, Italy

Program Committee

Michel Abdalla	ENS & CNRS, France
Vijay Atluri	Rutgers University, USA
Lucas Ballard	Google, USA
Paulo Barreto	University of São Paulo, Brazil
Lujo Bauer	Carnegie Mellon University, USA
Marina Blanton	University of Notre Dame, USA
Carlo Blundo	Universitá degli Studi di Salerno, Italy
Levente Buttyan	Budapest University of Technology and Economics, Hungary
Liqun Chen	Hewlett-Packard Laboratories, UK
Chen-Mou Cheng	National Taiwan University, Taiwan
Jung Hee Cheon	Seoul National University, Korea
Sherman S.M. Chow	University of Waterloo, Canada
S. De Capitani di Vimercati	Università degli Studi di Milano, Italy
Robert Deng	Singapore Management University, Singapore
Roberto Di Pietro	Università di Roma Tre, Italy
Xuhua Ding	Singapore Management University, Singapore
Wenliang Du	Syracuse University, USA

Organizing Chair

Shen-Tat Goh	Institute for Infocomm Research, Singapore

Publicity Chairs

Xinyi Huang	Fujian Normal University, China
Giovanni Livraga	Università degli Studi di Milano, Italy

Steering Committee

Yongfei Han	ONETS, China
Moti Yung	Google, USA
Jianying Zhou	Institute for Infocomm Research, Singapore

External Reviewers

Abu Rajab, Moheeb
Agudo, Isaac
Albanese, Massimiliano
Aliasgari, Mehrdad
Asano, Tomoyuki
Athanasopoulos, Elias
Au, Man Ho
Baek, Joonsang
Bandhakavi, Sruthi
Brzuska, Christina
Cao, Jin
Chen, Jiun-Ming
Cheng, Yueqiang
Cheong, Kai Yuen
Chu, Cheng-Kang
Coisel, Iwen
Cremers, Cas
Cui, Hui
D'Arco, Paolo
De Caro, Angelo
Ding, Ning
Dong, Xinshu
Drogkaris, Prokopios
Duan, Pu
Elkhiyaoui, Kaoutar
Fernandez, Gerardo
Fett, Daniel
Fiore, Dario
Fuchs, Andreas
Fuchsbauer, Georg
Galdi, Clemente
Gorantla, Choudary
Gu, Haihua
Guo, Fuchun
Hakuta, Keisuke
Han, Jinguang
Hanzlik, Lucjan
Hao, Hao
Henricksen, Matt
Huang, Qiong
Huang, Tao
Iovino, Vincenzo
Iskander, Marian
Jee, Kangkook
Jeske, Tobias
Jin, Xing
Kalabis, Lukas
Kalloniatis, Christos
Kemerlis, Vasileios P.
Kim, Hongtae
Kim, Myungsun
Kim, Sungwook
Kohlweiss, Markulf
Kolias, Constantinos
Kontaxis, Georgios
Krzywiecki, Lukasz
Kubiak, Przemyslaw
Kuo, Po-Chun
Künnemann, Robert
Lai, Junzuo
Laszka, Aron
Le, Meixing
Lee, Hyung Tae
Lee, Younho
Leontiadis, Iraklis
Li, Fagen
Li, Juanru
Li, Xiaolei
Li, Yen-Huan
Liang, Kaitai
Lin, Jingqiang
Liu, Joseph
Liu, Junrong
Liu, Ya
Liu, Zhen
Liu, Zhiqiang
Lu, Jiqiang
Luo, Tongbo
Luo, Weiliang
Majcher, Krzysztof
Morozov, Kirill

Najera, Pablo
Neven, Gregory
Nuñez, David
Ohtake, Go
Omote, Kazumasa
Önen, Melek
Oren, Yossef
Pala, Massimiliano
Pappas, Vasilis
Patil, Kailas
Pek, Gabor
Pelosi, Gerardo
Pereira, Geovandro
Petit, Christophe
Pointcheval, David
Polychronakis, Michalis
Portokalidis, Georgios
Quaglia, Elizabeth A.
Rangasamy, Jothi
Ratazzi, Paul
Raykova, Mariana
Rekleitis, Evangelos
Rizomiliotis, Panagiotis
Roman, Rodrigo
Roudier, Yves
Roy, Arnab
Ràfols, Carla
S. Shiva, Ashwathi
Sakiyama, Kazuo
Schröder, Dominique
Schuldt, Jacob
Schwabe, Peter
Schäge, Sven
Seo, Jae Hong
Shafiq, Basit
Shin, Seungwon
Simo, Hervais
Standaert, F.-X.
Steinebach, Martin
Stopczynski, Martin
Sun, Wenhai
Sun, Xiaoyan
Ta, Vinh Thonh
Tan, Xiao
Tang, Qiang
Tillich, Stefan
Triandopoulos, Nikos
Tuengerthal, Max
Tunstall, Michael
Tzouramanis, Theodoros
Verde, Nino Vincenzo
Villani, Antonio
Visconti, Ivan
Vogt, Andreas
Wang, Boyang
Wang, Guilin
Wang, Jun
Wang, Lusha
Wang, Yifei
Wikström, Douglas
Wolny, Kamil
Wu, Wei
Wu, Wenling
Xiong, Xi
Xu, Jia
Xu, Zhaoyan
Yang, Chao
Yang, Guomin
Yap, Wun-She
Yian, Chee Hoo
Ying, Jason
Yu, Ching-Hua
Yu, Yong
Yun, Aaram
Zhan, Zhenxin
Zhang, Bin
Zhang, Jialong
Zhang, Lei
Zhang, Mingwu
Zhang, Shengzhi
Zhang, Xiao
Zhang, Yinghui
Zhao, Mingyi
Zhao, Xingwen
Zheng, Qingji
Zhong, Chen
Zhu, Youwen

Table of Contents

Authentication

Key Management

Block Ciphers

Identity-Based Cryptography

Cryptographic Primitives

Cryptanalysis

Side Channel Attacks

Network Security

Web Security

Security and Privacy in Social Networks

Security and Privacy in RFID Systems

Security and Privacy in Cloud Systems

Security and Privacy in Smart Grids

Security Analysis of a Multi-factor Authenticated Key Exchange Protocol

Feng Hao and Dylan Clarke

School of Computing Science
Newcastle University
{feng.hao,dylan.clarke}@ncl.ac.uk

Abstract. This paper shows several security weaknesses of a Multi-Factor Authenticated Key Exchange (MK-AKE) protocol, proposed by Pointcheval and Zimmer at ACNS'08. The Pointcheval-Zimmer scheme was designed to combine three authentication factors in one system, including a password, a secure token (that stores a private key) and biometrics. In a formal model, Pointcheval and Zimmer formally proved that an attacker had to break all three factors to win. However, the formal model only considers the threat that an attacker may impersonate the client; it however does not discuss what will happen if the attacker impersonates the server. We fill the gap by analyzing the case of the server impersonation, which is a realistic threat in practice. We assume that an attacker has already compromised the password, and we then present two further attacks: in the first attack, an attacker is able to steal a fresh biometric sample from the victim without being noticed; in the second attack, he can discover the victim's private key based on the Chinese Remainder theorem. Both attacks have been experimentally verified. In summary, an attacker actually only needs to compromise a single password factor in order to break the entire system. We also discuss the deficiencies in the Pointcheval-Zimmer formal model and countermeasures to our attacks.

1 Introduction

Authenticated Key Exchange (AKE) is a fundamental security protocol for almost all secure communication systems. Depending on how the "authentication" is defined, AKE schemes are generally divided into two categories: Password-based Authenticated Key Exchange (PAKE) and PKI-based Authenticated Key Exchange [7]. In the former case, the authentication is based on the knowledge of a shared password, without requiring any Public Key Infrastructure (PKI). In the latter case, each party possesses a unique pair of the public and private keys. The authentication is based on the possession of the private key, which is usually stored in a tamper resistant device. A PKI is needed to securely distribute authentic public keys to all users [18].

While the above two AKE categories correspond to something you know (i.e., a password) and something you have (i.e., a secure token that stores a private key) respectively, there is a third authentication factor: namely, something you are

F. Bao, P. Samarati, and J. Zhou (Eds.): ACNS 2012, LNCS 7341, pp. 1–11, 2012.

(i.e., biometrics). Biometrics are an advanced authentication mechanism, which works by measuring a person's unique behavioral or physical characteristics [2]. The security of biometrics largely depends on whether a trusted path exists, which ensures the biometric sample is freshly obtained from the live subject. Such a trusted path can be realized, for example by enforcing supervision in a controlled environment (e.g., airport). In an unsupervised environment, the security will have to depend on the liveness detection features embedded with the biometric scanning equipment [18]. Assuming a trusted path already in place, researchers have made progress in designing biometrics-based AKE schemes [3, 5, 6].

So far all the above-mentioned AKE schemes are based on a single factor. In recent years, Multi-Factor Authenticated Key Exchange (MF-AKE) has emerged as an active research topic [10–14, 18–23]. The rationale is to improve single factor based AKE by combining two or even more factors in one system. This is a worthy goal, but extra caution should be taken. The past thirty years of research in the area of authenticated key exchange has proved that it is incredibly difficult to get even a single factor based AKE scheme right [4]. Designing a multi-factor AKE protocol can only be harder.

Many MF-AKE protocols have been proposed – and subsequently broken. For example, in 2010, Lee et al. proposed a two-factor AKE protocol that combines a smartcard and a password [11]. But a year later, their protocol was found insecure: the compromise of the smartcard factor breaks the entire scheme [21]. Xu et al. proposed a similar two-factor AKE protocol based on a smartcard and a password with "formal security proofs" in [23]. Within a year, their protocol was broken and a patched protocol was proposed in [19]. Yet the patched protocol was shortly found insecure [20]. Li and Hwang proposed a different type of two-factor AKE protocol, which consists of biometrics and a smart card [13]. In less than a year, their scheme was broken [14]. These examples show that MK-AKE is still a young field; more research is very much needed.

Recently at ACNS'08, Pointcheval and Zimmer proposed the first Multi-Factor Authenticated Key Exchange protocol that combines all three factors in one system: a password, a smartcard and biometrics [18]. Furthermore, the authors defined a formal model and formally proved that an adversary had to break all three factors in order to win. Unfortunately, we find their protocol vulnerable, as we explain below.

2 Pointcheval-Zimmer Protocol

In this section, we will describe Pointcheval-Zimmer's Multi-Factor Authenticated Key Exchange scheme. We will follow the original notations in [18] as closely as possible.

2.1 Notation

The client $\mathcal{C}$ owns a tuple $t_\mathcal{C} = (W'_\mathcal{C}, \mathrm{sk}_\mathcal{C} = \mathrm{x}_\mathcal{C}, \mathrm{pwd}_\mathcal{C})$ where $W'_\mathcal{C}$ is a biometric, $\mathrm{sk}_\mathcal{C}$ a private key and $\mathrm{pwd}_\mathcal{C}$ a password. The iris code is used in [18] as a specific example for biometrics.

The server $\mathcal{S}$ holds a list of tuples for each client $t_{\mathcal{S}} =< t_{\mathcal{S}}[\mathcal{C}] >$, where $t_{\mathcal{S}}[\mathcal{C}]$ is a transformed-tuple of $t_{\mathcal{C}}$. More specifically, $t_{\mathcal{S}}[\mathcal{C}]$ contains the following information about the client $\mathcal{C}$:

- The client's public key $h = g^{x_{\mathcal{C}}}$.
- An encrypted copy of the iris-code template that was enrolled during registration. The template is denoted as $W_{\mathcal{C}} = (W_i)_{i \leq N}$, where W_i is the i-th bit of $W_{\mathcal{C}}$ and N is the number of bits of an iris code. The ciphertext is obtained by using the El Gamal encryption algorithm; the result is $(g^{r_i}, h^{r_i} \cdot g^{W_i})_i$, where r_i is a random element in $\mathbb{Z}_q$.
- The client's password $\text{pwd}_{\mathcal{C}}$.

2.2 Description of Protocol

The protocol is based on a cyclic group with parameters (p, g, q). The p and q are big prime numbers, and $q \mid p - 1$. Let $\mathbb{G}_q$ be a subgroup in $\mathbb{Z}_p^*$ with prime order q, and g be its generator. In addition, the protocol defines two random elements in $\mathbb{G}_q$, namely u and v. Figure 1 specifies how the protocol works. The symbols used in the figure should be self-explanatory[1].

	Client		Server
	$\mathcal{C} : (W'_c = (W'_i)_i,\ sk_{\mathcal{C}} = x_{\mathcal{C}},\ \text{pwd}_{\mathcal{C}})$		$\mathcal{S} : ((g^{r_i}, h^{r_i} g^{W_i})_i,\ h = g^{x_{\mathcal{C}}},\ \text{pwd}_{\mathcal{C}},\ \mathcal{C})_{\mathcal{C}}$
1	$b \xleftarrow{\$} \mathbb{Z}_q$ and $B = g^b,\ B^* = B \cdot v^{\text{pwd}_{\mathcal{C}}}$	$\xrightarrow{\mathcal{C},\ B^*}$	For $1 \leq i \leq N$
2			$r'_i \xleftarrow{\$} \mathbb{Z}_q$ and compute
3			$g^{s_i} = g^{r'_i} \cdot g^{r_i},\ h^{s_i} g^{W_i} = h^{r'_i} \cdot h^{r_i} \cdot g^{W_i}$
4	For $1 \leq i \leq N$	$\xleftarrow{\mathcal{S},\ (g^{s_i})_i,\ A^*}$	$a \xleftarrow{\$} \mathbb{Z}_q$, $A = g^a,\ A^* = A \cdot u^{\text{pwd}_{\mathcal{C}}}$
5	compute $H(K'_i) = \alpha'_i \Vert \beta'_i \Vert k'_i$ with:		
6	$K_{\mathcal{C}} = (\frac{A^*}{u^{\text{pwd}_{\mathcal{C}}}})^b,\ K^i_{\mathcal{C}} = (g^{s_i})^{x_{\mathcal{C}}} \cdot g^{W'_i}$		
7	$K'_i = \mathcal{S} \Vert \mathcal{C} \Vert (g^{s_i})_i \Vert A^* \Vert B^* \Vert K^i_{\mathcal{C}} \Vert K_{\mathcal{C}} \Vert \text{pwd}_{\mathcal{C}} \Vert i$	$\xrightarrow{(\alpha'_i)_i}$	For $1 \leq i \leq N$, $H(K_i) = \alpha_i \Vert \beta_i \Vert k_i$ with:
8			$K_{\mathcal{S}} = (\frac{B^*}{v^{\text{pwd}_{\mathcal{C}}}})^a,\ K^i_{\mathcal{S}} = h^{s_i} \cdot g^{W_i}$
9			$K_i = \mathcal{S} \Vert \mathcal{C} \Vert (g^{s_i})_i \Vert A^* \Vert B^* \Vert K^i_{\mathcal{S}} \Vert K_{\mathcal{S}} \Vert \text{pwd}_{\mathcal{C}} \Vert i$
10			
11			If $\#\{i : \alpha_i \neq \alpha'_i\} \leq t$
12			Then acc $= 1$, $K = \text{lsb}_k(\Vert_{i:\alpha_i=\alpha'_i} k_i)$
13	If $\#\{i : \beta_i \neq \beta'_i\} \leq t$	$\xleftarrow{(\beta_i)_i}$	Else acc $= 0$, $K \xleftarrow{\$} \{0,1\}^k$, $\beta_i \xleftarrow{\$} \{0,1\}^l$
14	Then acc $= 1$, $K' = \text{lsb}_k(\Vert_{i:\beta_i=\beta'_i} k'_i)$		
15	Else acc $= 0$, $K' \xleftarrow{\$} \{0,1\}^k$		

Fig. 1. Pointcheval-Zimmer protocol

In [18], the authors also suggest practical parameters for a real-world implementation. They assume an iris scan has $N = 1024$ bits. A value of $t = 300$ is defined as the threshold, so two iris codes with less than t disparate bits

[1] The original specification in in [18] does not explicitly explain the meaning of lsb in Line 12 and 14. We assume lsb refers to "least significant bits" and we interpret it as a key derivation function that derives a session key from raw keying materials. This ambiguity does not affect our security analysis however.

(i.e., Hamming distance) are considered belonging to the same eye; otherwise, they are regarded from different eyes. Furthermore, the authors use l to denote the bit lengths of $\alpha, \alpha', \beta, \beta'$ (see line 5 and 7 in Figure 1). That is: $l = ||\alpha'|| = ||\beta'|| = ||\alpha|| = ||\beta||$.

The value of l is critical to the correctness of the protocol. For small values of l, say $l = 1$, then the protocol will be guaranteed to fail even between two honest players. To ensure a successful honest execution of the protocol, the value l must not be small. Pointcheval and Zimmer recommend $l = 24$, and they estimate that with this parameter, the probability for a successful execution between two honest players is $1 - 2 \cdot t \cdot 2^{-24} = 1 - 2^{-14}$.

Pointcheval and Zimmer also define a formal model to prove the security of the protocol [18]. The model assumes the adversary is able to corrupt a client $\mathcal{C}$ in the following ways: by compromising the password $\text{pwd}_{\mathcal{C}}$, by stealing the private key $\text{sk}_{\mathcal{C}}$, or by spoofing biometrics $W_{\mathcal{C}}$. Under this model, the authors formally prove that an attacker has to compromise all three factors in order to win. However, the formal model only considers the unilateral authentication from the client to the server. In other words, it implicitly assumes the server is honest. The authors of [18] acknowledge that "this can be seen as a strong limitation", but on the other hand, they argue "it is not in practice: if the password and the secret keys are compromised, an adversary can easily play the role of the server". We find this justification weak and unclear; in particular, the authors do not explicitly explain what will happen if the attacker is able to impersonate the server to the client (e.g., in a man-in-the-middle attack). We fill the gap by analyzing this threat in the following section.

3 Attacks

When the attacker is able to impersonate the server (which is a realistic threat in practice), we show the Pointcheval-Zimmer protocol is insecure. First of all, we assume that the attacker has compromised the client's password (say by phishing). Then, we show how the attacker is able to subsequently compromise the other two factors: the biometrics and private key, hence breaking the entire system.

3.1 Stealing Biometrics

First, we show how the compromise of the password factor will lead to the breach of the biometrics factor. Figure 2 shows how the attack works. As shown in Line 2, the attacker selects random values for s_i and computes g^{s_i} accordingly. This is a clear deviation from the original specification, because if the server had honestly followed the specification in computing $g^{s_i} = g^{r'_i} \cdot g^{r_i}$, it will not have knowledge of the exponent of g^{s_i}. But in this case, the server (attacker) knows the exponent (which allows carrying out the subsequent attack). This deviation is undetectable to the client.

One principle in designing robust security protocols is "never let yourself be used as an oracle by your opponent" [1]. Unfortunately, in this case, the client has

	Client (Victim)		Server (Attacker)
	$\mathcal{C} : (W'_c = (W'_i)_i,\ sk_C = x_C,\ \mathrm{pwd}_C)$		$\mathcal{M} : (h = g^{x_C},\ \mathrm{pwd}_C)$
1	$b \xleftarrow{\$} \mathbb{Z}_q$ and $B = g^b$, $B^* = B \cdot v^{\mathrm{pwd}_C}$	$\xrightarrow{\mathcal{C},\ B^*}$	For $1 \leq i \leq N$
2			$s_i \xleftarrow{\$} \mathbb{Z}_q$ and compute g^{s_i}
3	For $1 \leq i \leq N$	$\xleftarrow{\mathcal{S},\ (g^{s_i})_i,\ A^*}$	$a \xleftarrow{\$} \mathbb{Z}_q$, $A = g^a$, $A^* = A \cdot u^{\mathrm{pwd}_C}$
4	compute $H(K'_i) = \alpha'_i \| \beta'_i \| k'_i$ with:		
5	$K_C = (\frac{A^*}{u^{\mathrm{pwd}_C}})^b$, $K^i_C = (g^{s_i})^{x_C} \cdot g^{W'_i}$		
6	$K'_i = \mathcal{S} \| \mathcal{C} \| (g^{s_i})_i \| A^* \| B^* \| K^i_C \| K_C \| \mathrm{pwd}_C \| \mathrm{i}$	$\xrightarrow{(\alpha'_i)_i}$	
7		$\xleftarrow{(\beta_i)_i}$	$((\beta_i)_i,\ (k_i)_i,\ (\mathrm{W}''_\mathrm{i})_\mathrm{i}) \leftarrow \mathrm{SearchBiometrics}((\alpha')_i)$
8	If $\#\{i : \beta_i \neq \beta'_i\} \leq t$		Compute $K = \mathrm{lsb}_k(\|_i k_i)$
9	Then acc $= 1$, $K' = \mathrm{lsb}_\mathrm{k}(\|_{\mathrm{i}:\beta_\mathrm{i}=\beta'_\mathrm{i}} \mathrm{k}'_\mathrm{i})$		
10	Else acc $= 0$, $K' \xleftarrow{\$} \{0,1\}^k$		

Fig. 2. Attack 1: stealing fresh biometrics from the client without being detected

made itself an oracle to the attacker. After receiving the data from the attacker, the client proceeds to compute values of $(\alpha'_i)_i$ and sends them over to the server. Those values will allow the attacker to discover the biometric sample from the user, as described in Algorithm 1.

The recovered sample from Algorithm 1 is high-quality biometric data. By "high-quality", we mean the recovered sample $(W''_i)_i$ is extremely close to the client's sample $(W'_i)_i$, which is freshly acquired in a favorable supervised condition. The exact difference between $(W''_i)_i$ and $(W'_i)_i$ depends on the parameter l (which is the bit length of $\alpha, \alpha', \beta, \beta'$). If we take $l = 24$ as recommended in [18], the probability of each bit in W''_i being correct (i.e., it equals the bit in W_i) is $p = 1 - 2^{24}$. Hence, the probability of all $N = 1024$ bits in W''_i being correct is $p^N = 99.994\%$. With an almost identical biometric sample, it is trivial for the attacker to compute $(\beta_i)_i$ so that he can successfully finish the rest of the protocol (see Algorithm 1). In reality, having two identical biometric samples normally suggests a replay attack, so the attacker may randomly corrupt some (up to t) of the β_i values to artificially make the biometric matching look "fuzzy". The same attack also applies to the three-party extension of Pointcheval-Zimmer's protocol, which was proposed in [15].

3.2 Disclosing Private Key

In the second attack, we show how the attacker is able to recover the client's private key, based on a compromised password and stolen biometrics (obtained from the first attack).

The attack is possible because the client is not required to perform public key validation on the data received from the server[2] (Line 4-7 in Figure 1). Many protocols omit the step of public key validation in order to increase the protocol efficiency. But this is often done at the expense of security. One well-known example is HMQV, which "provably" drops public key validation based on

[2] In the three-party extension of the Pointcheval-Zimmer protocol [15], there is no public key validation either. So the same attack applies.

Algorithm 1. SearchBiometrics algorithm

Input: $(\alpha_i')_i$
Output: $(\beta_i)_i$, $(k_i)_i$, $(W_i'')_i$

for $i = 1, \ldots, N$ **do**
  Compute $H(K_i') = \alpha_i' || \beta_i' || k_i'$ with
  $K_C = (\frac{B^*}{v^{\text{pwd}_C}})^a$, $K_C^i = (g^{x_C})^{s_i}$
  $K_i' = \mathcal{S}||\mathcal{C}||(g^{s_i})_i||A^*||B^*||K_C^i||K_C||\text{pwd}_C||i$
  if $\alpha_i = \alpha_i'$ **then**
    $(W_i'')_i = 0$
    $(\beta_i)_i = \beta_i'$
    $(k_i)_i = k_i'$
  else
    $(W_i'')_i = 1$
    Recompute $H(K_i') = \alpha_i' || \beta_i' || k_i'$ with
    $K_i' = \mathcal{S}||\mathcal{C}||(g^{s_i})_i||A^*||B^*||K_C^i \cdot g||K_C||\text{pwd}_C||i$
    $(\beta_i)_i = \beta_i'$
    $(k_i)_i = k_i'$
  end if
end for

a formal model and security proofs [9]. However, Menenzes *et al.* subsequently pointed out the flaws in the formal model and attacks on HMQV [17]. Their paper highlights the importance of performing the public key validation. In [18], Pointcheval and Zimmer defined a formal model and provided formal security proofs for their MK-AKE protocol. However, the formal model in [18] implicitly assumes the server is *honest*, which is not a valid assumption.

The attack works as follows. Instead of computing g^{s_i} as in the original protocol (Line 3 in Figure 1), the attacker selects small subgroup elements in $\mathbb{Z}_p^*$ and sends them to the client. We use $\mathbb{G}_s$ to denote a small subgroup of prime order s. Let us take $i = 1$ as an example. Let b_1 be a generator of $\mathbb{G}_s$ (i.e., an arbitrary non-identity element). After receiving b_1, the client proceeds to compute α_1' as specified in the original protocol and sends it to the server. We know $W_1 = W_1'$ with a high probability (say 90%). For simplicity of illustration, we first assume $W_1 = W_1'$. The attacker knows all the concatenated items in K_i' (Line 7 in Figure 1), except K_C^i. Based on $W_1 = W_1'$ and the fact that $b_1^{x_C} \bmod p$ falls within a small range, the attacker can easily obtain the value of $a_1 = b_1^{x_C} \bmod p$ by exhaustive search (i.e., against the value of α'). In the subsequent step, the attacker can compute $x_C \bmod s$ – once again by exhaustive search (based on a_1, b_1 and p). Since b_1 is a generator in $\mathbb{G}_s$, $b_1^{x_C}$ also falls within the same small subgroup. Through exhaustive search, the attacker can obtain $x_C \bmod s$. By repeating the same procedure for different subgroups, the attacker can recover more secret bits of the private key. Depending on the group setting, it is possible to recover a full copy of the private key based on the Chinese Remainder theorem (e.g., see a concrete implementation of the attack in Appendix A).

In the above analysis, we have assumed $W_1 = W_1'$, but in reality the equality only holds for a probability say 90%. This problem can be easily addressed by exploiting the large amount of redundancies in an iris code. Instead of sending one b_i value, the attacker could send several b_i values from the same small subgroup $\mathbb{G}_s$. For example, with 30 b_i values, there will be on average $30 \times 0.9 = 27$ results that give the same $x_C \bmod s$. This removes any uncertainty due to the fuzzy nature of biometrics.

3.3 Combining Two Attacks

It is easy to combine the two attacks together. For that, we need to modify the first attack slightly: after successfully stealing a biometric sample, the adversary sends to the client random β_i' values to trigger a "rejection". Because the β_i' values are random, the matching at the client will fail. However, the failure will hardly raise any suspicion from the client, as any biometric system has a non-zero false rejection rate in the real-world operation[3]. Most likely, the user will be prompted to try again (and be ready to be more cooperative with the iris photo-taking). In the second attempt, the attacker will send small subgroup elements to discover the user's private key. Subsequently, the attacker will send correct β' values to trick the client into believing that the second attempt is successful.

Thus, with a stolen password, the attacker has successfully compromised the user's two other factors, hence breaking the entire system.

4 Discussion

Between the two attacks, the second attack might look more damaging, but it is easier to fix. We can address it by adding public key validation in the protocol. However, this will significantly decrease the computational efficiency of the original protocol, as it normally takes a full exponentiation to verify if the received public key is a valid element in the correct subgroup of $\mathbb{Z}_p^*$. Nonetheless, our attack shows that this step is necessary (more explanation about the importance of public key validation can be found in [17]).

The first attack indicates a more fundamental flaw with the protocol. The question concerns the exact role of biometric authentication in the Pointcheval-Zimmer protocol. Assume the attacker has compromised the password and the private key, but he does not have user's biometric data. What kind of security assurance can the remaining biometrics factor provide? Almost none – based on the following observations. First, the client cannot safeguard the biometrics. As shown in Section 3.1, the protocol allows a remote attacker to steal the user's freshly obtained biometric sample without the user's awareness at all. Second, the encryption of biometrics at the server does not really preserve the privacy of biometrics as claimed in [18]. The attacker will target the weaker part in

[3] Even with two perfectly matching biometric samples, a false reject may still occur in the Pointcheval-Zimmer scheme (due to the deficiency in the engineering design of the protocol) with a probability 2^{-14} for $l = 20$ (see [18]).

the system – the client – to steal a biometric sample in the plaintext. Third, although by design biometric matching is done at the server, the result of the matching cannot be securely sent back to the client. As demonstrated in Section 3.1, the attacker is able to arbitrarily manipulate the matching outcome. All these suggest a fundamental problem with the protocol, which seems not easily fixable with the current structural design of the protocol. More research on this is much needed (e.g., some techniques in biometrics-based AKE [3,5,6] may be applicable to address this issue).

The above flaw was missed by the formal analysis in [18], because the authors assume biometrics as "fully public". Their justification is that the "the opposite assumption is not reasonable in practice", which refers to the treatment of biometrics as "fully secret" in [3,5,6]. However, "biometric applications lie between the extremes of secret data and fully public data" [8]. Indeed, it is possible to steal users' fingerprints that were left on the keyboard, or surreptitiously photograph their irises using hidden cameras. But, samples stolen this way tend to be of poor-quality [2]. Getting high-quality biometric samples usually requires user cooperation and a favorable environment: e.g., proper posture, distance to the camera and lighting etc. Stealing such samples without being detected by the user is not easy. When designing biometrics-based protocols, it is prudent not to rely on the secrecy of biometrics, but on the other hand, it is obviously a security flaw to give away freshly acquired biometric samples to a remote attacker.

5 Conclusion

In this paper, we described two attacks on Pointcheval-Zimmer's Multi-Factor Authenticated Key Exchange (MF-AKE) protocol. In the first attack, we showed how an attacker could make use of a stolen password to subsequently compromise the biometrics factor without being detected. This violates the privacy of biometric data. In the second attack, we showed how an attacker could further discover the user's private key by exploiting the small subgroup confinements. This attack breaches the presumed tamper-resistance of a secure token that stores the private key. In summary, the attacker only needs to compromise a single password factor in order to compromise the three-factor AKE protocol. These attacks are rooted in the deficiency of Pointcheval-Zimmer's formal security model, and highlight the importance of making valid assumptions in the model.

Acknowledgement. We sincerely thank Pointcheval and Zimmer for quickly and frankly acknowledging our attacks and the deficiencies in their formal model, and also for confirming our suggested countermeasures.

Appendix A: An Implementation of Small Subgroup Confinement Attack on Pointcheval-Zimmer's Protocol

We provide implementation details about the second attack below. (We have also experimentally verified the first attack, but since that implementation is straightforward, we do not include the details in the paper.)

Group Parameters

Let us assume a cyclic group $\mathbb{Z}_p^*$ that has a subgroup of prime order q. Hence, $q \mid p-1$. Let g be a generator, $g^q \bmod p = 1$. As a specific example, we define a 512-bit p, 160-bit q, 512-bit g with the following values (in the hex format).

$$\begin{aligned}
p &= f95b8b2f45b3016efb6ec51d342931aea4a5f4516d15c4ed2cf79e4d318\ldots \\
&\quad e28837989bedcbe4ce8693f68de6b72b1f74c8e109bc9155f5d2d65e9f6d\ldots \\
&\quad 091e7f79b \\
q &= e80f99e4981ee1eac37d8f0bf707b2067f6fe8cf \\
g &= 33c65b25ad4c47ac067083b7f2acf53ed3a053dbe508acbabe179029dad\ldots \\
&\quad 77a04c0953c1dbce02ce2f8cf5b030a36de7868b7434194816dbe7da920\ldots \\
&\quad 13bc4696d
\end{aligned}$$

Note that this (artificial) example is for illustration only. In practice, the bit lengths of p and q are normally much longer. In addition, we assume $\mathbb{Z}_p^*$ has several small subgroups of primer order s_i, so $s_i \mid p-1$. The hex values of s_i are given in Table 1. Except s_1, all other s_i are 21-bit long.

Table 1. Orders of small subgroups of $\mathbb{Z}_p^*$

i	1	2	3	4	5	6	7	8	9
s_i	2	15a661	1182bb	12b357	1fa9e7	1f1c9f	1c58b7	16b6b3	1727c3

Small Subgroup Confinement Attack

To demonstrate how an adversary can recover the private key, we first define a random 160-bit private key in the range of $[0,\ q-1]$:

$$x = 538b2f452c20f9cd7e356455e2ae66e9924ddd5d$$

By exploiting the small subgroup confinement, the adversary can obtain $a_i = b_i^x \bmod p$ for each of the small subgroups with prime orders s_i. Furthermore, he can obtain $c_i = x \bmod s_i$ through exhaustive search. Table 2 shows the results of a_i and c_i for each of the small subgroups $\mathbb{G}_{s_i}$. On average, it takes about 68.5 seconds to find c_i using exhaustive search, on a 2.93 GHz desktop PC with 4 GB memory.

With the c_i values, the private key can be recovered based on the Chinese Remainder theorem [16]. For example, we could apply Gauss's algorithm to solve the simultaneous congruences problem as follows. Let $n = s_1 \cdot s_2 \cdot s_3 \cdot s_4 \cdot s_5 \cdot s_6 \cdot s_7 \cdot s_8 \cdot s_9$. Then $x = \sum_i c_i N_i M_i \bmod n$ where $N_i = n/s_i$ and $M_i = N_i^{-1} \bmod s_i$. In this specific example, n (169-bit) $>$ q (160-bit), so we are able to recover the full private key. It takes merely 1 millisecond to obtain the following result using Gauss's algorithm.

$$\begin{aligned}
x' &= 538b2f452c20f9cd7e356455e2ae66e9924ddd5d \bmod n \\
&= 538b2f452c20f9cd7e356455e2ae66e9924ddd5d \bmod q
\end{aligned}$$

Table 2. Exhaustive search among the small subgroup confinements

i	$a_i = b_i^x \bmod p$	$c_i = x \bmod s_i$	Time for exh search (ms)
1	f95b8b2f45b3016efb6ec51d342931aea4a5f4516d15c4ed2cf79e4d318e2883 7989bedcbe4ce8693f68de6b72b1f74c8e109bc9155f5d2d65e9f6d091e7f79a	1	0
2	791778359b242e573617fff6735703be986dd9a6271be8b413381f4211ffbc0cb 6cb2da17c880a115e223752cb431708c4a64d68cbd8109c7a2e31e434839682	122156	124854
3	133023e42630ff22b9f9fb3e6a1ceeddf2d7fc1014fcab33eeedc1af416951c2c9 099d8fef275408d1f82c7090cd72744a260685381a15cc8e58bdd052bb1928	517f1	31481
4	93358cdb4153463b93f0525d6819426b7b4c0fcb5905cf10db5f39eeea68a879 c8ad548cef72476671ff2805e1cbf8644f39f2d4fc71570522e4e89784a0a0aa	42cb2	25271
5	f517a5a96bd4a92af483727dd0c73c251ac159056ec5b2eea90854379ba8344 cc41e641d9e84ec319fd4ab545e038cba799972a2db93b2bf315fb62b08402cb	e0125	93050
6	e1895eb12f66645c2359ad6185d85fb02f39ddd3b2e80392c3f53ffc5ebcb23a9 81eb1d9cfd7a8eeea8a13e83af81a726280fcd7d545450fb6871786f9e2ebe6	15d34	7544
7	8dc566a29c84977ceb1e1a466321859fe3022f7ab3adae44ead9d8b7c2dc5461 b1ea9441b19425c13da5b7c998ea7fbe41aaea70177118b37438c5f36cad85d2	121427	121793
8	d73e16cf041ecd7c9c73d6414ad01b0ce85deaedbcf6834591f373091a51903 1cbb4aebe31fd56afab5750226584a762eaf7ffef0f5d7e2f940e27ea8d8b0d41	ef2c6	99664
9	dbdb75657da4a1e4312f17b7519eff2fe2c0b0b2fbe225482e3f78f73530c545a 3c8cd4d6fa0abd3f27058090cd1992263e72f16e47b7256f916d4a20201e5c0	10c700	112565

References

1. Anderson, R.J., Needham, R.: Robustness Principles for Public Key Protocols. In: Coppersmith, D. (ed.) CRYPTO 1995. LNCS, vol. 963, pp. 236–247. Springer, Heidelberg (1995)
2. Anderson, R.J.: Security Engineering: A Guide to Building Dependable Distributed Systems, 2nd edn. Wiley (2008)
3. Boyen, X., Dodis, Y., Katz, J., Ostrovsky, R., Smith, A.: Secure Remote Authentication Using Biometric Data. In: Cramer, R. (ed.) EUROCRYPT 2005. LNCS, vol. 3494, pp. 147–163. Springer, Heidelberg (2005); Lee, J.K., Ryu, S.R.: Fingerprint-based Remote User Authentication Scheme Using Smart Cards. Electronics Leters 38(12), 554–555 (2005)
4. Boyd, C., Mathuria, A.: Protocols for Authentication and Key Establishment. Springer (2003)
5. Boyen, X.: Reusable Cryptographic Fuzzy Extractors. In: ACM CCS 2004, pp. 82–91 (2004)
6. Dodis, Y., Reyzin, L., Smith, A.: Fuzzy Extractors: How to Generate Strong Keys from Biometrics and Other Noisy Data. In: Cachin, C., Camenisch, J.L. (eds.) EUROCRYPT 2004. LNCS, vol. 3027, pp. 523–540. Springer, Heidelberg (2004)
7. Hao, F.: On Robust Key Agreement Based on Public Key Authentication. In: Sion, R. (ed.) FC 2010. LNCS, vol. 6052, pp. 383–390. Springer, Heidelberg (2010)
8. Hao, F., Anderson, R., Daugman, J.: Combining crypto with biometrics effectively. IEEE Transactions on Computers 55(9), 1081–1088 (2006)
9. Krawczyk, H.: HMQV: A High-Performance Secure Diffie-Hellman Protocol. In: Shoup, V. (ed.) CRYPTO 2005. LNCS, vol. 3621, pp. 546–566. Springer, Heidelberg (2005)

10. Hwang, M., Chong, S., Chen, T.: DoS-Resistant ID-Based Password Authentication Scheme Using Smart Cards. Computer Journal of Systems and Software 7(50), 147–150 (2009)
11. Lee, Y., Kim, S., Won, D.: Enhancement of Two-Factor Authenticated Key Exchange Protocols in Public Wireless LANs. Computers and Electrical Engineering 36(1), 213–223 (2010)
12. Lian, I.E., Lee, C.C., Hwang, M.S.: A Password Authentication Scheme Over Insecure Networks. Journal of Computer System Sciences 72, 727–740 (2006)
13. Li, C.T., Hwang, M.S.: An Efficient Biometrics-Based Remote User Authentication Scheme Using Smart Cards. Journal of Network and Computer Applications 33(1), 1–5 (2010)
14. Li, X., Niu, J.W., Ma, J., Wang, W.D.: Cryptanalysis and Improvement of a Biometrics-Based Remote User Authentication Scheme Using Smart Cards. Journal of Network and Computer Applications 34(1), 73–79 (2011)
15. Liu, Y., Wei, F., Ma, C.: Multi-Factor Authenticated Key Exchange Protocol in the Three-Party Setting. In: Lai, X., Yung, M., Lin, D. (eds.) Inscrypt 2010. LNCS, vol. 6584, pp. 255–267. Springer, Heidelberg (2011)
16. Menezes, A.J., van Oorschot, P.C., Vanstone, S.A.: Handbook of Applied Cryptography. CRC Press (1996)
17. Menezes, A., Ustaoglu, B.: On the Importance of Public-Key Validation in the MQV and HMQV Key Agreement Protocols. In: Barua, R., Lange, T. (eds.) INDOCRYPT 2006. LNCS, vol. 4329, pp. 133–147. Springer, Heidelberg (2006)
18. Pointcheval, D., Zimmer, S.: Multi-factor Authenticated Key Exchange. In: Bellovin, S.M., Gennaro, R., Keromytis, A.D., Yung, M. (eds.) ACNS 2008. LNCS, vol. 5037, pp. 277–295. Springer, Heidelberg (2008)
19. Song, R.: Advanced Smart Card Based Password Authentication Protocol. Computer Standards & Interfaces 32, 321–325 (2010)
20. Tapiador, J.E., Hernandez-Castro, J.C., Peris-Lopez, P., Clark, J.A.: Cryptanalysis of Song's Advanced Smart Card Based Password Authentication Protocol (2011), Technical report available at `http://arxiv.org/pdf/1111.2744`
21. Wu, S., Zhu, Y.: Improved Two-Factor Authenticated Key Exchange Protocol. The International Arab Journal of Information Technology 8(4), 430–439 (2011)
22. Xiang, T., Wong, K., Liao, X.: Cryptanalysis of A Password Authentication Scheme Over Insecure Networks. Journal of Computer System Sciences 74, 657–661 (2008)
23. Xu, J., Zhu, W.T., Feng, D.G.: An Improved Smart Card Based Password Authentication Scheme with Provable Security. Computer Standards & Interfaces 31, 723–728 (2009)

Breaking an Animated CAPTCHA Scheme

Vu Duc Nguyen[1], Yang-Wai Chow[2], and Willy Susilo[1,*]

[1] Centre for Computer and Information Security Research
{dvn108,wsusilo,caseyc}@uow.edu.au
[2] Advanced Multimedia Research Laboratory
School of Computer Science and Software Engineering,
University of Wollongong, Australia

Abstract. CAPTCHAs have become a ubiquitous security countermeasure to protect online web-services against automated attacks. However, attackers have managed to successfully break many existing CAPTCHA schemes. Animated CAPTCHA schemes have been proposed as a method of producing CAPTCHAs that are more human usable and more secure. The addition of the time dimension is supposed to increase the robustness of animated CAPTCHAs. This paper investigates the robustness of HelloCaptcha, an animated text-based CAPTCHA scheme with a total of 84 different variations. In this paper, we show that simple techniques can be used to extract important information from the animation frames of an animated CAPTCHA. Our approach essentially reduces the animated CAPTCHA into a traditional single image CAPTCHA challenge. Furthermore, the methods presented in this paper can be generalized to break other animated CAPTCHAs.

Keywords: Animated CAPTCHA, character extraction, segmentation, optical character recognition.

1 Introduction

A CAPTCHA (Completely Automated Public Turing test to tell Computers and Humans Apart) is an automated challenge and response test to verify whether or not an online transaction is being carried out by a human [13]. At present, CAPTCHAs are used on many web-based services as a standard security mechanism for deterring automated attacks by bots or other malicious programs. Unfortunately, a large number of CAPTCHA schemes to date have been successfully broken. Many researchers and practitioners alike have shown that various design flaws can be exploited by automated programs to break these CAPTCHAs.

To increase the security strength and to confuse Optical Character Recognition (OCR) programs, traditional text-based CAPTCHAs rely on techniques like distorting the text and/or the overlaying of visual noise. However, this often makes the resulting CAPTCHA difficult for humans to use. A good CAPTCHA

* This work is supported by ARC Future Fellowship FT0991397.

F. Bao, P. Samarati, and J. Zhou (Eds.): ACNS 2012, LNCS 7341, pp. 12–29, 2012.

scheme must be both secure and human usable. It is for this reason that CAPTCHA developers have explored alternate paradigms in CAPTCHA design in an attempt to designing more secure and usable CAPTCHAs.

Animated CAPTCHAs have been proposed as a means of overcoming the limitations of traditional single image CAPTCHAs. One of the key principles behind the design of animated CAPTCHA schemes is that the information required to solve the CAPTCHA is not contained within a single image. As such, a human has to observe the animated CAPTCHA over its animation cycle in order to gather appropriate information to correctly solve the CAPTCHA challenge. This is assumed to be difficult for a computer because information is spread over multiple images. In addition, noise and other impediments can be added to the CAPTCHA challenge to deter automated attacks.

This paper addresses the question of whether animated CAPTCHAs really provide more security. In particular, this paper describes methods of breaking a representative animated text-based CAPTCHA scheme called HelloCaptcha [12], which at the time of writing has a total of 84 different variations. The results of this research show that we can successfully break all the variations of this animated CAPTCHA scheme with a high success rate using simple methods to extract information from the animation frames.

Our Contribution. In this paper we study the robustness of animated CAPTCHAs by breaking a representative animated CAPTCHA scheme named HelloCaptcha [12]. Our research shows that even though animated CAPTCHAs spread important information over multiple animation frames, it is possible to collect the relevant information from these frames in order to break the CAPTCHA. We present our approach of extracting key information from the animation frames and effectively reducing the animated CAPTCHA challenge into a single image. It is likely that the methods presented in this paper can be extended to break other animated CAPTCHA schemes. Our goal is to identify flaws in the design of animated CAPTCHAs that make them easy to break, so that future CAPTCHA schemes can be designed to avoid such pitfalls.

2 Related Work

2.1 Breaking CAPTCHAs

Over the year, many techniques have been proposed for breaking CAPTCHAs. These automated CAPTCHA solving methods are often based on pattern recognition, image processing, machine learning algorithms and so on. For example, Mori and Malik [9] developed an approach to break the Gimpy and EZ-Gimpy CAPTCHAs using object recognition techniques to identify words amidst background clutter. In their work, they presented a holistic approach of recognizing entire words at once, rather than attempting to identify individual characters in severe clutter. Li et al. [8] have also shown that image processing and pattern recognition algorithms, such as k-means clustering, digital image in-painting, character recognition based on cross-correlation, etc. have been successful in breaking a variety of e-Banking CAPTCHAs.

In a systematic study regarding the strengths and weaknesses of text-based CAPTCHAs, Bursztein et al. [3] observed that typical automated CAPTCHA solving processes can be divided into five generic steps, namely, pre-processing, segmentation, post-segmentation, recognition, and post-processing. While segmentation, the separation of a sequence of characters into individual characters, and recognition, the identification of those characters, are intuitive and generally understood, the additional pre-processing and post-processing steps are also included as part of a standard process. For example, pre-processing a CAPTCHA image can remove background patterns or eliminate other impediments that could interfere with segmentation, while a post-segmentation step can 'clean up' the segmentation output. After recognition, a post-processing step can used to improve accuracy by, for example, applying spell checking to any CAPTCHA that is based on actual dictionary words.

2.2 Segmentation Resistant

It is widely accepted that state of the art in text-based CAPTCHA design requires that a robust CAPTCHA be segmentation resistant. This segmentation resistant principle is based on the work by Chellapilla et al. [4] who have shown that automated programs can recognize single characters even better than humans. As such, if a text-based CAPTCHA can be segmented into its constituting characters, it is essentially broken.

This segmentation resistant principle required for robust CAPTCHA design, led a research team at Microsoft to develop a CAPTCHA scheme that was meant to be segmentation resistant. Unfortunately, it was shown that the Microsoft CAPTCHA could in fact be segmented, and thus broken, by a low-cost attack [15]. In addition, researchers have demonstrated the use of novel segmentation techniques to break various other CAPTCHAs [14]. Nevertheless, the success of these attacks does not negate the segmentation resistant principle in the design of robust CAPTCHAs. However, relying on segmentation alone does not provide reliable defense against automated attacks [3].

2.3 Animated CAPTCHAs

Animated CAPTCHAs have been proposed to overcome the limitations of traditional single image CAPTCHAs. Animated CAPTCHAs can be presented on webpages by means of three main formats; as an animated GIF (Graphics Interchange Format) image, a flash video or via video streaming. To date, animated CAPTCHA are not commonly used on websites yet. However, a number of animated CAPTCHA schemes have been proposed by the web community as well as the research community.

Athanasopoulos and Antonatos [2] contend that their proposed CAPTCHA, enhanced with animation technology, can resist sophisticated attacks better than standard CAPTCHAs based on static images with distorted text. Cui et al. [6] presented a sketch of an animated CAPTCHA approach based on moving letters amid a noisy background. An animated CAPTCHA method based on the idea of presenting distorted text on the face of a deforming surface was proposed by Fischer and Herfet [7]. Naumann et al. [10] suggest another animated CAPTCHA technique based on visual phenomena. The idea behind their approach is that by grouping different entities that move together, sketches or letters that are superimposed over a noisy background of the same color become visible once they are moving.

Chow and Susilo [5] devised an animated 3D CAPTCHA named AniCAP, that was designed with the segmentation resistant principle in mind. Their approach is based on motion parallax, the perception of depth through motion, where humans are supposed to distinguish the main characters located in the foreground from the background characters. NuCaptcha is another animated CAPTCHA designed to be segmentation resistant. The idea behind this CAPTCHA is that characters are joined together, but when seen to be moving, the user's mind sees the different parts and fills in the blanks; the parts that are moving together are grouped together, and user can clearly differentiate the letters. On the other hand, computers do not have this advantage and see a smear of pixels [11].

3 The Targeted Animated CAPTCHA Scheme

To investigate the robustness of animated text-based CAPTCHA schemes, the HelloCaptcha [12] was selected as a representative scheme. HelloCaptcha is an animated CAPTCHA scheme that has been made freely available via a web service with an easy to use embedding code. The developers of HelloCaptcha state that their CAPTCHA schemes are more readable, more secure and nice. The animation approach is intended to provide an additional time dimension in order to increase its difficulty against automated attacks [12]. However, whilst being usable and easy on the eyes, it is obvious that HelloCaptcha was not designed with the segmentation resistant principle in mind. Our purpose is to investigate whether this implementation of the time dimension alone really provides additional security to animated CAPTCHAs.

One of the key reasons why HelloCaptcha was selected as a representative animated CAPTCHA scheme for this study is because the developers have provided a variety of different variations to their scheme. At the time of writing, the developers of HelloCaptcha have provided 12 broad categories of different animated CAPTCHA schemes. Furthermore, each category has a number of different variations. In total, there are 84 different types, which have been grouped into the 12 broad categories. We have not found any other developer whom have provided a greater number of animated CAPTCHA types.

Please note that there too many HelloCaptcha variations to adequately describe in this paper. Thus, we encourage interested readers to visit the HelloCaptcha website at `http://hellocaptcha.com/` for further details and to view the different animated CAPTCHAs for themselves. In general, each animated CAPTCHA challenge provided by HelloCaptcha consists of a sequence of six letters and/or digits presented in an animated GIF image with the dimensions of 180×60 pixels. This paper will use the notation '*Category/Type*' to denote a specific HelloCaptcha ***type*** of a particular ***category***.

4 Breaking HelloCaptcha

In general, our attack can be divided into 4 stages. An overview of these stages is depicted in Fig. 1. At any time, only 1 of these 84 different types is randomly selected and presented to the user. Therefore, the first stage to breaking the CAPTCHA was to automate the process of identifying which of the 84 types is the current type being presented. This phase was called the type distinction stage. Upon successfully distinguishing the CAPTCHA type, the appropriate technique was then selected to extract a single image containing all the separate characters in the CAPTCHA challenge. The resulting image then underwent a pre-processing stage to remove noise and to increase the legibility of the characters. Subsequently, the final character recognition stage involves the image being passed through an OCR program to solve the CAPTCHA by recognizing the individual characters contained in the challenge.

4.1 Type Distinction

In order to automatically distinguish which of the 84 different types was currently being presented, a number of factors were considered. These factors are described as follows.

- **Number of frames.** Each HelloCaptcha challenge is displayed as an animated GIF that is constructed from a sequence of image frames. The total

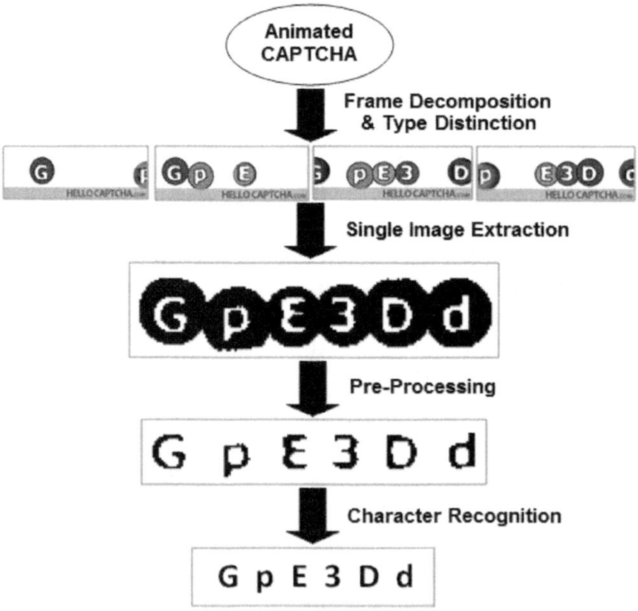

Fig. 1. Overview of the stages used to break HelloCaptcha

number of frames contained in the GIF file can be used to distinguish between some of the different types. For example, some HelloCaptcha types always contain a fixed number of frames (e.g. the types named *Noir Nois*, *Autumn Ant* and *Col Noi* of the *Noisy Mosaic* category are always fixed at 25 frames), while the total number of frames for other types may vary within a fixed range of frames (e.g. the number of frames in the *Contour* and *Redblue* types of the *TextFlood* category vary between 65 to 76 frames).

- **Maximum frame delay.** Animated GIFs display successive frames after a certain time duration, or frame delay time, which controls the speed of the animation. In other words, the frame delay informs the playback application how long to wait after displaying the current frame before it is to display the next frame in the animation. This information can be obtained from the GIF file and the maximum frame delay time can be used as a distinguishing factor between the different HelloCaptcha types. For instance, all types in the *Swapper* category consistently use a 100 ms frame delay whereas all type in the *Spring* category consistently use a value of 4 ms.
- **Number of blank frames.** We define blank frames as frames that contain only the background color. Some HelloCaptcha types always have a fixed number of blank frames over the entire animation. As such, this value can be used for type distinction. For example, there are always 17 blank frames in the *Smarties/Follower* type whereas the *Smarties/Negative* type always contains 7 blank frames.
- **Background color.** This is another value that is also very useful for type distinction. Many types are displayed using specific background colors. As an example, the background color used in the *Search Light/Stille Natch* type is made up of the following RGB color values Red=6, Green=38, Blue=66. The combination of these color values is unique in all the 84 types and it helps to easily distinguish the *Search Light/Stille Natch* type from other types. In practice, the background color can be ascertained by determining the most used pixel color in the first animation frame.

These simple pieces of information can easily be obtained from an animated GIF file. By using the different values of the number of frames, maximum frame delay, number of blank frames and background color, all the different HelloCaptcha types can be distinguished with a high degree of accuracy. Our experimental results show that we can accurately distinguish between the different types between 80%−100% of the time.

4.2 Single Image Extraction

One of the main security mechanisms provided by animated CAPTCHAs is that the information required to solve the CAPTCHA challenge is spread over multiple animation frames, unlike traditional CAPTCHAs where all information has to be presented in a single image. The same applies to the HelloCaptcha scheme, where the addition of this time dimension is intended to increase the difficulty of automated attacks [12].

In most animated CAPTCHAs, the challenge is obscured in individual frames where text characters may only be partially visible, joined together, overlapped with extra characters at random, and so on. While it is indeed impossible to solve most animated CAPTCHAs using individual animation frames, we show that it is possible to solve the CAPTCHA challenge by collecting information from multiple animation frames. Our strategy is to extract relevant information from the animation frames into a single image by using a set of simple techniques. We demonstrate that once this information is extracted, the animated CAPTCHA is in effect converted into a single image and the standard techniques used to break traditional CAPTCHAs can easily be applied to this resulting image.

Extraction by Pixel Delay Map (PDM). This is the main method used to extract relevant information to break HelloCaptcha. A total of 61 of the 84 HelloCaptcha types can be broken using this approach. For the purpose of CAPTCHA usability, text characters required to solve the challenge in a text-based CAPTCHA often needs to be emphasized somehow. This is because the human brain needs to be able to distinguish and recognize the characters required to solve the challenge.

In many HelloCaptcha types (e.g. *Flitter/Colorful*, *H-Mover/Street*, *Search Light/Cyber*, etc.), in order to get the human user's attention, the text characters required to solve the challenge are displayed at certain fixed locations for longer periods of time compared to the noise elements and other peripheral components. This allows us to build what we call a Pixel Delay Map (PDM) from the animation frames. The PDM is an image resulting from the accumulation of the total amount of time that a pixel gets displayed in a color that is different from the background color. Given an animated CAPTCHA, *AC*, the following is a depiction of the basic algorithm used to construct a PDM.

CalculatePDM(AC)

```
Initialize PDM(x,y) ← 0
begin
    for  i ← 1 to TotalFrames
       for  x ← 1 to TotalRows
          for  y ← 1 to TotalColumns
            if  Color(x,y) ≠ BackgroundColor  then
                PDM(x,y) ← PDM(x,y) + FrameDelay(i)
end
```

Fig. 2 shows examples of PDMs constructed from two different HelloCaptcha types, namely the *Flitter/Colorful* and the *H-Mover/Baller* types respectively. Fig. 2(a) and Fig. 2(b) depict six individual frames obtained at different times during the animation, with time increasing from left to right. In the *Flitter/Colorful* challenge that is shown in Fig. 2(a), small random colored squares gradually assemble to form the characters for a period of time, then disassemble

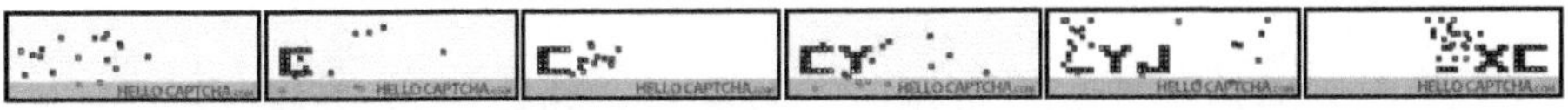

(a) Example frames from a *Flitter/Colorful* CAPTCHA challenge.

(b) Example frames from a *H-Mover/Baller* CAPTCHA challenge.

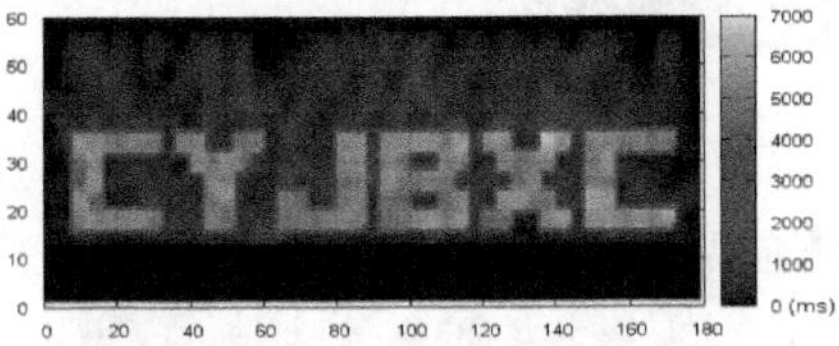

(c) PDM for challenge in Fig. 2(a).

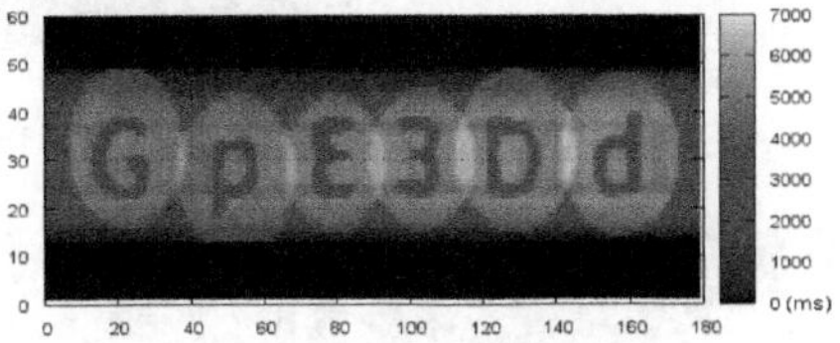

(d) PDM for challenge in Fig. 2(b).

(e) Extracted from PDM in Fig. 2(c).

(f) Extracted from PDM in Fig. 2(d).

Fig. 2. Examples of character extraction using the Pixel Delay Map (PDM) approach

as if blown apart. For the *H-Mover/Baller* challenge in Fig. 2(a), the characters gradually slide in from the right at random speeds, pause for a short while, before sliding out from the left.

The resulting PDM images are shown in Fig. 2(c) and Fig. 2(d). The x-axis and y-axis denote the pixel positions, whereas the different colors in the PDM image illustrate the length of time in which the animated CAPTCHA's pixels were displayed in a color that was distinct from the background color. In particular, a PDM pixel with a higher value means that the color at that pixel position was different from the background color for a longer period of time compared to PDM pixels with lower values.

The extracted binarized images that are shown in Fig. 2(e) and Fig. 2(f) were obtained by simply thresholding the PDM image by $\frac{1}{2}$ the highest PDM value (i.e. 3500 ms). In other words, pixels in the PDM that had a value lower than this threshold were set to white whereas pixels with a value higher than the threshold were set to black. One can see that the CAPTCHA challenges in these images can easily be solved.

In practice, the PDM technique is flexible and can be used in other ways. For example, if the basic PDM approach was used for the *Smarties/Smarties* type where the pixel times were accumulated over all the frames, the challenge characters would significantly overlap and other techniques would have to be used to separate the characters. In the *Smarties/Smarties* challenge, characters gradually fade in and fade out, one by one from left to right. Some sample frames depicting this are shown in Fig. 3(a). Fig. 3(b) in turn shows a PDM that was

(a) Example frames from a *Smarties/Smarties* CAPTCHA challenge.

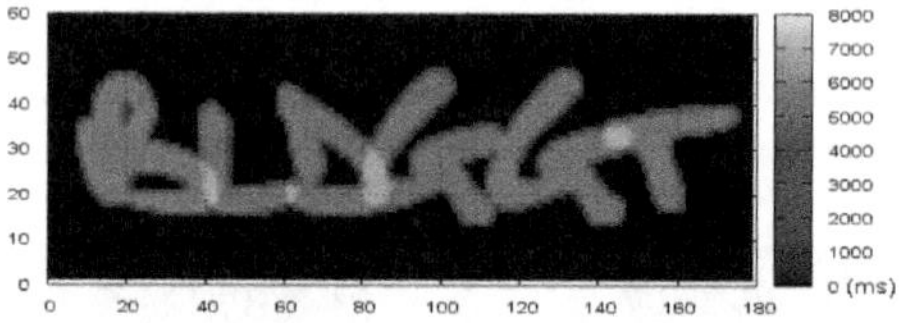

(b) Single PDM from all the frames.

(c) Three PDMs, each constructed from consecutive $\frac{1}{6}$ of the frames.

(d) Separate characters from six PDMs.

Fig. 3. Example of multiple PDMs from different portions of the animated frames

constructed using all the animation frames, one can see that the characters are joined together.

Nevertheless, in this situation we can construct six PDMs by splitting the animation frames into six distinct frame sequences. This is because challenges in HelloCaptcha always consist of six characters. In other words, a PDM is constructed by taking frames 0 to $\frac{1}{6}$ the total number of frames, another is constructed from frames $\frac{1}{6}$ to $\frac{2}{6}$ the total number of frames, another from frames $\frac{2}{6}$ to $\frac{3}{6}$ the total number of frames, and so on. The first three PDMs resulting from this approach are shown in Fig. 3(c). Consequently, the highest values from all the six resulting PDMs can be extracted giving the six separate characters that are depicted in Fig. 3(d).

For all the different HelloCaptcha types in the *Text Flood* and *Mass Flood* categories, a PDM can be constructed from the first $\frac{1}{3}$rd of the animation frames. This is because the relevant information to solve the challenge only appears during the first $\frac{1}{3}$rd of the animation then disappears and is no longer displayed again in the remaining frames.

Extraction by Catching Line (CL). The *Spring/Jumpers* HelloCaptcha type is an example of a type where the Catching Line (CL) approach can be used to extract a single image. In the *Spring/Jumpers* type, the characters constantly move in a vertical direction. This is depicted in Fig. 4(a), where the characters appear to 'spring' or 'jump' up and down. The security is based on the fact that

(a) Example frames from a *Spring/Jumpers* CAPTCHA challenge.

(b) Example of the virtual CL and 3 'caught' characters (in red).

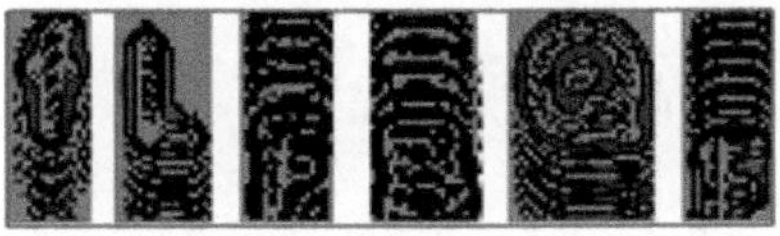

(c) Character vertical moving areas.

(d) 'Caught' characters.

Fig. 4. Example of character extraction using the Catching Line (CL) approach

during individual frames, the upper or lower parts of certain characters may be hidden from view. However, the full characters appear when moving in the central area of certain frames. The idea behind the CL approach is to 'catch' (get an image of) full characters. This can easily be done by getting an image of the full character the very first time the individual character 'touches' a virtual line which is located at a position $\frac{1}{10}$th from the upper boundary of the animated GIF, as portrayed in Fig. 4(b). Since all the characters only move in the vertical direction and do not overlap, one can easily determine the image area to capture based on the characters' moving areas, as shown in Fig. 4(c). Fig. 4(d) shows an example of an extracted image after catching all six characters.

Extraction by Color Selection (CS). Color can also be used to break a CAPTCHA scheme. The *Spring/Autumn Water* HelloCaptcha type is an example type where a Color Selection (CS) approach can be used to extract relevant information into a single image. The *Spring/Autumn Water* animated CAPTCHA challenge is composed of six moving characters on a white background, as depicted in Fig. 5(a). The characters move at random speeds and directions so they often overlap one another or are only partially visible if partly outside the image boundary. The color of each character is also random and they typically have different colors in a challenge.

We simply use the distinct colors to extract individual character images from a 'good' frame. A good frame is defined as one where the total number of pixels which make up the character is at a maximum. In other words, because the characters have distinct colors we can count the number of pixels of a particular color contained in each frame. A frame where the maximum number of pixels for a particular color is displayed, indicates that the character is not overlapped by other characters nor is it partially outside the image boundary. Hence, we can do this separately for the different colors, as shown in Fig. 5(b). The resulting extracted image is shown in Fig. 5(c).

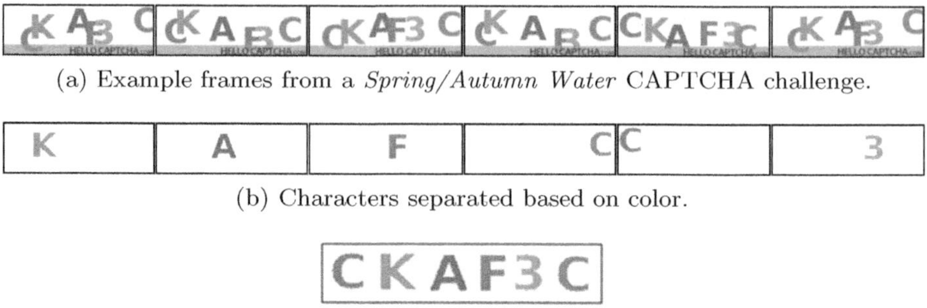

(a) Example frames from a *Spring/Autumn Water* CAPTCHA challenge.

(b) Characters separated based on color.

(c) Extracted characters.

Fig. 5. Example of character extraction using the Color Selection (CS) approach

Extraction by Frame Selection (FS). Unlike the *Spring/Autumn Water* type where each character has a different color, the *Spring/Default* type uses the same color for all the characters. Therefore, to extract the characters a different approach is required. In this Frame Selection (FS) method, the 'best' frame is chosen as one that satisfies all, or at least two, of the following conditions:

1. The characters can fully be separated into six distinct regions. To find this we use the Vertical Segmentation approach as described in Yan and Ahmad [15], in which a histogram representing the total number of pixels per column is constructed and the columns containing the characters can easily be identified.
2. None of the characters are in contact with the image boundaries. This condition is fulfilled as long as the border pixels all contain the background color.
3. Each character is made up of the maximum number of pixels. By counting the total number of connected character pixels, the total number of pixels that represents each of the six characters can be determined. The frame in which all characters are made up of the maximum number of pixels is selected.

An example of a sequence of frames from the *Spring/Default* type is provided in Fig. 6(a). Fig. 6(b) shows a frame where all three conditions are satisfied. On the other hand, Fig. 6(c) shows an example of a frame where only conditions 2 and 3 are met (condition 1 is not satisfied as the pixels of 'E' and 'T' overlap in the vertical direction). Nevertheless, it can be seen that the characters can still clearly be identified in that frame.

Extraction by Roller Selection (RS). The animation for all types in the *Roller* category is such that each character rotates around its center, starting from different rotation positions and rotating at different speeds. The *Roller/Rolling Clot* type is an example that uses this character rotation approach, as can be seen in Fig. 7(a). As the rotation progresses, at times characters may be connected or joined together. To obtain separate characters for

(a) Example frames from a *Spring/Default* CAPTCHA challenge.

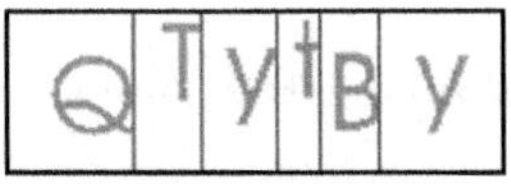

(b) All 3 conditions met.

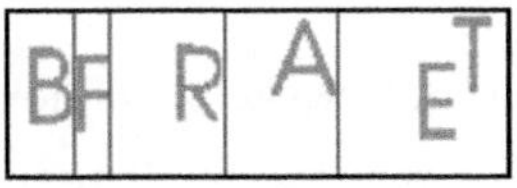

(c) 2 of 3 conditions met.

Fig. 6. Example of character extraction using the Frame Selection (FS) approach

(a) Example frames from a *Roller/Rolling Clot* CAPTCHA challenge.

(b) Extracted characters.

Fig. 7. Example of character extraction using the Roller Selection (RS) approach

these HelloCaptcha types, a Roller Selection (RS) method was used to extract the characters. The RS approach was done as follows:

1. Perform flood-fill starting from the center pixel of each the six characters (each character always rotates around its center). The flood-fill algorithm will give us the connected pixels of the same color, hence, we can obtain six individual characters.
2. Select the character when its highest pixel reaches its maximum height, as shown in Fig. 7(a). This was done to reduce the character recognition training set required by the OCR program, discussed later in section 4.4 below.

The end result is a set of six separate characters, at rotation angles that produce their maximum heights. This is shown in Fig. 7(b).

4.3 Pre-processing and Character Recognition

The image extraction stage produces a single image with all the characters. This image may contain noise and other impediments. Thus, to improve character recognition, a pre-processing stage is conducted to clean up the image. Depending on the HelloCaptcha type, different pre-processing steps may be used. The pre-processing methods that were used are described as follows.

- **Noise removal.** After using the PDM extraction method on types like the *Noisy Mosaic/Autumn Ant* type. The binarized image may contain unwanted noise, as can be seen in Fig. 8(a). This can be removed by performing

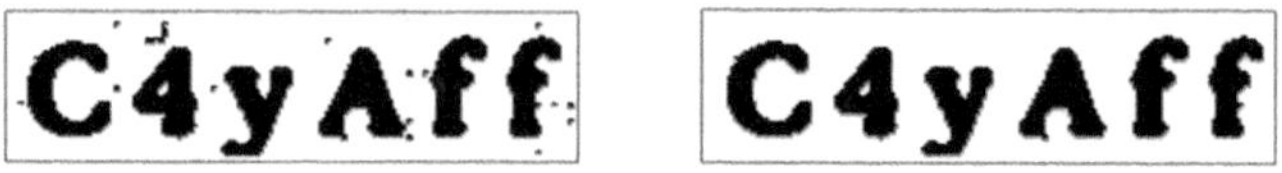

(a) Before noise removal. (b) After noise removal.

Fig. 8. Example of noise removal from extraction image

(a) Before circle removal. (b) Background filled in. (c) With colors inverted.

Fig. 9. Example of circle removal from extraction image

flood-fill on the different pixel clusters in the image and only saving the large areas. An example of the results can be seen in Fig. 8(b).

- **Circle removal.** In some types (e.g. *Roller/Cirque de Couleur*, *H-mover/Baller*, *Popup/Cirque*), the characters are encapsulated within circles, as shown in Fig. 9(a). We simple flood-fill the background starting from the image border, as depicted in Fig. 9(b) and invert the colors to obtain the image shown in Fig. 9(c).
- **Outline removal.** A similar process must be done for types where the outline of the characters are obtained, as shown in Fig. 10(a). The reason for this is that the OCR program typically achieves lower accuracy for characters displayed solely by their outlines. However, in these character outline cases care must be taken for characters such as the letter 'Q', in Fig. 10(a), which may contains a small internal white region. For these cases, we flood-fill the white internal regions of the characters to get the connected pixels and only save the large areas, resulting in the image shown in Fig. 10(b).
- **Refine by filling.** For certain types, the extracted image contains of characters with unwanted 'holes'. An example of this can be seen in Fig. 11(a). To remove these holes, we first obtain the character bounds by flood-filling the background (depicted in red in Fig. 11(b)), then flood-filling in the connected pixels within the characters which are less than 5 pixels from the character bounds (shown as green in Fig. 11(b)). This fills in the unwanted 'holes' whilst preserving the correct hollow regions for letters such as 'D' and digit '8' as can be seen in Fig. 11(c).

(a) Before outline removal. (b) Outlines removed.

Fig. 10. Example of removing the outlines surrounding the characters

(a) Before filling. (b) Removing 'holes'. (c) After filling.

Fig. 11. Example of refining the extracted image by filling in the unwanted holes

(a) Before shape removal. (b) Reference image. (c) After shape removal.

Fig. 12. Example of shape removal from the extracted image

- **Shape removal.** The *Popup/Love* type is an example of a HelloCaptcha type where each character is encapsulated within a heart shape, as can be seen in Fig. 12(a). To remove these shapes, we obtain a reference image containing the shapes, shown in Fig. 12(b), and subtract this from the extracted image. Fig. 12(c) shows the results with the shapes removed.

4.4 Character Recognition

After extracting the characters into a single image and cleaning up the image, the result is an image with six separate and distinct characters. All that remains is to perform character recognition. This is straightforward and can be done using any good OCR program. For this, we used the ABBYY FineReader 11 Professional Edition [1], which is one of the best OCR programs currently available on the market.

For more accurate results, the ABBYY FineReader uses a machine learning approach that can be trained from a training set of character samples. For some HelloCaptcha types, e.g. all the types in the *Flitter* category and for the *Search Light/Grun Ninja* type, we created a training set to be used in conjunction with the ABBYY FineReaders's existing embedded training database. For types such as types in the *Roller* category and *Swapper/3D*, we only used our own training set. Furthermore, for types that only used a selection of uppercase letters and digits, these were defined as the input language for the OCR.

5 Results and Discussion

An experiment was performed to test the accuracy of our methods. A total of 8,400 animated CAPTCHA samples were collected from the HelloCaptcha website [12] (i.e. 100 samples for each of the 84 different types). Overall, our approach achieves a high degree of accuracy. We can automatically distinguish between the different HelloCaptcha types between 80%−100% of the time. Most types can correctly be distinguished 100% of the time. In addition, the accuracy of breaking the different types (i.e. correctly recognizing all six characters in the animated CAPTCHA challenges) ranges between 16%−100% of the time. As stated in Bursztein et al. [3], a CAPTCHA scheme that can be broken more that 1% of the time is essentially broken. Our experiment was conducted on an Intel Core 2 Duo 3.33GHz PC and the average attack speed was around 4 seconds per challenge, which is well within the length of time that it would take for a normal human to solve the challenge.

5.1 Method of Attack

The PDM method that was presented in section 4.2, was the most commonly used approach to breaking HelloCaptcha. It was used to break 61 of the 84 different types. As CAPTCHAs have to be both usable and secure, designing a robust CAPTCHA scheme is a nontrivial task. From a usability standpoint, most CAPTCHAs are designed in a way where the important information is emphasized in order for the human brain to identify the main information required to solve the CAPTCHA. In a number of animated CAPTCHAs, the important information is emphasized by displaying it for longer periods of time. This is a fundamental problem that can be exploited using the PDM method. If one were to change the animated CAPTCHA design by displaying all information for equal lengths of time, this may severely impact the usability of the animated CAPTCHA, as no information really 'stands out' for a human user.

We observe that the PDM method can also be used to attack animated CAPTCHAs other than HelloCaptcha. Similarly the CL approach can also be used to break other animated CAPTCHA schemes which have properties whereby characters only move in the vertical direction. Other methods such as the CS, FS and RS methods, previously discussed in section 4.2, are possibly less general and may only be effective against specific animated CAPTCHAs.

5.2 Security Issues

One of the primary security issues in HelloCaptcha is that it was not designed with the segmentation resistant principle in mind. As such, individual characters can easily be extracted from the animation frames by exploiting a number of key features. Other than the segmentation resistant principle, several other security issues affect the overall robustness of a CAPTCHA scheme and should be considered when designing animated CAPTCHAs. We highlight a number of these issues in this section.

- **Number of characters and positions.** The PDM method is most effective for animated CAPTCHAs where the characters appear at fixed locations against a moving background/foreground. This is because the characters are displayed at relatively fixed locations for longer periods of time compared to the constantly changing noise. Furthermore, HelloCaptcha challenges always contain six characters which makes it predictable and easier to break. The issues of fixed length CAPTCHA challenges that appear at fixed locations are

(a) Frame 45. (b) Frame 82. (c) Combined image.

Fig. 13. Example where 2 halves of the challenge always appear in the specific frames

not unique to animated CAPTCHAs, it has previously been highlighted that this is a design issue that affects the robustness of CAPTCHAs in general [15].

- **The use of color.** The use of color is another common issue that has to be considered in the design CAPTCHAs, whether animated or otherwise. The incorrect use of color can negatively affect the security of a CAPTCHA, and may possible even affect the usability [16]. In a number of HelloCaptcha types, color played a negative role in the security of the CAPTCHA as important information could be obtained by identifying the different colors.
- **The number of frames.** The number of frames used in the animation is an issue that is unique to animated CAPTCHAs. A fixed number of frames can have a negative impact of the security of animated CAPTCHAs. In our study, we used the number of frames as one of the factors to differentiate between the various HelloCaptcha types. Furthermore, specific frames may be used to obtain certain pieces of information. For example, by evaluating 1,000 samples of the *H-mover/Default* type, we observed that it always uses a the total of 123 frames. In addition, frame number 45 always contains the first three characters and frame number 82 always contains the last three characters. This can be seen from the examples shown in Fig. 13(a) and Fig. 13(b) respectively. We can simply combine the first half of frame 45 with the second half of frame 82 to form the image shown in Fig. 13(c). Therefore, to improve the security of an animated CAPTCHA scheme, it is advisable to use a variable number of frames with information randomly distributed over the frames, even if this means increasing the complexity of generating the animated CAPTCHAs.
- **The frame delay.** If care is not taken, the frame delay can also be exploited to break animated CAPTCHAs. For example, in the *Swapper/Default* HelloCaptcha type, the correct solution is contained in the frame which has the longest delay time. Fig. 14(a) shows two frames with a frame delay time of 40 ms, note that the solution is only partially visible in these frames. However, Fig. 14(b) shows the frame with a frame delay time of 200 ms. One can see that the complete solution is clearly visible in this one frame. In such cases, the addition of the time dimension is completely redundant.
- **Method of delivery.** Animated GIF files are effective on the Web when used for small animation as such as animated CAPTCHAs because of its low implementation cost and it works on most browsers without the need for special plugin applications.However, it is also easy for attackers to download animated GIF files to extract and analyze all the animation frames. It has been argued that video streaming an animated CAPTCHA that is rendered on the client's browser is a more secure delivery mechanism [11].

(a) Examples of two 40 ms frames.

(b) A 200 ms frame.

Fig. 14. Example where frame delay time can be exploited

6 Conclusion

In this paper, we examined the robustness of an animated CAPTCHA scheme. The addition of the time dimension is meant to increase its robustness against automated attacks. However, this paper has demonstrated that if not designed well, the addition of the time dimension does not make the animated CAPTCHA more secure. We showed that simple but effective attacks can be used to break animated CAPTCHAs by exploiting certain characteristics in the CAPTCHA design. Our methods can successfully break all the different HelloCaptcha types with a high success rate. It is likely that our approach can be extended to break other animated CAPTCHA schemes.

References

1. ABBYY. ABBYY FineReader, `http://finereader.abbyy.com/`
2. Athanasopoulos, E., Antonatos, S.: Enhanced CAPTCHAs: Using Animation to Tell Humans and Computers Apart. In: Leitold, H., Markatos, E.P. (eds.) CMS 2006. LNCS, vol. 4237, pp. 97–108. Springer, Heidelberg (2006)
3. Bursztein, E., Martin, M., Mitchell, J.C.: Text-based CAPTCHA Strengths and Weaknesses. In: Chen, Y., Danezis, G., Shmatikov, V. (eds.) ACM Conference on Computer and Communications Security, pp. 125–138. ACM (2011)
4. Chellapilla, K., Larson, K., Simard, P.Y., Czerwinski, M.: Designing Human Friendly Human Interaction Proofs (HIPs). In: van der Veer, G.C., Gale, C. (eds.) CHI, pp. 711–720. ACM (2005)
5. Chow, Y.-W., Susilo, W.: AniCAP: An Animated 3D CAPTCHA Scheme Based on Motion Parallax. In: Lin, D., Tsudik, G., Wang, X. (eds.) CANS 2011. LNCS, vol. 7092, pp. 255–271. Springer, Heidelberg (2011)
6. Cui, J.-S., Mei, J.-T., Zhang, W.-Z., Wang, X., Zhang, D.: A CAPTCHA Implementation Based on Moving Objects Recognition Problem. In: ICEE, pp. 1277–1280. IEEE (2010)
7. Fischer, I., Herfet, T.: Visual CAPTCHAs for Document Authentication. In: 8th IEEE International Workshop on Multimedia Signal Processing (MMSP 2006), pp. 471–474 (2006)
8. Li, S., Shah, S.A.H., Khan, M.A.U., Khayam, S.A., Sadeghi, A.-R., Schmitz, R.: Breaking e-Banking CAPTCHAs. In: Proceedings of the 26th Annual Computer Security Applications Conference, ACSAC 2010, pp. 171–180. ACM, New York (2010)
9. Mori, G., Malik, J.: Recognizing Objects in Adversarial Clutter: Breaking a Visual CAPTCHA. In: CVPR (1), pp. 134–144 (2003)

10. Naumann, A.B., Franke, T., Bauckhage, C.: Investigating CAPTCHAs Based on Visual Phenomena. In: Gross, T., Gulliksen, J., Kotzé, P., Oestreicher, L., Palanque, P., Prates, R.O., Winckler, M. (eds.) INTERACT 2009. LNCS, vol. 5727, pp. 745–748. Springer, Heidelberg (2009)
11. NuCaptcha Inc., NuCaptcha, `http://www.nucaptcha.com/`
12. Program Produkt, HelloCAPTCHA, `http://www.hellocaptcha.com/`
13. von Ahn, L., Blum, M., Hopper, N.J., Langford, J.: CAPTCHA: Using Hard AI Problems for Security. In: Biham, E. (ed.) EUROCRYPT 2003. LNCS, vol. 2656, pp. 294–311. Springer, Heidelberg (2003)
14. Yan, J., Ahmad, A.S.E.: Breaking Visual CAPTCHAs with Naive Pattern Recognition Algorithms. In: ACSAC, pp. 279–291. IEEE Computer Society (2007)
15. Yan, J., Ahmad, A.S.E.: A Low-Cost Attack on a Microsoft CAPTCHA. In: ACM Conference on Computer and Communications Security, pp. 543–554 (2008)
16. Yan, J., Ahmad, A.S.E.: Usability of CAPTCHAs or Usability Issues in CAPTCHA Design. In: Cranor, L.F. (ed.) SOUPS. ACM International Conference Proceeding Series, pp. 44–52. ACM (2008)

Contextual OTP: Mitigating Emerging Man-in-the-Middle Attacks with Wireless Hardware Tokens

Assaf Ben-David[2,*], Omer Berkman[3,**], Yossi Matias[1], Sarvar Patel[1], Cem Paya[1], and Moti Yung[1,4]

[1] Google Inc.
[2] The Hebrew University, Jerusalem, Israel
[3] The Academic College of Tel Aviv Jaffa, Tel Aviv, Israel
[4] Columbia University, NY, USA

Abstract. OTP (One Time Password) devices are highly deployed trust enhancing (password entropy increasing) devices which are used to authenticate a user with a second factor (a pseudorandom sequence of digits produced by a device the user owns) and to cope with off-line phishing of password information. Wireless connection adds usability to OTP protocols in an obvious way: instead of the person copying the information between machines, the wireless (say, Bluetooth) mechanism can transfer the value directly. Indeed, OTP devices implemented in a smartphone and communicating with the browser over Bluetooth can act in usable fashion (and this extension was implemented in our organization and got very positive usability feedback). What we then noticed as a key observation is that this mode of OTP wireless transfer has turned the "man to machine" nature of the OTP tokens to a "(mobile) device to machine (the browser on the computer)" method, so we can now employ protocols between the two interacting computers. Thus, we asked what can this new mode contribute to security (rather than to usability only) and cope with increased set of attacks. Specifically, the question we are dealing with is whether wireless OTP devices (i.e., smartphones) can be hardened at a reasonable cost (i.e., without costly OTP infrastructural changes, public-key infrastructure/ operations, and with small modification to browsers) so as to be useful against one type of interesting and currently growing and highly publicized Man in the Middle (MITM) attacks. The work herein summarizes our study which is based on our proposed new notion of Contextual OTP (XOTP for short), which exploits session contexts to break the symmetry between the "user-MITM" and the "MITM-server" sessions.

1 Introduction

Users of Internet services are constantly under attack. For example, it is estimated that between 300 to 600 million phishing ([10]) emails are sent every day

* The work of this author was carried out during internship in Google.

** The work of this author was carried out during a sabbatical in Google.

F. Bao, P. Samarati, and J. Zhou (Eds.): ACNS 2012, LNCS 7341, pp. 30–47, 2012.

all over the world ([27]). Many of these phishing attacks are targeting users' accounts such as bank accounts and web services accounts (see the information on the increasing threat of phishing and other crimeware, such as the 2010 2d half report by the Anti-Phishing Working Group ([2]).

To mitigate such attacks and enhance security mechanisms in general, institutions from banks to universities alike, have required internal users to employ two-factor authentication schemes using one time passwords (OTPs). Unfortunately, phishers are adapting to the use of OTPs by using real-time man in the middle (MITM) attacks (e.g. [4,24,25]) and recently attacks of this type have intensified ([8]). In these attacks the phisher exploits the same phishing infrastructure as in offline attacks (so there is no need to build costly mechanisms and therefore it is the next natural step for an attacker). The phisher then gets the user's name and password as well as the user's current OTP over the web and use them in real time to access the user's account at the real server. Thus, while the OTP may guarantee to the server that the user is involved, it does not guarantee that the party it communicates with directly is indeed the user and not an MITM attacker.

Another set of MITM attacks which has been on the rise, involves corruption of "trusted" third parties, like Certification Authorities, obtaining faked certificates, and furthermore controlling and abusing the DNS infrastructure (the most recent such attack being "Operation Black Tulip" which was applied after compromising the Dutch DigiNotar CA [17] which went out of business). With such an attack, the user may employ the correct server's URL under HTTPS and still communicate with the MITM attacker rather than the real server. These attacks thus pose a real and significant threat to any transaction performed over the Web, and in particular to users living under repressive regimes (where the regime controls and manipulates the infrastructure). A very recent commercially motivated case is the CA Trustwave which sold companies "man in the middle" certificates(!!), so they can eavesdrop on their employees [15].

Our Goal: We aim at introducing a new notion, the XOTP, implementations of which we piggyback on and augment the OTP mechanism when run over wireless communication to mitigate MITM attacks. This demonstrates the added power of combining wireless communication and hardware tokens. Our approach is to minimally change the OTP mechanism/ function (and not change its infrastructure), which leads to elegant solutions, and, in turn, do not require much added investment.

The schemes we propose and their extensions are based on the basic notion that MITM type attacks can be mitigated by *cryptographically entangling the OTP with a mutually recognizable session context* (where context can be a combination of URL, server's certificate value, session key, etc. which can be utilized by the machines at both ends of the session), subject to this context being different when an attacker stands in the middle. Such a context can thus serve as a new type of authentication factor, and we refer to it as a **contextual factor** (since it represents the agreed-upon session's context to both session's parties); the resulting OTP is denoted **contextual OTP** or **XOTP**.

Ideally, the XOTP assures **(1) correctness:** When there is no attack, the XOTP computed by the client will match the one computed by the server; **(2) mitigating real-time phishing:** if a user is phished, the XOTP it computes will reflect the context of the user's session with the phisher, while the server expects an XOTP based on the context of its session with the phisher which is different.

Results: We first present the above notion in the abstract idealized setting in its most simple form. We then instantiate it in protocols with various combinations of actual contextual information available to web sessions with increasing degree of complexity. This demonstrates how as the combination of contextual factors gets stronger, more powerful attacks can be foiled: From the most common attack which tricks the user into mistaking the phisher site with that of the desired server, through an attacker that also controls the DNS infrastructure, and up to an attacker that additionally holds a fake server certificate. We note that in addition to breaching a CA as was done in the DigiNotar case, there are other means for an attacker to get the server certificate. Examples include CA issuance mistakes, CA colluding with the attacker (Trustwave) and CAs run by governments. Beyond our primary MITM attacks, to some extent, contextual factors even offer mitigation from simple forms of Cross Site Scripting (XSS) attacks. We finally show that further extensions of the scheme to authenticate transaction data further copes with malware and general XSS attacks.

Methodological Remarks: On a general methodological level, we note that XOTP demonstrates the power of designing for the specific channels employed. This enables solutions to problems that are typically unsolvable in an abstract channel setting (i.e., MITM attacks) due to the specific channel setting and the concrete technology advancements. We further note that while we define here the notion of contextual factor to describe binding of strong keyed cryptographic function at the OTP device to "an implicit context of the session/ transaction agreed by both sides," the general notion of *binding* variables representing the "current setting" to the cryptographic primitives has been used before in many cryptographic protocols, primarily to break symmetries. Perhaps password binding to the URL, presented in various works [28,7,22] is the closest application, but this was done for a different reason: mainly to diversify the password (in order to specialize the password to the URL and prevent password re-use across sites, which is a major source of password leaks), and with different cryptographic setting (key-less hashing techniques). Other such examples exists as well in two-party authentication protocols, etc. We also note that exploiting smartphone devices in order to cope with various phishing attacks has been suggested (e.g., [16,13]) but not as an extension of and on top of the OTP infrastructure, and not against an MITM attacker which uses fake server's certificates, and therefore not as a minimal elegant extension of the existing mechanism. We cover related work in more details in a special section.

2 The Contextual Factor Protocol

2.1 The Setting

The basic login scenario is where a client C wishes to login securely to a server S over session $Sess$ by using her browser. The scenario is such that protocols start with C providing to the server her claimed identity which is usually a user name and a password. Since any protocol starts like this, assume (w.l.o.g.) that S already knows the claimed name C of the client, and we only concern ourselves with the one-time password part of the protocol.

We assume that the user is equipped with a **smart device** containing (i) an application which implements the scheme; (ii) a secret key shared with the server which is used by the application to compute a pseudorandom function used in the scheme; and (iii) a communication capability between the browser and to the smart device. The system would also require provisioning procedures, establishing a new shared key and/or application (in case the key needs to be revoked, the smart device is lost, etc). Additionally the system needs a browser extension. We note that the system we suggest is generic and allows secure communication with multiple parties. We hope that the browser extension will eventually be an integral part of the browser; as such, the use of white-lists is not practical. Application and key setup as well as update have been discussed in similar scenarios before (e.g., [16,13]), and are already part of the OTP infrastructure; therefore we will not discuss them further.

We start by designing a protocol which captures an abstract idealized notion of the above scenario in which the user side is assumed to be a single entity and factors of the right form are assumed to be available . This allows us to ignore issues relating to secure flow of information within the client side (between the user, smart device and browser) as well as issues and attacks relating to the specific contextual factors used. We will deal with these issues later.

Suppose the client is being authenticated on session $Sess$ (by the server or by the MITM attacker masquerading as the server). We define some symbols to be used:

– $\mathbf{key_{S,C}}$ is a secret key shared between S and C (in C's device).

–$\mathbf{Chal_{S,C}}$ Synchronized, non-secret, non-repeating challenge between server S and client C (e.g., time and/or counter based challenge).

–$\mathbf{PRF(key, data)}$ a Pseudorandom Function (with truncation) algorithm applied to the given data using the given key. The result of the PRF is assumed to be hard to guess, i.e., we make the standard assumption that the PRF output is unpredictable. We assume that for any PRF input parameter that we employ throughout and given access to the history of previous outputs (plausible history of polynomial length) the probability of guessing the next output is bounded by a small probability ϵ (which we will employ throughout).

–**ID** stands for user name, password, account number or any other identity-related parameter or combination of such parameters. It may also be empty.

–**Nonce** Represents server and/or client nonce(s) which is part of data to the PRF. It may also be empty.

–"||" means concatenation (as an example of a one-to-one reversible binary function).

2.2 The Current OTP Protocol

The following exchange denoted as the **current OTP protocol**, generalizes many of the currently used OTP mechanisms.

- C computes $OTP = PRF(key_{S,C}, Chal_{S,C}||ID||nonce)$ and sends it to S (see remarks below).
- S verifies the received OTP by applying the same computation.

The Attack Considered: It is immediately obvious that a MITM attacker who impersonates as the real server, gets the OTP and can forward it to the real server, thus impersonating as the user.

2.3 The Contextual Factor Protocol

The goal of this section is to describe and motivate our notion and its simple yet powerful properties by considering an abstract simplified setting of idealized factors.

A contextual factor of a session $Sess$ is a string denoted XF_{Sess} which satisfies the following "factor requirements:"

1. **Recognition:** Both sides of the session $Sess$ recognize (the same) XF_{Sess}.
2. **Uniqueness:** For any two different sessions, $Sess1$ and $Sess2$, it holds that $XF_{Sess1} \neq XF_{Sess2}$.

Protocol:

1. C computes
   $XOTP = PRF(key_{S,C}, Chal_{S,C}||ID||nonce||XF_{Sess})$ and sends the computed $XOTP$ to S.
2. S verifies the received $XOTP$ by applying the same computation.

Correctness: In, both, the current protocol and the contextual factor protocol the challenge $CHAL_{S,C}$, the ID, the nonce and the contextual factor XF_{Sess} (in the contextual-factor protocol) are known to both parties: The challenge has been assumed to be synchronized, the ID can be inferred from the claimed user name, the nonce if exists is sent over, and the contextual factor is assumed to be known to both parties. Thus both sides apply the same deterministic computation employing the shared key which obviously produce the same result.

Security: As already noted, the standard protocol is susceptible to a relay attack by a MITM attacker P with probability 1.

On the other hand, in the contextual-factor protocol such relay attack will be rejected with high probability: Denote the session between the client C and the attacker P by $Sess1$ and the session between P and the server S by $Sess2$.

Let H be the history of XOTP's communications observed by P till these sessions. Denote the XOTP sent to the attacker by the client by $XOTP_1$ and the XOTP expected by the server by $XOTP_2$. P needs to compute $XOTP_2$ based on his current history $H||(XF_{Sess1}, XOTP_1)$. However, because of the uniqueness requirement, XF_{Sess1} as well as all contextual factors in H are different than XF_{Sess2}. Thus, by the PRF definition (following [12]), the probability of P to successfully produce/ guess $XOTP_2$ is bounded by ϵ.

Remarks:

(i) We do not require that the user be aware of the value of the contextual factor.

(ii) Instead of modifying the OTP algorithm, it is also possible to entangle the contextual factor with the resulting OTP by using an additional PRF computation within the smart device.

(iii) In our realization of the protocol, we do not need to truncate and format the PRF output, since user involvement is minimal.

(iv) When the user is involved (see discussion in Section 4.1) or if compatibility demands this at the server side, the PRF result is further processed and truncated. For example, from the 160-bit output of HMAC ([18]) which is used in the HOTP algorithm ([20]) between 6-8 decimal digits are extracted. It is assumed that the result of such processing and truncating gives a **pseudorandom output**, so that if one knows $Chal_{S,C}$, ID and the nonce, still calculating the response is hard (for an 8 digit pseudorandom value response we can assume $\epsilon = c/10^8$ for c being a constant very close to 1 in case of complete synchronization between the parties or a larger constant depending on the window of synchronization). The security definitions of PRF allows the adversary to first observe a polynomial size history of input output pairs, before guessing the value of a new input (in fact, before being able to tell this value from a randomly chosen one), see, e.g. [12].

(v) In HOTP the challenge is an 8-byte counter which is synchronized between the client and server (so it does not need to be sent between them) and no ID or nonce are involved. The standard mentions as possible extensions adding to the challenge identity-based parameters (such as address and PIN) or time. By adding ID and allowing $Chal_{S,C}$ to include both counter and time, our OTP definition (and the way we defined ϵ) include such extensions. We also allow using server and/or client nonce(s) (but in case a nonce is indeed used it needs to be sent to the other side).

(vi) In case the ID includes the user's password (or other secret information), one may consider requiring that the user enter it into the smart device rather than into the browser (as suggested in a similar context in, e.g., [13]). Note, however, that this hurts usability and does not add much security as the password can still be phished by clever wording on the browser.

Now that the idea has been motivated, the next two sections will demonstrate, first, the various ways to implement XOTP from concrete contextual factors, and then the robustness of the XOTP notion and how it can be used in conjunction with additional techniques (which are realizable on smartphones and are implementable as support and extensions of OTP software) to deal with extended set of attackers exploiting typical weaknesses of web technologies.

3 Possible Contextual Factors

We present several candidates for a contextual factor and the protection each of them gives against **Man-In-The-Network (MITN)** attacks, still relating to the client as a single entity. Because none of the candidates for contextual factors fully satisfy the second factor requirement, i.e., uniqueness, we distinguish between two types of MITN attacks (extending the discussion beyond the simplified real time attack considered above), according to the status of the contextual factor.

- In a **Different-XF** attack, the contextual factor in the session between the user and attacker is different than that in the session between the attacker and server.
- In a **Same-XF** attack, the contextual factor in the session between the user and attacker is identical to that in the session between the attacker and server.

Different-XF attacks represent the majority of MITM attacks. For example, when the contextual factor used is the URL (see below), the attack would be of the Different-XF type whenever the URL used by the attacker is different from that of the server - as is the case in most phishing attacks. As shown in Section 2.3, attacks of the Different-XF type, will fail when using the contextual factor protocol. We thus only consider the more advanced attacks - those of the Same-XF type. We start by discussing the use of the URL as a contextual factor. Although the URL turns out to be quite strong for dealing with several types of attacks, other types require further protection, and the respective contextual factors providing that protection will follow. Finally, we discuss the practically of using client and server IP's as contextual factors and mention some additional possible factors.

3.1 URL

We first note that the URL satisfies the first factor requirement - recognition - in the definition of a contextual factor (i.e., both sides to a session know the URL). Requirement 2 - uniqueness - is not always satisfied, e.g., by the attacker contaminating DNS entries so that the client uses the correct server's URL but is routed to the attacker.

Fortunately, such attacks will not work if the server requires its session with the client to be secure (HTTPS), which secure protocols should indeed require! The reason is that because in this case the browser uses the correct server's URL, it also uses the correct server's certificate. Thus, assuming the MITM attacker cannot break the server's private key but with probability $\delta << \epsilon$, he is with high probability no more than an observer of encrypted traffic so such an attack will indeed fail.

Because in virtually all implementations of a two-factor authentication scheme the server requires secure session by SSL/TLS, it follows that using the URL as a contextual factor in these server-certificate based sessions, fully mitigates both attack types mentioned above.

Remarks:

(i) Observe that although the URL is the only factor used in the XOTP computation, we implicitly use in this case, a contextual factor based on the URL and the session key wrapping it inside. We denote this implicit factor as a **nested URL** or in the general case a **nested contextual factor**.

(ii) It is important to note that insistence on a secure session would not help when OTP is used rather than XOTP since when the OTP is used, the MITM attacker will get the correct OTP from the user client (on a secure or insecure session between the client and attacker), and relay that OTP to the server on a secure session between the attacker and the server.

(iii) The URL is a good suggestion for a contextual factor since it is a built-in factor in the infrastructure which is extremely versatile as it may encode almost any additional information as may be required. For example, the server may incorporate into the URL transaction data, messages to the smart device, and more.

3.2 Server's Certificate

As with the URL, when the server's certificate is used as the contextual factor, recognition is satisfied but uniqueness is not in general satisfied since an attacker may present the server's certificate as well as contaminate DNS entries. Also, as with the URL, such attack won't work if the server requires its session with the client to be secure (HTTPS) because the scheme uses implicitly a nested contextual factor (the SSL/TLS session key wraps the server's certificate).

Additionally, using the server's certificate as the contextual factor foils an attack in which, in addition to contaminating DNS entries, the attacker uses a fake certificate of the server (issued unwittingly by a "trusted CA"). This is so since the real server expects the XOTP to be based on its own certificate rather than on the faked one (the probability of the attacker getting an employable faked certificate with the same public key as the original certificate, amounts to breaking the original public key, which is a negligible probability event).

It is important to note that a browser extension implementing such a scheme requires enforcing that the same certificate used to calculate the XOTP is also used to setup the SSL/TLS connection used to transfer the XOTP to the server. Without such an atomicity condition, an attack is possible where 1) the MITM attacker uses the server's real certificate on a first SSL/TLS connection to make the browser extension calculate the XOTP and then 2) switches to a different SSL/TLS connection using a fake server certificate, over which the XOTP would be sent to the attacker.

3.3 URL + Server's Certificate

The server's certificate can be used together with the URL which would allow also the inclusion of transaction data. In general, combining several contextual factors, gives the **accumulative benefits of all the factors**.

3.4 Session Key

Suppose any of the SSL/TLS session keys with the server (or a pseudo random function of it) is available to be used as a contextual factor in addition to the server's certificate (and possibly also the URL). This is an alternative way to foil the previous attack using a faked certificate because the XOTP created by the browser extension and sent to the attacker (using one SSL/TLS connection) will be different than the server expected XOTP for a second SSL/TLS connection used by the attacker to the real server.

In addition to foiling the previous attacks, this foils an additional attack where a malicious JavaScript attack on the page (see, e.g., [9]) steals the XOTP from the page and sends it to a malicious attacker which can then serve as a client and authenticate as the user. Such an attack is foiled because the server expects that the XOTP be based also on that fresh session key but the session key between the attacker and the server would be different (w.h.p). In Section 4.3 we show how to deal with such an attack when session keys are not available to be used as contextual factors. Note that using session key can somewhat increase failure probability because a SSL/TLS connection can be restarted with a different session key during the XOTP protocol for perfectly innocent reasons (violating the atomicity requirement). However, since the duration of the protocol is small we expect the increased failure risk to be small as well.

Note that we consider here XSS attacks on the authentication protocol and not attacks on already authenticated sessions [11] (but we consider such attacks later together with general malware attacks). Indeed, there may potentially be other ways in which a malicious script may abuse the session.

3.5 IP

In some cases client IP or server IP address (or both) may serve as contextual factors provided they can satisfy the first factor requirement of recognition. Recognition of the same IP address on both sides may be complicated by the fact that intermediary devices may be changing the IP address. The client IP can be used only if it is known that it has not been modified (by e.g., proxies). The server IP can usually be used, when it can be assured that the server knows its own IP as seen by the client which may not be the case when IP anycast or load balancers are used. Sometimes the whole IP address cannot be used but a part of it can (e.g. high bits).

In practice, a server's IP address is better to utilize since a server would generally know if its IP address (or its high bits) is known to a client and only then inclusion of the IP address in the XOTP calculation be requested of the client. Another alternative to using high bits, is to use server's location as known by mapping the server's IP address to an approximate location.

We note that when IP may not be in exact agreement between the parties, then it can be sent on the clear (like a nonce), to allow the other party to check that the IP used is approximately the agreed upon value (e.g., shares the same domain). This principle of sending values on the clear regarding parameters that are approximately in agreement applies generally.

To examine Requirement 2 - uniqueness - as well as attacks on such a system, let us assume that the Server IP is used as a contextual factor, and that the server only supports HTTPS sessions (or uses different IPs between HTTP and HTTPS sessions). First note that, as before, uniqueness is not satisfied in general, since the MITM attacker may subvert the routing infrastructure. Moreover, it turns out that because the server's IP address is not strongly correlated with the server's certificate, the MITM attacker can complete an attack against such a system: The attacker presents the client with a wrong URL (he cannot present the user with the correct server's HTTPS URL since this would cause the session to be secure). He also modifies DNS entries so that on client request, the server's IP would be returned (this is required, because the correct server's IP is associated with a server's HTTPS URL). Additionally, because the correct server's IP is used, the attacker must also subverts the routing infrastructure so that connections to the server's IP would be routed to the attacker. Clearly such an advanced resource-heavy MITM attack requires big efforts, and can be mitigated by, e.g., using DNSSEC ([19]).

3.6 Other Factors

Time and location and other "physical factors" suggested for relay attack prevention, can in principle (but generally not in practice) serve as contextual factors. See also section 5.

4 System Setting Considerations and Extended Attacks

In this section we extend our discussion to include the actual client environment (Section 4.1 where we employ tools available in modern smartphones) and consider additional attacks (Section 4.2 and Section 4.3). Finally the table in Section 4.4 summarizes our results.

4.1 Secure Communication of the XOTP in the Client's Environment

Recall that the contextual factor protocol in Section 2.3 and its analysis as well as the discussion of the URL as a contextual factor, both, assume that the client is a single entity and thus ignore issues related to the flow of information (specifically, the contextual factor and the XOTP) within the client environment. In this section we propose a technological solution to these issues. We assume throughout this section that the contextual factor is the URL and that the server requires that its authentication session with the client be secure (HTTPS). We also assume that the attacker has no control over the client environment (i.e., no XSS attack or other code breaches) and does not posses a fake server certificate (which as mentioned in Section 3.2 requires the use of more advanced contextual factors and further work by the browser extension). We discuss such attacks further in Section 4.3.

As mentioned in Section 2.1, we utilize a generic browser extension (which we hope to be eventually incorporated into the browser itself). The extension communicates with the smart device through Bluetooth (or any other secure channel). The extension adds itself into a page that contains a specific string (such as a field named "XOTP") waiting for a predefined event (clicking a predefined button added by the extension, clicking an XOTP field, submitting the page, etc.). When the event occurs, the extension sends the page URL to the smart device and copies the received XOTP into a specified field in the page. The extension can either submit the page or let the user (or code on the page) to submit it later.

In case of a Different-XF attack this XOTP would be sent to the attacker (and be exposed to the attacker's code) but, as argued before, it would be different from the value expected by the server in its session with the attacker. Because the attacker has no control over the user environment, he cannot trick the browser to request XOTP computation of a contextual factor of his choice. In the case of a Same-XF attack or simply a session with the genuine server, the page comes from the server and we assume that it is safe. (We ignore attacks which just spoil the XOTP computation since they only provide denial of service which a MITM attacker, given its location in the middle, can easily achieve by other means.)

We note that communication between the smartphone device and the browser (rather than a manual typing) neutralizes a (real-time) shoulder surfer attacks as well [23].

Below we discuss other methods for communicating the contextual factor and XOTP in the user environment that do not rely on direct communication between the browser and device. However, these methods involve the user and are thus vulnerable to direct phishing (social engineering) attacks on the user: In the **Manual** method the browser generates a (short) hash of the URL (so the XOTP would be based on the hash of the URL rather than on the URL itself), the user types it into the device, and then types the result back to the browser. With the **QR-code** method the browser displays a QR-code of the URL, the user scans it with the smart device, and types the result into the browser (this is similar to [14]).

Even without considering the user's last action of typing the XOTP to the browser, both the Manual and QR-code methods are much more susceptible to an attacker's Different-XF attack than the communication channel method since the user involvement is increased: Denote by URL_{PS} the URL in the attacker's session with the server and by URL_{PU} the URL in the attacker's session with the user client. The attacker may trick the user into initiating the authentication procedure on the user's session with the attacker. In the case of the Manual method the attacker could then display the hash of URL_{PS} (instead of the hash of URL_{PU}) and in the case of the QR-code method, the attacker can display the QR-code of URL_{PS} (instead of the QR-code of URL_{PU}). In both cases the XOTP that would be sent to the attacker allows him to then masquerade as the user. Thus, if these methods are employed, careful user guidelines are needed to bind the URL used and the page responded to in a clever way. Moreover, usability of both methods above (in particular of the Manual one) is inferior to that of using Bluetooth. On the other hand, Bluetooth requires a one-time pairing procedure between the user's machine and the smart device.

Reliable Communication Requirement: We note that the communication between the device and the computer has to be active which in modern computers is the case. In case there are communication problems another access method may apply, such as an OTP by SMS or by calling a help desk (where the parties are now aware they are exposed to potential MITM attack in that specific session). Note that similar issues can occur with legacy OTP devices. Such issues are sometimes handled by password reset type mechanisms (which we mention in Section 4.3 below).

4.2 Routing to a Different User Registered Website

A user is typically registered with multiple web sites and each may have its servers expecting an XOTP protocol to be satisfied before entrance is granted. It is always possible for a MITM attacker to switch a client's connection from one intended 'user-registered web site' to another simply by sending a redirect message or by acting as a MITM actually relaying all messages back and forth between the client and the server. However, this is not too damaging because the SSL connection would be extended from the client to the server and the MITM attacker would not be able to eavesdrop or effectively intervene in the communication. Secondly, the client should eventually realize that it is at a wrong web site and log off. In fact, such an attack is no different than an accident where the user logs into one web site and only afterwords realizing that she really meant to log in to another.

4.3 Additional Types of Extended Attacks

We consider two more types of attacks which are MITM-variants: (1) **Man-In-The-Browser (MITB)** in which the attacker has installed malware on the user's machine, and (2) a separate case of **Man-In-The-Device (MITD)** in which the attacker has malware in the user's smart device.

We start by describing extensions to the protocol. In Section 3.4 we have shown that using a session key with the server as the contextual factor prevents an XOTP stolen from the page and sent to the attacker by malicious JavaScript code, from being used to authenticate as the user (due to a fresh key being generated in each session). If, however, a session key with the server is not available (or the failure rate as discussed in Section 3.4 is too high), we can use instead a session key established between the server and the phone. Note that this variant of the protocol also solves the situation of SSL re-binding attacks [29].

More generally, we describe below two extensions to the protocol, noting that the session key between the server and phone can be taken from the result of the two-way authentication described first (which is an efficient way of implementing this; we are fully aware that other two party protocols can be employed):

- *Two-way authentication:* The contextual factor protocol allows the server to authenticate the client but does not prevent MITM attack in the strong sense. It only ensures that no matter what information was obtained by the attacker, this information cannot be used to masquerade the user to the (specific) server. When a smart device is used, it is easy to extend the scheme and authenticate the server as well (by a flow that carries a new returned-XOTP from the server to the device). The smart device could then warn the user if the server has not been authenticated within a very short period of time following client authentication.
- *Direct communication channel with the server:* The smart device (having, in fact, a general computer capability) may communicate directly with the server. For example, when using the URL as a contextual factor, the device may access the server directly via a second URL which is a default variation on the original URL used by the browser. We may cryptographically secure the original URL (using, e.g., PRF of the URL with the shared key) to ensure that the smart device only visits genuine server sites. Using this channel, (secured) transaction data or other information may be sent to the device without involving the browser. Note that the contextual factor **must be delivered to the smart device from the user's machine** and not from the server as it needs to represent the session in which the user client takes part.

Man-In-The-Browser (MITB) attacks:

- **MITB keyloggers:** When using Bluetooth, the scheme withstands the popular MITB keyloggers as the XOTP is not typed by the user.
- **General MITB:** The scheme as described does not protect against a **general attacker** that totally controls the browser or the machine since a MITB could eavesdrop or insert messages in the SSL session or issue requests. Further, an XSS attack post initial authentication can have this nature as well [11]. Because the scheme can additionally reflect context, it can, in addition to user initial authentication, provide **transaction authentication** by incorporating transaction data into the computation of the XOTP, and use this data for authentication of transaction steps (even user initial authentication can be handled as a random transaction, btw). The interaction will require a few XOTP computations by the device and the server to assure transaction safety; from the user point of view, she will have to check the display on the device and approve the transaction. For example, when transferring money between accounts, the account numbers and amounts involved

can be part of the data which is also integrated into the XOTP computation (in the right order and tagged with the field name to avoid "shuffling of information attack"). The flow of a transaction can go as following: (1) transaction data with initial XOTP should be sent to the server, and then (2) this information should be sent from the server to the device with server returned-XOTP (the XOTP is computed differently than the initial XOTP and the sending is encrypted end-to-end between the server and the device), (3) it should then be displayed on the device for user confirmation, then after confirmation (4) an additional final XOTP (computed differently than the other two above) is computed and sent to the server, closing the loop on (and effectively committing to) the transaction.

Man-In-The-Device (MITD) Attacks: The attacker can get access to the long term shared key between the device and the server and thus can create any XOTP it desires. Even without access to the long term key, the attacker can request an XOTP for any URL of its choice. However, the attacker will still not be able to complete the protocol without knowing the other factor (i.e., the secret password). Such an attack demonstrates that having a strong second factor does not diminish the importance of using also a strong first factor (i.e., choosing a password wisely). The principle of compartmentalizing the various factors is important whenever possible (it also prevents a spurious device from acting on its own) and is advocated here (this principle was not always kept and was given up for other advantages in various places, e.g., [13]; we believe it is important to keep it and our approach of minimalistic extension of the OTP leads to keeping this in a natural way).

Extended System Context: Finally we note that "user sign on" is a procedure within a larger system. In that system several authentication mechanisms may co-exist, and obviously the system can only be as strong as the weakest among them. Thus, our goal in designing the system should be that regardless of which information is "phished" and over which mechanism, the attacker will fail. In particular, it is important to verify that **"password reset mechanisms"** do not undermine the primary authentication scheme, namely that we insist on extra factors even in the reset of passwords procedure, and that we replace authentication devices in a highly secure fashion as well.

4.4 Summary of Results: Real-Time Impersonation and More

The following table summarizes our results. For each attack type (in increasing severity), we give the contextual factor that can be used to mitigate the attack and the technology required on the smart device for implementing it. The term *impersonation* below refers to a MITM attack (because the attacker impersonates as the server). It may be carried out by the attacker using a different URL than that of the server (first case) or the same URL (all other impersonation cases).

Attack type	Factor checked by server	Technology
Impersonation (different URL)	URL	Smart device
Impers. (same URL) + DNS poisoning	URL + HTTPS	As above
Impers. + DNS poisoning + Fake cert	URL + HTTPS + Server's cert	As above
Impers. + DNS poisoning + Fake cert + Simple XSS	URL + HTTPS + Server's cert + Session key	As above
Malware	URL + HTTPS + Transaction data + User confirmation on device to server	Smart device with input capabilities and display

5 Related Work

Although the conceptualization of the term "contextual factor" and its use as suggested in this paper is to the best of our knowledge novel, there is prior work that uses notions of similar flavor (i.e., cryptographic entanglement). Generally speaking, in authentication methods some binding of challenges and history to the response are common place and may be considered the motivation for our notion. Specifically in SSL/TLS [21] a connection is established based on a server certificate whose name must match (part of) the host name. In fact, an excellent way to mitigate many attacks would be to use SSL/TLS two-way authentication rather than just server authentication. However, client-side certificates are hardly ever used in consumer client-server authentication, and in any case cannot be used when the user logs in from an unknown computer (such as an Internet cafe). In this respect, see [16] (discussed below) which suggests to use PKI with client-side certificates on the user's mobile device. It should be noted that such a scheme is susceptible to an attack using a fake server certificate.

Preventing relay attacks: It was shown [3] that relay attacks on physical objects cannot be solved based solely on game theoretic techniques, and a solution was given based on measuring time. Recently, [26] suggested the use of unreliable channels to prevent relay attacks and observed that by measuring time [3] in fact uses time as such an unreleasable channel (other new channels presented are abstract ideas based on physical objects with no electronic realization). Although these results (and others such as [1]) concern physical objects and we are only interested in entity authentication in which cryptography can solve the problem (e.g., two-way SSL/TLS authentication), these works are relevant in the sense that they consider relay attacks as we do and the unreliable channel solutions they suggest are physical factors which are contextual in nature. We note that using time and location as contextual factors requires very refined method of measurement (not commonly available over the web).

Next we note that other mechanisms in past works bear some (at least superficial) similarity to our techniques.

(1) Smart mobiles: [16] and [13] each suggested an authentication scheme based on a smart mobile phone (or other smart device), though they cannot be considered as a minimal enhancement to OTP. In [13] the phone keeps the server's public key and uses it to encrypt the user name and password which the user types into the phone (instead of into the her computer). The encrypted password acts like an OTP. The scheme is susceptible to a MITM attack (it was designed against off line phishing): in case of such an attack, the protocol will end with a secure (SSL/TLS) session between the server and phisher, as well as between the phisher and the browser. Thus, security of further transactions relies on authenticating each transaction by using the session key established between the server and the mobile phone. Moreover, as mentioned above, the system is susceptible to the phisher tricking the user (e.g., to type her password into the browser). [16] basically implements an SSL/TLS two-way authentication protocol with a client certificate stored in the mobile phone. Because it relies on the SSL/TLS protocol it is susceptible to a MITM attack using a fake server certificate as we mentioned above. Additionally, a server in this case needs to remember clients' public keys or to employ the costly PKI operations (like probing revocation lists, or probing online certificate status servers), whereas an OTP based approach only implies a symmetric key per user, which in fact can be derived from a single long term stored master key and the user ID.

(2) URL-based hashed passwords: Entangling of the URL with the OTP mechanism (which is one of our basic contextual factors) resembles the way several publications entangled the URL with the user password (or other user data), though the earlier goals were completely different. In [6] and [5] this entangling is used to allow pseudonymous (anonymous, but still personalized) web browsing, as well as anti-spam, whereas in [28], [7] and [22], this entangling provides (among other things) a unique password per URL to prevent the usage of the same password everywhere (which is an important different concern than the ones in this work). The entanglement in these cases is by non-keyed cryptographic hashing (while we augment the OTP pseudorandom keyed function) and the entanglement does not prevent off-line dictionary based password cracking by an eavesdropper, say.

We note again, in light of the earlier works, that none has attempted the following experiment: move OTP to use wireless channel and seek minimal addition to the mechanism (essentially function code change) to cope with as many as possible rising threats. This approach forced us to consider the various changes to the OTP function incrementally, and to consider the change against the exact threat this hardening mitigates.

6 Concluding Remarks

OTP has been known to be superior to password-only login since it adds entropy to the login procedure, which is well understood by now. Here the principle of "added entropy" of the OTP devices is followed by a new principle of "added context" (i.e. a cryptographic computation does not only add entropy but is also capable of binding the factor to a specific context). The principle is strong enough to foil many MITM attacks (though, the scheme is not intended, all by itself, to stop all attacks defined in the very broad sense and combine rogue infrastructure, social engineering and so on). Indeed, we demonstrated how attackers with simultaneous control over numerous mechanisms and mounting quite sophisticated attacks, can be frustrated by the scheme when implemented correctly and with the right available wireless tools (while minimally enhancing the mechanism). For example, attacks like DNS contamination and tricking a CA to produce a faked cert, while attempting to employ the stateless nature of the SSL/TLS can nevertheless be dealt with, employing careful design and careful choice of context. This demonstrated the fact that with modern wireless tools, the security, practicality, and usability of the XOTP/ OTP tools can be substantially improved without much added cost, showing how added wireless channels can contribute much more than the obviously recognized user convenience aspect. More generally, we believe that the idea of adding contextual information to cryptographic computations can enhance the security of other protocols among devices and machines, especially in the emerging world of computing where users simultaneously operate numerous gadgets and devices.

Acknowledgments. We are grateful to Glenn Durfee and Úlfar Erlingsson for insightful comments and suggestions.

References

1. Alkassar, A., Stüble, C., Sadeghi, A.R.: Secure object identification - or: Solving the chess grandmaster problem. In: Proceedings of the Workshop on New Security Paradigms (2003)
2. APGW: Phishing activity trends report (2011), `http://www.antiphishing.org/reports/apwg_report_h2_2010.pdf`
3. Beth, T., Desmedt, Y.G.: Identification Tokens – or: Solving the Chess Grandmaster Problem. In: Menezes, A., Vanstone, S.A. (eds.) CRYPTO 1990. LNCS, vol. 537, pp. 169–176. Springer, Heidelberg (1991)
4. Eweek (2007), `http://www.eweek.com/c/a/Security/RSA-Catches-Financial-Phishing-Kit`
5. Gabber, E., Gibbons, P.B., Kristol, D.M., Matias, Y., Mayer, A.: On secure and pseudonymous client-relationships with multiple servers. ACM Transactions on Information and System Security (1999)
6. Gabber, E., Gibbons, P.B., Matias, Y., Mayer, A.: How to Make Personalized Web Browsing Simple, Secure, and Anonymous. In: Luby, M., Rolim, J.D.P., Serna, M. (eds.) FC 1997. LNCS, vol. 1318, pp. 17–31. Springer, Heidelberg (1997)
7. Halderman, J.A., Waters, B., Felten, E.W.: A convenient method for securely managing passwords. In: Proc. 14th International World Wide Web Conference (2005)
8. Higgins, K.J.: Researchers see real-time phishing jump (2010), `http://www.darkreading.com/authentication/security/attacks/s-howArticle.jhtml?artic%leID=228200550`
9. Jackson, C., Barth, A.: Beware of finer-grained origins. In: Proceedings of Web 2.0 Security and Privacy, W2SP 2008 (2008)

10. Jakobsson, M., Myers, S. (eds.): Phishing and Countermeasures: Understanding the Increasing Problem of Electronic Identity Theft. Wiley (2006)
11. Karlof, C., Tyger, J., Wagner, D., Shankar, U.: Dynamic pharming attacks and locked same-origin policies for web browsers. In: Computer and Communication Security, CCS (2007)
12. Luby, M.: Pseudorandomness and Cryptographic Applications. Princeton University Press, Princeton (1996)
13. Mannan, M.S., van Oorschot, P.C.: Using a Personal Device to Strengthen Password Authentication from an Untrusted Computer. In: Dietrich, S., Dhamija, R. (eds.) FC 2007 and USEC 2007. LNCS, vol. 4886, pp. 88–103. Springer, Heidelberg (2007)
14. McCune, J.M., Perrig, A., Reiter, M.K.: Seeing-is-believing: Using camera phones for human-verifiable authentication. In: IEEE Symposium on Security and Privacy (2005)
15. The H Security: Mozilla considers removing trustwave CA (2012), `http://www.h-online.com/security/news/item/Mozilla-considers-removing-Trustw%ave-CA-1430998.html`
16. Parno, B., Kuo, C., Perrig, A.: Phoolproof Phishing Prevention. In: Di Crescenzo, G., Rubin, A. (eds.) FC 2006. LNCS, vol. 4107, pp. 1–19. Springer, Heidelberg (2006)
17. Prins, J.: Diginotar certificate authority breach - operation black tulip (2011), `http://www.rijksoverheid.nl/documenten-en-publicaties/rapporten/2011/09/05/d%iginotar-public-report-version-1.html`
18. RFC-2104: HMAC: Keyed-hashing for message authentication
19. RFC-4033: DNS security introduction and requirements
20. RFC-4226: HOTP: An HMAC-based one-time password algorithm
21. RFC-5246: The transport layer security (TLS) protocol version 1.2
22. Ross, B., Jackson, C., Miyake, N., Boneh, D., Mitchell, J.C.: Stronger password authentication using browser extensions. In: Proceedings of the 14th Conference on USENIX Security (2005)
23. Roth, V., Richter, K., Freidinger, R.: A PIN-entry method resilient against shoulder surfing. In: Proceedings of the 13th ACM Conference on Computer and Communications Security (2004)
24. Schneier, B.: Two-factor authentication: Too little, too late. Communications of the ACM 4(4) (2005)
25. Schneier, B.: Hacking two-factor authentication (2009), `http://www.schneier.com/blog/archives/2009/09/hacking_two-fac.html`
26. Stajano, F., Wong, F.-L., Christianson, B.: Multichannel Protocols to Prevent Relay Attacks. In: Sion, R. (ed.) FC 2010. LNCS, vol. 6052, pp. 4–19. Springer, Heidelberg (2010)
27. Symantec (2010), `http://www.symantec.com/connect/blogs/beware-new-type-phishing-attack`
28. Yee, K., Sitaker, K.: Passpet: Convenient password management and phishing protection. In: Symposium On Usable Privacy and Security, SOUPS (2006)
29. Zusman, M., Sotirov, A.: Breaking the myths of extended validation SSL certificates. In: Black Hat (2009)

RIKE: Using Revocable Identities to Support Key Escrow in PKIs

Nan Zhang[1,2], Jingqiang Lin[1,*], Jiwu Jing[1], and Neng Gao[1]

[1] State Key Lab of Information Security, Institute of Information Engineering, Chinese Academy of Sciences, Beijing 100195, China
{zhangnan,linjq,jing,gaoneng}@lois.cn

[2] Graduate University of Chinese Academy of Sciences, Beijing 100049, China

Abstract. Public key infrastructures (PKIs) are proposed to provide various security services. Some security services such as confidentiality, require key escrow in certain scenarios; while some others such as non-repudiation, prohibit key escrow. Moreover, these two conflicting requirements can coexist for one user. The common solution in which each user has two certificates and an escrow authority backups all escrowed private keys for users, faces the problems of efficiency and scalability. In this paper, a novel key management infrastructure called *RIKE* is proposed to integrate the *inherent key escrow* of identity-based encryption (IBE) into PKIs. In RIKE, a user's PKI certificate also serves as a *revocable identity* to derive the user's IBE public key, and the revocation of its IBE key pair is achieved by the certificate revocation of PKIs. Therefore, the certificate binds the user with two key pairs, one of which is escrowed and the other is not. RIKE is an effective certificate-based solution and highly compatible with traditional PKIs.

Keywords: Certificate, identity-based encryption, key escrow, key management, public key infrastructure, revocation.

1 Introduction

Public key infrastructures (PKIs) are proposed to publish public keys. In PKIs, the public key of a user[1] is bound to its identity in a certificate, signed by a certification authority (CA). After querying and validating a certificate, everybody can use the contained public key for authentication, confidentiality, data integrity and non-repudiation.

Key escrow is required or prohibited in different security services, even for one user. On one hand, if a key pair is used for data encryption and decryption, key escrow is usually needed. A typical example is that, in a corporation, the

* Corresponding author: The authors were partially supported by National Natural Science Foundation of China grant 70890084/G021102, 61003273 and 61003274, and Strategy Pilot Project of Chinese Academy of Sciences sub-project XDA06010702.

[1] In this paper, "user" may refer to a person, a device, an application process or any equivalent entity.

F. Bao, P. Samarati, and J. Zhou (Eds.): ACNS 2012, LNCS 7341, pp. 48–65, 2012.

backups of all employees' private keys are stored in a trusted party, called the escrow authority (EA) or the key management center (KMC); so the corporation can decrypt all encrypted data in case the employee's private key is unavailable. On the other hand, if the key pair is used to sign and verify messages for non-repudiation, key escrow is usually prohibited. For example, key escrow is forbidden or unrecommended in the digital signatures laws and the guidelines of several countries and organizations [3,11,16,17].

To satisfy the conflicting requirements of key escrow, a common solution is to let each user hold two key pairs and accordingly two PKI certificates. The key pair and the certificate used for non-repudiation, don't support key escrow; while the others do. An instruction (called the key usage extension) is embedded in each certificate to indicate the purpose of the key pair [13]. In this two-certificate solution, every user generates its own key pair for non-repudiation and the EA generates the user's key pair for other purposes. The two public keys are contained separately in two certificates. As a result, more resources are needed to sign, publish and revoke certificates. Moreover, the EA faces the problem of scalability: as new key pairs are generated to replace expired certificates and new users join the system, the task of maintaining the great amount of escrowed private keys becomes very heavy.

Identity-based encryption (IBE) is a special type of public key algorithms, with the feature of inherent key escrow. In IBE, a trusted private key generator (PKG) initializes a secret master key and publishes the corresponding pubic parameters. A user's public key can be calculated from its identity (and the pubic parameters) by other users. Thus, certificates are not needed. When receiving messages encrypted by its public key (or its identity), the user asks the PKG to generate the private key corresponding to its identity. As for key recovery, the PKG only needs to hold and protect the secret master key, to regenerate private keys for all identities (or users). However, key revocation (when a private key is compromised) is a problem in IBE, since the public key are bound to identities automatically and users are not willing to change their identities. One the contrary, there are various certificate revocation mechanisms in PKIs, applicable to different scenarios.

The above observation that IBE and PKIs have complementary advantages gives us the motivation to integrate IBE and PKIs. In this paper, we propose RIKE, using Revocable Identities of IBE to support Key Escrow in PKIs. RIKE doesn't invent any new public key algorithm; instead, it is an innovative key management infrastructure assembling the advantages of both these two cryptosystems. Each RIKE user has one certificate and the public key in it is only used for the security services prohibiting (or not requiring) key escrow. The other security services requiring key escrow, are achieved through IBE; however, the IBE public key is derived from the user's certificate, not from its real identity. Compared with traditional PKIs, RIKE provides security services with the conflicting requirements of key escrow, while each user has only one certificate. Moreover, supporting key escrow in RIKE is much easier than that in PKIs, due to the inherent key escrow of IBE.

RIKE also solves the key revocation problem of IBE, by utilizing the certificate revocation mechanisms of PKIs. Unlike deriving the public key from the receiver's unchangeable identity in IBE, the RIKE user derives the IBE public key from the receiver's certificate after validating that certificate (including checking its revocation status). If the receiver's IBE private key is compromised, the CA will revoke its certificate and then the corresponding IBE public key shall not be used. If a new certificate is issued, a new IBE key pair is available automatically while the receiver's identity keeps constant. Therefore, in RIKE, a certificate works as a "revocable identity".

Another advantage of RIKE is the high compatibility with PKIs. RIKE is implemented on the prevailing X.509-based PKIs, with new-designed certificate extensions (see Section 4). So the existing PKIs can be smoothly transferred to RIKE. If some users receive a certificate contained such extensions and do not understand the extensions or support IBE, they just ignore these extensions and process it as a common X.509 PKI certificate.

This paper is organized as follows. Section 2 surveys the related work. Section 3 presents the detailed architecture and the analysis of RIKE. Section 4 discusses how to implement RIKE using X.509 PKI certificates. Finally, Section 5 draws the conclusion.

2 Related Work

How to perfectly support key escrow in PKIs is still an open problem. The common and prevailing approach is to store the backup of a user's private key in a centralized component [15,32]. This solution is compatible with traditional PKIs, but lacks of scalability because more and more private keys are stored as the number of users increases and certificates expire. Self-escrow PKIs (SE-PKIs) are proposed [9,31] to embed the public part of trapdoor information when users generate private keys. The key recovery agent (KRA), holding the secret part of trapdoor information, can recover users' private keys. However, the specifically-designed public key algorithms are not supported by most users and then obstruct the adoption of SE-PKIs. In this work, we try to provide a solution supporting efficient key escrow and compatible with traditional PKIs.

The concept of IBE originally proposed by Shamir [33]. Boneh *et al.* [7] and Cocks [12] invented secure IBE algorithms, in which an arbitrary bit-string can be used as a public key. Hierarchical IBE [20,21] is designed to reduce the workload of the centralized PKG. All these algorithms have two basic features: (*a*) a user's identity known by others are used as the user's public key, so the certificate is eliminated; and (*b*) IBE inherently supports key escrow because all users' private keys are generated by PKGs and can be recovered by PKGs.

The inherent key escrow is usually considered as a drawback of IBE in many scenarios, because of the risk that the users' private keys may be disclosed or maliciously used by the PKG. [18,23] proposed to generate the private keys by distributed PKGs, so that the keys are still secure when the PKGs are partially compromised. Alternatively, the privilege of PKGs can be constrained. In certificate-based encryption (CBE) [19], a user's non-escrowed secret key and a certificate from its CA are both needed to decrypt messages; and [34] designed a CBE-based proxy cryptosystem with revocable proxies. In certificateless public key cryptography (CL-PKC) [2], the private key is generated by a user and a key generating center (KGC) cooperatively. However, the feature of key escrow can be leveraged in the scenarios where key escrow is necessary, and the inherence makes the key-escrow design of RIKE concise and efficient.

The inherent binding of identities and public keys becomes a problem in IBE when the user's public key needs to be revoked (e.g., the private key is compromised), because identities are usually expected to remain constant. To deal with the revocation problem, the user's identity and a period of validity can be combined together to derive its public key [7]. Once the period expires, the key pair becomes invalid automatically; however, if the private key is compromised during the period, the key pair can not be revoked in this way. So, the period of validity shall be very short for high security, and the PKG shall generate and users shall apply for private keys very frequently. Another approach is the security mediator (SEM) architecture [5,6,26], in which an online SEM keeps a partition of each user's private key and every decryption operation requires the SEM's help. Revocation is achieved as long as the SEM stops helping the user to decrypt messages. But an always-online SEM service faces more risks, so more expensive protections are needed in practical deployment.

On the contrary, a public key (or a certificate) in PKIs is used only after its revocation status is checked. Lots of approaches are proposed to revoke PKI certificates, such as certificate revocation lists (CRLs) [13], redirect CRLs [1], the online certificate status protocol (OCSP) [29], NOVOMODO [27], certificate revocation trees (CRTs) [25] and authenticated dictionaries [30]. These revocation mechanisms have advantages in different environments, and all can be used in RIKE to revoke certificates. More detailed comparisons and evaluations of revocation mechanisms can be found in [22,28].

Some schemes are proposed to provide benefits similar to IBE, focusing on the compatibility and interoperability issues. In the RSA-based schemes [14,24], a user can encrypt messages to another user by its identity (without a certificate); or, the identity-to-key binding is implemented by online query [10]. Key escrow is not considered in these schemes. Our solution applies IBE to support key escrow in PKIs, keeping the complete compatibility with PKI certificates. For those users that do not support IBE algorithms, they can still use a certificates with the RIKE-parameter extension as a common PKI certificate.

3 RIKE: Supporting Key Escrow in PKIs

In this section, we firstly describes the background of PKIs and IBE, and the basic architecture of RIKE. Then, we extend this basic architecture to work with hierarchical PKIs and cross certification. Finally, we present the features of RIKE and the comparison with other schemes.

3.1 Background

PKIs are built based on traditional public key algorithm. A CA signs certificates, each of which binds a user's identity and its public key. Trusting the CA, a user verifies the CA's signature on a certificate and obtains another's identity and public key. Then, these certified information is used for various security services.

The notations and conceptions about PKIs are listed as follows. The superscript $\mathcal{P}$ indicates the key pairs in PKIs (to distinguish with those key pairs in IBE, for which the superscript $\mathcal{I}$ is used).

- ID_U, ID_{CA}: the identities of a user U and the CA, respectively.
- $PK^{\mathcal{P}}$: a public key of traditional public key algorithms.
- $SK^{\mathcal{P}}$: a private key of traditional public key algorithms.
- $Cert(U[,e]) = Sign_{SK^{\mathcal{P}}_{CA}}(ID_{CA}, ID_U|PK^{\mathcal{P}}[,e])$: the certificate signed by the CA, binding ID_U and $PK^{\mathcal{P}}$, with an *optional* extension e. The CA and U are called the *issuer* and the *subject* of $Cert(U[,e])$, respectively.

IBE is a special type of public key algorithms, where a public key is derived from the user's identity. A PKG firstly generates a master key and public parameters. The public parameters are publicly known, then a user can calculate another's public key based on its identity and the public parameters. Only the PKG can calculate the corresponding private key by the secret master key. The notations and conceptions about IBE are listed as follows:

- PM: the public parameters generated by the PKG.
- MK: the PKG's master key.
- $PK^{\mathcal{I}}(ID_U) = GenPK(ID_U, PM)$: the public key derived from the identity of U and PM. It can be calculated by any user.
- $SK^{\mathcal{I}}(ID_U) = GenSK(ID_U, MK)$: the private key generated by the PKG for U. Only the PKG can calculate it.

3.2 Basic RIKE

The basic idea of RIKE is to use (the hash value of) a user's PKI certificate as its identity to generate the IBE public key, used for the security services requiring key escrow. Then, each RIKE user has two key pairs but *only one* certificate. The basic architecture of RIKE is composed of a CA and an arbitrary number of users. The component responsible for signing certificates and generating escrowed private keys in RIKE, is still called the CA, to emphasizing its high compatibility with PKIs. Different from that in PKIs, the CA in RIKE owns a *PKG agent*, holding the IBE master key and generating escrowed private keys for users. Nevertheless, when RIKE is deployed, the PKG agent and the CA can be managed by two departments or implemented in one system.

We can easily implement the basic RIKE as follows, and $H(\cdot)$ is a collision-free hash function.

Initialization. The CA generates $(PK^{\mathcal{P}}_{CA}, SK^{\mathcal{P}}_{CA})$ and (PM, MK), and signs a *self-signed* certificate $Cert(CA, PM) = Sign_{SK^{\mathcal{P}}_{CA}}(ID_{CA}, ID_{CA}|PK^{\mathcal{P}}_{CA}, PM)$.

Then, the certificate is delivered to all users in out-of-band means, as it is done in traditional PKIs; while $SK^{\mathcal{P}}_{CA}$ and MK are known only to the CA. PM is embedded in the CA's certificate as an extension. See Section 4 for more details about the extension.

Certificate and Key Application. A user U generates $(PK^{\mathcal{P}}_U, SK^{\mathcal{P}}_U)$, and then applies $Cert(U) = Sign_{SK^{\mathcal{P}}_{CA}}(ID_{CA}, ID_U|PK^{\mathcal{P}}_{CA})$ and $SK^{\mathcal{I}}_U = SK^{\mathcal{I}}(H(Cert(U)))$ from the CA. $Cert(U)$ is published publicly by the CA, and the user keeps $SK^{\mathcal{P}}_U$ and $SK^{\mathcal{I}}_U$ secret. $(PK^{\mathcal{P}}_U, SK^{\mathcal{P}}_U)$ is called the *non-escrowed key pair* of U, and $(PK^{\mathcal{I}}_U, SK^{\mathcal{I}}_U)$ is called the *escrowed key pair* of U, where $PK^{\mathcal{I}}_U = PK^{\mathcal{I}}(H(Cert(U)))$.

Signing and Verification. U signs a message by $SK^{\mathcal{P}}_U$. Everybody can query and validate $Cert(U)$ to obtain $PK^{\mathcal{P}}_U$, and verify the signed message. Here, the validation of $Cert(U)$ includes checking the period of validity, the CA's signature on it, and its revocation status, as it does in PKIs.

Encryption and Decryption. Another user that wants to send encrypted data to U, firstly queries and validates $Cert(U)$, and calculates $PK^{\mathcal{I}}_U$ based on $Cert(U)$ and PM extracted from the CA's certificate. Then, data are encrypted by $PK^{\mathcal{I}}_U$ and sent to U. When receiving encrypted data, U decrypts them by $SK^{\mathcal{I}}_U$.

3.3 Certificate Renewal and Revocation in Basic RIKE

Before using either the public key in a certificate or the IBE public key derived from it, a user shall check the CA's signature, the certificate's period of validity and its revocation status. All certificate revocation mechanisms in PKIs can be used in RIKE. If the certificate expires or is revoked, the two public key shall not be used any more. The two key pairs may change or not after certificate renewal and revocation.

A user's certificate can be renewed when it expires. In the new certificate, the period of validity is updated, so the escrowed key pair changes automatically. But the non-escrowed key pair contained in the certificate will be either regenerated or kept unchanged (if it is still considered as secure).

A user's certificate is revoked and a new one with valid information is usually signed, when (a) the information in the certificate becomes invalid (e.g., its affiliation changes), or (b) the non-escrowed key pair or the escrowed key pair is (suspected to be) compromised.

- If a certificate is revoked due to the invalid information, the user's non-escrowed key pair may keep unchanged in the new certificate, but the escrowed key pair derived from the new certificate becomes different.
- If the certificate is revoked due to the compromise of the non-escrowed key pair, both the two key pairs will change even if the escrowed key pair is not compromised.
- If the certificate is revoked due to the compromise of the escrowed key pair, the user's non-escrowed key pair may keep unchanged in the new certificate (but at least one bit in the certificate shall be modified, e.g., the period of validity, so the escrowed key pair will change).

In summary, the non-escrowed key pair does not change unless it needs to change, while the escrowed key pair always changes once the certificate is replaced by a new one. But the sender who uses another user's IBE public key doesn't need to know whether the key pair has changed or not, it only validates the recipient's certificate (just as what they do in traditional PKIs) and derives the IBE public key from it. There is no extra burden for RIKE users to deal with certificate renewal and revocation.

3.4 Hierarchical RIKE

PKIs are usually built hierarchically [13], as a user applies its certificate from another user. For example, a CA (called the *root CA* in hierarchical PKIs) generates the self-signed certificate, and signs certificates for the 2nd-level users, some of which sign certificates for other users. The user that signs certificates for others, is also called a *subordinate CA*. In this paper, we call it a user or a CA alternatively according to the context. This structure can be easily extended to support more levels, where each user applies for a certificate from an upper-level user or the root CA.

The following new notations are used in a hierarchical PKI:

- $U_{j,k}$: the kth j^{th}-level user.
- $Cert(U_{j,k}) = Sign_{SK^{\mathcal{P}}_{U_{j-1,k'}}}(ID_{Isr}, ID_{U_{j,k}} | PK^{\mathcal{P}}_{U_{j,k}})$: the certificate of $U_{j,k}$ signed by Isr, which is another user $U_{j-1,k'}$ or the root CA (when $j = 2$).

In order to validate $Cert(U_{j,k})$, a vector of certificates (called the *certificate chain*) are needed: $\overrightarrow{Cert}(U_{j,k}) = (Cert(U_{2,k_2}), \cdots, Cert(U_{j-1,k_{j-1}}), Cert(U_{j,k}))$, where the issuer of each certificate is the subject of the preceding one and $Cert(U_{2,k_2})$ is signed by the root CA. Every user has been already configured with the root CA's self-signed certificate and uses it to validate $Cert(U_{2,k_2})$, and then repeatedly uses a validated certificate to validate the next one in the vector until $Cert(U_{j,k})$ is validated.

The hierarchical structure brings benefits. The root CA's workload is distributed among lower-level CAs, and users can apply for certificates locally. The root CA serves only a few users (or CAs) and then works off-line in most time to reduce attack risks. If a lower-level CA is compromised, only a limited number of users are impacted.

To build RIKE on hierarchical PKIs, we need to find a compatible way to manage IBE key pairs (or escrowed key pairs). Two intuitive and simple approaches are listed as follows:

- Only the root CA owns the PKG agent, which generates the escrowed private keys for all users. The IBE public parameters are embedded only in the root CA's self-signed certificate. Or,
- Each CA owns its PKG agent, which generates the escrowed private keys for its users only. Then, each CA's certificate contains the IBE public parameters of its own PKG agent.

However, neither of these simple approaches works well. In the first approach, the root CA takes the workload of generating escrowed private keys for all users, which violates the intentions of hierarchical PKIs. In the second one, an upper-level PKG agent cannot recover the private keys generated by the lower-level PKG agents, while sometimes centralized key escrow is needed.

We propose to build RIKE by combining hierarchical PKIs and hierarchical IBE [20,21]. Hierarchical IBE works as follows. For example, a PKG (called the *root PKG*) generates its master key and the public key parameters, and generates IBE private keys for the 2nd-level users, some of which generate IBE private keys for other users. The user that generates IBE private keys for others is called a *subordinated PKG*. In this paper, we call it a user or a PKG alternatively according to the context. The structure can be easily extended to support more levels, where each user applies its IBE private key from an upper-level user or the root PKG. Note that all users' private keys are generated by the secret master key directly or indirectly, so the root PKG can recover any user's private key.

In hierarchical IBE, each user's public key is derived from its *identity chain*, and the following new notations are introduced:

- $ID_{U_{j,k}}$: the identity of the kth j^{th}-level user $U_{j,k}$; particularly, the root PKG can be denoted as $U_{1,1}$ and its identity is null.
- $\overrightarrow{ID}_{U_{j,k}} = (\overrightarrow{ID}_{\star}, ID_{U_{j,k}})$: the identity chain of $U_{j,k}$, where $\overrightarrow{ID}_{\star}$ is (a) the identity chain of another user $U_{j-1,k'}$ if $U_{j,k}$ applies its IBE private key from $U_{j-1,k'}$, or (b) null if $U_{j,k}$ applies it from the root PKG (i.e, the root PKG's identity chain is also null).

Thus, the identity chain of $ID_{U_{j,k}}$ is $(ID_{U_{2,k_2}}, \cdots, ID_{U_{j-1,k_{j-1}}}, ID_{U_{j,k}})$, and its private key is generated by $U_{j-1,k_{j-1}}$ as follows:

- $PK^{\mathcal{I}}(\overrightarrow{ID}_{U_{j,k}}) = GenPK((ID_{U_{2,k_2}}, \cdots, ID_{U_{j-1,k_{j-1}}}, ID_{U_{j,k}}), PM)$: the public key of $U_{j,k}$ derived from $\overrightarrow{ID}_{U_{j,k}}$ and PM.
- $SK^{\mathcal{I}}(\overrightarrow{ID}_{U_{j,k}}) = GenSK(ID_{U_{j,k}}, SK^{\mathcal{I}}(\overrightarrow{ID}_{U_{j-1,k_{j-1}}}))$: the private key of $U_{j,k}$ generated by $U_{j-1,k_{j-1}}$. Note that $SK^{\mathcal{I}}(\overrightarrow{ID}_{U_{1,1}}) = MK$ is the root PKG's master key.

Finally, hierarchical RIKE is built on hierarchical PKIs, where each CA owns its PKG agent and these PKG agents work as the PKGs of hierarchical IBE. Only the root PKG agent publishes the IBE public parameters in the root CA's self-signed certificate (See Section 4 for details). With the same IBE public parameters, the hash values of a user's certificate chain are used to generate its IBE public key. Here, we firstly define $H(\overrightarrow{Cert}(U_{j,k})) =$
$(H(Cert(U_{2,k_2})), \cdots, H(Cert(U_{j-1,k_{j-1}})), H(Cert(U_{j,k})))$.

- The root CA generates $(PK^{\mathcal{P}}_{CA}, SK^{\mathcal{P}}_{CA})$ and (PM, MK), and signs a self-signed certificate $Cert(CA, PM) = Sign_{SK^{\mathcal{P}}_{CA}}(ID_{CA}, ID_{CA}|PK^{\mathcal{P}}_{CA}, PM)$.
- A 2nd-level user $U_{2,k}$ generates $(PK^{\mathcal{P}}_{U_{2,k}}, SK^{\mathcal{P}}_{U_{2,k}})$, and applies $Cert(U_{2,k}) = Sign_{SK^{\mathcal{P}}_{CA}}(ID_{CA}, ID_{U_{2,k}}|PK^{\mathcal{P}}_{U_{2,k}})$ and $SK^{\mathcal{I}}_{U_{2,k}} = SK^{\mathcal{I}}(H(\overrightarrow{Cert}(U_{2,k})))$ from

the root CA. $(PK^{\mathcal{P}}_{U_{2,k}}, SK^{\mathcal{P}}_{U_{2,k}})$ is the non-escrowed key pair of $U_{2,k}$, and $(PK^{\mathcal{I}}_{U_{2,k}}, SK^{\mathcal{I}}_{U_{2,k}})$ is the escrowed key pair of $U_{2,k}$, where $PK^{\mathcal{I}}_{U_{2,k}} = PK^{\mathcal{I}}(H(\overrightarrow{Cert}(U_{2,k})))$.
- A 3rd-level user $U_{3,k}$ generates $(PK^{\mathcal{P}}_{U_{3,k}}, SK^{\mathcal{P}}_{U_{3,k}})$, and applies $Cert(U_{3,k}) = Sign_{SK^{\mathcal{P}}_{U_{2,k'}}}(ID_{U_{2,k'}}, ID_{U_{3,k}}|PK^{\mathcal{P}}_{U_{3,k}})$ and $SK^{\mathcal{I}}_{U_{3,k}} = SK^{\mathcal{I}}(H(\overrightarrow{Cert}(U_{3,k})))$ from a 2nd-level user $U_{2,k'}$.
- Any user can follow the process above to generate its non-escrowed key pair and apply its certificate and escrowed key pair.

Hierarchical RIKE has both the features of distributed workload and centralized key escrow. Hierarchical RIKE distributes the workload among subordinate PKG agents of all levels. Each subordinate PKG agent is responsible for generating the escrowed private keys of only the users directly subordinated to it. At the same time, hierarchical RIKE gives the root PKG agent the ability to recover the escrowed key pairs of all users. Given a user's certificate chain, its IBE private key can be regenerated by the root PKG agent's master key.

3.5 Hierarchical RIKE with Cross Certification

Theoretically, one hierarchical PKI (and then hierarchical RIKE) with one root CA can serve all users in the world. However, there are lots of root CAs with different self-signed certificates in the real world. Usually, a user is configured with a certificate trust list (CTL), a limited set of self-signed certificates. Users apply for certificates from different root CAs (directly or indirectly) and have different CTLs. Thus, if a user U receives $\overrightarrow{Cert}(U')$ which is signed by a root CA not in the CTL of U, it cannot validate $Cert(U')$ and communicate with U' securely. Note that self-signed certificates shall be delivered in out-of-band means and be configured carefully, so a user doesn't change its CTL rashly.

Cross certification [13] helps a user validate certificates signed by a root CA not in its CTL. As shown in Figure 1, $Cert(U')$ is signed by $RootCA_2$ (indirectly), while the CTL of U contains the self-signed certificate of $RootCA_1$ only. To help U validate $Cert(U')$, $RootCA_1$ signs a *cross certificate* $CrsCert(CA') =$

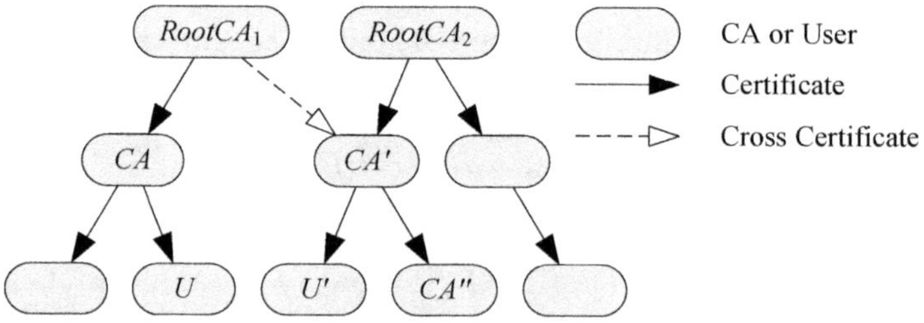

Fig. 1. Cross Certification and Cross Certificate

$Sign_{SK^{\mathcal{P}}_{RootCA_1}}(ID_{RootCA_1}, ID_{CA'}|PK^{\mathcal{P}}_{CA'})$, so U can validate $Cert(U')$ using the certificate chain $(CrsCert(CA'), Cert(U'))$.

A cross certificate is the same as a common certificate signed to a CA, except that the subject of the cross certificate is a CA which (*a*) already has a valid certificate and (*b*) has signed certificates for users. On receiving a certificate, a user cannot distinguish whether it is a cross certificate or not. Moreover, the subject (or the issuer) of a cross certificate may be a root or non-root CA, and cross certification may be bidirectional or not. For example, in Figure 1, CA' may also sign a cross certificate for $RootCA_1$ or not.

When cross certificates are issued, a user will have multiple certificate chains. For example, in Figure 1, U' has two certificate chains $(Cert(CA'), Cert(U'))$ and $(CrsCert(CA'), Cert(U'))$: one is validated by $Cert(RootCA_2)$ and the other is done by $Cert(RootCA_1)$.

The coexisting multiple certificate chains lead to two problems, if the method in Section 3.4 is directly used to derive a user's IBE public key. Firstly, different certificate chains result in different escrowed key pairs, all users (and CAs) subordinated to the subject of the cross certificate have to apply new IBE private keys after each cross certification happens. Secondly, given a user, some of its private keys are escrowed in PKI components managed by other organizations. The consequence is that the user's own organization can not recover those private keys. For example, in Figure 1, the private key that decrypts the encrypted data sent from U to U' will be escrowed only in the PKG agent of $RootCA_1$. Note that U' is a user of $RootCA_2$, but $RootCA_2$ can not recover this private key.

The solution is to always derive user's IBE public key from the same certificate chain. We choose the one in which there is no cross certificate and name it the *primary certificate chain* of a user. The user's escrowed private key is only generated based on the primary certificate chain.

To achieve this aim, the cross certificate carries the information of (*a*) the IBE public parameters of the root CA (or the PKG agent) in the subject's primary certificate chain and (*b*) the primary certificate chain of the subject (i.e., the hash values of certificates from the 2nd-level CA to the certificated subject, called the *ID-prefix* in this paper). The above information is embedded in the cross certificate[2] as an extension called the *RIKE-parameter* extension (see Section 4 for details). The verifier can use the above information to derive the same IBE public key as that from the primary certificate chain.

For example, in Figure 1, $(Cert(CA'), Cert(U'))$ is the primary certificate chain of U', and the cross certificate $CrsCert(CA')$ contains the IBE public parameters of $RootCA_2$ and the ID-prefix of CA' (i.e., $H(\overrightarrow{Cert}(CA'))$). So the IBE public key of U' is always derived from $(H(Cert(CA')), H(Cert(U')))$ and PM_{RootCA_2}, even if U validates the certificate chain $(CrsCert(CA'), Cert(U'))$ of U' by $Cert(RootCA_1)$, because $H(\overrightarrow{Cert}(CA'))$ and PM_{RootCA_2} are embedded in $CrsCert(CA')$ as a certificate extension.

[2] In RIKE, before a CA signs a cross certificate to another CA, it needs to query the primary certificate chain of the subject CA, to obtain the information.

In particular, Algorithm 1 is used to derive the IBE public key of U', when a verifier U receives a certificate chain $\overrightarrow{Cert}(U')$ different from the primary certificate chain of U'. U can reassemble (the hash values of) the primary certificate chain of U', when validating $\overrightarrow{Cert}(U')$. During this process, if $Cert_i$ is a cross certificate (containing the IBE public parameters PM and the ID-prefix), the IBE public parameters used by the verifier is substituted by PM and the ID-prefix is used to reassemble the identity chain of U'. Note that the algorithm works even when there are multiple cross certificates in $\overrightarrow{Cert}(U')$.

Algorithm 1. The Derivation of IBE Public Key

Input: The certificate chain of U', $\overrightarrow{Cert}(U') = (Cert_2, Cert_3, \cdots, Cert_j)$;
The self-signed certificate in the CTL of U, $Cert_1$;
Output: The IBE public key of U', $PK^{\mathcal{I}}_{U'}$;

```
PM = PMField(RikeParamExt(Cert_1));
// PM is set to the IBE public parameters in Cert_1.
ID⃗ = null;
IsrCert = Cert_1;
for (i = 2; i ≤ j; i++) do
    if Verify(IsrCert, Cert_i) then
        // Cert_i is verified by IsrCert.
        if e = RikeParamExt(Cert_i) then
            // Cert_i contains a RIKE parameter extension e.
            PM = PMField(e); // PM is set to the IBE public parameters in Cert_i.
            ID⃗ = IDPrefixField(e); // ID⃗ is set to the ID-prefix in Cert_i.
        else
            // No RIKE parameter extension in Cert_i.
            ID⃗ = AppendID(ID⃗, Hash(Cert_i));
        end if
        IsrCert = Cert_i; // To verify the next certificate.
    else
        return null;
    end if
end for
return PK^I_U' = GenPK(ID⃗, PM);
```

In Section 3.4, to guarantee the root CA (or the PKG agent) can recover all users' private keys, we required that the IBE public key parameters are only embedded in the root CA's self-signed certificate. However, with the RIKE-parameter extension in cross certificates, the IBE public key parameters may actually be obtained from a non-self-signed cross certificate. A new risk appears that some subordinate CA may maliciously embed different IBE public parameters when signing certificates for users. Note that a verifier cannot distinguish such certificates from a cross certificate. Thus, the root PKG agent can not recover the users' IBE private keys and the centralized key escrow is undermined.

Another certificate extension called the *RIKE-parameter-lock* extension, is proposed to avoid the above risk. Once this extension is set in a certificate, the subject CA and all its (directly and indirectly) subordinated CAs shall not issue a certificate with different IBE public parameters; otherwise, such a certificate is considered as invalid. Of course, those CAs are deprived of the privilege to issue cross certificates. It is reasonable, because the potentially malicious CAs should not be granted this privilege.

3.6 Certificate Renewal and Revocation in Hierarchical RIKE

In hierarchical RIKE, certificate renewal and revocation are somehow more complicated than those in basic RIKE. If the renewed (or revoked) certificate is held by a bottom-level user, the cases are the same as in basic RIKE discussed in Section 3.3.

However, if the renewed (or revoked) certificate is held by a CA, the two key pairs of the CA may change (as discussed in Section 3.3) and then impact the users subordinated to the CA. If the non-escrowed key pair of the CA (i.e., the key pair to sign and verify certificates) is changed, the CA needs to revoke all certificates it signed before. It is the same as in hierarchical PKIs, so we do not discuss this case here. The escrowed key pair of the CA may change, when the CA's certificate is renewed or revoked. Three cases are analyzed as follows:

Case 1. The renewed (or revoked) certificate is not a cross certificate, and neither the CA nor its (directly and indirectly) subordinate CAs hold cross certificates. The subordinate CAs' primary certificate chains changes because the CA's certificate changes. So all their escrowed key pairs are changed.

Case 2. The renewed (or revoked) certificate is not a cross certificate, but the CA or its (directly or indirectly) subordinate CA holds a cross certificate. In addition to all these CAs' escrowed key pairs are changed as in Case 1, the cross certificate shall be revoked and a new cross certificate shall be signed, because the ID-prefix in it needs to be updated.

Case 3. The renewed (or revoked) certificate is a cross certificate signed to the CA. Since the CA's is primary certificate chain doesn't depend on any cross certificate, its escrowed key pair is kept unchanged. So the escrowed keys of its subordinate CAs are also kept unchanged.

3.7 Features of RIKE

In RIKE, each user holds two key pairs but only one certificate: the non-escrowed key pair $(PK_U^{\mathcal{P}}, SK_U^{\mathcal{P}})$ is used in security services (e.g., non-repudiation) where key escrow is prohibited, and the escrowed key pair $(PK_U^{\mathcal{I}}, SK_U^{\mathcal{I}})$ is used in security services (e.g., confidentiality) where key escrow is required. We summarize the features of RIKE as follows.

Inherent Key Escrow. RIKE carries forward the inherent-key-escrow feature of IBE, in which all users' private keys are generated by the PKG with the secret master key. Therefore, unlike the EA in PKIs that stores all users's private keys, the CA (or the PKG agent) in RIKE only stores the secret master key itself and avoids the problem of scalability when the user amount becomes enormous.

Effective Certificate-Based Solution. RIKE extends traditional PKIs by creatively leveraging the PKI certificates as revocable identities to support both key escrow and non-repudiation without coming into conflict. For each user, $PK_U^{\mathcal{P}}$ and $PK_U^{\mathcal{I}}$ are published together in one certificate. These two public keys' integrity and validity are guaranteed simultaneously by validating this certificate. In this way, RIKE does not bring extra communications and validations to obtain $PK_U^{\mathcal{I}}$. Therefore, RIKE is an effective solution, especially in the environments where resources are limited.

Compatibility with Traditional PKIs. RIKE is completely compatible with all policies and procedures in traditional PKIs to create, manage, distribute, use, store and revoke certificates. As a result, for the security services without key escrow, RIKE and PKIs work with thorough interoperability. The transfer from a traditional PKI to RIKE is very simple and straightforward: the root CA (or the PKG agent) generates (PM, MK) and signs a new self-signed certificate containing the essential extensions (details in Section 4), and then generates escrowed key pairs for users and subordinate CAs.

Revocable Identities. Although RIKE borrows the key generation mechanism of IBE, RIKE does not suffer from the key revocation problem as IBE does. The reason is that $PK_U^{\mathcal{I}}$ is not derived from the user's real identity, but from the user's PKI certificate. When the certificate has been revoked and replaced by a new one, $PK_U^{\mathcal{I}}$ will change automatically. Therefore, the revocation of the escrowed key pair is implemented easily by certificate revocation, for which there are already abundant approaches. In this way, RIKE supports the "revocable identity" of the user (not the real identity, but the "identity" in the perspective of IBE).

Algorithm-Independency. RIKE combines the advantages of PKIs and IBE and does not pose any additional requirements on the cryptographic algorithms. Any algorithm applicable in traditional PKIs and IBE can be used in RIKE for signature and encryption, respectively. In a word, RIKE is an algorithm-independent framework and the algorithms can be adaptively chosen in different implementations.

3.8 Comparisons with other Schemes

RIKE vs. PKI. The most straightforward way to satisfy the conflicting key escrow requirements in PKIs is to use two key pairs and two certificates in parallel. One of the two key pairs is generated and escrowed by the EA, and can be recovered by it when needed. A key usage extension is embedded in each certificate to indicate the key pair's purpose.

We compare RIKE with PKIs in three aspects as follows, showing that RIKE is more efficient than PKIs in both client-side and server-side. Firstly, to satisfy the conflicting key escrow requirements, almost all the components in PKIs need to be scaled up. The CA shall have the ability to sign twice as many certificates as before. Resources of certificate distribution and revocation shall also increase twofold. In contrast, RIKE supports two key pairs implicitly in only one certificate. Almost no extra resource pressure is applied on PKI components and no additional certificate distribution is needed.

Secondly, in PKIs, both the two certificates of each user shall be obtained by (or transmitted to) other users. So the communication cost doubles, which impedes this solution in bandwidth-limited applications. In RIKE, no additional certificates of users are needed by encrypted-message senders.

Finally, the EA in PKIs needs to appropriately store all users' escrowed keys in a well-protected repository. In practise, besides currently valid keys, the EA also needs to store all historical keys (out-of-date keys and revoked ones), which are still useful to decrypt the old ciphertexts created when those keys were valid. As time goes on, more and more historical keys will be accumulated. Furthermore, all these private keys should be well protected with confidentiality. In RIKE, the PKG agent only stores the IBE master key in stead of all users' private keys. Whenever a specific private key shall be recovered, the PKG agent regenerates it by the IBE master key and the corresponding certificate chain. Since all the current and historical certificates are stored in plaintext, only one secret master key needs to protect, which is much simpler and more trustworthy than protecting a huge and accumulating set of private keys.

RIKE vs. SE-PKI. SE-PKIs enable PKIs to recover the private keys of all users. All the users' private keys are escrowed by the KRA which is independent from the CA. The KRA generates its own key pair (the private key SK_{KRA} and the pubic key PK_{KRA}) and provides PK_{KRA} as a parameter for users to generate their key pairs. In this way, a trapdoor is placed in the user key generation process. The user's private key can be calculated by the KRA with SK_{KRA}, so key escrow is achieved without storing and managing all users' private keys.

However, SE-PKIs also have limitations where RIKE has advantages. SE-PKIs escrow all users' private keys without differentiation, so it can not solve the conflict between key escrow and non-repudiation requirements. If SE-PKIs are adopted in traditional PKIs, each user still needs to apply for and hold two certificates. But each RIKE user only holds one certificate. Moreover, in a system with a huge amount of users, the centralized KRA in SE-PKIs undergoes a heavy workload, whereas the distributed PKGs in hierarchical RIKE (or hierarchical IBE) share the workload.

SE-PKIs depend on specially-designed algorithms, and the key generation and encryption algorithms are not compatible with the current widely-adopted PKIs'. On the contrary, RIKE is an algorithm-independent framework and could be smoothly transferred from the existing PKI systems.

Therefore, compared with SE-PKIs, RIKE is more suitable for large-scale cases and more compatible with legacy PKI systems.

RIKE vs. IBE As described above, the key idea of RIKE is the integration of PKIs and IBE by using the PKI certificate (actually, its hash value) as the "identity" in IBE. IBE provides the good feature that the sender can obtain the recipient's public key from the recipient's identity (an arbitrary bit-string, e.g. name or email address), without an online lookup. RIKE inherits this feature. So the sender obtains the recipient's escrowed public key from the recipient's PKI certificate, eliminating the additional certificate to carry it.

The major obstacle for IBE to become a fully-blown public key cryptosystem is its lack of key revocation mechanism which is necessary in practice. In IBE, the key pair is hard to revoke, because the public key is one-to-one bound with a user's identity and changing identity brings unacceptable inconvenience to the user. In contrast, RIKE supports key revocation without changing the user's identity. By leveraging the revocation mechanism of PKIs, RIKE converts PKI certificates into revocable identities.

Strictly speaking, RIKE borrows the IBE's spirit of using an arbitrary bit-string as a public key, and uses a PKI user's certificate as its identity. Email address is usually accepted as an IBE identity, because it is already commonly held and easy for human to remember and input. When RIKE is deployed for current PKI users, they have already held others' certificates and these certificates are used as IBE public keys automatically by the client applications. Therefore, although the revocable identity of RIKE is much longer than the identity of IBE, its usability is not reduced.

4 X.509-Based RIKE

In this section, we discuss how to use X.509 PKI certificates to implement RIKE, by defining two new certificate extensions, namely, the RIKE-parameter extension and the RIKE-parameter-lock extension.

The descriptions in ASN.1 syntax are as follows.

```
-- The RIKE-parameter extension
RIKEParameters ::= SEQUENCE {
    ibeAlgorithm              OBJECT IDENTIFIER,
    ibePublicParameterData    OCTET STRING,
    hashAlgorithm             OBJECT IDENTIFIER,
    idPrefix                  IDPrefix OPTIONAL }
IDPrefix ::= SEQUENCE SIZE (1..MAX) OF OCTET STRING

-- The RIKE-parameter-lock extension
RIKEParamLock ::= BOOLEAN
```

The RIKE-parameter extension `RIKEParameters` can be used in two kinds of certificates: self-signed certificates and cross certificates. The first three fields are mandatory. The field `ibeAlgorithm` and the filed `ibePublicParameterData` describe the IBE algorithm and its detailed public parameters, and the structure of `ibePublicParameterData` depends on which algorithm is used [4,8]. The field `hashAlgorithm` specifies the hash function to convert a certificate into a revocable identity. The field `idPrefix`, which is a sequence of hash values corresponding to the certificates in the subject's primary certificate chain, only exists in cross certificates.

The RIKE-parameter-lock extension `RIKEParamLock` is a boolean value to indicate whether the RIKE parameters are locked or not. When this extension is set to true in a certificate, any other certificate following the certificate in a certificate chain, is considered invalid if it has the RIKE-parameter extension (with different RIKE parameters).

With the above extensions, a PKI can be transferred to RIKE simply and smoothly. In particular, to deploy hierarchical RIKE, the root PKG agent generates (PM, MK) and signs a new self-signed certificate with a RIKE-parameter extension. After updating this certificate in their CTLs, all the users supporting IBE algorithms can encrypt messages by the recipients' IBE public keys derived from their existing PKI certificates.

5 Conclusions

In this paper, we integrate PKIs and IBE into a novel key management scheme called RIKE, which leverages revocable identities to support key escrow in PKIs. As an innovative key management infrastructure, RIKE satisfies the conflicting requirements of key escrow, and reduces the cost of managing key pairs and certificates. Each RIKE user holds two key pairs, one of which is escrowed and the other is non-escrowed, with only one certificate. This efficient scheme assembles the advantages of both the two cryptosystems and compatibly works with hierarchical PKIs.

References

1. Adams, C., Zuccherato, R.: A general, flexible approach to certificate revocation. Technical report, Entrust (1998)
2. Al-Riyami, S.S., Paterson, K.G.: Certificateless Public Key Cryptography. In: Laih, C.-S. (ed.) ASIACRYPT 2003. LNCS, vol. 2894, pp. 452–473. Springer, Heidelberg (2003)
3. Asia-Pacific Economic Cooperation (APEC). Guidelines for schemes to issue certificates capable of being used in cross jurisdiction ecommerce (2004)
4. Appenzeller, G., Martin, L.: IETF RFC 5408: Identity-based encryption architecture and supporting data structures (2009)
5. Baek, J., Zheng, Y.: Identity-Based Threshold Decryption. In: Bao, F., Deng, R., Zhou, J. (eds.) PKC 2004. LNCS, vol. 2947, pp. 262–276. Springer, Heidelberg (2004)
6. Boneh, D., Ding, X., Tsudik, G., Wong, M.: A method for fast revocation of public key certificates and security capabilities. In: 10th USENIX Security Symposium, pp. 297–308 (2001)
7. Boneh, D., Franklin, M.: Identity-Based Encryption from the Weil Pairing. In: Kilian, J. (ed.) CRYPTO 2001. LNCS, vol. 2139, pp. 213–229. Springer, Heidelberg (2001)

8. Boyen, X., Martin, L.: IETF RFC 5091: Identity-based cryptography standard (IBCS) #1: Supersingular curve implementations of the BF and BB1 cryptosystems (2007)
9. Brown, J., Gonzalez Nieto, J., Boyd, C.: Efficient and secure self-escrowed public-key infrastructures. In: 2nd ACM Symposium on Information, Computer and Communications Security, pp. 284–294 (2007)
10. Callas, J.: Identity-based encryption with conventional public-key infrastructure. In: 4th Annual PKI Workshop, pp. 98–111 (2005)
11. China. Electronic signature law (2004)
12. Cocks, C.: An Identity Based Encryption Scheme Based on Quadratic Residues. In: Honary, B. (ed.) Cryptography and Coding 2001. LNCS, vol. 2260, pp. 360–363. Springer, Heidelberg (2001)
13. Cooper, D., Santesson, S., Farrell, S., Boeyen, S., Housley, R., Polk, W.: IETF RFC 5280: Internet X.509 public key infrastructure certificate and certificate revocation list (CRL) profile (2008)
14. Ding, X., Tsudik, G.: Simple Identity-Based Cryptography with Mediated RSA. In: Joye, M. (ed.) CT-RSA 2003. LNCS, vol. 2612, pp. 193–210. Springer, Heidelberg (2003)
15. Entrust. Entrust authority digital certificate solution (2012)
16. European Telecommunications Standards Institute (ETSI). Policy requirements for certification authorities issuing qualified certificates (2000)
17. European Union (EU). Directive on a community framework for electronic signatures (1999)
18. Geisler, M., Smart, N.P.: Distributing the Key Distribution Centre in Sakai–Kasahara Based Systems. In: Parker, M.G. (ed.) Cryptography and Coding 2009. LNCS, vol. 5921, pp. 252–262. Springer, Heidelberg (2009)
19. Gentry, C.: Certificate-Based Encryption and the Certificate Revocation Problem. In: Biham, E. (ed.) EUROCRYPT 2003. LNCS, vol. 2656, pp. 272–293. Springer, Heidelberg (2003)
20. Gentry, C., Silverberg, A.: Hierarchical ID-Based Cryptography. In: Zheng, Y. (ed.) ASIACRYPT 2002. LNCS, vol. 2501, pp. 548–566. Springer, Heidelberg (2002)
21. Horwitz, J., Lynn, B.: Toward Hierarchical Identity-Based Encryption. In: Knudsen, L.R. (ed.) EUROCRYPT 2002. LNCS, vol. 2332, pp. 466–481. Springer, Heidelberg (2002)
22. Iliadis, J., Spinellis, D., Katsikas, S., Gritzalis, D., Preneel, B.: Evaluating certificate status information mechanisms. In: 7th ACM Conference on Computer and Communications Security, pp. 1–8 (2000)
23. Kate, A., Goldberg, I.: Distributed Private-Key Generators for Identity-Based Cryptography. In: Garay, J.A., De Prisco, R. (eds.) SCN 2010. LNCS, vol. 6280, pp. 436–453. Springer, Heidelberg (2010)
24. Khurana, H., Basney, J.: On the risks of IBE. In: International Workshop on Applied PKC, pp. 1–10 (2006)
25. Kocher, P.C.: On Certificate Revocation and Validation. In: Hirschfeld, R. (ed.) FC 1998. LNCS, vol. 1465, pp. 172–177. Springer, Heidelberg (1998)
26. Libert, B., Quisquater, J.-J.: Efficient revocation and threshold pairing based cryptosystems. In: 22nd Annual ACM Symposium on Principles of Distributed Computing, pp. 163–171 (2003)
27. Micali, S.: NOVOMODO: Scalable certificate validation and simplified PKI management. In: 1st Annual PKI Workshop, pp. 15–25 (2002)

28. Myers, M.: Revocation: Options and Challenges. In: Hirschfeld, R. (ed.) FC 1998. LNCS, vol. 1465, pp. 165–171. Springer, Heidelberg (1998)
29. Myers, M., Ankney, R., Malpani, A., Galperin, S., Adams, C.: IETF RFC 2560: X.509 Internet public key infrastructure online certificate status protocol - OCSP (1999)
30. Naor, M., Nissim, K.: Certificate revocation and certificate update. In: 7th USENIX Security Symposium, pp. 217–228 (1998)
31. Paillier, P., Yung, M.: Self-Escrowed Public-Key Infrastructures. In: Song, J.S. (ed.) ICISC 1999. LNCS, vol. 1787, pp. 257–268. Springer, Heidelberg (2000)
32. RSA, the security division of EMC. RSA digital certificate solution (2012)
33. Shamir, A.: Identity-Based Cryptosystems and Signature Schemes. In: Blakely, G.R., Chaum, D. (eds.) CRYPTO 1984. LNCS, vol. 196, pp. 47–53. Springer, Heidelberg (1985)
34. Wang, L., Shao, J., Cao, Z., Mambo, M., Yamamura, A.: A Certificate-Based Proxy Cryptosystem with Revocable Proxy Decryption Power. In: Srinathan, K., Rangan, C.P., Yung, M. (eds.) INDOCRYPT 2007. LNCS, vol. 4859, pp. 297–311. Springer, Heidelberg (2007)

TreVisor
OS-Independent Software-Based Full Disk Encryption Secure against Main Memory Attacks

Tilo Müller, Benjamin Taubmann, and Felix C. Freiling

Department of Computer Science
Friedrich-Alexander University of Erlangen-Nuremberg

Abstract. Software-based disk encryption techniques store necessary keys in main memory and are therefore vulnerable to DMA and cold boot attacks which can acquire keys from RAM. Recent research results have shown operating system dependent ways to overcome these attacks. For example, the TRESOR project patches Linux to store AES keys solely on the microprocessor. We present TreVisor, the first software-based and OS-independent solution for full disk encryption that is resistant to main memory attacks. It builds upon BitVisor, a thin virtual machine monitor which implements various security features. Roughly speaking, TreVisor adds the encryption facilities of TRESOR to BitVisor, i. e., we move TRESOR one layer below the operating system into the hypervisor such that secure disk encryption runs transparently for the guest OS. We have tested its compatibility with both Linux and Windows and show positive security and performance results.

1 Introduction

Why Disk Encryption Matters. Disk encryption is an increasingly popular method to protect sensitive data against physical loss and theft of nomadic computer systems. According to a Ponemon survey from 2010 [24], the majority of U. S. enterprises has an overall strategy for data protection from which full disk encryption (FDE) is the fastest growing favorite (59 %). Also the U. S. government recommends agencies to encrypt all data on mobile devices to compensate the lack of physical security outside their agency location [15].

However, widespread FDE solutions under Windows, such as BitLocker [17] and TrueCrypt [30], do *not* protect data effectively in all scenarios where an adversary has physical access to the computer.

Attacks against Disk Encryption. Since public weaknesses in common cryptographic primitives like AES are unknown, practical attacks against FDE often target the key management. Weak passphrases give rise to efficiently guessable keys, a fact that has been exploited by law enforcement authorities for many years. If strong passphrases are used, however, retrieving the cryptographic key can be circumvented by accessing the encrypted data through system subversion. Trojans and system level rootkits are usually sufficient to circumvent any kind of FDE and hard to prevent in general.

F. Bao, P. Samarati, and J. Zhou (Eds.): ACNS 2012, LNCS 7341, pp. 66–83, 2012.

But even if the user is not tricked into installing malicious software, physical access alone can be sufficient for system subversion. The master boot record (MBR) of an encrypted hard disk must necessarily be left unencrypted in software-based solutions for bootstrapping purposes. As a consequence, software keyloggers can always be placed in the master boot record of the disk. Such attacks have been called *evil maid attacks* [14] and typically require access to the target machine twice, once before, and once after the victim has entered the password.

However, evil maid attacks are based on system infiltration and are therefore not always applicable for lawful actions. Instead, physical memory access can be used to break FDE in many cases since widespread FDE products, including BitLocker and TrueCrypt, hold the encryption key in main memory. These attacks require access to the machine while it is running or suspended to RAM, a common scenario in the lawful seizure of evidence.

Historically, the first successful attacks on main memory used direct memory access (DMA). DMA allows devices to bypass the operating system (OS) and to access physical memory directly for performance reasons. Attacks exploiting this feature first came up in 2005 when an Apple Macintosh's system space was compromised by plugging in a malicious iPod via Firewire [3]. Later, similar attacks were used to unlock Windows Vista [21] and Windows 7 [4] even though BitLocker was active. Other interfaces, including PC Card [8,11], PCI Express [5], and Thunderbolt [25], are believed to have the same security issues as Firewire.

Another way to access main memory is to exploit the *remanence effect* of DRAM, i. e., the fact that RAM contents fade away gradually over time. Due to the remanence effect, encryption keys can be restored, e. g., after rebooting the system with a malicious USB flash drive. This type of attack became known as *cold boot* [10] in 2008. Cold boot attacks are generic and pose a threat to all current software-based FDE technologies, including Microsoft's BitLocker, Apple Macintosh's FileVault and Linux' dm-crypt.

Threat Model		Windows	Linux		OS-Independent	
Memory Attack	System Status	BitLocker/ TrueCrypt	dm-crypt/ TrueCrypt	TRESOR/ LoopAmnesia	BitVisor	TreVisor
cold boot	off or S4					
	run/locked	X	X		X	
	S3	X	X		X	
DMA	off or S4					
	run/locked	X	X	X		
	S3	X	X			

Fig. 1. Categories of our threat model. S3 and S4 stands for "suspended to RAM" and "disk", respectively. Vulnerable scenarios are marked with X.

Threat Model. Since it is impossible to protect FDE against system subversion by malware, we delegate this threat to the malware detection community and concentrate on memory attacks that rely on physical access. Such attacks can be classified according to the state of the system when an adversary gained access: (1) *switched off* or *suspended to disk*, (2) *running* but locked, and (3) *suspended to RAM*.

The resultant threat categories as well as countermeasures and vulnerable scenarios are depicted in Fig. 1. If a laptop was lost or got stolen, it was hopefully switched off or suspended to disk. In this category, all common FDE solutions protect the data on disk successfully. If the system was running (but the screen was locked[1]), widespread FDE systems for Windows and Linux are vulnerable to both cold boot and DMA attacks. The same holds for the third category: most machines that are suspended to RAM by the time an adversary gains access are vulnerable today.

Countermeasures. The current technological responses to cold boot attacks attempt to keep the key inside the CPU rather than in RAM: AESSE [18] stores the key inside SSE registers, TRESOR [19] inside debug registers, LoopAmnesia [23] inside MSRs, and FrozenCache [16] inside CPU caches. Besides AES keys, asymmetric keys are considered in the literature [22]. All of those systems treat registers and caches inside the CPU as more secure against attacks than RAM. As long as nobody finds a practical way to read out CPU contents, e. g., by injecting malicious code onto the bus, those systems are in fact more secure than conventional disk encryption systems. Unfortunately, these countermeasures require deep changes within the operating system. This is why all mentioned projects [18,19,23,16] are written for Linux and are not applicable to Windows.

The idea of keeping keys outside of RAM protects against DMA attacks on suspended machines, too, because in TRESOR and LoopAmnesia the key is irretrievably lost during suspend-to-RAM when the CPU is switched off. However, these solutions do not protect against DMA attacks on running machines, because DMA attacks generally allow to compromise the system space and consequently to read out key storage registers by executing code with ring 0 privileges. Currently, Intel's VT-d [1] technology can fully protect against such DMA attacks. Intel VT-d comprises an I/O Memory Mapping Unit (IOMMU) that enables address remapping for DMA data transfers. Just like traditional MMUs that translate virtual to physical addresses, the IOMMU translates device-visible addresses into physical ones. Hence, certain memory locations, e. g., the system space, can effectively be protected. However, Intel VT-d was introduced as virtualization technology and Windows 7 does not use it to protect against DMA attacks.

On the contrary, there exist hypervisor-based systems that do. BitVisor [28], for example, is a thin virtual machine monitor (VMM) which implements various security functionalities for a single guest, meaning that it exploits virtualization technology to enhance security and not to run multiple systems in parallel.

[1] If the screen was *not* locked but a privileged user was logged in, an adversary can access data trivially.

Among others, BitVisor supports full disk encryption and activates the IOMMU to enforce DMA security. But it does *not* protect against cold boot attacks.

Overall, we are not aware of any solution in the literature that (1) is operating system independent, and (2) secure against all threats listed in Fig. 1. Of course, another valid countermeasure is to revert to purely hardware-based FDE systems such as Intel's SSD 320 Series [13]. Since the key is held within the hard disk and cannot be read out, such approaches withstand attacks on main memory. But software-based solutions enjoy many advantages like reduced costs, vendor independence, configurability, and versatility; this is why we focus on software-based solutions in the remainder of this paper.

Contributions. In this paper, we present TreVisor. TreVisor is a software-based solution to FDE that is designed to be

- practically secure against the threats from Fig. 1 (i. e., cold boot and DMA),
- cryptographically secure by supporting the ciphers AES-128, -192, and -256,
- transparent to the operating system by using virtualization technology, and
- fast, in particular through Intel's new AES instruction set [9].

To this end, TreVisor unites two working projects (TRESOR [19] and BitVisor [28]) resulting in a free and robust FDE system that can be used by both Windows and Linux users. To the best of our knowledge, no other system exists that meets all these requirements.

We believe that integrating TRESOR's encryption routine into a hypervisor is a promising way for future FDE solutions, because using virtualization technology for disk encryption has several advantages:

- Hypervisors are isolated from the OS; even root users and local privilege escalations [2] cannot harm the encryption process or retrieve the key.
- All OSs are equally supported; it is possible to access the same encrypted partitions simultaneously from Windows and Linux.
- Hypervisors are small; the risk to have serious programming errors shrinks with the size of code.
- Building up a *full* disk encryption system is easy; only the hypervisor must be present unencrypted.
- DMA threats can centrally be counteracted by VT-d/IOMMU settings.

Altogether, TreVisor brings many of the advantages known from hardware-based FDE, such like resistance against memory attacks and transparency, into a software-based solution. As we show in this paper, TreVisor can achieve all these goals with a performance penalty of about one-third compared to unencrypted disks, which we find acceptable.

However, TreVisor requires both Intel's new AES instruction set (AES-NI) and support for VT-x/VT-d, which is currently available only with Intel Core i5 and Core i7 processors.

Outline. The remainder of our paper is structured as follows: first we give background information about TRESOR and BitVisor in Sect. 2. Then we introduce the implementation of TreVisor from a technical point of view in Sect. 3, followed by an evaluation regarding compatibility (Sect. 4), performance (Sect. 5), and, most notably, security (Sect. 6). Finally, we conclude in Sect. 7.

2 Background

We now briefly give the necessary background on TRESOR (Sect. 2.1) and BitVisor (Sect. 2.2). Readers familiar with these projects can safely skip this section.

2.1 TRESOR

TRESOR [19] runs encryption securely outside RAM; it is a Linux kernel patch for the x64 architecture designed to run AES resistant against cold boot attacks. TRESOR avoids RAM usage completely and runs the key management as well as the AES algorithm entirely on the microprocessor. To that end, some registers of the x64 architecture must permanently be used as cryptographic key storage and are not applicable for their intended purpose. The four breakpoint registers `dr0` to `dr3` have been chosen for that because they are (1) only accessible with ring 0 privileges, (2) large enough to store an AES-256 key, and (3) seldom used by end-users. Indeed, hardware breakpoints cannot be set by userland debuggers like GDB anymore. But given the fact that only developers and reverse engineers need to set breakpoints (and software breakpoints can still be set), the absence of these registers on end-user systems is acceptable as compared with the gain in security.

Other registers, like the SSE and general purpose registers (GPR), are used temporarily inside atomic sections, too. Before leaving the atomic section, these registers are reset to null and hence, they are safe to be swapped out to RAM during context switching. Only debug registers are not safe to be swapped out by context switching and cannot be used by other threads.

To overcome future cryptanalysis based on memory residues, TRESOR follows a strict security policy: no intermediate state of AES is ever going to RAM, meaning that nothing but the output block is written back to RAM after the input block has been read. Thereto, TRESOR's atomic sections encompass encryption (resp. decryption) of entire AES blocks.

AES uses a key schedule with 10-14 round keys based on the secret key. This key schedule is computed once and then stored inside RAM by all conventional AES implementations for performance reasons. But in TRESOR, the debug registers are fully occupied with the AES key itself and round keys cannot be stored permanently. Therefore, TRESOR uses a so called *on-the-fly key schedule*, meaning that each round key is re-computed inside the atomic section of each block. This contains a potential performance drawback; we get back on this in Sect. 5.

Despite this drawback, or because of it, TRESOR accelerates its AES computations by Intel's new AES instruction set (AES-NI) that implements AES efficiently in hardware. It is currently available to Intel Core i7 processors, most Core i5 and will be available to many upcoming x86 CPU, including AMD processors.

To sum up, TRESOR is a cold boot resistant implementation of AES, primarily designed for hard disk encryption. It is currently restricted to CPUs of the Core-i series, and so is TreVisor.

2.2 BitVisor

Traditional VMMs like Xen and VMware require numerous components to provide virtual hardware devices that can be shared among parallel guests. To the contrary, BitVisor [28] is a thin hypervisor architecture based on Intel VT-x (and AMD-V) which is designed to enforce I/O device security of single VMs. It is OS-independent, meaning that the VM can run unmodified versions of Windows, Linux, or any other x86 operating system.

BitVisor minimizes the overhead introduced by virtualization, leading to a so called *parapass-through* architecture: hardware is directly passed through to the guest except for a few administrated devices. Thereby, the need for most device drivers is eliminated inside the VMM and the guest OS handles devices directly. The exceptional case are devices which must explicitly be administrated in order to enforce security functionalities. In our disk encryption scenario, these are primarily hard disks. Thereto, BitVisor comes with its own set of *parapass-through drivers* for (S)ATA disks; these drivers know the specification of the target device and can handle intercepted I/O accesses correctly without fully virtualizing them. Extracted data, such as the sectors of a hard disk, can be manipulated by the VMM, e. g., for encryption and decryption.

In order to prevent attacks from malicious I/O devices against the memory region of the hypervisor, external DMA accesses are restricted by the IOMMU. From the operating system's point of view, I/O devices still have access to the system space. But since the disk encryption routine runs with ring -1 privileges, BitVisor's IOMMU settings guarantee resistance against read- and writeable DMA attacks.

Using the IOMMU to restrict direct memory access, no parapass-through drivers are required. For example, DMA attacks based on Firewire are successfully defeated while parapass-through drivers for Firewire can be excluded.

To sum up, BitVisor is a secure lightweight hypervisor which, by running only a single VM, eliminates the need for most components that are required to share system resources among VMs.

3 Design and Implementation

The use cases of TreVisor are mobile single-user systems, i. e., laptops, because these get frequently lost and stolen at public places like airports while running or being suspended, and left unattended in hotel rooms. In other words, TreVisor is designed to protect systems that are especially exposed to attacks based on physical access. Additionally, TreVisor should be available for Windows users, or, even better, be operating system independent.

The requirement to be OS-independent quickly brings hypervisor-based solutions into mind. A first idea is to move a solution like TRESOR [19] or LoopAmnesia [23] into a Linux-based hypervisor like Xen or VMware and to run Windows on top of that. Such a design would have had the advantage that TRESOR, which is a Linux kernel patch, could have been applied to the setup directly. However, we generally consider such a design as bad because running a second, full operating system introduces significant overhead. Instead, TreVisor is implemented as a thin hypervisor for single guests, meaning that no resource must be virtualized or shared among VMs, drastically facilitating the VMM implementation. Only hard disk accesses are intercepted in order to encrypt and decrypt them securely with TRESOR. Everything else, like the keyboard, mouse, printer, video, and sound card, is passed through to the guest without intervention.

In a nutshell, TreVisor is implemented as a patch for BitVisor and introduces the OS-independent parts of TRESOR to it. That is, to enforce security measures, TreVisor exploits hardware capabilities of modern Intel CPUs that are otherwise hardly used by end-users. Usually, end-user systems do neither utilize the debugging registers nor do they make use of the VT-d/IOMMU technology. TreVisor activates these, otherwise mostly idle, components to protect against cold boot and DMA attacks.

To sum up, TreVisor aims to be a disk encryption solution that prevents information leakages through main memory attacks while being transparent to the OS and to the end-user. Thereto it combines two technologies: BitVisor taken by itself is not resistant against cold boot attacks, and TRESOR does neither support Windows nor does it defeat DMA attacks on running machines.

In the following sections we describe technical challenges we faced when integrating TRESOR's encryption routines into the BitVisor code.

Key Storage Registers. TRESOR employs the debug registers `dr0` to `dr3` as cryptographic key storage, because those are (1) only accessible from ring 0, (2) large enough to store AES-256 keys, and (3) seldom used by end-users. While point two and three remain valid inside hypervisors, point one does not because we need to protect the key against ring 0 (the guest's kernel space) rather than ring 3 (the userland). In other words, in TreVisor we need a set of registers that is only accessible with ring -1 privileges.

As such a set of registers does not exist, we have to stay with the debug registers. Luckily, virtualization allows us to define sensitive events that lead the processor to switch context from the guest into the hypervisor, e. g., on executing privileged instructions or on accessing certain registers. Using VM execution control [12], we can effectively hide the debug registers from ring 0, i. e., from the guest OS, as follows:

- `cpuid` is an instruction that provides information about present CPU features. We hook into `cpuid` and forge its result by negating the DE (debugging extension) feature.

- `cr4` is an x86 control register that contains a flag to enable/disable debugging. We intercept `cr4` write accesses and deny all attempts to enable debugging.
- "MOV-DR exiting" is a virtualization control which causes the VM to exit on every `mov` instruction to or from debug registers. We enable this feature to ultimately prevent the guest from reading or overwriting the key.

All these measures cause unconditional VM exits and bring the hypervisor into action. The first measure is required to inform the OS about missing debugging capabilities. Since we generally disallow debugging in the second measure, an OS might react unexpectedly (e. g., crash) if it assumes debugging is present. The third measure is required to deny any read and write access to debug registers. Of course, a well programmed OS would never try to access a debug register if the CPUID negates its presence, but we want to be on the safe side for two reasons:

First, the OS (or any device driver) might be erroneous and write to debug registers despite their alleged absence. This immediately leads to data corruption on the encrypted hard disk because the key gets falsified. Second, an adversary who compromises the guest's system space could easily retrieve the secret key.

Summarizing, we used virtualization techniques to effectively change the privilege level of the debug registers from 0 to -1.

Key Management. As pointed out in the last section, we can get exclusive control over debug registers inside the hypervisor. We now take this exclusive access for granted and examine how to securely read the key from a user prompt into those registers. Due to numerous encryption features, BitVisor already comes with some kind of key management. Unfortunately, this key management is inadequate for our purpose as it stores passwords and keys in RAM (where they are vulnerable to cold boot attacks). Thus, we have to replace BitVisor's key management with a more secure one. To this end, we display a TreVisor specific password prompt at boot time.

The password is transformed into a 256-bit key by multiple SHA-256 iterations and then copied into the debug registers. We make sure that all residues of both the password and the key are erased from RAM thoroughly. Therefore we do *not* send the password or key into untrusted libraries; instead we use a custom implementation of the SHA-256 algorithm.

To support multicore CPUs, we have to write the secret key into the debug registers of all CPUs because we cannot assume that the TreVisor code is always executed on CPU0. To the contrary, hypervisor code is generally executed on the CPU which led to the VM-exit. And since process migration is a costly task, we distribute the secret key among all cores.

Writing the secret key into different cores turned out to be more complicated than we expected. The problem actually is that BitVisor itself does not fully initialize the APs (application processors) but runs on the BSP (boot strap processor) until the OS boots. In other words, BitVisor leaves it to the guest OS to set up remaining cores. As a solution we hook into the first inter-processor interrupt (IPI) of each CPU; IPIs are used by the OS to signal events between processors. TreVisor intercepts the start-up signals in order to initialize the debug registers with keybits before they may be used for encryption. When a guest OS does not activate all cores, the key will only be present in those that have been activated.

The mechanism we implemented requires that the keybits are temporarily copied from the BSP into RAM and then further into the APs. Unfortunately, it is impossible to copy data between processors directly. That is, our solution might not be absolutely secure against cold boot attacks for the time of initializing a new CPU. But all OSs that we are aware of initialize CPU cores during boot-up or never and thus, we consider the risk as negligible.

Disk Encryption. TreVisor supports disk encryption with AES-128, AES-192, and AES-256. During startup, we always copy 256 keybits into the debug registers; if AES-128 or AES-192 are used, superfluous bits are just ignored.

In order to encrypt the disk, we have to hook into the guest's HDD activities. Fortunately, BitVisor already provides most of the necessary functionality, including parapass-through drivers for ATA and USB disks. To put it simply, we only have to replace the standard AES routines in BitVisor by TRESOR. To this end, we add TRESOR to an internal crypto API and activate it within the configuration.

Comparable to context switches inside an OS, hypervisors must switch the context between the guest and itself. Thereto Intel VT-x supports virtual machine control structures (VMCS) to hold guest states. These structures encompass all necessary registers, e. g., the GPRs and potentially the SSE and debug registers, too. On multicore systems, each processor has its own guest context.

Since TRESOR uses GPR and SSE registers, we have to run disk encryption routines inside a critical section where interrupts and context switches are disabled. Only by running TRESOR atomically, we can guarantee that no intermediate state of AES or the key schedule is ever going to the VMCS structures in RAM. Before leaving the atomic section, we zero-fill the GPR and SSE registers so that they are safe to be swapped out.

Additionally we have to take care of the SSE task switch bit (TS) and make sure that we save/restore the guest's SSE state before/after encrypting a disk block. The SSE registers are quite large (4 kilobits in total) and since a hypervisor usually does not use them, they are not saved by default for performance reasons.

Suspend to RAM. Lastly, we spent considerable effort to support ACPI suspend modes in TreVisor. Since the CPU is switched off during suspend modes, we have to re-read the key upon wakeup. Normally, all CPU registers are backed up in RAM during suspend modes, but in TreVisor the debug registers are naturally prevented from being stored inside RAM.

The ACPI mode S3 (suspend-to-RAM) is basically supported by BitVisor since release 1.2 (October 2011). We consider S3 as an important feature for mobile users as it speeds up the boot process and reduces power consumption. It just enhances the usefulness of mobile environments. Above that, it is especially important to the cold boot scenario since the key of conventional FDE systems is *not* lost during S3.

Therefore we want to support S3 in TreVisor, and "all" we have to do is to re-read the key upon wakeup. This may sound easy, but in practice, video cards continuously fail to get re-initialized after S3 and the screen stays blank (until we return to the OS which re-initializes the video card smoothly thanks to the parapass-through architecture). We ended up with a workaround where the user must re-enter the password "blind" and thus, we must consider S3 support in TreVisor still as work in progress.

4 Compatibility

We analyzed TreVisor regarding its compatibility with userland programs, operating systems, encrypted partitions and hardware components.

Userland Programs. Generally, userland programs are supported unless they use one of the hardware components which are occupied by TreVisor; these are VT-x/VT-d and the debugging extensions. Although the TRESOR encryption routines additionally use multimedia (SSE) and general purpose registers, other programs are fully supported, including office and internet applications, 3D games, simulations, and more.

Debuggers are *not* fully supported. In particular hardware breakpoints cannot be set as we have verified with GDB in Linux and OllyDBG in Windows. Although the running debugger does not crash, setting a hardware breakpoint has just no effect (since TreVisor effectively prevents overwriting the debug registers). However, setting software breakpoints works fine and software breakpoints are the default breakpoints in most debuggers.

Virtualization software like VirtualBox and VMware is not fully supported either. Again, these programs do not crash but they run less efficiently as they cannot make use of Intel's VT-x (it is already used by TreVisor). The problem could be solved in the future because nested VT-x is generally possible when supported by the lowest hypervisor.

Operating Systems. One of the most important advantages of TreVisor compared to TRESOR is its capability to run an unmodified version of basically every x86 OS, including Windows 7, Linux, BSD variants, and more. We have verified this for the most important OSs, in particular for the 32- and 64-bit variants of Windows and Linux.

During development, the Windows operating system and its device drivers caused somehow more trouble than Linux. For example, booting up the 64-bit variant of Windows 7 fails with a blue screen when the CPUID negates debugging extensions. The problem can magically be solved by not changing the return value of CPUID but still disabling the debugging extensions (which is in fact a wrong configuration).

Encrypted Partitions. TreVisor supports full AES, including the 128-, 192-, and 256-bit variant; other ciphers are not supported.

Indeed, TreVisor allows to encrypt several partitions, but all of those must be encrypted with the same secret key because no space is left to store additional keys in debug registers. A possible solution to this problem is to store a keyring in RAM that is encrypted with a *master key*. As a downside, this solution induces a significant performance drawback because the key has to be decrypted before scrambling a block. Hence, and since TreVisor is primarily designed for single-user systems, we decided against it.

Partitions that are encrypted with TreVisor can be used by both Windows and Linux. Using partitions that were encrypted with another disk encryption software is not possible (at least not without re-encryption).

Hardware Components. Since we utilize recent Intel technologies, we had to write TreVisor close to the hardware and consequently, there *are* incompatibilities with many existing systems. It is almost impossible to support older CPUs because hardware virtualization simplifies the implementation of hypervisors considerably.

From current Intel CPUs, only the Core i5 and i7 series are compatible with TreVisor as these are the only 64-bit CPUs that support VT-x/VT-d as well as the AES instructions. We have tested TreVisor successfully with both, the i5 and i7. Most notably, CPU frequency scaling features can, unfortunately, not be used from within the guest at the time of this writing.

As mentioned above, we are also aware of problems with some video cards and ACPI wakeup. After TreVisor resumes from suspend-to-RAM, users must re-enter the password "blind" because the password prompt regularly fails to get displayed.

Tu sum up, guaranteeing the support of a wide range of commercial hardware components, including CPUs, video cards, and ACPI, is a difficult task. In this sense, TreVisor must be seen as an academic prototype, not as a market-ready product.

5 Performance

We now present TreVisor disk encryption benchmarks. We have evaluated the performance of TreVisor on two different systems: Windows 7 (64-bit) and Ubuntu Linux (64-bit, kernel 3.0). On both systems our tests revealed practical benchmark data; the performance drawback compared to unencrypted disks, and compared to disks encrypted with AES-NI, is about one-third.

Table 1 illustrates the encryption and decryption speed of TreVisor in megabytes per seconds. We list the encryption speed of all three TRESOR variants (TRESOR-128, -192, and -256) and compared TRESOR-256 with reference values of plain disk access (No-VMM), BitVisor without encryption (No-Crypto), BitVisor with our own, memory-based AES-NI implementation (AES-NI/256), and BitVisor with its default, OpenSSL-based AES variant (StdAES/256).

Table 1. Basic throughput data (a) of TreVisor in MB/s. The penalty of the reference values (b) is in comparison to the throughput of TRESOR-256.

	TRESOR-128	TRESOR-192	TRESOR-256
Linux/write	59.3	54.4	56.0
Linux/read	38.8	36.3	32.5
Windows	43.4	39.4	35.3

(a)

	No-VMM		No-Crypto		AES-NI/256		StdAES/256	
Linux/write	60.7	7.74 %	59.9	6.51 %	59.2	5.41 %	57.6	2.78 %
Linux/read	63.7	48.98 %	63.2	48.58 %	62.9	48.33 %	37.8	14.02 %
Windows	54.1	34.75 %	53.9	34.51 %	53.4	33.90 %	41.4	14.73 %

(b)

Under Linux, we evaluated the encryption speed (write to an encrypted partition) as well as the decryption speed (read from an encrypted partition) with the `dd` utility. To minimize the effect of disk caching, we copied large files (10 gigabytes) that do not fit into RAM and averaged over several test runs. Under Windows we used the PC analysis software SiSoft Sandra 2011 to create combined disk benchmarks (read and write). It remains unclear how the values are determined by SiSoft Sandra in detail and thus, the inferior throughput compared to Linux does not necessarily mean that hard disks under Windows work slower.

We took the disk speed of BitVisor without encryption (No-Crypto) into account to investigate the performance drawback that arises from hooking into disk writes without modifying them. As shown by Table 1, this effect is minor. Additionally we added our own, memory-based implementation of AES to BitVisor (AES-NI/256), because the default implementation of BitVisor does not utilize the AES instruction set. To investigate the exact performance drawback caused by TRESOR's on-the-fly key schedule, we patched TRESOR's encryption routine to store all round keys persistently in RAM. We were surprised by the outstanding acceleration attributed to AES-NI; the difference between AES-NI and no encryption is nominal.

As shown in Table 1, the performance drawback of TRESOR-256 compared to standard AES-256 is only about 3 % for writing but about 14 % for reading. This effect stems from the on-the-fly key schedule: decryption round keys are derived from encryption round keys by an additional, costly operation (namely `aesimc`, *inverse mix columns*). As the key schedule must be recomputed for each input block, this extra operation becomes noticeable. In comparison with plain (not encrypted) disks the effect intensifies and the performance decreases to almost 50 % for decryption/reading. Writing, on the other hand, is not affected by the extra operation and the performance decrease compared to all reference values is below 10 %.

In summary, TreVisor can practically be employed without a noteable performance drawback for encryption, but with a performance drawback of up to 50 % for decryption. However, compared to the gain in security, we consider this performance loss as acceptable. Furthermore, we want to point to the combined read/write benchmarks under Windows, revealing a decrease of about one-third on average. Lastly, we want to note that the decrease relative to traditional, non AES-NI implementations is below 15 %; and those systems were practically deployed for many years until AES-NI CPUs came up recently.

6 Security

First we show that TreVisor is, above all, secure against cold boot attacks and DMA attacks. Furthermore it can be configured in a way that is mostly secure against evil maid attacks. Last we discuss timing attacks and attacks by privileged users.

Cold Boot Attacks. The major development object of TreVisor is to offer Windows users the opportunity to encrypt their hard disks resistant to cold boot attacks. All existent, software-based FDE systems for Windows store necessary keys in RAM. As shown by a recent study about the practicalness of cold boot attacks [6], this problem *must* be taken seriously.

To defeat cold boot attacks we consistently treat RAM as insecure and adopted TRESOR's techniques to hold keys in CPU registers. But the correctness of TRESOR does not necessarily imply the correctness of TreVisor. We have to show that no programming error slipped into our implementation that, for example, leads the debug registers to be written out into RAM during context switching.

To prove resistance against cold boot attacks we analyzed the RAM of a TreVisor system in order to show that (1) debug registers are never swapped out into VMCS structures, and (2) GPR and SSE registers are not swapped out inside critical sections. The most comfortable way to analyze the memory of a TreVisor system is to run TreVisor itself inside a VM and examine its memory from outside. Unfortunately, nested VT-x does not work in current virtualization software; either it is not supported at all, or, as in VMware (4.0.2), it is officially supported but still causes TreVisor to crash.

Hence, we had to examine memory via cold boot attacks. We booted a mini OS from USB as well as over network (PXE) and dumped everything that was left in memory. Additionally we used BitVisor's debug console to examine the entire main memory at runtime and patched TreVisor to go through the crucial VMCS structures. Last but not least, we implemented log messages on access to debug registers.

Instead of using `aeskeyfind` [10], which is based on the AES key schedule that is not stored in TreVisor at all, we had advantage over real attackers by searching for known keybits. Despite this advantage, we were not able to find any match of the key that exceeds three bytes, a finding which can be explained by chance alone. On the other hand, when we explicitly wrote keybits into RAM, we *were* able to recover them.

Summarizing, we were not able to recover keybits, the key schedule or any part of them despite strenuous efforts.

DMA Attacks. Read-only DMA attacks are prevented by TreVisor as explained in the previous section – since the TreVisor key never enters RAM, it cannot be retrieved by reading from RAM. However, writeable DMA attacks may still be harmful because they can compromise the system space and execute privileged code, e. g., code that displays the debug registers.

Writeable DMA attacks can only be circumvented by restricting the memory space that is available to DMA capable I/O ports like Firewire. Such restrictions can effectively be implemented via the IOMMU of Intel's VT-d technology.

Normally, hypervisors use the IOMMU to protect themselves from access of untrusted guest OSs. However, the same mechanism can be used to improve security against DMA attacks, and so does BitVisor. BitVisor verifies that addresses, which are specified by guest DMA descriptors, do not point into hypervisor regions.

It should be mentioned that BitVisor protects only its own memory regions; the system space of the guest is *not* protected and still vulnerable to DMA attacks. But since the disk encryption logic of TreVisor resides in the hypervisor, the key cannot be accessed from the guest.

Summarizing, BitVisor implements an effective approach to protect its memory space against malicious I/O devices and hence, TreVisor is secure against DMA attacks.

Evil Maid Attacks. As a matter of fact, software-based FDE systems do not enforce *full* disk encryption due to bootstrapping reasons. At least the MBR and a small decryption routine must be stored unencrypted in order to launch the remaining system. This weakness affects most existent FDE solutions, including TrueCrypt, and allows for so called evil maid attacks: unencrypted MBRs can be infiltrated with bootkits (which, for example, may have keylogging functionality).

To overcome such attacks, we successfully tested a configuration of TreVisor that enables true FDE, meaning that the master boot record of the disk must not be left unencrypted. For this reason we store both the bootloader and the hypervisor onto an external USB flash drive which is required to be plugged in during boot time. (Additionally, reconfiguring the BIOS should be protected by a password.)

Although this countermeasure thwarts the most evident MBR attacks, it is not perfect. First, the USB device must be handled like a physical key, meaning that loss of the device threatens the protection mechanism. Second, only the integrity of the bootloader can be verified, not the integrity of the BIOS (e. g., BIOS kits [26]). Therefore, we are working on the support of trusted platform modules in future versions of TreVisor as it would be both more convenient and more secure. BitLocker, for example, already supports such a TPM configuration.

Timing Attacks. We want to mention briefly that TRESOR is resistant to another kind of side channel attacks – timing and cache-based attacks [20].

However, this is not special because all disk encryption systems which build upon Intel's AES-NI, including BitLocker and TrueCrypt, *are* resistant to timing attacks. Intel itself states that, beyond improving performance, the AES instruction set provides security benefits by running in data-independent time [27].

Privileged User Attacks. In TreVisor, an adversary who could gain root or administrator privileges *cannot* compromise the key. Of course, such an adversary is mostly able to read out the disk, but the encryption key itself cannot be accessed. This again is an improvement compared to BitLocker and TrueCrypt where the key resides in system space – always available to attacks by the superuser. In TreVisor the key does not reside in system space but ring -1 privileges are required to access it. TreVisor does not grant access from the untrusted guest OS and hence, it is impossible for an adversary to read the key without breaking out of the VM.

Preventing such VM escapes was one of the design goals of BitVisor. The authors argue that BitVisor comprises only about 20 KLOC (kilo lines of code) which is quite small compared to other VMMs. Reduced code size effectively increases the reliability of a hypervisor as it reduces the risk of serious programming errors. For example, local privilege escalations for the Linux kernel appear regularly (e. g., CVE-2009-2692, CVE-2010-3081, and CVE 2012-0056 [2]). Thus, even when using TRESOR, which is a Linux kernel patch, the secret key cannot effectively be protected against local attacks because once a user compromised system space, the key can easily be retrieved. In contrast, a TreVisor user would additionally have to break out of the VM – which is at least a further obstacle.

7 Conclusions and Future Work

In this paper we described TreVisor, a disk encryption scheme that is primarily designed to resist main memory attacks. Our proposal goes along with a prototype implementation that runs reliably in practice but allows for many improvements in future work, too.

Conclusions. Software based disk encryption solutions like BitLocker and TrueCrypt are designed to preserve confidentiality and integrity in the case of physical loss. But in many practical cases they do *not* fulfill these requirements as it has been shown by several known attacks: cold boot, DMA, and evil maid attacks.

Hence, there is practical need for a disk encryption solution that finally integrates countermeasures to all these threats [7]. TreVisor, which claims to be such a solution, has several advantages as compared with conventional disk encryption software:

- As only a hypervisor must be present unencrypted, truly full disk encryption can easily be enforced. For example, the small hypervisor can be stored on external bootable devices, improving protection against evil maid attacks.
- As VT-x technology is utilized to run below the operating system, TreVisor encrypts transparently for the OS. That is, TreVisor permits, for example, Linux and Windows to access the same encrypted partition simultaneously.

– As it is based on TRESOR and VT-d/IOMMU technology, TreVisor secures against attacks on main memory, namely cold boot and DMA attacks.

We believe that exploiting otherwise hardly used components, i. e., the debug registers and VT-x/VT-d, is a reasonable way to deploy more secure disk encryption in mobile end-user systems.

To conclude, TreVisor substantially increases the security of disk encryption systems. It is the first system which is secure against known main memory attacks, in particular against cold boot and DMA attacks. Before TreVisor, countermeasures against these attacks were spread over small, academic projects and serious disk encryption systems still do not implement them.

Future Work. We believe that utilizing virtualization technology in order to enforce OS-independent disk encryption is the most promising software-alternative to compete with hardware-FDE, which becomes increasingly popular. At the time of this writing, however, TreVisor can only be treated as a prototype of what *could* be done in future. To be deployable on the mass market, many issues have to be solved, mainly regarding usability and compatibility:

– Easy installation, particularly for Windows users. Currently, TreVisor must be compiled under Linux and manually be started from within GRUB before booting the OS.
– User-friendly suspend-to-RAM support. At the moment, suspend-to-RAM must be considered as "experimental" since TreVisor fails to display a visual password prompt.
– CPU frequency scaling from within the guest. This is especially important to save battery life of notebooks.

Above that, we are continuously working on further security improvements, primarily on the integration of trusted platform modules into the boot process. Proving the integrity of boot components by means of the TPM would be a secure and convenient add-on to TreVisor in the future.

Acknowledgments. We would like to thank *Richard Mäckl*, *Johannes Stüttgen*, and *Stefan Vömel* as well as the anonymous reviewers for reading a prior version of this paper and giving us valuable suggestions for improving it.

Availability. TreVisor is free software which is published under the GPL v2 [29]. Its source code is publicly available at `www1.cs.fau.de/trevisor`

References

1. Abramson, D., Jackson, J., Muthrasanallur, S., Neiger, G., Regnier, G., Sankaran, R., Schoinas, I., Uhlig, R., Vembu, B., Wieger, J.: Intel Virtualization Technology for Directed I/O. Intel Technology Journal 10 (August 2006)
2. Aedla, J.: Linux Kernel CVE-2012-0056 Local Privilege Escalation Vulnerability (January 2012); Common Vulnerabilities and Exposures, `http://www.securityfocus.com/bid/51625/`

3. Becher, M., Dornseif, M., Klein, C.N.: FireWire - All Your Memory Are Belong To Us. In: Proceedings of the Annual CanSecWest Applied Security Conference, Vancouver, British Columbia, Canada. Laboratory for Dependable Distributed Systems, RWTH Aachen University (2005)
4. Böck, B.: Firewire-based Physical Security Attacks on Windows 7, EFS and BitLocker. Secure Business Austria Research Lab (August 2009)
5. Carrier, B.D., Spafford, E.H.: Getting Physical with the Digital Investigation Process. IJDE 2(2) (2003)
6. Carbone, Bean, Salois: An in-depth analysis of the cold boot attack. Technical report, DRDC Valcartier, Defence Research and Development, Canada, Technical Memorandum (January 2011)
7. Cardwell, M.: Protecting a Laptop from Simple and Sophisticated Attacks (August 2011), `https://grepular.com/Protecting_a_Laptop_from_Simple_and_Sophisticated_Attacks`
8. Devine, C., Vissian, G.: Compromission physique par le bus PCI. In: Proceedings of SSTIC 2009. Thales Security Systems (June 2009)
9. Gueron, S.: Intel's New AES Instructions for Enhanced Performance and Security. In: Dunkelman, O. (ed.) FSE 2009. LNCS, vol. 5665, pp. 51–66. Springer, Heidelberg (2009)
10. Halderman, J.A., Schoen, S.D., Heninger, N., Clarkson, W., Paul, W., Calandrino, J.A., Feldman, A.J., Appelbaum, J., Felten, E.W.: Lest We Remember: Cold Boot Attacks on Encryptions Keys. In: Proceedings of the 17th USENIX Security Symposium, San Jose, CA, pp. 45–60. Princeton University, USENIX Association (2008)
11. Hulton, D.: Cardbus Bus-Mastering: 0wning the Laptop. In: Proceedings of ShmooCon 2006, Washington DC, USA (January 2006)
12. Intel Corporation. Intel 64 and IA-32 Architectures Developer's Manual, Combined Volumes: 1, 2A, 2B, 2C, 3A, 3B and 3C edition (December 2011)
13. Intel Corporation. Solid-State Drive 320 Series (2011), `http://www.intel.com/content/www/us/en/solid-state-drives/solid-state-drives-320-series.html`
14. Rutkowska, J.: Evil Maid goes after TrueCrypt. The Invisible Things Lab (October 2009), `http://theinvisiblethings.blogspot.com/2009/10/evil-maid-goes-after-truecrypt.html`
15. Johnson, C.: Protection of Sensitive Agency Information. U.S. Executive Office of the President, Washington, D.C. 20503 (June 2006)
16. Pabel, J.: Frozen Cache (January 2009), `http://frozenchache.blogspot.com/`
17. Microsoft Corporation. Windows BitLocker Drive Encryption: Technical Overview. Microsoft (July 2009)
18. Müller, T., Dewald, A., Freiling, F.: AESSE: A Cold-Boot Resistant Implementation of AES. In: Proceedings of the Third European Workshop on System Security (EUROSEC), Paris, France, pp. 42–47. RWTH Aachen / Mannheim University, ACM (April 2010)
19. Müller, T., Freiling, F., Dewald, A.: TRESOR Runs Encryption Securely Outside RAM. In: 20th USENIX Security Symposium, San Francisco, California. University of Erlangen-Nuremberg, USENIX Association (August 2011)
20. Osvik, D.A., Shamir, A., Tromer, E.: Cache Attacks and Countermeasures: The Case of AES. In: Pointcheval, D. (ed.) CT-RSA 2006. LNCS, vol. 3860, pp. 1–20. Springer, Heidelberg (2006)

21. Panholzer, P.: Physical Security Attacks on Windows Vista. Technical report. SEC Consult Vulnerability Lab, Vienna (May 2008)
22. Parker, T.P., Xu, S.: A Method for Safekeeping Cryptographic Keys from Memory Disclosure Attacks. In: Chen, L., Yung, M. (eds.) INTRUST 2009. LNCS, vol. 6163, pp. 39–59. Springer, Heidelberg (2010)
23. Simmons, P.: Security Through Amnesia: A Software-Based Solution to the Cold Boot Attack on Disk Encryption. CoRR, abs/1104.4843. University of Illinois at Urbana-Champaign (2011)
24. Ponemon, L.: 2010 Annual Study: U.S. Enterprise Encryption Trends. Ponemon Institute, Symantec (2010)
25. Graham, R.D.: Thunderbolt: Introducing a new way to hack Macs. Errata Security, `http://erratasec.blogspot.com/2011/02/thunderbolt-introducing-new-way-to-hack.html` (February 2011)
26. Sacco, A.L., Ortega, A.A.: Persistent BIOS Infection: The early bird catches the worm. In: Proceedings of the Annual CanSecWest Applied Security Conference, Vancouver, British Columbia, Canada. Core Security Technologies (2009)
27. Gueron, S.: Intel Advanced Encryption Standard (AES) Instruction Set White Paper. Intel Corporation, Rev. 3.0 edn. Intel Mobility Group, Israel Development Center (January 2010)
28. Shinagawa, T., Eiraku, H., Omote, K., Hasegawa, S., Hirano, M., Kourai, K., Oyama, Y., Kawai, E., Kono, K., Chiba, S., Shinjo, Y., Kato, K.: In: International Conference on Virtual Execution Environments, Washington, DC, USA. University of Tsukuba (March 2009)
29. Richard Stallman and Jerry Cohen. GNU General Public License Version 2. Free Software Foundation (June 1991)
30. TrueCrypt Foundation. TrueCrypt: Free Open-Source Disk Encryption Software for Windows, Mac OS and Linux (2010), `http://www.truecrypt.org/`

Authenticated Encryption: How Reordering Can Impact Performance*

Basel Alomair

Computer Research Institute (CRI)
King Abdulaziz City for Science and Technology (KACST)
alomair@kacst.edu.sa

Abstract. In this work, we look at authenticated encryption schemes from a new perspective. As opposed to analyzing the *security* of different methods of constructing authenticated encryption schemes, we investigate the effect of the method used to construct an authenticated encryption scheme on the *performance* of the construction. We show that, by performing the authentication operation before the encryption operation, the security requirements on the authentication operation can be relaxed, leading to more efficient constructions, without affecting the security of the overall construction.

Keywords: Universal hash-function families, pseudorandom permutations, authenticated encryption, provable security.

1 Introduction

There are three different methods to generically compose an authenticated encryption scheme by combining an encryption algorithm with a MAC algorithm: Encrypt-and-MAC ($E\&M$), Encrypt-then-MAC (EtM), or MAC-then-Encrypt (MtE). Although significant efforts have been devoted to analyzing the security implications of different generic compositions (see, e.g., [10, 29]), little effort has been devoted to the study of the performance implications of different generic compositions [2]. Of particular interest to this work is the performance aspect of generic compositions when the encryption algorithm is block cipher based and the MAC algorithm is universal hash-function family based. (We focus on such constructions since block ciphers are the recommended building block for secure encryption [25] and since universal hash families based MACs are the fastest method for message authentication [39].)

In a typical EtM composition, the plaintext is broken into blocks. Each block is processed with a block cipher, resulting in a ciphertext block. The resulting ciphertext blocks are then authenticated using a MAC based on a universal hash-function family (in the Carter-Wegman style [15]). One of the most recent authenticated encryption schemes is the Carter-Wegman Counter (CWC) block cipher mode of authenticated encryption proposed by Kohno et al. in [27]. (The

* A more complete version of this paper can be found in [3].

F. Bao, P. Samarati, and J. Zhou (Eds.): ACNS 2012, LNCS 7341, pp. 84–99, 2012.

National Institute of Standards and Technology (NIST) has adopted the CWC mode of operation in the standardized Galois/Counter Mode (GCM) of authenticated encryption [17].) The CWC mode of operation gives high-performance authenticated encryption by combining the counter mode of encryption with a Wegman-Carter universal hash-function family for authentication.

The OKH Solution. In this work, we investigate the performance implications of the order in which the two operations, encryption and authentication, are performed. We describe the Odd Key Hashing (OKH) mode of authenticated encryption. The OKH mode is motivated by the CWC mode of authenticated encryption proposed by Kohno et al. [27]. However, unlike the CWC and the GCM schemes, the order of encrypt-then-authenticate is reversed in the OKH mode. That is, as opposed to applying the hashing operation on the ciphertext, it is applied on the plaintext, before block cipher encryption. The main result of this study is to show that, while the hash family used to construct a MAC in the *EtM* composition must be universal, this need not be the case in the *MtE* composition.[1] The performance implication of this result is that, since the hash family need not be universal, it can be computed faster than the fastest universal hash family in the cryptographic literature. The theoretical significance of this result is that relaxing the security requirements on the MAC algorithm does not affect the provable security of the overall authenticated encryption composition.

Background and Related Work. The notion of authenticated encryption was introduced independently by Katz and Yung in [26], and by Bellare and Rogaway in [11]. Since then, many authenticated encryption schemes have been proposed, such as, RPC of Katz and Yung [26], XECB of Gligor and Donescu [20], IAPM of Jutla [24], OCB of Rogaway et al. [37], EAX of Bellare et al. [12], and CWC of Kohno et al. [27]. Alomair and Poovendran showed that one can utilize the *E&M* composition to eliminate redundant computations in the MAC algorithm in order to come up with more efficient generic constructions [5,6]. Stream cipher based authenticated encryption primitives have appeared in [19,40]. However, these stream cipher based proposals have been analyzed and shown to be vulnerable to attacks [31,33,34,41].

The use of universal hash-function families to construct MAC algorithms is due to Carter and Wegman [15]. Compared to block cipher based MACs, such as [9,16], and cryptographic hash function based MACs, such as [7,35], universal hashing based MACs lead to faster message authentication [14,21,28,36]. The speed of a universal hash family based MAC relies mainly on the speed of the used universal hash family. The security of MACs based on universal hashing has been extensively studied. In [22], key recovery attacks against universal hash functions was introduced. In [4], it was shown that the security of universal hashing based on integer arithmetic is proportional to the smallest prime factor of the used modulus.

[1] Although the same result can be shown for the *E&M* composition, we restrict our discussion to the *MtE* composition for brevity.

There has been a significant effort for the design of fast universal hash families. In [28], Krawczyk introduced the cryptographic CRC which hashes in about 6 cycles/byte, as shown by Shoup in [38]. In [36], Rogaway proposed the bucket hashing which was the first hash family explicitly targeted for fast software implementation; it runs in about $1.5 - 2.5$ cycles/byte [14]. In [23], Johansson described bucket hashing with smaller key size. In [21], Halevi and Krawczyk proposed the MMH family, which hashes at about $1.2 - 3$ cycles/byte. In [18], Etzel et al. proposed the square hash, an MMH-variant that can be more efficient than MMH in certain settings [14]. In [13], Bernstein proposed floating-point arithmetic based hash function that achieves a peak speed of 2.4 cycles/byte. In [1], Afanassiev et al. described an application of hashing based on polynomial evaluation over finite fields. In [32], Nevelsteen and Preneel study the performance of several universal hash functions proposed for MACs. The speed champion of universal hash functions directed for software implementation is the NH family of Black et al. proposed in [14]. The NH family is an extension to the MMH family of [21]. The speed improvement comes from eliminating the non-trivial modular reduction required by the MMH family. The novelty of NH family is that it uses arithmetic modulo powers of two, i.e., "computations that computers like to do [14]." The NH family hashes at about 0.34 cycles/byte for 2^{-32} probability of message collision.

Organization. In Section 2, we give some preliminaries. In Section 3, we describe the OK hash family. In Section 4, we describe the construction of the proposed OKH authenticated encryption scheme. In Section 5, we state and prove the authenticity and privacy theorems of the proposed scheme. In Section 6, we summarize the basic ideas behind the proposed mode of operation and provide performance comparisons. In Section 7, we conclude the paper.

2 Notations and Preliminaries

2.1 Notations

- If s is a binary string, $|s|$ denotes the length of s in bits.
- For a positive integer β, $\{0,1\}^\beta$ denotes a binary string of length β-bits, and $\{0,1\}^*$ denotes a binary string of arbitrary length.
- If b is a bit and β is a positive integer, we denote by b^β the concatenation of b with itself β times.
- For a non-empty set $\mathcal{H}$, we denote by $h \xleftarrow{\$} \mathcal{H}$ the selection of a member of $\mathcal{H}$ uniformly at random and assigning it to h.
- If x and n are positive integers so that $0 \le x < 2^n$, we denote by $\mathsf{tostr}(x, n)$ the binary representation of x as an n-bit string (in a big-endian format).
- If s is a binary string, we denote by $\mathsf{toint}(s)$ the unsigned integer representation of s (in a big-endian format).
- If s is a binary string and ℓ is a positive integer, we denote by $\mathsf{setlen}(s, \ell)$ the truncation of s into its ℓ most significant bits. If $|s| < \ell$, then $\mathsf{setlen}(s, \ell)$ denotes the ℓ-bit long string $s||0^{\ell-|s|}$.

2.2 Universal Hash-Function Families

A family of hash functions $\mathcal{H}$ is specified by a finite set of keys $\mathcal{K}$. Each key $k \in \mathcal{K}$ defines a member of the family $\mathcal{H}_k \in \mathcal{H}$. As opposed to thinking of $\mathcal{H}$ as a set of functions from $\mathcal{D}$ to $\mathcal{R}$, it can be viewed as a single function $\mathcal{H} : \mathcal{K} \times \mathcal{D} \rightarrow \mathcal{R}$, whose first argument is usually written as a subscript. A random element $h \xleftarrow{\$} \mathcal{H}$ is determined by selecting a $k \xleftarrow{\$} \mathcal{K}$ uniformly at random and setting $h \leftarrow \mathcal{H}_k$. There are many classes of universal hash families, depending on their probability of message collision, we give below a formal definition of one class of universal hash families called ϵ-almost universal.

Definition 1. *Let $\mathcal{H} = \{h : \mathcal{D} \rightarrow \mathcal{R}\}$ be a family of hash functions and let $\epsilon \geq 0$ be a real number. $\mathcal{H}$ is said to be ϵ-almost universal, denoted ϵ-AU, if for all distinct $M, M' \in \mathcal{D}$, we have that $\Pr_{h \leftarrow \mathcal{H}}\left[h(M) = h(M')\right] \leq \epsilon$. $\mathcal{H}$ is said to be ϵ-almost universal on equal-length strings if for all distinct, equal-length strings $M, M' \in \mathcal{D}$, we have that $\Pr_{h \leftarrow \mathcal{H}}\left[h(M) = h(M')\right] \leq \epsilon$.*

2.3 Block Ciphers

A block cipher mapping ℓ-bit strings to ℓ-bit strings is a family of permutations, $\mathcal{F}$, specified by a finite set of keys, $\mathcal{K}_e$. Each key $K \in \mathcal{K}_e$ defines a member of the family $\mathcal{F}_K \in \mathcal{F}$. As opposed to thinking of $\mathcal{F}$ as a set of functions mapping elements from $\{0,1\}^\ell$ to elements in $\{0,1\}^\ell$, it can be viewed as a single function $\mathcal{F} : \mathcal{K}_e \times \{0,1\}^\ell \rightarrow \{0,1\}^\ell$, whose first argument is usually written as a subscript. A random element $f \xleftarrow{\$} \mathcal{F}$ is determined by selecting a $K \xleftarrow{\$} \mathcal{K}_e$ uniformly at random and setting $f \leftarrow \mathcal{F}_K$.

As in [27], we adopt the notion of security for block ciphers introduced in [30] and adopted for the concrete setting in [8]. Let $\mathcal{F} : \{0,1\}^L \times \{0,1\}^\ell \rightarrow \{0,1\}^\ell$, where L is the key length and ℓ is the block size of the block cipher, be a block cipher and let $\mathsf{Perm}(\ell)$ denote the set of all permutations on $\{0,1\}^\ell$. Let $\mathcal{A}$ be an adversary with access to an oracle and that returns a bit. Then,

$$\mathsf{Adv}^{\mathrm{prp}}_{\mathcal{F}}(\mathcal{A}) = \Pr\left[f \xleftarrow{\$} \mathcal{F} : \mathcal{A}^{f(\cdot)} = 1\right] - \Pr\left[\pi \xleftarrow{\$} \mathsf{Perm}(\ell) : \mathcal{A}^{\pi(\cdot)} = 1\right] \qquad (1)$$

denotes the prp-advantage of $\mathcal{A}$ in distinguishing a random instance of $\mathcal{F}$ from a random permutation. Intuitively, we say that $\mathcal{F}$ is a secure PRP, or a secure block cipher, if the prp-advantages of all adversaries using reasonable resources is small.

A block cipher is said to be strong pseudorandom permutation (sprp) if it is indistinguishable from a random permutation even if the adversary is given an oracle access to the inverse function. Then,

$$\mathsf{Adv}^{\mathrm{sprp}}_{\mathcal{F}}(\mathcal{A}) = \Pr\left[f \xleftarrow{\$} \mathcal{F} : \mathcal{A}^{f(\cdot), f^{-1}(\cdot)} = 1\right] - \Pr\left[\pi \xleftarrow{\$} \mathsf{Perm}(\ell) : \mathcal{A}^{\pi(\cdot), \pi^{-1}(\cdot)} = 1\right] \qquad (2)$$

denotes the sprp-advantage of $\mathcal{A}$ in distinguishing a random instance of $\mathcal{F}$ from a random permutation. Modern block ciphers, such as AES, are believed to be secure SPRPs.

2.4 Authenticated Encryption Schemes

The authenticated encryption model that we use is similar to the one in [27,37]. A nonce-using, symmetric authenticated encryption scheme $\mathcal{AE} = (\mathcal{K}, \mathcal{SE}, \mathcal{VD})$ consists of three algorithms: the key generation algorithm ($\mathcal{K}$), the signed encryption algorithm ($\mathcal{SE}$), and the verified decryption algorithm ($\mathcal{VD}$). $\mathcal{AE}$ is defined over some key space KeySp, some nonce space $\mathsf{NonceSp} = \{0,1\}^{\mathsf{nl}}$, for a positive integer nl, and some message space $\mathsf{MsgSp} = \{0,1\}^*$. We require that membership in MsgSp can be efficiently tested and that if M, M' are two strings such that $M \in \mathsf{MsgSp}$ and $|M| = |M'|$, then $M' \in \mathsf{MsgSp}$.

The randomized key generation algorithm $\mathcal{K}$ returns a key $K \in \mathsf{KeySp}$. The deterministic signed encryption algorithm $\mathcal{SE}$ takes as input a key $K \in \mathsf{KeySp}$, a nonce $N \in \mathsf{NonceSp}$, and a payload message $M \in \mathsf{MsgSp}$, and returns a ciphertext $\sigma \in \{0,1\}^*$. The deterministic verified decryption algorithm $\mathcal{VD}$ takes as input a key $K \in \mathsf{KeySp}$, a nonce $N \in \mathsf{NonceSp}$, a string $\sigma \in \{0,1\}^*$, and outputs a message $M \in \mathsf{MsgSp}$ or the special symbol $\mathsf{INVALID}$ on error. We ask for the basic validity requirement that if $\sigma = \mathcal{SE}_K(N, M)$ then it must be the case that $\mathcal{VD}_K(N, \sigma) = M$.

2.5 Adversarial Model

We adopt the standard adversarial model used in authenticated encryption schemes. The adversary is given oracle access to the signed encryption algorithm $\mathcal{SE}_K(N, M)$. The adversary can call the $\mathcal{SE}$ oracle on nonce-message pairs (N, M) of her choice and observing the outputs. After calling the $\mathcal{SE}$ oracle for q times, the adversary attempts a forgery by calling the verified decryption algorithm $\mathcal{VD}_K(N, \sigma)$ for an (N, σ) pair of her choice. Note that the adversary does not see the secret key K. If the verified decryption oracle returns the $\mathsf{INVALID}$ symbol, the adversary is considered unsuccessful; otherwise, the forgery attempt is said to be successful.

A standard assumption in authenticated encryption schemes is that the adversary is nonce-respecting. An adversary is said to be nonce-respecting if she never repeats a nonce. That is, after calling the signed encryption oracle on (N, M), the adversary never asks its oracle a query (N, M'), regardless of the oracle responses. We emphasize, however, that the nonce used in the forgery attempt may coincide with a nonce used in one of the adversary's queries.

2.6 Properties of Odd Integers

We state here two lemmas about odd integers in the finite integer ring $\mathbb{Z}_{2^n}$ that will be used for the rest of the paper.

Lemma 1. *For any nonzero integers α and β in $\mathbb{Z}_{2^n}$, 2^n divides $\alpha\beta$ only if both α and β are even integers. Formally, the following one-way implication must hold:*

$$\alpha\beta \equiv 0 \mod 2^n \quad \Rightarrow \quad \alpha \equiv \beta \equiv 0 \mod 2. \tag{3}$$

Lemma 2. *Let $\boldsymbol{X_1}$ be the random variable representing the experiment of drawing a number x_1 from the set of integers $\{0, 1, \cdots, 2^n - 1\}$ uniformly at random. Then, for any odd integer $k \in \mathbb{Z}_{2^n}$, the random variable $\boldsymbol{Y_1} = k \cdot \boldsymbol{X_1} \mod 2^n$ is uniformly distributed over the set $\{0, 1, \cdots, 2^n - 1\}$.*

Lemmas 1 and 2, along with the fact that there is a one-to-one correspondence between n-bit strings and the integer ring $\mathbb{Z}_{2^n}$, will be used to establish the results of this paper. The proofs of the two lemmas can be found in [3].

3 The Odd Key Hash Family

In this section, we give a description of the OK hash family that will be used in the construction of our OKH authenticated encryption. Fix an integer $n \geq 1$ (the "block size") and an integer $b \geq 1$ (the "number of blocks"). We define the family of functions $\mathsf{OK}[n, b]$ as follows. The domain is $\mathcal{D} = \{0,1\}^n \cup \{0,1\}^{2n} \cup \cdots \cup \{0,1\}^{bn}$ and the range is $\mathcal{R} = \{0,1\}^n$. Each function in $\mathsf{OK}[n, b]$ is defined by the b-tuple $K = (k_1, \cdots, k_b)$, where $k_i \in \mathbb{Z}^*_{2^n}$ for $i = 1, \cdots, b$. A random function in $\mathsf{OK}[n, b]$ is given by drawing the k_i's at random from the multiplicative group $\mathbb{Z}^*_{2^n}$. The function determined by K is written as $\mathsf{OK}_K(\cdot)$.

For an input message $M \in \mathcal{D}$, view M as a sequence of n-bit blocks, i.e., $M = m_1 \cdots m_\ell$, where $\ell \leq b$, and write each block in its unsigned integer representation in $\mathbb{Z}_{2^n}$ (in a big-endian format). Then, the compressed image of M is given by

$$\mathsf{OK}_K(M) = \sum_{i=1}^{\ell} k_i m_i \mod 2^n. \tag{4}$$

When the values of n and b are known, we will write OK instead of $\mathsf{OK}[n, b]$ to simplify the notations.

4 Description of the OKH Authenticated Encryption

As mentioned earlier, the key idea allowing for more efficient authentication over the CWC mode of operation is advancing the hashing phase to be applied on the plaintext instead of the ciphertext. The mode of operation used for encryption is similar to the counter mode (CTR) but with the the requirement that the plaintext is to be processed by the block cipher, as illustrated in Figure 1.

As in previous authenticated encryption schemes, we require the block cipher BC to be a strong pseudorandom permutation. Let BC: $\{0,1\}^{\mathsf{kl}} \times \{0,1\}^{\mathsf{bs}} \rightarrow \{0,1\}^{\mathsf{bs}}$ be the used block cipher, where kl and bs are the key length and block size of the block cipher, respectively. The authenticated encryption using BC for

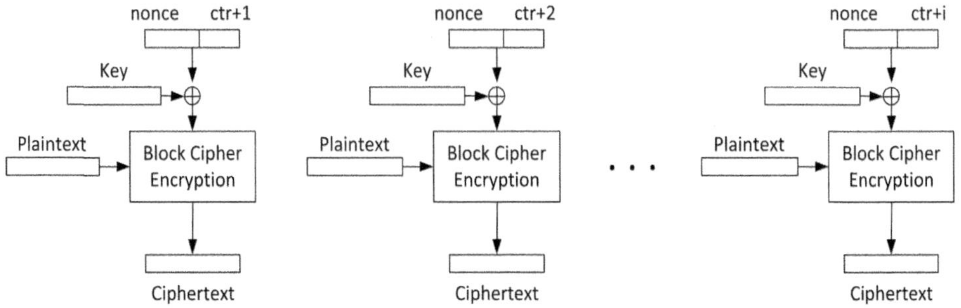

Fig. 1. The used mode of encryption to construct the OKH authenticated encryption. Note that the nonce-counter concatenation is XORed with the key, not the plaintext block. Therefore, given a nonce-respecting adversary, the encryption key of each block is different than all other blocks of the same message and different than all other blocks of different messages due to the use of nonce-counter concatenation. This observation is critical for the security of the proposed scheme.

encryption and the OK family for hashing, OKH $= (\mathcal{K}, \mathcal{SE}, \mathcal{VD})$, is defined as follows. The message spaces are

$$\mathsf{MsgSp} = \Big\{M \in \{0,1\}^* : |M| \leq \mathsf{MaxLen}\Big\}, \tag{5}$$

$$\mathsf{KeySp} = \Big\{(K_e, K_h) \in \{0,1\}^{\mathsf{kl}} \times \{0,1\}^{\mathsf{MaxLen}}\Big\}, \tag{6}$$

$$\mathsf{NonceSp} = \Big\{N \in \{0,1\}^* : |N| \leq \mathsf{kl} - \log_2 \mathsf{MaxLen}\Big\}, \tag{7}$$

where MaxLen is the message maximum length. The size of the counter, CtrLen, in the mode of operation of Figure 1 is at least $\log_2 \mathsf{MaxLen}$. The concatenation of the nonce and the counter is of length kl-bits.

Informally speaking, the authentication tag is computed by dividing the plaintext message into blocks of n-bit long, hash it according to equation (4) using a member of the OK family, and encrypt the resulting n-bit hashed image (as shown in the MAC algorithm below). The size of the hashing images, n, is less than or equal to the block cipher size, bs. The encryption part is done the natural way (as shown in the $\mathcal{E}$ algorithm below).

Remark 1. There are two important points to note about the OK family. First, the OK family is defined over the domain $\mathcal{D}$ only. This issue, however, can be easily solved with an appropriate padding (e.g. with zeros as we do in this paper). Second, and more important, as in universal hash families, the OK family as described in Section 3 can only be used to authenticate equal-length messages. For example, a message block consisting of all zeros will not contribute to the value of the hashed image. Hence, it is easy to come up with two distinct messages that collide and, eventually, achieve a successful forgery. However, there are known techniques to make the hash function applicable to arbitrary-length messages. For instance, in [14] the authors proposed appending the length of

Algorithm $\mathcal{K}$	Algorithm $\mathcal{SE}_K(N, M)$	Algorithm $\mathcal{VD}_K(N, \sigma)$
$K_e \overset{\$}{\leftarrow} \{0,1\}^{\mathsf{kl}}$	$C \longleftarrow \mathcal{E}_{K_e}(N, M)$	if $\lvert\sigma\rvert \leq \mathsf{tl}$ then return INVALID
$K_h \leftarrow$ KeyGenOK	$\tau \leftarrow \mathsf{MAC}_{K_h}(N, M)$	parse σ as $C \| \tau$ where $\lvert\tau\rvert = \mathsf{tl}$
return $K = K_e \parallel K_h$	return $\sigma = C \parallel \tau$	if $C \notin$ MsgSp then return INVALID
		$M \longleftarrow \mathcal{D}_{K_e}(N, C)$
		if $\tau \neq \mathsf{MAC}_{K_h}(N, M)$ then return INVALID
		return M

Fig. 2. Pseudocodes of the key generation ($\mathcal{K}$), signed encryption ($\mathcal{SE}$), and verified decryption ($\mathcal{VD}$) algorithms

the message at the end. In our case, it suffices to append the bit '1' at the end of the message since this will guarantee that changing the message length will change the hashed image in an unpredictable manner depending on the hashing key corresponding to the last message block.

Another important remark is related to the message length. For a message that is longer than MaxLen, it is treated as multiple chunks of length MaxLen or less and the corresponding tags are concatenated; that is, arbitrary long messages can be authenticated using the same fixed-length hashing key (this is actually the case for any universal hashing based MAC, not just the proposed one [14]). In typical settings, however, one need not worry about such messages since MaxLen is often sufficiently long. For instance, typical AES key-lengths are 128, 192, or 256. Assuming the shortest key-length of $\mathsf{kl} = 128$ bits, and setting $|N| = 88$ and $\mathsf{CtrLen} = 40$ as can be found in [27], MaxLen can be more than Tera bits long, while one is often dealing with much shorter messages (about third of the messages on the backbone of the Internet, for instance, are only 43 bytes [37]). For the rest of the paper, we will assume messages of length MaxLen or less for simplicity.

The OKH's key generation ($\mathcal{K}$), signed encryption ($\mathcal{SE}$), and verified decryption ($\mathcal{VD}$) algorithms are defined as in Figure 2. The rest of the algorithms (KeyGenOK, $\mathcal{E}$, $\mathcal{D}$, MAC, OK-HASH) are defined in Figure 3. Algorithm KeyGenOK handles the generation of the key that defines the used member of the OK hashing family. Algorithms $\mathcal{E}$ and $\mathcal{D}$ handle the encryption and decryption corresponding to the mode of operation depicted in Figure 1. Algorithm MAC handles the generation of authentication tags, which calls algorithm OK-HASH to compress the message.

5 Theorem Statements

In this section, we give the main security statements of the proposed scheme, formal security proofs can be found in the full version [3].

5.1 Security of Authentication

Fix an authenticated encryption scheme OKH $= (\mathcal{K}, \mathcal{SE}, \mathcal{VD})$ and run an adversary $\mathcal{A}$ with an oracle $\mathcal{SE}_K(\cdot,\cdot)$ for some key K. The adversary $\mathcal{A}$

Algorithm KeyGenOK
- $\mathsf{Key} \xleftarrow{\$} \{0,1\}^{\mathsf{MaxLen}}$
- $\alpha \leftarrow |\mathsf{Key}|/\mathsf{tl}$
- break Key into tl-bit chunks K_i's
- for $i = 1$ to α do
  - $K_i \leftarrow K_i \vee \mathsf{tostr}(1, \mathsf{tl})$
- return $K_h = K_1 || \cdots || K_\alpha$

Algorithm $\mathcal{E}_{K_e}(N, M)$
- $\ell \leftarrow$ min int so that bs divides $|M||0^\ell|$
- $M \leftarrow M||1||0^{\ell-1}$
- $\alpha \leftarrow |M|/\mathsf{bs}$
- break M into bs-bit chunks M_i's
- for $i = 1$ to α do
  - $K_{e_i} \leftarrow K_e \oplus (N||\mathsf{tostr}(i, \mathsf{CtrLen}))$
  - $c_i \leftarrow \mathsf{BC}_{K_{e_i}}(M_i)$
- return $C = c_1 || \cdots || c_\alpha$

Algorithm $\mathcal{D}_{K_e}(N, C)$
- $\alpha \leftarrow |C|/\mathsf{bs}$
- break C into bs-bit chunks c_i's
- for $i = 1$ to α do
  - $K_{e_i} \leftarrow K_e \oplus (N||\mathsf{tostr}(i, \mathsf{CtrLen}))$
  - $M_i \leftarrow \mathsf{BC}^{-1}_{K_{e_i}}(c_i)$
- return $M = M_1 || \cdots || M_\alpha$

Algorithm $\mathsf{MAC}_{K_h}(N, M)$
- $\gamma' \leftarrow \mathsf{OK\text{-}HASH}_{K_h}(M)$
- $\ell \leftarrow \mathsf{bs} - n$
- $\gamma \leftarrow \gamma' || 0^\ell$
- $\tau' \leftarrow \mathsf{BC}_{K_e}(\gamma \oplus \mathsf{setlen}(N, \mathsf{bs}))$
- $\tau \leftarrow \mathsf{setlen}(\tau', \mathsf{tl})$
- return τ

Algorithm $\mathsf{OK\text{-}HASH}_{K_h}(M)$
- $\ell \leftarrow$ min int so that n divides $|M||0^\ell|$
- $M \leftarrow M||1||0^{\ell-1}$
- $\alpha \leftarrow |M|/n$
- break M into n-bit chunks M_i's
- for $i = 1$ to α do
  - $m_i \leftarrow \mathsf{toint}(M_i)$
  - $k_i \leftarrow \mathsf{toint}(K_i)$
- $\gamma \leftarrow \sum_{i=1}^{\alpha} k_i m_i \mod 2^n$
- return γ

Fig. 3. Pseudocodes of the KeyGenOK, $\mathcal{E}$, $\mathcal{D}$, MAC, OK-HASH

successfully forges in this run if $\mathcal{A}$ is nonce-respecting, $\mathcal{A}$ outputs a pair (N, σ) where $\mathcal{VD}_K(N, \sigma) \neq \mathsf{INVALID}$, and $\mathcal{A}$ made no earlier query (N, M) which resulted in the response σ. Let $\mathsf{Adv}^{\mathrm{auth}}_{\mathrm{OKH}[\mathsf{BC},\mathsf{OK}]}(\mathcal{A}) = \Pr\left[K \xleftarrow{\$} \mathcal{K} : \mathcal{A}^{\mathcal{SE}_K(\cdot,\cdot)} \text{ forges}\right]$ be the adversary's advantage of successful forgery against the scheme OKH that uses BC as a block cipher for encryption and the OK family for hashing.

We give here information-theoretic bounds on the authenticity of the scheme of Section 4 assuming the use of a true random permutation, $\mathsf{Perm}(\ell)$, for encryption.

Theorem 1. *Fix an* $\mathsf{OK}[n, b]$ *hash family and let* $\mathsf{Perm}(\ell) : \{0,1\}^\ell \to \{0,1\}^\ell$ *be a true random permutation and let* tl *be the desired tag length. Let* $\mathcal{A}$ *be a nonce-respecting adversary that asks* q *queries and then makes its forgery attempt against the* OKH *of Section 4. Then,* $\mathcal{A}$*'s advantage of successful forgery is bounded by*

$$\mathsf{Adv}^{\mathrm{auth}}_{\mathrm{OKH}[\mathsf{Perm}(\ell),\mathsf{OK}_{[n,b]}]}(\mathcal{A}) \leq 2^{-n} + 2^{-\mathsf{tl}}.$$

It is standard to pass a complexity-theoretic analog of Theorem 1, but in doing this one will need access to a BC^{-1} oracle in order to verify a forgery attempt, which translates into needing the strong pseudorandom permutation assumption. One gets the following. Fix an $\mathsf{OK}[n, b]$ hash family and a block cipher $\mathsf{BC} : \mathcal{K} \times \{0,1\}^\ell \to \{0,1\}^\ell$. Let $\mathcal{A}$ be a nonce-respecting adversary that asks q queries totaling at most λ bits of payload and then makes its forgery attempt. Then, there is an adversary $\mathcal{B}$ attacking the block cipher in which

$$\mathsf{Adv}^{\mathrm{auth}}_{\mathrm{OKH}[\mathsf{BC},\mathsf{OK}]}(\mathcal{A}) \leq \mathsf{Adv}^{\mathrm{sprp}}_{\mathsf{BC}}(\mathcal{B}) + 2^{-n} + 2^{-\mathsf{tl}}.$$

Furthermore, adversary $\mathcal{B}$ takes the same time adversary $\mathcal{A}$ takes and makes at most $\lambda/\ell + q$ oracle queries.

Before we proceed with the security analysis, we give the intuition behind the choices of the OK hash family and the mode of encryption in Figure 1. First, observe that the OK family is not a universal hash family. To see this, let the OK family be defined over the ring $\mathbb{Z}_{2^n}$. Let $M = m_1||m_2$ be a message of two n-bit blocks (for the rest of the paper, we overload m_i to denote both the n-bit binary string of the ith message block and its integer representation as an element of $\mathbb{Z}_{2^n}$ in a big-endian format; the distinction between the two representations will be omitted as long as it is clear from the context). Consider now the message $M' = m'_1||m'_2 = m_1 + 2^{n-1}||m_2 + 2^{n-1} \neq M$. Then,

$$\begin{aligned} k_1 m'_1 + k_2 m'_2 &= k_1\left(m_1 + 2^{n-1}\right) + k_2\left(m_2 + 2^{n-1}\right) \\ &\equiv k_1 m_1 + k_2 m_2 \mod 2^n, \end{aligned} \tag{8}$$

where equation (8) holds since both k_1 and k_2 are odd integers by design.

Remark 2. Equation (8) implies that the OK family, unlike universal hash families, cannot be used to construct standard MACs since forgers can easily find two colliding messages. However, one key idea of the proposed OKH is that finding two messages that collide does not translate into a successful forgery (since the adversary must also predict the correct ciphertext corresponding to the colliding messages). The other key idea behind the design of OKH is that the effect of modifying an observed ciphertext block will result in modifying its corresponding plaintext block randomly (assuming the block cipher is a strong pseudorandom permutation). For that, the following lemma addresses the adversary's chances of causing a collision in the hashing phase by modifying the ciphertext.

Lemma 3. *Let $C \neq C'$ be any two distinct ciphertexts and let $M \neq M'$ be the plaintext messages corresponding to C and C', respectively. Then, assuming the use of a random permutation for block encryption,* $\Pr_{h \leftarrow \mathsf{OK}[n,b]}\left[h(M) = h(M')\right] \leq 2^{-n}$.

Proof. Let c_i and c'_i denote the ith blocks of C and C', respectively. Similarly, let m_i and m'_i denote the ith blocks of M and M', respectively. Since $M \neq M'$, they must be different in at least one block. Assume M and M' differ in a single block only. Without loss of generality, let $m'_1 = m_1 + \epsilon \neq m_1$ and the rest of the blocks are the same. Then, since k_1 is an odd integer, by Lemma 1, $k_1 \cdot \epsilon \not\equiv 0 \mod 2^n$ and the probability $\Pr_{h \leftarrow \mathsf{OK}[n,b]}\left[h(M) = h(M')\right] = 0$.

Assume now that M and M' differ by more than one block. Write $m'_i = m_i + \epsilon_i \neq m_i$ for each block i in which the two messages differ. Since a random permutation is used for encryption, even if c'_i differs with c_i by a known constant,

ϵ_i will be a random element of $\mathbb{Z}_{2^n}$ (for any user with no knowledge of the encryption key). Therefore, by Lemma 2, $\Pr_{h \leftarrow \mathsf{OK}[n,b]}\left[h(M) = h(M')\right] = 2^{-n}$ and the lemma follows. □

Remark 3. Lemma 3 illustrates the significance of restricting the hashing keys to the set of odd integers. To see this, let the keys be drawn from $\mathbb{Z}_{2^n}$ as opposed to $\mathbb{Z}^*_{2^n}$. Assume that k_i, for some i, happened to be equal to 2^{n-1}. Then, using the fact that the used block cipher can be modeled as a strong pseudorandom permutation, any modification of the ith ciphertext block will go undetected with a probability 1/2 (this is because $\epsilon \cdot 2^{n-1}$ is congruent to zero modulo 2^n for any even ϵ). In general, if k_i is equal to $2^{n-\ell}$, for any positive integer $\ell < n$, then any modification of the ith ciphertext block will go undetected with a probability $1/2^\ell$.

With Remark 2 and Lemma 3 in mind, one can proceed with the formal proof of Theorem 1, which can be found in [3].

5.2 Security of Encryption

In this section, we show that the privacy of the proposed scheme is provably secure assuming the used block cipher is a secure pseudorandom permutation. Consider an adversary $\mathcal{A}$ who has one of two types of oracles: a real encryption oracle and a fake encryption oracle. The real encryption oracle $\mathcal{E}_K(\cdot,\cdot)$ takes as input a pair (N, M) and returns a ciphertext $C \overset{\$}{\leftarrow} \mathcal{E}_K(N, M)$. Assume that the length of the ciphertext depends only on the length of the plaintext, that is, $|C| = l(|M|)$. The fake encryption oracle, $\$(\cdot,\cdot)$, takes as input a pair (N, M) and returns a random string $C \overset{\$}{\leftarrow} \{0,1\}^{l(|M|)}$. Given adversary $\mathcal{A}$ and authenticated encryption scheme OKH $= (\mathcal{K}, \mathcal{SE}, \mathcal{VD})$, define

$$\mathsf{Adv}^{\text{priv}}_{\text{OKH}[\mathsf{BC},\mathsf{OK}]}(\mathcal{A}) = \Pr\left[K \overset{\$}{\leftarrow} \mathcal{K} : \mathcal{A}^{\mathcal{SE}_K(\cdot,\cdot)} = 1\right] - \Pr\left[\mathcal{A}^{\$(\cdot,\cdot)} = 1\right]$$

to be $\mathcal{A}$'s advantage of breaking the privacy of the authenticated encryption scheme using BC as a block cipher and OK for hashing. That is, as in previous authenticated encryption proposals (e.g., [27, 37]), the strong model of distinguishing the ciphertext from a random string is used to model the privacy of encryption.

Theorem 2. *Let* OKH[BC;OK] *be the authenticated encryption scheme described in Section 4 using the* OK *hash family for compression and the block cipher* BC *for encryption. Then given a nonce-respecting adversary, $\mathcal{A}$, against* OKH[BC;OK], *one can construct an adversary $\mathcal{B}$ against* BC *such that*

$$\mathsf{Adv}^{\text{priv}}_{\text{OKH}[\mathsf{BC;OK}]}(\mathcal{A}) \leq \mathsf{Adv}^{\text{prp}}_{\mathsf{BC}}(\mathcal{B}). \tag{9}$$

Furthermore, the experiment for $\mathcal{B}$ takes the same time as the experiment for $\mathcal{A}$ and, if $\mathcal{A}$ makes at most q oracle queries totaling at most μ bits of payload data, then $\mathcal{B}$ makes at most $\mu/\ell + q$ oracle queries.

Table 1. Performance comparison of the MMH hash family of Halevi and Krawczyk [21], the polynomial-evaluation (POLY) hash family of Bernstein [13], the NH hash family of Black et al. [14], and the proposed OK family.

	MMH family	POLY family	NH family	OK family
Collision probability	2^{-30}	2^{-96}	2^{-64}	2^{-64}
Hashed image (bits)	32	128	128	64
Speed (cycles/byte)	1.2	2.4	0.84	0.27

Theorem 2 states that, if BC is a secure pseudorandom permutation, then the proposed authenticated encryption scheme provides data privacy; the formal proof can be found in [3].

6 Design and Performance Discussions

In this section, we discuss the main design ideas behind the proposed OKH scheme and compare its performance to other authentication codes in the cryptographic literature.

First, note that there is a one-to-one correspondence between the set of n-bit sequences and the integer ring $\mathbb{Z}_{2^n}$. Hence, the integer ring $\mathbb{Z}_{2^n}$ is the natural choice when performing arithmetic on binary sequences. From a computational-efficiency point of view, the integer ring $\mathbb{Z}_{2^n}$ has an advantage over other finite integer rings in that modular reduction can simply be realized by truncating what is beyond the nth bit (no nontrivial modular reduction is required). From a mathematical point of view, the integer ring $\mathbb{Z}_{2^n}$ possesses the unique property that an element $a \in \mathbb{Z}_{2^n}$ is invertible if and only if it is an odd integer. That is, the multiplicative group $\mathbb{Z}_{2^n}^*$ consists of all odd integers less than 2^n, and nothing else. Consequently, for a random number ϵ drawn uniformly from $\mathbb{Z}_{2^n}$ and an odd key k, the value of $\epsilon \cdot k \mod 2^n$ is uniformly distributed over $\mathbb{Z}_{2^n}$ (by Lemma 2). This fact, along with the fact that block ciphers can be realized as strong pseudorandom permutations, are the main principles behind the design of the OKH authenticated encryption composition. That is, by advancing the hashing phase to be applied to the plaintext, before block cipher encryption, and restricting the hashing keys to the set of odd integers, one can show that a successful forgery by causing a collision in the hashing phase can occur with a negligible probability (by Theorem 1).

In the literature, without advancing the hashing phase to be performed before block cipher encryption, the objective of guaranteeing that a forgery attempt by causing a collision in the hashing phase can succeed with a negligible probability has been achieved by restricting the hash function to be universal. Intuitively, removing such a restriction on the hash function should only increase its speed.

In what follows, we give a detailed performance comparison between the OK family and the NH family of Black et al. [14], the fastest universal hash family in the cryptographic literature for software implementations. Comparison with other known hash families is summarized in Table 1. As before, let M be a message to be authenticated and write M as a sequence of n-bit strings; i.e.,

$M = m_1||\cdots||m_\ell$, where $|m_i| = n$. Similarly, let the key $K = k_1||\cdots||k_\ell$ be the hashing key. Let NH_K be a random member of the NH family determined by the key K. Then, the compressed image of M is computed as

$$\mathsf{NH}_K(M) = \sum_{i=1}^{\ell/2} \Big((k_{2i-1} + m_{2i-1} \mod 2^n) \cdot (k_{2i} + m_{2i} \mod 2^n) \Big) \mod 2^{2n}. \tag{10}$$

On the other hand, the hash image of M as computed by the OK family is

$$\mathsf{OK}_K(M) = \sum_{i=1}^{\ell} k_i \cdot m_i \mod 2^n. \tag{11}$$

That is, when the block size is n, the OK computations are performed over $\mathbb{Z}_{2^n}$ while the NH computations are performed over the larger integer ring $\mathbb{Z}_{2^{2n}}$.

To give a numerical example, let $n = 64$ bits. Then, the OK family requires 64-bit computations while the NH family requires 128-bit computations. When using a 64-bit machine, this implies that NH computations must be split over two registers, while OK computations are performed using a single register. Splitting operations over two registers can slow down the speed by about 63%. More importantly, in standard compilers, there is no integer data type of size 128-bit. Therefore, to multiply two 64-bit integers, one needs to split each integer into two 32-bit parts and multiply with appropriate shifts. Using a 64-bit machine with 3.00GHz Intel(R) Xeon(TM) CPU running on UNIX operating system, the NH family runs at 0.87 cycles/byte while the OK family runs at 0.27 cycles/byte (All codes are written in C).

7 Conclusion

In this paper, we proposed the OKH authenticated encryption scheme. By advancing the hashing phase before block cipher encryption, we showed how a hashing function that is not universal can be used without affecting security of authentication. Since the hash function does not have to be universal, it can be computed faster than universal hash function in the literature of cryptography.

References

1. Afanassiev, V., Gehrmann, C., Smeets, B.: Fast Message Authentication Using Efficient Polynomial Evaluation. In: Biham, E. (ed.) FSE 1997. LNCS, vol. 1267, pp. 190–204. Springer, Heidelberg (1997)
2. Alomair, B.: Towards Authenticated and Private Computer and Wireless Communications. PhD thesis, University of Washington (2011)
3. Alomair, B.: Authenticated Encryption: How Reordering Can Impact Performance. In: Bao, F., Samarati, P., Zhou, J. (eds.) ACNS 2012. LNCS, vol. 7341, pp. 84–99. Springer, Heidelberg (2012), `http://cri.kacst.edu.sa/goto.php?link=alomair`

4. Alomair, B., Clark, A., Poovendran, R.: The power of primes: security of authentication based on a universal hash-function family. Journal of Mathematical Cryptology 4(2), 121–147 (2010)
5. Alomair, B., Poovendran, R.: Efficient Authentication for Mobile and Pervasive Computing. In: Soriano, M., Qing, S., López, J. (eds.) ICICS 2010. LNCS, vol. 6476, pp. 186–202. Springer, Heidelberg (2010)
6. Alomair, B., Poovendran, R.: $\mathcal{E}$-MACs: Towards More Secure and More Efficient Constructions of Secure Channels. In: Rhee, K.-H., Nyang, D. (eds.) ICISC 2010. LNCS, vol. 6829, pp. 292–310. Springer, Heidelberg (2011)
7. Bellare, M., Canetti, R., Krawczyk, H.: Keying Hash Functions for Message Authentication. In: Koblitz, N. (ed.) CRYPTO 1996. LNCS, vol. 1109, pp. 1–15. Springer, Heidelberg (1996)
8. Bellare, M., Desai, A., Jokipii, E., Rogaway, P.: A concrete security treatment of symmetric encryption. In: Proceedings of the 38th IEEE Symposium on Foundations of Computer Science – FOCS 1997, pp. 394–403. IEEE Computer Society Press (1997)
9. Bellare, M., Guérin, R., Rogaway, P.: XOR MACs: New Methods for Message Authentication Using Finite Pseudorandom Functions. In: Coppersmith, D. (ed.) CRYPTO 1995. LNCS, vol. 963, pp. 15–28. Springer, Heidelberg (1995)
10. Bellare, M., Namprempre, C.: Authenticated encryption: Relations among notions and analysis of the generic composition paradigm. Journal of Cryptology 21(4), 469–491 (2008)
11. Bellare, M., Rogaway, P.: Encode-Then-Encipher Encryption: How to Exploit Nonces or Redundancy in Plaintexts for Efficient Cryptography. In: Okamoto, T. (ed.) ASIACRYPT 2000. LNCS, vol. 1976, pp. 317–330. Springer, Heidelberg (2000)
12. Bellare, M., Rogaway, P., Wagner, D.: The EAX Mode of Operation. In: Roy, B., Meier, W. (eds.) FSE 2004. LNCS, vol. 3017, pp. 389–407. Springer, Heidelberg (2004)
13. Bernstein, D.: Floating-point arithmetic and message authentication (2004) (unpublished manuscript), `http://cr.yp.to/hash127.html`
14. Black, J., Halevi, S., Krawczyk, H., Krovetz, T., Rogaway, P.: UMAC: Fast and Secure Message Authentication. In: Wiener, M. (ed.) CRYPTO 1999. LNCS, vol. 1666, pp. 216–499. Springer, Heidelberg (1999)
15. Carter, J., Wegman, M.: Universal classes of hash functions. In: Proceedings of the 9th ACM Symposium on Theory of Computing – STOC 1977, pp. 106–112. ACM SIGACT (1977)
16. Dworkin, M.: Recommendation for block cipher modes of operation: The CMAC mode for authentication. National Institute of Standards and Technology (NIST) Special Publication 800-38B (2005)
17. Dworkin, M.: Recommendation for Block Cipher Modes of Operation: Galois/Counter Mode (GCM) and GMAC. National Institute for Standards and Technology (NIST) Special Publication 800-38D (2007)
18. Etzel, M., Patel, S., Ramzan, Z.: SQUARE HASH: Fast Message Authentication via Optimized Universal Hash Functions. In: Wiener, M. (ed.) CRYPTO 1999. LNCS, vol. 1666, pp. 234–251. Springer, Heidelberg (1999)
19. Ferguson, N., Whiting, D., Schneier, B., Kelsey, J., Lucks, S., Kohno, T.: Helix: Fast Encryption and Authentication in a Single Cryptographic Primitive. In: Johansson, T. (ed.) FSE 2003. LNCS, vol. 2887, pp. 330–346. Springer, Heidelberg (2003)

20. Gligor, V.D., Donescu, P.: Fast Encryption and Authentication: XCBC Encryption and XECB Authentication Modes. In: Matsui, M. (ed.) FSE 2001. LNCS, vol. 2355, pp. 92–108. Springer, Heidelberg (2002)
21. Halevi, S., Krawczyk, H.: MMH: Software Message Authentication in the Gbit/Second Rates. In: Biham, E. (ed.) FSE 1997. LNCS, vol. 1267, pp. 172–189. Springer, Heidelberg (1997)
22. Handschuh, H., Preneel, B.: Key-Recovery Attacks on Universal Hash Function Based MAC Algorithms. In: Wagner, D. (ed.) CRYPTO 2008. LNCS, vol. 5157, pp. 144–161. Springer, Heidelberg (2008)
23. Johansson, T.: Bucket Hashing with a Small Key Size. In: Fumy, W. (ed.) EUROCRYPT 1997. LNCS, vol. 1233, pp. 149–162. Springer, Heidelberg (1997)
24. Jutla, C.: Encryption modes with almost free message integrity. Journal of Cryptology 21(4), 547–578 (2008)
25. Katz, J., Lindell, Y.: Introduction to modern cryptography. Chapman & Hall/CRC (2008)
26. Katz, J., Yung, M.: Unforgeable Encryption and Chosen Ciphertext Secure Modes of Operation. In: Schneier, B. (ed.) FSE 2000. LNCS, vol. 1978, pp. 284–299. Springer, Heidelberg (2001)
27. Kohno, T., Viega, J., Whiting, D.: CWC: A High-Performance Conventional Authenticated Encryption Mode. In: Roy, B., Meier, W. (eds.) FSE 2004. LNCS, vol. 3017, pp. 408–426. Springer, Heidelberg (2004)
28. Krawczyk, H.: LFSR-Based Hashing and Authentication. In: Desmedt, Y.G. (ed.) CRYPTO 1994. LNCS, vol. 839, pp. 129–139. Springer, Heidelberg (1994)
29. Krawczyk, H.: The Order of Encryption and Authentication for Protecting Communications (or: How Secure Is SSL?). In: Kilian, J. (ed.) CRYPTO 2001. LNCS, vol. 2139, pp. 310–331. Springer, Heidelberg (2001)
30. Luby, M., Rackoff, C.: How to Construct Pseudorandom Permutations from Pseudorandom Functions. SIAM Journal on Computing 17(2), 373–386 (1988)
31. Muller, F.: Differential Attacks against the Helix Stream Cipher. In: Roy, B., Meier, W. (eds.) FSE 2004. LNCS, vol. 3017, pp. 94–108. Springer, Heidelberg (2004)
32. Nevelsteen, W., Preneel, B.: Software Performance of Universal Hash Functions. In: Stern, J. (ed.) EUROCRYPT 1999. LNCS, vol. 1592, pp. 24–41. Springer, Heidelberg (1999)
33. Paul, S., Preneel, B.: Near Optimal Algorithms for Solving Differential Equations of Addition with Batch Queries. In: Maitra, S., Veni Madhavan, C.E., Venkatesan, R. (eds.) INDOCRYPT 2005. LNCS, vol. 3797, pp. 90–103. Springer, Heidelberg (2005)
34. Paul, S., Preneel, B.: Solving Systems of Differential Equations of Addition. In: Boyd, C., González Nieto, J.M. (eds.) ACISP 2005. LNCS, vol. 3574, pp. 75–88. Springer, Heidelberg (2005)
35. Preneel, B., van Oorschot, P.C.: MDx-MAC and Building Fast MACs from Hash Functions. In: Coppersmith, D. (ed.) CRYPTO 1995. LNCS, vol. 963, pp. 1–14. Springer, Heidelberg (1995)
36. Rogaway, P.: Bucket hashing and its application to fast message authentication. Journal of Cryptology 12(2), 91–115 (1999)
37. Rogaway, P., Bellare, M., Black, J., Krovetz, T.: OCB: A block-cipher mode of operation for efficient authenticated encryption. In: Proceedings of the 8th ACM Conference on Computer and Communications Security – CCS 2001, pp. 196–205. ACM SIGSAC (2001)

38. Shoup, V.: On Fast and Provably Secure Message Authentication Based on Universal Hashing. In: Koblitz, N. (ed.) CRYPTO 1996. LNCS, vol. 1109, pp. 313–328. Springer, Heidelberg (1996)
39. van Tilborg, H.: Encyclopedia of cryptography and security. Springer (2005)
40. Whiting, D., Schneier, B., Lucks, S., Muller, F.: Phelix – fast encryption and authentication in a single cryptographic primitive. ECRYPT Stream Cipher Project, Report 2005/020 (2005), `http://www.ecrypt.eu.org/stream`
41. Wu, H., Preneel, B.: Differential-Linear Attacks Against the Stream Cipher Phelix. In: Biryukov, A. (ed.) FSE 2007. LNCS, vol. 4593, pp. 87–100. Springer, Heidelberg (2007)

Length-Doubling Ciphers and Tweakable Ciphers

Haibin Zhang

Dept. of Computer Science, University of California, Davis, USA
hbzhang@cs.ucdavis.edu

Abstract. We motivate and describe a mode of operation HEM (resp., THEM) that turns a n-bit blockcipher into a variable-input-length cipher (resp., tweakable cipher) that acts on strings of $[n..2n-1]$ bits. Both HEM and THEM are simple and intuitive and use only *two* blockcipher calls, while prior work at least takes *three*. We prove them secure in the sense of strong PRP and tweakable strong PRP, assuming the underlying blockcipher is a strong PRP.

Keywords: ciphers, tweakable ciphers, deterministic encryption, enciphering scheme, symmetric encryption, universal hash function.

1 Introduction

Designing modes of operation from a blockcipher (e.g., AES) is one of the central tasks in shared-key cryptography. They allow the repeated use of a blockcipher to achieve confidentiality or authentication for variable-input-length (VIL) messages. Many confidentiality applications like disk-sector encryption require that the encryption be *length-preserving*. The requirement entails the usage of *ciphers* [19]. A cipher is a family of keyed length-preserving permutations. Such a primitive is also called an *enciphering scheme*, or *deterministic encryption*. The conventional security notions for a cipher are "pseudorandom permutation" (PRP) and "strong pseudorandom permutation" (SPRP) [17]. Tweakable cipher [18] is an extension of conventional cipher which takes a "tweak" (or "associated data", that does not have to be encrypted) as an additional input. Correspondingly, the security notions are "tweakable PRP" and "tweakable SPRP".

Compared to many other cryptographic primitives like signatures, MACs, and encryption schemes, it is not easy to build a VIL cipher from a fixed-input-length (FIL) cipher (e.g., blockcipher), where techniques such as "padding" and "tainting" fail to work. Indeed, to this end, a very large number of wide blocksize ciphers [1, 3, 4, 6, 10–13, 27, 28] are proposed (though not all of them can handle settings where the messages need not be a multiple of n bits).

This work mainly considers a special case of this problem—on how to turn a blockcipher of size n into a VIL cipher and a VIL tweakable cipher with the message space $\bigcup_{i=n}^{2n-1} \{0,1\}^i$, which basically "doubles" the domain of a blockcipher in the VIL sense. First, this length-doubling problem is of *historical* interest. Luby and Rackoff's Feistel construction [17] can be viewed as the first attempt

F. Bao, P. Samarati, and J. Zhou (Eds.): ACNS 2012, LNCS 7341, pp. 100–116, 2012.

to double the domain of a cipher for fixed-input-length (FIL) messages, while our work deals with the problem in the VIL setting and aims to improve its efficiency. Besides its theoretical value, it is particularly applicable to many useful settings. For instance, short headers in the Internet are usually of the targeted lengths that we study. If messages to be encrypted are of such short lengths, one should be careful about the cipher used since otherwise it can influence the efficiency *by percentage*. And, perhaps more importantly, length-doubling ciphers are useful tools to build high-level protocols. In particular, Rogaway and Zhang show how to turn such a VIL length-doubling tweakable cipher into an arbitrary-length-input online cipher [24]. Last, length-doubling cipher seems to be the "right" method of dealing with the incomplete final block for IEEE P1619 standard [21]. This standard is applicable to cases like disk-sector encryption, which cannot afford the extra hardware or latency required by a two-pass wide blocksize encryption mode. Current standard XTS-AES makes use of *ciphertext stealing* [19] to handle the issue. Though P1619 does not provide a formal definition for security property (which is somewhat unfortunate), an intuitive one may be that each block should be enciphered by an independent SPRP (indexed by a tweak), and the "long" final block (i.e., the second last complete block and the partial final block) should be enciphered by an independent length-doubling tweakable SPRP secure in the VIL setting.[1] It is easy to verify that ciphertext stealing construction described in [21] and its possible variants (see, e.g., [25]) do *not* satisfy such a definition of security even in the PRP sense.[2] We believe that it is necessary to reconsider the problem and find efficient alternatives.

On the other hand, the above-mentioned general wide blocksize ciphers, if being restricted to our targeted domain, do not give very efficient length-doubling schemes. For instance, EME2 cipher of Halevi [10] uses five blockcipher calls to achieve a VIL length-doubling tweakable SPRP. It takes at least four blockcipher calls to use unbalanced Feistel networks [26]. The currently best solution is obtained from the XLS construction by Ristenpart and Rogaway [23] which essentially uses three blockcipher calls and little extra work. In fact, *none* of them are designed specifically for the length-doubling problem—a problem that we aim to address in this paper.

OUR METHOD. At the heart of our motivation is how to construct *efficient* VIL (tweakable) ciphers using *only two* blockcipher calls. The question itself has both theoretical and practical interest. The blockcipher implementation is still the most expensive one in most of the platforms. Trimming one blockcipher call would be highly likely to result in a considerable improvement in efficiency.[3] But this goal is not as easy as it looks. (From a purely theoretical view—without considering efficiency, the problem can be well solved with good provable-security.)

---

[1] Another possible definition is to ask the last block to be "short", but this will not give a tight security reduction for very short messages.

[2] We note that Liskov and Minematsu [16] provide a "proof" to justify the security of ciphertext stealing in XTS-AES. However, one can barely argue anything from such a proof since there is no security notion.

[3] Other examples of this kind include [9, 15].

We extend the idea of Naor and Reingold [20] to construct an efficient VIL length-doubling cipher and tweakable cipher. The overhead for the VIL cipher construction is about two blockcipher calls and two AXU hash function calls and little additional work. We name the VIL cipher construction "HEM" to indicate that we use hash function, blockcipher encryption, and mixing function [23] as components.

We begin by describing a mode FHEM (fixed-length HEM) for a fixed length $n+s$ where n is the blocksize and $s \in [1..n-1]$. The mode is depicted in Figure 1. Given an input M of length $n+s$, it first parses M as M_1 and M_2 such that $|M_1| = n$ and $|M_2| = s$. The algorithm takes four "rounds". The first and last rounds use AXU hash functions, and the second and third rounds use regular blockciphers. The overall structure can be viewed as following the framework of Feistel networks, but is neither exactly like Feistel nor unbalanced Feistel networks. The input and output of the hash function are both simply from $\{0,1\}^n$. This is *crucial* for our construction—the efficiency of the hash functions usually decreases rapidly if the input size gets larger, while the security would lose if the output size gets smaller. Furthermore, we use a third tool, a mixing function [23], to "fix" the consequence of not exactly following the four rounds Feistel. This makes our construction a little less elegant but does not essentially hurt efficiency. Besides, different from Naor-Reingold construction, the security assumption needed for the underlying blockcipher is SPRP rather than PRP. Moreover, the round two and three functions must be permutations. Clearly, these two requirements are insignificant to the implementation since one usually chooses to use AES anyway.

It may still seem hard to make a VIL cipher, since intuitively it needs an AXU hash function for VIL messages. We circumvent the problem by applying the same AXU hash function (with an independent key) to the length of M_2. See Figure 3 for our construction. We stress that we can pre-compute all values for the additional hash functions since there are only n of them where n is the underlying blocksize (e.g., 128). Concretely, the additional operations needed compared to FHEM are just two xors. Thus, we can practically make the VIL cipher from the FIL construction with "no" extra work. We comment that this basically uses an idea similar to that used in [22] yet in a more efficient way.

We go on to present two constructions of VIL length-doubling tweakable ciphers. One of them gets better provable security, while the other is more concise.

To instantiate the modes, we can use many ready solutions for AXU hash functions. One notable construction is the GF(2^n) multiply. The efficiency of AXU hash functions may vary due to software and hardware support and other factors, and thus we do not give a specific recommendation. For the mixing function, we recommend using the more efficient one in [23] that takes only three xors and a single one-bit circular rotation. Another immediate consequence of our results is that any progress in the area of AXU hash functions will result in improved VIL length-doubling (tweakable) ciphers.

FURTHER RELATED WORK. Luby and Rackoff [17] showed the classical result that four rounds of Feistel suffice to construct a length-doubling SPRP in the

FIL sense. Naor and Reingold [20] revisited four rounds Feistel construction and showed that the first and fourth layer blockciphers can be well replaced with two pairwise independent permutations. (In particular, they showed that a weaker almost XOR universal (AXU) hash function is sufficient.) This change improves the efficiency and enables a simpler proof. Patel, Ramzan, and Sundaram [22] constructed a VIL cipher achieving SPRP security for a larger domain $\bigcup_{i \geq 2n} \{0,1\}^i$, by combining unbalanced Feistel networks [26] and pairwise independent permutations. It is not clear how to extend their idea to design even a FIL cipher for our target domain, say, a cipher of length $3n/2$, with a *tight* reduction. Cook, Yung, and Keromytis [5] designed "from scratch" the elastic blockcipher to solve the same length-doubling problem. Their construction is not designed from the perspective of provably secure mode of operation. The XLS mode of operation by Ristenpart and Rogaway [23] in essence solved a more general problem that turns a m-bit-size cipher and a n-bit-size cipher to a cipher that acts on strings of $[m..m+n-1]$ bits. Goldenberg *et al.* addressed the question on how to directly incorporate a tweak on Luby-Rackoff blockciphers [7].

Discussion. Intel released AES-New Instructions (AES-NI) [8], starting from Westmere, in order to more efficiently implement AES. The gap about the efficiency between a well-chosen and carefully-implemented AXU hash function and AES becomes less obvious. However, this change only appears for the recent Intel and AMD architectures. For other platforms, especially for specified hardware-based ones, our scheme outperforms other schemes notably.

Our results also answer an interesting question regarding how to construct *efficient* length-doubling (tweakable) ciphers in the VIL sense using *only two* blockcipher calls, which may be a more important contribution. In fact, the problems studied in our work can be understood from a broader perspective: How do we achieve an *efficient* VIL cipher for messages with the domain $\bigcup_{i \geq n} \{0,1\}^i$ using the *least* blockcipher calls? Of course, this question only makes sense if there exists a lower bound for the number of blockcipher calls for an *efficient* construction. We conjecture on this informal question that it needs at least $\lceil i/n \rceil$ blockcipher calls.

2 Preliminaries

Notation. A *string* is a member of $\{0,1\}^*$. If $A, B \in \{0,1\}^*$ then $A \,\|\, B$ or $A\,B$ denotes their concatenation. If X is a string then $|X|$ denotes its length. The *empty string* is denoted ε. Throughout this work, we fix a number n called the *blocksize.*

Ciphers, blockciphers and tweakable ciphers. A map $f\colon \mathcal{X} \to \mathcal{X}$ for $\mathcal{X} \subseteq \{0,1\}^*$ is a *length-preserving function* if $|f(x)| = |x|$ for all $x \in \{0,1\}^*$. It is a length-preserving *permutation* if it is also a permutation. A *cipher* is a map $\mathcal{E}\colon \mathcal{K} \times \mathcal{M} \to \mathcal{M}$ where $\mathcal{K}$ is a nonempty set, $\mathcal{M} \subseteq \{0,1\}^*$ is a nonempty set, and $\mathcal{E}_K = \mathcal{E}(K, \cdot)$ is a length-preserving permutation for all $K \in \mathcal{K}$. The set $\mathcal{K}$ is called the *key space* and $\mathcal{M}$ is called the *message space.* If $\mathcal{E}\colon \mathcal{K} \times$

$\mathcal{M} \to \mathcal{M}$ is a cipher then its *inverse* is the cipher $\mathcal{E}^{-1}: \mathcal{K} \times \mathcal{M} \to \mathcal{M}$ defined by $\mathcal{E}^{-1}(K, Y) = \mathcal{E}_K^{-1}(Y)$ being the unique point X such that $\mathcal{E}_K(X) = Y$. A *blockcipher* is a map $E: \mathcal{K} \times \{0,1\}^n \to \{0,1\}^n$ where $\mathcal{K}$ is a finite nonempty set and $E_K(\cdot) = E(K, \cdot)$ is a permutation on $\{0,1\}^n$ for every $K \in \mathcal{K}$. Equivalently, a blockcipher is a cipher with message space $\mathcal{M} = \{0,1\}^n$. A *tweakable cipher* is a map $\widetilde{\mathcal{E}}: \mathcal{K} \times \mathcal{T} \times \mathcal{M} \to \mathcal{M}$ where $\mathcal{K}$ is a finite nonempty set and $\mathcal{T}$ is a nonempty set (the *tweak space*) and $\mathcal{M}$ is a nonempty set (the message space) and $\widetilde{\mathcal{E}}_K^T(\cdot) = \widetilde{\mathcal{E}}(K, T, \cdot)$ is a permutation on $\mathcal{M}$ for every $K \in \mathcal{K}$, $T \in \mathcal{T}$.

Let Perm(n) be the set of all permutations on n bits, Perm$(\mathcal{M})$ be the set of all length-preserving permutations on the finite set $\mathcal{M} \subseteq \{0,1\}^*$, and Perm$(\mathcal{T}, \mathcal{M})$ be the set of all functions $\pi: \mathcal{T} \times \mathcal{M} \to \mathcal{M}$ where $\pi_T(\cdot) = \pi(T, \cdot)$ is a permutation for each $T \in \mathcal{T}$. We may regard Perm(n), Perm$(\mathcal{M})$, and Perm$(\mathcal{T}, \mathcal{M})$ as blockciphers, ciphers, and tweakable ciphers, respectively; they are the ideal blockcipher on n bits, the ideal cipher on $\mathcal{M}$, and the ideal tweakable cipher on message space $\mathcal{M}$ and tweak space $\mathcal{T}$. When an adversary $\mathcal{A}$ is run with an oracle $\mathcal{O}$ we let $\mathcal{A}^{\mathcal{O}} \Rightarrow 1$ denote the event that $\mathcal{A}$ outputs 1. Define the $\pm$prp (i.e., SPRP) and $\pm\widetilde{\text{prp}}$ (i.e., tweakable SPRP) *advantage* of $\mathcal{A}$ against E, $\mathcal{E}$ or $\widetilde{\mathcal{E}}$ by:

$$\mathbf{Adv}_E^{\pm\text{prp}}(\mathcal{A}) = \Pr[K \xleftarrow{\$} \mathcal{K}:\ \mathcal{A}^{E_K, E_K^{-1}} \Rightarrow 1] - \Pr[\pi \xleftarrow{\$} \text{Perm}(n):\ \mathcal{A}^{\pi, \pi^{-1}} \Rightarrow 1]$$
$$\mathbf{Adv}_{\mathcal{E}}^{\pm\text{prp}}(\mathcal{A}) = \Pr[K \xleftarrow{\$} \mathcal{K}:\ \mathcal{A}^{\mathcal{E}_K, \mathcal{E}_K^{-1}} \Rightarrow 1] - \Pr[\pi \xleftarrow{\$} \text{Perm}(\mathcal{M}):\ \mathcal{A}^{\pi, \pi^{-1}} \Rightarrow 1]$$
$$\mathbf{Adv}_{\widetilde{\mathcal{E}}}^{\pm\widetilde{\text{prp}}}(\mathcal{A}) = \Pr[K \xleftarrow{\$} \mathcal{K}:\ \mathcal{A}^{\widetilde{\mathcal{E}}_K, \widetilde{\mathcal{E}}_K^{-1}} \Rightarrow 1] - \Pr[\pi \xleftarrow{\$} \text{Perm}(\mathcal{T}, \mathcal{M}):\ \mathcal{A}^{\pi, \pi^{-1}} \Rightarrow 1]$$

ALMOST XOR UNIVERSAL HASH FUNCTION. We recall the definition of ϵ-*almost XOR universal* (ϵ-AXU) *hash function* [14]. A hash function $H: \mathcal{K} \times \mathcal{X} \to \{0,1\}^n$ is called ϵ-AXU, if for all distinct $X, X' \in \mathcal{X}$ and all $C \in \{0,1\}^n$, we have that $\Pr[K \xleftarrow{\$} \mathcal{K}:\ H_K(X) \oplus H_K(X') = C] \le \epsilon$. There are many efficient AXU hash functions candidates. For concreteness, we review one such function for $\mathcal{X} = \{0,1\}^n$—multiplication in *Galois Field* GF(2^n) (i.e., $H_K(X) = K \cdot X$ where $K, X \in \{0,1\}^n$), which achieves 2^{-n} for ϵ—the minimum value one can hope for. Assume that a, b are strings of $\{0,1\}^n$ where $a = a_{n-1} \cdots a_1 a_0$ and $b = b_{n-1} \cdots b_1 b_0$. The Galois Field addition is defined as their bitwise xor. To multiply $a, b \in \text{GF}(2^n)$, write them as polynomials $a(\mathtt{x}) = a_{n-1}\mathtt{x}^{n-1} + \cdots + a_1\mathtt{x} + a_0$ and $b(\mathtt{x}) = b_{n-1}\mathtt{x}^{n-1} + \cdots + b_1\mathtt{x} + b_0$, compute $c(\mathtt{x}) = a(\mathtt{x}) \cdot b(\mathtt{x}) \bmod p(\mathtt{x})$ where $p(\mathtt{x})$ is a fixed irreducible polynomial over GF(2^n), and then return the binary representation of $c(\mathtt{x})$ as output. For the AXU hash function input string $X \in \{0,1\}^*$ and $|X| < n$, we let $\mathsf{pad}(X)$ be the string $X \| 0^{n-|X|}$. (Namely, the minimal number of zero-bits are padded on the right such that $\mathsf{pad}(X)$ is a complete block.) Note that for example $\mathsf{pad}(Y) = \mathsf{pad}(Y\|0)$ for some string Y where $|Y| < n$. This all zero padding method can be applied to other AXU hash functions.

MIXING FUNCTION. We review the definition of the mixing function formally defined and studied by Ristenpart and Rogaway [23]. We define a mixing function $\mathsf{mix}: \mathcal{S}^2 \to \mathcal{S}^2$ ($\mathcal{S} \supseteq \bigcup_{i=1}^{n-1}\{0,1\}^i$) such that $\mathsf{mix}_\mathsf{L}(\cdot,\cdot)$ and $\mathsf{mix}_\mathsf{R}(\cdot,\cdot)$ are the left projection and right projection of mix function. Ideally, we want such a primitive to have the property of a *multipermutation*: namely, for any $A \in \mathcal{S}$, $\mathsf{mix_L}(A,\cdot)$, $\mathsf{mix_L}(\cdot,A)$, $\mathsf{mix_R}(A,\cdot)$, and $\mathsf{mix_R}(\cdot,A)$ are all permutations. Like the construction in [23], a relaxed notion of mixing function can work almost as well for our schemes. We say that mix is an $\epsilon(s)$-*good mixing function*, if for all s such that $\{0,1\}^s \in \mathcal{S}$ and all $A,B,C \in \{0,1\}^s$, we have both $\mathsf{mix_L}(A,\cdot)$ and $\mathsf{mix_R}(\cdot,B)$ are permutations, and $\Pr[R \xleftarrow{\$} \{0,1\}^s : C = \mathsf{mix_L}(R,B)]$ and $\Pr[R \xleftarrow{\$} \{0,1\}^s : C = \mathsf{mix_R}(A,R)]$ are both less than $\epsilon(s)$. In their work, two efficient mixing functions are given. The more efficient one with $\epsilon(s) = 2^{1-s}$ only takes three xors and a one-bit circular rotation.

3 A Fix-Input-Length Cipher

In this section, we provide a cipher for a fixed length $n+s$ where n is the blocksize and $s \in [1..n-1]$. In other words, the cipher we shall describe is secure against adversaries who are only allowed to ask queries of length $n+s$. The construction only works in the FIL setting, but it serves as the basis for constructing VIL (tweakable) ciphers.

Let $E\colon \mathcal{K}_1 \times \{0,1\}^n \to \{0,1\}^n$ be a blockcipher and let $H\colon \mathcal{K}_2 \times \{0,1\}^n \to \{0,1\}^n$ be a 2^{-n}-AXU hash function family. Define an $\epsilon(s)$-good mixing function $\mathsf{mix}\colon \mathcal{S}^2 \to \mathcal{S}^2$ where $\mathcal{S} \supseteq \bigcup_{i=1}^{n-1}\{0,1\}^i$. We define a cipher $\mathcal{E} = \mathrm{FHEM}[H, E, \mathsf{mix}]$ with key space $\mathcal{K}_1^2 \times \mathcal{K}_2^2$. See Figure 1 for the construction. We claim that this cipher is $\pm$prp-secure for fixed-input-length $n+s$.

The intuition of the proof is as follows. Like the Naor-Reingold construction, by using AXU-hash function, for any two different messages M^i and M^j of the same length, the probability that M_3^i and M_3^j "collide" is negligible. After applying a random function, the output $M_4||M_5$ is now uniformly distributed. This perfectly hides the complete block M_1, but the partial block M_2 remains unprotected. A mixing function is used to force the output of $M_5||M_2$ to inherit the distribution of M_5. An independent random function is then employed to further hide part of the mixing function output. The overall construction should be made "symmetrizing" in order to achieve strong PRP security. Also note that unlike Naor-Reingold and subsequent work, our constructions (i.e., one in this section and the following extensions) ask the underlying blockcipher to be reversible and the complexity assumption for it is SPRP.

The following theorem establishes the security of FHEM.

Theorem 1. *Let* $\mathcal{E} = \mathrm{FHEM}[H, \mathrm{Perm}(n), \mathsf{mix}]$ *with message space* $\{0,1\}^{n+s}$. *If* $\mathcal{A}$ *asks at most* q *queries then* $\mathbf{Adv}_{\mathcal{E}}^{\pm\mathrm{prp}}(\mathcal{A}) \le 1.5\,q^2/2^n + 0.5\,q^2\frac{\epsilon(s)}{2^{n-s}} + 0.5\,q^2/2^{n+s}$, *and if we use a mixing function with* $\epsilon(s) = 2^{1-s}$ *then we have that* $\mathbf{Adv}_{\mathcal{E}}^{\pm\mathrm{prp}}(\mathcal{A}) \le 3\,q^2/2^n$. ∎

Proof. We assume without loss of generality that $\mathcal{A}$ is deterministic and makes q queries from $\{0,1\}^{n+s}$. We further assume that it does not ask "pointless" queries: it never repeats an encipher query, never repeats a decipher query, never asks a decipher query of a value that it earlier received from an encipher query, and never makes an encipher query of a value that it earlier received from a decipher query.

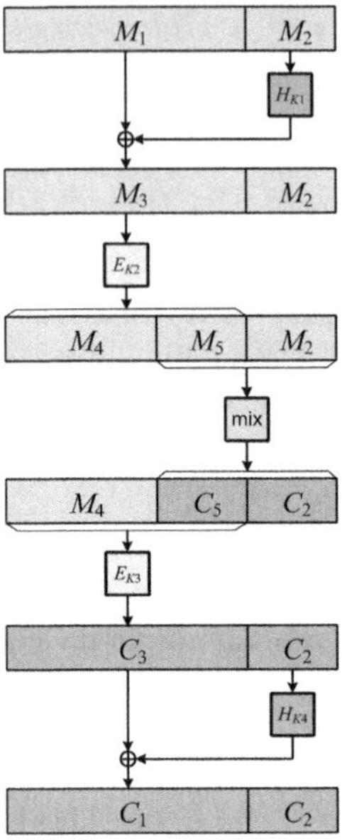


```
algorithm 𝓔_K(M)   where K = K_1||K_2||K_3||K_4
if |M| ≠ n + s then return ⊥
M_1||M_2 ← M   where |M_1| = n and |M_2| = s
M_3 ← M_1 ⊕ H_{K_1}(pad(M_2))
M_4||M_5 ← E_{K_2}(M_3)   where |M_4| = n − s and |M_5| = s
C_5||C_2 ← mix(M_5||M_2)   where |C_5| = |C_2| = s
C_3 ← E_{K_3}(M_4||C_5)
C_1 ← C_3 ⊕ H_{K_4}(pad(C_2))
C ← C_1||C_2
return C
```

Fig. 1. Mode FHEM. Each input M is parsed as a complete block M_1 and a partial block M_2. We should pad M_2 to a complete block before applying the AXU hash function. Similar operations should be carried out for the deciphering algorithm.

We use the code-based games [2] in Figure 2. Variable *bad* is initialized to false. A functions π is initialized to everywhere undefined. Its current domain and range are denoted domain(π) and range(π), while their complements relative to $\{0,1\}^n$ are denoted codomain(π) and corange(π).

We begin with game G_1, which precisely describes the FHEM construction with the ideal blockcipher π_1 and π_2. Game G_6 always outputs random values, simulating a pair of random functions. Let p denote the probability that $\mathcal{A}$ outputs 1 in the game simulating a random permutation and its inverse. The difference between p and $\Pr[\,G_6^{\mathcal{A}} \Rightarrow 1\,]$ is at most $0.5\,q^2/2^{n+s}$, due to the PRP/PRF switching lemma. We must bound $\Pr[\,G_1^{\mathcal{A}} \Rightarrow 1\,] - \Pr[\,G_6^{\mathcal{A}} \Rightarrow 1\,]$.

procedure $E(M)$
$j \leftarrow j+1;\ M^j \leftarrow M$
$M_1^j \| M_2^j \leftarrow M^j$
$M_3^j \leftarrow M_1^j \oplus H_{K_1}(\mathsf{pad}(M_2^j))$
$M_4^j \| M_5^j \leftarrow \pi_1(M_3^j)$
$C_5^j \| C_2^j \leftarrow \mathsf{mix}(M_5^j \| M_2^j)$
$C_3^j \leftarrow \pi_2(M_4^j \| C_5^j)$
$C_1^j \leftarrow C_3^j \oplus H_{K_4}(\mathsf{pad}(C_2^j))$
$C^j \leftarrow C_1^j \| C_2^j$
return C^j

procedure $D(C)$
$j \leftarrow j+1;\ C^j \leftarrow C$
$C_1^j \| C_2^j \leftarrow C^j$
$C_3^j \leftarrow C_1^j \oplus H_{K_1}(\mathsf{pad}(C_2^j))$
$M_4^j \| C_5^j \leftarrow \pi_2^{-1}(C_3^j)$
$M_5^j \| M_2^j \leftarrow \mathsf{mix}(C_5^j \| C_2^j)$
$M_3^j \leftarrow \pi_1^{-1}(M_4^j \| M_5^j)$
$M_1^j \leftarrow M_3^j \oplus H_{K_1}(\mathsf{pad}(M_2^j))$
$M^j \leftarrow M_1^j \| M_2^j$
return M^j

Game G_1

procedure $E(M)$
$j \leftarrow j+1;\ M^j \leftarrow M$
$M_1^j \| M_2^j \leftarrow M^j$
$M_3^j \leftarrow M_1^j \oplus H_{K_1}(\mathsf{pad}(M_2^j))$
if $M_3^j \notin \mathrm{domain}(\pi_1)$ **then**
  $Y_1 \xleftarrow{\$} \{0,1\}^n$
  if $Y_1 \in \mathrm{range}(\pi_1)$ **then**
    $bad \leftarrow \mathsf{true}$; [$Y_1 \xleftarrow{\$} \mathrm{corange}(\pi_1)$]
  $\pi_1(M_3^j) \leftarrow Y_1$
$M_4^j \| M_5^j \leftarrow \pi_1(M_3^j)$
$C_5^j \| C_2^j \leftarrow \mathsf{mix}(M_5^j \| M_2^j)$
if $M_4^j \| C_5^j \notin \mathrm{domain}(\pi_2)$ **then**
  $Y_2 \xleftarrow{\$} \{0,1\}^n$
  if $Y_2 \in \mathrm{range}(\pi_2)$ **then**
    $bad \leftarrow \mathsf{true}$; [$Y_2 \xleftarrow{\$} \mathrm{corange}(\pi_2)$]
  $\pi_2(M_4^j \| C_5^j) \leftarrow Y_2$
$C_3^j \leftarrow \pi_2(M_4^j \| C_5^j)$
$C_1^j \leftarrow C_3^j \oplus H_{K_4}(\mathsf{pad}(C_2^j))$
$C^j \leftarrow C_1^j \| C_2^j$
return C^j

procedure $D(C)$
$j \leftarrow j+1;\ C^j \leftarrow C$
$C_1^j \| C_2^j \leftarrow C^j$
$C_3^j \leftarrow C_1^j \oplus H_{K_4}(\mathsf{pad}(C_2^j))$
if $C_3^j \notin \mathrm{range}(\pi_2)$ **then**
  $X_2 \xleftarrow{\$} \{0,1\}^n$
  if $X_2 \in \mathrm{domain}(\pi_2)$ **then**
    $bad \leftarrow \mathsf{true}$; [$X_2 \xleftarrow{\$} \mathrm{codomain}(\pi_2)$]
  $\pi_2^{-1}(C_3^j) \leftarrow X_2$
$M_4^j \| C_5^j \leftarrow \pi_2^{-1}(C_3^j)$
$M_5^j \| M_2^j \leftarrow \mathsf{mix}(C_5^j \| C_2^j)$;
if $M_4^j \| M_5^j \notin \mathrm{range}(\pi_1)$ **then**
  $X_1 \xleftarrow{\$} \{0,1\}^n$
  if $X_1 \in \mathrm{domain}(\pi_1)$ **then**
    $bad \leftarrow \mathsf{true}$; [$X_1 \xleftarrow{\$} \mathrm{codomain}(\pi_1)$]
  $\pi_1^{-1}(M_4^j \| M_5^j) \leftarrow X_1$
$M_3^j \leftarrow \pi_1^{-1}(M_4^j \| M_5^j)$
$M_1^j \leftarrow M_3^j \oplus H_{K_1}(\mathsf{pad}(M_2^j))$
$M^j \leftarrow M_1^j \| M_2^j$
return M^j

[Game G_2]
Game G_3

procedure $E(M)$
$j \leftarrow j+1;\ M^j \leftarrow M$
$M_1^j \| M_2^j \leftarrow M^j$
$M_3^j \leftarrow M_1^j \oplus H_{K_1}(\mathsf{pad}(M_2^j))$
$Y_1 \xleftarrow{\$} \{0,1\}^n$
if $M_3^j \in \mathrm{domain}(\pi_1)$ **then**
  $bad \leftarrow \mathsf{true}$; [$Y_1 \leftarrow \pi_1(M_3^j)$] **else**
$\pi_1(M_3^j) \leftarrow Y_1$
$M_4^j \| M_5^j \leftarrow \pi_1(M_3^j)$
$C_5^j \| C_2^j \leftarrow \mathsf{mix}(M_5^j \| M_2^j)$
$Y_2 \xleftarrow{\$} \{0,1\}^n$
if $M_4^j \| C_5^j \in \mathrm{domain}(\pi_2)$ **then**
  $bad \leftarrow \mathsf{true}$; [$Y_2 \leftarrow \pi_2(M_4^j \| C_5^j)$] **else**
$\pi_2(M_4^j \| C_5^j) \leftarrow Y_2$
$C_3^j \leftarrow \pi_2(M_4^j \| C_5^j)$
$C_1^j \leftarrow C_3^j \oplus H_{K_4}(\mathsf{pad}(C_2^j))$
$C^j \leftarrow C_1^j \| C_2^j$
return C^j

procedure $D(C)$
$j \leftarrow j+1;\ C^j \leftarrow C$
$C_1^j \| C_2^j \leftarrow C^j$
$C_3^j \leftarrow C_1^j \oplus H_{K_4}(\mathsf{pad}(C_2^j))$
$X_2 \xleftarrow{\$} \{0,1\}^n$
if $C_3^j \in \mathrm{range}(\pi_2)$ **then**
  $bad \leftarrow \mathsf{true}$; [$X_2 \leftarrow \pi_2^{-1}(C_3^j)$] **else**
$\pi_2^{-1}(C_3^j) \leftarrow X_2$
$M_4^j \| C_5^j \leftarrow \pi_2^{-1}(C_3^j)$
$M_5^j \| M_2^j \leftarrow \mathsf{mix}(C_5^j \| C_2^j)$
$X_1 \xleftarrow{\$} \{0,1\}^n$
if $M_4^j \| M_5^j \in \mathrm{range}(\pi_1)$ **then**
  $bad \leftarrow \mathsf{true}$; [$X_1 \leftarrow \pi_1^{-1}(M_4^j \| C_5^j)$] **else**
$\pi_1^{-1}(M_4^j \| C_5^j) \leftarrow X_1$
$M_3^j \leftarrow \pi_1^{-1}(M_4^j \| M_5^j)$
$M_1^j \leftarrow M_3^j \oplus H_{K_1}(\mathsf{pad}(M_2^j))$
$M^j \leftarrow M_1^j \| M_2^j$
return M^j

[Game G_4]
Game G_5

procedure $E(M)$
$j \leftarrow j+1;\ M^j \leftarrow M$
$C^j \xleftarrow{\$} \{0,1\}^{n+s}$; **return** C^j

procedure Finalize
$M_1^j \| M_2^j \leftarrow M^j$
$C_1^j \| C_2^j \leftarrow C^j$
$M_3^j \leftarrow M_1^j \oplus H_{K_1}(\mathsf{pad}(M_2^j))$
$M_4^j \xleftarrow{\$} \{0,1\}^{n-s}$
$M_5^j \leftarrow \mathsf{mix}_{\mathsf{R}}^{-1}(C_2^j \| M_2^j)$
$C_5^j \leftarrow \mathsf{mix}_{\mathsf{L}}(M_5^j \| M_2^j)$
$C_3^j \leftarrow C_1^j \oplus H_{K_4}(\mathsf{pad}(C_2^j))$
$bad \leftarrow (M_3^j = M_3^i)$ **or**
  $(M_4^j \| C_5^j = M_4^i \| C_5^i)$, for some $i < j$

Game G_6

procedure $D(C)$
$j \leftarrow j+1;\ C^j \leftarrow C$
$M^j \xleftarrow{\$} \{0,1\}^{n+s}$; **return** M^j

procedure Finalize
$C_1^j \| C_2^j \leftarrow C^j$
$M_1^j \| M_2^j \leftarrow M^j$
$C_3^j \leftarrow C_1^j \oplus H_{K_4}(\mathsf{pad}(C_2^j))$
$M_4^j \xleftarrow{\$} \{0,1\}^{n-s}$
$C_5^j \leftarrow \mathsf{mix}_{\mathsf{R}}^{-1}(M_2^j \| C_2^j)$
$M_5^j \leftarrow \mathsf{mix}_{\mathsf{L}}(C_5^j \| C_2^j)$
$M_3^j \leftarrow M_1^j \oplus H_{K_1}(\mathsf{pad}(M_2^j))$
$bad \leftarrow (C_3^j = C_3^i)$ **or**
  $(M_4^j \| M_5^j = M_4^i \| M_5^i)$, for some $i < j$

Fig. 2. Games used in the proof of Theorem 1. Game G_2 includes the bracketed statements while game G_3 does not. Similarly, game G_4 includes the bracketed statements while game G_5 does not. In game G_6, encipher and decipher queries are answered by random values.

Game G_2 rewrites G_1 using lazy sampling [2] and these two games are adversarially indistinguishable. The probability that *bad* gets set to true in G_3 is bounded by the PRP/PRF switching lemma; $\Pr[G_2^{\mathcal{A}} \Rightarrow 1] - \Pr[G_3^{\mathcal{A}} \Rightarrow 1] \leq 2 \times 0.5\, q^2/2^n$. Game G_4 makes several trivial modifications compared to G_3; they are adversarially indistinguishable. $\Pr[G_4^{\mathcal{A}} \Rightarrow 1] - \Pr[G_5^{\mathcal{A}} \Rightarrow 1]$ is at most the probability that $\mathcal{A}$ manages to set *bad* in game G_5. Game G_6 simply changes the order of many random choices; it is adversarially indistinguishable from game G_5. In game G_6, the encipher and decipher queries are answered by random values over $\{0,1\}^{n+s}$. It remains to bound the probability that *bad* gets set to true in this game.

In game G_6, we let $\mathsf{mix}_{\mathsf{R}}^{-1}(\cdot, B)$ denote the inverse of $\mathsf{mix}_{\mathsf{R}}(\cdot, B)$. We first analyze the circumstance where the j-th query is an encipher query. If the j-th and i-th queries M^j and M^i are both encipher queries and $i < j$, then $M^j \neq M^i$ since encipher queries are not repeated. If $M_2^j \neq M_2^i$ then $\mathsf{pad}(M_2^j) \neq \mathsf{pad}(M_2^i)$. By the definition of AXU hash function the probability that $M_1^j \oplus H_{K_1}(\mathsf{pad}(M_2^j)) = M_1^i \oplus H_{K_1}(\mathsf{pad}(M_2^i))$ is at most 2^{-n}. Otherwise, we have that $M_2^j = M_2^i$ and $M_1^j \neq M_1^i$, and the probability that $M_1^j \oplus H_{K_1}(\mathsf{pad}(M_2^j)) = M_1^i \oplus H_{K_1}(\mathsf{pad}(M_2^i))$ is zero. Thus we have that the probability that $M_3^j = M_3^i$ is at most 2^{-n}. If the j-th query is an encipher query and the i-th query is a decipher query, then we still have $M^j \neq M^i$ because $\mathcal{A}$ never makes an encipher query of a value that it earlier received from a decipher query. Similarly, in this case, the probability that $M_3^j = M_3^i$ is at most 2^{-n}.

On the other hand, we claim that the probability that $M_4^j || C_5^j = M_4^i || C_5^i$ (for some $i < j$) is at most $\frac{\epsilon(s)}{2^{n-s}}$. This can be justified as follows. First, M_4^j is freshly chosen at random and thus the probability that $M_4^j = M_4^i$ is 2^{s-n}. Second, by $M_5^j = \mathsf{mix}_{\mathsf{R}}^{-1}(C_2^j || M_2^j)$, we have that M_5^j is uniformly distributed since C_2^j is independently chosen at random. By the definition of $\epsilon(s)$-good mixing function and by $C_5^j = \mathsf{mix}_{\mathsf{L}}(M_5^j || M_2^j)$, we have that the probability that $C_5^j = C_5^i$ is at most $\epsilon(s)$. Therefore, the probability that $M_4^j || C_5^j$ equals $M_4^i || C_5^i$ (for some $i < j$) is at most $\frac{\epsilon(s)}{2^{n-s}}$.

The same probability results hold for the case where the j-th query is a decipher query with a proof symmetric to the above one. Since there are at most $q^2/2$ possible collisions at both Line 620/670 and Line 621/671, the probability that $\mathcal{A}$ manages to set *bad* in this game is at most $0.5\, q^2/2^n + 0.5\, q^2 \frac{\epsilon(s)}{2^{n-s}}$. This completes the proof of the claim. ▮

It is easy to pass from the information-theoretic setting to complexity-theoretic one.

Insecurity of FHEM against VIL adversaries. FHEM is designed against FIL adversaries. It is not a PRP secure cipher with respect to VIL attacks. A simple attack is illustrated as follows. The adversary simply makes two encipher queries 0^{n+1} and 0^{n+2}, and gets two replies from the oracle $C_1 || C_2$ and $C_1' || C_2'$ where $|C_1| = |C_1'| = n$. If $C_1 = C_1'$, the adversary returns 1; otherwise, it returns 0. If the adversary is given a FHEM oracle, one can check that the probability

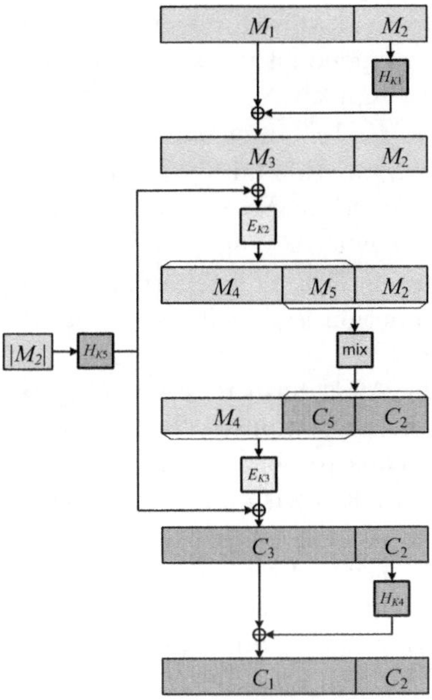

algorithm $\mathcal{E}_K(M)$ **where** $K = K_1||K_2||K_3||K_4||K_5$

if $M \notin \bigcup_{i=n+1}^{2n-1} \{0,1\}^i$ **then return** $\perp$

$M_1||M_2 \leftarrow M$ **where** $|M_1| = n$ and $|M_2| = s$

$M_3 \leftarrow M_1 \oplus H_{K_1}(\mathsf{pad}(M_2))$

$M_4||M_5 \leftarrow E_{K_2}(M_3 \oplus H_{K_5}(\mathsf{pad}(|M_2|)))$ **where** $|M_4| = n - s$ and $|M_5| = s$

$C_5||C_2 \leftarrow \mathsf{mix}(M_5||M_2)$ **where** $|C_5| = |C_2| = s$

$C_3 \leftarrow E_{K_3}(M_4||C_5) \oplus H_{K_5}(\mathsf{pad}(|M_2|))$

$C_1 \leftarrow C_3 \oplus H_{K_4}(\mathsf{pad}(C_2))$

$C \leftarrow C_1||C_2$

return C

Fig. 3. Mode HEM. The input for the AXU hash functions H_{K_1}, H_{K_4}, and H_{K_5} should be all padded to a complete block. In particular, the first $\log n$ bits of input for H_{K_5} is the length encoding of the partial block M_2, while the remaining are $n - \log n$ zero-bits.

that $C_1 = C_1'$ is quite high; otherwise, it is about 2^{-n}. Such an adversary can thus attack the PRP security of FHEM.

4 A Length-Doubling VIL Cipher

We now show how to make a VIL cipher based on the FIL one in the above section. The basic idea is to replace the AXU hash function with a length related AXU hash function. Namely, we want a primitive that enjoys the AXU hash function property even for variable length input. We do not design such a primitive from scratch. Instead, this can be achieved by applying the same AXU hash function (with an independently and uniformly chosen key) to the length of incomplete block of the input.

Let $E\colon \mathcal{K}_1 \times \{0,1\}^n \rightarrow \{0,1\}^n$ be a blockcipher and let $H\colon \mathcal{K}_2 \times \{0,1\}^n \rightarrow \{0,1\}^n$ be a 2^{-n}-AXU hash function family. Let $\mathsf{mix}\colon \mathcal{S}^2 \rightarrow \mathcal{S}^2$ ($\mathcal{S} \supseteq \bigcup_{i=1}^{n-1}\{0,1\}^i$) be an $\epsilon(s)$-good mixing function. From these building blocks we define a cipher $\mathcal{E} = \mathrm{HEM}[H, E, \mathsf{mix}]$ with key space $\mathcal{K}_1^2 \times \mathcal{K}_2^3$ and message space $\mathcal{M} = \bigcup_{i=n+1}^{2n-1}\{0,1\}^i$. See Figure 3. The construction is ±prp-secure for VIL adversaries. The AXU-hash function H_{K_5} can be replaced with a blockcipher E_{K_5} and the security remains.

We emphasize that the AXU hash function H_{K_5} taking as input the length of incomplete block can be precomputed. It only needs n (typically, 128) invocations of hash function calls (or blockcipher calls). This thus yields a highly efficient implementation of HEM with a few preprocessed operations and a little additional storage.

To make this cipher also secure for queries of length n, one can choose an independent blockcipher to encipher all n-bit messages. The complexity assumption used is SPRP. We give the security analysis for the scheme with message space $\bigcup_{i=n+1}^{2n-1}\{0,1\}^i$, using AXU hash function H_{K_5}.

Theorem 2. *Let* $\mathcal{E} = \mathrm{HEM}[H, \mathrm{Perm}(n), \mathsf{mix}]$. *If* $\mathcal{A}$ *asks at most* q *queries then* $\mathbf{Adv}_{\mathcal{E}}^{\pm\mathrm{prp}}(\mathcal{A}) \leq 2\,q^2/2^n + 0.5\,q^2\frac{\epsilon(s)}{2^{n-s}}$, *and if we use a* 2^{1-s}*-good mixing function then we have that* $\mathbf{Adv}_{\mathcal{E}}^{\pm\mathrm{prp}}(\mathcal{A}) \leq 3\,q^2/2^n$. ∎

Proof. The proof follows an analogous line to the one of Theorem 1. We begin with game G_1, which precisely describes the HEM construction with the ideal blockcipher π_1 and π_2. Game G_6 simulates a pair of random functions. We let p' denote the probability that $\mathcal{A}$ outputs 1 in the game simulating a random permutation and its inverse. The difference between p' and $\Pr[\,G_6^{\mathcal{A}} \Rightarrow 1\,]$ is at most $0.5\,q^2/2^n$, again due to the PRP/PRF switching lemma. We have to bound $\Pr[\,G_1^{\mathcal{A}} \Rightarrow 1\,] - \Pr[\,G_6^{\mathcal{A}} \Rightarrow 1\,]$.

Game G_2 modifies G_1 using lazy sampling and they are adversarially indistinguishable. By the PRP/PRF switching lemma, $\Pr[\,G_2^{\mathcal{A}} \Rightarrow 1\,] - \Pr[\,G_3^{\mathcal{A}} \Rightarrow 1\,] \leq 2 \times 0.5\,q^2/2^n$. Game G_4 and G_3 are easily seen to be adversarially indistinguishable. $\Pr[\,G_4^{\mathcal{A}} \Rightarrow 1\,] - \Pr[\,G_5^{\mathcal{A}} \Rightarrow 1\,]$ is at most the probability that $\mathcal{A}$ can set *bad* in game G_5. We delay the calculation of the probability in an adversarially indistinguishable game G_6 where the encipher and decipher queries are answered by random values from $\bigcup_{i=n+1}^{2n-1}\{0,1\}^i$.

It remains to bound the probability that $\mathcal{A}$ can set *bad* in game G_6. We only give the analysis for the circumstance where the j-th query is an encipher query. (The case where j-th query is a decipher query is symmetric.)

Consider the j-th and i-th queries M^j and M^i (where $i < j$). We have that $M^j \neq M^i$ since encipher queries are not repeated and $\mathcal{A}$ never makes an encipher query of a value that it earlier received from a decipher query. We have several cases to consider:

```
procedure E(M)
j ← j + 1;  M^j ← M
M_1^j || M_2^j ← M^j
M_3^j ← M_1^j ⊕ H_{K_1}(pad(|M_2^j|))
M_4^j || M_5^j ← π_1(M_3^j ⊕ H_{K_5}(pad|M_2^j|))
C_5^j || C_2^j ← mix(M_5^j || M_2^j)
C_3^j ← π_2(M_4^j || C_5^j) ⊕ H_{K_5}(pad(|M_2^j|))
C_1^j ← C_3^j ⊕ H_{K_4}(pad(C_2^j))
C^j ← C_1^j || C_2^j
return C^j

procedure D(C)
j ← j + 1;  C^j ← C
C_1^j || C_2^j ← C^j
C_3^j ← C_1^j ⊕ H_{K_1}(pad(C_2^j))
M_4^j || C_5^j ← π_2^{-1}(C_3^j ⊕ H_{K_5}(pad(|C_2^j|)))
M_5^j || M_2^j ← mix(C_5^j || C_2^j)
M_3^j ← π_1^{-1}(M_4^j || M_5^j) ⊕ H_{K_5}(pad(|C_2^j|))
M_1^j ← M_3^j ⊕ H_{K_1}(pad(M_2^j))
M^j ← M_1^j || M_2^j
return M^j
                                                  Game G_1
```

```
procedure E(M)
j ← j + 1;  M^j ← M
M_1^j || M_2^j ← M^j
M_3^j ← M_1^j ⊕ H_{K_1}(pad(M_2^j))
if M_3^j ⊕ H_{K_5}(pad(|M_2^j|)) ∉ domain(π_1) then
    Y_1 ←$ {0,1}^n
    if Y_1 ∈ range(π_1) then
        bad ← true;  [ Y_1 ←$ corange(π_1) ]
    π_1(M_3^j ⊕ H_{K_5}(pad(|M_2^j|))) ← Y_1
M_4^j || M_5^j ← π_1(M_3^j ⊕ H_{K_5}(pad(|M_2^j|)))
C_5^j || C_2^j ← mix(M_5^j || M_2^j)
if M_4^j || C_5^j ∉ domain(π_2) then
    Y_2 ←$ {0,1}^n
    if Y_2 ∈ range(π_2) then
        bad ← true;  [ Y_2 ←$ corange(π_2) ]
    π_2(M_4^j || C_5^j) ← Y_2
C_3^j ← π_2(M_4^j || C_5^j) ⊕ H_{K_5}(pad(|M_2^j|))
C_1^j ← C_3^j ⊕ H_{K_4}(pad(C_2^j))
C^j ← C_1^j || C_2^j
return C^j

procedure D(C)
j ← j + 1;  C^j ← C
C_1^j || C_2^j ← C^j
C_3^j ← C_1^j ⊕ H_{K_4}(pad(C_2^j))
if C_3^j ⊕ H_{K_5}(pad(|C_2^j|)) ∉ range(π_2) then
    X_2 ←$ {0,1}^n
    if X_2 ∈ domain(π_2) then
        bad ← true;  [ X_2 ←$ codomain(π_2) ]
    π_2^{-1}(C_3^j ⊕ H_{K_5}(pad(|C_2^j|))) ← X_2
M_4^j || C_5^j ← π_2^{-1}(C_3^j ⊕ H_{K_5}(pad(|C_2^j|)))
M_5^j || M_2^j ← mix(C_5^j || C_2^j)
if M_4^j || M_5^j ∉ range(π_1) then
    X_1 ←$ {0,1}^n
    if X_1 ∈ domain(π_1) then
        bad ← true;  [ X_1 ←$ codomain(π_1) ]
    π_1^{-1}(M_4^j || M_5^j) ← X_1
M_3^j ← π_1^{-1}(M_4^j || M_5^j) ⊕ H_{K_5}(pad(|C_2^j|))
M_1^j ← M_3^j ⊕ H_{K_1}(pad(M_2^j))
M^j ← M_1^j || M_2^j                              [ Game G_2 ]
return M^j                                          Game G_3
```

```
procedure E(M)
j ← j + 1;  M^j ← M
M_1^j || M_2^j ← M^j
M_3^j ← M_1^j ⊕ H_{K_1}(pad(M_2^j))
Y_1 ←$ {0,1}^n
if M_3^j ⊕ H_{K_5}(pad(|M_2^j|)) ∈ domain(π_1) then
    bad ← true;
    [ Y_1 ← π_1(M_3^j ⊕ H_{K_5}(pad(|M_2^j|))) ] else
π_1(M_3^j ⊕ H_{K_5}(pad(|M_2^j|))) ← Y_1
M_4^j || M_5^j ← π_1(M_3^j)
C_5^j || C_2^j ← mix(M_5^j || M_2^j)
Y_2 ←$ {0,1}^n
if M_4^j || C_5^j ∈ domain(π_2) then
    bad ← true;  [ Y_2 ← π_2(M_4^j || C_5^j) ] else
π_2(M_4^j || C_5^j) ← Y_2
C_3^j ← π_2(M_4^j || C_5^j) ⊕ H_{K_5}(pad(|M_2^j|))
C_1^j ← C_3^j ⊕ H_{K_4}(pad(C_2^j))
C^j ← C_1^j || C_2^j
return C^j

procedure D(C)
j ← j + 1;  C^j ← C
C_1^j || C_2^j ← C^j
C_3^j ← C_1^j ⊕ H_{K_4}(pad(C_2^j))
X_2 ←$ {0,1}^n
if C_3^j ⊕ H_{K_5}(pad(|C_2^j|)) ∈ range(π_2) then
    bad ← true;
    [ X_2 ← π_2^{-1}(C_3^j ⊕ H_{K_5}(pad(|C_2^j|))) ] else
π_2^{-1}(C_3^j ⊕ H_{K_5}(pad(|C_2^j|))) ← X_2
M_4^j || C_5^j ← π_2^{-1}(C_3^j)
M_5^j || M_2^j ← mix(C_5^j || C_2^j)
X_1 ←$ {0,1}^n
if M_4^j || M_5^j ∈ range(π_1) then
    bad ← true;  [ X_1 ← π_1^{-1}(M_4^j || M_5^j) ] else
π_1^{-1}(M_4^j || M_5^j) ← X_1
M_3^j ← π_1^{-1}(M_4^j || M_5^j) ⊕ H_{K_5}(pad(|C_2^j|))
M_1^j ← M_3^j ⊕ H_{K_1}(pad(M_2^j))
M^j ← M_1^j || M_2^j                              [ Game G_4 ]
return M^j                                          Game G_5
```

```
procedure E(M)
j ← j + 1;  M^j ← M
C^j ←$ {0,1}^{n+s};  return C^j

procedure Finalize
M_1^j || M_2^j ← M^j
C_1^j || C_2^j ← C^j
X^j ← M_1^j ⊕ H_{K_1}(pad(M_2^j)) ⊕ H_{K_5}(pad(|M_2^j|))
M_4^j ←$ {0,1}^{n-s}
M_5^j ← mix_R^{-1}(C_2^j || M_2^j)
C_5^j ← mix_L(M_5^j || M_2^j)
C_3^j ← C_1^j ⊕ H_{K_4}(pad(C_2^j)) ⊕ H_{K_5}(pad(|M_2^j|))
bad ← (X^j = X^i) or
      (M_4^j || C_5^j = M_4^i || C_5^i), for some i < j

procedure D(C)                                      Game G_6
j ← j + 1;  C^j ← C
M^j ←$ {0,1}^{n+s};  return M^j

procedure Finalize
C_1^j || C_2^j ← C^j
M_1^j || M_2^j ← M^j
Y_j ← C_1^j ⊕ H_{K_4}(pad(C_2^j)) ⊕ H_{K_5}(pad(|C_2^j|))
M_4^j ←$ {0,1}^{n-s}
C_5^j ← mix_R^{-1}(M_2^j || C_2^j)
M_5^j ← mix_L(C_5^j || C_2^j)
M_3^j ← M_1^j ⊕ H_{K_1}(pad(M_2^j)) ⊕ H_{K_5}(pad(|C_2^j|))
bad ← (Y^j = Y^i) or
      (M_4^j || M_5^j = M_4^i || M_5^i), for some i < j
```

Fig. 4. Games used in the proof of Theorem 2. In game G_6, encipher and decipher queries are answered by random values.

- If $|M_2^j| \neq |M_2^i|$ then by the definition of AXU hash function (for H_{K_5}) the probability that $M_1^j \oplus H_{K_1}(\mathsf{pad}(M_2^j)) \oplus H_{K_5}(\mathsf{pad}(|M_2^j|)) = M_1^i \oplus H_{K_1}(\mathsf{pad}(M_2^i)) \oplus H_{K_5}(\mathsf{pad}(|M_2^i|))$ is at most 2^{-n}. In other words, we have that $\Pr[\, X^j = X^i \,] \leq 2^{-n}$.
- If $|M_2^j| = |M_2^i|$ and $M_2^j \neq M_2^i$ then again by the definition of AXU hash functions (for H_{K_1}) we have $\Pr[\, X^j = X^i \,] \leq 2^{-n}$.
- If $|M_2^j| = |M_2^i|$ and $|M_2^j| = |M_2^i|$ then we immediately have that $M_1^j \neq M_1^i$. The probability that X^j equals X^i is zero.

In any case, we have $\Pr[\, X^j = X^i \,] \leq 2^{-n}$.

We now bound the probability that $M_4^j||C_5^j = M_4^i||C_5^i$ (for some $i < j$). M_4^j is freshly chosen at random, and the probability that $M_4^j = M_4^i$ is 2^{s-n}. Moreover, we have $M_5^j = \mathsf{mix}_\mathsf{R}^{-1}(C_2^j||M_2^j)$, and thus M_5^j is uniformly distributed since C_2^j is independently chosen at random. We also have $C_5^j = \mathsf{mix}_\mathsf{L}(M_5^j||M_2^j)$; by the definition of mixing function we have that the probability that $C_5^j = C_5^i$ is at most $\epsilon(s)$. The probability that $M_4^j||C_5^j = M_4^i||C_5^i$ (for some $i < j$) is at most $\frac{\epsilon(s)}{2^{n-s}}$.

There are at most $q^2/2$ pairs of possible collisions at both Line 620/670 and Line 621/671, the probability that $\mathcal{A}$ manages to set *bad* in this game is at most $0.5\, q^2/2^n + 0.5\, q^2 \frac{\epsilon(s)}{2^{n-s}}$. The theorem now follows. ∎

It is straightforward to pass from the information-theoretic setting to complexity-theoretic one. For completeness, we show it as follows.

Corollary 1. *Let E be a blockcipher, let H be a 2^{-n}-AXU hash function family, and let mix be an $\epsilon(s)$-good mixing function. Let $\mathcal{E} = \mathrm{HEM}[H, E, \mathsf{mix}]$ and let $\mathcal{A}$ be an adversary that asks at most q queries. Then there exist adversaries $\mathcal{B}$ and $\mathcal{C}$ such that $\mathbf{Adv}_{\mathcal{E}}^{\pm\mathrm{prp}}(\mathcal{A}) \leq \mathbf{Adv}_{E}^{\pm\mathrm{prp}}(\mathcal{B}) + \mathbf{Adv}_{E}^{\pm\mathrm{prp}}(\mathcal{C}) + 2\, q^2/2^n + 0.5\, q^2 \frac{\epsilon(s)}{2^{n-s}}$ for any $s \in [n-1]$. Specifically, if we use a mixing function with $\epsilon = 2^{1-s}$ then we have $\mathbf{Adv}_{\mathcal{E}}^{\pm\mathrm{prp}}(\mathcal{A}) \leq \mathbf{Adv}_{E}^{\pm\mathrm{prp}}(\mathcal{B}) + \mathbf{Adv}_{E}^{\pm\mathrm{prp}}(\mathcal{C}) + 3\, q^2/2^n$.* ∎

5 Length-Doubling VIL Tweakable Ciphers

In this section, we present a VIL tweakable cipher over $\bigcup_{i=n+1}^{2n-1} \{0,1\}^i$ with tweak space $\{0,1\}^n$. It is easy to modify the scheme to support larger tweak space. We also give a variant with a slightly more succinct structure.

Let $E\colon \mathcal{K}_1 \times \{0,1\}^n \to \{0,1\}^n$ be a blockcipher and let $H\colon \mathcal{K}_2 \times \{0,1\}^n \to \{0,1\}^n$ be a 2^{-n}-AXU hash function family. Let $\mathsf{mix}\colon \mathcal{S}^2 \to \mathcal{S}^2$ ($\mathcal{S} \supseteq \bigcup_{i=1}^{n-1}\{0,1\}^i$) be an $\epsilon(s)$-good mixing function. Define from the above primitives a VIL tweakable cipher $\widetilde{\mathcal{E}} = \mathrm{THEM}[H, E, \mathsf{mix}]$ with key space $\mathcal{K}_1^2 \times \mathcal{K}_2^4$ and tweak space $\mathcal{T} = \{0,1\}^n$ and message space $\mathcal{M} = \bigcup_{i=n+1}^{2n-1} \{0,1\}^i$. See Figure 5. The construction is $\pm\widetilde{\mathrm{prp}}$-secure for VIL adversaries. To extend the domain of THEM to include $\{0,1\}^n$, we can choose an independent tweakable blockcipher to encipher the n-bit messages. The following theorem establishes the security of THEM.

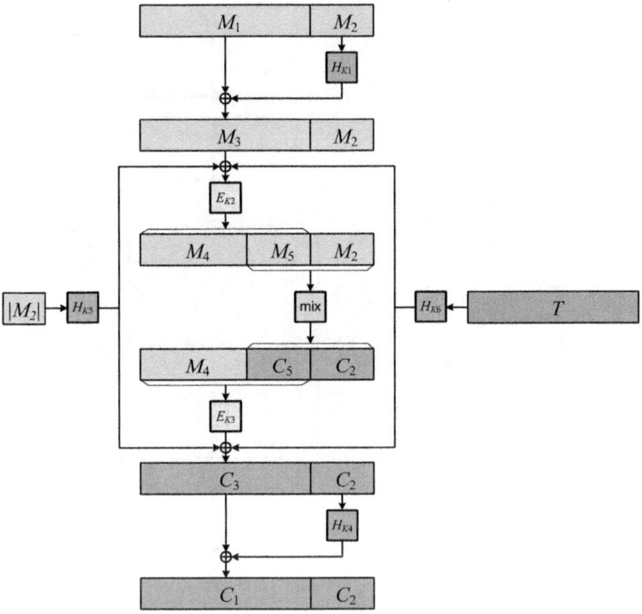

algorithm $\widetilde{\mathcal{E}}_K^T(M)$ **where** $K = K_1||K_2||K_3||K_4||K_5||K_6$

if $M \notin \bigcup_{i=n+1}^{2n-1} \{0,1\}^i$ **then return** $\perp$

$M_1||M_2 \leftarrow M$ **where** $|M_1| = n$ and $|M_2| = s$

$M_3 \leftarrow M_1 \oplus H_{K_1}(\mathsf{pad}(M_2))$

$M_4||M_5 \leftarrow E_{K_2}(M_3 \oplus H_{K_5}(\mathsf{pad}(|M_2|)) \oplus H_{K_6}(T))$ **where** $|M_4| = n - s$, $|M_5| = s$

$C_5||C_2 \leftarrow \mathsf{mix}(M_5||M_2)$ **where** $|C_5| = |C_2| = s$

$C_3 \leftarrow E_{K_3}(M_4||C_5) \oplus H_{K_5}(\mathsf{pad}|M_2|) \oplus H_{K_6}(T)$

$C_1 \leftarrow C_3 \oplus H_{K_4}(\mathsf{pad}(C_2))$

$C \leftarrow C_1||C_2$

return C

Fig. 5. Mode THEM. Compared to HEM mode, FHEM takes an additional tweak T as input. For simplicity, we can assume that the tweak space is $\{0,1\}^n$. Of course, it is trivial to handle larger tweak space by selecting an AXU hash function that supports longer input.

Theorem 3. *Let* $\widetilde{\mathcal{E}} = \mathrm{THEM}[H, \mathrm{Perm}(n), \mathsf{mix}]$. *If* $\mathcal{A}$ *asks at most* q *queries then* $\mathbf{Adv}_{\widetilde{\mathcal{E}}}^{\pm\widetilde{\mathrm{prp}}}(\mathcal{A}) \leq 2\,q^2/2^n + 0.5\,q^2 \frac{\epsilon(s)}{2^{n-s}}$, *and if we use a* 2^{1-s}*-good mixing function then we have that* $\mathbf{Adv}_{\widetilde{\mathcal{E}}}^{\pm\widetilde{\mathrm{prp}}}(\mathcal{A}) \leq 3\,q^2/2^n$. ∎

The proof of the above theorem largely resembles the previous ones, and is thus omitted.

An Alternative Design–Tweak Stealing. A more compact variant using the idea of "tweak stealing" is depicted in Figure 6. This algorithm causes a small decrease in tweak space to $\{0,1\}^{n-\log n}$ (if we insist using a AXU hash function from $\{0,1\}^n$ to $\{0,1\}^n$ for the tweak input), and leads to a slight security loss.

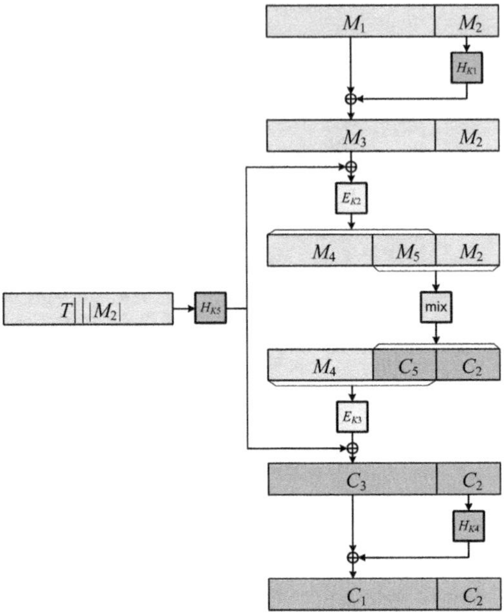

algorithm $\widetilde{\mathcal{E}}_K^T(M)$ **where** $K = K_1||K_2||K_3||K_4||K_5$
if $M \notin \bigcup_{i=n+1}^{2n-1} \{0,1\}^i$ **then return** $\perp$
$M_1||M_2 \leftarrow M$ **where** $|M_1| = n$ and $|M_2| = s$
$M_3 \leftarrow M_1 \oplus H_{K_1}(\mathsf{pad}(M_2))$
$M_4||M_5 \leftarrow E_{K_2}(M_3 \oplus H_{K_5}(T\ |M_2|))$ **where** $|M_4| = n - s$ and $|M_5| = s$
$C_5||C_2 \leftarrow \mathsf{mix}(M_5||M_2)$ **where** $|C_5| = |C_2| = s$
$C_3 \leftarrow E_{K_3}(M_4||C_5) \oplus H_{K_5}(T\ |M_2|)$
$C_1 \leftarrow C_3 \oplus H_{K_4}(\mathsf{pad}(C_2))$
$C \leftarrow C_1||C_2$
return C

Fig. 6. An Alternative Mode—"Tweak Stealing". This mode is specified to support tweak space $\mathcal{T} = \{0,1\}^{n-\log n}$. The input for AXU hash function H_{K_5} is a tweak $T \in \mathcal{T}$ concatenating $\log n$ bits encoding of the length of partial input M_2.

However, this does not necessarily restrict its usage. For instance, it suffices for constructing arbitrary-input-length online ciphers [24]: the *stolen* tweak does not impair the encipher and decipher algorithms. We comment that in spite of its structural simplicity, the variant does not seem to give a notable improvement of efficiency if we consider pre-computation.

References

1. Bellare, M., Rogaway, P.: On the Construction of Variable-Input-Length Ciphers. In: Knudsen, L.R. (ed.) FSE 1999. LNCS, vol. 1636, pp. 231–244. Springer, Heidelberg (1999)

2. Bellare, M., Rogaway, P.: The Security of Triple Encryption and a Framework for Code-Based Game-Playing Proofs. In: Vaudenay, S. (ed.) EUROCRYPT 2006. LNCS, vol. 4004, pp. 409–426. Springer, Heidelberg (2006)
3. Chakraborty, D., Sarkar, P.: HCH: A New Tweakable Enciphering Scheme Using the Hash-Encrypt-Hash Approach. In: Barua, R., Lange, T. (eds.) INDOCRYPT 2006. LNCS, vol. 4329, pp. 287–302. Springer, Heidelberg (2006)
4. Chakraborty, D., Sarkar, P.: A New Mode of Encryption Providing a Tweakable Strong Pseudo-random Permutation. In: Robshaw, M. (ed.) FSE 2006. LNCS, vol. 4047, pp. 293–309. Springer, Heidelberg (2006)
5. Cook, D., Yung, M., Keromytis, A.: Elastic block ciphers: method, security and instantiations. Int. J. Inf. Sec. 8(3), 211–231 (2009)
6. McGrew, D.A., Fluhrer, S.R.: The Security of the Extended Codebook (XCB) Mode of Operation. In: Adams, C., Miri, A., Wiener, M. (eds.) SAC 2007. LNCS, vol. 4876, pp. 311–327. Springer, Heidelberg (2007)
7. Goldenberg, D., Hohenberger, S., Liskov, M., Schwartz, E.C., Seyalioglu, H.: On Tweaking Luby-Rackoff Blockciphers. In: Kurosawa, K. (ed.) ASIACRYPT 2007. LNCS, vol. 4833, pp. 342–356. Springer, Heidelberg (2007)
8. Gueron, S.: Intel's New AES Instructions for Enhanced Performance and Security. In: Dunkelman, O. (ed.) FSE 2009. LNCS, vol. 5665, pp. 51–66. Springer, Heidelberg (2009)
9. Halevi, S.: An observation regarding Jutla's modes of operation. Cryptology ePrint report 2001/015 (April 2, 2001)
10. Halevi, S.: EME*: Extending EME to Handle Arbitrary-Length Messages with Associated Data. In: Canteaut, A., Viswanathan, K. (eds.) INDOCRYPT 2004. LNCS, vol. 3348, pp. 315–327. Springer, Heidelberg (2004)
11. Halevi, S.: Invertible Universal Hashing and the TET Encryption Mode. In: Menezes, A. (ed.) CRYPTO 2007. LNCS, vol. 4622, pp. 412–429. Springer, Heidelberg (2007)
12. Halevi, S., Rogaway, P.: A Parallelizable Enciphering Mode. In: Okamoto, T. (ed.) CT-RSA 2004. LNCS, vol. 2964, pp. 292–304. Springer, Heidelberg (2004)
13. Halevi, S., Rogaway, P.: A Tweakable Enciphering Mode. In: Boneh, D. (ed.) CRYPTO 2003. LNCS, vol. 2729, pp. 482–499. Springer, Heidelberg (2003)
14. Krawczyk, H.: LFSR-Based Hashing and Authentication. In: Desmedt, Y.G. (ed.) CRYPTO 1994. LNCS, vol. 839, pp. 129–139. Springer, Heidelberg (1994)
15. Krovetz, T., Rogaway, P.: The Software Performance of Authenticated-Encryption Modes. In: Joux, A. (ed.) FSE 2011. LNCS, vol. 6733, pp. 306–327. Springer, Heidelberg (2011)
16. Liskov, M., Minematsu, K.: Comments on XTS-AES (September 2008), `http://csrc.nist.gov/`
17. Luby, M., Rackoff, C.: How to construct pseudorandom permutations from pseudorandom functions. SIAM Journal of Computing 17(2), 373–386 (1988)
18. Liskov, M., Rivest, R.L., Wagner, D.: Tweakable Block Ciphers. In: Yung, M. (ed.) CRYPTO 2002. LNCS, vol. 2442, pp. 31–46. Springer, Heidelberg (2002)
19. Meyer, C., Matyas, M.: Cryptography: A new dimension in data security. John Wiley & Sons, New York (1982)
20. Naor, M., Reingold, O.: On the construction of pseudorandom permutations: Luby-Rackoff revisited. Journal of Cryptology 12(1), 29–66 (1999)
21. IEEE P 1619. IEEE Standard for Cryptographic Protection of Data on Block-Oriented Storage Devices

22. Patel, S., Ramzan, Z., Sundaram, G.S.: Efficient Constructions of Variable-Input-Length Block Ciphers. In: Handschuh, H., Hasan, M.A. (eds.) SAC 2004. LNCS, vol. 3357, pp. 326–340. Springer, Heidelberg (2004)
23. Ristenpart, T., Rogaway, P.: How to Enrich the Message Space of a Cipher. In: Biryukov, A. (ed.) FSE 2007. LNCS, vol. 4593, pp. 101–118. Springer, Heidelberg (2007)
24. Rogaway, P., Zhang, H.: Online Ciphers from Tweakable Blockciphers. In: Kiayias, A. (ed.) CT-RSA 2011. LNCS, vol. 6558, pp. 237–249. Springer, Heidelberg (2011)
25. Rogaway, P., Wooding, M., Zhang, H.: The security of ciphertext stealing. In: FSE 2012. LNCS. Springer (2012)
26. Schneier, B., Kelsey, J.: Unbalanced Feistel Networks and Block Cipher Design. In: Gollmann, D. (ed.) FSE 1996. LNCS, vol. 1039, pp. 121–144. Springer, Heidelberg (1996)
27. Wang, P., Feng, D., Wu, W.: HCTR: A Variable-Input-Length Enciphering Mode. In: Feng, D., Lin, D., Yung, M. (eds.) CISC 2005. LNCS, vol. 3822, pp. 175–188. Springer, Heidelberg (2005)
28. Zheng, Y., Matsumoto, T., Imai, H.: On the Construction of Block Ciphers Provably Secure and Not Relying on Any Unproved Hypotheses. In: Brassard, G. (ed.) CRYPTO 1989. LNCS, vol. 435, pp. 461–480. Springer, Heidelberg (1990)

Extending Higher-Order Integral: An Efficient Unified Algorithm of Constructing Integral Distinguishers for Block Ciphers

Wentao Zhang[1], Bozhan Su[2], Wenling Wu[1], Dengguo Feng[2], and Chuankun Wu[1]

[1] State Key Laboratory of Information Security, Institute of Information Engineering, Chinese Academy of Sciences, Beijing, P.R. China
[2] Institute of Software, Chinese Academy of Sciences, Beijing, P.R. China
zhangwt06@yahoo.com, {subozhan,wwl,feng}@is.iscas.ac.cn, chuankun.wu@gmail.com

Abstract. In this paper, we give an extension of the concept of higher-order integral, which can make us design better higher-order integral distinguishers for some block ciphers (structures). Using the new extension, we present a unified algorithm of searching for the best possible higher-order integral distinguishers for block ciphers. We adopt the inside-out approach, trying to predict the behavior of a set of carefully chosen data, not only along encryption direction, but also along decryption direction. Applying the unified algorithm, we search for the best possible higher-order integral distinguishers of Gen-SMS4 structure, Gen-Fourcell structure and Present. For Gen-SMS4 structure and Present, the best higher-order integral distinguishers given by our algorithm are better than the best results known so far. For Gen-Fourcell structure, the best higher-order integral distinguishers given by our algorithm are the same as the best results known so far. We expect that the inside-out method is helpful to understand higher-order integral of block ciphers better, and the unified algorithm presented in this paper can be used as a tool for efficiently evaluating the security of a block cipher against integral cryptanalysis.

Keywords: block cipher, integral cryptanalysis, higher-order integral, integral distinguisher, generalized Feistel structure, Present.

1 Introduction

Integral cryptanalysis [15] is originally proposed by L.R.Knudsen and D.Wagner as a dedicated attack against Square block cipher [8], so is firstly known as "Square attack". Afterwards, the original idea used in Square attack has been extended and given different names, including saturation attack [18], collision attack [12], multiset attack [5] and integral cryptanalysis [15].

Integral cryptanalysis is of particular significance for its applicability to AES. AES is designed to be resistant to differential cryptanalysis and linear cryptanalysis, and very successful in this aspect, only 6-round AES can be resistant to differential cryptanalysis and linear cryptanalysis. However, 6-round AES can be broken using integral cryptanalysis, only with $6 \cdot 2^{32}$ chosen plaintexts

F. Bao, P. Samarati, and J. Zhou (Eds.): ACNS 2012, LNCS 7341, pp. 117–134, 2012.

and 2^{44} time [11]. Up to now, integral cryptanalysis is one of the most effective attacks for round-reduced AES [11,12] and round-reduced IDEA block cipher [4].

Integral cryptanalysis is a chosen-plaintext attack, which considers the propagation of sums of many values. The goal of an attacker is to derive information about the secret key using integral distinguishers. Assume a block cipher has n data subblocks, each data subblock has a length of m bits. When mounting an integral attack, the attacker typically chooses one or several specific subblocks, assume he chooses d subblocks. Then, the attacker chooses $2^{d \times m}$ plaintexts, which take on all possible values in the d subblocks, and have constant values in the other subblocks. The attacker considers these $2^{d \times m}$ chosen plaintexts at a time, trying to predict the properties in some subblock(s) after a certain number of encryption rounds. Customarily, the following four properties are considered:

(1)Constant: The state of a subblock is called "constant" if every data in this subblock has the same constant value.

(2)Active: The state of a subblock is called "active" either if the data in this subblock are all different and have constant values in the other subblocks, or if the data can be divided into some pairwise disjoint subsets and the following condition holds for each subset: the data in this subblock are all different and have constant values in the other subblocks.

(3)Balanced: The state of a subblock is called "balanced" if the XOR of all values is zero.

(4)Unkown: The state of a subblock is called "unknown" if no information is known.

We collectively call the above four states as integral states. Notice that some of the properties are implied by others. For example, a constant or active state is automatically balanced.

The security of a block cipher against integral cryptanalysis depends on several factors, including the length of integral distinguishers, specific input/output forms, the strength of one-round encryption/decryption. Among them, the design of integral distinguishers is the most important. In spite of a long time study of integral cryptanalysis on block ciphers, integral distinguishers have often been designed based on ad hoc approaches and the experience of cryptanalysts. There is no common method of designing integral distinguishers so far.

In this paper, we give an extension of the concept of higher-order integral. Furthermore, based on the new extension, we present an efficient unified algorithm to the design of higher-order integral distinguishers using the method of symbol calculation. The main ideas and contributions are as follows:

- The actual value of a constant state has no influence on the attack, thus all constant states can be denoted as a single letter "C"; A balanced state is usually a sum of some active states. Hence, the state of any subblock can be expressed either as "C", or a sum of some active states and some unknown states. Note that an unknown state is a sum of 0 active state and 1 unknown state. Compared with the customary description, the above expression is more accurate, thus makes the information kept as undamaged as possible.

• Traditionally, integral distinguishers are designed from top to bottom, an attacker tries to predict the behavior of a set of carefully chosen plaintexts after a certain number of encryption rounds. By contrast, we adopt the inside-out approach, trying to predict the behavior of a set of carefully intermediate data, not only after a certain number of encryption rounds, but also after a certain number of decryption rounds. Consequently, we make an extension of the concept of higher-order integral, which can make us design more effective integral distinguishers for some block ciphers (structures).

• Using the matrix method introduced in [13, 14], we propose an efficient unified algorithm of designing the best possible integral distinguishers for block ciphers (structures). The algorithm can be applied widely, not only for byte-oriented block ciphers and some generalized Feistel structures such as AES, Camellia [2], Gen-SMS4 structure [3] and Gen-Fourcell structure [7], but also for bit-oriented block ciphers such as Noekeon [9], Serpent [1] and Present [6]. For Camellia, Gen-SMS4, Noekeon, Serpent and Present, the best integral distinguishers given by our algorithm are better than the best results known so far. For AES and Gen-Fourcell, the best integral distinguishers given by our algorithm are the same as the best results known so far. Hence, we believe that the unified algorithm presented in this paper can be used as a tool for efficiently evaluating the security of block ciphers against integral cryptanalysis.

Due to the length limitation of this paper, we only use Gen-SMS4 structure, Gen-Fourcell structure and Present as 3 typical examples. More examples will be presented in the extended paper.

The focus of this paper is the construction of integral distinguishers for block ciphers. How to design an attack algorithm using these integral distinguishers is out of the scope of this paper, and we leave it for further work.

2 Preliminaries

Throughout this paper, we always assume that: (1) A block cipher structure $\mathbb{S}$ has n data subblocks; (2) The round functions F of $\mathbb{S}$ are all bijective; (3) The operation to connect a subblock with another one is $\oplus$, thus the sum in integral cryptanalysis considered in this paper is referred to as "$\oplus$". Although some block ciphers do not satisfy all the above conditions, e.g., IDEA and RC6, yet we believe that the similar idea can also be applied, with some modifications.

2.1 Higher-Order Integral

The concept of higher-order integral is proposed by L.R.Knudsen and D.Wagner [15]. Consider a set of 2^m elements (representing a set of plaintexts), which differ only in one particular subblock, such that each of the 2^m possible values for this particular subblock occurs exactly once, the sum over the elements of this set is called a **first-order integral**. Consider next a set of $2^{d \times m}$ elements, which differ in d subblocks, such that each of the $2^{d \times m}$ possible values for the d-tuple of values from these subblocks occurs exactly once, the sum of this set is called a ***dth*-order integral**, and **integral** for short. A *dth*-order integral is called a **higher-order integral** when $d \geq 2$.

Consider a set $\overline{S} = S_1 \cup \cdots \cup S_s$ composed of s sets, where each S_i forms an integral. Then, clearly, if one can determine the sum of the elements of S_i for each i, then one can also determine the sum of all elements in $\overline{S}$. This fact is the key point for understanding higher-order integral.

2.2 Matrix Characterization of a Block Cipher Structure

Modern block ciphers are designed by iterating a round function certain times. The following gives a matrix characterization of one round of a block cipher.

Definition 1. [14] ***(Encryption/Decryption Characteristic Matrix)*** *For a block cipher structure $\mathbb{S}$, let $(X_0, X_1, \cdots, X_{n-1})$ and $(Y_0, Y_1, \cdots, Y_{n-1})$ respectively denote the input and output of one-round encryption, then the $n \times n$ encryption/decryption characteristic matrix are defined as follows:*
(1)Encryption characteristic matrix *$\mathcal{E}_{n\times n}$: If $Y_j = X_i \oplus R$, where R is some value* [1]*, the (i,j) entry of $\mathcal{E}$ is set to 1; If Y_j is nonlinearly affected by X_i, the (i,j) entry of $\mathcal{E}$ is set to 2; If Y_j is not affected by X_i, the (i,j) entry of $\mathcal{E}$ is set to 0.*
(2)Decryption characteristic matrix *$\mathcal{D}_{n\times n}$: If $X_j = Y_i \oplus T$, where T is some value, the (i,j) entry of $\mathcal{D}$ is set to 1; If X_j is nonlinearly affected by $F(Y_i)$, the (i,j) entry of $\mathcal{D}$ is set to 2; If X_j is not affected by Y_i, the (i,j) entry of $\mathcal{D}$ is set to 0.*

In Definition 1, each entry of the encryption/decryption characteristic matrix has only one of the three values: 0, 1, or 2. For byte-oriented block ciphers, such as AES and Camellia, the length of a subblock is chosen to be 8 bits; For bit-oriented block ciphers, such as Noekeon, Serpent and Present, the length of a subblock is chosen to be 1 bit. For some block ciphers, one characteristic matrix is sufficient to describe one-round encryption (decryption); While for some other block ciphers, it needs a composition of two characteristic matrices, i.e., firstly the first matrix E_1 (D_1), then the second matrix E_2 (D_2). In the following, we can see that most popular block ciphers can be represented by one characteristic matrix or a composition of two characteristic matrices.

3 A Unified Approach for the Design of Integral Distinguishers

3.1 A New Representation of the 4 kinds of Integral States

The following observations are very important to our new representation:

(1) For a constant subblock state, it is sufficient to know that it is a constant state, and ignoring its exact value. Thus, we can label all constant subblock states with a single letter "C".

[1] Note that it is a formal expression, R can either be independent of X_i, or have a nonlinear relation with X_i.

(2) Generally, a balanced subblock state is produced by an XOR sum of some active states. Thus, we can express a balanced subblock state as $\bigoplus_{i \in I_A} A_i$, where A_i denotes an active subblock state, I_A is the index set, note that the constant monomial is ignored. Compared with a single letter "B", it is more accurate to label a balanced state as an XOR sum of some active states.

(3) For an unknown subblock state, we can express it as $(\bigoplus_{i \in I_A} A_i) \oplus (\bigoplus_{j \in I_?} ?_j)$, where A_i denotes an active state, $?_j$ denotes an unknown state, I_A and $I_?$ is the index set respectively, $I_?$ is not empty, similarly the constant monomial is ignored. Compared with a single letter "?", such an expression is more accurate.

Example 1. Assume a block cipher has 2 subblocks, the state is $(?_0, ?_0 \oplus A_0)$ at some point, where $?_0$ denotes an unknown state, A_0 an active state. Considering XOR sum of the values in the two subblocks, we can get that $?_0 \oplus (?_0 \oplus A_0) = A_0$. However, if we just express the state as $(?, ?)$, we can get nothing. □

Based on the above 3 observations, an integral state is not limited to the 4 types: constant, active, balanced or unknown. An integral state may have much more types, it can be either "C", or an XOR sum of some active states and some unknown states. The following gives a formal description.

Definition 2. ***(Integral Form in Subblock)****For a given set of plaintexts or intermediate data blocks, fixing a subblock, define integral form in the subblock as*

$$<< A.set, A.maxs >, < U.set, U.maxs >>$$

where A.set is a set consisting of some active subblock states, U.set is a set consisting of some unknown subblock states. A.maxs is defined as the maximum subscript in A.set plus 1 (i.e., let h be the maximum subscript among all the elements of A.set, then $A.maxs \equiv h + 1$), especially $\emptyset.maxs \equiv 0$ for an empty set $\emptyset$. Similarly, U.maxs is defined.

Notice that $A.maxs$ ($U.maxs$) is necessary in Definition 2 for the expression of a newly-produced active (unknown) subblock state.

Example 2. (**Integral form in sublock**) For a constant subblock state, its integral form is $<< \emptyset, 0 >, < \emptyset, 0 >>$; For an active subblock state A_0, its integral form is $<< \{A_0\}, 1 >, < \emptyset, 0 >>$; For an integral subblock state $A_0 \oplus A_2 \oplus ?_1 \oplus ?_5$, its integral form is $<< \{A_0, A_2\}, 3 >, < \{?_1, ?_5\}, 6 >>$. □

On the other hand, let $<< A.set, A.maxs >, < U.set, U.maxs >>$ denote an integral form in subblock, define $Unionset \equiv A.set \cup U.set$, where "∪" is the operation of set union, then the integral state in this subblock is just the XOR sum of all elements in $Unionset$, we will use this representation together with integral form in subblock defined in Definition 2 interchangeably in the following.

Assume a block cipher has n data subblocks, naturally, we can define integral form in block.

Definition 3. *(Integral Form in Block) For a given set of plaintexts or intermediate data blocks, define its integral form as* $(\alpha_0, \alpha_1, \cdots, \alpha_{n-1})$*, where* α_i *is the integral form in subblock corresponding to the i-th subblock,* $0 \le i \le n-1$.

We will simply write "integral form" instead of "integral form in (sub)block", when the context is clear.

3.2 Rules for Applying Encryption/Decryption Characteristic Matrix to An Integral Form in Block

For a given set of plaintexts or intermediate data blocks, we can determine its integral form in block. Next, we need to define rules to calculate the integral form after one-round encryption/decryption. In the following, we only focus on encryption process, since decryption process can be treated similarly.

Firstly, we define an operator "$\uplus$" between two integral forms in subblock, this operator is something like adding mod 2.

Definition 4. *Let* μ *and* ν *be two integral forms in subblock, let*

$$\mu = << (A.set)_\mu, (A.maxs)_\mu >, < (U.set)_\mu, (U.maxs)_\mu >>,$$
$$\nu = << (A.set)_\nu, (A.maxs)_\nu >, < (U.set)_\nu, (U.maxs)_\nu >>$$

Define $\omega = \mu \uplus \nu = << (A.set)_\omega\, (A.maxs)_\omega >, < (U.set)_\omega, (U.maxs)_\omega >>$, *thereinto,*

$$(A.set)_\omega \equiv ((A.set)_\mu \setminus (A.set)_\nu) \cup ((A.set)_\nu \setminus (A.set)_\mu),$$
$$(U.set)_\omega \equiv ((U.set)_\mu \setminus (U.set)_\nu) \cup ((U.set)_\nu \setminus (U.set)_\mu),$$
$$(A.maxs)_\omega \equiv Maxsubscript((A.set)_\omega) + 1,$$
$$(U.maxs)_\omega \equiv Maxsubscript((U.set)_\omega) + 1$$

where "$\setminus$*" is the operation of set minus, "*$\cup$*" is the operation of set union, and* $Maxsubscript(X)$ *function returns the maximum subscript in* X.

Example 3. Here is an example of set minus "$\setminus$". Let $(A.set)_\mu = \{A_0, A_1, A_3\}$ and $(A.set)_\nu = \{A_1, A_2, A_4\}$. Then, $(A.set)_\mu \setminus (A.set)_\nu = \{A_0, A_3\}$, $(A.set)_\nu \setminus (A.set)_\mu = \{A_2, A_4\}$. □

Now, we are ready to present the rules.

Definition 5. *Let* $\mathcal{E}_{n\times n} = [e_{ij}]_{n\times n}$ *be the encryption characteristic matrix of a block cipher. For a given set of plaintexts or intermediate data blocks, let* $\alpha = (\alpha_0, \alpha_1, \cdots, \alpha_{n-1})$ *be its integral form. Let* $\gamma = \mathcal{E}_{n\times n}(\alpha) = (\gamma_0, \gamma_1, \cdots, \gamma_{n-1})$ *be the integral form of the outputs after one-round encryption, then* γ_i *is defined as* $\gamma_i = \biguplus_{j=1}^{n} e_{ij}(\alpha_j)$, *where* $e_{ij}(\alpha_j)$ *means applying the transformation* e_{ij} *to* α_j.

The entry e_{ij} has 3 possible values, 0, 1, or 2. Thereinto, 0 is the zero transformation, which transforms any integral form x to $<< \emptyset, 0 >, < \emptyset, 0 >>$; 1 is the identical transformation, which transforms any integral form x to x ; 2 is a bijective transformation, which transforms a constant state to a constant state, an active state to a new active state, and any other state to a new unknown

Table 1. Rules for Applying 0, 1, 2 to an Integral Form in Subblock–Along Encryption Direction

Trans.	input	output
0	x	$<< \emptyset, 0 >, < \emptyset, 0 >>$
1	x	x
2	C A_i otherwise	C $A_{emaxsA+1}$ $U_{emaxsU+1}$

state. Table 1 summarizes the above rules, where $emaxsA$ ($emaxsU$) denotes the maximum subscript of all active(unknown) states brought forth so far, along encryption direction.

Example 4. The structure of SMS4 [3] is a kind of 4-branch generalized Feistel structure (denoted as Gen-SMS4), one round encryption of Gen-SMS4 is described as follows:

$$Y_0 = X_1, Y_1 = X_2, Y_2 = X_3, Y_3 = X_0 \oplus F(X_1 \oplus X_2 \oplus X_3)$$

It needs two characteristic matrices to describe one-round encryption (decryption) of Gen-SMS4, i.e., firstly the first matrix $E1_{GenSMS4}$ ($D1_{GenSMS4}$), then the second matrix $E2_{GenSMS4}$ ($D2_{GenSMS4}$). The uppermost row is the 0-th row, the leftmost column is the 0-th column. The encryption and decryption characteristic matrices of Gen-SMS4 are as follows:

$$E1_{GenSMS4} = \begin{pmatrix} 0,1,1,1 \\ 1,0,0,0 \\ 0,1,0,0 \\ 0,0,1,0 \end{pmatrix}, \qquad E2_{GenSMS4} = \begin{pmatrix} 0,0,1,0 \\ 0,0,0,1 \\ 1,0,1,1 \\ 2,1,0,0 \end{pmatrix}$$

$$D1_{GenSMS4} = \begin{pmatrix} 1,1,1,0 \\ 0,1,1,0 \\ 1,0,1,0 \\ 0,0,0,1 \end{pmatrix}, \qquad D2_{GenSMS4} = \begin{pmatrix} 2,0,0,1 \\ 1,1,0,0 \\ 1,0,1,0 \\ 1,1,1,0 \end{pmatrix}$$

Assume an attacker chooses a set of 2^m data, which has the form of $\{(c_0, x \oplus c_1, x \oplus c_2, x \oplus c_3)\}$, where x takes on all the 2^m possible values, c_0, c_1, c_2 and c_3 are 4 constants. The integral form of the data set is $\alpha^0 = (C, A_0, A_0, A_0)$. Applying $E1_{GenSMS4}$ to α^0, using Def. 5 and the rules in Table 1, we can get:

$$\begin{pmatrix} 0,1,1,1 \\ 1,0,0,0 \\ 0,1,0,0 \\ 0,0,1,0 \end{pmatrix} \begin{pmatrix} C \\ A_0 \\ A_0 \\ A_0 \end{pmatrix} = \begin{pmatrix} A_0 \\ C \\ A_0 \\ A_0 \end{pmatrix}$$

Next, applying $E2_{GenSMS4}$ to (A_0, C, A_0, A_0), we can get:

$$\begin{pmatrix} 0,0,1,0 \\ 0,0,0,1 \\ 1,0,1,1 \\ 2,1,0,0 \end{pmatrix} \begin{pmatrix} A_0 \\ C \\ A_0 \\ A_0 \end{pmatrix} = \begin{pmatrix} A_0 \\ A_0 \\ A_0 \\ A_1 \end{pmatrix}$$

Hence the integral form of the outputs after one-round encryption is $\alpha^1 = (A_0, A_0, A_0, A_1)$.

Along the encryption direction, let $\alpha^i = (\alpha_0^i, \alpha_1^i, \cdots, \alpha_{n-1}^i)$ denote the integral form of the outputs after i-round encryption, $i = 1, 2, \cdots$. Let $\beta^0 = \alpha^0$, along the decryption direction, let $\beta^j = (\beta_0^j, \beta_1^j, \cdots, \beta_{n-1}^j)$ denote the integral form of the outputs after j-round decryption. Similarly, we can calculate α^i for $i = 2, 3, \cdots$, and β^j for $j = 1, 2, 3, \cdots$. Table 2 presents the results.

Table 2. An Example of Gen-SMS4: Application of Def. 5 and Rules in Table 1

χ	χ_0	χ_1	χ_2	χ_3
β^5	$A(0,2)$	A_1	A_0	A_0
β^4	A_1	A_0	A_0	A_0
β^3	A_0	A_0	A_0	C
β^2	A_0	A_0	C	A_0
β^1	A_0	C	A_0	A_0
$\alpha^0 = \beta^0$	C	A_0	A_0	A_0
α^1	A_0	A_0	A_0	A_1
α^2	A_0	A_0	A_1	$A(0,2)$
α^3	A_0	A_1	$A(0,2)$	$A_0 \oplus ?_0$
α^4	A_1	$A(0,2)$	$A_0 \oplus ?_0$	$A_0 \oplus ?_1$
α^5	$A(0,2)$	$A_0 \oplus ?_0$	$A_0 \oplus ?_1$	$A_1 \oplus ?_2$

χ_k : integral form in the k-th subblock of α^i or β^j, $k = 0, \cdots, n-1$;
$A(i, j, \cdots, k)$: a simplified expression for $A_i \oplus A_j \oplus \cdots \oplus A_k$;
$?(i, j, \cdots, k)$: a simplified expression for $?_i \oplus ?_j \oplus \cdots \oplus ?_k$.

In fact, Table 2 presents a 10-round integral distinguisher for Gen-SMS4, we will give more explainations in the following sections. □

3.3 Finishing Conditions for Calculus and an Extension of Higher-Order Integral

For a given set of plaintexts or intermediate data blocks, Def. 5 and Table 1 show that we can calculate the integral form of the outputs after one-round encryption. Decryption process can be treated similarly. Theoretically, such a process can be iterated for arbitrary number of rounds, either along encryption direction, or along decryption direction. However, we must give some restrictions to terminate the process for deriving useful integral distinguishers.

Finishing Condition along Encryption Direction. Let $(\alpha_0, \alpha_1, \cdots, \alpha_{n-1})$ be an integral form in block. If there exists a subset $I^* \subseteq \{0, 1, \cdots, n-1\}$, such that the *U.set* of $\biguplus_{i \in I^*} \alpha_i$ is empty, that means, $\biguplus_{i \in I^*} \alpha_i$ is either a constant state or an XOR sum of some active states. In either case, the attacker can derive useful information from the corresponding set of data.

On the other hand, if the *U.set* of $\biguplus_{i \in I} \alpha_i$ is non-empty for every subset $I \subseteq \{0, 1, \cdots, n-1\}$, then the attacker can derive nothing from the corresponding set of data. The following gives a formal definition.

Definition 6. ***(Integral-Nothing)*** *An integral form in block $(\alpha_0, \alpha_1, \cdots, \alpha_{n-1})$ is called integral-nothing, if the U.set of $\biguplus_{i \in I} \alpha_i$ is non-empty for every subset $I \subseteq \{0, 1, \cdots, n-1\}$.*

If an integral form is integral-nothing, the attacker can derive nothing from the corresponding set of data. Hence, along encryption direction, when the integral form becomes integral-nothing, the attacker should terminate the process.

Example 5. This example comes from a 18-round integral distinguisher of Gen-Fourcell structure, see Table 3 for more details.

We have $\alpha^{15} = (\alpha_0^{15}, \alpha_1^{15}, \alpha_2^{15}, \alpha_3^{15}) = (A(0,1) \oplus ?(0,2), A(0,1,2) \oplus ?(0,1,2,3), A(2,3) \oplus ?(1,3,4), A(3,4) \oplus ?_4)$. Every component of α^{15} has unkown ingredients, but we have $\alpha_0^{15} \uplus \alpha_1^{15} \uplus \alpha_2^{15} \uplus \alpha_3^{15} = A_4$, thus α^{15} is not integral-nothing.

Applying the encryption matrices of Gen-Fourcell to α_{15}, we get $\alpha^{16} = (\alpha_0^{16}, \alpha_1^{16}, \alpha_2^{16}, \alpha_3^{16}) = (A(0,1,2) \oplus ?(0,1,2,3), A(2,3) \oplus ?(1,3,4), A(3,4) \oplus ?_4, A(0,1,4) \oplus ?(0,2,5))$, we can verify that α^{16} is integral-nothing. □

Finishing Condition along Decryption Direction and an Extension of Higher-Order Integral. The finishing condition is different along decryption direction.

In the original definition [15] (also refer to section 2.1), dth-order integral is related to a set of $2^{d \times m}$ elements, which differ only in d subblocks. However, we argue that the linear relations among different subblocks should be taken into account. In the following, we give an extension of higher-order integral, a dth-order integral is related to $2^{d \times m}$ elements, but they can differ in d^* subblocks where $d^* \geq d$. To do this, we should firstly define "integral order of an integral form", which takes the linear relations among different subblocks into account.

Definition 7. ***(Integral Order of an Integral Form)*** *Given a set of plaintexts or intermediate data blocks, let $\beta = (\beta_0, \beta_1, \cdots, \beta_{n-1})$ be its integral form, and let*

$$\beta_i = << (A.set)_i, (A.maxs)_i >, < (U.set)_i, (U.maxs)_i >>$$

for $i = 0, 1, \cdots, n-1$. Let dmaxsA (respectively dmaxsU) denote the maximum subscript among all active (respectively unknown) states brought forth so far along decryption direction (If no unknown state is brought forth, then $dmaxsU \equiv -1$), note that they are irrelevant with the process along encryption direction.

Denote $w \equiv (dmaxsA+1)+(dmaxsU+1)$*, construct a* $n\times w$ *matrix* $\mathcal{G}_{n\times w} = (g_{ij})$ *as follows: Each element* g_{ij} *is firstly initialized as* 0, $0 \leq i \leq n-1, 0 \leq j \leq w-1$*. Next, if* $A_j \in (A.set)_i$*, then* g_{ij} *is modified to 1, for* $j = 0, 1, \cdots, dmaxsA$*; If* $U_j \in (U.set)_i$*, then* $g_{i(j+dmaxsA+1)}$ *is modified to 1, for* $j = 0, 1, \cdots, dmaxsU$*. Notice that the i-th row is completely determined by* $\beta_i (i = 0, 1, \cdots, n-1)$*, and the first* $(dmaxsA + 1)$ *columns are corresponding to the active ingredients, the last* $(dmaxsU + 1)$ *columns corresponding to the unknown ingredients. Define the* ***integral order*** *of* β *as* $d = rank(\mathcal{G}_{n\times w})$*, here* $rank(\mathcal{G}_{n\times w})$ *is the rank of* $\mathcal{G}_{n\times w}$*, where* $\mathcal{G}$ *is regarded as a matrix over* $GF(2)$*.*

Integral cryptanalysis is a kind of chosen-plaintext attack, the attacker can choose a priori a set of plaintexts and obtain the corresponding ciphertexts. For a successful attack, the amount of the chosen plaintexts must be less than 2^l, where $l\ (= n\times m)$ is the block length. An integral form with an integral order d is corresponding to $2^{d\times m}$ data (accordingly, corresponding to $2^{d\times m}$ plaintexts), hence d must satisfy that $d \leq n-1$. Therefore, when the integral order d of the integral form satisfies that $d = n$ after some decryption rounds (since there are n subblocks in total, integral order can not be larger than n), the attacker should terminate the process.

Along decryption direction, let d_j denote the integral order of the integral form of the outputs after j-round decryption, $j = 1, 2, \cdots$. Due to diffusion of the block cipher (structure) along decryption direction, d_j will increase or keep unchanged as j increases. Hence, there must exist a unique t which satisfies that $d_t \leq n-1$ and $d_{t+1} = n$. That means, the attacker should terminate the process after $(t+1)$ rounds along decryption direction. We call d_t the integral order of the corresponding integral distinguisher, and the distinguisher is called a $d_t th$-order integral distinguisher. The following gives a formal description.

Definition 8. ***(Integral Order of an Integral Distinguisher)*** *Let Dis be an integral distinguisher, which is constructed by the above inside-out approach. Let* d_j *denote the integral order of the integral form of the outputs after* j*-round decryption, then there must exist a unique integer* t *satisfying* $d_t \leq n-1$ *and* $d_{t+1} = n$*.* d_t *is called the integral order of Dis, and Dis is called a* $d_t th$*-order integral distinguisher.*

Example 6. Considering the integral distinguisher of Gen-SMS4 in Table 2.

We have $\beta^5 = (A_0 \oplus A_2, A_1, A_0, A_0)$, the corresponding matrix $\mathcal{G}_{\beta^5}$ is :

$$\mathcal{G}_{\beta^5} = \begin{pmatrix} 1,0,1 \\ 0,1,0 \\ 1,0,0 \\ 1,0,0 \end{pmatrix}$$

The rank of $\mathcal{G}_{\beta^5}$ is 3, thus the integral order of β^5 is 3.

Applying the composition of $D1_{GenSMS4}$ and $D2_{GenSMS4}$ to β^5, we get $\beta^6 = (A_0 \oplus ?_0, A_0 \oplus A_2, A_1, A_0)$, the corresponding matrix $\mathcal{G}_{\beta^6}$ is :

$$\mathcal{G}_{\beta^6} = \begin{pmatrix} 1,0,0,1 \\ 1,0,1,0 \\ 0,1,0,0 \\ 1,0,0,0 \end{pmatrix}$$

The first 3 columns of $\mathcal{G}_{\beta^6}$ are corresponding to active ingredients, and the last column to unknown ingredients. The rank of $\mathcal{G}_{\beta^6}$ is 4, i.e., the integral order of β^6 is 4, which is equal to the number of subblocks. Hence, the attacker should terminate the process aftre 6-round decryption along decryption direction. The integral order of this distinguisher is equal to the integral order of β^5, i.e., 3. □

In Section 4, we will see that better higher-order integral distinguishers can be constructed using our new extension of higher-order integral, including Gen-SMS4 structure and Present.

Let *Dis* be an integral distinguisher for a block cipher (structure), which is constructed according to the rules and finishing conditions in Section 3.1-3.3. Assume *Dis* has w rounds along encryption direction, t rounds along decryption direction, let β_j denote the integral form of the outputs after j-round decryption along decryption direction, and let d_j denote the integral order of β_j. Now, we present some details about the part of *Dis* along decryption direction. We will see that *Dis* is indeed a $(w + t)$-round integral distinguisher.

Note the following facts:

(1.) Firstly, considering the outputs after $(j+1)$-round decryption. The attacker can choose d_{j+1} independent subblocks, which take on all possible values (corresponding to $2^{d_{j+1} \times m}$ data); For each of the other $(n - d_{j+1})$ subblocks, the state is either constant, or the value in this subblock can be linearly determined by the values in the chosen d_{j+1} subblocks. Thus, the attacker chooses a set of $2^{d_{j+1} \times m}$ data blocks, this set is denoted as Ω^{j+1}.

(2.) Secondly, considering the set of the 1-round encryption outputs of all of the elements of Ω^{j+1}, we will get a new set of $2^{d_{j+1} \times m}$ data blocks, which is denoted as Ω^j. Since β^{j+1} and β_j are correlated by the decryption characteristic matrices and the calculus rules, also Ω^{j+1} and Ω^j are correlated by the one-round encryption function, the $2^{d_{j+1} \times m}$ elements of Ω^j can be separated into d_{j+1}/d_i groups, which satisfy the following condition: each group has $2^{d_j \times m}$ elements with an integral form of β_j, thus each group is corresponding to a $(w + j)$-round integral distinguisher. Hence, if the sum of the outputs after $(w + j)$-round decryption is zero (corresponding to $2^{d_j \times m}$ data blocks), then the sum of the outputs after $(w+j+1)$-round decryption is also zero (corresponding to $2^{d_{j+1} \times m}$ data blocks), since the XOR of many zeros is also zero.

(3.) For *Dis*, the w-round part along encryption direction can be regarded as a traditional integral distinguisher. Then, exucute one-round decryption, and applying the above induction to $j = 0$, we get a $(w + 1)$-round integral distinguisher. The induction can be applied iteratively along decryption direction for $j = 0, 1, \cdots, t-1$, totally t times. Finally, we get a $(w+t)$-round integral distinguisher, that is to say, Dis is indeed a $(w+t)$-round integral distinguisher.

Example 7. Considering the integral distinguisher of Gen-SMS4 in Table 2. We have that $\beta^5 = (A_0 \oplus A_2, A_1, A_0, A_0)$ and $\beta^4 = (A_1, A_0, A_0, A_0)$. Based on β^5, the attacker chooses a set of $2^{3\times m}$ elements, which has the form of $(u_0 \oplus c_0, u_1 \oplus c_1, u_2 \oplus c_2, u_2 \oplus c_3)$, for each possible $(u_0, u_1, u_2) \in (GF(2^m))^3$, and c_0, c_1, c_2 and c_3 are m-bit constants. For simplicity, let $c_0 = c_1 = c_2 = c_3 = 0$. Then, applying the encryption characteristic matrices to β^5, we can get that the set of the $2^{3\times m}$ outputs after one-round encryption (this set is denoted by Ω) have the form of $(u_1, u_2, u_2, u_0 \oplus F(u_1 \oplus c_4))$, where c_4 is a new constant (which depends on the key). In the following, we will show that the $2^{3\times m}$ elements of Ω can be divided into 2^m groups, each group has $2^{2\times m}$ elements, satisfying that the integral form of each group is equal to β^4.

Let $const$ denote a m-bit constant, let $u_0 = u_2 \oplus F(u_1 \oplus c_4) \oplus const$, then we get a subset of $2^{2\times m}$ elements of Ω, which have the form of $(u_1, u_2, u_2, u_2 \oplus const)$, for each possible $(u_1, u_2) \in (GF(2^m))^2$, we use $Group_{const}$ to denote this group. It is easy to see that the integral form of $Group_{const}$ is equal to β^4. There are 2^m possible values of $const$, thus there are 2^m disjoint groups, and the union of these 2^m groups will cover every element of Ω. □

If an integral distinguisher has an integral order d_t, then it is corresponding to $2^{d_t \times m}$ plaintexts. Hence, the integral order of an integral distinguisher reflects the amount of data blocks needed by this integral distinguisher.

3.4 A Unified Algorithm of Constructing Integral Distinguishers

Based on the results of Section 3.1-3.3, we are now ready to present a unified algorithm of constructing integral distinguishers for block ciphers.

For a block cipher (structure), let $\{\mathcal{E}_{n\times n}\}/\{\mathcal{D}_{n\times n}\}$ denote its encryption/de -cryption characteristic matrices. Choose a set of data blocks, let $\alpha^0 = (\alpha_0, \alpha_1, \cdots,$
$\alpha_{n-1})$ denote its integral form. Along encryption direction, let $\alpha^i = (\alpha_0^i, \alpha_1^i, \cdots, \alpha_{n-1}^i)$ denote the integral form of the outputs after i-round encryption, $i = 1, 2, \cdots$. Let $\beta^0 = \alpha^0$, along decryption direction, let $\beta^j = (\beta_0^j, \beta_1^j, \cdots, \beta_{n-1}^j)$ denote the integral form of the outputs after j-round decryption, and let d_j denote the integral order of β^j, $j = 1, 2, \cdots$.

In integral cryptanalysis, an attacker is usually intended to derive the longest distinguishers. In the following, we present an algorithm to calculate the length of the longest possible integral distinguishers.

Once we get the length of the longest possible integral distinguishers using Algorithm 1, we can backtrack to derive the corresponding distinguishers. Note that there are usually many longest integral distinguishers using Algorithm 1.

In Algorithm 1, Step 2 needs to enumerate all the cases of α^0. For some block ciphers (structures), it is impossible to enumerate all the cases due to the computing limitation. We will discuss the selection of initial integral forms $\alpha^0 (= \beta^0)$ for different block ciphers(structures) in Section 4.4.

Algorithm 1. Compute the Length of the Longest Possible Integral Distinguishers

Input: Encryption characteristic matrices $\{\mathcal{E}_{n\times n}\}$, decryption characteristic matrices $\{\mathcal{D}_{n\times n}\}$.
Output: The length of the longest possible integral distinguishers, denoted by r, and r is initialized to be 0.

Step1. For a chosen integral form $\alpha^0 = \beta^0$, do the following:
- **(1)** Find the largest integer s such that α^{s+1} is integral-nothing and α^s is not integral-nothing.
- **(2)** Find the largest integer t such that $d_{t+1} = n$ and $d_t < n$.
- **(3)** Calculate $h = s + t$, then h is the length of the longest integral distinguisher corresponding to α^0. If $h > r$, let $r \leftarrow h$.

Step2. Repeat Step 1 until all the cases of α^0 are enumerated.
Step3. Output r.

---

4 Experimental Results – Application to Gen-SMS4, Gen-Fourcell and Present

In this section, we present experimental results of applying Algorithm 1 to Gen-SMS4 structure [3], Gen-Fourcell structure [7] and Present [6].

Assume the round subkey is XORed with the state, and the sum in integral cryptanalysis considered is XOR sum. Thus, subkey addition has no effect on the design of integral distinguishers, we will omit it. For a given block cipher (structure), there are usually many longest possible integral distinguishers applying Algorithm 1. Although some are the same or equivalent, we will not tell them apart.

In the following, let χ_k denote the integral form in the k-th subblock of α^i (or β^j), where $i = 0, 1, \cdots,\ \ j = 0, 1, \cdots$ and $k = 0, \cdots, n-1$.

4.1 Gen-SMS4

Using Algorithm 1, we found 256 10-round integral distinguishers. Table 2 presents one: 5 rounds along encryption direction, and 5 rounds along decryption direction. From β^5, it is a $3rd$-order integral distinguisher; The attacker chooses a set of $2^{3\times m}$ elements, which has the form of $(u_0 \oplus c_0, u_1 \oplus c_1, u_2 \oplus c_2, u_2 \oplus c_3)$, for each possible $(u_0, u_1, u_2) \in (GF(2^m))^3$, and c_0, c_1, c_2 and c_3 are m-bit constants. Then, considering the outputs after 10-round encryption. From $\alpha_0^5 = A(0, 2)$, we can get that the XOR sum of all the $2^{3\times m}$ values in the 0th subblock (corresponding to χ_0) is zero. A 8-round integral distinguisher of Gen-SMS4 is given in [17], the 10-round distinguisher in Table 2 truncated from the 3rd round to the 10th round is equivalent to the 8-round distinguisher in [17].

4.2 Gen-Fourcell

The structure of Fourcell [7] is also a kind of 4-branch generalized Feistel structure (denoted as Gen-Fourcell), one round of Gen-Fourcell is described as follows:

Table 3. A 18-round Integral Distinguisher for Gen-Fourcell

χ	χ_0	χ_1	χ_2	χ_3
β^3	$?_0$	A_2	A_1	C
β^2	A_2	A_1	C	C
β^1	A_1	C	C	C
$\alpha^0 = \beta^0$	C	C	C	A_0
α^1	C	C	A_0	A_0
α^2	C	A_0	A_0	C
α^3	A_0	A_0	C	C
α^4	A_0	C	C	$A(0,1)$
α^5	C	C	$A(0,1)$	$A(0,1,2)$
α^6	C	$A(0,1)$	$A(0,1,2)$	A_2
α^7	$A(0,1)$	$A(0,1,2)$	A_2	C
α^8	$A(0,1,2)$	A_2	C	$A(0,1)\oplus ?_0$
α^9	A_2	C	$A(0,1)\oplus ?_0$	$A(0,1,2)\oplus ?(0,1)$
α^{10}	C	$A(0,1)\oplus ?_0$	$A(0,1,2)\oplus ?(0,1)$	$A(2,3)\oplus ?_1$
α^{11}	$A(0,1)\oplus ?_0$	$A(0,1,2)\oplus ?(0,1)$	$A(2,3)\oplus ?_1$	A_3
α^{12}	$A(0,1,2)\oplus ?(0,1)$	$A(2,3)\oplus ?_1$	A_3	$A(0,1)\oplus ?(0,2)$
α^{13}	$A(2,3)\oplus ?_1$	A_3	$A(0,1)\oplus ?(0,2)$	$A(0,1,2)\oplus ?(0,1,2,3)$
α^{14}	A_3	$A(0,1)\oplus ?(0,2)$	$A(0,1,2)\oplus ?(0,1,2,3)$	$A(2,3)\oplus ?(1,3,4)$
α^{15}	$A(0,1)\oplus ?(0,2)$	$A(0,1,2)\oplus ?(0,1,2,3)$	$A(2,3)\oplus ?(1,3,4)$	$A(3,4)\oplus ?_4$

$$Y_0 = X_1, Y_1 = X_2, Y_2 = X_3, Y_3 = F(X_0) \oplus X_1 \oplus X_2 \oplus X_3$$

The encryption and decryption characteristic matrices of Gen-Fourcell are as follows:

$$E_{GenFourcell} = \begin{pmatrix} 0,1,0,0 \\ 0,0,1,0 \\ 0,0,0,1 \\ 2,1,1,1 \end{pmatrix}$$

$$D1_{GenFourcell} = \begin{pmatrix} 1,1,1,1 \\ 0,1,1,1 \\ 1,0,1,1 \\ 1,1,0,1 \end{pmatrix}, \qquad D2_{GenFourcell} = \begin{pmatrix} 2,0,0,0 \\ 1,1,0,0 \\ 1,0,1,0 \\ 1,0,0,1 \end{pmatrix}$$

Note: the uppermost row is the 0-th row, the leftmost column is the 0-th column.

Using Algorithm 1, we found 56 18-round integral distinguishers. Table 3 presents one: 15 rounds along encryption direction, and 3 rounds along decryption direction. From β^3, it is a $3rd$-order integral distinguisher; The attacker chooses a set of $2^{3\times m}$ elements, which has the form of $(u_0 \oplus c_0, u_1 \oplus c_1, u_2 \oplus c_2, c_3)$, for each possible $(u_0, u_1, u_2) \in (GF(2^m))^3$, c_0, c_1, c_2 and c_3 are m-bit constants. Then, considering the outputs after 18-round encryption. Based on α^{15}, considering $\alpha_0^{15} \uplus \alpha_1^{15} \uplus \alpha_2^{15} \uplus \alpha_3^{15}$, we have $(A(0,1) \oplus ?(0,2)) \uplus (A(0,1,2) \oplus ?(0,1,2,3)) \uplus (A(2,3) \oplus ?(1,3,4)) \uplus (A(3,4) \oplus ?_4) = A_4$. Hence the XOR sum of all the $2^{3\times m}$ values of the XOR sum of the 4 subblocks is zero. In [16], a 18-round integral distinguisher of Gen-Fourcell is given. The distinguisher in Table 3 is equivalent to that in [16].

4.3 Present

Present [6] is a SP-network block cipher, the block length is 64. Since it is bit-oriented, we will treat a bit as a data subblock, then 64 subblocks in total, i.e., $m = 1$ and $n = 64$. Figure 1 gives the bit indexing of a 64-bit data block.

One round of Present [6] is described as $Y = Theta \circ Gamma(X)$, where $Gamma$ is the S-box layer, and $Theta$ is a linear transformation. $Gamma$ operates independently on 16 4-tuple of bits, the first S-box takes bits 0-3 as input, the next S-box takes bits 4-7 as input, and so on. Let a_i denote the i-th bit of a, $i = 0, 1, \cdots, 63$, then $Theta(a_i) = a_j$, where $j = 16 \times (i \, mod \, 4) + \lfloor i/4 \rfloor$, $\lfloor x \rfloor$ is the integer portion of x.

Present uses a 4×4 S-box. Let $x = x_3x_2x_1x_0$, where x_i is the i-th bit of x, $i = 0, 1, 2, 3$. Let $(\Delta x \rightarrow \Delta y)$ denote a differential with input difference Δx and output difference Δy. For the S-box of Present, there are 3 truncated differentials with probability 1:

$$(1001 \rightarrow ***0), \quad (0001 \rightarrow ***1), \quad (1000 \rightarrow ***1)$$

where "*" denotes an unknown bit.

For Present, the size of the characteristic matrices is 64×64. It is impractical to use Definition 6 as the finishing condition along encryption direction. However, Present is a SP-network cipher, the outputs of different S-boxes can be regarded as being independent. Thus, Definition 6 can be revised as follows, without any effect on the design of the best possible integral distinguishers for SP-network block ciphers.

Definition 6' (Integral-Nothing for SP-network Ciphers). For a SP-network block cipher (structure), an integral form $(\alpha_0, \alpha_1, \cdots, \alpha_{n-1})$ is integral-nothing, if the $U.set$ of α_i is non-empty for every $i \in \{0, 1, \cdots, n-1\}$.

60	56	52	48	44	40	36	32	28	24	20	16	12	8	4	0
61	57	53	49	45	41	37	33	29	25	21	17	13	9	5	1
62	58	54	50	46	42	38	34	30	26	22	18	14	10	6	2
63	59	55	51	47	43	40	35	31	27	23	19	15	11	7	3

Fig. 1. 4×16 Bit Indexing of a 64-bit Data Block

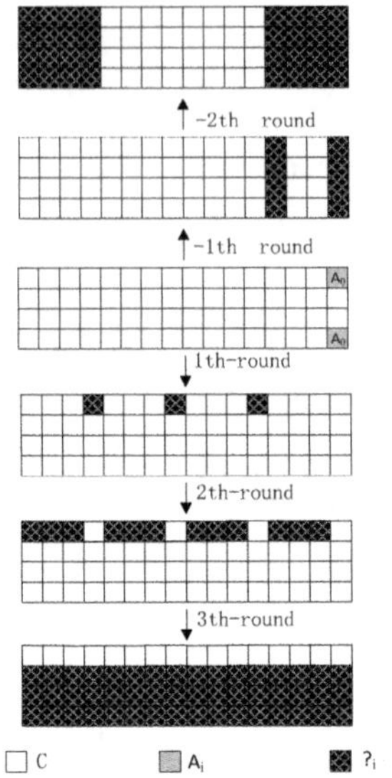

Fig. 2. The 5-round integral distinguisher for Present (In the initial integral form, we label A_0 both in bit 0 and bit 3, which means that the two bits are linearly dependent, the integral order of the integral form is 1, instead of 2

Using Algorithm 1, we found many 5-round integral distinguishers of Present. Figure 2 illustrates one of the best, which uses $Prob_{Sbox}(1001 \rightarrow ***0) = 1$, 3 rounds along encryption direction, and 2 rounds along decryption direction. It is a $32th$-order integral distinguisher. The attacker chooses a set of 2^{32} plaintexts, which have constant values in the following 32 bits: 0-15 and 48-63, while taking all possible values in the other 32 bits. Then, considering the outputs after 5-round encryption, the XOR sum of all the 2^{32} values in each of the following 16 bits is zero: 0, 4, 8, 12, 16, 20, 24, 28, 32, 36, 40, 44, 48, 52, 56 and 60. Note that this distinguisher uses the new extension concept of higher-order integral, the initial integral form has an integral order of 1, while it is related to 2 bits.

In [20], a 3-round integral distinguisher of Present is given, which has an integral order of 4. The distinguisher in Figure 2 is a 5-round integral distinguisher, furthermore, the truncated 3-round distinguisher of the last 3 rounds has an integral order of 2. Thus, our distinguisher (as illustrated in Figure 2) is more better than that in [20].

4.4 Selection of Initial Integral Forms

Let $(\chi_0, \chi_1, \chi_2, \cdots, \chi_{n-1})$ denote an initial integral form in block.

For Gen-Fourcell and Gen-SMS4, χ_i can be any element or an XOR combination of the elements in the set $\{C, A_0, A_1, A_2\}$, which have 2^4 possibilities for each subblock. There are 4 subblocks, thus $2^{16} - 1$ possibilities in total. In our experiments, we have tried them exhaustively.

For Present, the integral order of an initial integral form can be $1, 2, \cdots, 63$, it is impractical to search for each case. However, Present is a SP-network cipher, and the diffusion layer is very simple. Our experiments also show that the smaller the integral order of an initial integral form, the better the integral distinguishers. Hence, we only considered 1st-order initial integral forms.

5 Discussion and Conclusion

Our work in this paper is originally inspired by the work of J.Kim et al [13, 14]. In [13, 14], the authors proposed a general tool for finding impossible differentials of block cipher structures using matrix method and meet-in-the-middle approach, and applied their tool to some block ciphers (structures). They also pointed out that the matrix method can be converted into a tool for the Square attack. However, they only considered the 1st-order integral, not considering higher-order integral. In [19], Y.Y.Luo et al. greatly improved the results of [13, 14].

In this paper, we adopted inside-out approach to construct integral distinguishers for block ciphers, and extended the concept of higher-order integral by considering the linear relations among different subblocks. Furthermore, we presented an efficient unified algorithm to the design of the longest possible integral distinguishers for block ciphers. We applied the algorithm to many block ciphers (structures), the experiments showed: For Gen-SMS4 structure and Present, the best integral distinguishers given by our algorithm are better than the best results known so far; For Gen-Fourcell structure, the best integral distinguishers given by our algorithm are the same as the best results known so far.

To sum up, we believe that the inside-out method for designing integral distinguishers and the new extension of higher-order integral are helpful to better understand integral cryptanalysis of block ciphers. Also, the unified algorithm in this paper can be useful as a tool for efficiently evaluating the security of block ciphers against integral cryptanalysis.

Acknowledgment. We would like to thank anonymous referees for their helpful comments and suggestions. The research presented in this paper is supported by the National Natural Science Foundation of China (No.60903212), the "Strategic Priority Research Program" of the Chinese Academy of Sciences (No.XDA06010701), and the Knowledge Innovation Project of the Chinese Academy of Sciences.

References

1. Anderson, R., Biham, E., Knudsen, L.R.: Serpent: A Proposal for the Advanced Encryption Standard. In: NIST AES Proposal (1998), http://www.cl.cam.ac.uk/
2. Aoki, K., Ichikawa, T., Kanda, M., Matsui, M., Moriai, S., Nakajima, J., Tokita, T.: *Camellia*: A 128-Bit Block Cipher Suitable for Multiple Platforms - Design and Analysis. In: Stinson, D.R., Tavares, S. (eds.) SAC 2000. LNCS, vol. 2012, pp. 39–56. Springer, Heidelberg (2001)
3. SMS4 Encryption Algorithm for Wireless Networks, English translation of the Chinese document, http://eprint.iacr.org/2008/329.pdf

4. Biham, E., Dunkelman, O., Keller, N.: A New Attack on 6-Round IDEA. In: Biryukov, A. (ed.) FSE 2007. LNCS, vol. 4593, pp. 211–224. Springer, Heidelberg (2007)
5. Biryukov, A., Shamir, A.: Structural Cryptanalysis of SASAS. In: Pfitzmann, B. (ed.) EUROCRYPT 2001. LNCS, vol. 2045, pp. 394–405. Springer, Heidelberg (2001)
6. Bogdanov, A., Knudsen, L.R., Leander, G., Paar, C., Poschmann, A., Robshaw, M.J.B., Seurin, Y., Vikkelsoe, C.: PRESENT: An Ultra-Lightweight Block Cipher. In: Paillier, P., Verbauwhede, I. (eds.) CHES 2007. LNCS, vol. 4727, pp. 450–466. Springer, Heidelberg (2007)
7. Choy, J., Chew, G., Khoo, K., Yap, H.: Cryptographic Properties and Application of a Generalized Unbalanced Feistel Network Structure. In: Boyd, C., González Nieto, J. (eds.) ACISP 2009. LNCS, vol. 5594, pp. 73–89. Springer, Heidelberg (2009)
8. Daemen, J., Knudsen, L.R., Rijmen, V.: The Block Cipher SQUARE. In: Biham, E. (ed.) FSE 1997. LNCS, vol. 1267, pp. 149–165. Springer, Heidelberg (1997)
9. Daemen, J., Peeters, M., Assche, G.V., Rijmen, V.: Nessie Proposal: NOEKEON. In: First Open NESSIE Workshop (2000), `http://gro.noekeon.org/`
10. Daemen, J., Rijnmen, V.: AES Proposal: Rijndael, `http://csrc.nist.gov/encryption/aes/rijndael`
11. Ferguson, N., Kelsey, J., Lucks, S., Schneier, B., Stay, M., Wagner, D., Whiting, D.L.: Improved Cryptanalysis of Rijndael. In: Schneier, B. (ed.) FSE 2000. LNCS, vol. 1978, pp. 213–230. Springer, Heidelberg (2001)
12. Gilbert, H., Minier, M.: A collision attack on seven rounds of Rijndael. In: Proceedings of the Third AES Candidate Conference, pp. 230–241
13. Kim, J., Hong, S., Sung, J., Lee, S., Lim, J.: Impossible Differential Cryptanalysis for Block Cipher Structures. In: Johansson, T., Maitra, S. (eds.) INDOCRYPT 2003. LNCS, vol. 2904, pp. 82–96. Springer, Heidelberg (2003)
14. Kim, J., Hong, S., Lim, J.: Impossible differential cryptanalysis using matrix method. Discrete Mathematics 310(5), 988–1002 (2010)
15. Knudsen, L.R., Wagner, D.: Integral Cryptanalysis (Extended Abstract). In: Daemen, J., Rijmen, V. (eds.) FSE 2002. LNCS, vol. 2365, pp. 112–127. Springer, Heidelberg (2002)
16. Li, R., Sun, B., Li, C., Qu, L.: Cryptanalysis of a Generalized Unbalanced Feistel Network Structure. In: Steinfeld, R., Hawkes, P. (eds.) ACISP 2010. LNCS, vol. 6168, pp. 1–18. Springer, Heidelberg (2010)
17. Liu, F., Ji, W., Hu, L., Ding, J., Lv, S., Pyshkin, A., Weinmann, R.-P.: Analysis of the SMS4 Block Cipher. In: Pieprzyk, J., Ghodosi, H., Dawson, E. (eds.) ACISP 2007. LNCS, vol. 4586, pp. 158–170. Springer, Heidelberg (2007)
18. Lucks, S.: Attacking seven rounds of Rijndael under 192-bit and 256-bit keys. In: in Proc. 3rd AES Candidate Conf., pp. 215–229 (2000)
19. Luo, Y., Wu, Z., Lai, X., Gong, G.: A Unified Method for Finding Impossible Differentials of Block Cipher Structures. Cryptology ePrint Archive: Report 2009/627
20. Z'aba, M.R., Raddum, H., Henricksen, M., Dawson, E.: Bit-Pattern Based Integral Attack. In: Nyberg, K. (ed.) FSE 2008. LNCS, vol. 5086, pp. 363–381. Springer, Heidelberg (2008)

Security Enhancements by OR-Proof in Identity-Based Identification

Atsushi Fujioka[1], Taiichi Saito[2], and Keita Xagawa[1]

[1] NTT Secure Platform Laboratories
{fujioka.atsushi,xagawa.keita}@lab.ntt.co.jp
[2] Tokyo Denki University
taiichi@c.dendai.ac.jp

Abstract. We investigate three security enhancement transformations, based on the well known OR-proof technique, in identity-based identification (IBI) protocols and show a required condition of the underlying IBI protocols. The transformations can convert an IBI protocol, which satisfies a property similar to the Σ-protocol and is secure against impersonation under passive attacks, to one secure against impersonation under concurrent attacks in both adaptive and weak selective identity attack models. In addition, we argue that enhancing the security in the static identity attack model with two of the transformations seem to be difficult; however, we prove that the other can convert an IBI protocol, which satisfies another property, in the model.

Keywords: identity-based identification, OR-proof, impersonation under concurrent attacks.

1 Introduction

Identification is an important research topic in cryptography and is formulated as a protocol between a *prover* and a *verifier*, where the prover wants to prove his/her identity to the verifier. In normal settings, the verifier needs to know the identity and public-key of the prover. The security of identification protocols is defined through an experiment with an adversary who acts as a verifier to gather much knowledge in the *learning phase* and then acts as a prover to impersonate an entity in the *challenge phase*. We say that the protocol is secure when the probability that the adversary succeeds in the impersonation is negligible.

A strong security model of identification protocols has been established as *security against impersonation under concurrent attacks* (imp-ca model) [2]. Roughly speaking, an adversary is allowed to concurrently access entities who can prove their identities, even in the challenge phase. On the other hand, in the security model against impersonation under passive attacks (called imp-pa

F. Bao, P. Samarati, and J. Zhou (Eds.): ACNS 2012, LNCS 7341, pp. 135–152, 2012.

model) [2], an adversary is only allowed to eavesdrop on identification communications in the learning phase.[1]

Identification using public-keys (i.e., identification in the public key infrastructure (PKI) model) is called *standard identification* (SI), and many investigations on SI protocols have been done. On the other hand, after the proposal of identity-base cryptography [10], identification in the identity-base setting, called *identity-base identification* (IBI), has also been investigated. In an IBI protocol, the existence of a *key generation center* (KGC) is assumed as in other identity-based cryptographic schemes. The KGC publishes a master public key, generates a secret key corresponding to the identity of an entity, and gives the secret key to the entity. Thus, IBI has three phases: SETUP, EXTRACT, and IDENTIFICATION. In this setting, the verifier needs to know the master public key and the identity of the prover only.

The first security formulation for IBI was given by Kurosawa and Heng [7]. They also proposed the first generic construction of IBI protocols from digital signature schemes. The construction generates an imp-pa secure IBI protocol from a digital signature scheme *existentially unforgeable against chosen message attacks* (euf-cma secure) and its Σ-protocol [5] for signature possession. Bellare, Namprempre, and Neven provided formal security definitions of IBI protocols, the imp-pa security and the imp-ca security [1].[2]

For IBI protocols, we have three types of attack models regarding the selection of identities. One is called *security against impersonation under adaptive identity attacks* (imp-atk security) [1,9], the second one is called *security against impersonation under static identity attacks* (stat-id-imp-atk security) [9], and the third one is called *security against impersonation under weak selective identity attacks* (wsid-imp-atk security) [11], where atk denotes a type of attack such that atk $\in$ {pa, ca}. The wsid-imp-atk security implies a weak version of the imp-atk security such that an adversary in the wsid-imp-atk security declares the identities of all entities to be in queries or challenged only at the beginning of the learning phase, and the stat-id-imp-atk security implies a weak version of the imp-atk security such that an adversary in the stat-id-imp-atk security requests secret keys of identities only at the beginning of the learning phase. Throughout this paper, we denote security against impersonation under adaptive identity attacks as adapt-id-imp-atk security (not imp-atk). Note that the wsid-imp-atk security is weaker than the stat-id-imp-atk security since a stat-id-imp-atk adversary can be constructed from a wsid-imp-atk adversary (see **Appendix** A).

Security Enhancement Transformations of Identity-Based Identification. It is well known that the OR-proof technique [6] enhances the security of SI protocols from the imp-pa security to the imp-ca security. Thus, we expect that the OR-proof technique can be applied to IBI protocols to enhance

[1] There is another security model against impersonation under active attacks called imp-aa *model* [2]. An adversary in the model is allowed to *sequentially* access entities who can prove their identities; thus, it is stronger than the imp-pa model but weaker than the imp-ca model. Throughout this paper, we do not consider this security model since the imp-ca model is the strongest among them.

[2] Note that the imp-aa security model is also defined in [1].

security. However, to the best of our knowledge, no formal description of the OR-proof technique for IBI protocols has been provided. In addition, since the security notions for IBI protocols, differing in how and when an adversary declares the target identity, have been presented in various ways, as shown in the above [11,1,9], it is not clear how (and how much) the OR-proof technique can enhance the security of IBI protocols.

We formulate three types of security enhancement transformations based on the OR-proof technique for IBI protocols, *dual-identity (DI)*, *master-identity (MI)*, and *double-parameter (DP)*.

In an IBI protocol by the DI transformation, an entity with an identity id is given a secret key corresponding to either imaginary identity $(id, 0)$ or $(id, 1)$ for the underlying IBI protocol, and shows his/her identity by proving that the imaginary identity is either $(id, 0)$ or $(id, 1)$. It is easy to have a variant of DI transformation where each entity of id is given both secret keys corresponding to imaginary identities $(id, 0)$ and $(id, 1)$, and shows that his/her imaginary identity is $(id, 0)$ or $(id, 1)$ with either secret key. We call this double-key variant *DIdk*. Single-key DI is called *DIsk* if necessary to distinguish with DIdk.

In an IBI protocol by the MI transformation, an entity with an identity id is simply given a secret key corresponding to the identity id for the underlying IBI protocol, and the entity of the identity id proves that his/her identity is either id or an (imaginary) master identity.

In an IBI protocol by the DP transformation, an entity of an identity id is given a secret key corresponding to the identity id based on either of two master public keys in the underlying IBI protocol, and proves that his/her identity is id under either master public key. It is easy to have a variant of the DP transformation where an entity of an identity id is given two secret keys based on both master public keys, and shows that his/her identity is id either based on either master public key. This variant is used to construct an adapt-id-imp-ca secure IBI protocol proposed by Kurosawa and Heng [8]. We call this double-key variant *DPdk*. Single-key DP is called *DPsk* if necessary to distinguish with DPdk.

Our Contributions. We introduce two properties of IBI protocols similar to Σ-protocols [5] in the proof-of-knowledge protocols, called *Σ^+-type* and *Σ^*-type*, and show that the DI, MI, and DP transformations require underlying IBI protocols to have either property. Many existing IBI protocols satisfy these properties. For example, all IBI protocols produced by the generic constructions in [7,11] are Σ^+-type, and a large proportion of them seem to be Σ^*-type.

Next, we formally prove that both DI and MI transformations can convert an adapt-id-imp-pa secure Σ^+-type IBI protocol to an adapt-id-imp-ca secure one, and we show that they can also convert a wsid-imp-pa secure Σ^+-type IBI protocol to a wsid-imp-ca secure one. In addition, we show that the DP transformation can convert an adapt-id-imp-pa (resp. stat-id-imp-pa and wsid-imp-pa) secure Σ^*-type IBI protocol to an adapt-id-imp-ca (resp. stat-id-imp-ca and wsid-imp-ca) secure one.

We argue that it seem to be difficult to enhance passive security of IBI protocol with the DI and MI transformations in the static identity attack model. In addition, we discuss that the DI, MI, and DP transformations can be applied to identification protocols in the hierarchical identity setting.

Organization. We gives a definition of identity-based identification and related notions in **Section 2**. The security enhancement transformations are provided in **Section 3**. An additional security discussion and security relations are given in **Appendix**.

2 Definitions

In this section, we define identity-based identification (IBI) protocols and introduce a property similar to Σ-protocols [5].

Identity-Based Identification. We adopt the definition of IBI protocols in [1]. Let $\mathsf{IBI} = (\mathsf{SetUp}, \mathsf{KG}, \mathsf{P}, \mathsf{V})$ be an IBI protocol, where SetUp is the master-key-generation algorithm that on input 1^κ outputs mpk and msk, KG is the user-key-generation algorithm that on input (msk, id) outputs sk_{id}, P is the prover algorithm that, taking inputs mpk, id and sk_{id}, interacts with V, and V is the verifier algorithm that, taking inputs mpk and id, interacts with P and finally outputs $dec \in \{accept, reject\}$.

We describe the formal definitions of the security of IBI based on the following experiment $\mathbf{Exp}^{\mathsf{adapt\text{-}id\text{-}imp\text{-}atk}}_{\mathsf{IBI},\mathcal{I}}(\kappa)$ between a challenger and an impersonator $\mathcal{I} = (\mathsf{CV}, \mathsf{CP})$, where atk denotes a type of attack such that $\mathsf{atk} \in \{\mathsf{pa}, \mathsf{ca}\}$.

Experiment $\mathbf{Exp}^{\mathsf{adapt\text{-}id\text{-}imp\text{-}atk}}_{\mathsf{IBI},\mathcal{I}}(\kappa)$**:**

Setup Phase: The challenger obtains $(mpk, msk) \leftarrow \mathsf{SetUp}(1^\kappa)$ and initializes HU, CU, TU, $PS \leftarrow \emptyset$, where HU, CU, and TU denote the sets of honest users, corrupted users, and target users, respectively, and PS denotes the set of provers' sessions. The impersonator CV is given the security parameter 1^κ and the master public key mpk.

Learning Phase: CV can query the oracles INIT, CORR, and CONV when $\mathsf{atk} = \mathsf{pa}$, and also to PROV when $\mathsf{atk} = \mathsf{ca}$. Note that $id \notin HU \setminus TU$ means that id is a target user, corrupted user, or non-initiated user.

- The oracle INIT receives input id. If $id \in HU \cup CU \cup TU$, then it returns $\perp$. Otherwise, it obtains $sk_{id} \leftarrow \mathsf{KG}(msk, id)$, adds id to HU, and provides CV with id.
- The oracle CORR receives input id. If $id \notin HU \setminus TU$, then it returns $\perp$. Otherwise, it adds id to CU, deletes id in HU, and returns sk_{id} to CV.
- The oracle CONV receives input id. If $id \notin HU$, then it returns $\perp$. Otherwise it returns a transcript of a transaction between the prover with identity id and a verifier.
- (only when $\mathsf{atk} = \mathsf{ca}$) The oracle PROV receives inputs id, s, and M_{in}. If $id \notin HU \setminus TU$, then it returns $\perp$. If $(id, s) \notin PS$, then it adds (id, s) to PS, picks a random coin ρ, and sets a state of the prover $st_{\mathsf{P}}[(id, s)] \leftarrow (mpk, sk_{id}, \rho)$. Next, it obtains $(M_{out}, st_{\mathsf{P}}[(id, s)]) \leftarrow \mathsf{P}(M_{in}, st_{\mathsf{P}}[(id, s)])$. Finally, it returns M_{out}.

Challenge Phase: CV outputs a target identity id^* and state information st_{CP}. If id^* is not in HU, then the challenger outputs *reject* and halts. Otherwise, the challenger sets $TU \leftarrow \{id^*\}$, and gives st_{CP} to CP. CP can query INIT, CORR, and CONV, (and PROV when $\mathsf{atk} = \mathsf{ca}$) as in the learning phase. Finally, the challenger obtains $(tr, dec) \leftarrow \mathbf{Run}[\mathsf{CP}(st_{\mathsf{CP}})^{\text{INIT,CORR,CONV(,PROV)}} \leftrightarrow \mathsf{V}(mpk, id^*)]$ and outputs dec.

In these experiments, the adversary is allowed to obtain a secret key of an adaptively chosen identity and a transcript of a transaction between the prover of an adaptively chosen identity and a verifier. In the case of $\mathsf{atk} = \mathsf{ca}$, it is allowed to adaptively and concurrently access entities who can prove their identities.

Definition 2.1. *Let* $\mathsf{IBI} = (\mathsf{SetUp}, \mathsf{KG}, \mathsf{P}, \mathsf{V})$ *be an IBI protocol and* $\mathcal{I} = (\mathsf{CV}, \mathsf{CP})$ *an impersonator. Let* κ *be a security parameter. The advantage of* $\mathcal{I}$ *in attacking* IBI *is defined as*

$$\mathbf{Adv}_{\mathsf{IBI},\mathcal{I}}^{\mathsf{adapt\text{-}id\text{-}imp\text{-}atk}}(\kappa) := \Pr\left[\ \mathbf{Exp}_{\mathsf{IBI},\mathcal{I}}^{\mathsf{adapt\text{-}id\text{-}imp\text{-}atk}}(\kappa) = accept\ \right].$$

We say that IBI *is* secure against impersonation under adaptive identity and concurrent attacks (adapt-id-imp-ca secure) *if* $\mathbf{Adv}_{\mathsf{IBI},\mathcal{I}}^{\mathsf{adapt\text{-}id\text{-}imp\text{-}ca}}(\kappa)$ *is negligible for every polynomial-time* $\mathcal{I}$ *and is* secure against impersonation under adaptive identity and passive attacks (adapt-id-imp-pa secure) *if* $\mathbf{Adv}_{\mathsf{IBI},\mathcal{I}}^{\mathsf{adapt\text{-}id\text{-}imp\text{-}pa}}(\kappa)$ *is negligible for every polynomial-time* $\mathcal{I}$.

According to [9] and [11], we describe two other security definitions, which are weaker than the adapt-id-imp-atk security, of IBI based on the following experiments, $\mathbf{Exp}_{\mathsf{IBI},\mathcal{I}}^{\mathsf{stat\text{-}id\text{-}imp\text{-}atk}}(\kappa)$ and $\mathbf{Exp}_{\mathsf{IBI},\mathcal{I}}^{\mathsf{wsid\text{-}imp\text{-}atk}}(\kappa)$ ($\mathsf{atk} \in \{\mathsf{pa}, \mathsf{ca}\}$), between a challenger and an impersonator $\mathcal{I} = (\mathsf{CV}, \mathsf{CP})$.

Experiment $\mathbf{Exp}_{\mathsf{IBI},\mathcal{I}}^{\mathsf{stat\text{-}id\text{-}imp\text{-}atk}}(\kappa)$**:**

Setup Phase: At the beginning of this phase, CV on input 1^κ issues a single corrupt query $(id_1, \ldots, id_t)$ to the challenger before receiving the master public key. The challenger is given the security parameter 1^κ, obtains $(mpk, msk) \leftarrow \mathsf{SetUp}(1^\kappa)$, and computes $sk_{id_i} \leftarrow \mathsf{KG}(msk, id_i)$ $(1 \leq i \leq t)$. It sets $CU \leftarrow \{id_1, id_2, \ldots, id_t\}$ then returns $(sk_{id_1}, \ldots, sk_{id_t})$ to CV. The challenger initializes HU, TU, $PS \leftarrow \emptyset$. CV is given the master public key mpk.

Learning and Challenge Phases: The learning and challenge phases are defined the same as those in experiment $\mathbf{Exp}_{\mathsf{IBI},\mathcal{I}}^{\mathsf{adapt\text{-}id\text{-}imp\text{-}atk}}(\kappa)$, except that impersonator $\mathcal{I}$ is not allowed additional queries to CORR during these phases.

Experiment $\mathbf{Exp}_{\mathsf{IBI},\mathcal{I}}^{\mathsf{wsid\text{-}imp\text{-}atk}}(\kappa)$**:**

Setup Phase: At the beginning of this phase, CV on input 1^κ issues a single initialization query $(id_1, \ldots, id_t)$ to the challenger before receiving the master public key. The challenger is given the security parameter 1^κ and obtains $(mpk, msk) \leftarrow \mathsf{SetUp}(1^\kappa)$. It sets $HU \leftarrow \{id_1, id_2, \ldots, id_t\}$ and provides CV with $(id_1, \ldots, id_t)$. The challenger initializes CU, TU, $PS \leftarrow \emptyset$. CV is given the master public key mpk.

Learning and Challenge Phases: The learning and challenge phases are defined the same as those in experiment $\mathbf{Exp}_{\mathsf{IBI},\mathcal{I}}^{\mathsf{adapt\text{-}id\text{-}imp\text{-}atk}}(\kappa)$, except that $\mathcal{I}$ is not allowed additional queries to the INIT oracle during these phases.

In the stat-id-imp-atk experiment, the adversary has to choose all identities that it wants to corrupt at the beginning of the experiment. After that, it is allowed to access oracles except for CORR. In the wsid-imp-atk experiment, the adversary has to select all identities that it wants to initialize at the beginning of the experiment. Then, it is allowed to send queries of only the identities chosen at the beginning,

Let $\mathsf{IBI} = (\mathsf{SetUp}, \mathsf{KG}, \mathsf{P}, \mathsf{V})$ be an IBI protocol and $\mathcal{I} = (\mathsf{CV}, \mathsf{CP})$ an impersonator. Let κ be a security parameter. The advantages of $\mathcal{I}$ in attacking IBI are defined as $\mathbf{Adv}_{\mathsf{IBI},\mathcal{I}}^{\mathsf{stat\text{-}id\text{-}imp\text{-}atk}}(\kappa) := \Pr\left[\ \mathbf{Exp}_{\mathsf{IBI},\mathcal{I}}^{\mathsf{stat\text{-}id\text{-}imp\text{-}atk}}(\kappa) = accept\ \right]$ and $\mathbf{Adv}_{\mathsf{IBI},\mathcal{I}}^{\mathsf{wsid\text{-}imp\text{-}atk}}(\kappa) := \Pr\left[\ \mathbf{Exp}_{\mathsf{IBI},\mathcal{I}}^{\mathsf{wsid\text{-}imp\text{-}atk}}(\kappa) = accept\ \right]$. We say that IBI is *secure against impersonation under static (resp. weak selective) identity and concurrent attacks (*stat-id-imp-ca *(resp.* wsid-imp-ca*) secure)* if $\mathbf{Adv}_{\mathsf{IBI},\mathcal{I}}^{\mathsf{stat\text{-}id\text{-}imp\text{-}ca}}(\kappa)$ (resp. $\mathbf{Adv}_{\mathsf{IBI},\mathcal{I}}^{\mathsf{wsid\text{-}imp\text{-}ca}}(\kappa)$) is negligible for every polynomial-time $\mathcal{I}$ and is *secure against impersonation under static (resp. weak selective) identity and passive attacks (*stat-id-imp-pa *(resp.* wsid-imp-ca*) secure)* if $\mathbf{Adv}_{\mathsf{IBI},\mathcal{I}}^{\mathsf{stat\text{-}id\text{-}imp\text{-}pa}}(\kappa)$ (resp. $\mathbf{Adv}_{\mathsf{IBI},\mathcal{I}}^{\mathsf{wsid\text{-}imp\text{-}pa}}(\kappa)$) is negligible for every polynomial-time $\mathcal{I}$.

The relations between the security notions are given in **Appendix** A.

Σ^+- and Σ^*-type IBI Protocols. We define two analogues of Σ-protocols [5] in the context of IBI protocols. Let $\mathsf{IBI} = (\mathsf{SetUp}, \mathsf{KG}, \mathsf{P}, \mathsf{V})$ be an identity-based identification protocol.

Suppose that P and V interact by using four probabilistic polynomial time algorithms $(\Sigma_{\mathsf{ibi\text{-}com}}, \Sigma_{\mathsf{ibi\text{-}ch}}, \Sigma_{\mathsf{ibi\text{-}res}}, \Sigma_{\mathsf{ibi\text{-}vrfy}})$ as follows:

$\mathsf{P} \to \mathsf{V}$: P computes $(a, st) \leftarrow \Sigma_{\mathsf{ibi\text{-}com}}(mpk, id, sk_{id})$ and sends a to V.
$\mathsf{V} \to \mathsf{P}$: V computes $c \leftarrow \Sigma_{\mathsf{ibi\text{-}ch}}(mpk, id)$ and sends c to P.
$\mathsf{P} \to \mathsf{V}$: P computes $z \leftarrow \Sigma_{\mathsf{ibi\text{-}res}}(mpk, id, sk_{id}, a, c, st)$ and sends z to V.
V: V computes $dec \leftarrow \Sigma_{\mathsf{ibi\text{-}vrfy}}(mpk, id, a, c, z)$ and outputs $dec \in \{accept, reject\}$.

We call these types of three-move IBI protocols *canonical* [2]. We also call an IBI protocol IBI *Σ^+-type* if it is canonical and satisfies the following three properties: *special zero-knowledge*, *special soundness*, and *special challenge*:

Special Zero-knowledge: We can obtain an accepting transcript from a challenge c, mpk, and id. That is, there is a probabilistic polynomial time algorithm $\Sigma_{\mathsf{ibi\text{-}sim}}$ that takes on input mpk, id and c such that $c \leftarrow \Sigma_{\mathsf{ibi\text{-}ch}}(mpk, id)$, and outputs (a, z) such that $accept = \Sigma_{\mathsf{ibi\text{-}vrfy}}(mpk, id, a, c, z)$. The distribution of transcripts generated by $\Sigma_{\mathsf{ibi\text{-}ch}}$ and $\Sigma_{\mathsf{ibi\text{-}sim}}$ is indistinguishable from those of real transcripts.

Special Soundness: We can compute the user secret key sk_{id} for an identity id from mpk, id, and two accepting transcripts (a, c, z) and (a, c', z') such that

$c \neq c'$. That it, there is a probabilistic polynomial time algorithm $\Sigma_{\mathsf{ibi\text{-}ext}}$ that takes as input mpk, id, and two transcripts (a, c, z) and (a, c', z') satisfying $accept = \Sigma_{\mathsf{ibi\text{-}vrfy}}(mpk, id, a, c, z) = \Sigma_{\mathsf{ibi\text{-}vrfy}}(mpk, id, a, c', z')$ and $c \neq c'$, and outputs sk_{id}.

Special Challenge: $\Sigma_{\mathsf{ibi\text{-}ch}}$ depends only on mpk, not on (mpk, id), and the output c is uniformly distributed over a commutative group $\mathbb{G}$. In addition, the group operation $+$ is computable in polynomial time. $\mathbb{G}$ is determined only by mpk, not by (mpk, id). That it, there is a probabilistic polynomial time algorithm $\Sigma^{+}_{\mathsf{ibi\text{-}ch}}$ such that it takes as input mpk (without id) and outputs c, and c is uniformly distributed over $\mathbb{G}$.

We call an IBI protocol IBI *Σ^*-type* if the special challenge property is replaced with the following property:

Strongly Special Challenge: $\Sigma_{\mathsf{ibi\text{-}ch}}$ depends only on 1^κ, not on mpk, and the output c is uniformly distributed over $\mathbb{G}$. In addition, $+$ is computable in polynomial time. $\mathbb{G}$ is determined only by 1^κ, not by mpk. That it, there is a probabilistic polynomial time algorithm $\Sigma^{*}_{\mathsf{ibi\text{-}ch}}$ such that it takes as input 1^κ (not mpk) and outputs c, and c is uniformly distributed over $\mathbb{G}$.

For example, all IBI protocols by the generic constructions in [7] and [11] are Σ^+-type. The Chin-Heng-Goi IBI protocol [3] and the protocols in [8] can be seen as Σ^*-type.

3 Security Enhancement Transformations

Let $\mathsf{IBI}' = (\mathsf{SetUp}', \mathsf{KG}', \mathsf{P}', \mathsf{V}')$ be a Σ^+-type (Σ^*-type) IBI protocol in which $(\mathsf{P}', \mathsf{V}')$ use four probabilistic polynomial time algorithms $\Sigma_{\mathsf{ibi\text{-}com}}$, $\Sigma^{+}_{\mathsf{ibi\text{-}ch}}$ ($\Sigma^{*}_{\mathsf{ibi\text{-}ch}}$), $\Sigma_{\mathsf{ibi\text{-}res}}$, and $\Sigma_{\mathsf{ibi\text{-}vrfy}}$ and have the special zero-knowledge property with a probabilistic polynomial time algorithm $\Sigma_{\mathsf{ibi\text{-}sim}}$ and the special soundness with a probabilistic polynomial time algorithm $\Sigma_{\mathsf{ibi\text{-}ext}}$.

3.1 Dual-Identity Transformation

We show a security enhancement transformation based on the OR-proof, applicable to the Σ^+-type IBI protocol. We call this transformation *DI transformation.*

We describe an IBI protocol $\mathsf{IBI} = (\mathsf{SetUp}, \mathsf{KG}, \mathsf{P}, \mathsf{V})$ produced by applying the DI transformation to IBI' in **Fig. 1**.

Due to the special challenge property, c is an element in $\mathbb{G}$ determined by mpk', so are c_0 and c_1 since $c = c_0 + c_1$ and the operation $+$ is defined in $\mathbb{G}$. Then, c_0 and c_1 are possible challenges under mpk'.

It is easy to have a variant of the DI transformation, the DIdk transformation such that each entity is given both secret keys of identities $(id, 0)$ and $(id, 1)$, and the entity shows that it has either the secret key of $(id, 0)$ or $(id, 1)$. However, this variant requires double sized secret keys for each user.

SETUP

$\mathsf{SetUp}(1^\kappa)$

$(mpk', msk') \leftarrow \mathsf{SetUp}'(1^\kappa)$
output $(mpk, msk) = (mpk', msk')$

EXTRACT

$\mathsf{KG}(msk, id)$

$msk = msk'$
$b_{id} \leftarrow \{0,1\}$
$sk'_{(id,b_{id})} \leftarrow \mathsf{KG}'(msk', (id, b_{id}))$
output $sk_{id} = (sk'_{(id,b_{id})}, b_{id})$

IDENTIFICATION

$\mathsf{P}(mpk, id, sk_{id})$		$\mathsf{V}(mpk, id)$
$mpk = mpk'$		$mpk = mpk'$
$sk_{id} = (sk'_{(id,b_{id})}, b_{id})$		
$(a_{b_{id}}, st) \leftarrow \Sigma_{\mathsf{ibi\text{-}com}}(mpk', (id, b_{id}), sk'_{(id,b_{id})})$		
$c_{\bar{b}_{id}} \leftarrow \Sigma^+_{\mathsf{ibi\text{-}ch}}(mpk')$		
$(a_{\bar{b}_{id}}, z_{\bar{b}_{id}}) \leftarrow \Sigma_{\mathsf{ibi\text{-}sim}}(mpk', (id, \bar{b}_{id}), c_{\bar{b}_{id}})$	(a_0, a_1) $\longrightarrow$	$c \leftarrow \Sigma^+_{\mathsf{ibi\text{-}ch}}(mpk')$
$c_{b_{id}} = c - c_{\bar{b}_{id}}$	c $\longleftarrow$	
$z_{b_{id}} \leftarrow \Sigma_{\mathsf{ibi\text{-}res}}(mpk', (id, b_{id}), sk'_{(id,b_{id})}, a_{b_{id}}, c_{b_{id}}, st)$	(c_0, z_0, z_1) $\longrightarrow$	$c_1 = c - c_0$
		$dec_0 \leftarrow \Sigma_{\mathsf{ibi\text{-}vrfy}}(mpk', (id, 0), a_0, c_0, z_0)$
		$dec_1 \leftarrow \Sigma_{\mathsf{ibi\text{-}vrfy}}(mpk', (id, 1), a_1, c_1, z_1)$
		output *accept* if $dec_0 = dec_1 = accept$; otherwise, output *reject*

Fig. 1. DI Transformation

3.2 Master-Identity Transformation

We show another security enhancement transformation based on the OR-proof, applicable to the Σ^+-type IBI protocol. We call this transformation *MI transformation.*

We describe an IBI protocol $\mathsf{IBI} = (\mathsf{SetUp}, \mathsf{KG}, \mathsf{P}, \mathsf{V})$ produced by applying the MI transformation to IBI' in **Fig. 2**.

Due to the special challenge property, c is an element in $\mathbb{G}$ determined by mpk' so are c_0 and c_1 since $c = c_0 + c_1$ and the operation $+$ is defined in $\mathbb{G}$. Then, c_0 and c_1 are possible challenges under mpk'.

Here, id_{master} is randomly chosen from the set of identities, but does not coincide with any identities of real entities. It is clear that an adversary should not be allowed to obtain the secret key of id_{master}. In the construction of KG, $\mathsf{KG}(msk, id)$ outputs $\perp$ if $id = id_{master}$. This means that the space of all possible identities of entities does not include id_{master}.

Note that the secret key size in the MI transformation is one bit smaller than that in the DI one.

3.3 Double-Parameter Transformation

We show the other security enhancement transformation based on the OR-proof, applicable to the Σ^*-type IBI protocol. We call this transformation *DP transformation.*

SETUP

$\mathsf{SetUp}(1^\kappa)$
$(mpk', msk') \leftarrow \mathsf{SetUp}'(1^\kappa)$
choose a master identity id_{master}
output $(mpk, msk) = ((mpk', id_{master}), msk')$

EXTRACT

$\mathsf{KG}(msk, id)$
output $\perp$ if $id = id_{master}$
$msk = msk'$
$sk'_{id} \leftarrow \mathsf{KG}'(msk', id)$
output $sk_{id} = sk'_{id}$

IDENTIFICATION

$\mathsf{P}(mpk, id, sk_{id})$		$\mathsf{V}(mpk, id)$
$mpk = (mpk', id_{master})$		$mpk = (mpk', id_{master})$
$sk_{id} = sk'_{id}$		
$(a_0, st) \leftarrow \Sigma_{\mathsf{ibi\text{-}com}}(mpk', id, sk'_{id})$		
$c_1 \leftarrow \Sigma^+_{\mathsf{ibi\text{-}ch}}(mpk')$		
$(a_1, z_1) \leftarrow \Sigma_{\mathsf{ibi\text{-}sim}}(mpk', id_{master}, c_1)$	(a_0, a_1) $\longrightarrow$	$c \leftarrow \Sigma^+_{\mathsf{ibi\text{-}ch}}(mpk')$
$c_0 = c - c_1$	c $\longleftarrow$	
$z_0 \leftarrow \Sigma_{\mathsf{ibi\text{-}res}}(mpk', id, sk'_{id}, a_0, c_0, st)$	(c_0, z_0, z_1) $\longrightarrow$	$c_1 = c - c_0$
		$dec_0 \leftarrow \Sigma_{\mathsf{ibi\text{-}vrfy}}(mpk', id, a_0, c_0, z_0)$
		$dec_1 \leftarrow \Sigma_{\mathsf{ibi\text{-}vrfy}}(mpk', id_{master}, a_1, c_1, z_1)$
		output *accept* if $dec_0 = dec_1 = accepts$; otherwise, output *reject*

Fig. 2. MI Transformation

We describe an IBI protocol $\mathsf{IBI} = (\mathsf{SetUp}, \mathsf{KG}, \mathsf{P}, \mathsf{V})$ produced by applying the DI transformation to IBI' in **Fig. 3**.

Due to the strongly special challenge property, c is an element in $\mathbb{G}$ determined only by 1^κ so are c_0 and c_1 since $c = c_0 + c_1$ and the operation $+$ is defined in $\mathbb{G}$. Then, c_0 and c_1 are possible challenges under mpk'_0 and mpk'_1, respectively.

It is easy to have a variant of the DP transformation such that each entity is given both secret keys based on mpk'_0 and mpk'_1, and the entity shows that it has either a secret key in mpk'_0 or mpk'_1. This variant is used to construct the adapt-id-imp-ca secure IBI protocol developed by Kurosawa and Heng [8]. However, the variant requires double sized secret keys for each user.

Although the DP transformation requires two master public keys and is less efficient than the DI and MI transformations, the transformation can enhance the security of a Σ^*-type IBI protocol even in the static identity attack model (also see **Section 3.5**).

3.4 Security of DI, MI and DP Transformations

We formally prove that the DI, MI and DP transformations can convert an adapt-id-imp-pa secure IBI protocol to an adapt-id-imp-ca secure one.

Proposition 3.1. *The DI transformation converts an* adapt-id-imp-pa *secure Σ^+-type IBI protocol into an* adapt-id-imp-ca *secure one.*

Setup

$\mathsf{SetUp}(1^\kappa)$
$(mpk'_0, msk'_0) \leftarrow \mathsf{SetUp}'(1^\kappa)$
$(mpk'_1, msk'_1) \leftarrow \mathsf{SetUp}'(1^\kappa)$
output $(mpk, msk) = ((1^\kappa, mpk'_0, mpk'_1), (msk'_0, msk'_1))$

Extract

$\mathsf{KG}(msk, id)$
$msk = (msk'_0, msk'_1)$
$b_{id} \leftarrow \{0,1\}$
$sk'_{(id,b_{id})} \leftarrow \mathsf{KG}'(msk'_{b_{id}}, id)$
output $sk_{id} = (sk'_{(id,b_{id})}, b_{id})$

Identification

$\mathsf{P}(mpk, id, sk_{id})$		$\mathsf{V}(mpk, id)$
$mpk = (1^\kappa, mpk'_0, mpk'_1)$		$mpk = (1^\kappa, mpk'_0, mpk'_1)$
$sk_{id} = (sk'_{(id,b_{id})}, b_{id})$		
$(a_{b_{id}}, st) \leftarrow \Sigma_{\mathsf{ibi\text{-}com}}(mpk'_{b_{id}}, id, sk'_{(id,b_{id})})$		
$c_{\bar{b}_{id}} \leftarrow \Sigma^*_{\mathsf{ibi\text{-}ch}}(1^\kappa)$		
$(a_{\bar{b}_{id}}, z_{\bar{b}_{id}}) \leftarrow \Sigma_{\mathsf{ibi\text{-}sim}}(mpk'_{\bar{b}_{id}}, id, c_{\bar{b}_{id}})$	(a_0, a_1) $\longrightarrow$	$c \leftarrow \Sigma^*_{\mathsf{ibi\text{-}ch}}(1^\kappa)$
	c $\longleftarrow$	
$c_{b_{id}} = c - c_{\bar{b}_{id}}$		
$z_{b_{id}} \leftarrow \Sigma_{\mathsf{ibi\text{-}res}}(mpk'_{b_{id}}, id, sk'_{(id,b_{id})}, a_{b_{id}}, c_{b_{id}}, st)$	(c_0, z_0, z_1) $\longrightarrow$	$c_1 = c - c_0$
		$dec_0 \leftarrow \Sigma_{\mathsf{ibi\text{-}vrfy}}(mpk'_0, id, a_0, c_0, z_0)$
		$dec_1 \leftarrow \Sigma_{\mathsf{ibi\text{-}vrfy}}(mpk'_1, id, a_1, c_1, z_1)$
		output *accept* if $dec_0 = dec_1 = accepts$; otherwise, output *reject*

Fig. 3. DP Transformation

Proof. We prove this by contradiction, i.e., we show that if there exists an adapt-id-imp-ca attacker $\mathcal{I} = (\mathsf{CV}, \mathsf{CP})$ for the resulting IBI protocol IBI, then we can construct an adapt-id-imp-pa adversary $\mathcal{I}' = (\mathsf{CV}', \mathsf{CP}')$ for the underlying IBI protocol IBI'.

Reduction from adapt-id-imp-ca to adapt-id-imp-pa:

Setup Phase: The adapt-id-imp-pa adversary CV' is given the security parameter 1^κ and the master public key mpk'.

Learning Phase: CV' initializes HU, CU, TU, PS, $SK \leftarrow \emptyset$, where SK denotes the set of secret keys, and gives the security parameter 1^κ and the master public key mpk to the adapt-id-imp-ca impersonator CV, where $mpk = mpk'$. CV' simulates the oracles for CV as follows:

- (Simulation of Init) CV' receives a query id: If $id \in HU \cup CU \cup TU$, then CV' returns $\perp$ to CV. Otherwise, CV' sends $(id, 0)$ and $(id, 1)$ to the external Init oracle, generates $b_{id} \leftarrow \{0,1\}$, sends (id, b_{id}) the external Corr oracle to obtain $sk'_{(id,b_{id})}$, adds id and $(id, b_{id}, sk'_{(id,b_{id})})$ to HU and SK, respectively, and provides CV with id.
- (Simulation of Corr) CV' receives a query id: If $id \notin HU \setminus TU$, then CV' returns $\perp$ to CV. Otherwise, CV' adds id to CU, deletes id in HU, retrieves $(id, b_{id}, sk'_{(id,b_{id})})$ from SK, sets $sk_{id} = (sk'_{(id,b_{id})}, b_{id})$, and returns sk_{id} to CV.

- (Simulation of CONV) CV' receives a query id: If $id \notin HU$, then CV' returns $\perp$ to CV. Otherwise, CV' sends $(id, 0)$ and $(id, 1)$ to the external CONV oracle to obtain (a_0, c_0, z_0) and (a_1, c_1, z_1), respectively, and returns $((a_0, a_1), c, (c_0, z_0, z_1))$ to CV where $c = c_0 + c_1$.
- (Simulation of PROV) CV' receives a query (id, s, M_{in}): If $id \notin HU \setminus TU$, then CV' returns $\perp$ to CV. If $(id, s) \notin PS$, then CV' adds (id, s) to PS, selects a random coin ρ, retrieves $(id, b_{id}, sk'_{(id,b_{id})})$ from SK, and sets a state of the prover $st_{\mathsf{P}}[(id, s)] \leftarrow (mpk, sk_{id}, \rho)$, where $sk_{id} = (sk'_{(id,b_{id})}, b_{id})$. Next, CV' computes M_{out} depending on M_{in}: If M_{in} is a null string (i.e., CV' is requested to generate the first message), CV' sends $(id, \bar{b}_{id})$ to the external CONV oracle to obtain $(a_{\bar{b}_{id}}, c_{\bar{b}_{id}}, z_{\bar{b}_{id}})$, runs $(a_{b_{id}}, st) \leftarrow \Sigma_{\mathsf{ibi\text{-}com}}(mpk', (id, b_{id}), sk'_{(id,b_{id})})$, sets $M_{out} = (a_0, a_1)$, and adds $(st, c_{\bar{b}_{id}}, z_{\bar{b}_{id}})$ to $st_{\mathsf{P}}[(id, s)]$. If M_{in} is c (i.e., CV' is requested to generate the last message), CV' computes $c_{b_{id}} = c - c_{\bar{b}_{id}}$, runs $z_{b_{id}} \leftarrow \Sigma_{\mathsf{ibi\text{-}res}}(mpk', (id, b_{id}), sk'_{(id,b_{id})}, a_{b_{id}}, c_{b_{id}}, st)$, and sets $M_{out} = (c_0, z_0, z_1)$. Finally, CV' returns M_{out}.

CV can query the oracles INIT, CORR, CONV, and PROV, as shown above. Then, CV outputs a target identity id^* and state information st_{CP}. If id^* is not in HU, then CV' outputs *reject* and halts. Otherwise, CV' sets $TU \leftarrow \{id^*\}$, and gives st_{CP} to CP. At some point, CV' obtains $(tr, dec) \leftarrow \mathbf{Run}[\mathsf{CP}(st_{\mathsf{CP}})^{\text{INIT,CORR,CONV,PROV}} \leftrightarrow \mathsf{V}(mpk, id^*)]$ acting as V, where $tr = ((a_0, a_1), c, (c_0, z_0, z_1))$. Then, CV' reruns CP to obtain $(tr', dec') \leftarrow \mathbf{Run}[\mathsf{CP}(st_{\mathsf{CP}})^{\text{INIT,CORR,CONV,PROV}} \leftrightarrow \mathsf{V}(mpk, id^*)]$, where $tr' = ((a_0, a_1), c', (c'_0, z'_0, z'_1))$. The special soundness property of IBI' implies that $sk_{(id^*,b)}$ ($b = 0$ or 1) can be computed from tr and tr'. CV' finally outputs a target identity $(id^*, \bar{b}_{id^*})$ and state information $(sk'_{(id^*,b)}, id^*, mpk')$ to the challenger.

Challenge Phase: After the challenger returns $(sk'_{(id^*,b)}, id^*, mpk')$ to CP', CP' acts as P' with the extracted secret key $sk'_{(id^*,b)}$ to impersonate $(id^*, \bar{b}_{id^*})$.

In the simulation of the CONV oracle, CV' generates two queries from the received query, sends the two queries to the external CONV oracles, and constructs an answer from two replies from the CONV oracle. For PROV queries, CV' uses the secret key obtained in the INIT oracle simulation, and simulates the PROV oracle.

It is clear that $\mathcal{I}'$ succeeds in impersonating $(id^*, \bar{b}_{id^*})$ if b coincides with $\bar{b}_{id^*}$. Since the resulting IBI protocol is witness indistinguishable [6] and $\mathcal{I}'$ has the secret key of either $(id, 0)$ or $(id, 1)$, the PROV oracle simulation by $\mathcal{I}'$ can be perfect. Thus, we have $\mathbf{Adv}^{\mathsf{adapt\text{-}id\text{-}imp\text{-}pa}}_{\mathsf{IBI},\mathcal{I}'}(\kappa) \geq \frac{1}{2}(\mathbf{Adv}^{\mathsf{adapt\text{-}id\text{-}imp\text{-}ca}}_{\mathsf{IBI},\mathcal{I}}(\kappa) - \frac{1}{|\mathbb{G}|})^2$ since the underlying IBI protocol is canonical and the Reset Lemma [2] is applicable, where $\mathbb{G}$ is a commutative group over which the output challenge is uniformly distributed.

Note that two transcripts, $(((a_0, a_1), c, (c_0, z_0, z_1)), ((a_0, a_1), c', (c'_0, z'_0, z'_1)))$, where $c \neq c'$, implies either $((a_0, c_0, z_0), (a_0, c'_0, z'_0))$ where $c_0 \neq c'_0$ or $((a_1, c_1, z_1), (a_1, c'_1, z'_1))$ where $c_1 \neq c'_1$.

A detailed probability analysis is given in the final version of this paper. □

Proposition 3.2. *The MI transformation converts an* adapt-id-imp-pa *secure* Σ^+*-type IBI protocol into an* adapt-id-imp-ca *secure one.*

Due to page limitation, we only show an outline of the proof.

We assume two types of impersonators and show that there exist reductions from each impersonator to $\mathcal{I}'$. We let $\mathcal{I}_{master}$ be an impersonator from which $\mathcal{I}'$ derives the secret key corresponding to id_{master}, and $\mathcal{I}_{user}$ be the other impersonator from which $\mathcal{I}'$ derives a secret key of a user. In the reduction from $\mathcal{I}_{master}$, $\mathcal{I}'$ can perfectly simulate the PROV oracle by obtaining secret keys of users from the external CORR oracle. Thus, $\mathcal{I}'$ can extract a secret key of id_{master} from $\mathcal{I}_{master}$ (by using the Reset Lemma), and can impersonate id_{master}. In the reduction from $\mathcal{I}_{user}$, $\mathcal{I}'$ can perfectly simulate the PROV oracle with a secret key of id_{master} obtained from the external CORR oracle, and extract a secret key of the target identity id^* from $\mathcal{I}_{user}$ (by using the Reset Lemma). Thus, it can impersonate id^*.

A full proof will be given in the final version of this paper.

Proposition 3.3. *The DP transformation converts an* adapt-id-imp-pa *secure* Σ^**-type IBI protocol into an* adapt-id-imp-ca *secure one.*

Due to page limitation, we only show an outline of the proof.

After $\mathcal{I}'$ receives mpk' from the challenger, $\mathcal{I}'$ internally generates another key pair (mpk'_*, msk'_*). $\mathcal{I}'$ can perfectly simulate all oracles since $\mathcal{I}'$ obtains the secret keys of all users with this msk'_*. Thus, $\mathcal{I}'$ can extract from $\mathcal{I}$ a secret key of the target identity id^* either for mpk' or mpk'_* (by using the Reset Lemma). If $\mathcal{I}'$ obtains secret key for id^* in mpk', $\mathcal{I}'$ can impersonate id^* in mpk'.

A full proof will be given in the final version of this paper.

Similarly to the above propositions, the security enhancement transformations can be applied to wsid-imp-pa secure IBI protocols. Thus, the following theorems hold.

Theorem 3.1. *The DI transformation converts a* wsid-imp-pa *secure* Σ^+*-type IBI protocol into a* wsid-imp-ca *secure one.*

We describe an outline of the proof.

In the Setup phase, the adversary who breaks the wsid-imp-ca security issues $(id_1, \ldots, id_t)$ to the challenger. Then, the adversary who breaks the wsid-imp-pa security issues $((id_1, 0), (id_1, 1), \ldots, (id_t, 0), (id_t, 1))$ to the external challenger. After receiving mpk', the wsid-imp-pa adversary sends (id_i, b_i) to the external CORR where b_i is a random bit $(1 \leq i \leq t)$, and receives the keys $sk_{(id_i, b_i)}$. The wsid-imp-pa adversary simulates the oracles the same as in the proof of **Proposition 3.1**.

A full proof will be given in the final version of this paper.

Theorem 3.2. *The MI transformation converts a* wsid-imp-pa *secure* Σ^+*-type IBI protocol into a* wsid-imp-ca *secure one.*

Theorem 3.3. *The DP transformation converts a* wsid-imp-pa *secure* Σ^**-type IBI protocol into a* wsid-imp-ca *secure one.*

The strategy of proving **Theorem 3.2** (resp. **Theorem 3.3**) is the same as **Proposition 3.2** (resp. **Proposition 3.3**) and **Theorem 3.1**. Full proofs will be given in the final version of this paper.

On the other hand, the DP transformation can enhance passive security to a concurrent one also in the static identity attack model.

Theorem 3.4. *The DP transformation converts a* stat-id-imp-pa *secure* Σ^**-type IBI protocol into a* stat-id-imp-ca *secure one.*

Due to page limitation, we only show an outline of the proof.

For a single CORR query on $(id_1, \ldots, id_t)$ from the adversary, $\mathcal{I}$, who breaks the stat-id-imp-ca security, the adversary, $\mathcal{I}'$, who breaks the stat-id-imp-pa security makes a single external CORR call $(id_1, \ldots, id_t)$, obtains $(sk'_{id_1}, \ldots, sk'_{id_t})$, and is given the master public key mpk'. $\mathcal{I}'$ randomly selects $b^* \leftarrow \{0,1\}$, sets $mpk'_{\bar{b}^*} = mpk'$ and $sk_{(id_i,\bar{b}^*)} = sk'_{id_i}$ $(1 \leq i \leq t)$, runs $(mpk'_{b^*}, msk'_{b^*}) \leftarrow \mathsf{SetUp}'(1^\kappa)$ and $sk_{(id_i,b^*)} \leftarrow \mathsf{KG}'(msk'_{b^*}, id_i)$ $(1 \leq i \leq t)$, randomly selects $b_{id_i} \leftarrow \{0,1\}$ $(1 \leq i \leq t)$, sets $mpk = (1^\kappa, mpk'_0, mpk'_1)$, and provides $\mathcal{I}$ with $(sk_{(id_1,b_{id})}, \ldots, sk_{(id_t,b_{id})})$ and the master public key mpk. $\mathcal{I}'$ can perfectly simulate all oracles since $\mathcal{I}'$ obtains the secret keys of all users with msk_{b^*}. Thus, $\mathcal{I}'$ can extract from $\mathcal{I}$ a secret key of the target identity id^* either for mpk' or msk'_{b^*} (by using the Reset Lemma). If $\mathcal{I}'$ obtains a secret key for id^* in mpk', $\mathcal{I}'$ can impersonate id^* in mpk'.

A full proof will be given in the final version of this paper.

3.5 Discussions

While the DP transformation can convert a stat-id-imp-pa secure Σ^*-type IBI protocol to a stat-id-imp-ca secure one, the DI and MI transformations seem not to be able to do so. In the DP transformation, two master public keys in the underlying IBI protocol, mpk'_0 and mpk'_1, compose a master public key in the resulting IBI protocol, and the secret key of each entity in the resulting IBI protocol is computed with either the master secret keys, msk'_0 or msk'_1, in the underlying IBI protocol. Even in the stat-id-imp-atk security model, since the simulator has a master secret key msk'_{b^*} for the master public key mpk'_{b^*}, it can generate a secret key for any entity and then simulate the PROV oracle.

On the other hand, we consider the DI and MI transformations. In the DI transformation, the master public key in the resulting IBI protocol is set with the master public key in the underlying IBI protocol, mpk', and a secret key of an entity corresponding to identity id in the resulting IBI protocol is a secret key corresponding to either identity, $(id, 0)$ or $(id, 1)$, in the underlying IBI protocol. The secret key is computed with the master secret key, msk', corresponding to mpk'. Since the simulator does not have msk' in the proof for the stat-id-imp-atk security, it can not obtain secret keys for identities queried to the PROV oracle and fail to simulate it.

Table 1. Applicability of Transformations

	adapt-id	stat-id	wsid
DIsk	✓		✓
DIdk	✓		✓
MI	✓		✓
DPsk	✓	✓	✓
DPdk	✓	✓	✓

In the MI transformation, the master public and secret keys and the secret keys of entities in the resulting IBI protocol are the same as those in the underlying IBI protocol. In the proof for the stat-id-imp-atk security, the simulator might obtain the secret key for the master identity, id_{master}, at the Setup phase and simulate the CONV oracle. In the challenge phase, however, if the secret key extracted from transcripts coincides with the key for id_{master}, the simulator would not obtain the non-trivial secret key and the impersonation would fail.

Next, we consider identification protocols in the hierarchical identity setting, called *hierarchical identity-based identification* (HIBI) [4], and discuss the possibility for converting passively secure HIBI protocols to concurrently secure ones.

Since each intermediate KGC is given both secret keys based on the two independent master public keys in the DPdk transformation, it can issue two corresponding secret keys for child KGCs or users. Thus, the DPdk transformation can convert an adapt-id-imp-pa secure HIBI protocol into an adapt-id-imp-ca secure one.

In the DIdk transformation, the OR-proof can be accomplished when we adopt $(0, id_1, id_2, \ldots, id_\ell)$ and $(1, id_1, id_2, \ldots, id_\ell)$ as two imaginary identities corresponding to an identity $(id_1, \ldots, id_\ell)$ $(\ell \geq 1)$. The DIdk transformation can convert an adapt-id-imp-pa secure HIBI protocol into an adapt-id-imp-ca secure one.

It is clear that the MI transformation can convert an adapt-id-imp-pa secure HIBI protocol into an adapt-id-imp-ca secure one since each entity is assigned the secret key as usual and proves his/her identity by showing the possession of his/her own key or the secret key for id_{master}. Note that we need a restriction in which the adversary is not allowed to make CORR queries whose prefixes are id_{master}.

3.6 Comparisons

In the previous subsections, we saw that the DI and MI transformations are applicable to the Σ^+-type IBI protocol, while the DP transformation is applicable to Σ^*-type. In this section, we compare the IBI protocols produced with these transformations.

Table 2 compares computational costs, where $|\cdot|$ denotes the computational cost of the algorithm. Here, $|\mathsf{SetUp}'|$, $|\mathsf{KG}'|$, $|\mathsf{P}'|$ and $|\mathsf{V}'|$ are the computational costs of the underlying IBI protocol. Thus, $|\mathsf{P}'|$ and $|\mathsf{V}'|$ can be expressed as

Table 2. Computational Cost

	$\lvert\mathsf{SetUp}\rvert$	$\lvert\mathsf{KG}\rvert$	$\lvert\mathsf{P}\rvert$	$\lvert\mathsf{V}\rvert$
DIsk	$\lvert\mathsf{SetUp}'\rvert$	$\lvert\mathsf{KG}'\rvert + \lvert\mathsf{b}\rvert$	$\lvert\mathsf{P}'\rvert + \lvert\Sigma_{\mathsf{ibi\text{-}ch}}\rvert + \lvert\Sigma_{\mathsf{ibi\text{-}sim}}\rvert$	$\lvert\mathsf{V}'\rvert + \lvert\Sigma_{\mathsf{ibi\text{-}vrfy}}\rvert$
DIdk	$\lvert\mathsf{SetUp}'\rvert$	$2\lvert\mathsf{KG}'\rvert$	$\lvert\mathsf{P}'\rvert + \lvert\Sigma_{\mathsf{ibi\text{-}ch}}\rvert + \lvert\Sigma_{\mathsf{ibi\text{-}sim}}\rvert + \lvert\mathsf{b}\rvert$	$\lvert\mathsf{V}'\rvert + \lvert\Sigma_{\mathsf{ibi\text{-}vrfy}}\rvert$
MI	$\lvert\mathsf{SetUp}'\rvert$	$\lvert\mathsf{KG}'\rvert + \lvert\mathsf{id}\rvert$	$\lvert\mathsf{P}'\rvert + \lvert\Sigma_{\mathsf{ibi\text{-}ch}}\rvert + \lvert\Sigma_{\mathsf{ibi\text{-}sim}}\rvert$	$\lvert\mathsf{V}'\rvert + \lvert\Sigma_{\mathsf{ibi\text{-}vrfy}}\rvert$
DPsk	$2\lvert\mathsf{SetUp}'\rvert$	$\lvert\mathsf{KG}'\rvert + \lvert\mathsf{b}\rvert$	$\lvert\mathsf{P}'\rvert + \lvert\Sigma_{\mathsf{ibi\text{-}ch}}\rvert + \lvert\Sigma_{\mathsf{ibi\text{-}sim}}\rvert$	$\lvert\mathsf{V}'\rvert + \lvert\Sigma_{\mathsf{ibi\text{-}vrfy}}\rvert$
DPdk	$2\lvert\mathsf{SetUp}'\rvert$	$2\lvert\mathsf{KG}'\rvert$	$\lvert\mathsf{P}'\rvert + \lvert\Sigma_{\mathsf{ibi\text{-}ch}}\rvert + \lvert\Sigma_{\mathsf{ibi\text{-}sim}}\rvert + \lvert\mathsf{b}\rvert$	$\lvert\mathsf{V}'\rvert + \lvert\Sigma_{\mathsf{ibi\text{-}vrfy}}\rvert$
IBI'	$\lvert\mathsf{SetUp}'\rvert$	$\lvert\mathsf{KG}'\rvert$	$\lvert\mathsf{P}'\rvert$	$\lvert\mathsf{V}'\rvert$

$\lvert\mathsf{b}\rvert$ and $\lvert\mathsf{id}\rvert$ are the costs for selecting one bit and id_{master}, respectively.

$\lvert\Sigma_{\mathsf{ibi\text{-}com}}\rvert + \lvert\Sigma_{\mathsf{ibi\text{-}res}}\rvert$ and $\lvert\Sigma_{\mathsf{ibi\text{-}ch}}\rvert + \lvert\Sigma_{\mathsf{ibi\text{-}vrfy}}\rvert$, respectively. Note that $\Sigma_{\mathsf{ibi\text{-}com}}$ is $\Sigma^{+}_{\mathsf{ibi\text{-}com}}$ or $\Sigma^{*}_{\mathsf{ibi\text{-}com}}$ depending on the type.

The DP transformation requires double computational costs in the SETUP phase, and the double-key variants (DIdk and DPdk) require double computational costs in the EXTRACT phase. In all transformation, the computational costs of $\Sigma_{\mathsf{ibi\text{-}ch}}$ and $\Sigma_{\mathsf{ibi\text{-}sim}}$ are required for a prover, and that of $\Sigma_{\mathsf{ibi\text{-}vrfy}}$ for a verifier. We may ignore $\lvert\mathsf{b}\rvert$ and $\lvert\mathsf{id}\rvert$ since they are comparably smaller than others.

If a transcript of a transaction of the underlying IBI protocol has the form (a, c, z), in all transformations that of the resulting protocol has the form $((a_0, a_1), c, (c_0, z_0, z_1))$. Therefore, the communication costs for all transformations are the same, and they require additional elements $(|a| + |z| + |c|)$ for a prover than the underlying IBI protocol. The communication costs for a verifier of the resulting IBI protocol are the same with that of the underlying one. Here, $|\cdot|$ denotes the bit length of the variable.

Table 3. Key Size

	$\lvert mpk\rvert$	$\lvert msk\rvert$	$\lvert sk\rvert$
DIsk	$\lvert mpk'\rvert$	$\lvert msk'\rvert$	$\lvert sk'\rvert + 1$
DIdk	$\lvert mpk'\rvert$	$\lvert msk'\rvert$	$2\lvert sk'\rvert$
MI	$\lvert mpk'\rvert + \lvert id\rvert$	$\lvert msk'\rvert$	$\lvert sk'\rvert$
DPsk	$2\lvert mpk'\rvert + \kappa$	$2\lvert msk'\rvert$	$\lvert sk'\rvert + 1$
DPdk	$2\lvert mpk'\rvert + \kappa$	$2\lvert msk'\rvert$	$2\lvert sk'\rvert$
IBI'	$\lvert mpk'\rvert$	$\lvert msk'\rvert$	$\lvert sk'\rvert$

κ is the security parameter.

Table 3 compares key sizes. The DP transformation requires double sized master public and secret keys in addition to the security parameter, and the double-key variants (DIdk and DPdk) requires double sized user secret key. The DIsk and DPsk transformations require one more bit for the user secret key.

4 Conclusion

We introduced two properties of IBI protocols, Σ^{+}-type and Σ^{*}-type, similar to Σ-protocols. We then proved that both DI and MI transformation can convert an adapt-id-imp-pa (and wsid-imp-pa) secure Σ^{+}-type IBI protocol to an adapt-id-imp-ca (and wsid-imp-ca) secure one, respectively. We also showed that the DP transformation can convert an adapt-id-imp-pa, stat-id-imp-pa, and wsid-imp-pa secure Σ^{*}-type IBI protocol to an adapt-id-imp-ca, stat-id-imp-ca, and wsid-imp-ca secure one, respectively.

Converting an IBI protocol in the static identity attack model seems to be difficult with the DI and MI transformations. The DI and MI transformations require only a single master public key, while the DP transformation requires two. Now we give an open problem related to the number of master public keys. **Open problem:** Determine whether an OR-proof security enhancement transformation exists, based on a single master public key, that converts stat-id-imp-pa IBI protocol to stat-id-imp-ca one.

References

1. Bellare, M., Namprempre, C., Neven, G.: Security proofs for identity-based identification and signature schemes. Journal of Cryptology 22(1), 1–61 (2009); Preliminary version In: EUROCRYPT 2004 (2004)
2. Bellare, M., Palacio, A.: GQ and Schnorr Identification Schemes: Proofs of Security against Impersonation under Active and Concurrent Attacks. In: Yung, M. (ed.) CRYPTO 2002. LNCS, vol. 2442, pp. 162–177. Springer, Heidelberg (2002)
3. Chin, J.-J., Heng, S.-H., Goi, B.-M.: An Efficient and Provable Secure Identity-Based Identification Scheme in the Standard Model. In: Mjølsnes, S.F., Mauw, S., Katsikas, S.K. (eds.) EuroPKI 2008. LNCS, vol. 5057, pp. 60–73. Springer, Heidelberg (2008)
4. Chin, J.-J., Heng, S.-H., Goi, B.-M.: Hierarchical Identity-Based Identification Schemes. In: Ślęzak, D., Kim, T.-h., Fang, W.-C., Arnett, K.P. (eds.) SecTech 2009. CCIS, vol. 58, pp. 93–99. Springer, Heidelberg (2009)
5. Cramer, R.: Modular Design of Secure, yet Practical Cryptographic Protocols. PhD thesis, University of Amsterdam (1996)
6. Feige, U., Shamir, A.: Witness indistinguishable and witness hiding protocols. In: STOC 1990, pp. 416–426. ACM (1990)
7. Kurosawa, K., Heng, S.-H.: From Digital Signature to ID-based Identification/Signature. In: Bao, F., Deng, R., Zhou, J. (eds.) PKC 2004. LNCS, vol. 2947, pp. 248–261. Springer, Heidelberg (2004)
8. Kurosawa, K., Heng, S.-H.: Identity-Based Identification Without Random Oracles. In: Gervasi, O., Gavrilova, M.L., Kumar, V., Laganá, A., Lee, H.P., Mun, Y., Taniar, D., Tan, C.J.K. (eds.) ICCSA 2005. LNCS, vol. 3481, pp. 603–613. Springer, Heidelberg (2005)
9. Rückert, M.: Adaptively Secure Identity-Based Identification from Lattices without Random Oracles. In: Garay, J.A., De Prisco, R. (eds.) SCN 2010. LNCS, vol. 6280, pp. 345–362. Springer, Heidelberg (2010)
10. Shamir, A.: Identity-Based Cryptosystems and Signature Schemes. In: Blakely, G.R., Chaum, D. (eds.) CRYPTO 1984. LNCS, vol. 196, pp. 47–53. Springer, Heidelberg (1985)
11. Yang, G., Chen, J., Wong, D.S., Deng, X., Wang, D.: A new framework for the design and analysis of identity-based identification schemes. Theoretical Computer Science 407(1-3), 370–388 (2008); Preliminary version In: ACNS 2007 (2007)

A Relations between Security Notions

A.1 Relation between x-id-imp-pa Security and x-id-imp-ca Security

If an adversary who breaks the x-id-imp-pa security exists, then it immediately implies that an adversary who breaks the x-id-imp-ca security exists, where x denotes a type of attack such that $\mathsf{x} \in \{\mathsf{ws}, \mathsf{stat}, \mathsf{adapt}\}$. Thus, the x-id-imp-ca security implies the x-id-imp-pa security.

A.2 Relation between stat-id-imp-atk Security and adapt-id-imp-atk Security

If an adversary who breaks the stat-id-imp-atk security exists, then it immediately implies that an adversary who breaks the adapt-id-imp-atk security exists, where atk denotes a type of attack such that $\mathsf{atk} \in \{\mathsf{pa}, \mathsf{aa}, \mathsf{ca}\}$. Thus, the adapt-id-imp-atk security implies the stat-id-imp-atk security.

There is a gap between the stat-id-imp-atk and adapt-id-imp-atk security. Assume that there is a stat-id-imp-atk secure IBI protocol. We modify it to an IBI protocol such that the master public key contains a special identity, and the user-key-generation algorithm outputs an additional identity and a corresponding secret key together with a secret key of the special identity if that key is required. Then, the resulting IBI protocol is still stat-id-imp-atk secure but not adapt-id-imp-atk secure.

A.3 Relation between wsid-imp-atk Security and stat-id-imp-atk Security

Assume that an adversary who breaks the wsid-imp-atk security exists. Then, we can construct an adversary who breaks the stat-id-imp-atk security, where atk denotes a type of attack such that $\mathsf{atk} \in \{\mathsf{pa}, \mathsf{aa}, \mathsf{ca}\}$.

The wsid-imp-atk adversary initially issues a single INIT query that consists of identities, which only are allowed to be used in queries during the wsid-imp-atk experiment. The stat-id-imp-atk adversary randomly selects a target identity among them, and makes a single external CORR query that consists of all identities but the target identity, to obtain their secret keys. When the wsid-imp-atk adversary makes a CORR query on an identity, the stat-id-imp-atk adversary has already obtained it by the initial CORR query. At the end, if the wsid-imp-atk adversary succeeds in impersonating the target identity, the stat-id-imp-atk adversary can also impersonate the target identity by using the messages output by the wsid-imp-atk adversary exactly as they are. Thus, the stat-id-imp-atk security implies the wsid-imp-atk security. Note that the success probability decreases in proportion to a fraction of users issued in the initial INIT query.

There is a gap between the wsid-imp-atk and stat-id-imp-atk security. Assume that there is a wsid-imp-atk secure IBI protocol. We modify it to an IBI protocol such that the master public key contains an additional identity and a corresponding secret key. Then, the resulting IBI protocol is still wsid-imp-atk secure but not stat-id-imp-atk secure.

A.4 Relations among Security Notions

Fig. 4 summarizes the relations among the security notions for IBI protocols.

$$\begin{array}{ccccc} \mathsf{wsid\text{-}imp\text{-}ca} & \Leftarrow & \mathsf{stat\text{-}id\text{-}imp\text{-}ca} & \Leftarrow & \mathsf{adapt\text{-}id\text{-}imp\text{-}ca} \\ \Downarrow & & \Downarrow & & \Downarrow \\ \mathsf{wsid\text{-}imp\text{-}pa} & \Leftarrow & \mathsf{stat\text{-}id\text{-}imp\text{-}pa} & \Leftarrow & \mathsf{adapt\text{-}id\text{-}imp\text{-}pa} \end{array}$$

Fig. 4. Diagram of Security Notions

The symbol $\Rightarrow$ indicates "imply", and $A \Rightarrow B$ means that if an IBI protocol is A secure, then the protocol is B secure.

Identity-Based Extractable Hash Proofs and Their Applications

Yu Chen[1], Zongyang Zhang[2], Dongdai Lin[1], and Zhenfu Cao[2,*]

[1] State Key Laboratory of Information Security (SKLOIS), Institute of Information Engineering, Chinese Academy of Sciences, China
{chenyu,ddlin}@iie.ac.cn

[2] Department of Computer Science and Engineering, Shanghai Jiao Tong University, China
{zongyangzhang,zfcao}@sjtu.edu.cn

Abstract. In this paper, we introduce a general paradigm called identity-based extractable hash proof system (IB-EHPS), which is an extension of extractable hash proof system (EHPS) proposed by Wee (CRYPTO '10). We show how to construct identity-based encryption (IBE) scheme from IB-EHPS in a simple and modular fashion. Our construction provides a generic method of building and interpreting CCA-secure IBE schemes based on computational assumptions. As instantiations, we realize IB-EHPS from the bilinear Diffie-Hellman assumption and the modified bilinear Diffie-Hellman assumption, respectively.

1 Introduction

Security against adaptive chosen-ciphertext attack (CCA-security) [27] is now accepted as the standard security notion for public-key encryption (PKE) schemes as well as identity-based encryption (IBE) schemes. In contrast to security against adaptive chosen-plaintext attack (CPA-security) [25], CCA-security captures the immunity against an active adversary who is given access to a decryption oracle that allows it to obtain the decryptions of ciphertexts of its choice.

On the other hand, in most cases related to cryptography, decisional assumptions form a much stronger class of assumptions than the corresponding search (computational) assumptions[1]. As such, cryptosystems based on search problems are generally preferred to those based on decisional assumptions. From now on, we will use the term *computational* and *search* interchangeablely.

Up to now, only a handful of IBE schemes [11, 14, 19] have been proven to be CCA-secure from computational assumptions in the standard model. Besides, there seems no overarching concept explaining these constructions. Inspired by the notion of extractable hash proof system [31] in the public key setting, we introduce a new notion named identity-based extractable hash proof system and show how to construct CCA-secure IBE schemes from it.

* Corresponding author.

[1] Unless the decisional assumption can be proved equivalent to its computational counterpart, as it is the case with cryptosystems based on the problem of "leaning with error" (LWE) [26].

F. Bao, P. Samarati, and J. Zhou (Eds.): ACNS 2012, LNCS 7341, pp. 153–170, 2012.

1.1 Background

The concept of identity-based encryption (IBE) was introduced by Shamir [28] in 1984. Boneh and Franklin [6] proposed the first practical IBE scheme whose security is based on the computational bilinear Diffie-Hellman (CBDH) assumption. Cocks [12] described another IBE scheme based on the decisional quadratic residues (DQR) assumption modulo a composite. Both of them are proven secure under the random oracle model [2]. However, a proof in the random oracle model can only serve as a heuristic argument and possibly lead to insecure schemes in the standard model. This posed an interesting problem of constructing IBE schemes in the standard model.

First, Canetti, Halevi, and Katz [8] made the breakthrough by giving a solution in the standard model, but under a weaker notion named "selective-identity" where the attacker must declare the target identity id^* before seeing the public parameters. Boneh and Boyen [4] then provided two efficient selective-identity CPA-secure IBE schemes known as BB_1-IBE and BB_2-IBE. The former is based on the decisional bilinear Diffie-Hellman (DBDH) assumption while the latter is based the decisional q-BDHI assumption. Subsequently, Waters [29] proposed an efficient and adaptive-identity CPA-secure IBE scheme (Waters-IBE) in the standard model which is also based on the DBDH assumption by employing Waters hash in place of Boneh-Boyen hash used in BB_1-IBE. One drawback is that it suffers from large public parameter size. Gentry [15] proposed an IBE scheme (Gentry-IBE) which enjoys short public parameters and tight security reduction. Although Gentry-IBE achieves adaptive-identity CCA-security in the standard model, it did so at the cost of relying a non-standard and non-static assumption called the decisional q-ABHDE assumption. Waters [30] then introduced dual system encryption methodology and proposed an adaptive-identity CPA-secure IBE scheme based on the DBDH assumption and the decisional linear (DLIN) assumption in the standard model. Recently, Gentry *et al.* [16] proposed an IBE scheme based on the LWE assumption in the random oracle model. Cash *et al.* [9] and Agrawal *et al.* [1] showed how to construct IBE schemes based on the LWE assumption in the standard model.

As previously stated, CCA-security is the *de facto* level of security required for IBE schemes used in practice. Unfortunately, constructing CCA-secure IBE scheme without resorting to random oracle heuristic turns out to be difficult. Boneh, Canetti, Halevi, and Katz [5] proposed a generic transformation (known as the BCHK transformation) from any CPA-secure 2-level HIBE scheme to a CCA-secure IBE scheme, which is the only generic approach known for constructing efficient CCA-secure IBE in the standard model.

1.2 Motivation

As we have already mentioned, a decisional assumption is generally stronger than its computational counterpart. From both theoretical and practical perspective, it is more desirable to reduce the security of cryptographic schemes to computational assumptions. Considering an IBE scheme obtained from the BCHK transformation, its CCA-security relies on the CPA-security of the underlying 2-level HIBE scheme and the security of one-time signature or MAC. Hence its assumption cannot be directly counted as computational or decisional assumption. However, the indistinguishability against CPA-attack is of decisional flavor, thus it is arguably closer to decisional assumptions.

Haralambiev *et al.* [19] proposed several efficient PKE schemes in the standard model. They also sketched that one of their PKE schemes can be extended to a BB_1-style identity-based key encapsulation mechanism (IB-KEM). Galindo [14] gave an IB-KEM from the PKE scheme due to Hanaoka and Kurosawa [18]. Chen *et al.* [11] proposed another BB_1-style IB-KEM. All the above IB-KEMs are proven to be selective-identity CCA-secure based on the CBDH assumption in the standard model. All of them fall outside of the BCHK [5] methodology. While the IB-KEMs due to [19] and [11] are similar, it seems that the IB-KEM [14] relies on different techniques to achieve CCA-security. So far, there is no overarching framework explaining these constructions.

Recently, several CCA-secure PKE schemes from various computational assumptions emerged, such as [10, 18–20]. Inspired in part by hash proof system (HPS) [13], Wee [31] introduced the notion of extractable hash proof system (EHPS) and showed how to derive efficient CCA-secure PKE via EHPS. Roughly speaking, EHPS resembles hash proof system (HPS) [13] in that both of them are essentially a special kind of non-interactive zero-knowledge proof, except that EHPS replaces the soundness requirement with a *proof of knowledge property* [27]. The framework of EHPS does not only encompass a series of CCA-secure PKE schemes [21,22] based on decisional assumptions, but also can explain a series of CCA-secure PKE schemes [19, 20] based on computational assumptions in a unified way, which is the most appealing advantage of EHPS.

Although the realm of IBE and PKE are inherently different, the techniques are sometimes interchangeable. Motivated by the above discussion, we find the following intriguing question:

Does there exist a general framework for the construction of identity-based encryption from computational assumptions in the standard model?

1.3 Our Contributions

EHPSs and their benefits are confined to the realm of public-key setting. In this paper we bring them to the identity-based setting, defining identity-based extractable hash proof system (IB-EHPS). Using IB-EHPS, we obtain new insights into the construction of CCA-secure IBE schemes. In particular, we show that this notion unifies many seemingly unrelated IBE constructions under a single framework. We summarize our main contributions as follows.

Identity-Based Extractable Hash Proof Systems. We introduce the notion of IB-EHPS by tailoring EHPS to the identity-based setting. We show that IB-EHPS instantly yields adaptive-identity CPA-secure IBE. However, the basic IB-EHPS is too generic to encompass more applications. To resolve this problem, we further propose the notion of all-but-one (ABO) IB-EHPS, which can in turn be used to construct adaptive-identity CCA-secure IBE.

Practical CCA-secure IBE from IB-EHPS. We present two ABO IB-EHPSs from the CBDH assumption and the modified CBDH assumption, respectively. As a result, we obtain two efficient adaptive-identity CCA-secure IBE schemes based on computational assumptions in the standard model.

2 Preliminaries

2.1 Definitions

For a positive integer n, we use $[n]$ to denote the set $[n] = \{1, \ldots, n\}$. For a finite set X, we use $x \stackrel{R}{\leftarrow} X$ to denote that x is sampled from X uniformly at random. The main security parameter through this paper is κ, and all algorithms are implicitly given κ as input. We use standard asymptotic notation O and o to denote the growth of functions. Let $\mathsf{poly}(\kappa)$ denote an unspecified function $f(\kappa) = O(\kappa^c)$ for some constant c. Let $\mathsf{negl}(\kappa)$ denote an unspecified function $f(\kappa)$ such that $f = o(\kappa^{-c})$ for every constant c. We say that a probability is overwhelming if it is $1 - \mathsf{negl}(\kappa)$. A probabilistic polynomial-time (PPT) algorithm is a randomized algorithm that runs in time $\mathsf{poly}(\kappa)$. If $\mathcal{A}$ is a randomized algorithm, we write $z \leftarrow \mathcal{A}(x_1, \ldots, x_n; r)$ to indicate that $\mathcal{A}$ outputs z on inputs $(x_1, \ldots, x_n)$ and random coins r. We will omit r and write $z \leftarrow \mathcal{A}(x_1, \ldots, x_n)$ when it is not necessary to make explicit the randomness $\mathcal{A}$ uses. We assume that an algorithm returns $\perp$ if any of its inputs is $\perp$.

2.2 Identity-Based Key Encapsulation Mechanisms

Instead of providing the full functionality of an IBE scheme, in many applications it is sufficient to allow sender and receiver to agree on a common random session key. This can be accomplished by *identity-based key encapsulation mechanism* (IB-KEM) as formalized in [3]. Considering there are many practical reasons to prefer an IB-KEM over an IBE scheme, we define IBE schemes as IB-KEM in this paper. An IB-KEM consists of four PPT algorithms as follows:

- $\mathsf{Setup}(\kappa)$: takes as input a security parameter κ, outputs the master public key mpk and the master secret key msk. mpk will be used as an implicit input by all other algorithms KeyGen, Encap, Decap. Let I, C, and K be the identity space, ciphertext space, and the key space (for DEM), respectively.
- $\mathsf{KeyGen}(msk, id)$: takes as input msk and an identity $id \in I$, outputs a private key sk of id.
- $\mathsf{Encap}(id)$: takes as input an identity $id \in I$, outputs a ciphertext $c \in C$ and a DEM key $k \in K$.
- $\mathsf{Decap}(sk, c)$: takes as input a private key sk of identity id and a ciphertext c, outputs a DEM key $k \in K$ or an distinguished symbol $\perp$ (which is not in K) indicating that c is not consistent under id. Here we say that a ciphertext is *consistent* or *well-formed* or *valid* if it can be "honestly generated" by the encryption algorithm. For a PKE or IBE scheme, if anyone can do the "consistency check", we say that it is *public verifiable*. Otherwise, we say that it is *private verifiable*.

We refer to [23] for formal security definition of IB-KEM. For correctness, we require that for any $(mpk, msk) \leftarrow \mathsf{Setup}(\kappa)$, any $(c, k) \leftarrow \mathsf{Encap}(id)$, and any $sk \leftarrow \mathsf{KeyGen}(msk, id)$, we have $\Pr[\mathsf{Decap}(sk, c) = k] = 1$.

2.3 Bilinear Diffie-Hellman Assumption

Let $(p, \mathbb{G}, \mathbb{G}_T, e) \leftarrow \mathsf{GroupGen}(1^\kappa)$, where $\mathsf{GroupGen}(\cdot)$ is a bilinear groups parameters generator [7]. Let g be a random generator of $\mathbb{G}$. Define $\mathrm{bdh}(A, B, C) := T$, where $A = g^a$, $B = g^b$, $C = g^c$, and $T = e(g, g)^{abc}$. The computational bilinear Diffie-Hellman (CBDH) problem is computing $\mathrm{bdh}(A, B, C)$ given random $A, B, C \in \mathbb{G}$. The CBDH assumption asserts that the CBDH problem is hard, that is, $\Pr[\mathcal{A}(A, B, C) = \mathrm{bdh}(A, B, C)] \leq \mathsf{negl}(\kappa)$ for all PPT algorithms $\mathcal{A}$.

In the bilinear setting, the Goldreich-Levin theorem [17] gives us the following lemma for a Goldreich-Levin hardcore predicate $f_{\mathrm{gl}} : \mathbb{G}_T \times \{0,1\}^u \to \{0,1\}$.

Lemma 2.1 *Let $A, B, C \xleftarrow{R} \mathbb{G}$, $R \xleftarrow{R} \{0,1\}^u$, $K = f_{\mathrm{gl}}(\mathrm{bdh}(A, B, C), R)$, and $U \xleftarrow{R} \{0,1\}$. Suppose there exists a PPT algorithm $\mathcal{B}$ distinguishing the distributions $\Delta_{\mathrm{bdh}} = (g, A, B, C, K, R)$ and $\Delta_{\mathrm{rand}} = (g, A, B, C, U, R)$ with non-negligible advantage. Then there exists a PPT algorithm solving the CBDH problem with non-negligible correct probability.*

The modified computational bilinear Diffie-Hellman (mCBDH) problem [24] is similar to the CBDH problem except that an additional point $B' = g^{b^2}$ is given. We can prove a similar lemma regarding mCBDH problem as Lemma 2.1.

2.4 Binary Relations for Search Problems

A search problem $\mathbf{S} = (S_\kappa)_{\kappa \geq 0}$ is a collection of distributions. For every value of $\kappa \geq 0$, an instance of S_κ specifies two finite, non-empty sets X and W, public parameter PP, and a binary relation $\mathsf{R}_{\mathrm{pp}} \in X \times W$. A search problem also provides two algorithms, namely SampS and SampR. SampS takes as input a security parameter κ, and outputs an instance of S_κ. We write $(X, W, \mathrm{PP}, \mathsf{R}_{\mathrm{pp}}) \leftarrow \mathsf{SampS}(\kappa; \mathrm{SP})$, where SP is the random coins used in SampS. SampR takes as input PP, and outputs a tuple (x, w) belong to R_{pp}. We write $(x, w) \leftarrow \mathsf{SampR}(\mathrm{PP}; r)$, where r is the random coins used in SampR. Note that PP is often assumed to be an implicit input and it is useful to make the random coins explicitly in SampR algorithm, thus we often write $\mathsf{SampR}(r)$ henceforth whenever the context is clear. Different to the requirement in EHPS [31], we do not require that R_{pp} can be efficiently verifiable in IB-EHPS.

Intuitively, the relation R_{pp} corresponds to a hard search problem, that is, given a random element $x \in X$, it is hard to find $w \in W$ such $(x, w) \in \mathsf{R}_{\mathrm{pp}}$. More formally, we say that a binary relation R_{pp} is one-way if:

- with overwhelming probability over PP, for any $x \in X$, there exists at most one $w \in W$ such that $(x, w) \in \mathsf{R}_{\mathrm{pp}}$ (we say that w is a *witness* for x); and
- there is an efficiently computable function F from W to $\{0,1\}^l$ for some positive integer l such that given x, $\mathsf{F}(w)$ is pseudo-random over $\{0,1\}^l$ where $(x, w) \leftarrow \mathsf{SampR}(\mathrm{PP})$.

For relations where computing w given x is hard on average, we may derive a function GL with a one-bit output via the Goldreich-Levin hardcore predicate f_{gl}. Note that GL is an instantiation of the above function F.

Bilinear Diffie-Hellman Relation. Let $(p, \mathbb{G}, \mathbb{G}_T, e) \leftarrow \mathsf{GroupGen}(\kappa)$. The public parameter PP is given by (g, g^a, g^b) for a random $g \in \mathbb{G}$ and random $a, b \xleftarrow{R} \mathbb{Z}_p$[2]. We consider the bilinear Diffie-Hellman relation over $\mathbb{G} \times \mathbb{G}_T$:

$$\mathsf{R}^{\mathsf{bdh}}_{\mathrm{pp}} = \left\{(x, w) \in \mathbb{G} \times \mathbb{G}_T : w = e(g, x)^{ab}\right\}$$

The associated SampR picks $r \xleftarrow{R} \mathbb{Z}_p$ and outputs $(g^r, e(g^a, g^b)^r)$. Lemma 2.1 shows that we may extract a single hardcore bit from w using $\mathsf{GL}(w)$ for relation $\mathsf{R}^{\mathsf{bdh}}_{\mathrm{pp}}$. The modified BDH relation $\mathsf{R}^{\mathsf{mbdh}}_{\mathrm{pp}}$ can be defined analogously.

2.5 General Hashing

Let X, I, and Y be finite, non-empty sets. Let $\mathsf{H} = (\mathsf{H}_{mpk})_{mpk \in MPK}$ be a collection of functions indexed by MPK, so that for every $mpk \in MPK$, H_{mpk} is a function from $I \times X$ into Y. We call $\mathbf{H} = (\mathsf{H}, MPK, I, X, Y)$ a hash family.

3 Identity-Based Extractable Hash Proofs

An IB-EHPS $\mathbf{P}$ for $\mathbf{S}$ associating with each instance $(X, W, \mathrm{PP}, \mathsf{R}_{\mathrm{pp}}) \leftarrow \mathsf{SampS}(\kappa)$ of S_κ and an identity space I and a hash family $\mathbf{H} = (\mathsf{H}, MPK, I, X, Y)$, is a tuple of algorithms (SetupExt, SetupHash, KeyGen, KeyGen*, Pub, Priv, Ext). Loosely speaking, an IB-EHPS can behave in one of two modes, namely the extraction mode and the hashing mode. We will rely on the extraction mode for the normal functionality of the resulting IBE scheme, and on the hashing mode for the proof of security.

Extraction Mode

- $\mathsf{SetupExt}(\mathrm{PP}, \mathrm{SP})$: takes as input (PP, SP), outputs the master public key mpk and the master secret key msk.
- $\mathsf{Pub}(mpk, id, r)$: takes as input mpk, an identity $id \in I$ and random coins r, outputs $y \in Y$ such that $y = \mathsf{H}_{mpk}(id, x)$ where $(x, w) \leftarrow \mathsf{SampR}(r)$. This is the *public evaluation algorithm*.
- $\mathsf{KeyGen}(msk, id)$: takes as input msk and an identity $id \in I$, outputs a private key sk for id.
- $\mathsf{Ext}(sk, x, y)$: takes as input a private key sk of identity $id \in I$, $x \in X$ and $y \in Y$, outputs $w \in W$.

For the correctness of extraction mode, we require that for any $(mpk, msk) \leftarrow \mathsf{SetupExt}(\mathrm{PP}, \mathrm{SP})$ and any $id \in I$ and any $sk \leftarrow \mathsf{KeyGen}(msk, id)$, we have $y = \mathsf{H}_{mpk}(id, x) \Longrightarrow (x, \mathsf{Ext}(sk, x, y)) \in \mathsf{R}_{\mathrm{pp}}$.

[2] We assume PP also includes p and the descriptions of $(e, \mathbb{G}, \mathbb{G}_T)$.

Hashing Mode

- $\mathsf{SetupHash}(\mathrm{PP})$: takes PP as input, outputs the master public key mpk and the master secret key msk^*. msk^* implicitly splits the whole identity space I into two orthogonal subspaces I_1 and I_2, namely $I = I_1 \cup I_2$ and $I_1 \cap I_2 = \emptyset$.
- $\mathsf{Priv}(msk^*, id, x)$: takes as input msk^* and an identity $id \in I$, if $id \in I_2$ outputs $y \in Y$, else outputs $\perp$. This is the *private evaluation algorithm.*
- $\mathsf{KeyGen}^*(msk^*, id)$: takes as input msk^* and an identity $id \in I$, if $id \in I_1$ outputs a private key sk for id, else outputs $\perp$.

For the correctness of hashing mode, we require that for any $(mpk, msk) \leftarrow \mathsf{SetupHash}(\mathrm{PP})$ and any $id \in I_2$, we have $\mathsf{Priv}(msk^*, id, x) = \mathsf{H}_{mpk}(id, x)$.

Indistinguishability. We require the first output (mpk) of $\mathsf{SetupExt}(\mathrm{PP}, \mathrm{SP})$ and $\mathsf{SetupHash}(\mathrm{PP})$ are statistically indistinguishable. For any mpk and any identity $id \in I_1$, we require the output of $\mathsf{KeyGen}(msk, id)$ and $\mathsf{KeyGen}^*(msk^*, id)$ are statistically indistinguishable.
Well Partition. We now set a property that is sufficient for the existence of an efficient transformation that we will use to obtain CPA-secure IB-KEM. Intuitively, this property guarantees that no PPT adversary can distinguish the CPA-security games simulated by operating IB-EHPS in extraction mode and hashing mode with non-negligible probability. We formally define this property via the following game played between a PPT adversary $\mathcal{A}$ and a challenger $\mathcal{CH}$.

Given PP and SP, $\mathcal{CH}$ picks $b \xleftarrow{R} \{0,1\}$, and plays Sub-Game b with $\mathcal{A}$.

Sub-Game 0. $\mathcal{CH}$ interacts with $\mathcal{A}$ by operating IB-EHPS in extraction mode.
Setup: $\mathcal{CH}$ generates $(mpk, msk) \leftarrow \mathsf{SetupExt}(\mathrm{PP}, \mathrm{SP})$ and gives mpk to $\mathcal{A}$. $\mathcal{CH}$ also samples $(x^*, w^*) \leftarrow \mathsf{SampR}(r^*)$ and records them for latter use.
Phase 1 - Private key queries: When $\mathcal{A}$ submits an private key query $\langle id \rangle$, $\mathcal{CH}$ responds with $\mathsf{KeyGen}(msk, id)$.
Phase Middle: When $\mathcal{A}$ submits an identity $id^* \in I$ on the condition that id^* did not appear in any private key query in Phase 1, $\mathcal{CH}$ obtains $y^* = \mathsf{H}_{mpk}(id^*, x^*)$ by evaluating $\mathsf{Pub}(mpk, id^*, r^*)$, then sets $k_0^* = \mathsf{F}(w^*)$ and $k_1^* \xleftarrow{R} \{0,1\}^l$. $\mathcal{CH}$ picks a random bit $\beta \in \{0,1\}$ and returns (x^*, y^*, k_β^*) to $\mathcal{A}$.
Phase 2 - Private key queries: Same as Phase 1 except that the private key query $\langle id^* \rangle$ is not allowed.

Sub-Game 1. $\mathcal{CH}$ interacts with $\mathcal{A}$ by operating IB-EHPS in hashing mode.
Setup: $\mathcal{CH}$ generates $(mpk, msk^*) \leftarrow \mathsf{SetupHash}(\mathrm{PP})$ and gives mpk to $\mathcal{A}$. $\mathcal{CH}$ also samples $(x^*, w^*) \leftarrow \mathsf{SampR}(r^*)$ and records them for latter use.
Phase 1 - Private key queries: When $\mathcal{A}$ submits a private key query $\langle id \rangle$, $\mathcal{CH}$ responds with $\mathsf{KeyGen}(msk^*, id)$.
Phase Middle: When $\mathcal{A}$ submits an identity $id^* \in I$ on the condition that id^* did not appear in any private key query in Phase 1, $\mathcal{CH}$ computes $y^* = \mathsf{H}_{mpk}(id, x^*)$ via $\mathsf{Priv}(msk^*, id^*, x^*)$, and sets $k_0^* = \mathsf{F}(w^*)$ and $k_1^* \xleftarrow{R} \{0,1\}^l$. $\mathcal{CH}$ picks a random bit $\beta \in \{0,1\}$ and returns (x^*, y^*, k_β^*) to $\mathcal{A}$.

Phase 2 - Private key queries: Same as Phase 1 except that the private key query $\langle id^* \rangle$ is not allowed.

At the end of the game, $\mathcal{A}$ outputs its guess b' for b and wins the game if $b = b'$. Let $\Pr[\mathcal{A}\text{ wins}]$ be the probability that $\mathcal{A}$ wins the game, where the probability space is over the random coins consumed by $\mathcal{CH}$. Let δ be a real number in $[0, 1]$. It is straightforward to see that if $\Pr[\mathcal{A}\text{ wins}] \leq 1 - \frac{1}{2}\delta$, then the probability that the $\mathcal{A}$'s view in Sub-Game 1 is identical to Sub-Game 0 is at least δ. Let Q_e be the number of private key queries. We say such an IB-EHPS is (Q_e, δ)-well-partition.

All-But-One Identity-Based Extractable Hash Proofs. For our applications, it is convenient to work with a richer abstraction. More precisely, an ABO IB-EHPS is a tuple of algorithms (SetupExt, SetupABO, Pub, Priv, Verify, Verify*, KeyGen, KeyGen*, Ext, Ext*).

Extraction Mode

- The algorithms SetupExt, Pub, and KeyGen related to the extraction mode are identical to that in IB-EHPS.
- Verify(id, sk, x, y): takes as input an identity $id \in I$, a private key sk for id, $x \in X$ and $y \in Y$, if $y = \mathsf{H}_{mpk}(id, x)$ returns 1, else returns 0. Particularly, when sk is not necessary, we say H_{mpk} is *public verifiable.*
- Ext(sk, x, y): takes as input a private key sk for identity $id \in I$, $x \in X$ and $y \in Y$, if Verify$(id, sk, x, y) = 1$ then outputs $w \in W$, else outputs $\perp$.

For the correctness of extraction mode, we require that for any $(mpk, msk) \leftarrow$ SetupExt$(\mathrm{PP}, \mathrm{SP})$, any $id \in I$ and any $sk \leftarrow$ KeyGen(msk, id), we have:

$$y = \mathsf{H}_{mpk}(id, x) \Longrightarrow (x, \mathsf{Ext}(sk, x, y)) \in \mathsf{R}_{\mathrm{pp}} \tag{1}$$

ABO Hashing Mode

- SetupABO(PP, x^*): similar to SetupHash(PP) in IB-EHPS except taking an extra input $x^* \in X$.
- KeyGen$^*(msk^*, id)$: same as KeyGen* in IB-EHPS.
- Priv(msk^*, id, x): takes as input msk^*, $id \in I$, and $x \in X$, if $id \in I_2$ and $x = x^*$ outputs $y \in Y$, else outputs $\perp$.
- Verify$^*(id, msk^*, x, y)$: takes as input an identity $id \in I$, msk^*, $x \in X$ and $y \in Y$, if $y = \mathsf{H}_{mpk}(id, x)$ returns 1 else returns 0. When H_{mpk} is public verifiable, msk^* is not necessary.
- Ext$^*(msk^*, x, y)$: takes as input msk^*, $x \in X$, and $y \in Y$, if $x \neq x^*$ and Verify$^*(id, msk^*, x, y) = 1$ then outputs $w \in W$, else outputs $\perp$.

For the correctness of ABO hashing mode, we require for any $x^* \in X$ and any $(mpk, msk^*) \leftarrow$ SetupABO(SP, x^*) and any $id \in I_2$, we have Priv$(msk^*, id, x^*) =$ $\mathsf{H}_{mpk}(id, x^*)$, and for any $id \in I$ if $x \neq x^*$ we have:

$$y = \mathsf{H}_{mpk}(id, x) \Longrightarrow (x, \mathsf{Ext}^*(msk^*, x, y)) \in \mathsf{R}_{\mathrm{pp}} \tag{2}$$

INDISTINGUISHABILITY. We require that the similar indistinguishable properties hold as that for IB-EHPS, namely for any $x^* \in X$ the first output of $\mathsf{SetupExt}(\mathrm{PP}, \mathrm{SP})$ and $\mathsf{SetupHash}(\mathrm{PP}, x^*)$ are statistically indistinguishable. For any mpk and any identity $id \in I_1$, we require that the output of $\mathsf{KeyGen}(msk, id)$ and $\mathsf{KeyGen}^*(msk^*, id)$ are statistically indistinguishable.
WELL PARTITION. This property for ABO IB-EHPS is defined analogously as that for IB-EHPS. We formally defined it via the following game played between a PPT adversary $\mathcal{A}$ and a challenger $\mathcal{CH}$.

Given PP and SP, $\mathcal{CH}$ picks $b \xleftarrow{R} \{0,1\}$, and plays Sub-Game b with $\mathcal{A}$.

Sub-Game 0. $\mathcal{CH}$ interacts with $\mathcal{A}$ by operating ABO IB-EHPS in extraction mode.
Setup: Same as Sub-Game 0 in IB-EHPS.
Phase 1 - Private key queries: Same as Sub-Game 0 in IB-EHPS.
Phase 1 - Decapsulation queries: When $\mathcal{A}$ submits a query $\langle id, x, y \rangle$, if $x = x^*$, $\mathcal{CH}$ directly returns $\perp$. Otherwise $\mathcal{CH}$ responds with $\mathsf{F}(\mathsf{Ext}(sk, x, y))$.
Phase Middle: Same as Sub-Game 0 in IB-EHPS.
Phase 2 - Private key queries: Same as Sub-Game 0 in IB-EHPS.
Phase 2 - Decapsulation queries: When $\mathcal{A}$ submits a query $\langle id, x, y \rangle$, $\mathcal{CH}$ computes $sk \leftarrow \mathsf{KeyGen}(msk, id)$ and responds with $\mathsf{F}(\mathsf{Ext}(sk, x, y))$. The query $\langle id^*, x^*, y^* \rangle$ is not allowed.

Sub-Game 1. $\mathcal{CH}$ interacts with $\mathcal{A}$ by operating ABO IB-EHPS in ABO hashing mode.
Setup: $\mathcal{CH}$ generates $(mpk, msk^*) \leftarrow \mathsf{SetupABO}(\mathrm{PP}, x^*)$ and gives mpk to $\mathcal{A}$. $\mathcal{CH}$ also samples $(x^*, w^*) \leftarrow \mathsf{SampR}(r^*)$ and records it for latter use.
Phase 1 - Private key queries: Same as Sub-Game 1 in IB-EHPS.
Phase 1 - Decapsulation queries: When $\mathcal{A}$ submits a query $\langle id, x, y \rangle$, if $x = x^*$, $\mathcal{CH}$ returns $\perp$. Otherwise if $id \in I_1$, $\mathcal{CH}$ extracts $sk = \mathsf{KeyGen}^*(msk^*, id)$ and responds with $\mathsf{F}(\mathsf{Ext}(sk, x, y))$, else responds with $\mathsf{F}(\mathsf{Ext}^*(msk^*, x, y))$.
Phase Middle: Same as Sub-Game 1 in IB-EHPS.
Phase 2 - Private key queries: Same as Sub-Game 1 in IB-EHPS.
Phase 2 - Decapsulation queries: When $\mathcal{A}$ submits a query $\langle id, x, y \rangle$, if $id \in I_1$, $\mathcal{CH}$ computes $sk = \mathsf{KeyGen}^*(msk^*, id)$ and responds with $\mathsf{F}(\mathsf{Ext}(sk, x, y))$, else responds with $\mathsf{F}(\mathsf{Ext}^*(msk^*, x, y))$. The extraction query $\langle id^*, x^*, y^* \rangle$ is not allowed.

At the end of the game, $\mathcal{A}$ outputs its guess b' for b and wins the game if $b = b'$. Similar to the analysis we have done before, if $\Pr[\mathcal{A} \text{ wins}] \leq 1 - \frac{1}{2}\delta$, then the probability that the $\mathcal{A}$'s view in Sub-Game 1 is identical to Sub-Game 0 is at least δ. Let Q_e and Q_d be the number of private key queries and extraction queries, respectively. We say such an ABO IB-EHPS is (Q_e, Q_d, δ)-well-partition.

In ABO IB-EHPS, property (1) for the extraction mode ensures the functionality of the resulting IB-KEM while the property (2) for the ABO hashing mode ensures the correctness of simulation. The crux to achieve CCA-security is to make sure that the decryption oracle does not help the adversary to distinguish w^* from random; in other words, the output of the decryption algorithm should contain no knowledge of w^* related to x^* when the input ciphertext (x^*, y) is not consistent. In line of this, to yield CCA-secure IBE, the ABO IB-EHPS should also have the following two properties:

$$y \neq \mathsf{H}_{mpk}(id, x) \Longrightarrow (x, \mathsf{Ext}(sk, x, y)) \notin \mathsf{R}_{\mathrm{PP}} \tag{3}$$

$$y \neq \mathsf{H}_{mpk}(id, x) \Longrightarrow (x, \mathsf{Ext}^*(msk^*, x, y)) \notin \mathsf{R}_{\mathrm{PP}} \tag{4}$$

We achieve properties (3) and (4) by equipping the ABO IB-EHPS with algorithms Verify and Verify* which can determine if $y = \mathsf{H}_{mpk}(id, x)$, and algorithms Ext and Ext* returns a distinguished symbol $\perp$ when $y \neq \mathsf{H}_{mpk}(id, x)$. We note that it is also possible to achieve properties (3) and (4) without requiring algorithms Verify and Verify* available. The trick is for certain relation R we may re-design algorithms Ext and Ext* smartly using the "implicit rejection" idea [21, 24], namely for $y \neq \mathsf{H}_{mpk}(id, x)$, $\mathsf{Ext}(sk, x, y)$ and $\mathsf{Ext}^*(msk^*, x, y)$ returns a random value $w \in W$ which is independent of x. Thus the properties (3) and (4) will hold with overwhelming probability.

Combining properties (3) and (4) with (1) and (2), the ABO IB-EHPS in fact has the following stronger properties: $y = \mathsf{H}_{mpk}(id, x) \Longleftrightarrow (x, \mathsf{Ext}(sk, x, y)) \in \mathsf{R}_{\mathrm{PP}}$ for the extraction mode and $y = \mathsf{H}_{mpk}(id, x) \Longleftrightarrow (x, \mathsf{Ext}(msk^*, x, y)) \in \mathsf{R}_{\mathrm{PP}}$ for the ABO mode (when $x \neq x^*$), which is reminiscent of ABO EHPS [31]. The key difference is that [31] achieves properties (3) and (4) by requiring the relation R_{PP} can be efficiently verifiable, which may make it too stringent to cover many known CCA-secure IBE schemes, such as [11, 19, 23].

3.1 Relation to Extractable Hash Proof System

IB-EHPS is the corresponding notion of EHPS in the IBE setting. However, we stress that the extension is not straightforward for the following main differences.

1. The (ABO) hashing mode for (ABO) IB-EHPS is defined in partitioning style. More precisely, the setup algorithm generates (mpk, msk^*) and implicitly splits the whole identity space I into two orthogonal subspaces, — 1) I_1: identities for which KeyGen* can generate private keys; and 2) I_2: identities for which Priv can evaluate the hash value. We note that (ABO) IB-EHPS inherently relies on the partitioning strategy. Suppose that there is an identity id belongs to the intersection of I_1 and I_2, then given (PP, x) one can compute the corresponding w such that $(x, w) \in \mathsf{R}_{\mathrm{PP}}$ by itself as follows: first computes $y = \mathsf{H}_{mpk}(id, x)$ via $\mathsf{Priv}(msk^*, id, x)$, then obtains a private key sk of id via $\mathsf{KeyGen}(msk^*, id)$ and uses it to extract w via $\mathsf{Ext}(sk, x, y)$. This contradicts the one-wayness of R_{PP}. This feature of IB-EHPS makes it particularly well-suited to yield IBE schemes whose provable security follows the partitioning strategy [15, 30].
2. In ABO EHPS, the ABO hashing mode is defined with respect to a tag t^*, which in turn is the hash value of x^* for some target collision resistant (TCR)

hash function. Hence the correctness of the ABO hashing mode is related to the TCR hash function. In our case, we define the ABO hashing mode directly with respect to x^*. We do so out of two reasons. One is that for an abstract paradigm it is more preferable to minimize the dependence on other primitives, while the other is that the proof for the transformation from IB-EHPS to CCA-secure IBE would be rather clean and simple. Nevertheless, TCR hash function turns out to be a useful tool when instantiating EHPS/IB-EHPS from concrete number-theoretic assumptions.

4 Generic Constructions from Identity-Based Extractable Hash Proofs

In this section, we present the generic constructions of IBE from (ABO) IB-EHPS. As a warm up, we first show the transformation from IB-EHPS to adaptive-identity CPA-secure IBE, then the transformation from ABO IB-EHPS to adaptive-identity CCA-secure IBE. Before going into details, we first give an intuitive explanation of the constructions from IB-EHPS to IBE with respect to the underlying relation. Suppose that the binary relation of an IB-EHPS is R_{pp} and (x, w) is a tuple that belongs to R_{pp}. The overall construction is: first encrypt (or commit to) a fresh DEM key (the corresponding witness is w) which is in turn used to encrypt the actual message, and then provide an identity-based extractable hash proof $y = \mathsf{H}_{mpk}(id, x)$ (which is also zero-knowledge) of the key. The ciphertext is of the form (x, y). In fact, such an approach was used implicitly in the PKE schemes based on computational assumptions and its connection to the Rackoff-Simon paradigm [27] was made explicit in [31]. Here we make its link to the underlying relation R clear. It is useful to note the distinguished feature in the construction from IB-EHPS to IBE that the value w (used to compute the session key) is uniquely determined by PP and the random coins used by SampR. This explains why IB-EHPS cannot encompass the IBE schemes whose session keys are related to the identity, e.g. Boneh-Franklin IBE [7].

4.1 IND-ID-CPA Secure IBE

Starting from an IB-EHPS (SetupExt, SetupHash, Pub, Priv, Ext, KeyGen, KeyGen*) associating with a one-way relation instance $(X, W, \mathrm{PP}, \mathsf{R}_{\mathrm{pp}})$ and a hash family $\mathbf{H} = (\mathsf{H}, MPK, I, X, Y)$, we construct an IB-KEM as follows:

- $\mathsf{Setup}(\kappa)$: same as $\mathsf{SetupExt}(\mathrm{PP})$ in IB-EHPS.
- $\mathsf{KeyGen}(msk, id)$: same as $\mathsf{KeyGen}(msk, id)$ in IB-EHPS.
- $\mathsf{Encap}(id)$: samples $(x, w) \leftarrow \mathsf{SampR}(r)$, computes $y = \mathsf{Pub}(mpk, id, r)$, and returns a ciphertext $c = (x, y)$ and a DEM key $k = \mathsf{F}(w)$,
- $\mathsf{Decap}(sk, c)$: parses c as (x, y), and returns $\mathsf{F}(\mathsf{Ext}(sk, x, y))$.

The functionality of the above IB-KEM follows readily from the correctness of the extraction mode. For the security, we have the following theorem whose proof appears in the full version of this paper.

Theorem 4.1 *If* R_{pp} *is a one-way relation and the IB-EHPS is* (Q_e, δ)*-well-partition, then the above IB-KEM is IND-ID-CPA secure as long as* δ *is non-negligible.*

4.2 IND-ID-CCA Secure IBE

Starting from an ABO IB-EHPS (SetupExt, SetupABO, Pub, Priv, Verify, Verify*, Ext, Ext*, KeyGen, KeyGen*) for a one-way relation instance $(X, W, \mathrm{PP}, \mathsf{R}_{\mathrm{pp}})$ and a hash family $\mathbf{H} = (\mathsf{H}, MPK, I, X, Y)$, we construct an IB-KEM as follows:

- $\mathsf{Setup}(\kappa)$: same as $\mathsf{SetupExt}(\mathrm{PP}, \mathrm{SP})$ in ABO IB-EHPS.
- $\mathsf{KeyGen}(msk, id)$: same as $\mathsf{KeyGen}(msk, id)$ in ABO IB-EHPS.
- $\mathsf{Encap}(id)$: samples $(x, w) \leftarrow \mathsf{SampR}(r)$, computes $y = \mathsf{Pub}(mpk, id, r)$, and returns a ciphertext $c = (x, y)$ and a associated DEM key $k = \mathsf{F}(w)$.
- $\mathsf{Decap}(sk, c)$: parses c as (x, y), and returns $\mathsf{F}(\mathsf{Ext}(sk, x, y))$.

The functionality of the above IB-KEM follows readily from the correctness of the extraction mode. For the security, we have the following theorem.

Theorem 4.2 *If* R_{pp} *is a one-way relation and the ABO IB-EHPS is* (Q_e, Q_d, δ)*-well-partition, then the above IB-KEM is IND-ID-CCA secure as long as* δ *is non-negligible.*

Proof. To establish the IND-ID-CCA security based on the one-wayness of relation R_{pp}, we proceed via a sequence of games. Let A be the event that $\mathcal{A}$ wins in Game CCA, and A_i be the event that $\mathcal{A}$ wins in Game i.

Game CCA. Given PP and SP, $\mathcal{CH}$ plays with $\mathcal{A}$ in the following game.
Setup: $\mathcal{CH}$ generates $(mpk, msk) \rightarrow \mathsf{SetupExt}(\mathrm{PP}, \mathrm{SP})$ and gives mpk to $\mathcal{A}$.
Phase 1 - Private key queries: When $\mathcal{A}$ submits a private key query $\langle id \rangle$, $\mathcal{CH}$ responds with $\mathsf{KeyGen}(msk, id)$.
Phase 1 - Decapsulation queries: When $\mathcal{A}$ submits a decapsulation query $\langle id, c = (x, y) \rangle$, $\mathcal{CH}$ extracts $sk = \mathsf{KeyGen}(msk, id)$ and responds with $\mathsf{Ext}(sk, x, y)$.
Challenge: When $\mathcal{A}$ submits a target identity id^* such that id^* did not appear in any private key query in Phase 1, $\mathcal{CH}$ samples $(x^*, w^*) \leftarrow \mathsf{SampR}(r^*)$ and computes $y^* = \mathsf{H}_{mpk}(id^*, x^*)$ via $\mathsf{Pub}(mpk, id^*, r^*)$, then sets $k_0^* = \mathsf{F}(w^*)$ and $k_1^* \xleftarrow{R} \{0,1\}^l$. $\mathcal{CH}$ picks $\beta \xleftarrow{R} \{0,1\}$ and returns (x^*, y^*, k_β^*) to $\mathcal{A}$ as the challenge.
Phase 2 - Private key queries: Same as in Phase 1 except that the query $\langle id^* \rangle$ is not allowed.
Phase 2 - Decapsulation queries: Same as in Phase 1 except that the query $\langle id^*, x^*, y^* \rangle$ is not allowed.
Guess: $\mathcal{A}$ outputs its guess β' for β and wins if $\beta' = \beta$.

$\mathcal{A}$'s view in Game CCA is identical to the standard IND-ID-CCA game, thus

$$\Pr[A] = 1/2 + \mathsf{Adv}_{\mathcal{A}}^{\mathrm{CCA}}(\kappa) \tag{5}$$

Game 0. Given PP and PP, $\mathcal{CH}$ plays with $\mathcal{A}$ in the following game.
Setup: Same as in Sub-Game 0 for ABO IB-EHPS.
Phase 1 - Private key queries: Same as in Sub-Game 0 for ABO IB-EHPS.
Phase 1 - Decapsulation queries: Same as in Sub-Game 0 for ABO IB-EHPS.
Challenge. Same as the Phase Middle in Sub-Game 0 for ABO IB-EHPS.
Phase 2 - Private key queries: Same as in Sub-Game 0 for ABO IB-EHPS.
Phase 2 - Decapsulation queries: Same as in Sub-Game 0 for ABO IB-EHPS.
Guess: $\mathcal{A}$ outputs its guess β' for β and wins if $\beta = \beta'$.

Observe that $\mathcal{A}$'s view in Game 0 is essentially the same as in Sub-Game 0. There are two differences between Game 0 and Game CCA: 1) in Game 0 the challenger samples (x^*, w^*) at the setup phase while in Game CCA the challenger samples (x^*, w^*) at the challenge phase. It is easy to see that this difference is invisible in $\mathcal{A}$'s view. 2) in Game 0 the challenger will return $\perp$ when encountering a decapsulation query with $x = x^*$ in Phase 1. We conclude that $\mathcal{A}$'s view in Game 0 is identical to Game CCA if the event that in Phase 1 $\mathcal{A}$ submits a decapsulation query with $x = x^*$ does not happen, whose probability is at most $Q_d/|X|$. Thus we have $|\Pr[A_0] - \Pr[A]| \le Q_d/|X|$. Since $Q_d = \mathsf{poly}(\kappa)$, we have that $Q_d/|X| = \mathsf{negl}(\kappa)$ and hence $\Pr[A_0] \approx \Pr[A]$. We claim that $\mathsf{Adv}_{\mathcal{A}}^{\mathrm{CCA}} = \mathsf{negl}(\kappa)$ based on the one-wayness of R_{pp}. Suppose that there exists an algorithm $\mathcal{A}$ whose advantage against the CCA-security of IB-KEM is not negligible in κ, then we can construct an adversary $\mathcal{B}$ breaking the pseudo-randomness of F, which is sufficient to prove CCA-security under the one-wayness of R_{pp}.

Game 1. $\mathcal{B}$ receives a challenge instance (PP, x^*, k^*), where x^* is picked from the tuple $(x^*, w^*) \in \mathsf{R}_{\mathrm{pp}}$ generated by $\mathsf{SampR}(r^*)$ and k^* is either $\mathsf{F}(w^*)$ or randomly picked from $\{0,1\}^l$. $\mathcal{B}$ is asked to determine $k^* = \mathsf{F}(w^*)$ or $k^* \xleftarrow{R} \{0,1\}^l$. $\mathcal{B}$ plays with $\mathcal{A}$ in the following game.
Setup: $\mathcal{B}$ operates as $\mathcal{CH}$ does in Sub-Game 1 for ABO IB-EHPS except that $\mathcal{B}$ skips the sampling step.
Phase 1 - Private key queries: $\mathcal{B}$ operates as $\mathcal{CH}$ does in Sub-Game 1 for ABO IB-EHPS.
Phase 1 - Decapsulation queries: $\mathcal{B}$ operates as $\mathcal{CH}$ processes the decapsulation queries in Sub-Game 1 for ABO IB-EHPS.
Challenge: When $\mathcal{A}$ submits a target identity id^* on the condition that id^* did not appear in any private key query in Phase 1, $\mathcal{B}$ computes $y^* = \mathsf{H}_{mpk}(id^*, x^*)$ via $\mathsf{Priv}(msk^*, id^*, x^*)$, then instead of creating the challenge by explicitly generating a random bit β, it sends (x^*, y^*, k^*) to $\mathcal{A}$ as the challenge.
Phase 2 - Private key queries: $\mathcal{B}$ operates as $\mathcal{CH}$ does in Sub-Game 1 for ABO IB-EHPS.
Phase 2 - Decapsulation queries: $\mathcal{B}$ operates as $\mathcal{CH}$ processes the decapsulation queries in Sub-Game 1 for ABO IB-EHPS.
Guess: $\mathcal{A}$ outputs its guess β' for β and $\mathcal{B}$ forwards β' to its own challenger.

Observe that $\mathcal{A}$'s view in Game 1 is essentially the same as Sub-Game 1. Since the underlying ABO IB-EHPS is (Q_e, Q_d, δ)-well-partition, then we conclude that $\mathcal{B}$ can break the pseudo-randomness of F with advantage:

$$\mathsf{Adv}_{\mathcal{B}} = |(1-\delta)/2 + \delta \cdot \Pr[A_0] - 1/2| = \delta \cdot |\Pr[A_0] - 1/2| \approx \delta \cdot \mathsf{Adv}_{\mathcal{A}}^{\mathrm{CCA}}$$

If δ is non-negligible, then the above IB-KEM is IND-ID-CCA secure based on the one-wayness of R_{pp}. This proves the theorem. □

5 Instantiations of IB-EHPS

ABO IB-EHPS for the BDH Relation

We first run $\mathsf{SampS}(\kappa)$ to generate an instance $(X, W, \mathrm{PP}, \mathsf{R}_{\mathrm{pp}})$ of the BDH relation defined in Section 2.4, where $X = \mathbb{G}$, $W = \mathbb{G}_T$, $\mathrm{PP} = (g, g^a, g^b)$. $\mathbb{G}$ and $\mathbb{G}$ are two groups of prime order p and equipped with bilinear map $e : \mathbb{G} \times \mathbb{G} \to \mathbb{G}_T$, and g is a random generator of $\mathbb{G}$. The random coins SP consumed by SampS consists of $(a, b) \in \mathbb{Z}_p^2$ and the randomness used to pick g. For the choice of $\mathbf{H} = (\mathsf{H}, MPK, I, X, Y)$, let $MPK = \mathbb{G}^{5+n}$ for some integer n, $I = \{0,1\}^n$, $Y = \mathbb{G}^2$. We write $\overline{u}$ for a n-length vector $(u_1, \ldots, u_n)$ hereafter. We also need a TCR hash function TCR from $\mathbb{G}$ to $\mathbb{Z}_p$. For $mpk = (g, g_1', g_1, g_2, u_0, \overline{u}) \in MPK$, we define:

$$\mathsf{H}_{mpk}(id, x) = (y_1, y_2) := ((g_1^t g_1')^r, F(id)^r)$$

Here $x = g^r$, $t = \mathsf{TCR}(x)$, and $F(id) = u_0 \prod_{i=1}^n u_i^{id_i}$ (id_i denotes the i-th bit of identity id) is known as Waters-hash.

Extraction Mode

- $\mathsf{SetupExt}(\mathrm{PP}, \mathrm{SP})$: sets $g = g$, $g_1 = g^a$, $g_2 = g^b$ (from PP), picks $g_1', u_0 \xleftarrow{R} \mathbb{G}$, $\overline{u} \xleftarrow{R} \mathbb{G}^n$, and returns $mpk = (g, g_1', g_1, g_2, u_0, \overline{u})$, $msk = a$ (from SP).
- $\mathsf{Pub}(mpk, id, r)$: returns $(y_1, y_2) = ((g_1^t g_1')^r, F(id)^r)$ where $t = \mathsf{TCR}(g^r)$.
- $\mathsf{KeyGen}(msk, id)$: picks $s \xleftarrow{R} \mathbb{Z}_p$, and returns $sk = (g_2^a F(id)^s, g^s)$.
- $\mathsf{Verify}(id, sk, x, y)$: parses y as (y_1, y_2), computes $t = \mathsf{TCR}(x)$, if $e(x, g_1^t g_1') = e(g, y_1)$ and $e(x, F(id)) = e(g, y_2)$ returns 1, else returns 0.
- $\mathsf{Ext}(sk, x, y)$: parses sk as (sk_1, sk_2) and y as (y_1, y_2), if $\mathsf{Verify}(id, sk, x, y) = 1$ then returns $e(x, sk_1)/e(y_2, sk_2)$, else returns $\perp$.

The correctness of extraction follows from the following simple calculation:

$$y = ((g_1^t g_1')^r, F(I)^r) = \mathsf{H}_{mpk}(I, u) \Longrightarrow e(x, g_2^a F(I)^s)/e(y_2, g^s) = e(g_1, g_2)^r$$

ABO Hashing Mode

- $\mathsf{SetupABO}(\mathrm{PP}, x^*)$: sets $g = g$, $g_1 = g^a$, $g_2 = g^b$ (from PP), picks $d \xleftarrow{R} \mathbb{Z}_p$, computes $t^* = \mathsf{TCR}(x^*)$, sets $g_1' = g_1^{-t^*} g^d$; sets $m = 2(Q_e + Q_d)$, and chooses $k \xleftarrow{R} [n+1]$; picks $\alpha' \xleftarrow{R} \mathbb{Z}_m$, $\overline{\alpha} \xleftarrow{R} \mathbb{Z}_m^n$, $\beta' \xleftarrow{R} \mathbb{Z}_p$, $\overline{\beta} \xleftarrow{R} \mathbb{Z}_p^n$, sets $u_0 = g_2^{p-km+\alpha'} g^{\beta'}$ and $u_i = g_2^{\alpha_i} g^{\beta_i}$ for $1 \le i \le n$; returns $mpk = (g, g_1', g_1, g_2, u_0, \overline{u})$, $msk^* = (t^*, d, \alpha', \overline{\alpha}, \beta', \overline{\beta})$. For ease of narration we define two functions, namely $J(id) = (p - mk) + \alpha' + \sum \alpha_i id_i$ and $K(id) = \beta' + \sum \beta_i id_i$. Hence $F(id)$ is essentially of the form $g_2^{J(id)} g^{K(id)}$. The structure of mpk implicitly splits the whole identity space I into I_1 and I_2. For an identity $id \in I$, if $J(id) \neq p$ it belongs to I_1, otherwise it belongs to I_2.

- $\mathsf{Priv}(msk^*, id, x)$: if $id \in I_2$ and $x = x^*$ returns $(y_1, y_2) = ((x^*)^d, (x^*)^{K(id)})$, else returns $\perp$.
- $\mathsf{Verify}^*(id, msk^*, x, y)$: same as Verify.
- $\mathsf{KeyGen}^*(msk^*, id)$: if $id \notin I_1$ returns $\perp$, else picks $s \xleftarrow{R} \mathbb{Z}_p$ and returns

$$sk = (sk_1, sk_2) = \left(g_1^{\frac{-K(id)}{J(id)}} F(id)^s, g_1^{\frac{-1}{J(id)}} g^s \right)$$

- $\mathsf{Ext}^*(msk^*, x, y)$: parses y as (y_1, y_2), if $\mathsf{Verify}^*(id, msk^*, x, y) = 1$ and $t \neq t^*$ then returns $e((y_1/x^d)^{1/(t-t^*)}, g_2)$ where $t = \mathsf{TCR}(x)$, else returns $\perp$.

The correctness of ABO hashing mode follows from the following two facts:

1. If $id \in I_2$ and $x = x^*$, we have $\mathsf{Priv}(msk^*, id, x^*) = ((x^*)^d, (x^*)^{K(id)}) = ((g^d)^{r^*}, (g^{K(id)})^{r^*}) = ((g_1^{t^*} g_1')^{r^*}, F(id)^{r^*}) = \mathsf{H}_{mpk}(id, x^*)$.
2. If $x \neq x^*$, then $((g_1^t g_1')^r, F(id)^r) = \mathsf{H}_{mpk}(id, x) \Longrightarrow e((y_1/x^d)^{1/(t-t^*)}, g_2) = e(g_1, g_2)^r)$, where $t = \mathsf{TCR}(x)$. The property of TCR ensures that $t = t^*$ holds with overwhelming probability when $x = x^*$.

The indistinguishability is established from the following two facts:

1. The distribution of mpk in both modes are identical.
2. For any mpk and any identity $id \in I_1$, the output of $\mathsf{KeyGen}(msk, id)$ and $\mathsf{KeyGen}^*(msk^*, id)$ are statistically indistinguishable. To see this, let $\tilde{s} = s - a/J(id)$, we have

$$sk_1 = g_1^{\frac{-K(id)}{J(id)}} F(id)^s = g_1^{\frac{-K(id)}{J(id)}} (g_2^{J(id)} g^{K(id)})^s = g_2^a F(id)^{s - \frac{a}{J(id)}} = g_2^a F(id)^{\tilde{s}}$$

$$sk_2 = g_1^{\frac{-1}{J(id)}} g^s = g^{s - \frac{a}{J(id)}} = g^{\tilde{s}}$$

Since s is uniform in $\mathbb{Z}_p$, then $\tilde{s}$ is also uniform in $\mathbb{Z}_p$. Thereby the distribution of $\mathsf{KeyGen}(msk, id)$ and $\mathsf{KeyGen}^*(msk^*, id)$ are identical.

Follow the same analysis in [23], the above IB-EHPS is (Q_e, Q_d, δ)-well-partition, where $\delta \geq \frac{1}{8(n+1)(Q_e+Q_d)}$. Applying the transformation in Section 4.2 to this ABO IB-EHPS, we obtain an IB-KEM (see Fig. 1), which can be viewed as a variant of the IB-KEM in [23]. Combining theorem 4.2, we conclude that this IB-KEM is IND-ID-CCA secure based the CBDH assumption.

ABO IB-EHPS for the mBDH Relation

Based on the modified bilinear Diffie-Hellman relation $\mathsf{R}_{\mathsf{pp}}^{\mathsf{mbdh}}$, we can create an ABO IB-EHPS whose Ext and Ext^* algorithms implement the "implicitly rejection" idea. Applying the transformation from Section 4.2 to the ABO IB-EHPS, we obtain a CCA-secure IB-KEM based on the mBDH assumption (see Fig. 2), which is a variant of the IB-KEM in [24].

Setup(κ):	Extract(msk, I)
$g, g_1', g_2, u_0 \xleftarrow{R} \mathbb{G}, \overline{u} \xleftarrow{R} \mathbb{G}^n$; $a \xleftarrow{R} \mathbb{Z}_p$	$s \xleftarrow{R} \mathbb{Z}_p$
$F(id) = u_0 \prod_{i=1}^{n} u_i^{id_i}$	$sk = (g_2^a F(id)^s, g^s)$
$mpk = (g, g_1 = g^a, g_1', g_2, u_0, \overline{u})$; $msk = a$	return sk
return (mpk, msk)	
Encap(id)	Decap(sk, c)
$r \xleftarrow{R} \mathbb{Z}_p$, $x \leftarrow g^r$	parse sk as (sk_1, sk_2), c as (x, y_1, y_2)
$t \leftarrow \mathsf{TCR}(x)$	$t = \mathsf{TCR}(x)$
$y_1 = (g_1^t g_1')^r$, $y_2 = F(id)^r$	If $e(x, g_1^t g_1') \neq e(g, y_1)$ or
$k \leftarrow \mathsf{GL}(e(g_1, g_2)^r)$	$e(x, F(id)) \neq e(g, y_2)$, then return $\perp$
return $c = (x, y_1, y_2)$	else return $\mathsf{GL}(e(x, sk_1)/e(y_2, sk_2))$

Fig. 1. An IND-ID-CCA secure IB-KEM based on BDH (variant of [23])

Setup(κ):	Extract(msk, I)
$g, g_2, u_0 \xleftarrow{R} \mathbb{G}, \overline{u} \xleftarrow{R} \mathbb{G}^n$; $a \xleftarrow{R} \mathbb{Z}_p$	$s \xleftarrow{R} \mathbb{Z}_p$
$F(id) = u_0 \prod_{i=1}^{n} u_i^{id_i}$	$sk = (g_2^a F(id)^s, g^{-s}, g_2^s)$
$mpk = (g, g_1 = g^a, g_2, u_0, \overline{u})$; $msk = a$	return sk
return (mpk, msk)	
Encap(id)	Decap(sk, c)
$r \xleftarrow{R} \mathbb{Z}_p$, $x \leftarrow g^r$, $t \leftarrow \mathsf{TCR}(x)$	parse sk as (sk_1, sk_2, sk_3), c as (x, y)
$y = (F(id) g_2^t)^r$, $k \leftarrow \mathsf{GL}(e(g_1, g_2)^r)$	$t = \mathsf{TCR}(x)$
return $c = (x, y)$	return $\mathsf{GL}(e(x, sk_1 \cdot sk_3^t) \cdot e(y, sk_2))$

Fig. 2. An IND-ID-CCA secure IB-KEM based on mBDH (variant of [24])

6 Extension

We also put forward the notion of dual ABO IB-EHPS, which can be viewed as a special case of ABO IB-EHPS whose I_2 contains a single point id^*. The term "dual ABO" reflects that the algorithm Priv returns $\mathsf{H}_{mpk}(id, x)$ only on the point that $id = id^*$ and $x = x^*$. The dual ABO IB-EHPS turns out to be a useful paradigm for constructing selective-identity CCA-secure IB-KEM. In particular, the instantiation of dual ABO IB-EHPS from the BDH relation serves as a clarification of all the known selective-identity CCA-secure IB-KEMs [11,14,19] based on the CBDH assumption. Due to space limit, we include this part in the full version of this paper.

Acknowledgment. The authors greatly thank the anonymous ACNS 2012 reviewers for many helpful and insightful comments. This work is supported by the Foundation of SKLOIS of Chinese Academy of Sciences, the Strategic Priority Research Program of Chinese Academy of Sciences under Grant No. XDA06010701, the National 973 Program of China under Grant No. 2011CB302400, and the National Natural Science Foundation of China under Grant No. 60970152. The second and the fourth authors are partially supported by the National Natural Science Foundation of China under Grant Nos. 61033014, 60970110, 60972034 and Asia 3 Foresight Program under Grant No. 61161140320.

References

1. Agrawal, S., Boneh, D., Boyen, X.: Efficient Lattice (H)IBE in the Standard Model. In: Gilbert, H. (ed.) EUROCRYPT 2010. LNCS, vol. 6110, pp. 553–572. Springer, Heidelberg (2010)
2. Bellare, M., Rogaway, P.: Random oracles are practical: A paradigm for designing efficient protocols. In: ACM Conference on Computers and Communication Security, pp. 62–73 (1995)
3. Bentahar, K., Farshim, P., Malone-Lee, J., Smart, N.P.: Generic constructions of identity-based and certificateless kems. Journal of Cryptology 21(2), 178–199 (2008)
4. Boneh, D., Boyen, X.: Efficient Selective-ID Secure Identity-Based Encryption Without Random Oracles. In: Cachin, C., Camenisch, J.L. (eds.) EUROCRYPT 2004. LNCS, vol. 3027, pp. 223–238. Springer, Heidelberg (2004)
5. Boneh, D., Canetti, R., Halevi, S., Katz, J.: Chosen-ciphertext security from identity-based encryption. SIAM J. Comput. 36(5), 1301–1328 (2007)
6. Boneh, D., Franklin, M.: Identity-Based Encryption from the Weil Pairing. In: Kilian, J. (ed.) CRYPTO 2001. LNCS, vol. 2139, pp. 213–229. Springer, Heidelberg (2001)
7. Boneh, D., Franklin, M.K.: Identity-based encryption from the weil pairing. SIAM Journal on Computation 32, 586–615 (2003)
8. Canetti, R., Halevi, S., Katz, J.: Chosen-Ciphertext Security from Identity-Based Encryption. In: Cachin, C., Camenisch, J.L. (eds.) EUROCRYPT 2004. LNCS, vol. 3027, pp. 207–222. Springer, Heidelberg (2004)
9. Cash, D., Hofheinz, D., Kiltz, E., Peikert, C.: Bonsai Trees, or How to Delegate a Lattice Basis. In: Gilbert, H. (ed.) EUROCRYPT 2010. LNCS, vol. 6110, pp. 523–552. Springer, Heidelberg (2010)
10. Cash, D., Kiltz, E., Shoup, V.: The Twin Diffie-Hellman Problem and Applications. In: Smart, N.P. (ed.) EUROCRYPT 2008. LNCS, vol. 4965, pp. 127–145. Springer, Heidelberg (2008)
11. Chen, Y., Chen, L., Zhang, Z.: CCA secure IB-KEM from the computational bilinear diffie-hellman assumption in the standard model (2011), `http://eprint.iacr.org/2011/593`
12. Cocks, C.: An Identity Based Encryption Scheme Based on Quadratic Residues. In: Honary, B. (ed.) Cryptography and Coding 2001. LNCS, vol. 2260, pp. 360–363. Springer, Heidelberg (2001)
13. Cramer, R., Shoup, V.: Universal Hash Proofs and a Paradigm for Adaptive Chosen Ciphertext Secure Public-Key Encryption. In: Knudsen, L.R. (ed.) EUROCRYPT 2002. LNCS, vol. 2332, pp. 45–64. Springer, Heidelberg (2002)
14. Galindo, D.: Chosen-Ciphertext Secure Identity-Based Encryption from Computational Bilinear Diffie-Hellman. In: Joye, M., Miyaji, A., Otsuka, A. (eds.) Pairing 2010. LNCS, vol. 6487, pp. 367–376. Springer, Heidelberg (2010)
15. Gentry, C.: Practical Identity-Based Encryption Without Random Oracles. In: Vaudenay, S. (ed.) EUROCRYPT 2006. LNCS, vol. 4004, pp. 445–464. Springer, Heidelberg (2006)

16. Gentry, C., Peikert, C., Vaikuntanathan, V.: Trapdoors for hard lattices and new cryptographic constructions. In: Proceedings of the 40th Annual ACM Symposium on Theory of Computing, STOC, pp. 197–206. ACM (2008)
17. Goldreich, O., Levin, L.A.: A hard-core predicate for all one-way functions. In: Proceedings of the Twenty-First Annual ACM Symposium on Theory of Computing, STOC, pp. 25–32. ACM (1989)
18. Hanaoka, G., Kurosawa, K.: Efficient Chosen Ciphertext Secure Public Key Encryption under the Computational Diffie-Hellman Assumption. In: Pieprzyk, J. (ed.) ASIACRYPT 2008. LNCS, vol. 5350, pp. 308–325. Springer, Heidelberg (2008)
19. Haralambiev, K., Jager, T., Kiltz, E., Shoup, V.: Simple and Efficient Public-Key Encryption from Computational Diffie-Hellman in the Standard Model. In: Nguyen, P.Q., Pointcheval, D. (eds.) PKC 2010. LNCS, vol. 6056, pp. 1–18. Springer, Heidelberg (2010)
20. Hofheinz, D., Kiltz, E.: Practical Chosen Ciphertext Secure Encryption from Factoring. In: Joux, A. (ed.) EUROCRYPT 2009. LNCS, vol. 5479, pp. 313–332. Springer, Heidelberg (2009)
21. Kiltz, E.: Chosen-Ciphertext Security from Tag-Based Encryption. In: Halevi, S., Rabin, T. (eds.) TCC 2006. LNCS, vol. 3876, pp. 581–600. Springer, Heidelberg (2006)
22. Kiltz, E.: Chosen-Ciphertext Secure Key-Encapsulation Based on Gap Hashed Diffie-Hellman. In: Okamoto, T., Wang, X. (eds.) PKC 2007. LNCS, vol. 4450, pp. 282–297. Springer, Heidelberg (2007)
23. Kiltz, E., Galindo, D.: Direct Chosen-Ciphertext Secure Identity-Based Key Encapsulation Without Random Oracles. In: Batten, L.M., Safavi-Naini, R. (eds.) ACISP 2006. LNCS, vol. 4058, pp. 336–347. Springer, Heidelberg (2006)
24. Kiltz, E., Vahlis, Y.: CCA2 Secure IBE: Standard Model Efficiency through Authenticated Symmetric Encryption. In: Malkin, T. (ed.) CT-RSA 2008. LNCS, vol. 4964, pp. 221–238. Springer, Heidelberg (2008)
25. Micali, S., Rackoff, C., Sloan, B.: The notion of security for probabilistic cryptosystems. SIAM J. Comput. 17(2), 412–426 (1988)
26. Peikert, C., Waters, B.: Lossy trapdoor functions and their applications. In: STOC 2008, pp. 187–196 (2008)
27. Rackoff, C., Simon, D.R.: Non-interactive Zero-Knowledge Proof of Knowledge and Chosen Ciphertext Attack. In: Feigenbaum, J. (ed.) CRYPTO 1991. LNCS, vol. 576, pp. 433–444. Springer, Heidelberg (1992)
28. Shamir, A.: Identity-Based Cryptosystems and Signature Schemes. In: Blakely, G.R., Chaum, D. (eds.) CRYPTO 1984. LNCS, vol. 196, pp. 47–53. Springer, Heidelberg (1985)
29. Waters, B.: Efficient Identity-Based Encryption Without Random Oracles. In: Cramer, R. (ed.) EUROCRYPT 2005. LNCS, vol. 3494, pp. 114–127. Springer, Heidelberg (2005)
30. Waters, B.: Dual System Encryption: Realizing Fully Secure IBE and HIBE under Simple Assumptions. In: Halevi, S. (ed.) CRYPTO 2009. LNCS, vol. 5677, pp. 619–636. Springer, Heidelberg (2009)
31. Wee, H.: Efficient Chosen-Ciphertext Security via Extractable Hash Proofs. In: Rabin, T. (ed.) CRYPTO 2010. LNCS, vol. 6223, pp. 314–332. Springer, Heidelberg (2010)

On Structural Signatures for Tree Data Structures

Kai Samelin[1,*], Henrich C. Pöhls[2,**], Arne Bilzhause[2], Joachim Posegga[2], and Hermann de Meer[1]

[1] Chair of Computer Networks and Computer Communication
[2] Chair of IT-Security
Institute of IT-Security and Security Law (ISL), University of Passau, Germany
{ks,hcp,ab,jp}@sec.uni-passau.de, demeer@uni-passau.de

Abstract. In this paper, we present new attacks on the redactable signature scheme introduced by *Kundu* and *Bertino* at VLDB '08. This extends the work done by *Brzuska* et al. at ACNS '10 and *Samelin* et al. at ISPEC '12. The attacks address unforgeability, transparency and privacy. Based on the ideas of *Kundu* and *Bertino*, we introduce a new provably secure construction. The corresponding security model is more flexible than the one introduced by *Brzuska* et al. Moreover, we have implemented the schemes introduced by *Brzuska* et al. and *Kundu* and *Bertino*. The practical evaluation shows that schemes with a quadratic complexity become unuseable very fast.

1 Introduction

A redactable signature scheme (RSS) allows a third party, we name *sanitizer*, to redact contents of a signed document $m = m[1]||\dots||m[n]$ to generate a sanitized version of the document, which signature is still valid. This action can be performed without involvement of the original signer and without knowing any private keys. In more detail, a RSS allows to replace a block $m[i] \in m$ with $\emptyset$. Thus, a redaction leaves a blinded document m', where $m' = m \setminus m[i]$. Still, a third party is able to verify that all received blocks and their ordering are authentic. *Kundu* and *Bertino* were the first to apply this paradigm to tree-structured data [12]. In this paper, we present new attacks on the scheme introduced by *Kundu* and *Bertino* [12]. This extends the work done by *Brzuska* et al. [5] and *Samelin* et al. [20]. We introduce a provably secure scheme based on *Kundu* and *Bertino*'s initial idea. We have implemented the first version of *Kundu* and *Bertino*'s scheme [12] and the secure scheme by *Brzuska* et al. [5] and provide a detailed performance analysis. Our construction's performance is comparable to the one introduced in [12] by *Kundu* and *Bertino*: the runtime and storage complexity is in $\mathcal{O}(n)$, if one uses a RSS for lists, that is also in $\mathcal{O}(n)$ for both

* Is supported by "Regionale Wettbewerbsfähigkeit und Beschäftigung", Bayern, 2007-2013 (EFRE) as part of the SECBIT project. (http://www.secbit.de)
** Is funded by BMBF (FKZ:13N10966) and ANR as part of the ReSCUeIT project.

F. Bao, P. Samarati, and J. Zhou (Eds.): ACNS 2012, LNCS 7341, pp. 171–187, 2012.

metrics. The used RSS must protect the order among its elements. As an example, the scheme introduced in [19] achieves this. Other existing provably secure schemes for trees have a runtime and storage complexity of $\mathcal{O}(n^2)$ [5].

State of the Art. The term RSS was coined in 2002 by *Johnson* et al. in [10]. In the same year, *Steinfeld* and *Bull* introduced the same concept as "Content Extraction Signatures" [21]. Since then, RSSs have been extended to tree-structured data [5,12,13] and to arbitrary graphs [15]. *Samelin* et al. introduced the concept of independently redactable structure in [20].

A related concept are Sanitizable Signature Schemes (SSS), introduced by *Ateniese* et al. in [3]. In a SSS, the sanitizer does not delete blocks, but can modify $m[i]$ into an arbitrary string $m[i]' \in \{0,1\}^*$ [6]. It is possible to restrict sanitizers in SSSs to certain values. This is a well-known field; refer to [7,11] for related work. Approaches to merge SSSs and tree-structured data have also been published, e.g., in [18]. *Pöhls* et al. show in [18] that RSSs for lists are not suitable for tree-structured documents in certain scenarios. We only discuss RSSs in this paper.

Brzuska et al. defined and formalized a set of desired properties for redactable tree-structured documents in [5]. *Kundu* and *Bertino* showed in [13], that non-private RSS can be attacked using "side-channel" information, i.e., if the position of a redacted block is visible, one may be able to reconstruct some information. They name these type of vulnerability "inference attacks" [13]. Consider the following example clarifying this statement: in a tree-based patient record of a hospital, which has exactly two wards, i.e., cancer and surgery, a sub-tree represents the treatments in each ward. If a patient has been treated in both wards, and one subtree has been removed, an adversary sees $\emptyset$. It can deduce that the patient has been treated in both wards, using the knowledge that two wards exist and the information about the structure of the patient record. This impacts on privacy and is not acceptable in certain scenarios. The scheme introduced in [12] by *Kundu* and *Bertino* was originally proposed to address these problems. However, two attacks on this scheme have been published already: one attacking transparency and privacy [5] and one attacking structural integrity [20].

Our Contribution. The scheme introduced in [12] by *Kundu* and *Bertino* has some very useful properties. In particular, it has a low runtime and storage complexity, i.e., both are $\mathcal{O}(n)$, where n is the number of vertices in a tree $T = (V, E)$. This makes their scheme the fastest one known to the authors for both metrics. Other provably secure and transparent schemes proposed, e.g., [5], have a complexity of $\mathcal{O}(n^2)$, are only useable for lists [8,20], or are only able to quote substrings [2]. However, the scheme introduced in [2] fulfills a very strong privacy notation, i.e., strong context-hiding. This notation prohibits even the signer from deciding if two quotes have been derived from the same source. Our goal is a secure scheme not limited to quoting for trees. The worst-case approximation of the scheme introduced by *Brzuska* et al. in [5] depends on the branching factor, which in the case of a tree with height 1, is $\mathcal{O}(n)$, where n is the number of leafs. As the branching factor is not a constant factor anymore, the growth will be quadratic in n. All other current schemes in $\mathcal{O}(n)$, e.g., the ones introduced in [9] or [13], are susceptible to the attack on privacy introduced by *Brzuska* et al.'s for the scheme by *Kundu* and *Bertino* in [5]. This is due to fact that appending *ordered* random numbers cannot be sufficient to hide redactions [5]. We use the ideas of *Kundu* and *Bertino* [12,13] to derive a new construction which is provably secure and does not inherit their flaws. We contribute by adding new attacks, breaking unforgeability, transparency and privacy. We derive a provably secure construction based on their ideas. Moreover, we have implemented the schemes by *Brzuska* et al. [5] and the scheme introduced in [12] and show that schemes with a quadratic runtime complexity become unuseable very fast.

Outline of the Paper. The rest of the paper is structured as follows. In Sect. 2, we extend the existing definitions required to understand the schemes presented. We will shortly restate the scheme by *Kundu* and *Bertino* [12], along with the new attack vectors. Our new scheme is presented in Sect. 3. We extend the new scheme to add additional useful properties, as proposed in [17] and [20], in Sect. 4. We present our implementations along with the corresponding performance analysis in Sect. 5. Finally, Sect. 6 concludes our work. All formal proofs can be found in the appendices.

2 Preliminaries, the Scheme by *Kundu* and *Bertino* and the Extended Security Model

We start this section by defining the algorithms of an RSS in general. Our notation is inspired by *Brzuska* et al. [5], but extended to allow redaction of non-leafs. The redaction of non-leafs is allowed in the schemes by *Kundu* and *Bertino* and offers more flexibility [12,13]. However, securely allowing this flexibility requires a more sophistacted approach. A thorough discussion is given in Sect. 2.3.

Definition 1 (Redactable Signature Scheme). *A RSS for trees $\mathcal{RSS}_T$ consists of four PPT algorithms: $\mathcal{RSS}_T :=$ (KeyGen, TSign, TVerify, TShare).*

KeyGen. *The key generation algorithm* $(\mathrm{sk}, \mathrm{pk}) \leftarrow \mathsf{KeyGen}(1^\lambda)$ *outputs a key pair consisting of the private key* sk *and the public key* pk*:* $(\mathrm{sk}, \mathrm{pk}) \leftarrow \mathsf{KeyGen}(1^\lambda)$*, on input of a security parameter* λ.

TSign. *The signing algorithm* $\mathsf{TSign}(\mathrm{sk}, T, r, i)$ *takes as input the secret key* sk*, the tree T, a flag $r \in \{0,1\}$ indicating, if the root is allowed to be redacted, and a flag $i = \{0,1\}$ which indicates, if non-leafs can be redacted. We define that a 1 indicates that the corresponding action is allowed. It outputs a signature σ_T over the tree T:* $(T, \sigma_T) \leftarrow \mathsf{TSign}(\mathrm{sk}, T, r, i)$

TVerify. *The algorithm* $\mathsf{TVerify}(T, \mathrm{pk}, \sigma_T)$ *takes as input a tree T, a public key* pk *and the signature σ_T. It outputs a bit $d \in \{0,1\}$, which indicates, if σ_T is a valid ($d = 1$) signature on T under the public key* pk*:* $d \leftarrow \mathsf{TVerify}(T, \mathrm{pk}, \sigma_T)$

TShare. *The algorithm* $\mathsf{TShare}(T, pk, \sigma_T, \mathcal{N})$ *takes as input a tree T, a public key* pk *and signature σ_T, as well as a set of nodes $\mathcal{N} \subseteq T$ to redact. It returns a new tree $T' \leftarrow \textsc{mod}(T, \mathcal{N})$, along with a new signature σ'_T resp. $\perp$ on error, where* MOD *is the function modifying the tree T w.r.t. $\mathcal{N}$, i.e.,*

$T' = T \setminus \mathcal{N}$. Hence, TShare *outputs* $(T', \sigma'_T) \leftarrow \mathsf{TShare}(T, \mathrm{pk}, \sigma_T, \mathcal{N})$*, resp.* $\perp$ *on error. We do not consider, if and how the edges between the nodes are treated, if a redaction takes place. This is the concern of an instantiation of a* $\mathcal{RSS}_T$*. In particular, it depends on the concrete use case, if non-leafs are allowed to be redacted. We want to leave this choice to the signer.*

All algorithms must fulfill the usual correctness requirements. In particular, all genuinely signed trees and trees created from them by TShare must verify. This also implies that TShare always outputs a valid tree. How to actually ensure this and what a valid tree in terms of the signature is, depends on the algorithms and use cases. Note, if any result of the algorithms given is not a tree, the result is treated as $\perp$. We state the extended security model next.

The Extended Security Model. *Brzuska* et al. introduced and formalized some security properties in [5]. Their model was the first rigid approach. It is restrictive by allowing the cutting of leafs of the tree only. We extend their model, since the extended functionality of non-leaf redaction requires an adjustment. We denote the transitive closure of T, w.r.t. TShare, as $span_{\vdash}(T)$, following [8] and [20]. $b \xleftarrow{\$} \{0,1\}$ denotes that b is a random bit, drawn from a uniform distribution. The following security properties have been derived from [5]. As in the original model, the following definitions cater only for the information an adversary can derive from the signature. If obvious redactions took place, that are detectable or reversible using side-channel information, it may be trivial to decide whether something has been redacted.

1. *Unforgeability:* No one should be able to compute a valid signature on a tree T^* verifying under pk outside $span_{\vdash}(T)$, without access to the corresponding secret key sk. This is analogous to the standard unforgeability requirement for signature schemes, as already noted in [5]. A scheme $\mathcal{RSS}_T$ is unforgeable, iff for any efficient (PPT) adversary $\mathcal{A}$, the probability that the game depicted in Fig. 1 returns 1, is negligible (as a function of λ). In this game, the adversary has access to a signing oracle.
2. *Privacy:* No one should be able to gain any knowledge about the unmodified tree from a redacted version without having access to the original. This is similar to the standard indistinguishability notation for encryption schemes [5]. We say that a scheme $\mathcal{RSS}_T$ is private, iff for any efficient (PPT) adversary $\mathcal{A}$, the probability that the game shown in Fig. 3 returns 1, is negligibly close to $\frac{1}{2}$ (as a function of λ). In this game, the adversary has to figure out, which input has been used by the LoR-Oracle.
3. *Transparency:* A third party should not be able to decide whether a signature σ_T of a tree T has been created from scratch or through TShare. In other words, a party who receives a signed tree T cannot tell whether he received a freshly signed tree or a tree which has potentially been modified [5]. We say that a scheme $\mathcal{RSS}_T$ is transparent, iff for any efficient (PPT) adversary $\mathcal{A}$, the probability that the game shown in Fig. 2 returns 1, is negligibly close to $\frac{1}{2}$ (as a function of λ). In this game, the adversary has to figure out, if

Experiment $\mathsf{Unforgeability}_{\mathcal{A}}^{\mathsf{RSS}_T}(\lambda)$
$(pk, sk) \leftarrow \mathsf{KeyGen}(1^\lambda)$
$(T^*, \sigma_T^*) \leftarrow \mathcal{A}^{\mathsf{TSign}(sk,\dots)}(pk)$
let $i = 1, 2, \dots, q$ index the queries to the signing oracle
return 1 iff
$\mathsf{TVerify}(T^*, pk, \sigma_T^*) = 1$ and
$\forall i : 0 < i \leq q,\ T^* \notin span_{\vdash}(T_i)$

Fig. 1. Game for Unforgeability

Experiment $\mathsf{Transparency}_{\mathcal{A}}^{\mathsf{RSS}_T}(\lambda)$
$(pk, sk) \leftarrow \mathsf{KeyGen}(1^\lambda)$
$b \xleftarrow{\$} \{0, 1\}$
$d \leftarrow \mathcal{A}^{\mathsf{TSign}(sk,\cdot),\mathsf{Adapt/Sign}(\dots,sk,b)}(pk)$
where oracle Adapt/Sign for input $T, \mathcal{N}$:
if $\mathcal{N} \not\subseteq T$, return $\perp$
if $b = 0$: $(T, \sigma_T) \leftarrow \mathsf{TSign}(sk, T, r, i)$,
$(T', \sigma'_T) \leftarrow \mathsf{TShare}(T, pk, \sigma_T, \mathcal{N})$
if $b = 1$: $T' \leftarrow \textsc{mod}(T, \mathcal{N})$
$(T', \sigma'_T) \leftarrow \mathsf{TSign}(sk, T', r, i)$,
return (T', σ'_T).
return 1 iff $b = d$

Fig. 2. Game for Transparency

Experiment $\mathsf{Privacy}_{\mathcal{A}}^{\mathsf{RSS}_T}(\lambda)$
$(pk, sk) \leftarrow \mathsf{KeyGen}(1^\lambda)$
$b \xleftarrow{\$} \{0, 1\}$
$d \leftarrow \mathcal{A}^{\mathsf{TSign}(sk,\cdot),\mathsf{LoRAdapt}(\dots,sk,b)}(pk)$
return 1 iff $b = d$

Fig. 3. Game for Privacy

$\mathsf{LoRAdapt}(T_{j,0}, \mathcal{N}_{j,0}, T_{j,1}, \mathcal{N}_{j,1}, r, i, sk, b)$
if $\textsc{mod}(T_{j,0}, \mathcal{N}_{j,0}) \neq \textsc{mod}(T_{j,1}, \mathcal{N}_{j,1})$ return $\perp$
$(T_j, \sigma_{T_j}) \leftarrow \mathsf{TSign}(sk, T_{j,b}, r, i)$
$(T'_j, \sigma'_{T_j}) \leftarrow \mathsf{TShare}(T_{j,b}, pk, \sigma_{T_{j,b}}, \mathcal{N}_{j,b})$
return $(T'_{j,b}, \sigma'_{T,b})$

Fig. 4. LoRAdapt Oracle for Privacy

the signature has been created through TSign or TShare. This encapsulated by the $\mathsf{Adapt/Sign}$-Oracle.

Implications and Seperations. The implications given by *Brzuska* et al. in [5] and [6] do not change: transparency $\Longrightarrow$ privacy, privacy $\not\Rightarrow$ transparency, and unforgeability is independent. To avoid duplicate work, we omit the proofs, as only minor adjustments are required.

2.1 Aggregate Signatures and Bilinear Pairings

Aggregate signatures ($\mathcal{AGG}$) have been introduced by *Boneh* et al. in [4]. The basic idea is as follows: given ℓ signatures, i.e., $\{\sigma_i \mid 0 < i \leq \ell\}$, a $\mathcal{AGG}$ constructs one compressed signature σ_c which contains all signatures σ_i. This allows verifying all given signatures σ_i by verifying σ_c. To construct such a scheme, let $\mathbb{G}_1$ be a cyclic multiplicative group with prime order q, generated by g, i.e., $\mathbb{G}_1 = \langle g \rangle$. Further, let $\mathbb{G}_T$ denote a cyclic multiplicative group with the same prime order q. Let $\hat{e} : \mathbb{G}_1 \times \mathbb{G}_1 \rightarrow \mathbb{G}_T$, where $\mathbb{G}_T = \langle \hat{e}(g, g) \rangle$, be a bilinear map such that:

1. Bilinearity: $\forall u, v \in \mathbb{G}_1 : \forall a, b \in \mathbb{Z}/q\mathbb{Z} : \hat{e}(u^a, v^b) = \hat{e}(u, v)^{ab}$
2. Non-degeneracy: $\exists u, v \in \mathbb{G}_1 : \hat{e}(u, v) \neq 1$
3. Computability: There is an efficient algorithm $\mathcal{A}_{bimap}$ that calculates the mapping $\hat{e}$ for all $u, v \in \mathbb{G}_1$

Definition 2 (The BGLS-Scheme). *The $\mathcal{AGG}$ by* Boneh *et al. [4] (BGLS-Scheme) with public aggregation consists of five efficient algorithms. We will only use one public key, Q, which allows a performance improvement, while making sure that just one signing key is used. Next, we define:*

$$\mathcal{AGG} := (\mathit{AKeyGen}, \mathit{ASign}, \mathit{AVerf}, \mathit{AAgg}, \mathit{AAggVerf})$$

AKeyGen. *The algorithm* KeyGen *outputs the public and private key of the signer, i.e.,* (pk, sk). $\mathrm{sk} \xleftarrow{\$} \mathbb{Z}/q\mathbb{Z}$ *denote the signer's private key. Additionally, let* $\mathcal{H}_k : \{0,1\}^* \to \mathbb{G}_1$ *be an ordinary cryptographic hash-function from the family* $\mathcal{H}_K$*, modeled as a random oracle, and set* $Q \leftarrow g^{\mathrm{sk}}$*, where* $\langle g \rangle = \mathbb{G}_1$*. Set the public parameters and key* $\mathrm{pk} \leftarrow (g, Q, \mathbb{G}_1, \mathbb{G}_T, \mathcal{H}_k, \hat{e})$.

ASign. *The algorithm* ASign *outputs the signature* σ_i *on input of the secret key* sk *and a single string* $m_i \in \{0,1\}^*$: $\sigma_i \leftarrow (\mathcal{H}_k(m_i))^{\mathrm{sk}}$.

AVerf. *To verify a signature* σ_i*, check, if the following equation holds:* $\hat{e}(\sigma_i, g) \stackrel{?}{=} \hat{e}(\mathcal{H}_k(m_i), Q)$.

AAgg. *To aggregate* ℓ *signatures* σ_i *into an aggregated signature* σ_c*, the aggregator computes* $\sigma_c \leftarrow \prod_{i=1}^{\ell} \sigma_i$*, denoted as* AAgg(pk, $\mathcal{S}$)*, where* $\mathcal{S}$ *is the set of* ℓ *signatures signed using the same public parameters contained in* pk. AAgg *can be used by untrusted parties and without knowing the private key.*

AAggVerf. *To verify an aggregated signature* σ_c*, check whether* $\hat{e}(\sigma, g) \stackrel{?}{=} \prod_{i=1}^{\ell} \hat{e}(\mathcal{H}_k(m_i), Q)$ *holds, on input of* σ_c*,* pk *and a list of all signed* m_i*. To improve efficiency, the right side can be rewritten as* $\hat{e}(\prod_{i=1}^{\ell} \mathcal{H}_k(m_i), Q)$*, due to the use of only one public key. We denote the algorithm as* $d \leftarrow$ AAggVerf(pk, σ, $\{m_i\}_{0<i\leq\ell}$).

The usual correctness requirements must hold, which have been formally proven in [4]. Moreover, we require the expected security properties to hold, i.e., unforgeability under chosen message attacks (UNF-CMA) and the k-element extraction assumption. The proofs and formal definitions can also be found in [4].

2.2 The Scheme by *Kundu* and *Bertino*

In this section, we shortly restate the scheme by *Kundu* and *Bertino* [12]. Afterwards, we describe the attacks. We use the following notations: n denotes the number of nodes (i.e., $|V|$); a node i is denoted as n_i; c_i denotes the label or content of n_i. A family of cryptographic hash-functions is denoted as $\mathcal{H}_\mathcal{S}$, where $\mathcal{S}$ denotes the key space of the hash function family. The following algorithm is the original one introduced in [12]. In their scheme, the input flags r and i are both fixed to 1. This indicates, that both, root and non-leaf node redaction, is always allowed. If nodes have to be ordered, we assume that the ordering algorithm used is known to every party involved, e.g., pre-order traversal. "||" denotes a concatenation which is uniquely reversible.

Construction 1 (The Kundu-Scheme.) *The scheme by Kundu and Bertino* $\mathcal{KS}$ *consists of four efficient algorithms. In particular* $\mathcal{KS}$:= (KeyGen, TSign, TVerify, TShare).

KeyGen. *The key generation algorithm* (sk, pk) $\leftarrow$ KeyGen(1^λ) *outputs a key pair of an aggregate signature scheme. It outputs* (sk, pk) $\leftarrow$ $\mathcal{AGG}$.AKeyGen(1^λ)

TSign. *The signing algorithm* $(T, \sigma_T) \leftarrow$ TSign(sk, T, r, i) *takes the secret key* sk *and the tree* T*. The flags* r *and* i *are defined to be* 1*. To sign the tree, perform the following steps:*

1. *Choose a cryptographic hash-function* $\mathcal{H}_s \in \mathcal{H}_\mathcal{S}$
2. *Get the pre-order traversal numbers of each node* $n_i \in T$*, denoted* l_i
3. *Get the post-order traversal numbers of each node* $n_i \in T$*, denoted* r_i
4. *Apply the randomizing, but order-preserving function* θ *to both lists of traversal numbers, using some distribution* Δ*. The randomized counterparts is denoted as* l_i^r *resp.* r_i^r*. How to calculate this randomization step is not important; the required order-preserving behaviour is enough to undermine transparency and privacy (See Sect. 2.3)*
5. *Let* $\rho_i := (l_i^r, r_i^r)$*. Set* $G_T \leftarrow \mathcal{H}_s(\rho_1||c_1||\dots||\rho_n||c_n)$
6. *For* $1 < i \leq n$*, compute* $\sigma_i \leftarrow \mathcal{AGG}.\mathsf{ASign}(\mathrm{sk}, \mathcal{H}_s(G_T||\rho_i||c_i))$
7. *Compress all resulting signatures* σ_i *into a compressed signature* σ_c*, i.e.,* $\sigma_c \leftarrow \mathcal{AGG}.\mathsf{AAgg}(\mathrm{pk}, \{\sigma_i\}_{0<i\leq n})$.
8. *Output* (T, σ_T)*, where* $\sigma_T = (\mathrm{pk}, \sigma_c, G_T, s, \{(\sigma_i, \rho_i)\}_{0<i\leq n})$

TVerify. *The algorithm* $\mathsf{TVerify}(T, \mathrm{pk}, \sigma_T)$ *takes as input the received tree* T*, a public key* pk *and the received signature* σ_T*. It outputs a bit* $d \in \{0, 1\}$*, which indicates, if the received* σ_T *is a valid and correct signature on* T *under the public key pk. It performs the following steps to do so:*

1. *Compute* $d \leftarrow \mathcal{AGG}.\mathsf{AAggVerf}(\mathrm{pk}, \sigma_c, \{\mathcal{H}_s(G_T||\rho_i||c_i)\}_{0<i\leq n})$.
   If $d = 0$*, output 0, else continue*
2. *Check, if each node is positioned correctly using both traversal numbers derived from* ρ_i*. If the positions are correct, output* 1*, else output* 0

TShare. *The algorithm* $\mathsf{TShare}(T, \mathrm{pk}, \sigma_T, \mathcal{N})$ *takes as input a tree* T*, a public key pk and the tree's signature* σ_T*, as well as a subset* $\mathcal{N} \subseteq T$ *of nodes*

1. *Check the validity of* σ_T *using* $\mathsf{TVerify}$
2. *Set* $T' \leftarrow T \setminus \mathcal{N}$*, and also remove the edge(s). If non-leaf nodes are subject to redaction, implicit edges are introduced: the redacted node is skipped in terms of the edge. See Sect. 2.4 for more details*
3. *Compute* $\sigma_c' \leftarrow \mathcal{AGG}.\mathsf{AAgg}(\mathrm{pk}, \{\sigma_i\}_{0<i\leq n'})$*, where* $n' = |V'|$*. Note, we have rearranged the indices to account for the redaction*
4. *Output* (T', σ_T')*, where* $\sigma_T' = (\mathrm{pk}, \sigma_c', G_T, s, \{(\sigma_i, \rho_i)\}_{0<i\leq n'})$

2.3 Attacks on *Kundu*'s Transparency and Privacy

Transparency and privacy can actually be attacked in many ways. We go through all known attacks. Some attacks can just be mounted on special revisions of the schemes by *Kundu* and *Bertino*. We state this for each attack. To have a complete list of attacks against the new scheme is shown to be resilient against, we restate all attacks known yet.

Randomized Traversal Numbers. *Kundu* and *Bertino* propose three ways to randomize the traversal numbers: [13]

1. **Sorted Random Numbers**: Generate $|V|$ random numbers and sort them.
2. **Order-Preserving Encryption**: Apply an order-preserving encryption scheme, e.g. [1], to the traversal numbers.
3. **Addition of Random Numbers**: Assign the numbers to the nodes by taking the previous traversal number and adding a random offset.

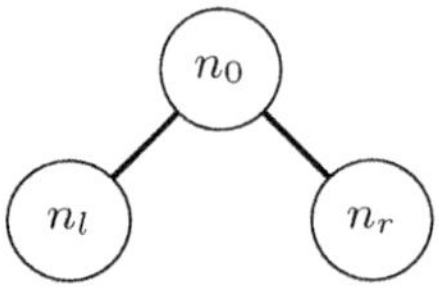

Fig. 5. Sample Tree I

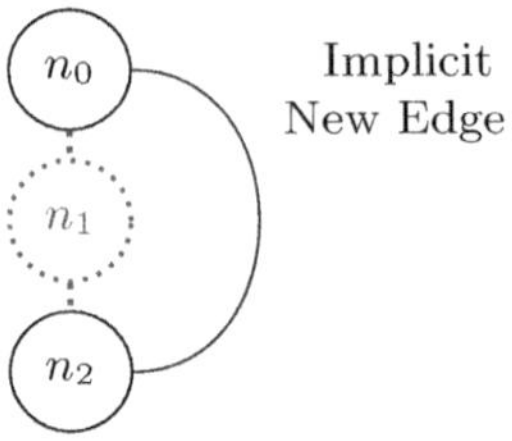

Fig. 6. Sample Tree II

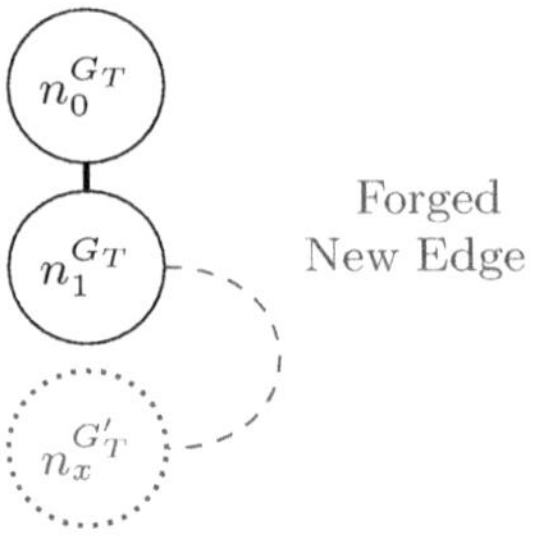

Fig. 7. Mix and Match

The following attack has been discovered by *Brzuska* et al. in [5]: Consider a tree with three nodes, i.e., the tree as depicted in Fig. 5. The algorithm θ outputs a tuple $(r_l, r_r, r_0) \leftarrow \theta(T, \Delta)$ on input of the depicted tree T in Fig. 5 and a distribution Δ. Assume that Δ is the uniform distribution and let $\mu = \mathrm{E}(r_l)$ be the expected random number associated to n_r. Furthermore, assume that $\Pr[r_l = \mu] = 0$ for a sufficiently large space. Hence, the probabilities are $\Pr[r_l < \mu] = \Pr[r_l > \mu] = 0.5$. Therefore, we obtain $\Pr[r_r > \mu] \geq 0.75$, since r_r is the largest random number and for $r_r < \mu$, both r_l and r_0 must be smaller than μ. Transparency can then be broken for the sample tree depicted in Fig. 5 as follows: transparency has been defined as a game where the adversary must guess a bit b. The game either returns a signature just containing n_0 and n_r or a signature which represents the whole tree where n_l has been cut off [5]. To win the game, the adversary just outputs 1, if $r_r \leq \mu$ and 0 otherwise. Note: $\Pr[r_r > \mu] \geq 0.75$ for a sanitized tree, while for a freshly signed one the probability changes: $\Pr[r_r > \mu] = 0.5$. Hence, b can be guessed with a non-negligible probability. Therefore, the scheme by *Kundu* and *Bertino* is neither transparent [5] nor private, since the leaked information allows to make statements about the original message. This attack works for other distributions as well, while it is required that the adversary knows the distribution Δ, which can be derived from the signing algorithm.

Leaking Structural Signature G_T. In the original paper [12], *Kundu* and *Bertino* introduce an additional digest which they name "structural signature", denoted as G_T. This "signature" is calculated as $G_T \leftarrow \mathcal{H}_s(\rho_1||c_1||\ldots||\rho_n||c_n)$. It is part of the calculation of each σ_i, i.e., $\sigma_i \leftarrow \mathcal{AGG}.\mathsf{ASign}(sk, G_T||\rho_i||c_i)$. Obviously, to check the signature σ_T, G_T needs to be available to the verifier. The verifier can calculate a digest G'_T on input of the tree T'. In particular, $G'_T \leftarrow \mathcal{H}_s(\rho_1||c_1||\ldots||\rho_{n'}||c_{n'})$, where $n' = |V'|$. Afterwards, the following equation is checked: $G'_T \stackrel{?}{=} G_T$. Following our definitions, this also destroys privacy. This problem has partially been circumvented in [13] by salting G_T; in particular the signer calculates: $G_T \leftarrow \mathcal{H}_s(\omega||\rho_1||c_1||\ldots||\rho_n||c_n)$, where ω is a nonce. However, this choice requires that $\mathcal{H}_s$ is modeled as a random oracle to ensure privacy [13]. In the revision introduced in [16], no structural signature is present, thus not inheriting this problem. However, omitting the tag renders that revision of the scheme forgeable, as we see next.

2.4 Attacks on *Kundu*'s Unforgeability and Structural Integrity

Besides the attacks on transparency and privacy, unforgeability can be attacked as well. In this section, two attacks are given. The first is adapted from [20]. The second attack is new.

Level Promotion. We shortly restate the observation which has recently been discovered by *Samelin* et al. in [20]. They show how to alter the semantic meaning of tree T by removing nodes which are not leafs. They name non-leafs "intermediate nodes". We adapt their nomenclature. Removing intermediate nodes allows introducing new implicit edges in certain scenarios. A sample tree, where this is possible, is depicted in Fig. 6. For this tree the traversal numbers are $(1, 2, 3)$ (pre-order) resp. $(3, 2, 1)$ (post-order). The randomization step can be omitted due to the order-preserving behaviour. If n_1 gets redacted, the traversal numbers of the remaining nodes are $(1, 3)$ (pre-order) resp $(3, 1)$, which, in terms of the signature, is valid. In particular, a sanitizer can introduce an edge $e_{n_0,n_2} \notin E$ which has not been signed by the signer. Strictly speaking, this destroys structural integrity, and therefore has a negative impact on unforgeability. Furthermore, this attack is possible in all versions of the schemes by *Kundu* and *Bertino* [12,13,16]. One might argue, that it depends on the definition of integrity, if this behavior can be considered as an attack. However, we think that such powerful possibilities must be under the sole control of the signer to avoid unwanted side-effects. This is in accordance with [20]. Our secured scheme allows such a level-promotion, but only after the signer explicitly allowed this behaviour during generation of the signature. Note, additionally the attack from [20] is also applicable, if one is able to redact parent nodes to allow distributing subtrees of a given tree. In our example, this would be the tree $T' = (\{n_1, n_2\}, \{e_{1,2}\})$. Obviously, this also leads to the same problem. This has not been mentioned in [20]. One may argue that intermediate node redaction lacks application scenarios. This is not true. Consider the following example: a tree's structure is implicitly describing the hierarchy within a company. Allowing to remove intermediate nodes now permits only to remove the department head. This leaves a list of employees. Without intermediate node redaction, this is not possible. Hence, there exist use cases, where removing leafs only is not sufficient.

Match-and-Mix Attacks. In the older revisions of the schemes by *Kundu* and *Bertino*, i.e., as proposed in [12,13], each node $n_x^{G_T}$ is bound to the tree T^{G_T} due to the structural signature G_T, which is part of each signature. Hence, a node $n_x^{G_T}$ cannot be merged into another tree $T^{G'_T}$, which nodes $n^{G'_T}$ have been signed with a different ***tag*** G'_T. However, in the newest revision [16], this tag is not present anymore. It is just aggregated onto the signature, but does not bind nodes to the tag, since it is not part of the signature generation of each node. In particular, $\sigma_i \leftarrow \mathcal{AGG}.\mathsf{ASign}(sk, G_T||\rho_i||c_i)$, where $G_T := \emptyset$. Hence, an adversary $\mathcal{A}$ can merge nodes of different trees into a new tree $T^{\mathcal{A}}$, which contains nodes originally signed for $T^{G'_T}$ and T^{G_T}. As an example, we show how the tree $T^{\mathcal{A}}$, as depicted in Fig. 7, can be constructed. W.l.o.g. let

$\rho_{n_0^{G_T}} = (1,3)$, $\rho_{n_1^{G_T}} = (2,2)$ be the nodes of tree T^{G_T} and $\rho_{n_x^{G'_T}} = (3,1)$ be the only node of $T^{G'_T}$. The adversary $\mathcal{A}$ has access to each of the individual signatures $\sigma_{n_0}, \sigma_{n_1}, \sigma_{n_x}$ and the "blinding signatures" σ_{ω_T} and $\sigma_{\omega'_T}$.[1] Hence, the adversary can construct a signature $\sigma_{\mathcal{A}}$ on the tree $T^{\mathcal{A}}$ by calculating $\sigma_{\mathcal{A}} \leftarrow \mathcal{AGG}.\mathsf{AAgg}(pk, \{\sigma_{n_0}, \sigma_{n_1}, \sigma_{n_x}, \sigma_{\omega_T}\})$ resp. $\sigma'_{\mathcal{A}} \leftarrow \mathcal{AGG}.\mathsf{AAgg}(pk, \{\sigma_{n_0}, \sigma_{n_1}, \sigma_{n_x}, \sigma_{\omega'_T}\})$, where $T^{\mathcal{A}}$ consists of all three nodes mentioned. Running $\mathsf{TVerify}(T^{\mathcal{A}}, pk, \sigma_{\mathcal{A}})$ outputs the bit 1. As a result, the attacker has successfully forged a signature for $T^{\mathcal{A}}$. This new attack can be applied as soon as two trees are signed with the same private key. Hence, the newest revision of the scheme by *Kundu* and *Bertino* is forgeable. Again, the older revisions do not have this problem, as every node $n_x^{G_T}$ is bound to the tree T^{G_T} due to G_T, which is unique for each tree signed. We conclude, that all revisions, i.e. [12,13,16], are not secure in our model.

3 Our New Secure Construction

We present a construction based on *Kundu* and *Bertino*'s ideas, but without inheriting the previously mentioned weaknesses. We use a transparent and unforgeable RSS for standard linear documents, which is denoted as Π in this section. In particular, let $\Pi = (\mathsf{KeyGen}, \mathsf{Sign}, \mathsf{Redact}, \mathsf{Verify})$ be a secure RSS for lists. Moreover, we require transparency, privacy and unforgeability. Please refer to [5,8,20] for the formal definitions. The existing secure schemes have a runtime of $\mathcal{O}(n^2)$ or $\mathcal{O}(n \cdot log(n))$ resp. [2], whereas the latter just allows quoting of substrings, hence not suitable for arbitrary redactions. Their scheme fulfills a strong privacy notation, strong context-hiding. We are aware of the fact that *Kundu* et al. introduced a new scheme in [14]. However, their new scheme also has a worst case complexity of $\mathcal{O}(n^2)$, only allows to remove leafs and therefore offers **no** advantage to the one introduced in [5] by *Brzuska* et al. Additionally, [14] relies on a different idea, based on signing binary relations, similar to [5,8,20]. Hence, we see their scheme as a completely new construction not related to the original idea. Our scheme makes use of a RSS for lists. Utilizing a RSS in $\mathcal{O}(n)$, our new scheme is considerably faster than the existing ones. Such a RSS has been proposed by *Pöhls* et al. in [19].

Sketch of Our Secure Construction. We first sketch our construction to increase readability of the algorithmic descriptions. The basic idea is to use the fact that a tree is uniquely determined by its pre- and postorder traversal numbers. However, as *Brzuska* et al. show in [5], ordering random numbers leads to insecure schemes. Hence, we need to find a way to sign the ordering of the traversal numbers without explicitly ordering them. We achieve this by using a secure transparent RSS, which protects the order of the signed parts. We use this order preserving RSS to sign two lists. Each list contains $|V|$ uniformly distributed random nonces, which are pairwise unique. We use one list for pre- and another one for post-order traversal numbers. We annotate a node n_i with

[1] An adversary can always build an inverse of all signatures received.

the nonce at the positions in the lists corresponding to the node's traversal numbers. The ordering of each list secured by signing it with the RSS. Note, the nonces in the lists are not ordered. When a node n_i is redacted, the two nonces used to annotate n_i are removed from the respective list using the RSS, while leaving each list of remaining nonces still uniformly distributed. Additionally, the signer can decide, if a redaction of the root or redaction of intermediate nodes is allowed, as the signer annotates the nodes accordingly. This keeps the signer in control, while giving more flexibility. This must be done by annotating nodes accordingly. We cannot use the content of the nodes in the lists, as they may not be unique, making the reconstruction ambiguous and therefore forgeable.

Construction 2 (The New Flexible Construction.) *We give an algorithmic explanation of our scheme. The security proofs can be found in the appendices, but from the algorithmic description the new scheme's resilience against the mentioned attacks are visible. The aggregate signature is used to improve the performance of the verification process and to achieve consecutive redaction control, as we show in Sect. 4.1.*

KeyGen. *The key generation algorithm outputs two key pairs: (1) A key pair for an aggregate signature scheme $\mathcal{AGG}$, i.e., $(sk_{\mathcal{AGG}}, pk_{\mathcal{AGG}}) \leftarrow \mathcal{AGG}.\mathsf{AKeyGen}(1^\lambda)$. (2) A key pair for Π, i.e., $(sk_\Pi, pk_\Pi) \leftarrow \Pi.\mathsf{KeyGen}(1^\lambda)$. Output* $(\mathrm{sk}, \mathrm{pk}) = ((\mathrm{sk}_\Pi, \mathrm{sk}_{\mathcal{AGG}}), (\mathrm{pk}_\Pi, \mathrm{pk}_{\mathcal{AGG}}))$

TSign. *The signing algorithm $(T, \sigma_T) \leftarrow \mathsf{TSign}(\mathrm{sk}, T, r, i)$ takes all secret keys, the tree T to sign and the flags r and i. To sign T, perform:*

1. *Generate two lists, $\mathcal{L}$ and $\mathcal{M}$, each containing $n = |V|$ pairwise distinct uniformly distributed random numbers $\in \{0,1\}^\lambda$. The elements of the list are addressed by an index, i.e., $\mathcal{L}_i$*
2. *Let i be the index of a pre-order traversal $c_i \leftarrow \mathcal{L}_i || c_i$*
3. *Let j be the index of a post-order traveral. Set $c_i \leftarrow \mathcal{M}_j || c_i$*
4. *Choose a random tag $\tau \xleftarrow{\$} \{0,1\}^\lambda$, which must be unique for each tree T*
5. *If $r = 0$, set $\tau \leftarrow \tau || noroot$ and annotate the root: $c_r \leftarrow c_r || theroot$, while all for all other nodes: $c_i \leftarrow c_i || nottheroot$. Otherwise, set $\tau \leftarrow \tau || root$ and annote all nodes: $c_i \leftarrow c_i || nottheroot$. Sign τ: $\sigma_\tau \leftarrow \mathcal{AGG}.\mathsf{ASign}(\mathrm{sk}_{\mathcal{AGG}}, \tau)$*
6. *Draw a nonce d, $d \xleftarrow{\$} \{0,1\}^\lambda$. For each node n_i: if $i = 0$, set $c_i \leftarrow (c_i || d + \mathrm{depth}(n_i))$, otherwise $c_i \leftarrow (c_i || -1)$. The function $\mathrm{depth} : V \rightarrow \mathbb{N}$ returns the distance from the root to the given node $n_i \in T$*
7. *For each node n_i sign $c_i || \tau$: $\sigma_i \leftarrow \mathcal{AGG}.\mathsf{ASign}(\mathrm{sk}_{\mathcal{AGG}}, c_i || \tau)$*
8. *Compress all resulting signatures into the aggregate signature σ_c, along with the signature of τ, i.e., $\sigma_c \leftarrow \mathcal{AGG}.\mathsf{AAgg}(\mathrm{pk}_{\mathcal{AGG}}, \sigma_\tau \cup \{\sigma_i\}_{0<i\leq n})$*
9. *Sign $\mathcal{L}$ using Π, i.e., $(\mathcal{L}, \sigma_\mathcal{L}) \leftarrow \Pi.\mathsf{Sign}(\mathrm{sk}_\Pi, \mathcal{L})$*
10. *Sign $\mathcal{M}$ using Π, i.e., $(\mathcal{M}, \sigma_\mathcal{M}) \leftarrow \Pi.\mathsf{Sign}(\mathrm{sk}_\Pi, \mathcal{M})$*
11. *Output (T, σ_T), where $\sigma_T = (\sigma_c, (\sigma_i)_{0<i\leq n}, \sigma_\tau, \mathcal{L}, \mathcal{M}, \sigma_\mathcal{L}, \sigma_\mathcal{M}, \mathrm{pk}_{\mathcal{AGG}}, \mathrm{pk}_\Pi, \tau)$*

TVerify. *Verifying the signature is similar to generating the signature.*

1. *Check the validity of $\sigma_\mathcal{L}/\mathcal{L}$ and $\sigma_\mathcal{M}/\mathcal{M}$ using $\Pi.\mathsf{Verify}$*
2. *For each node $n_i \in T$, parse c_i as $(m_i, l_i, e_i, a_i, d_i)$*

3. *Traverse* T *via pre-order:* $\forall n_i \in T$*, check if* $\mathcal{L}_i = l_i$*. If not, abort and return* 0
4. *Traverse* T *via pre-order:* $\forall n_j \in T$*, check if* $\mathcal{M}_j = m_j$*. If not, abort and return* 0
5. *Compute* $v \leftarrow \mathcal{AGG}.\mathsf{AAggVerf}(\mathrm{pk}, \sigma_T, \{\tau\} \cup \{m_i||l_i||e_i||d_i||a_i||\tau\}_{0<i\leq n})$*. If* $v = 0$*, abort and return* 0
6. *Let* r *denote the root of the tree* T*. If* $d_r \neq -1$*, check for all nodes* $n_i \in T \setminus r$*, if* $\mathrm{depth}(\mathrm{parent}(n_i) = d_i - 1)$*, where* $\mathrm{parent} : V \to V$ *returns the only parent of a given node* n_i*. If not, abort and return* 0
7. *Parse* τ *as* (τ, h)*. If* $h =$ *noroot, check, if the received root's annotion* r_a *equals "theroot". If so, return* 1*, otherwise return* 0

TShare. *The algorithm* $\mathsf{TShare}(T, \mathrm{pk}, \sigma_T, \mathcal{N})$ *takes as input the tree* T*, all public keys* pk *and the signature* σ_T*, as well as a set of nodes* $\mathcal{N} \subseteq T$*.*

1. *Remove nodes by setting* $T' \leftarrow T \setminus \mathcal{N}$
2. *If intermediate nodes have been redacted, adjust the edges of the intermediate nodes' successors. In particular, for each node* $n_i \in T' \setminus r$ *not having a parent, introduce the edge* $e_{i,j}$*, where* n_j *is the closest ancestor node not redacted. If the result is not a tree, return* $\perp$
3. *For each* $n_i \in \mathcal{N}$*, adjust both lists of nonces:* $\sigma'_{\mathcal{L}} \leftarrow \Pi.\mathsf{Redact}(\mathrm{pk}_\Pi, \mathcal{L}, i)$*, where* i *is the pre-order number of* n_i*. And* $\sigma'_{\mathcal{M}} \leftarrow \Pi.\mathsf{Redact}(pk_\Pi, \mathcal{M}, j)$*, where* j *is the post-order number of* n_i
4. *Set* $\sigma'_c \leftarrow \mathcal{AGG}.\mathsf{AAgg}(\mathrm{pk}_{\mathcal{AGG}}, \sigma_\tau \cup \{\sigma_i\}_{0<i\leq n'}\})$
5. *Construct* (T', σ'_T)*, where* $\sigma'_T = (\sigma'_c, (\sigma_i)_{0<i\leq n'}, \sigma_\tau, \mathcal{L}', \mathcal{M}', \sigma'_{\mathcal{L}}, \sigma'_{\mathcal{M}}, \mathrm{pk}_{\mathcal{AGG}}, \mathrm{pk}_\Pi, \tau)$
6. *Verify* σ_T*: If* $\mathsf{TVerify}(T', \mathrm{pk}, \sigma'_T)$ *outputs* 0*, abort and return* $\perp$
7. *Output* (T', σ'_T)

3.1 Security

Theorem 1 (The Construction is Secure). *Our construction is secure, if the used* RSS *is transparent and unforgeable, while the used* $\mathcal{AGG}$ *is unforgeable.*

Proof. Relegated to App. A.

3.2 Complexity Analysis

For signing our scheme requires $\mathcal{O}(n)$ steps. Each step "involves" each node $n_i \in T$ four times: we assign two random values and we generate two digests. Afterwards, all digests are compressed using $\mathcal{AGG}$, while $\mathcal{L}$ and $\mathcal{M}$ are signed using a order preserving secure RSS Π. We assume that drawing nonces is in $\mathcal{O}(\lambda)$ and therefore constant. The steps performed by Π are not considered in this approximation, as they depend on the actual RSS used, which can be exchanged. Verification "involves" each node in two operations to calculate the digest. Hence, verification is also in $\mathcal{O}(n)$. However, to verify the random values, Π.Verify is called twice. This is also the case during redaction: we simply delete the nodes, while redacting the random values from $\mathcal{L}$ and $\mathcal{M}$ involves two invocations of Π again. Hence, the asymptotic runtime depends on Π, while being at least in $\mathcal{O}(n)$.

4 Modifications to Our Scheme

4.1 Consecutive Redaction Control

To prohibit redaction of several nodes, the ideas introduced in [17] by *Miyazaki* et al. can be applied. Depending on the $\mathcal{AGG}$ used, it is possible to remove a signature from the compressed one. In particular, the *BGLS*-Scheme [4] allows such calculations due to its group-theoretic structure. Hence, to prohibit redaction, the signatures for these nodes are not delivered to the sanitizer. σ_τ is not given to the sanitizer by the signer in the first place. Additional proofs and a more detailed discussion can be found in [17] and [20].

4.2 Restricting to Sanitizers and Accountability

All proposed schemes allow everybody to redact nodes. To limit redaction to explicitly denoted sanitizers the signature σ_T is extended to hold an additional value d. Let $d \leftarrow \text{SIGN}(sk, \tau || \mathcal{CH}(\langle T \rangle))$, where $\langle T \rangle$ is a suitable binary representation of the signed tree T and $\mathcal{CH}$ is a chameleon hash. The values required for $\mathcal{CH}$ need to be delivered with σ_T. Hence, only sanitizers who possess the secret for $\mathcal{CH}$ can alter T without invalidating the signature. This can be enriched further to achieve sanitizer and signer accountability [6]: $\mathcal{CH}$ could be replaced with a tag-based chameleon-hash $\mathcal{CH}_{TAG}$, i.e., the construction of *Brzuska* et al. [6].

5 Implementations and Performance Analysis

We have implemented the scheme by *Kundu* and *Bertino* in the original version [12] and the scheme by *Brzuska* et al. [5]. We did not implement our scheme, since it only differs by a constant factor from the scheme by *Kundu* and *Bertino* in the original version [12], i.e., the underlying RSS Π. The source code used for this evaluation will be made available on request. The tests were performed on a *Lenovo Thinkpad T61* with an *Intel* T8300 Dual Core `@2.40` Ghz and 4 GiB of RAM. We ran *Ubuntu* Version 10.04 LTS (64 Bit) and Java version `1.6.0_26-b03`. We used 2048-Bit RSA-Keys. The trees signed are n-ary balanced ones with height h. Tab. 1 and 2 show the results for high trees with a low branching factors. Tab. 3 and 4 show the results for flat trees with a high branching factor. This gives a good impression for different use cases and allows determine the specific pros and cons of the schemes.

As shown in the Tables 1-4, all schemes have a comparable runtime for small trees. For binary and ternary trees of increasing depth the scheme by *Brzuska* et al. suffers from the quadratic runtime and becomes unusable, denoted by "-slow-", very fast. Measuring was aborted, if the runtime was greater than 20 minutes. We conclude, that a linear complexity is required to yield useable schemes, as waiting a few minutes to generate σ_T is not acceptable, even if signatures are normally more often verified than generated.

Table 1. *Brzuska* et al.: Low Branching Factor n; Median Runtime in ms

h / n	Generate σ_T 10	50	100	Verify σ_T 10	50	100
2	721	14,319	55,625	24	525	1,910
3	6,856	-slow-	-slow-	241	-slow-	-slow-

Table 2. *Kundu* and *Bertino*: Low Branching Factor n; Median Runtime in ms

h / n	Generate σ_T 10	50	100	Verify σ_T 10	50	100
2	544	2,247	5,293	5	8	10
3	4,319	106,694	369,854	8	165	319

Table 3. *Brzuska* et al.: High Branching Factor n; Median Runtime in ms

h / n	Generate σ_T 2	3	4	Verify σ_T 2	3	4
5	605	2,804	9,607	19	95	333
10	17,195	666,530	-slow-	614	18,838	-slow-

Table 4. *Kundu* and *Bertino*: High branching factor n; Median Runtime in ms

h / n	Generate σ_T 2	3	4	Verify σ_T 2	3	4
5	1,657	5,065	14,132	4	12	22
10	44,756	1,228,738	-slow-	111	864	-slow-

6 Conclusion and Open Questions

After several new attacks presented in this paper, the scheme by *Kundu* and *Bertino* has been found to be insecure with respect to all RSS security properties. Building on the fact that removing an element from a uniformly distributed list of random numbers preserves their distribution and a secure order preserving transparent RSS, the new construction given in this paper can reuse the idea of *Kundu* and *Bertino* that a node's position in a tree is specifiable only by his post- and pre-order traversal numbers. Moreover, our scheme is the first which allows that the signer can decide, if it is allowed to redact parent or intermediate nodes. The new construction is secure against the existing attacks against the scheme by *Kundu* and *Bertino* as given by *Brzuska* et al. [5] and *Samelin* et al. in [20], as well as the two new attacks described in this paper. The paper offers formal proofs of the new construction's security. While all existing secure schemes have a runtime overhead of $\mathcal{O}(n^2)$, our new construction has only an overhead of $\mathcal{O}(n)$, if the underlying order-preserving RSS is in $\mathcal{O}(n)$, i.e., [19]. The work demonstrates that the underlying idea of using traversal numbers to transparently redact nodes combined with a order-preserving RSS can be facilitated to build a simplistic and enhanceable redactable scheme, which is provably secure in terms of transparency, privacy and unforgeability, while being highly efficient and very flexible.

References

1. Agrawal, R., Kiernan, J., Srikant, R., Xu, Y.: Order-preserving encryption for numeric data. In: SIGMOD Conference, pp. 563–574 (2004)
2. Ahn, J.H., Boneh, D., Camenisch, J., Hohenberger, S., Shelat, A., Waters, B.: Computing on authenticated data. Cryptology ePrint Archive, Report 2011/096 (2011), `http://eprint.iacr.org/`

3. Ateniese, G., Chou, D.H., de Medeiros, B., Tsudik, G.: Sanitizable Signatures. In: de Capitani di Vimercati, S., Syverson, P.F., Gollmann, D. (eds.) ESORICS 2005. LNCS, vol. 3679, pp. 159–177. Springer, Heidelberg (2005)
4. Boneh, D., Gentry, C., Lynn, B., Shacham, H.: Aggregate and Verifiably Encrypted Signatures from Bilinear Maps. In: Biham, E. (ed.) EUROCRYPT 2003. LNCS, vol. 2656, pp. 416–432. Springer, Heidelberg (2003)
5. Brzuska, C., Busch, H., Dagdelen, O., Fischlin, M., Franz, M., Katzenbeisser, S., Manulis, M., Onete, C., Peter, A., Poettering, B., Schröder, D.: Redactable Signatures for Tree-Structured Data: Definitions and Constructions. In: Zhou, J., Yung, M. (eds.) ACNS 2010. LNCS, vol. 6123, pp. 87–104. Springer, Heidelberg (2010)
6. Brzuska, C., Fischlin, M., Freudenreich, T., Lehmann, A., Page, M., Schelbert, J., Schröder, D., Volk, F.: Security of Sanitizable Signatures Revisited. In: Jarecki, S., Tsudik, G. (eds.) PKC 2009. LNCS, vol. 5443, pp. 317–336. Springer, Heidelberg (2009)
7. Canard, S., Jambert, A.: On Extended Sanitizable Signature Schemes. In: Pieprzyk, J. (ed.) CT-RSA 2010. LNCS, vol. 5985, pp. 179–194. Springer, Heidelberg (2010)
8. Chang, E.-C., Lim, C.L., Xu, J.: Short Redactable Signatures Using Random Trees. In: Fischlin, M. (ed.) CT-RSA 2009. LNCS, vol. 5473, pp. 133–147. Springer, Heidelberg (2009)
9. Izu, T., Kunihiro, N., Ohta, K., Sano, M., Takenaka, M.: Sanitizable and Deletable Signature. In: Chung, K.-I., Sohn, K., Yung, M. (eds.) WISA 2008. LNCS, vol. 5379, pp. 130–144. Springer, Heidelberg (2009)
10. Johnson, R., Molnar, D., Song, D., Wagner, D.: Homomorphic Signature Schemes. In: Preneel, B. (ed.) CT-RSA 2002. LNCS, vol. 2271, pp. 244–262. Springer, Heidelberg (2002)
11. Klonowski, M., Lauks, A.: Extended Sanitizable Signatures. In: Rhee, M.S., Lee, B. (eds.) ICISC 2006. LNCS, vol. 4296, pp. 343–355. Springer, Heidelberg (2006)
12. Kundu, A., Bertino, E.: Structural Signatures for Tree Data Structures. In: Proc. of PVLDB 2008, New Zealand. ACM (2008)
13. Kundu, A., Bertino, E.: CERIAS Tech Report 2009-1 Leakage-Free Integrity Assurance for Tree Data Structures (2009)
14. Kundu, A., Atallah, M.J., Bertino, E.: Leakage-free redactable signatures. In: CODASPY, pp. 307–316 (2012)
15. Kundu, A., Bertino, E.: How to authenticate graphs without leaking. In: EDBT, pp. 609–620 (2010)
16. Kundu, A., Bertino, E.: Structural signatures: How to authenticate trees without leaking. Technical report, Purdue University (June 2010)
17. Miyazaki, K., Hanaoka, G., Imai, H.: Digitally signed document sanitizing scheme based on bilinear maps. In: Proceedings of the 2006 ACM Symposium on Information, Computer and Communications Security, ASIACCS 2006, pp. 343–354. ACM, New York (2006)
18. Pöhls, H.C., Samelin, K., Posegga, J.: Sanitizable Signatures in XML Signature — Performance, Mixing Properties, and Revisiting the Property of Transparency. In: Lopez, J., Tsudik, G. (eds.) ACNS 2011. LNCS, vol. 6715, pp. 166–182. Springer, Heidelberg (2011)
19. Pöhls, H.C., Samelin, K., Posegga, J., de Meer, H.: Length-hiding redactable signatures from one-way accumulators in $\mathcal{O}(n)$ (mip-1201). Technical report, University of Passau (April 2012)

20. Samelin, K., Pöhls, H.C., Bilzhause, A., Posegga, J., de Meer, H.: Redactable Signatures for Independent Removal of Structure and Content. In: Ryan, M.D., Smyth, B., Wang, G. (eds.) ISPEC 2012. LNCS, vol. 7232, pp. 17–33. Springer, Heidelberg (2012)
21. Steinfeld, R., Bull, L., Zheng, Y.: Content Extraction Signatures. In: Kim, K.-c. (ed.) ICISC 2001. LNCS, vol. 2288, pp. 285–304. Springer, Heidelberg (2002)

A Proofs

Theorem 2 (The Construction is Transparent). *Our construction is transparent and therefore also private [5].*

Proof. $\mathcal{L}$, $\mathcal{M}$, the public keys $pk_{\mathcal{AGG}}$ and pk_{Π} and the tag τ do not leak any information about the tree T. The lists $\mathcal{L}$ and $\mathcal{M}$ contain uniformly distributed random numbers. Even on redaction of a tree node, we only remove elements from a uniformly distributed list of randoms, which still results in a uniformly distributed list of randoms. If the tree is annotated with d, nothing can be derived either, since d is chosen from a uniform distribution and the tree grows at most linearly in $|V|$. Hence, we only need to show that $\sigma_{\mathcal{L}/\mathcal{M}}$ and σ_c together imply transparency. A successful attack on transparency would show that either Π and/or the aggregate signature scheme $\mathcal{AGG}$ is not transparent. To avoid duplicate work, we relegate the reader to [17] and [20], where the proofs for transparency of σ_c can be found. For $\sigma_{\mathcal{L}/\mathcal{M}}$, we show how an adversary would also break the transparency of the underlying RSS Π: Assume an efficient adversary $\mathcal{A}_{tra}$ which wins the transparency game of our scheme. Using $\mathcal{A}_{tra}$ we construct another adversary $\mathcal{B}_{tra}$ to break the transparency of Π in the following way: For any calls to $\mathcal{O}^{\mathsf{TSign}}$ resp. $\mathcal{O}^{\mathsf{Adapt/Sign}}$, $\mathcal{B}_{tra}$ genuinely returns the answers of its own oracle. Eventually, $\mathcal{A}_{tra}$ outputs its guess b^*. b^* is then outputted by $\mathcal{A}_{tra}$ as its own guess. We now need to argue that, due to the fact that $\mathcal{AGG}$ is information-theoretically transparent, $\mathcal{B}_{tra}$'s probability of success equals the one of $\mathcal{A}_{tra}$. Only Π could have leaked this information, as the lists contain just uniformly distributed nonces and redactions of elements in that list again lead to a uniformly distributed list. Hence, $\mathcal{B}_{tra}$ wins with the same probability as $\mathcal{A}_{tra}$.

Theorem 3 (The Construction is Unforgeable). *Our construction is unforgeable.*

Proof. We assume that no tag-collisions occur, winning the unforgeability game in a trivial manner. Note, that we do not need an induction over the tree, as we transform it into two lists, which are uniquely determined. Let the algorithm winning the unforgeability game be denoted as $\mathcal{A}$. Our scheme's security relies upon the security of $\mathcal{AGG}$ and Π. Given the game in Fig. 1, we can derive that a forgery must fall in at least one of the following cases:

Case 1: A value protected by σ_c has never been signed by the oracle
Case 2: The value protected by σ_c has been signed, but $T^* \not\subseteq \text{span}_{\vdash}(T)$. In other words: The tree T^* protected by σ_T is not in the transitive closure of any tree for which a signature was queried. This case has to be divided as well:

Case 2a: $T \notin \text{span}_{\vdash}(T^*)$
Case 2b: $T \in \text{span}_{\vdash}(T^*)$

We can construct an adversary $\mathcal{B}$, which breaks the unforgeability of the BGLS-Scheme, if an adversary $\mathcal{A}$ with a non-negligible advantage ϵ exists, winning our unforgeability game. To do so, $\mathcal{B}$ uses $\mathcal{A}$ as a black box. For every signature query $\mathcal{A}$ requests, $\mathcal{B}$ forwards the queries to its signing oracle $\mathcal{O}^{TSign}$ and genuinely returns the answers to $\mathcal{A}$. Eventually, $\mathcal{A}$ outputs (T^*, σ_T^*), where $T^* = (\sigma_c^*, \{\sigma_i\}_{0<i\leq n^*}, \sigma_T^*, \mathcal{L}^*, \mathcal{M}^*, \sigma_{\mathcal{L}}^*, \sigma_{\mathcal{M}}^*, pk_{\mathcal{AGG}}^*, pk_{\Pi}, \tau^*)$. Given the transcript of the simulation, $\mathcal{B}$ checks, if the outputted tuple is a trivial "forgery", i.e., just an allowed redaction. If so, $\mathcal{B}$ aborts the simulation. If $\mathcal{B}$ does not abort, we have to consider the following constellations: If σ_c^* contains messages never signed, $\mathcal{B}$ outputs (σ_c^*) along with forged strings, which can easily be extracted. If $T \notin \text{span}_{\vdash}(T^*)$, we have to consider two cases: (1) the values protected by $\sigma_{\mathcal{L}}$ resp. $\sigma_{\mathcal{M}}$ have been altered or (2) the strings protected by σ_T were modified. In either case, this allows to break the unforgeability of either Π or $\mathcal{AGG}$. The corresponding forgeries can easily be extracted. If $T \in \text{span}_{\vdash}(T^*)$, both Π and $\mathcal{AGG}$ must have been forged, since new elements are now contained in the aggregate or the RSS. As before, the forgeries can be extracted given the transcript.

Inner-Product Lossy Trapdoor Functions and Applications

Xiang Xie[1,2,⋆], Rui Xue[3], and Rui Zhang[3,⋆⋆]

[1] Institute of Software, Chinese Academy of Sciences
[2] Graduate University of Chinese Academy of Sciences
[3] The State Key Laboratory of Information Security
Institute of Information Engineering, Chinese Academy of Sciences
xiexiang@is.iscas.ac.cn, {rxue,r-zhang}@iie.ac.cn

Abstract. In this paper, we propose a new cryptographic primitive called inner-product lossy trapdoor function (IPLTDF). We give a formal definition, and a concrete construction from lattices. We then show this primitive is useful to obtain efficient chosen-plaintext secure inner-product encryption (IPE) schemes. The resulting IPE scheme has almost the same public key size for multi-bit encryption compared with a recent IPE scheme proposed by Agrawal, Freeman and Vaikuntanathan [2] for single-bit encryption. Unfortunately, our IPE scheme only supports attribute vectors with logarithmic length. On the positive side, our basic IPE scheme can be extended to achieve chosen-ciphertext (CCA) security. As far as we are aware, this is the first CCA-secure IPE scheme based on lattices.

Keywords: inner-product encryption, inner-product lossy trapdoor functions, lattices.

1 Introduction

In a traditional public key encryption system, data encrypted under a public key can only be decrypted by an receiver with the corresponding secret key. Inspired by a seminal work by Sahai and Waters [29], researchers have focused on more fine-grained encryption schemes, which led to the notion of *functional encryption* [13]. In a functional encryption, dedicated secret keys allow users to learn functions of encrypted data. Functional encryption is a broad framework and has many concrete expressions, among which, *predicate encryption* (PE) [19] is an important one. In a PE system, the secret key sk_f corresponding to a predicate f can be used to decrypt a ciphertext associated with attribute I if and only if $f(I) = 1$. A useful set of predicates for PE is called *inner-product*

⋆ Supported by the Fund of the National Natural Science Foundation of China under Grants No. 61170280.

⋆⋆ Supported by the One Hundred Person Project of the Chinese Academy of Sciences and the Fund of the National Natural Science Foundation of China under Grants No. 61100225.

F. Bao, P. Samarati, and J. Zhou (Eds.): ACNS 2012, LNCS 7341, pp. 188–205, 2012.

predicates. In an inner-product encryption (IPE) system, an attribute of inner-product predicates is expressed as vector $\mathbf{x}$ and predicate $f_{\mathbf{v}}$ is associated with vector $\mathbf{v}$, where $f_{\mathbf{v}}(\mathbf{x}) = 1$ iff $\langle \mathbf{x}, \mathbf{v} \rangle = 0$.

Katz, Sahai, and Waters defined the notion of predicate encryption and gave the first construction of IPE. However, their construction was based on complicated assumptions. Subsequently, Okamoto and Takashima [23] showed how to construct hierarchical IPE schemes. All the previous constructions are secure under selective adversaries until [20]. Lewko et al. [20] gave the first adaptively secure IPE scheme, which was further improved in [24,25]. All the above constructions made use of bilinear pairings. Very recently, Agrawal, Freeman and Vaikuntanathan [2] proposed the first IPE scheme under the *worst-case* lattice assumption.

In this work, we seek for a different way to construct IPE. We introduce a new primitive called inner-product lossy trapdoor functions (IPLTDFs). We define and construct IPLTDFs. Thanks to its lossiness, we can easily obtain IPE schemes from IPLTDFs via hardcore bits.

In a high level, for chosen public parameters pp and master secret key, an inner-product trapdoor function (IPTDF) F associated with any attribute vector $\mathbf{x}$ is an injective, deterministic map $F_{pp,\mathbf{x}}$ which can be inverted given a secret key derivable from a predicate vector $\mathbf{v}$ via the master secret key if and only if $\langle \mathbf{x}, \mathbf{v} \rangle = 0$. Suppose there is an another algorithm that generates "fake" public parameters pp^*, such that, for an adversary-specified challenge attribute vector $\mathbf{x}^*$, $F_{pp^*,\mathbf{x}^*}$ is no longer injective but has image much smaller than its domain. Moreover, given public parameters, one should not be able to tell whether it is real or fake. Importantly, as in inner-product encryption, this must hold even when the adversary may obtain, via a key-derivation oracle, an inversion key for predicates $\mathbf{v}$ with $\langle \mathbf{v}, \mathbf{x}^* \rangle \neq 0$. As with IPE, security may be selective (the adversary must specify $\mathbf{x}^*$ before seeing pp) or adaptive (no such restriction). In this paper, we only consider the selective security.

In order to build secure IPTDFs, an intuitive idea is to apply the matrix-based framework of [27] and encrypt each matrix entry with an IPE scheme. For already complicated IPE schemes this method brings us more complicated analysis. Alternatively, we derive one-way IPTDFs by applying the ideas from [2], and then show how to make it lossy which is non-trivial. Our solution shows that secure inner-product lossy trapdoor function (IPLTDF) can be achieved in principle, which was not clear prior to our work.

1.1 Our Contributions

In this work, we define the notion of inner-product lossy trapdoor functions, and we give a concrete construction of IPTDF based on lattices. However, to make the scheme correct and lossy simultaneously, our IPTDF only supports attribute vectors with logarithmic length. we then use it to construct a chosen-plaintext secure IPE scheme for multi-bit encryption based on lattices with public key size almost the same as the scheme presented by Agrawal, Freeman, and Vaikuntanathan [2] for single-bit encryption. Unfortunately, in order to invert correctly, we have to append the attribute vector after the function value of our IPTDFs, which limits our IPE scheme only achieves payload hiding security. We leave it as a future work to construct IPTDFs whose attribute information is hidden in the function value using our methodology.

As an interesting observation, we note that the information of the lossy attribute actually is hidden in the public parameters of the lossy IPTDFs. This property is also satisfied in identity-based trapdoor functions (IBTDFs). I.e., the public parameters of lossy IBTDFs hide the information of the lossy identity. Under the framework presented by Peikert and Waters [27], we obtain the first chosen-ciphertext secure IPE from lattices. As a by-product, we also observe that lossy IBTDFs are actually All-But-One (ABO) trapdoor functions [27]. Combine our IPTDF and the concrete construction of lossy IBTDF in [7] from lattices. We get a chosen-ciphertext secure IPE scheme with public size almost twice as our chosen-plaintext secure IPE scheme.

1.2 Related Works

Many encryption schemes of different types can be included in the framework of inner-product encryption. Identity-based encryption (IBE) [30,10,11] and hidden-vector encryption (HVE) [14] can be viewed as a special case of inner-product predicates encryption with equality-test predicates. Attribute-based encryption (ABE) [29,18,8] where policies are given by CNF or DNF can be implemented by inner-product encryption.

Inner-product encryption was introduced by Katz, Sahai, and Waters [19], however, their scheme only achieves selective security without delegatability (see [23]). Okamoto and Takashima [23] introduced a notion called dual pairing vector spaces (DPVS) and proposed a hierarchical IPE scheme based on DPVS, but again, only selective security is proven. To achieve adaptive security, Lewko et al. [20] adapted the dual system encryption methodology [31], and obtained the first adaptively secure IPE and hierarchical IPE schemes. Later, Okamoto and Takashima [24,25] proposed adaptively secure IPE and hierarchical IPE schemes under simpler assumptions. All these previous constructions use bilinear pairings except a recent scheme proposed by Agrawal, Freeman and Vaikuntanathan [2], which is the first IPE scheme under the *worst-case* lattice assumption. We note that the scheme in [2] seems difficult to be improved into a hierarchical IPE scheme.

The notion of lossy trapdoor functions (LTDFs) was first explicitly presented in [27]. A trapdoor function F specifies, for each public key pk, an injective, *deterministic* map F_{pk} that can be inverted given an associated trapdoor. There is an algorithm that generates a "fake" public key pk^* indistinguishable from the real one, such that F_{pk^*} has image much smaller than its domain. Peikert and Waters [27] call such a trapdoor function lossy. LTDFs was shown to be a powerful tool. Peikert and Waters [27] showed that LTDFs provided very natural constructions of many cryptographic primitives, including chosen-ciphertext secure public key encryptions, pseudo-random generators, collision-resistant hash functions, and oblivious transfer. Besides the original work of Peikert and Waters, Many other applications of LTDFs were discovered, these include deterministic public key encryption [9], hedged public key encryption [5] and selective-opening secure public key encryption [6].

Another notion related to inner-product trapdoor function is identity-based trapdoor functions (IBTDFs) [7], which can be viewed as an identity-based version of LTDFs. In an IBTDF, each encryption function is associated with an identity and anyone has the secret key corresponding to the same identity can invert. Bellare et al. gave two constructions of IBTDFs and described two applications of IBTDFs in [7]: deterministic identity-based encryption schemes and hedged identity-based encryption schemes. Actually, IBTDFs can be viewed as a special case of our IPTDFs. More specifically, when we use the 2-dimensional attribute vector $\mathbf{x} = (id, -1)$ and predicate vector $\mathbf{v} = (1, id')$ in our IPTDFs, this is exactly the case of IBTDFs, since $\langle \mathbf{x}, \mathbf{v} \rangle = 0$ if and only if $id = id'$.

2 Notations

If x is a string, $|x|$ denotes its length. If S is a set, $|S|$ denotes its size. If S is a set then $s \leftarrow S$ denotes the operation of picking an element s of S uniformly at random. We write $z \leftarrow \mathcal{A}^{\mathcal{O}}(x)$ to indicate that $\mathcal{A}$ is an algorithm with input x and access to oracle $\mathcal{O}$ and output z. We say a function $f(n)$ is *negligible* if $f(n) < 1/n^c$ for any $c > 0$ and all sufficiently large n, denoted as $\mathrm{negl}(n)$. Let X and Y be two random variables over set S. The *statistic distance* between X and Y is defined as $\Delta(X, Y) = \frac{1}{2}\sum_{s \in S} \big| \Pr[X = s] - \Pr[Y = s] \big|$. We say X and Y are statistically indistinguishable if $\Delta(X, Y)$ is negligible.

We use bold capital letters (e.g. $\mathbf{A}$) to denote matrices, and use $\mathbf{I}_n$ to denote the identity matrix with dimension n. We use bold lowercase letters (e.g. $\mathbf{x}$) to denote vectors. $\mathbf{A}^t$ denotes the transpose of the matrix $\mathbf{A}$. When we say a matrix defined over $\mathbb{Z}_q$ has full rank, we mean that it has full rank modulo q. If $\mathbf{A}_1$ is an $n \times m$ matrix and $\mathbf{A}_2$ is an $n \times m'$ matrix, then $[\mathbf{A}_1 | \mathbf{A}_2]$ denotes the $n \times (m + m')$ matrix formed by concatenating $\mathbf{A}_1$ and $\mathbf{A}_2$. If $\mathbf{x}_1$ is a length m vector and $\mathbf{x}_2$ is a length m' vector, then we let $[\mathbf{x}_1 | \mathbf{x}_2]$ denote the length $m + m'$ vector formed by concatenating $\mathbf{x}_1$ and $\mathbf{x}_2$. When doing matrix-vector multiplication, we always view vectors as column vectors.

3 Inner-Product Lossy Trapdoor Functions

In this section, we define the notion of *inner-product trapdoor functions*. In inner-product trapdoor functions, each function value is associated with an attribute $\mathbf{x}$ and each secret key sk_f corresponds to an inner-product predicate f. A user with sk_f can invert the function value if and only if $f(\mathbf{x}) = 1$. An inner-product trapdoor function consists of four algorithms (IPTDF.Pg, IPTDF.Kg, IPTDF.Ev, IPTDF.Inv) associated with input space InSp, output space OutSp, a class of inner-product predicate functions F, and a set of attributes Σ.

IPTDF.Pg(λ) takes as input a security parameter λ. It returns public parameters PP and a master secret key msk.

$\texttt{IPTDF.Kg}(PP, f, msk)$ takes as input public parameters PP, a predicate $f \in \texttt{F}$, and a master secret key msk. It returns an inversion key sk_f for f.

$\texttt{IPTDF.Ev}(PP, \mathbf{x}, \cdot)$ which is an injective function, takes as input public parameters PP, an attribute $\mathbf{x} \in \Sigma$, and a value in $\texttt{InSp}$. It returns a function value in $\texttt{OutSp}$.

$\texttt{IPTDF.Inv}(PP, sk_f, \cdot)$ takes as input public parameters PP, a secret key sk_f for f, and a function value in $\texttt{OutSp}$. It returns a value in $\texttt{InSp}$.

For correctness, we require that $\forall (PP, msk) \leftarrow \texttt{IPTDF.Pg}(\lambda)$, $\forall f \in \texttt{F}$, $\forall sk_f \leftarrow \texttt{IPTDF.Kg}(PP, f, msk)$ and $\forall \mathbf{x} \in \Sigma$, if $C_\mathbf{x} \leftarrow \texttt{IPTDF.Ev}(PP, \mathbf{x}, m)$, where $m \in \texttt{InSp}$,

- If $f(\mathbf{x}) = 1$ then $\texttt{IPTDF.Inv}(PP, sk_f, C_\mathbf{x}) = m$.
- If $f(\mathbf{x}) = 0$ then $\texttt{IPTDF.Inv}(PP, sk_f, C_\mathbf{x}) = \perp$ with all but negligible probability.

An inner-product trapdoor function is associated with a sibling. An ℓ-lossy sibling L-IPTDF=(L-IPTDF.Pg, L-IPTDF.Kg, L-IPTDF.Ev, L-IPTDF.Inv) differs from IPTDF in the following sense:

1. $\texttt{L-IPTDF.Pg}(\lambda, \mathbf{x}^*)$ takes as input a security parameter λ and an attribute $\mathbf{x}^*$. It returns public parameters PP and a master secret key msk. We call $\mathbf{x}^*$ a lossy attribute.
2. $\texttt{L-IPTDF.Kg}(PP, f, msk)$ takes as input public parameters PP, a predicate f, and a master secret key msk. It returns an inversion key sk_f for all f with the requirement that $f(\mathbf{x}^*) = 0$.
3. For any $\mathbf{x} \neq \mathbf{x}^*$, $\texttt{L-IPTDF.Ev}(PP, \mathbf{x}, \cdot)$ computes an injective function over $\texttt{InSp}$, and $\texttt{L-IPTDF.Inv}(PP, sk_f, \cdot)$ computes its inversion if $f(\mathbf{x}) = 1$. Additionally, $\texttt{L-IPTDF.Ev}(PP, \mathbf{x}^*, \cdot)$ computes a function such that $\frac{|\texttt{OutSp}|}{|\texttt{InSp}|} \leq 2^\ell$.

We say IPTDF is ℓ-lossy with sibling L-IPTDF, if for any probabilistic polynomial-time (PPT) adversary $\mathcal{A}$, the advantage of the following game is negligible.

$$\begin{aligned}
&\textbf{Experiment } \mathbf{Exp}^{sAtt\text{-}lossy}_{\texttt{IPTDF},\texttt{L-IPTDF},\mathcal{A}}(\lambda)\\
&\quad \mathbf{x}^* \leftarrow \mathcal{A}(\lambda);\\
&\quad b \leftarrow \{0,1\}, \text{ if } b = 0,\ (PP_0, msk_0) \leftarrow \texttt{IPTDF.Pg}(\lambda);\\
&\quad \text{if } b = 1,\ (PP_1, msk_1) \leftarrow \texttt{L-IPTDF.Pg}(\lambda, \mathbf{x}^*);\\
&\quad b' \leftarrow \mathcal{A}^{\mathcal{O}(b,\cdot)}(PP_b);\\
&\quad \text{if } b = b' \text{ return } 1, \text{ otherwise } 0.
\end{aligned}$$

Where oracle $\mathcal{O}(b, f)$ returns $sk_f \leftarrow \texttt{L-IPTDF.Kg}(PP_1, f, msk_1)$ when $b = 1$, and returns $sk_f \leftarrow \texttt{IPTDF.Kg}(PP_0, f, msk_0)$ when $b = 0$ with the restriction that $\mathcal{A}$ is not allowed to query f such that $f(\mathbf{x}^*) = 1$. We define the advantage of $\mathcal{A}$ in the above experiment as

$$\mathbf{Adv}^{sAtt\text{-}lossy}_{\texttt{IPTDF},\texttt{L-IPTDF},\mathcal{A}}(\lambda) = \left| \Pr[\mathbf{Exp}^{sAtt\text{-}lossy}_{\texttt{IPTDF},\texttt{L-IPTDF},\mathcal{A}}(\lambda) = 1] - \frac{1}{2} \right|.$$

3.1 Lossy Attribute Hiding

We observe that the sibling L-IPTDF enjoys an interesting property. We call it *lossy attribute hiding* property. Informally, the public parameters of the L-IPTDF generated from any distinct lossy attributes are indistinguishable, even given access to obtain the inversion key of predicates that the lossy attributes do not satisfy. For any PPT adversary $\mathcal{A}$ associated with the following game:

$$
\begin{array}{l}
\textbf{Experiment } \mathbf{Exp}^{sAtt\text{-}lah}_{\texttt{L-IPTDF},\mathcal{A}}(\lambda) \\
\quad \mathbf{x}_0, \mathbf{x}_1 \leftarrow \mathcal{A}(\lambda); \\
\quad b \leftarrow \{0,1\},\ (PP_b, msk_b) \leftarrow \texttt{L-IPTDF.Pg}(\lambda, \mathbf{x}_b); \\
\quad b' \leftarrow \mathcal{A}^{\mathcal{O}(b,\cdot)}(PP_b); \\
\quad \text{if } b = b' \text{ return 1, otherwise 0.}
\end{array}
$$

Where oracle $\mathcal{O}(b, f)$ returns $sk_f \leftarrow \texttt{L-IPTDF.Kg}(PP_b, f, msk_b)$ with the restriction that $\mathcal{A}$ is not allowed to query f such that $f(\mathbf{x}_0) = 1$ or $f(\mathbf{x}_1) = 1$. We define the advantage of $\mathcal{A}$ in the above experiment as

$$
\mathbf{Adv}^{sAtt\text{-}lah}_{\texttt{L-IPTDF},\mathcal{A}}(\lambda) = \left| \Pr[\mathbf{Exp}^{sAtt\text{-}lah}_{\texttt{L-IPTDF},\mathcal{A}}(\lambda) = 1] - \frac{1}{2} \right|.
$$

We say L-IPTDF is lossy attribute hiding, if for any PPT adversary $\mathcal{A}$, the above advantage is negligible.

Next, we show that the lossiness of inner-product trapdoor functions implies the lossy attribute hiding property of the corresponding sibling functions.

Lemma 1. *Let* IPTDF *be an inner-product trapdoor function, and* L-IPTDF *be its sibling. If* IPTDF *is ℓ-lossy, then* L-IPTDF *is lossy attribute hiding.*

Proof. Considering the lossy attribute hiding game that the challenger generates the public parameters and master key under $b = 0$, we denote this game as Game_0. Since IPTDF is ℓ-lossy and L-IPTDF is its sibling, no PPT adversary can tell differences with non-negligible probability if the public parameters and master key are changed from IPTDF.Pg(λ), as long as the adversary do not query f with $f(\mathbf{x}_0) = 1$. Analogously, we denote the lossy attribute hiding game as Game_1 when the challenger generates the public parameters and master key under $b = 1$. No PPT adversary can tell differences with non-negligible probability if the public parameters and master key are changed from IPTDF.Pg(λ), as long as the adversary do not query f with $f(\mathbf{x}_1) = 1$. We then conclude that no PPT adversary can distinguish the public parameters between Game_0 and Game_1 with non-negligible probability, with restriction that the adversary do not query f such that $f(\mathbf{x}_0) = 1$ or $f(\mathbf{x}_1) = 1$. This completes the proof. □

Remark. We can similarly define the *lossy identity hiding* property of lossy IBTDFs. This means that the information of the identity will be hidden in the public parameters of lossy IBTDFs. Analogously, the lossiness of IBTDF implies the lossy identity hiding property.

4 Inner-Product Trapdoor Functions from Lattices

Background. A full-rank m-dimensional integer lattice $\Lambda \subseteq \mathbb{Z}^m$ is a discrete additive subgroup whose linear span is $\mathbb{R}^m$. Every lattice is generated as the $\mathbb{Z}$-linear combination of some basis of linearly independent vectors i.e.,$\Lambda = \{\sum_{i=1}^{m} z_i \mathbf{b}_i : z_i \in \mathbb{Z}\}$. In this work we deal exclusively with "q-ary" lattices, where for simplicity we always take q =poly(n) to be prime. For a matrix $\mathbf{A} \in \mathbb{Z}_q^{n\times m}$, define the integer lattice

$$\Lambda^{\perp}(\mathbf{A}) = \{\mathbf{z} \in \mathbb{Z}^m : \mathbf{A}\mathbf{z} = 0 \mod q\}.$$

Let $\mathbf{S} = \{\mathbf{s}_1, ..., \mathbf{s}_k\}$ be a set of vectors in $\mathbb{R}^m$. We use $\widetilde{\mathbf{S}} = \{\widetilde{\mathbf{s}_1}, ..., \widetilde{\mathbf{s}_k}\}$ to denote the Gram-Schmidt orthogonalization of the vectors $\mathbf{s}_1, ..., \mathbf{s}_k$. We use $\|\mathbf{S}\|$ to denote the length of the longest vector in $\mathbf{S}$. For a real-valued matrix $\mathbf{R}$, we let $s_1(\mathbf{R})$ denote the largest singular value of $\mathbf{R}$, i.e. $s_1(\mathbf{R})=\max_{\|\mathbf{u}\|=1}\|\mathbf{R}\mathbf{u}\|$.

Let Λ be a discrete subset of $\mathbb{Z}^m$. For any vector $\mathbf{c} \in \mathbb{R}^m$ and any positive parameter $\sigma \in \mathbb{R}_{>0}$, let $\rho_{\sigma,\mathbf{c}}(\mathbf{x}) = \exp(-\pi\|\mathbf{x} - \mathbf{c}\|^2/\sigma^2)$ be the Gaussian function on $\mathbb{R}^m$ with center $\mathbf{c}$ and parameter σ. Let $\rho_{\sigma,\mathbf{c}}(\Lambda) = \sum_{\mathbf{x}\in\Lambda} \rho_{\sigma,\mathbf{c}}(\mathbf{x})$ be the discrete integral of $\rho_{\sigma,\mathbf{c}}$ over Λ, and let $\mathcal{D}_{\Lambda,\sigma,\mathbf{c}}$ be the discrete Gaussian distribution over Λ with center $\mathbf{c}$ and parameter σ. Specifically, for all $\mathbf{y} \in \Lambda$, we have $\mathcal{D}_{\Lambda,\sigma,\mathbf{c}}(\mathbf{y}) = \frac{\rho_{\sigma,\mathbf{c}}(\mathbf{y})}{\rho_{\sigma,\mathbf{c}}(\Lambda)}$. For notional convenience, $\rho_{\sigma,\mathbf{0}}$ and $\mathcal{D}_{\Lambda,\sigma,\mathbf{0}}$ are abbreviated as ρ_σ and $\mathcal{D}_{\Lambda,\sigma}$, respectively.

Security of our construction reduces to the learning with errors (LWE) problem, a classic hard problem on lattices defined by Regev [28]. The (decisional) learning with errors problem in dimension n with error rate $\alpha \in (0, 1)$, stated in matrix form, is: given an input $(\mathbf{A}, \mathbf{b})$ where $\mathbf{A} \in \mathbb{Z}_q^{n\times m}$ for any m=poly(n) is uniformly random and $\mathbf{b} \in \mathbb{Z}_q^m$ is either of the form $\mathbf{b} = [\mathbf{I}_m|\mathbf{A}^t]\mathbf{x} \mod q$ for $\mathbf{x} \leftarrow \mathcal{D}_{\mathbb{Z},\alpha q}^{m+n}$, or is uniformly random (and independent of $\mathbf{A}$), distinguish which is the case, with non-negligible advantage.[1] By standard hybrid argument, replacing $\mathbf{x}$ with a matrix $\mathbf{X} \in \mathbb{Z}_q^{(m+n)\times\omega}$ (ω = poly(n)) whose each column sampled independently from $\mathcal{D}_{\mathbb{Z},\alpha q}^{m+n}$, and replacing $\mathbf{b}$ with either $\mathbf{B} = [\mathbf{I}_m|\mathbf{A}^t]\mathbf{X} \mod q$ or uniformly random $\mathbf{B}$ of the same dimension, yields an equivalent problem (up to a ω factor in the adversary's advantage). It is known that when $\alpha q \geq 2\sqrt{n}$, this decisional problem is at least as hard as approximating several problems on n-dimensional lattices in the *worst-case* to within $\widetilde{O}(n/\alpha)$ factors with a quantum computer [28] or on a classical computer for a subset of these problems [26]. We give some useful facts for our construction.

Lemma 2 ([22]). *Let Λ be an n-dimensional lattice, let $\mathbf{T}$ be a basis for Λ, and suppose $\sigma \geq \|\widetilde{\mathbf{T}}\| \cdot \omega(\sqrt{\log n})$. Then for any $\mathbf{c} \in \mathbb{R}^n$ we have*

$$\Pr[\|\mathbf{x} - \mathbf{c}\| > \sigma\sqrt{n} : \mathbf{x} \leftarrow \mathcal{D}_{\Lambda,\sigma,\mathbf{c}}] \leq \text{negl}(n).$$

[1] This is actually the "normal form" of the LWE problem, which is equivalent to the one from [28] in which the portion of $\mathbf{x}$ that is multiplied by $\mathbf{A}^t$ is uniformly random in $\mathbb{Z}_q^n$. The equivalence is shown in [4].

Lemma 3 **([17,21]).** *For prime q and integer $b \geq 2$. Let $m \geq n \log_b q + \omega(\log n)$. For $\mathbf{A} \leftarrow \mathbb{Z}_q^{n\times m}$ and $\mathbf{R} \leftarrow \mathcal{D}^{m\times m}_{\mathbb{Z}, b\cdot\omega(\sqrt{\log n})}$. Then $(\mathbf{A}, \mathbf{AR})$ is statistically close to uniform in $\mathbb{Z}_q^{n\times m} \times \mathbb{Z}_q^{n\times m}$.*

Lemma 4 **([3,21]).** *Let q, n, m, b be positive integers with $b \geq 2$ and $m \geq 6n \log_b q$. There is a probabilistic polynomial-time algorithm* $\mathtt{TrapGen}(q, n, m, b)$ *that outputs a pair $(\mathbf{A}, \mathbf{T}) \in \mathbb{Z}_q^{n\times m} \times \mathbb{Z}^{m\times m}$ such that $\mathbf{A}$ is statistically close to uniform in $\mathbb{Z}_q^{n\times m}$ and $\mathbf{T}$ is a basis for $\Lambda^{\perp}(A)$, satisfying $\|\widetilde{\mathbf{T}}\| \leq O(b \cdot \sqrt{n \log_b q})$.*

Lemma 5 **([17,21]).** *Let $\mathbf{A} \in \mathbb{Z}_q^{n\times m}$ be full-rank. Given $\mathbf{A}$ and any basis $\mathbf{T} \in \mathbb{Z}^{m\times m}$ of $\Lambda^{\perp}(\mathbf{A})$, one can efficiently recover $\mathbf{x} = [\mathbf{x}_1|\mathbf{x}_2] \in \mathbb{Z}_q^{m+n}$ from $[\mathbf{I}_m|\mathbf{A}^t] \cdot [\mathbf{x}_1|\mathbf{x}_2] \mod q = \mathbf{A}^t\mathbf{x}_2 + \mathbf{x}_1 \mod q$, as long as $\|\mathbf{x}_1\| \leq q/(2\|\widetilde{\mathbf{T}}\|)$.*

Lemma 6 **([1,16]).** *Let $q > 2, m > n$, $\mathbf{A}, \mathbf{B} \in \mathbb{Z}_q^{n\times m}$, $\mathbf{T_A}$ be a basis of $\Lambda^{\perp}(A)$, and $\sigma \geq \|\widetilde{\mathbf{T_A}}\| \cdot \omega(\sqrt{\log m})$. There exists an efficient randomized algorithm* `SampleLeft` *that, takes as inputs $\mathbf{A}, \mathbf{B}, \mathbf{T_A}, \sigma$, and outputs a basis $\mathbf{S}$ of $\Lambda^{\perp}(\mathbf{U})$ for $\mathbf{U} = [\mathbf{A}|\mathbf{B}]$ with $\|\widetilde{\mathbf{S}}\| \leq \sigma \cdot \sqrt{2m}$ whose distribution depends on $\mathbf{U}, \sigma$.*

Lemma 7 **([1]).** *Let $q > 2, m > n$, $\mathbf{A}, \mathbf{B} \in \mathbb{Z}_q^{n\times m}$, $\mathbf{B}$ be full-rank, $\mathbf{R} \in \mathbb{Z}^{m\times m}$, $\mathbf{T_B}$ be a basis of $\Lambda^{\perp}(\mathbf{B})$, and $\sigma \geq \|\widetilde{\mathbf{T_B}}\| \cdot s_1(\mathbf{R}) \cdot \omega(\sqrt{\log m})$. There exists an efficient randomized algorithm* `SampleRight` *that, takes as inputs $\mathbf{A}, \mathbf{B}, \mathbf{T_B}, \mathbf{R}, \sigma$, and outputs a basis $\mathbf{S}$ of $\Lambda^{\perp}(\mathbf{U})$ for $\mathbf{U} = [\mathbf{A}|\mathbf{AR} + \mathbf{B}]$ with $\|\widetilde{\mathbf{S}}\| \leq \sigma \cdot \sqrt{2m}$ whose distribution depends on $\mathbf{U}, \sigma$.*

4.1 An Inner-Product Trapdoor Function from Lattices

In this subsection, we present a concrete inner-product trapdoor function from lattices. In our construction, each inversion key will be associated with a predicate vector $\mathbf{a} = (a_1, ..., a_\ell) \in \mathbb{Z}_q^\ell$ for some fixed $\ell \geq 2$ and each function value will be associated with an attribute vector $\mathbf{b} = (b_1, ..., b_\ell) \in \mathbb{Z}_q^\ell$. Inversion should succeed if and only if $\langle \mathbf{a}, \mathbf{b} \rangle = 0 \mod q$. Hence the predicate associated with the inversion key is defined as $f_{\mathbf{a}}(\mathbf{b}) = 1$ if $\langle \mathbf{a}, \mathbf{b} \rangle = 0 \mod q$, and $f_{\mathbf{a}}(\mathbf{b}) = 0$ otherwise. We note that we have to append the attribute vector $\mathbf{b}$ after the function value in order to invert it. This restriction limits our `IPTDF` only enables *payload hiding* `IPE` scheme (see Sec. 5), however, this will not harm the lossy attribute hiding property of lossy `IPTDF`s, since the public parameters will still hide the information of the lossy attribute.

Let $c > 1$ be an positive integer to be determined later. Let $n = \lambda$ be a security parameter and ℓ be the length of predicate and attribute vectors. Let $q = \text{poly}(n)$ be a prime, $b = b(n, \ell) \geq 2$ be an integer, $\hat{n} = cn$, and $m = \Theta(\hat{n} \log_b q)$. Let $r = r(n, \ell) \geq 2$ be an integer and define $k = \lfloor \log_r q \rfloor$. Define $D_\beta = \{0, 1, ..., \beta - 1\}$ and D_γ similarly for some positive integers $\beta \geq \gamma$ to be determined later. Let σ, α be positive real Gaussian parameters. Our inner-product trapdoor function has domain `InSp`$=D_\beta^{(\ell(k+1)+1)m+n} \times D_\gamma^{\hat{n}-n}$.

`IPTDF.Pg`(n, ℓ) takes as input a security parameter n and a parameter ℓ, denoting the length of predicate and attribute vectors,

1. Use the algorithm of Lemma 4 to generate a (nearly uniform) $\mathbf{A} \in \mathbb{Z}_q^{\hat{n}\times m}$, together with a basis $\mathbf{T_A}$ for lattice $\Lambda^{\perp}(\mathbf{A})$ such that $\|\widetilde{\mathbf{T_A}}\| = O(b \cdot \sqrt{\hat{n}\log_b q})$.
2. Choose $\ell \cdot (k+1)$ uniformly random matrices $\mathbf{A}_{i,j} \in \mathbb{Z}_q^{\hat{n}\times m}$ for $1 \le i \le \ell$ and $0 \le j \le k$. Choose a uniformly random matrix $\mathbf{B} \in \mathbb{Z}_q^{\hat{n}\times m}$.

Output $PP = (\mathbf{A}, \mathbf{B}, \{\mathbf{A}_{i,j}\}_{1\le i\le \ell, 0\le j\le k})$ and $msk = \mathbf{T_A}$.

`IPTDF.Kg`$(PP, \mathbf{a}, msk)$ takes as input public parameters PP, a predicate vector $\mathbf{a} = (a_1, ..., a_\ell) \in \mathbb{Z}_q^\ell$, and a master secret key msk,

1. For $i = 1, ..., \ell$, write the r-ary decomposition of $a_i \in \mathbb{Z}_q$ as

$$a_i = \sum_{j=0}^{k} a_{i,j} \cdot r^j, \quad \text{where} \quad a_{i,j} \in [0, ..., r-1].$$

2. Define the matrix $\mathbf{U_a} = [\mathbf{A}| \sum_{i=1}^{\ell} \sum_{j=0}^{k} a_{i,j}\mathbf{A}_{i,j}]$.
3. Use the `SampleLeft` algorithm in Lemma 6 to generate a basis $\mathbf{S_a}$ of $\Lambda^{\perp}(\mathbf{U_a})$ with $\|\widetilde{\mathbf{S_a}}\| \le \sigma\sqrt{2m}$.

Output the inversion key $sk_{\mathbf{a}} = \mathbf{S_a}$.

`IPTDF.Ev`$(PP, \mathbf{b}, \mathbf{m})$ takes as input a public parameters PP, an attribute vector $\mathbf{b} = (b_1, ..., b_\ell) \in \mathbb{Z}_q^\ell$, and a message $\mathbf{m} = [\mathbf{x}_0|\mathbf{x}_{1,0}|...|\mathbf{x}_{i,j}|...|\mathbf{x}_{\ell,k}|\mathbf{x}] \in D_\beta^{(\ell(k+1)+1)m+n} \times D_\gamma^{\hat{n}-n}$, where $\mathbf{x}_0, \mathbf{x}_{i,j} \in D_\beta^m$ for $1 \le i \le \ell$, $0 \le j \le k$, and $\mathbf{x} \in D_\beta^n \times D_\gamma^{\hat{n}-n}$,

1. Define the matrix

$$\mathbf{F_b} = [\mathbf{A}|\mathbf{A}_{1,0} + r^0 b_1\mathbf{B}|\cdots|\mathbf{A}_{i,j} + r^j b_i \mathbf{B}|\cdots|\mathbf{A}_{\ell,k} + r^k b_\ell \mathbf{B}].$$

2. Compute $C_{\mathbf{b}} = [\mathbf{I}_{(\ell(k+1)+1)m}|\mathbf{F}_{\mathbf{b}}^t] \cdot \mathbf{m} \mod q$.

Output $C_{\mathbf{b}}$ together with the attribute vector $\mathbf{b}$.

`IPTDF.Inv`$(PP, sk_{\mathbf{a}}, (C_{\mathbf{b}}, \mathbf{b}))$ takes as input public parameters PP, an inversion key $sk_{\mathbf{a}}$ for predicate $\mathbf{a}$, and a function value $(C_{\mathbf{b}}, \mathbf{b})$ for attribute vector $\mathbf{b}$,

1. Parse $C_{\mathbf{b}}$ into $\mathbf{c}_0$, $\mathbf{c}_{i,j}$ for $1 \le i \le \ell, 0 \le j \le k$, where $\mathbf{c}_0 = \mathbf{A}^t\mathbf{x} + \mathbf{x}_0$, $\mathbf{c}_{i,j} = (\mathbf{A}_{i,j} + r^j b_i \mathbf{B})^t\mathbf{x} + \mathbf{x}_{i,j}$.
2. Compute $\widetilde{\mathbf{c}} = \sum_{i=1}^{\ell}\sum_{j=0}^{k} a_{i,j}\mathbf{c}_{i,j} = \sum_{i=1}^{\ell}\sum_{j=0}^{k} a_{i,j}\mathbf{A}_{i,j}^t\mathbf{x} + \langle \mathbf{a}, \mathbf{b}\rangle \mathbf{B}^t\mathbf{x} + \sum_{i=1}^{\ell}\sum_{j=0}^{k} a_{i,j}\mathbf{x}_{i,j}$.
3. Note that $[\mathbf{c}_0|\widetilde{\mathbf{c}}] = [\mathbf{A}| \sum_{i=1}^{\ell}\sum_{j=0}^{k} a_{i,j}\mathbf{A}_{i,j} + \langle \mathbf{a}, \mathbf{b}\rangle \mathbf{B}]^t\mathbf{x} + [\mathbf{x}_0| \sum_{i=1}^{\ell}\sum_{j=0}^{k} a_{i,j}\mathbf{x}_{i,j}]$. If $\langle \mathbf{a}, \mathbf{b}\rangle = 0 \mod q$, use $sk_{\mathbf{a}}$ and the inversion algorithm of Lemma 5 to compute $[\widetilde{\mathbf{x}}|\mathbf{x}]$ from $[\mathbf{c}_0|\widetilde{\mathbf{c}}]$, where $\widetilde{\mathbf{x}} = [\mathbf{x}_0| \sum_{i=1}^{\ell}\sum_{j=0}^{k} a_{i,j}\mathbf{x}_{i,j}]$. Then recover all $\mathbf{x}_{i,j}$ from $\mathbf{c}_{i,j}$ by using $\mathbf{x}$ and the attribute vector $\mathbf{b}$. Finally, It outputs $\mathbf{m}$ if $\langle \mathbf{a}, \mathbf{b}\rangle = 0 \mod q$.

We now describe the sibling L-IPTDF. The evaluation and inversion algorithms are those in IPTDF. We give the parameter generation and inversion key generation algorithms of L-IPTDF as follows.

L-IPTDF.Pg$(n, \ell, \mathbf{b}^*)$ takes as input a security parameter n, a parameter ℓ denoting the length of predicate and attribute vectors, and an attribute vector $\mathbf{b}^* = (b_1^*, ..., b_\ell^*) \in \mathbb{Z}_q^\ell$,

1. Use the algorithm of Lemma 4 to generate a (nearly uniform) $\mathbf{B} \in \mathbb{Z}_q^{\hat{n}\times m}$, together with a basis $\mathbf{T_B}$ for lattice $\Lambda^\perp(\mathbf{B})$ such that $\|\widetilde{\mathbf{T_B}}\| = O(b \cdot \sqrt{\hat{n}\log_b q})$.
2. Choose a uniformly random matrix $\bar{\mathbf{A}} \in \mathbb{Z}_q^{n\times m}$, and $\mathbf{E} \leftarrow \mathcal{D}_{\mathbb{Z},\alpha q}^{(m+n)\times(\hat{n}-n)}$ where $\alpha q = \Theta(\sqrt{n})$, and pairwise independent $\mathbf{R}_{i,j} \leftarrow \mathcal{D}_{\mathbb{Z},b\cdot\omega(\sqrt{\log n})}^{m\times m}$ for $1 \le i \le \ell, 0 \le j \le k$.
3. Set $\mathbf{A}^t = \left[\bar{\mathbf{A}}^t \middle| [\mathbf{I}_m|\bar{\mathbf{A}}^t]\cdot\mathbf{E} \mod q\right]$, and let $\mathbf{A}_{i,j} = \mathbf{A}\mathbf{R}_{i,j} - r^j b_i^* \mathbf{B}$ for $1 \le i \le \ell, 0 \le j \le k$.

Output $PP = (\mathbf{A}, \mathbf{B}, \{\mathbf{A}_{i,j}\}_{1\le i\le\ell,0\le j\le k})$ and msk=$(\mathbf{T_B}, \{\mathbf{R}_{i,j}\}_{1\le i\le\ell,0\le j\le k})$.

L-IPTDF.Kg(PP, f, msk) takes as input public parameters PP, a master secret key msk, and a predicate vector $\mathbf{a} = (a_1, ..., a_\ell) \in \mathbb{Z}_q^\ell$,

1. Define the matrix

$$\mathbf{U_a} = [\mathbf{A}|\sum_{i=1}^{\ell}\sum_{j=0}^{k} a_{i,j}\mathbf{A}_{i,j}] = [\mathbf{A}|\mathbf{A}(\sum_{i=1}^{\ell}\sum_{j=0}^{k} a_{i,j}\mathbf{R}_{i,j}) - \langle \mathbf{a}, \mathbf{b}^*\rangle\mathbf{B}].$$

2. If $\langle \mathbf{a}, \mathbf{b}^*\rangle \neq 0 \mod q$, use the SampleRight algorithm in Lemma 7 to generate a basis $\mathbf{S_a}$ of $\Lambda^\perp(\mathbf{U_a})$ with $\|\widetilde{\mathbf{S_a}}\| \le \sigma\sqrt{2m}$, else abort.

4.2 Correctness

We now show that for certain parameter choices, the inversion algrithms of IPTDF and L-IPTDF work correctly with overwhelming probability, and the evaluation of L-IPTDF.Ev$(PP, \mathbf{b}^*, \cdot)$ is lossy.

Lemma 8. *Suppose the parameters γ, β, b satisfy*

$$\gamma^{c-1} \ge 2^{\Omega((\ell(k+1)+1)m/n)}, \text{ and } \gamma\cdot\widetilde{\Omega}(bm\sqrt{n}) \le \beta \le \frac{q}{2\sqrt{2}\sigma m((r-1)\ell(k+1)+1)}.$$

We have:

1. *If $\langle \mathbf{a}, \mathbf{b}\rangle = 0 \mod q$, then with overwhelming probability* IPTDF.Inv *invert correctly, and* L-IPTDF.Inv *invert correctly with $\mathbf{b} \neq \mathbf{b}^*$.*
2. L-IPTDF.Ev$(PP, \mathbf{b}^*, \cdot)$ *is a lossy function with lossiness $\Omega((\ell(k+1)+1)m)$.*

Proof. During the second step of IPTDF.Inv($PP, \mathbf{S_a}, (C_\mathbf{b}, \mathbf{b})$), we compute $\widetilde{\mathbf{c}}$, which is equal to

$$\widetilde{\mathbf{c}} = \sum_{i=1}^{\ell}\sum_{j=0}^{k} a_{i,j}\mathbf{c}_{i,j} = \sum_{i=1}^{\ell}\sum_{j=0}^{k} a_{i,j}\mathbf{A}_{i,j}^t\mathbf{x} + \langle \mathbf{a}, \mathbf{b}\rangle \mathbf{B}^t\mathbf{x} + \sum_{i=1}^{\ell}\sum_{j=0}^{k} a_{i,j}\mathbf{x}_{i,j}.$$

If $\langle \mathbf{a}, \mathbf{b}\rangle = 0 \mod q$, then the middle term disappears, leaving $\widetilde{\mathbf{c}} = \sum_{i=1}^{\ell}\sum_{j=0}^{k} a_{i,j}\mathbf{A}_{i,j}^t\mathbf{x} + \sum_{i=1}^{\ell}\sum_{j=0}^{k} a_{i,j}\mathbf{x}_{i,j}$. It follows that

$$[\mathbf{c}_0|\widetilde{\mathbf{c}}] = [\mathbf{A}|\sum_{i=1}^{\ell}\sum_{j=0}^{k} a_{i,j}\mathbf{A}_{i,j}]^t \cdot \mathbf{x} + \widetilde{\mathbf{x}} = [\mathbf{I}_{2m}|\mathbf{U}_\mathbf{a}^t] \cdot [\widetilde{\mathbf{x}}|\mathbf{x}] \mod q,$$

where $\widetilde{\mathbf{x}} = [\mathbf{x}_0|\sum_{i=1}^{\ell}\sum_{j=0}^{k} a_{i,j}\mathbf{x}_{i,j}]$. Since $\mathbf{S_a}$ is a basis of $\Lambda^\perp(\mathbf{U_a})$, according to Lemma 5, if $\|\widetilde{\mathbf{x}}\| \le q/(2\|\widetilde{\mathbf{S_a}}\|)$, one can recover $[\widetilde{\mathbf{x}}|\mathbf{x}]$ by using $\mathbf{S_a}$. From the $\mathbf{c}_{i,j}$, the attribute vector $\mathbf{b}$ and $\mathbf{x}$, one can obtain $\mathbf{x}_{i,j}$ for $1 \le i \le \ell, 0 \le j \le k$. By Lemma 6 and 7 we have $\|\widetilde{\mathbf{S_a}}\| \le \sigma\sqrt{2m}$ with overwhelming probability. Since $\mathbf{m} \in D_\beta^{(\ell(k+1)+1)m+n} \times D_\gamma^{\hat{n}-n}$, and $a_{i,j} \in \{0, ..., r-1\}$, by the triangle inequality, we have

$$\begin{aligned}\|\widetilde{\mathbf{x}}\| \le \|\mathbf{x}_0\| + \|\sum_{i=1}^{\ell}\sum_{j=0}^{k} a_{i,j}\mathbf{x}_{i,j}\| &\le \beta\sqrt{m} + (r-1)\ell(k+1)\beta\sqrt{m} \\ &= (1 + (r-1)\ell(k+1))\beta\sqrt{m}.\end{aligned}$$

For β as in the lemma statement, $\beta \le \frac{q}{2\sqrt{2}\sigma m((r-1)\ell(k+1)+1)}$ is sufficient to recover $\mathbf{m}$, as desired.

In the third step of L-IPTDF.Pg, $\mathbf{A}^t = \left[\bar{\mathbf{A}}^t \middle| [\mathbf{I}_m|\bar{\mathbf{A}}^t] \cdot \mathbf{E} \mod q\right]$ and $\mathbf{A}_{i,j} = \mathbf{A}\mathbf{R}_{i,j} - r^j b_i^* \mathbf{B}$ for $1 \le i \le \ell, 0 \le j \le k$, then in L-IPTDF.Ev($PP, \mathbf{b}^*, \mathbf{m}$), we compute

$$\begin{aligned}\mathbf{F}_{\mathbf{b}^*} &= [\mathbf{A}|\mathbf{A}_{1,0} + r^0 b_1^*\mathbf{B}|\cdots|\mathbf{A}_{i,j} + r^j b_i^*\mathbf{B}|\cdots|\mathbf{A}_{\ell,k} + r^k b_\ell^*\mathbf{B}] \\ &= [\mathbf{A}|\mathbf{A}\mathbf{R}_{1,0}|\cdots|\mathbf{A}\mathbf{R}_{i,j}|\cdots|\mathbf{A}\mathbf{R}_{\ell,k}] = \mathbf{A}[\mathbf{I}_m|\mathbf{R}_{1,0}|\cdots|\mathbf{R}_{i,j}|\cdots|\mathbf{R}_{\ell,k}].\end{aligned}$$

Let $\mathbf{R} = [\mathbf{I}_m|\mathbf{R}_{1,0}|\cdots|\mathbf{R}_{i,j}|\cdots|\mathbf{R}_{\ell,k}]$, therefore,

$$\begin{aligned}\mathbf{F}_{\mathbf{b}^*}^t = \mathbf{R}^t \cdot \left[\bar{\mathbf{A}}^t \middle| [\mathbf{I}_m|\bar{\mathbf{A}}^t] \cdot \mathbf{E}\right] &= \left[(\bar{\mathbf{A}}\mathbf{R})^t \middle| [\mathbf{R}^t|(\bar{\mathbf{A}}\mathbf{R})^t] \cdot \mathbf{E}\right] \\ &= \left[(\bar{\mathbf{A}}\mathbf{R})^t \middle| [\mathbf{I}_{(\ell(k+1)+1)m}|(\bar{\mathbf{A}}\mathbf{R})^t] \begin{bmatrix}\mathbf{R}^t & \mathbf{0} \\ \mathbf{0} & \mathbf{I}_n\end{bmatrix} \cdot \mathbf{E}\right] \\ &= \left[(\bar{\mathbf{A}}\mathbf{R})^t \middle| [\mathbf{I}_{(\ell(k+1)+1)m}|(\bar{\mathbf{A}}\mathbf{R})^t] \cdot \mathbf{E}'\right],\end{aligned}$$

where, $\mathbf{E}' = \begin{bmatrix}\mathbf{R}^t & \mathbf{0} \\ \mathbf{0} & \mathbf{I}_n\end{bmatrix} \mathbf{E}$. Note that

$$s_1(\mathbf{E}') \le s_1(\mathbf{R}^t)s_1(\mathbf{E}) \le \widetilde{O}(b\sqrt{m}) \cdot O(\sqrt{mn}) \le \widetilde{O}(bm\sqrt{n}).$$

We have the following,

$$\begin{aligned} C_{\mathbf{b}^*} &= [\mathbf{I}_{(\ell(k+1)+1)m}|\mathbf{F}^t_{\mathbf{b}^*}] \cdot \mathbf{m} \\ &= [\mathbf{I}_{(\ell(k+1)+1)m}|(\bar{\mathbf{A}}\mathbf{R})^t] \cdot ([\mathbf{I}_{(\ell(k+1)+1)m+n}|\mathbf{E}'] \cdot \mathbf{m}) \mod q. \end{aligned}$$

Therefore, it suffices to bound the number of possible values of the form

$$[\mathbf{I}_{(\ell(k+1)+1)m+n}|\mathbf{E}'] \cdot \mathbf{m} \mod q.$$

Define $N_d(s)$ to be the number of integer points in an d-dimensional ball of radius s. For $r \geq \sqrt{d}$, from the volume of the ball and Stirling's approximation, we have $N_d(s) = O(s/\sqrt{d})^d$. Therefore, the number of possible values of $[\mathbf{I}_{(\ell(k+1)+1)m+n}|\mathbf{E}'] \cdot \mathbf{m}$ is at most $N_{(\ell(k+1)+1)m+n}(\|[\mathbf{I}_{(\ell(k+1)+1)m+n}|\mathbf{E}'] \cdot \mathbf{m}\|)$. Since $\mathbf{m} = [\mathbf{x}_0|\mathbf{x}_{1,0}|...|\mathbf{x}_{i,j}|...|\mathbf{x}_{\ell,k}|\mathbf{x}] \in D_\beta^{(\ell(k+1)+1)m+n} \times D_\gamma^{\hat{n}-n}$, we have

$$\begin{aligned} \|[\mathbf{I}_{(\ell(k+1)+1)m+n}|\mathbf{E}'] \cdot \mathbf{m}\| &\leq \beta \cdot \sqrt{(\ell(k+1)+1)m+n} + s_1(\mathbf{E}') \cdot \gamma\sqrt{\hat{n}-n} \\ &\leq \sqrt{(\ell(k+1)+1)m+n} \cdot (\beta + \gamma \cdot s_1(\mathbf{E}')). \end{aligned}$$

Therefore, the number of possible values of $[\mathbf{I}_{(\ell(k+1)+1)m+n}|\mathbf{E}'] \cdot \mathbf{m}$ is at most $O(\beta + \gamma \cdot s_1(\mathbf{E}'))^{(\ell(k+1)+1)m+n}$. For lossiness, observe that the base-2 logarithm of the domain size of L-IPTDF.Ev$(PP, \mathbf{b}^*, \cdot)$ is

$$((\ell(k+1)+1)m+n)\log\beta + (c-1)n\log\gamma.$$

Whereas by the above, and for $\beta \geq \gamma \cdot s_1(\mathbf{E}')$, the base-2 logarithm of the image size of L-IPTDF.Ev$(PP, \mathbf{b}^*, \cdot)$ is at most

$$\begin{aligned} &((\ell(k+1)+1)m+n)\log(O(\beta + \gamma \cdot s_1(\mathbf{E}'))) \\ &\quad \leq ((\ell(k+1)+1)m+n)\log\beta + O((\ell(k+1)+1)m). \end{aligned}$$

Let $\gamma^{c-1} \geq 2^{\Omega((\ell(k+1)+1)m/n)}$, and for sufficient large constant in $\Omega(\cdot)$, the two quantities above differ by at least $\Omega((\ell(k+1)+1)m)$ as desired. □

4.3 Security

In this subsection, we show that the inner-product function described above is $\Omega((\ell(k+1)+1)m)$-lossy under selective attribute attacker.

Theorem 1. *Suppose β, γ as in Lemma 8. If decisional LWE problem is infeasible with error rate α, then* IPTDF *described above is $\Omega((\ell(k+1)+1)m)$-lossy with sibling* L-IPTDF *described above, under selective attribute adversaries.*

Proof. We define a series of games (Game$_0$,...,Game$_3$) where in Game$_0$, an adversary $\mathcal{A}$ is against IPTDF, that is, the public parameter generation, key generation algorithms are from IPTDF. While in Game$_3$, $\mathcal{A}$ is against L-IPTDF, that is, the parameter generation, key generation algorithms are from L-IPTDF. We show that the adversary's views in the first game and the last game are indistinguishable.

Game$_0$: $\mathcal{A}$ submits a challenge attribute vector $\mathbf{b}^* = (b_1^*, ..., b_\ell^*)$ before setup. The challenger uses the algorithm of Lemma 4 to obtain $(\mathbf{A}, \mathbf{T_A}) \in \mathbb{Z}_q^{\hat{n}\times m} \times \mathbb{Z}^{m\times m}$, and chooses uniformly random matrices $\mathbf{B}, \mathbf{A}_{i,j} \in \mathbb{Z}_q^{\hat{n}\times m}$ for $1 \le i \le \ell, 0 \le j \le k$. The challenger gives $(\mathbf{A}, \mathbf{B}, \{\mathbf{A}_{i,j}\}_{1\le i\le \ell, 0\le j\le k})$ to $\mathcal{A}$, and keeps $\mathbf{T_A}$ as the master secret key. When $\mathcal{A}$ queries an inversion key for a predicate vector $\mathbf{a}$ with $\langle \mathbf{a}, \mathbf{b}^*\rangle \neq 0 \mod q$, the challenger uses $\mathbf{T_A}$ to respond an inversion key $\mathbf{S_a}$ by invoking algorithm `SampleLeft` from Lemma 6, the distribution of $\mathbf{S_a}$ depends on $\mathbf{U_a}$ and σ .

Game$_1$: This game is identical to Game$_0$ except that, the challenger changes the way to generate $\mathbf{A}_{i,j}$ for $1 \le i \le \ell, 0 \le j \le k$. Instead, the challenger first chooses pairwise independent $\mathbf{R}_{i,j} \in \mathcal{D}^{m\times m}_{\mathbb{Z}, b\cdot\omega(\sqrt{\log n})}$ for $1 \le i \le \ell, 0 \le j \le k$. Let $\mathbf{A}_{i,j} = \mathbf{A}\mathbf{R}_{i,j} - r^j b_i^* \mathbf{B}$.

Game$_2$: This game is identical to Game$_1$ except that the challenger changes the way to generate the master secret key and respond the key-extraction query. Instead, the challenger uses the algorithm of Lemma 4 to obtain $(\mathbf{B}, \mathbf{T_B})$, and chooses $\mathbf{A} \leftarrow \mathbb{Z}_q^{\hat{n}\times m}$ uniformly at random. When $\mathcal{A}$ queries an inversion key for a predicate vector $\mathbf{a}$ with $\langle \mathbf{a}, \mathbf{b}^*\rangle \neq 0 \mod q$, the challenger uses $\mathbf{T_B}$ and $\mathbf{R}_{i,j}$ for $1 \le i \le \ell, 0 \le j \le k$ to respond an inversion key $\mathbf{S_a}$ by invoking algorithm `SampleRight` from Lemma 7, the distribution of $\mathbf{S_a}$ depends on $\mathbf{U_a}$ and σ.

Game$_3$: This game is identical to Game$_2$ except that the challenger changes the way to generate $\mathbf{A}$. Instead, the challenger chooses a uniformly random matrix $\bar{\mathbf{A}} \in \mathbb{Z}_q^{n\times m}$, and $\mathbf{E} \leftarrow \mathcal{D}^{(m+n)\times(\hat{n}-n)}_{\mathbb{Z},\alpha q}$ where $\alpha q = \Theta(\sqrt{n})$. Set $\mathbf{A}^t = \left[\bar{\mathbf{A}}^t \middle| [\mathbf{I}_m|\bar{\mathbf{A}}^t]\cdot \mathbf{E} \mod q\right]$.

It's obvious that Game$_0$ is the `IPTDF` security definition and Game$_3$ is the `L-IPTDF` security definition. We now show that the adversary's views between the adjacent games are indistinguishable.

The only difference of Game$_0$ of Game$_1$ is the way $\mathbf{A}_{i,j}$ generated. In Game$_1$, $\mathbf{A}$ is uniform, therefore $\mathbf{A}, \mathbf{A}\mathbf{R}_{1,0}, ..., \mathbf{A}\mathbf{R}_{\ell,k}$ is statistically close to a uniform string of the same size by Lemma 3, and so is $\mathbf{A}, \mathbf{A}\mathbf{R}_{1,0} - r^0 b_1^*\mathbf{B}, ..., \mathbf{A}\mathbf{R}_{\ell,k} - r^k b_\ell^* \mathbf{B}$. Thus, the adversary's views between Game$_0$ and Game$_1$ are statistically indistinguishable.

The differences of Game$_1$ and Game$_2$ are the way $\mathbf{A}, \mathbf{B}$ generated and the way to answer key-extraction queries. By Lemma 4, the distributions of $\mathbf{A}, \mathbf{B}$ in Game$_1$ and Game$_2$ are statistically close. Therefore, the distributions of $\mathbf{U_a}$ in Game$_1$ and Game$_2$ are statistically close. In Game$_2$, note that $\mathbf{U_a} = [\mathbf{A}|\mathbf{A}(\sum_{i=1}^{\ell}\sum_{j=0}^{k} a_{i,j}\mathbf{R}_{i,j}) - \langle \mathbf{a}, \mathbf{b}^*\rangle \mathbf{B}]$, it can invoke the algorithm `SampleRight` by Lemma 7 as long as $\langle \mathbf{a}, \mathbf{b}^*\rangle \neq 0 \mod q$. By Lemma 6 and Lemma 7, for sufficiently large σ, the distribution of $\mathbf{S_a}$ in Game$_1$ and Game$_2$ are statistically close.

Apparently, matrix $\mathbf{A}$ in Game$_2$ and Game$_3$ is computational indistinguishable if the decisional LWE problem is infeasible. Summarize the above discussions, we complete the proof $\square$

4.4 Parameter Selection

We can extract from the above description the parameters required for correctness and security of the system. By Lemma 8 we require

$$\gamma^{c-1} \geq 2^{\Omega((\ell(k+1)+1)m/n)}, \text{ and } \gamma\cdot\widetilde{\Omega}(bm\sqrt{n}) \leq \beta \leq \frac{q}{2\sqrt{2}\sigma m((r-1)\ell(k+1)+1)}.$$

For security, we require $\alpha q = \Theta(\sqrt{n})$. The constants c and γ depend on the relationship of ℓ, m, and n. We need $\gamma^{c-1} \geq 2^{\Omega((\ell(k+1)+1)m/n)}$. In order to generate $\mathbf{A}$ with a trapdoor, we have $m = \Theta(\hat{n}\log_b q)$, so we need $\gamma > q^{\Theta((\ell(k+1)+1)/\log b)\cdot c/c-1}$. For any desired constant $C > 1$ and $\ell = O(\log n)$, we can choose constants $k > 1, c > 1$ and choose $b = \Theta(n)$ such that $\gamma \leq q^{1/C}$. The additional constraints imposed by our security reduction are as follows. From the description of `IPTDF.Pg`, we have $\|\widetilde{\mathbf{T}_\mathbf{A}}\| = O(b\sqrt{\hat{n}\log_b q})$ by Lemma 4, in order to respond the key-extraction queries by `SampleLeft`, σ subjects to the requirement that

$$\sigma \geq \|\widetilde{\mathbf{T}_\mathbf{A}}\| \cdot \omega(\sqrt{\log m}) = O(b\sqrt{\hat{n}\log_b q}) \cdot \omega(\sqrt{\log m}).$$

From the description of `L-IPTDF.Pg`, we have $\|\widetilde{\mathbf{T}_\mathbf{B}}\| = O(b\sqrt{\hat{n}\log_b q})$ by Lemma 4, in order to respond the key-extraction queries by `SampleRight`, σ subjects to the requirement that

$$\sigma \geq \|\widetilde{\mathbf{T}_\mathbf{B}}\| \cdot s_1(\sum_{i=1}^{\ell}\sum_{j=0}^{k} a_{i,j}\mathbf{R}_{i,j}) \cdot \omega(\sqrt{\log m}).$$

Since $\mathbf{R}_{i,j}$ are chosen from $\mathcal{D}^{m\times m}_{\mathbb{Z},b\omega(\sqrt{\log n})}$, and $a_{i,j} \in \{0, ..., r-1\}$, it follows that $s_1(\sum_{i=1}^{\ell}\sum_{j=0}^{k} a_{i,j}\mathbf{R}_{i,j}) \leq \widetilde{O}((r-1)\ell(k+1)\cdot b\sqrt{m})$ with overwhelming probability. We see that it suffices to choose

$$\sigma \geq O(b\sqrt{\hat{n}\log_b q}) \cdot \widetilde{O}((r-1)\ell(k+1)b\sqrt{m}) \cdot \omega(\sqrt{\log m}).$$

To satisfy the more stringent of the above two conditions, we set $\sigma = \Theta(r\ell k n^{3.5})$. For correctness and lossiness, it suffices to take

$$q^{1/C} \cdot \widetilde{\Omega}(bm\sqrt{n}) \leq \beta \leq q/(2\sqrt{2}\sigma m(r-1)(\ell(k+1)+1)).$$

In order to satisfy all the constraints, it is sufficient to set $r = q^{1/C'}$ for some constant $C' > 1$ (therefore $k = C'$ is a constant), and sufficiently large q such that

$$q^{1-1/C-2/C'} \geq \widetilde{\Omega}(m^2 n^5 \ell^2 k^2) = \widetilde{\Omega}(n^7 \ell^2 k^2).$$

The following selection of parameters satisfies all of these constrains. For a given $\ell = O(\log n)$ and constant C, C', set

$$m = 20cn, \qquad \beta = n^{3+\delta}, \qquad \gamma = n^{\delta},$$
$$\sigma = \lceil r\ell k n^{3.5}\rceil, \quad \alpha = (n^{7+\delta}\ell^2 k^2)^{-1}, \quad q = \text{the prime nearest to } \lceil n^{7.5+\delta}\ell^2 k^2\rceil,$$

where $n^{\delta} = \lceil q^{1/C+2/C'}\rceil$, and constant c is set as in the analysis, and $b = n$. Observe that the above setting of parameters satisfies all the constrains, the security of the scheme can be based on the hardness of approximating SIVP and GapSvp to within a factor of $\widetilde{O}(n/\alpha) = \widetilde{O}(n^{8+\delta}\ell^2 k^2)$ in the worst case by quantum algorithms.

5 Applications

In this section, we describe some applications of our IPTDF. These applications include chosen-plaintext secure IPE schemes and chosen-ciphertext secure IPE scheme. Katz, Sahai, and Waters [19] introduce two basic security notions of IPE: *payload hiding* and *attribute hiding*. Payload hiding guarantees that no efficient adversary can obtain any information about the encrypted message, but allows information about attributes to be revealed. Attribute hiding is a stronger notion which guarantees in addition that no efficient adversary can obtain any information about the attribute associated with a ciphertext.

Chosen-Plaintext Secure Inner-Product Encryption. A straightforward application of IPTDF is for inner-product encryption (IPE). By the lossiness of IPTDFs, it is easy for us to obtain a payload hiding IPE scheme (via hardcore bits) under selectively chosen-plaintext adversaries. However, as mentioned in Sec. 4.1 we have to append the attribute vector after the function value, therefore, anyone can learn the information of the attribute from the function value. In this case we can not achieve attribute hiding IPE schemes using our IPTDF, we leave it as a future work to construct IPTDFs whose attribute information is hidden in the function value.

Due to its lossiness, our IPTDF together with a pairwise independent hash function h imply an IPE scheme for multi-bit messages (with length $O(\ell n)$). The ciphertext of the IPE scheme consists of $c = (\texttt{IPTDF.Ev}(PP, \mathbf{b}, x), h(x) \oplus m)$, where x is randomly chosen from the domain of IPTDF and h, and m is the message. Our concrete construction of IPTDF is inspired by [2], then the efficiency of our IPE scheme is almost the same as the one in [2] except that our scheme can encrypt multi-bit messages simultaneously. However, our scheme only supports attribute vectors with logarithmic length, while the scheme in [2] supports attribute vectors with polynomial length.

Chosen-Ciphertext Secure Inner-Product Encryption. Peikert and Waters [27] gave a framework to construct chosen-ciphertext secure public key encryption schemes from lossy trapdoor functions. Our inner-product lossy trapdoor function also works in this framework. Following the framework in [27], to obtain a payload hiding IPE scheme under selectively chosen-ciphertext adversaries, one can combine an IPTDF and an All-But-One (ABO) [27] trapdoor function with a strongly unforgeable one-time signature. The ciphertext of the IPE scheme consists of $c = (vk, \texttt{IPTDF.Ev}(PP, \mathbf{b}, x), \texttt{ABO-TDF.Ev}(vk, x), h(x) \oplus m, \sigma)$, where x is randomly chosen from the domain of IPTDF and ABO-TDF, h is a pairwise independent hash function, m is the message, and σ is the one-time signature of $\texttt{IPTDF.Ev}(PP, \mathbf{b}, x)$, $\texttt{ABO-TDF.Ev}(vk, x)$, and $h(x) \oplus m$ under the signing key associated to vk.

In order to base the resulting IPE scheme on lattices, we need to provide an ABO trapdoor function based on lattices.[2] We have two ways to address this problem. The first one is to use the original lattice based ABO trapdoor function presented in [27]. However, this construction brings large public key size. We prefer to the second one, and as a by-product, we give a generic construction for ABO trapdoor functions. We observe that the lossy sibling of IBTDF is actually an ABO lossy trapdoor function if we view an identity as a *branch*. The lossy identity hiding property is exactly the hidden branch property of ABO trapdoor functions (even the adversary can access an oracle to obtain other inversion keys). Bellare et al. presented a "direct" construction of IBTDF from lattices. Use this ABO trapdoor function and our IPTDF scheme, we obtain the first chosen-ciphertext secure IPE scheme based on lattices.

Note that a generic method to construct chosen-ciphertext secure IPE scheme is the CHK/BK [15,12] transform. The CHK/BK transform transforms a 2-level chosen-plaintext secure hierarchical IPE scheme into chosen-ciphertext secure IPE scheme. However, the only lattice based IPE scheme in [2] seems difficult to be extended into a hierarchical IPE scheme. Therefore, it seems difficult to obtain a chosen-ciphertext secure IPE scheme from the scheme in [2] using CHK/BK transform.

Acknowledgements. Xiang Xie would like to thank Chris Peikert for helpful discussions. The authors would like to thank the anonymous reviewers for their useful comments.

References

1. Agrawal, S., Boneh, D., Boyen, X.: Efficient Lattice (H)IBE in the Standard Model. In: Gilbert, H. (ed.) EUROCRYPT 2010. LNCS, vol. 6110, pp. 553–572. Springer, Heidelberg (2010)
2. Agrawal, S., Freeman, D.M., Vaikuntanathan, V.: Functional Encryption for Inner Product Predicates from Learning with Errors. In: Lee, D.H. (ed.) ASIACRYPT 2011. LNCS, vol. 7073, pp. 21–40. Springer, Heidelberg (2011)
3. Alwen, J., Peikert, C.: Generating shorter bases for hard random lattices. Theory of Computing Systems, 1–19 (2011)
4. Applebaum, B., Cash, D., Peikert, C., Sahai, A.: Fast Cryptographic Primitives and Circular-Secure Encryption Based on Hard Learning Problems. In: Halevi, S. (ed.) CRYPTO 2009. LNCS, vol. 5677, pp. 595–618. Springer, Heidelberg (2009)

[2] It is easy to get strongly unforgeable one-time signatures from lattices.

5. Bellare, M., Brakerski, Z., Naor, M., Ristenpart, T., Segev, G., Shacham, H., Yilek, S.: Hedged Public-Key Encryption: How to Protect against Bad Randomness. In: Matsui, M. (ed.) ASIACRYPT 2009. LNCS, vol. 5912, pp. 232–249. Springer, Heidelberg (2009)
6. Bellare, M., Hofheinz, D., Yilek, S.: Possibility and Impossibility Results for Encryption and Commitment Secure under Selective Opening. In: Joux, A. (ed.) EUROCRYPT 2009. LNCS, vol. 5479, pp. 1–35. Springer, Heidelberg (2009)
7. Bellare, M., Kiltz, E., Peikert, C., Waters, B.: Identity-Based (Lossy) Trapdoor Functions and Applications. In: Pointcheval, D., Johansson, T. (eds.) EUROCRYPT 2012. LNCS, vol. 7237, pp. 228–245. Springer, Heidelberg (2012)
8. Bethencourt, J., Sahai, A., Waters, B.: Ciphertext-policy attribute-based encryption. In: IEEE Symposium on Security and Privacy, SP 2007, pp. 321–334. IEEE (2007)
9. Boldyreva, A., Fehr, S., O'Neill, A.: On Notions of Security for Deterministic Encryption, and Efficient Constructions without Random Oracles. In: Wagner, D. (ed.) CRYPTO 2008. LNCS, vol. 5157, pp. 335–359. Springer, Heidelberg (2008)
10. Boneh, D., Franklin, M.: Identity-Based Encryption from the Weil Pairing. In: Kilian, J. (ed.) CRYPTO 2001. LNCS, vol. 2139, pp. 213–229. Springer, Heidelberg (2001)
11. Boneh, D., Franklin, M.: Identity-based encryption from the Weil pairing. SIAM J. of Computing 32(3), 586–615 (2003)
12. Boneh, D., Katz, J.: Improved Efficiency for CCA-Secure Cryptosystems Built Using Identity-Based Encryption. In: Menezes, A. (ed.) CT-RSA 2005. LNCS, vol. 3376, pp. 87–103. Springer, Heidelberg (2005)
13. Boneh, D., Sahai, A., Waters, B.: Functional encryption: Definitions and challenges. Theory of Cryptography, 253–273 (2011)
14. Boneh, D., Waters, B.: Conjunctive, subset, and range queries on encrypted data. Theory of Cryptography, 535–554 (2007)
15. Canetti, R., Halevi, S., Katz, J.: Chosen-Ciphertext Security from Identity-Based Encryption. In: Cachin, C., Camenisch, J.L. (eds.) EUROCRYPT 2004. LNCS, vol. 3027, pp. 207–222. Springer, Heidelberg (2004)
16. Cash, D., Hofheinz, D., Kiltz, E., Peikert, C.: Bonsai Trees, or How to Delegate a Lattice Basis. In: Gilbert, H. (ed.) EUROCRYPT 2010. LNCS, vol. 6110, pp. 523–552. Springer, Heidelberg (2010)
17. Gentry, C., Peikert, C., Vaikuntanathan, V.: Trapdoors for hard lattices and new cryptographic constructions. In: Proceedings of the 40th Annual ACM Symposium on Theory of Computing, pp. 197–206. ACM (2008)
18. Goyal, V., Pandey, O., Sahai, A., Waters, B.: Attribute-based encryption for fine-grained access control of encrypted data. In: Proceedings of the 13th ACM Conference on Computer and Communications Security, pp. 89–98. ACM (2006)
19. Katz, J., Sahai, A., Waters, B.: Predicate Encryption Supporting Disjunctions, Polynomial Equations, and Inner Products. In: Smart, N.P. (ed.) EUROCRYPT 2008. LNCS, vol. 4965, pp. 146–162. Springer, Heidelberg (2008)
20. Lewko, A., Okamoto, T., Sahai, A., Takashima, K., Waters, B.: Fully Secure Functional Encryption: Attribute-Based Encryption and (Hierarchical) Inner Product Encryption. In: Gilbert, H. (ed.) EUROCRYPT 2010. LNCS, vol. 6110, pp. 62–91. Springer, Heidelberg (2010)
21. Micciancio, D., Peikert, C.: Trapdoors for Lattices: Simpler, Tighter, Faster, Smaller. In: Pointcheval, D., Johansson, T. (eds.) EUROCRYPT 2012. LNCS, vol. 7237, pp. 700–718. Springer, Heidelberg (2012)

22. Micciancio, D., Regev, O.: Worst-case to average-case reductions based on gaussian measures. SIAM Journal on Computing 37(1), 267 (2007)
23. Okamoto, T., Takashima, K.: Hierarchical Predicate Encryption for Inner-Products. In: Matsui, M. (ed.) ASIACRYPT 2009. LNCS, vol. 5912, pp. 214–231. Springer, Heidelberg (2009)
24. Okamoto, T., Takashima, K.: Fully Secure Functional Encryption with General Relations from the Decisional Linear Assumption. In: Rabin, T. (ed.) CRYPTO 2010. LNCS, vol. 6223, pp. 191–208. Springer, Heidelberg (2010)
25. Okamoto, T., Takashima, K.: Adaptively Attribute-Hiding (Hierarchical) Inner Product Encryption. In: Pointcheval, D., Johansson, T. (eds.) EUROCRYPT 2012. LNCS, vol. 7237, pp. 591–608. Springer, Heidelberg (2012)
26. Peikert, C.: Public-key cryptosystems from the worst-case shortest vector problem: extended abstract. In: Proceedings of the 41st Annual ACM Symposium on Theory of Computing, pp. 333–342. ACM (2009)
27. Peikert, C., Waters, B.: Lossy trapdoor functions and their applications. In: Proceedings of the 40th Annual ACM Symposium on Theory of Computing, pp. 187–196. ACM (2008)
28. Regev, O.: On lattices, learning with errors, random linear codes, and cryptography. In: Proceedings of the Thirty-Seventh Annual ACM Symposium on Theory of Computing, pp. 84–93. ACM (2005)
29. Sahai, A., Waters, B.: Fuzzy Identity-Based Encryption. In: Cramer, R. (ed.) EUROCRYPT 2005. LNCS, vol. 3494, pp. 457–473. Springer, Heidelberg (2005)
30. Shamir, A.: Identity-Based Cryptosystems and Signature Schemes. In: Blakely, G.R., Chaum, D. (eds.) CRYPTO 1984. LNCS, vol. 196, pp. 47–53. Springer, Heidelberg (1985)
31. Waters, B.: Dual System Encryption: Realizing Fully Secure IBE and HIBE under Simple Assumptions. In: Halevi, S. (ed.) CRYPTO 2009. LNCS, vol. 5677, pp. 619–636. Springer, Heidelberg (2009)

On the Joint Security of Signature and Encryption Schemes under Randomness Reuse: Efficiency and Security Amplification

Afonso Arriaga[1], Manuel Barbosa[1], and Pooya Farshim[2]

[1] HASLab/INESC TEC, Universidade do Minho, Braga, Portugal
[2] Department of Computer Science, Darmstadt University of Technology, Germany
{arriaga,mbb}@di.uminho.pt, farshim@cased.de

Abstract. We extend the work of Bellare, Boldyreva and Staddon on the systematic analysis of randomness reuse to construct multi-recipient encryption schemes to the case where randomness is reused across different cryptographic primitives. We find that through the additional binding introduced through randomness reuse, one can actually obtain a *security amplification* with respect to the standard black-box compositions, and achieve a stronger level of security. We introduce stronger notions of security for encryption and signatures, where challenge messages can depend in a restricted way on the random coins used in encryption, and show that two variants of the KEM/DEM paradigm give rise to encryption schemes that meet this enhanced notion of security. We obtain the most efficient signcryption scheme to date that is secure against insider attackers without random oracles.

Keywords: Signcryption, Insider Security, Randomness Reuse.

1 Introduction

Signcryption is a cryptographic primitive that aims to simultaneously provide the guarantees of public-key encryption and signature schemes [17], i.e., confidentiality, integrity, authentication and possibly non-repudiation, whilst offering efficiency gains. One trivial way to obtain the signcryption functionality—if one is not interested in saving computational power or bandwidth—is to use a black-box combination of the two primitives. This approach was systematically studied by An, Dodis and Rabin [2], by looking at Encrypt-then-Sign (EtS), Sign-then-Encrypt (StE) and Encrypt-and-Sign (EaS) compositions. The former two constructions are natural sequential compositions of the two primitives, whereas EaS is a parallel composition using a commitment scheme to enforce the necessary binding. A well-known, albeit surprising, result in [2] is that the interaction between the signature and encryption primitives can work *against* the security of the composition, making it impossible to achieve the strongest levels of security, even when the underlying encryption and signature schemes are themselves strongly secure. For example, in an StE construction an attacker knowing the secret key of the receiver is always able to forge valid signcryptions simply by decrypting and re-encrypting the contents of a legitimately generated ciphertext, regardless of the security guarantees provided by the underlying signature scheme: this translates into a trivial break of unforgeability against insider attackers in the signcryption setting.

F. Bao, P. Samarati, and J. Zhou (Eds.): ACNS 2012, LNCS 7341, pp. 206–223, 2012.

One general approach to obtaining efficiency gains in cryptography is to reuse randomness across instantiations of various cryptographic algorithms. This technique can allow for significant savings in processing load and bandwidth, as partial results (and even ciphertext elements) can be shared between multiple instances of cryptographic algorithms. For this reason, randomness reuse is frequently used in the context of batch encryption operations where (possibly different) messages are encrypted to multiple recipients, as recognized by Kurosawa [11] in the construction of multi-recipient encryption schemes. Furthermore, randomness reuse is also used as an optimization technique, in an ad-hoc way, in the construction of signcryption schemes [17,16]. Nevertheless, this avenue must be pursued with caution, since randomness reuse may, of course, hinder the security of the resulting cryptographic schemes.

Bellare et al. [3], building on the work of Kurosawa [11], systematically study the problem of reusing randomness in multi-recipient encryption. The authors consider the particular case of constructing such schemes by running multiple instances of a public-key encryption (PKE) scheme, whilst sharing randomness across them. An interesting result in this work is a general method for identifying PKE schemes that are secure when used in this scenario. Schemes which satisfy the so-called *reproducibility test* permit establishing the security for multiple recipients with randomness reuse through a variant of the hybrid argument.

OUR CONTRIBUTIONS. In this paper we extend the work of Bellare et al. [3] to the case where randomness is reused across *different* cryptographic primitives, and analyze the security of signcryption schemes constructed by composing encryption and signature schemes under randomness reuse. More in detail, our contributions are the following:

- We define a compatibility notion that establishes classes of signature and encryption schemes that can be composed under randomness reuse to obtain correct signcryption schemes. We then identify security properties that are sufficient for the EtS and StE compositions with randomness reuse to result in secure signcryption schemes. In particular, we introduce the notion of *randomness-dependent security* for both signatures and encryption schemes. Intuitively, security must be preserved when the messages chosen by attackers are allowed to depend (in a restricted way) on the implicit randomness input to the underlying cryptographic algorithms. We believe these security notions may be of independent interest in the study of the role of randomness in cryptographic security and, particularly, in the generic analysis of randomness reuse optimizations for scenarios where multiple (possibly heterogenous) cryptographic operations are carried out in a batch procedure (e.g., optimizing the overall performance of a server continuously carrying out key agreement, signature and encryption operations).
- We find that through the additional binding that is established via the reuse of randomness, it is possible to achieve *full* insider security. Our results hold in the dynamic multi-user setting, although in some cases we require adversaries to register the full key pairs of all users created for the attack. This is usually called the *registered key model* [13] and it captures natural restrictions in many PKI settings. This is a *security amplification* with respect to the equivalent compositions without randomness reuse, in which it is *not* possible to achieve this level of security, even starting from underlying schemes providing the same security guarantees we

require for our results. In other words, our results depend in an essential way on reusing randomness, and it is *not* the case that a standard composition of randomness-dependent secure signature and encryption schemes trivially yields a comparable result. In this respect, our work generalizes independent work in the same direction presented in [13], and that we contextualize in Section 2.
- We identify a set of simple and natural properties of KEMs and DEMs that suffice to ensure that PKE schemes constructed from *both* variants of the KEM/DEM composition paradigm proposed in [9,10] fall within our framework. As a particular case, when the Kurosawa–Desmedt [12] encryption scheme is composed with the Boneh–Boyen signature scheme [6] in the StE construction, we obtain the most efficient signcryption scheme to be proven insider secure in the standard model. One caveat is that our results hold only in the registered key model. In compensation, our scheme offers non-repudiation, inherited from the StE construction, and a combination of computational and communication (bandwidth) efficiency that outperforms previous solutions.

STRUCTURE OF THE PAPER. In the next section we review the related work in more detail (a more extensive discussion can be found in the full version of the paper). Then, in Section 3 we settle notation by introducing the standard syntax, correctness and security definitions for signature, encryption and signcryption schemes. In Section 4 we describe the properties that are sufficient for the EtS and StE compositions to yield secure signcryption schemes under randomness reuse, and prove the corresponding composition theorems. Finally, in Section 5 we describe the potential instantiations of our framework and compare our scheme with existing results.

2 Related Work

Matsuda et al. [13] and Chiba et al. [8] systematically study the construction of signcryption schemes using compositions of standard cryptographic primitives, aiming to obtain levels of efficiency and security that are comparable to the best concrete schemes in the literature via generic constructions. Independently of our work, Matsuda et al. [13] show how to perform compositions of tag-based KEMs and signature schemes to obtain efficiency gains in an StE-like construction via randomness reuse. They also describe a series of schemes that can be used to instantiate these constructions. The resulting compositions are efficient and achieve full insider security, with the caveat that strong unforgeability can only be proven in a slightly weaker model, where the adversary must register the secret keys for the public keys it chooses to query to the signcryption oracle. Our results have the same limitation.

The main differences between our work and the approach in [13] are the following. Our results are more general in that they consider the composition of encryption schemes and signature schemes under randomness reuse, rather than lower level primitives. On one hand, this sets our results as natural extensions to the work by An et al. [2] on signature and encryption compositions, and also of the work of Bellare et al. [3], allowing us to establish a connection between the two results. On the other hand, our results capture the ones included in [13] on randomness reuse for the construction of signcryption schemes as particular cases, and cover a broader class of constructions. More precisely, our compatibility framework and security results apply to general encryption schemes, rather than those specifically constructed from tag-based KEMs. This allows us to capture not only schemes constructed using a specific flavor of tag-KEMs [13], but also encryption schemes constructed from other known variants of the KEM/DEM paradigm [9,10], and even schemes that do not follow this paradigm.

Chiba et al. [8] propose the first fully secure signcryption schemes in the standard model by using a variant of the StE construction that relies on a chosen-ciphertext-secure tag-based KEM, a chosen-ciphertext-secure DEM that has a "one-to-one" property, and a strongly unforgeable signature scheme. Such schemes are less efficient than the one we propose, but are proven secure without the key registration requirement.

3 Preliminaries

NOTATION. We write $a \leftarrow b$ to denote the algorithmic action of assigning the value of b to the variable a. We use $\perp \notin \{0,1\}^\star$ to denote special failure symbol. If S is a set, we write $a \leftarrow_\$ \mathsf{S}$ for sampling a from S uniformly at random. If $\mathcal{A}$ is a probabilistic algorithm we write $a \leftarrow_\$ \mathcal{A}(i_1, i_2, \ldots, i_n)$ for the action of running $\mathcal{A}$ on inputs $i_1, i_2, \ldots, i_n$ with random coins, and assigning the result to a. Sometimes we run $\mathcal{A}$ on specific coins r and write $a \leftarrow \mathcal{A}(i_1, i_2, \ldots, i_n; \mathsf{r})$.

GAMES. In this paper we use the code-based game-playing language [4]. Each game has an **Initialize** and a **Finalize** procedure. It also has specifications of procedures to respond to an adversary's various queries. A game is run with an adversary $\mathcal{A}$ as follows. First **Initialize** runs and its outputs are passed to $\mathcal{A}$. Then $\mathcal{A}$ runs and its oracle queries are answered by the procedures of the game. When $\mathcal{A}$ terminates, its output is passed to **Finalize** which returns the outcome of the game. In each game, we restrict attention to legitimate adversaries, which is defined specifically for each game. We use lists as data structures to keep relevant state in the games. The empty list is represented by square brackets [ ]. We denote by $\mathsf{List} \leftarrow a : \mathsf{List}$ the action of appending element a to the head of a list List.

PUBLIC-KEY ENCRYPTION. A public-key encryption scheme $\mathcal{E} = (\mathsf{EGen}, \mathsf{Enc}, \mathsf{Dec})$ is specified by three polynomial-time algorithms (in the length of their inputs) associated with a message space $\mathcal{M}$ and a randomness space $\mathcal{R}$.

- $\mathsf{EGen}(1^\lambda)$ is the probabilistic key-generation algorithm, taking as input the security parameter and returning a secret key sk and a public key pk.
- $\mathsf{Enc}(\mathsf{m}, \mathsf{pk}; \mathsf{r})$ is the probabilistic encryption algorithm. On input a message $\mathsf{m} \in \mathcal{M}$, a public key pk, and possibly some random coins $\mathsf{r} \in \mathcal{R}$, this algorithm outputs a ciphertext c.
- $\mathsf{Dec}(\mathsf{c}, \mathsf{sk})$ is the deterministic decryption algorithm. On input of a ciphertext c and a key sk, this algorithm outputs a message m or failure symbol $\perp$.

The correctness of a public-key encryption scheme requires that for any $\lambda \in \mathbb{N}$, any $(\mathsf{sk}, \mathsf{pk}) \leftarrow_\$ \mathsf{EGen}(1^\lambda)$, any $\mathsf{m} \in \mathcal{M}$, and any random coins $\mathsf{r} \in \mathcal{R}$, we have that $\mathsf{Dec}(\mathsf{Enc}(\mathsf{m}, \mathsf{pk}; \mathsf{r}), \mathsf{sk}) = \mathsf{m}$.

The standard notion of security for a public-key encryption scheme considered here is indistinguishability under chosen-ciphertext attacks (IND-CCA). We refer the interested reader to the full version of the paper for a formal definition.

DIGITAL SIGNATURE. A signature scheme $\mathcal{S} = (\mathsf{SGen}, \mathsf{Sign}, \mathsf{Verify})$ is specified by three polynomial-time algorithms with a randomness space $\mathcal{R}$ and a message space $\mathcal{M}$.

- $\mathsf{SGen}(1^\lambda)$ is the probabilistic key-generation algorithm which takes as input the security parameter and returns a secret key sk and a public key pk.
- $\mathsf{Sign}(\mathsf{m}, \mathsf{sk}; \mathsf{r})$ is the probabilistic signature generation algorithm. On input a message m, a secret key sk, and possibly some random coins $\mathsf{r} \in \mathcal{R}$, this algorithm outputs a signature σ.
- $\mathsf{Verify}(\mathsf{m}, \sigma, \mathsf{pk})$ is the deterministic signature verification algorithm. On input of a signature σ, a message m and a public key pk, this algorithm outputs a boolean value T or F.

The correctness of a signature scheme requires that for any $\lambda \in \mathbb{N}$, any $\mathsf{m} \in \{0,1\}^*$, any $(\mathsf{sk}, \mathsf{pk}) \leftarrow_\$ \mathsf{SGen}(1^\lambda)$, and any $\mathsf{r} \in \mathcal{R}$, we have that $\mathsf{Verify}(\mathsf{Sign}(\mathsf{m}, \mathsf{sk}; \mathsf{r}), \mathsf{m}, \mathsf{pk}) = \mathsf{T}$.

The standard notion of security for a digital signature scheme considered in this paper is strong existential unforgeability under chosen-message attacks (sUF-CMA). We refer the interested reader to the full version of the paper for a formal definition.

SIGNCRYPTION. A signcryption scheme $\mathcal{SC} = (\mathsf{Gen}, \mathsf{Signcrypt}, \mathsf{Unsigncrypt})$ is specified by three polynomial-time algorithms associated with a message space $\mathcal{M}$ and a randomness space $\mathcal{R}$.

- $\mathsf{Gen}(1^\lambda)$ is the probabilistic key-generation algorithm which takes as input the security parameter and returns a secret key sk and a matching public key pk. Unless one wishes to signcrypt a message to oneself, two key pairs are required to signcrypt and unsigncrypt.
- $\mathsf{Signcrypt}(\mathsf{m}, \mathsf{sk}_S, \mathsf{pk}_R; \mathsf{r})$ is the probabilistic signcryption algorithm. On input a message $\mathsf{m} \in \mathcal{M}$, the sender's secret key sk_S, the receiver's public key pk_R, and possibly some random coins $\mathsf{r} \in \mathcal{R}$, this algorithm outputs a signcryption c.
- $\mathsf{Unsigncrypt}(\mathsf{c}, \mathsf{pk}_S, \mathsf{sk}_R)$ is the deterministic unsigncryption algorithm. On input a signcryption c, the sender's public key pk_S, and the receiver's secret key sk_R, this algorithm outputs a message m or failure symbol $\perp$.

The correctness of a signcryption scheme requires that for any $\mathsf{m} \in \mathcal{M}$, any $\lambda \in \mathbb{N}$, any $(\mathsf{sk}_S, \mathsf{pk}_S) \leftarrow_\$ \mathsf{Gen}(1^\lambda)$, any $(\mathsf{sk}_R, \mathsf{pk}_R) \leftarrow_\$ \mathsf{Gen}(1^\lambda)$, and any random coins $\mathsf{r} \in \mathcal{R}$, we have $\mathsf{Unsigncrypt}(\mathsf{Signcrypt}(\mathsf{m}, \mathsf{sk}_S, \mathsf{pk}_R; \mathsf{r}), \mathsf{pk}_S, \mathsf{sk}_R) = \mathsf{m}$. We consider here the strong notion of confidentiality, introduced by [16], in which the adversary is allowed to choose without restrictions pk_S to query to the **Unsigncrypt** oracle. The adversary may also choose the challenge key pair $(\mathsf{sk}_S, \mathsf{pk}_S)$, but the key pair is required to be valid. Analogously to IND-CCA for encryption, **LoR** oracle can only be called once. We refer to this model as dynamic multi-user indistinguishability against insider chosen-ciphertext attacks (IND-iCCA).

Definition 1. *A signcryption scheme is* IND-iCCA *secure if, for every legitimate PPT adversary $\mathcal{A}$, the following definition of advantage is negligible in λ*

$$\mathbf{Adv}^{\mathsf{IND\text{-}iCCA}}_{\mathcal{SC},\mathcal{A}}(\lambda) := 2 \cdot \Pr[\mathsf{IND\text{-}iCCA}_{\mathcal{SC},\mathcal{A}}(1^\lambda) \Rightarrow \mathsf{T}] - 1\,,$$

where game $\mathsf{IND\text{-}iCCA}_{\mathcal{SC},\mathcal{A}}$ described in Figure 1.

```
procedure Initialize(1^λ):              procedure LoR(m_0, m_1, (sk_S, pk_S)):
(sk_R, pk_R) ←$ Gen(1^λ)                c ←$ Signcrypt(m_b, sk_S, pk_R)
b ←$ {0, 1}                             List ← (c, pk_S) : List
List ← []                               Return c
Return (pk_R)
                                        procedure Unsigncrypt(c, pk_S):
procedure Finalize(b'):                 m ← Unsigncrypt(c, pk_S, sk_R)
Return (b = b')                         Return m
```

Fig. 1. Game IND-iCCA for a signcryption $\mathcal{SC} = (\mathsf{Gen}, \mathsf{Signcrypt}, \mathsf{Unsigncrypt})$. An adversary $\mathcal{A}$ is legitimate if: 1) it calls **LoR** once, with $\mathsf{m}_0, \mathsf{m}_1 \in \mathcal{M}$ and $|\mathsf{m}_0| = |\mathsf{m}_1|$, and a valid key pair $(\mathsf{sk}_S, \mathsf{pk}_S)$; and 2) it does not query **Unsigncrypt** with $(\mathsf{c}, \mathsf{pk}_S) \in \mathsf{List}$.

We also define dynamic multi-user strong existential unforgeability against insider chosen message attacks for authenticity, but in a slightly weaker model that obliges the adversary to register a key pair $(\mathsf{sk}_R, \mathsf{pk}_R)$ before querying the **Signcrypt** oracle or **Finalize** with pk_R. For this purpose, a **Key-Reg** oracle is also available. This model is called sUF-iCMA, for short.

Definition 2. *A signcryption scheme is* sUF-iCMA *secure if, for every legitimate PPT adversary $\mathcal{A}$, the following definition of advantage is negligible in λ*

$$\mathbf{Adv}^{\mathsf{sUF\text{-}iCMA}}_{\mathcal{SC},\mathcal{A}}(\lambda) := \Pr[\mathsf{sUF\text{-}iCMA}_{\mathcal{SC},\mathcal{A}}(1^\lambda) \Rightarrow \mathsf{T}] ,$$

where game $\mathsf{sUF\text{-}iCMA}_{\mathcal{SC},\mathcal{A}}$ *described in Figure 2.*

```
procedure Initialize(1^λ):              procedure Key-Reg(sk, pk):
(sk_S, pk_S) ←$ Gen(1^λ)                If isValid(sk, pk)
List ← []                                   List' ← (sk, pk) : List'
List' ← []                                  Return T
Return pk_S                             Else Return F

procedure Signcrypt(m, pk_R):           procedure Finalize(c, pk_R):
If (*, pk_R) ∈ List'                    If (c, pk_R) ∈ List Return F
    c ←$ Signcrypt(m, sk_S, pk_R)       If (sk_R, pk_R) ∈ List'
    List ← (c, pk_R) : List                 m ← Unsigncrypt(c, pk_S, sk_R)
    Return c                                If m ≠⊥ Return T
Else Return ⊥                           Return F
```

Fig. 2. Game sUF-iCMA for a signcryption $\mathcal{SC} = (\mathsf{Gen}, \mathsf{Signcrypt}, \mathsf{Unsigncrypt})$

REMARK. We assume one can confirm the validity of a key pair using an efficient algorithm isValid, which is usually the case for practical schemes. Under this assumption, one could omit the key pair validity restriction in the adversary legitimacy definition in the IND-iCCA security model, and require the signcryption algorithm to internally check for sender key pair validity. This does not apply to the key registration oracle in the unforgeability game, as this is conditioning the adversary to provide valid key pairs for the *receivers* (note that this check cannot be done internally by the signcryption algorithm). However, one could remove the validity check restriction in **Finalize**, and require that the unsigncryption algorithm does check for receiver key pair validity.

4 Compositions with Randomness Reuse

In this section we look at black-box compositions of signature and encryption under randomness reuse. We describe properties that are sufficient for the encrypt-then-sign and sign-then-encrypt constructions with shared randomness to yield secure signcryption schemes, and prove the corresponding composition theorems. Our proposed framework gives rise to signcryption schemes that attain full insider security in dynamic multi-user models. We defer a discussion on instantiability to Section 5.

4.1 Composition-Enabling Properties

PARTITIONED SCHEMES, COMPATIBILITY, AND CONDITIONAL INJECTIVITY. The notion of joint signature and encryption in the public-key setting with randomness reuse implies that the signature and encryption algorithms share the same randomness space. In order to clarify the concept and simplify the security proofs, we will restrict our attention to *partitioned* schemes [7]. Furthermore, to enable composition under randomness reuse, we also require the signature and encryption schemes to be *compatible*. We formalize these notions next.

Definition 3 (Partitioned schemes). *We say a signature scheme is* partitioned, *if its signature space is composed of pairs* (σ, R), *where the signature generation algorithm calculates* R *independently of the input message and keys. More precisely, we require that experiment* $\mathbf{Indep}_{\mathcal{S}}$ *in Figure 3 returns* T *with probability* 1 *for all messages* m_0 *and* m_1 *in the appropriate space. Similarly, an encryption scheme is* partitioned, *if its ciphertext space is composed of pairs* (c, R) *and experiment* $\mathbf{Indep}_{\mathcal{E}}$ *in Figure 3 returns* T *with probability* 1 *for all messages* m_0 *and* m_1 *in the appropriate space.*

Definition 4 (Compatibility). *A signature scheme* $\mathcal{S}$ *and an encryption scheme* $\mathcal{E}$ *are compatible if they are partitioned, share the same random space* $\mathcal{R}$, *and the experiment* $\mathbf{Compatibility}$ *in Figure 3 returns* T *with probability* 1 *for any messages* m_0 *and* m_1 *in the appropriate spaces.*

test $\mathbf{Indep}_{\mathcal{S}}(\mathsf{m}_0, \mathsf{m}_1)$:	**test** $\mathbf{Indep}_{\mathcal{E}}(\mathsf{m}_0, \mathsf{m}_1)$:	**test** $\mathbf{Compatibility}_{\mathcal{S},\mathcal{E}}(\mathsf{m}_0, \mathsf{m}_1)$:
$(\mathsf{sk}_0, \mathsf{pk}_0) \leftarrow_\$ \mathsf{SGen}(1^\lambda)$	$(\mathsf{sk}_0, \mathsf{pk}_0) \leftarrow_\$ \mathsf{EGen}(1^\lambda)$	$(\mathsf{sk}_0, \mathsf{pk}_0) \leftarrow_\$ \mathsf{SGen}(1^\lambda)$
$(\mathsf{sk}_1, \mathsf{pk}_1) \leftarrow_\$ \mathsf{SGen}(1^\lambda)$	$(\mathsf{sk}_1, \mathsf{pk}_1) \leftarrow_\$ \mathsf{EGen}(1^\lambda)$	$(\mathsf{sk}_1, \mathsf{pk}_1) \leftarrow_\$ \mathsf{EGen}(1^\lambda)$
$\mathsf{r} \leftarrow_\$ \mathcal{R}$	$\mathsf{r} \leftarrow_\$ \mathcal{R}$	$\mathsf{r} \leftarrow_\$ \mathcal{R}$
$(\sigma_0, \mathsf{R}_0) \leftarrow \mathsf{Sign}(\mathsf{m}_0, \mathsf{sk}_0; \mathsf{r})$	$(\mathsf{c}_0, \mathsf{R}_0) \leftarrow \mathsf{Enc}(\mathsf{m}_0, \mathsf{pk}_0; \mathsf{r})$	$(\sigma, \mathsf{R}_0) \leftarrow \mathsf{Sign}(\mathsf{m}_0, \mathsf{sk}_0; \mathsf{r})$
$(\sigma_1, \mathsf{R}_1) \leftarrow \mathsf{Sign}(\mathsf{m}_1, \mathsf{sk}_1; \mathsf{r})$	$(\mathsf{c}_1, \mathsf{R}_1) \leftarrow \mathsf{Enc}(\mathsf{m}_1, \mathsf{pk}_1; \mathsf{r})$	$(\mathsf{c}, \mathsf{R}_1) \leftarrow \mathsf{Enc}(\mathsf{m}_1, \mathsf{pk}_1; \mathsf{r})$
Return $(\mathsf{R}_0 = \mathsf{R}_1)$	Return $(\mathsf{R}_0 = \mathsf{R}_1)$	Return $(\mathsf{R}_0 = \mathsf{R}_1)$

Fig. 3. Partitioning and compatibility tests for a partitioned signature $\mathcal{S} = (\mathsf{SGen}, \mathsf{Sign}, \mathsf{Verify})$, and a partitioned public-key encryption $\mathcal{E} = (\mathsf{EGen}, \mathsf{Enc}, \mathsf{Dec})$

Finally, we also require the following injectivity properties in partitioned schemes, which essentially state that, once the randomness dependent component R is fixed, and for any fixed key pair, the signature generation and encryption algorithms become injective mappings from the message space onto the signature and ciphertext spaces, respectively. We observe that these properties can be relaxed to computational hardness assumptions, and all our results still go through.

Definition 5 (Conditional injectivity). *We say a partitioned signature scheme is* conditionally injective *if for all key pairs* $(\mathsf{sk}, \mathsf{pk})$*, all messages* m *and signatures* (σ, R) *in the appropriate spaces, it holds that:*

$$\mathsf{Sign}(\mathsf{m}, \mathsf{sk}) = (\sigma, \mathsf{R}) \wedge \sigma \neq \sigma' \Rightarrow \mathsf{Verify}(\mathsf{m}, (\sigma', \mathsf{R}), \mathsf{pk}) = \mathsf{F}.$$

We say a partitioned encryption scheme is conditionally injective *if for all key pairs* $(\mathsf{sk}, \mathsf{pk})$*, messages* m *and ciphertexts* (c, R) *in the appropriate spaces, it holds that:*

$$\mathsf{Enc}(\mathsf{m}, \mathsf{pk}) = (\mathsf{c}, \mathsf{R}) \wedge \mathsf{c} \neq \mathsf{c}' \Rightarrow \mathsf{Dec}((\mathsf{c}', \mathsf{R}), \mathsf{sk}) \neq \mathsf{m}.$$

REPRODUCIBILITY. Following the approach of Bellare et al. [3], we introduce new notions of *reproducibility* that allow identifying encryption and signature schemes for which it is possible to prove that randomness reuse does not hurt the security of compositions.

Definition 6 (Reproducibility). *We say that a signature scheme is* reproducible *if there exists a deterministic polynomial-time reproduction algorithm* $\mathsf{Rep}_{\mathcal{S}}$ *(resp.* $\mathsf{Rep}_{\mathcal{E}}$*) taking a message, a secret key, and a value* R *such that experiment* $\mathbf{Rep}_{\mathcal{S}}$ *(resp.* $\mathbf{Rep}_{\mathcal{E}}$*) in Figure 4 returns* T *with overwhelming probability for all messages* m *in the appropriate space.*

test $\mathbf{Rep}_{\mathcal{S}}(\mathsf{m})$:	**test** $\mathbf{Rep}_{\mathcal{E}}(\mathsf{m})$:
$(\mathsf{sk}, \mathsf{pk}) \leftarrow_{\$} \mathsf{SGen}(1^{\lambda})$	$(\mathsf{sk}, \mathsf{pk}) \leftarrow_{\$} \mathsf{EGen}(1^{\lambda})$
$\mathsf{r} \leftarrow_{\$} \mathcal{R}$	$\mathsf{r} \leftarrow_{\$} \mathcal{R}$
$(\sigma, \mathsf{R}) \leftarrow \mathsf{Sign}(\mathsf{m}, \mathsf{sk}; \mathsf{r})$	$(\mathsf{c}, \mathsf{R}) \leftarrow \mathsf{Enc}(\mathsf{m}, \mathsf{pk}; \mathsf{r})$
$\sigma' \leftarrow_{\$} \mathsf{Rep}_{\mathcal{S}}(\mathsf{m}, \mathsf{sk}, \mathsf{R})$	$\mathsf{c}' \leftarrow_{\$} \mathsf{Rep}_{\mathcal{E}}(\mathsf{m}, \mathsf{sk}, \mathsf{R})$
Return $(\sigma = \sigma')$	Return $(\mathsf{c} = \mathsf{c}')$

Fig. 4. Reproducibility test for a partitioned signature $\mathcal{S} = (\mathsf{SGen}, \mathsf{Sign}, \mathsf{Verify})$ with reproducibility algorithm $\mathsf{Rep}_{\mathcal{S}}$, and a partitioned public-key encryption $\mathcal{E} = (\mathsf{EGen}, \mathsf{Enc}, \mathsf{Dec})$ with reproducibility algorithm $\mathsf{Rep}_{\mathcal{E}}$

Intuitively, the schemes are reproducible if it is possible to reconstruct a valid signature (resp. ciphertext) without having explicit access to the random coins, but instead having access to the secret key. We note that this property seems natural for encryption schemes, where knowledge of the secret key may "compensate" for the lack of knowledge of the implicit randomness. As for signature schemes, this property may seem less natural, as the reproducibility algorithm should be able to produce valid signatures, while having access to apparently less information than the signature generation algorithm itself. However, one can easily see that if $\mathsf{R} = \mathsf{r}$, then a signature scheme is trivially reproducible. Furthermore, Matsuda et al. [13] present various (standard model) signature schemes that, not having this characteristic, are shown to be reproducible. We note that our formalization defines reproducibility as a property of a single scheme, and not as a property of a pair of schemes. We see this as an important definitional choice in ensuring that our framework can be extended to reason about randomness reuse between other cryptographic primitives.

4.2 Security under Randomness-Dependent Attacks

We introduce two new attack models, one for encryption and one for digital signatures. In a nutshell, these models allow messages queried by the adversaries to the relevant oracles to depend on the randomness component R, so this is provided to the adversary in advance. These models are specific for partitioned schemes and aimed at proving security under randomness reuse. We defer considerations on the feasibility of achieving this level of security to the following section.

SECURITY OF ENCRYPTION UNDER RANDOMNESS-DEPENDENT ATTACKS. We define a new security model for encryption, which we call "indistinguishability under randomness-dependent chosen-ciphertext attacks" (IND-RDA). This new model is similar to IND-CCA except that the adversary receives the R component for the challenge in the beginning of the game. To capture this notion of security, rather than partitioning the encryption algorithm, we simply encrypt the fixed all-zeros message at the beginning of the game, in order to obtain a pair (r, R). Note that, since R is guaranteed not to depend on the message, we have that reusing r to produce the challenge ciphertext will yield a consistent security game definition.

Definition 7. *A public-key encryption scheme is* IND-RDA *secure if, for every legitimate PPT adversary $\mathcal{A}$, the following definition of advantage is negligible in λ*

$$\mathbf{Adv}^{\mathsf{IND\text{-}RDA}}_{\mathcal{E},\mathcal{A}}(\lambda) := 2 \cdot \Pr[\mathsf{IND\text{-}RDA}_{\mathcal{E},\mathcal{A}}(1^\lambda) \Rightarrow \mathsf{T}] - 1\,,$$

where game $\mathsf{IND\text{-}RDA}_{\mathcal{E},\mathcal{A}}$ *described in Figure 5.*

procedure Initialize(1^λ):	procedure LoR$(\mathsf{m}_0, \mathsf{m}_1)$:	procedure Dec(c, R):
$b \leftarrow_\$ \{0,1\}$	$(\mathsf{c}, \mathsf{R}) \leftarrow \mathsf{Enc}(\mathsf{m}_b, \mathsf{pk}; \mathsf{r})$	$\mathsf{m} \leftarrow \mathsf{Dec}((\mathsf{c}, \mathsf{R}), \mathsf{sk})$
$\mathsf{List} \leftarrow [\,]$	$\mathsf{List} \leftarrow (\mathsf{c}, \mathsf{R}) : \mathsf{List}$	Return m
$(\mathsf{sk}, \mathsf{pk}) \leftarrow_\$ \mathsf{EGen}(1^\lambda)$	Return (c, R)	
$\mathsf{r} \leftarrow_\$ \mathcal{R}$		procedure Finalize(b'):
$(\mathsf{c}, \mathsf{R}) \leftarrow \mathsf{Enc}(0, \mathsf{pk}; \mathsf{r})$		Return $(b = b')$
Return $(\mathsf{pk}, \mathsf{R})$		

Fig. 5. Game IND-RDA for a partitioned public-key encryption $\mathcal{E} = (\mathsf{EGen}, \mathsf{Enc}, \mathsf{Dec})$. An adversary $\mathcal{A}$ is legitimate if: 1) it calls **LoR** once, with $\mathsf{m}_0, \mathsf{m}_1 \in \mathcal{M}$ and $|\mathsf{m}_0| = |\mathsf{m}_1|$; and 2) it never calls **Dec** on $(\mathsf{c}, \mathsf{R}) \in \mathsf{List}$.

SECURITY OF SIGNATURES UNDER RANDOMNESS-DEPENDENT ATTACKS. We also introduce a new security notion for partitioned signature schemes, which we call "strong unforgeability under randomness-dependent chosen message attacks" (sUF-RDA). This new model is similar to sUF-CMA, with the caveat that calls to the **Sign** oracle are done in two steps. On a first interaction, the adversary obtains the randomness component for the signature scheme, and in the next step it provides the message on which the full signature is generated.

Definition 8. *A digital signature scheme is* sUF-RDA *secure if, for every legitimate PPT adversary $\mathcal{A}$, the following definition of advantage is negligible in λ*

$$\mathbf{Adv}^{\mathsf{sUF\text{-}RDA}}_{\mathcal{S},\mathcal{A}}(\lambda) := \Pr[\mathsf{sUF\text{-}RDA}_{\mathcal{S},\mathcal{A}}(1^\lambda) \Rightarrow \mathsf{T}]\,,$$

where game $\mathsf{sUF\text{-}RDA}_{\mathcal{S},\mathcal{A}}$ *described in Figure 6.*

procedure Initialize(1^λ):
$(\mathsf{sk}, \mathsf{pk}) \leftarrow_\$ \mathsf{SGen}(1^\lambda)$
$\mathsf{List} \leftarrow []$
$\mathsf{flag} \leftarrow \mathsf{F}$
Return (pk_S)

procedure Sign(m):
If $\mathsf{flag} = \mathsf{T}$
  $(\sigma, \mathsf{R}) \leftarrow \mathsf{Sign}(\mathsf{m}, \mathsf{sk}; \mathsf{r})$
  $\mathsf{List} \leftarrow (\mathsf{m}, (\sigma, \mathsf{R})) : \mathsf{List}$
  $\mathsf{flag} \leftarrow \mathsf{F}$
  Return (σ, R)
Else
  $\mathsf{r} \leftarrow_\$ \mathcal{R}$
  $(\sigma, \mathsf{R}) \leftarrow \mathsf{Sign}(0, \mathsf{sk}; \mathsf{r})$
  $\mathsf{flag} \leftarrow \mathsf{T}$
  Return $(\perp, \mathsf{R})$

procedure Finalize$(\mathsf{m}, (\sigma, \mathsf{R}))$:
If $(\mathsf{m}, (\sigma, \mathsf{R})) \notin \mathsf{List} \wedge \mathsf{Verify}(\mathsf{m}, (\sigma, \mathsf{R}), \mathsf{pk})$
  Return T
Else Return F

Fig. 6. Game sUF-RDA for a partitioned signature $\mathcal{S} = (\mathsf{SGen}, \mathsf{Sign}, \mathsf{Verify})$

It is clear that the security notion IND-RDA implies IND-CCA, and that sUF-RDA implies sUF-CMA. On the other hand, it is easy to find counterexamples showing that IND-CCA does not imply IND-RDA, nor does sUF-CMA imply sUF-RDA: simply construct a scheme based on an encryption/signature algorithm that returns the secret key when the input message is a fixed function of (e.g., equal to) the randomness component. We note that such counterexamples can be constructed even if the underlying schemes are reproducible, which shows reproducibility is not sufficient to imply randomness-dependent security.

4.3 Secure Compositions under Randomness Reuse

Let a digital signature $\mathcal{S}$ and a public-key encryption $\mathcal{E}$ be two *compatible* schemes, with randomness space $\mathcal{R}$. Our first construction, denoted EtS and described in Figure 7, produces a signcryption scheme from $\mathcal{E}$ and $\mathcal{S}$ in an encrypt-then-sign composition with randomness reuse. Conversely, in the StE construction, $\mathcal{E}$ and $\mathcal{S}$ are used in a sign-then-encrypt composition as shown in Figure 8, also with randomness reuse. We observe that we adopt the strategy proposed by An et al. [2] to achieve security in the multi-user model, by always including the receiver's public key in the signed data, and always including the sender's public key in the encrypted payload, so that it can be checked for consistency upon decryption[1].

Gen(1^λ):
$(\mathsf{sk}_1, \mathsf{pk}_1) \leftarrow_\$ \mathsf{SGen}(1^\lambda)$
$(\mathsf{sk}_2, \mathsf{pk}_2) \leftarrow_\$ \mathsf{EGen}(1^\lambda)$
$(\mathsf{sk}, \mathsf{pk}) \leftarrow ((\mathsf{sk}_1, \mathsf{sk}_2), (\mathsf{pk}_1, \mathsf{pk}_2))$
Return $(\mathsf{sk}, \mathsf{pk})$

Signcrypt$(\mathsf{m}, \mathsf{sk}_S, \mathsf{pk}_S, \mathsf{pk}_R)$:
$(\mathsf{sk}_1, \mathsf{sk}_2) \leftarrow \mathsf{sk}_S$
$(\mathsf{pk}_1, \mathsf{pk}_2) \leftarrow \mathsf{pk}_R$
$\mathsf{r} \leftarrow_\$ \mathcal{R}$
$(\mathsf{c}, \mathsf{R}) \leftarrow \mathsf{Enc}((\mathsf{m}, \mathsf{pk}_S), \mathsf{pk}_2; \mathsf{r})$
$(\sigma, \mathsf{R}) \leftarrow \mathsf{Sign}((\mathsf{c}, \mathsf{R}, \mathsf{pk}_R), \mathsf{sk}_1; \mathsf{r})$
$\hat{\mathsf{c}} \leftarrow (\mathsf{c}, \sigma, \mathsf{R})$
Return $\hat{\mathsf{c}}$

Unsigncrypt$(\hat{\mathsf{c}}, \mathsf{pk}_S, \mathsf{sk}_R, \mathsf{pk}_R)$:
$(\mathsf{pk}_1, \mathsf{pk}_2) \leftarrow \mathsf{pk}_S$
$(\mathsf{sk}_1, \mathsf{sk}_2) \leftarrow \mathsf{sk}_R$
$(\mathsf{c}, \sigma, \mathsf{R}) \leftarrow \hat{\mathsf{c}}$
$(\mathsf{m}, \mathsf{pk}'_S) \leftarrow \mathsf{Dec}((\mathsf{c}, \mathsf{R}), \mathsf{sk}_2)$
If $\mathsf{Verify}((\mathsf{c}, \mathsf{R}, \mathsf{pk}_R), (\sigma, \mathsf{R}), \mathsf{pk}_1) \wedge$
  $\mathsf{pk}_S = \mathsf{pk}'_S$ Return m
Return $\perp$

Fig. 7. EtS construction with randomness reuse

The following theorems state the security guarantees provided by these constructions. The proofs can be found in the full version of this paper.

[1] The overhead of encrypting the public key can be greatly reduced by encrypting its image under a collision-resistant hash function, or using an efficient tag-based PKE as proposed in [13].

$\mathbf{Gen}(1^\lambda)$:	$\mathbf{Signcrypt}(\mathsf{m}, \mathsf{sk}_S, \mathsf{pk}_S, \mathsf{pk}_R)$:	$\mathbf{Unsigncrypt}(\hat{\mathsf{c}}, \mathsf{pk}_S, \mathsf{sk}_R, \mathsf{pk}_R)$:
$(\mathsf{sk}_1, \mathsf{pk}_1) \leftarrow_\$ \mathsf{SGen}(1^\lambda)$	$(\mathsf{sk}_1, \mathsf{sk}_2) \leftarrow \mathsf{sk}_S$	$(\mathsf{pk}_1, \mathsf{pk}_2) \leftarrow \mathsf{pk}_S$
$(\mathsf{sk}_2, \mathsf{pk}_2) \leftarrow_\$ \mathsf{EGen}(1^\lambda)$	$(\mathsf{pk}_1, \mathsf{pk}_2) \leftarrow \mathsf{pk}_R$	$(\mathsf{sk}_1, \mathsf{sk}_2) \leftarrow \mathsf{sk}_R$
$(\mathsf{sk}, \mathsf{pk}) \leftarrow ((\mathsf{sk}_1, \mathsf{sk}_2), (\mathsf{pk}_1, \mathsf{pk}_2))$	$\mathsf{r} \leftarrow_\$ \mathcal{R}$	$(\mathsf{c}, \mathsf{R}) \leftarrow \hat{\mathsf{c}}$
Return $(\mathsf{sk}, \mathsf{pk})$	$(\sigma, \mathsf{R}) \leftarrow \mathsf{Sign}((\mathsf{m}, \mathsf{pk}_R), \mathsf{sk}_1; \mathsf{r})$	$(\mathsf{m}, \sigma, \mathsf{pk}'_S) \leftarrow \mathsf{Dec}((\mathsf{c}, \mathsf{R}), \mathsf{sk}_2)$
	Return $\mathsf{Enc}((\mathsf{m}, \sigma, \mathsf{pk}_S), \mathsf{pk}_2; \mathsf{r})$	If $\mathsf{pk}_S = \mathsf{pk}'_S \wedge$
		$\mathsf{Verify}(\mathsf{m}, (\sigma, \mathsf{R}), \mathsf{pk}_1)$
		Return m
		Else Return $\perp$

Fig. 8. StE construction with randomness reuse

Theorem 1 (Security of the EtS **construction).** *Suppose signature scheme $\mathcal{S}$ and encryption scheme $\mathcal{E}$ are compatible and that $\mathcal{S}$ is conditionally injective. Then the following hold:*
1) If $\mathcal{E}$ is reproducible and $\mathcal{S}$ is sUF-RDA *secure, then the resulting* EtS *construction is* sUF-iCMA *secure.*
2) If $\mathcal{S}$ is reproducible and $\mathcal{E}$ is IND-CCA *secure, then the resulting* EtS *construction is* IND-iCCA *secure.*

Theorem 2 (Security of the StE **construction).** *Suppose signature scheme $\mathcal{S}$ and encryption scheme $\mathcal{E}$ are compatible and that $\mathcal{E}$ is conditionally injective. Then the following hold:*
1) If $\mathcal{S}$ is reproducible and $\mathcal{E}$ is IND-RDA *secure, then the resulting* StE *construction is* IND-iCCA *secure.*
2) If $\mathcal{E}$ is reproducible, and $\mathcal{S}$ is sUF-CMA *secure, then the resulting* StE *construction is* sUF-iCMA *secure.*

We note that we obtain chosen-ciphertext security and strong unforgeability, both against insider attackers, even though this could not be achieved simultaneously by plain sequential composition without randomness reuse. The intuition behind the proofs of both theorems is the following. All proofs require simulating challenge signcryptions with shared randomness across encryption and signatures, and such signcryptions must embed a challenge from a signature or encryption security game. If one tried to reduce directly to the standard notions of security for signature and encryption, this proof strategy would fail, as one needs to commit to challenge messages before having access to the randomness associated with the challenge. For example, this means that one would not be able to request a signature on a ciphertext which shares the same randomness, as this randomness is totally hidden from us. The randomness-dependent attack models fix this problem by allowing adversaries to have access to the randomness components R in challenge encryptions and signatures before committing to a challenge message. Having access to R, one can simulate signatures and encryptions using the reproducibility properties of the schemes. For example, when proving that the StE construction is IND-iCCA secure by reducing to the IND-RDA property of the encryption scheme, one constructs the challenge signcryption as follows. When the adversary provides two challenge messages $(\mathsf{m}_0, \mathsf{m}_1)$, one first obtains R and then uses the reproducibility property of $\mathcal{S}$ to simulate signatures σ_0 and σ_1 on the challenge messages. One then queries the **LoR** oracle on the resulting message/signature pairs $(\mathsf{m}_0||\sigma_0, \mathsf{m}_1||\sigma_1)$. The challenge will then be guaranteed to be correctly simulated with randomness reuse. We note that key registration is required for the unforgeability proofs, as the secret keys required to run the reproducibility algorithms must be provided by the adversary. This is not an issue in the chosen-ciphertext security proofs, since the sender's secret key must always be provided to the **LoR** oracle (i.e., this is the case even in the standard dynamic multi-user model for signcryption).

REMARK. The proofs of the theorems actually establish a slightly stronger result than that stated in the theorems. Indeed, the results would still go through if the randomness-dependent security models are modified in line with the weaker notions of generalized chosen-ciphertext security [2] and existential unforgeability. We have chosen not to include the details in the presentation for the sake of clarity.

REMARK. Our constructions aim to minimize the overhead of the resulting signcryption scheme. For this reason, StE construction does *not* include the full signature inside the ciphertext — notice that R is not included inside the ciphertext. We remark that by including the full signature we could relax the security requirements of the signature to *weak* unforgeability, whilst still achieving *strong* unforgeability for the resulting composed signcryption scheme. This security amplification is accomplished by combining the extra binding provided by the randomness sharing with the conditional injectivity of the algorithms.

REMARK. The combination of randomness-dependent security and reproducibility for encryption schemes may be of independent interest in the design of multi-recipient encryption schemes with randomness reuse. Indeed, it is straightforward to show that for schemes displaying both properties the techniques proposed by Bellare et al. [3] can be adapted to prove security under a stronger model than that originally adopted. Recall that in [3] the adversary can place parallel challenge queries of n message pairs to the challenge oracle, and this will return n ciphertexts under n different public keys. The returned ciphertexts share the same encryption randomness. Applying our techniques, one can give extra power to the adversary in that it need not be restricted to making parallel challenge queries, but may choose challenge messages adaptively after seeing ciphertexts that share the same randomness. Note that security can even be proven in an analogue of the key registration model, in which the adversary can choose some keys maliciously.

5 Instantiating the Constructions

5.1 Security under Randomness-Dependent Attacks

One interesting aspect of our results is the requirement for a stronger security guarantee from the underlying signature and encryption components, in order to obtain security under randomness reuse. Concretely, this translates into the randomness-dependent attack models we have introduced in the previous section and raises the obvious question of how likely it is that off-the-shelf public-key encryption or signature schemes meet this level of security. Although we have no positive results for signature schemes in the standard model, we will show in this section that the class of encryption schemes achieving randomness-dependent security is potentially large, simply by looking at

KEM/DEM paradigms for constructing PKEs. In fact, the Kurosawa–Desmedt [12] appears as a notably efficient example that falls within our general framework. This observation allows to go beyond the efficiency levels both in terms of computational load and bandwidth of the previously most efficient standard model constructions.

RDA-SECURE SIGNATURE SCHEMES. For signature schemes, and restricting our attention to constructions whose security does not rely on random oracles, we found that current signature schemes do not meet this level of security. The typical problem, which occurs for example in the Boneh–Boyen signature scheme, is that the security proof critically relies on the ability to postpone the release of the randomness-dependent signature component until after the adversary has provided the message to be signed. This is a possible explanation for the lack of EtS-like constructions with randomness reuse in the standard model. If we admit random oracles, then one can consider any deterministic signature scheme, and randomness reuse no longer makes much sense as an optimization. Luckily, a RDA-secure signature is only required for the EtS construction. We therefore concentrate our attention on StE compositions, where the randomness-dependent security requirement applies only to the underlying encryption scheme.

RDA-SECURE ENCRYPTION FROM THE KEM/DEM PARADIGM. The first formalization of a KEM/DEM composition theorem was presented by Cramer and Shoup in their seminal paper on chosen-ciphertext-secure public-key encryption [9]. To simplify our discussion, we will restrict our attention to KEMs where the ciphertext is public-key independent, i.e., where the user-specific components of the public key passed to encapsulation are not used to calculate the ciphertext c, but only in the calculation of the secret key[2] k. We observe that PKE constructed from KEM/DEM schemes such as the ones we consider are naturally partitioned, and that the KEM ciphertext can be seen as the R component of the PKE ciphertext.

The KEM/DEM composition theorem in [9] roughly goes as follows. One performs a single game hop, modifying the IND-CCA game so that, rather than using the secret key output by the KEM in the challenge ciphertext generation, one uses a random secret key as input to the DEM. The definition of the decryption oracle is also modified consistently with this change. The transition between the two games can then be reduced to the KEM security assumption. The adversary's advantage in the second game can finally be reduced to the security of the DEM, as the secret key being used for data encapsulation is totally unrelated to that output by the KEM.

The same proof strategy can easily be adapted to show that KEM/DEM composition yields a RDA-secure PKE. To see this, observe that the KEM adversary constructed in the proof outlined above is able to obtain the challenge ciphertext (i.e., the R component in the PKE ciphertext) right at the beginning of the game, independently of the PKE adversary's actions. Furthermore, the DEM attacker constructed in the final step of the proof can generate the KEM ciphertext for the challenge right at the beginning. We can therefore conclude that the KEM/DEM construction initially proposed by Cramer and Shoup achieves randomness dependent chosen-ciphertext security without any modifi-

[2] Such schemes are common, and include those originally proposed by Cramer and Shoup [9]. Our results could be generalized to schemes that do not meet this constraint, by introducing a notion of partitioned KEM schemes.

cation. This result shows how our framework generalizes the results published in [13], in which the authors define reproducibility over KEMs, and then prove security of a signcryption scheme constructed from a KEM, a DEM and a signature scheme, in a StE construction with randomness reuse across the KEM and the signature schemes.

REMARK. The authors in [13] actually present their results based on a notion of a tag-based KEM that allows them to bind the sender's public key to the KEM ciphertext, rather than encrypting it together with the payload, but the KEM/DEM composition theorem they rely on does not take advantage of this binding and is a particular case of the one we describe above. Indeed, it is interesting that the tag-KEM/DEM composition paradigm proposed in [1] does *not* immediately yield RDA-secure schemes. The problem here is that the tag-KEM ciphertext can only be obtained after the tag has been defined, and this depends on the encrypted message in the hybrid construction of [1].

RDA-SECURE ENCRYPTION FROM WEAKENED KEY ENCAPSULATION. Hofheinz and Kiltz [10] propose an alternative KEM/DEM composition framework in which the security of the KEM can be weakened, as long as the DEM scheme satisfies a stronger notion of security known as one-time authenticated encryption. Such schemes can be constructed using the encrypt-then-mac approach, but no length-preserving solutions exist [10]. Interestingly this hybrid encryption paradigm preserves the independence between KEM and DEM components that allowed our extension to randomness-dependent attacks to go through. Indeed, the proof for the composition theorem in [10] follows a similar structure as that described above. This means that restricting our attention to (weak) KEM schemes where ciphertexts are public key independent, we immediately obtain partitioned and randomness-dependent chosen-ciphertext secure PKEs that can be used to instantiate our signcryption constructions. Notably, the weak KEM that is used in the very efficient Kurosawa–Desmedt encryption scheme [12] has this property.

5.2 Compatibility, Reproducibility, and Conditional Injectivity

Matsuda et al. [13] presented an extensive description of schemes that meet compatibility, reproducibility and conditional injectivity properties as required by the generic constructions using a tag-based KEM, a signature and a DEM with randomness reuse. Although the presentation is slightly different, all the schemes used to instantiate their constructions can be used to instantiate our own. However, the Boneh–Boyen signature scheme [6] was not considered by [13] as a candidate for signcryption schemes constructed under randomness reuse. We present a modified version of this signature scheme in the following subsection that displays the necessary properties, which enables us to use it in the instantiation of our construction. Additionally, the KEM/DEM compositions we have described above, when using a public-key independent KEM and a deterministic DEM which is one-to-one over the messages, also give suitable encryption schemes for instantiation.

5.3 An Efficient Instantiation

In this section we present a concrete instantiation of our results that, to the best of our knowledge, is the most efficient signcryption providing full insider security without random oracles. The scheme instantiates our StE construction with randomness reuse with the Kurosawa–Desmedt encryption scheme [12] and the Boneh–Boyen signature scheme [6]. On the negative side, the scheme's strong unforgeability is only proven under the key registration restriction. On the other hand, the scheme offers non-repudiation for free, which is inherited from the StE construction: the receiver obtains a valid signature on the recovered message.

THE KUROSAWA–DESMEDT ENCRYPTION SCHEME. We recall the encryption scheme in [12]. Here, $\mathbb{G}$ is a cyclic group of prime order q in which the DDH assumption holds, and $g_1, g_2 \in \mathbb{G}$ are two random distinct generators. Also, SKE is a one-time authenticated symmetric-key encryption scheme. As referred in the previous section, SKE cannot be assumed to be length-preserving, so we will assume a minimum overhead of size $|\mathsf{MAC}|$, corresponding to a MAC tag. The scheme also requires two hash functions $H_1 : \mathbb{G} \to \{0,1\}^k$ and $H_2 : \mathbb{G} \times \mathbb{G} \to \mathbb{Z}_q$, where the former must be a secure key-derivation function (i.e., entropy smoothing), and the latter must be target collision resistant. We have shown in the previous section that the Kurosawa–Desmedt encryption scheme is partitioned, that it is randomness-dependent chosen-ciphertext-secure and that, when instantiated with a deterministic and one-to-one DEM it is conditionally injective.

To be used in our constructions, we further require the scheme to be reproducible. It is straightforward to show that the scheme satisfies this property. Given a ciphertext (c, R) under an arbitrary public key, a secret key sk and a message m, the reproducibility algorithm produces a randomness reusing encryption of m as follows. It first takes the $\mathsf{R} = (\mathsf{R}_1, \mathsf{R}_2)$ and calculates a secret key k precisely as this is done in the decryption algorithm using sk. It then encrypts the m under the DEM using k to obtain the required ciphertext (c, R).

algorithm Gen:	**algorithm Enc**(m, pk):	**algorithm Dec**((c, R), sk):
$w \leftarrow_\$ \mathbb{Z}_q, x \leftarrow_\$ \mathbb{Z}_q$	$(a, b) \leftarrow \mathsf{pk}$	$(w, x, y, z) \leftarrow \mathsf{sk}$
$y \leftarrow_\$ \mathbb{Z}_q, z \leftarrow_\$ \mathbb{Z}_q$	$\mathsf{r} \leftarrow_\$ \mathbb{Z}_q$	$(\mathsf{R}_1, \mathsf{R}_2) \leftarrow \mathsf{R}$
$a \leftarrow g_1^w g_2^x, b \leftarrow g_1^y g_2^z$	$\mathsf{R}_1 \leftarrow g_1^{\mathsf{r}}, \mathsf{R}_2 \leftarrow g_2^{\mathsf{r}}$	$s \leftarrow H_2(\mathsf{R}_1, \mathsf{R}_2)$
$\mathsf{sk} \leftarrow (w, x, y, z)$	$s \leftarrow H_2(\mathsf{R}_1, \mathsf{R}_2)$	$\mathsf{K} \leftarrow H_1(\mathsf{R}_1^{w+ys} \cdot \mathsf{R}_2^{x+zs})$
$\mathsf{pk} \leftarrow (a, b)$	$\mathsf{K} \leftarrow H_1(a^{\mathsf{r}} b^{s\mathsf{r}})$	$\mathsf{m} \leftarrow \mathsf{SKE.Dec}(\mathsf{K}, \mathsf{c})$
Return (sk, pk)	$\mathsf{c} \leftarrow \mathsf{SKE.Enc}(\mathsf{K}, \mathsf{m})$	Return m
	$\mathsf{R} \leftarrow (\mathsf{R}_1, \mathsf{R}_2)$	
	Return (c, R)	

Fig. 9. The Kurosawa–Desmedt encryption scheme [12]

THE BONEH–BOYEN SIGNATURE SCHEME. The Boneh–Boyen signature scheme [6] is strongly unforgeable in the standard model. It relies on bilinear groups, and so we briefly recall this notion below.

Definition 9. *A bilinear group description Γ is a tuple $(p, \mathbb{G}_1, \mathbb{G}_2, \mathbb{G}_T, e, g_3, g_4)$ where $\mathbb{G}_1$, $\mathbb{G}_2$ and $\mathbb{G}_T$ are groups of order p with efficiently computable group laws; g_3 and g_4 are generators of $\mathbb{G}_1$ and $\mathbb{G}_2$, respectively; and e is a bilinear pairing $e : \mathbb{G}_1 \times \mathbb{G}_2 \to \mathbb{G}_T$ satisfying the usual properties of bilinearity and non-degeneracy.*

We present the signature scheme of [6] in Figure 10. Observe that we slightly modified the signature and verification algorithms to make the scheme compatible with

Kurosawa–Desmedt encryption [12], i.e., so that signatures present the same R component. Intuitively, we replace the randomness generation operation in the signature algorithm so that, rather than sampling s directly, we obtain it from the R component in a Kurosawa–Desmedt ciphertext. We therefore consider a group $\mathbb{G}$ of order q as described by the Kurosawa–Desmedt encryption scheme, with two distinct generators $g_1, g_2 \in \mathbb{G}$.

We require an encoding function Map that takes a random element in group $\mathbb{G}$ onto an element in the randomness space of the Boneh–Boyen signature scheme.[3] This encoding function is fed with the first element in the Kurosawa–Desmedt R component g_1^{r}. The second element g_2^{r} is simply included as part of the signed message; we use the standard approach of extending the Boneh–Boyen signature scheme to messages of arbitrary length, introducing a collision-resistant hash function $H : \{0,1\}^{\star} \to \mathbb{Z}_p$. We note that the apparent loss in efficiency in the signature scheme disappears when one uses this version of the scheme in our StE construction. Also note that the signature scheme is reproducible. The reproduction algorithm proceeds identically to the signature algorithm, except it skips the steps where $\mathsf{r} \leftarrow_{\$} \mathbb{Z}_q$ and R are computed.

algorithm Gen:	**algorithm Sign**$(\mathsf{m}, \mathsf{sk})$:	**algorithm Verify**$(\mathsf{m}, (\sigma, \mathsf{R}), \mathsf{pk})$:
$x \leftarrow_{\$} \mathbb{Z}_p, y \leftarrow_{\$} \mathbb{Z}_p$	$(x, y) \leftarrow \mathsf{sk}$	$(u, v, z) \leftarrow \mathsf{pk}$
$u \leftarrow g_4^x, v \leftarrow g_4^y$	$\mathsf{r} \leftarrow_{\$} \mathbb{Z}_q$	$(\mathsf{R}_1, \mathsf{R}_2) \leftarrow \mathsf{R}$
$z \leftarrow e(g_3, g_4)$	$\mathsf{R}_1 \leftarrow g_1^{\mathsf{r}}, \mathsf{R}_2 \leftarrow g_2^{\mathsf{r}}$	$h \leftarrow H(\mathsf{m}, \mathsf{R}_2)$
$\mathsf{sk} \leftarrow (x, y)$	$s \leftarrow \mathsf{Map}(\mathsf{R}_1)$	$s \leftarrow \mathsf{Map}(\mathsf{R}_1)$
$\mathsf{pk} \leftarrow (u, v, z)$	$h \leftarrow H(\mathsf{m}, \mathsf{R}_2)$	If $e(\sigma, u \cdot g_4^h \cdot v^s) = z$
Return $(\mathsf{sk}, \mathsf{pk})$	$a \leftarrow 1/(x + h + ys) \bmod p$	Return T
	$\sigma \leftarrow g_3^a$	Else Return F
	$\mathsf{R} \leftarrow (\mathsf{R}_1, \mathsf{R}_2)$	
	Return (σ, R)	

Fig. 10. The Boneh–Boyen signature scheme [6] modified to be compatible with the Kurosawa–Desmedt encryption scheme

We now discuss the security of the modified Boneh–Boyen signature scheme. It is straightforward to show that this scheme remains strongly unforgeable provided that the DDH problem is hard in group $\mathbb{G}$ and that Map is a one-to-one and efficiently invertible mapping from $\mathbb{G}$ to $\mathbb{Z}_p$ (the inversion algorithm is only used in the proof of security). A closer look at the proof reveals that even weaker properties on Map suffice. Indeed, the function only needs to be injective, efficiently invertible, and map $\mathbb{G}$ to a sufficiently large fraction of $\mathbb{Z}_p$ elements. To meet these requirements, we may instantiate $\mathbb{G}$ as the group of points on an elliptic curve, where the DDH problem is assumed to be hard. Standard point compression techniques [5] allow us to instantiate Map with an injective encoding whose image corresponds to a sufficiently large fraction of $\mathbb{Z}_p$ values. More precisely, for carefully chosen elliptic curves, there exist injective and efficiently invertible mappings from curve points into bit strings of length l, where l is approximately the logarithm of the order of the group. Such encodings will have the property we require when p is chosen to be sufficiently close to 2^l.

[3] As in the original scheme, in the unlikely event that $s = -(x + h)/y$, we simply sample a new randomness. We omit this in Figure 10 for readability.

The security proof of the modified Boneh–Boyen signature scheme can be found in the full version of this paper. Intuitively, to reduce the security of the modified scheme to the original version, one simulates signature queries by repeatedly querying the signature oracle, until one obtains a signature where the randomness value can be inverted back into $\mathbb{G}$. Furthermore, a valid forgery on the modified scheme will still constitute a valid forgery on the original scheme.

COMPARISON. We present in Table 1 a comparison of our StE construction with randomness reuse, when instantiated with the Kurosawa–Desmedt encryption scheme and the Boneh–Boyen signature scheme, with previous signcryption constructions in various relevant parameters. We consider only signcryption schemes offering full insider security in dynamic multi-user models, and not relying on random oracles. We present results for the 80-bit security level. In addition to efficiency considerations, we also present the underlying computational assumptions, whether key registration is required, and whether the scheme offers non-repudiation by providing receiver's with valid signatures on the recovered messages.

For computational efficiency, we compare the number of exponentiations, multi-exponentiations, and pairing computations (in this order), both in the signcryption and unsigncryption operations. Clearly the new scheme matches the previously computationally more efficient solution from [14]. We also include the size of the random coins required for the signcryption operation. Here, our scheme displays a saving of 50% over previous solutions, due to the randomness reuse optimization. Finally, in terms of overhead (i.e., the difference between ciphertext length and message length), our scheme compares favorably with other solutions. The 160-bit overhead with respect to the solutions in [8,13] can be explained by including a digest of the sender's public key in the payload, which must be calculated using a collision-resistant hash function. This might be avoided by considering a tag-based variant of the encryption scheme as in [8,13], although we have not considered this possibility.

Table 1. Comparison with signcryption schemes in the literature. We consider [14,15] also instantiated with the BB signature scheme. We take $|\mathbb{G}| = |\mathbb{Z}_p| = |\mathsf{H}| = 160$ and $|\mathsf{MAC}| = 80$ bits.

Scheme	Assumptions	Key Reg.	Non-Rep.	Computations sc.	usc.	Randomness (bits)	Overhead (bits)
[8]	DBDH, q-SDH	No	Yes	$[4,0,0]$	$[1,1,2]$	320	640
[8]	DBDH, q-SDH	No	Yes	$[3,1,0]$	$[1,1,2]$	320	720
[13]	DBDH, co-CDH	Yes	Yes	$[4,1,0]$	$[1,1,3]$	320	640
[16]	DBDH, q-SDH	Yes	No	$[3,2,0]$	$[3,1,4]$	480	800
[14]	DDH, q-SDH	No	No	$[3,1,0]$	$[0,2,1]$	320	720
[15]	DDH, q-SDH	No	No	$[4,1,0]$	$[1,2,1]$	320	800
New scheme	DDH, q-SDH	Yes	Yes	$[3,1,0]$	$[0,2,1]$	160	720

Acknowledgements. The authors would like to thank Nigel Smart and the anonymous ACNS'12 reviewers for helping to improve the quality of the paper. A. Arriaga and M. Barbosa were supported by project SMART (ref. 120224), financed by National Funds through the Portuguese Foundation for Science and Technology and by ENIAC Joint Undertaking.

References

1. Abe, M., Gennaro, R., Kurosawa, K.: Tag-KEM/DEM: A new framework for hybrid encryption. Journal of Cryptology 21, 97–130 (2008)
2. An, J.H., Dodis, Y., Rabin, T.: On the Security of Joint Signature and Encryption. In: Knudsen, L.R. (ed.) EUROCRYPT 2002. LNCS, vol. 2332, pp. 83–107. Springer, Heidelberg (2002)
3. Bellare, M., Boldyreva, A., Staddon, J.: Randomness Re-use in Multi-recipient Encryption Schemeas. In: Desmedt, Y.G. (ed.) PKC 2003. LNCS, vol. 2567, pp. 85–99. Springer, Heidelberg (2002)
4. Bellare, M., Rogaway, P.: The Security of Triple Encryption and a Framework for Code-Based Game-Playing Proofs. In: Vaudenay, S. (ed.) EUROCRYPT 2006. LNCS, vol. 4004, pp. 409–426. Springer, Heidelberg (2006)
5. Blake, I., Seroussi, G., Smart, N.: Elliptic Curves in Cryptography. London Mathematical Society Lecture Note Series, vol. 265. Cambridge University Press (1999)
6. Boneh, D., Boyen, X.: Short signatures without random oracles and the SDH assumption in bilinear groups. Journal of Cryptology 21, 149–177 (2008)
7. Boneh, D., Shen, E., Waters, B.: Strongly Unforgeable Signatures Based on Computational Diffie-Hellman. In: Yung, M., Dodis, Y., Kiayias, A., Malkin, T. (eds.) PKC 2006. LNCS, vol. 3958, pp. 229–240. Springer, Heidelberg (2006)
8. Chiba, D., Matsuda, T., Schuldt, J.C.N., Matsuura, K.: Efficient Generic Constructions of Signcryption with Insider Security in the Multi-user Setting. In: Lopez, J., Tsudik, G. (eds.) ACNS 2011. LNCS, vol. 6715, pp. 220–237. Springer, Heidelberg (2011)
9. Cramer, R., Shoup, V.: A Practical Public Key Cryptosystem Provably Secure against Adaptive Chosen Ciphertext Attack. In: Krawczyk, H. (ed.) CRYPTO 1998. LNCS, vol. 1462, pp. 13–25. Springer, Heidelberg (1998)
10. Hofheinz, D., Kiltz, E.: Secure Hybrid Encryption from Weakened Key Encapsulation. In: Menezes, A. (ed.) CRYPTO 2007. LNCS, vol. 4622, pp. 553–571. Springer, Heidelberg (2007)
11. Kurosawa, K.: Multi-recipient Public-Key Encryption with Shortened Ciphertext. In: Naccache, D., Paillier, P. (eds.) PKC 2002. LNCS, vol. 2274, pp. 7–38. Springer, Heidelberg (2002)
12. Kurosawa, K., Desmedt, Y.: A New Paradigm of Hybrid Encryption Scheme. In: Franklin, M. (ed.) CRYPTO 2004. LNCS, vol. 3152, pp. 426–442. Springer, Heidelberg (2004)
13. Matsuda, T., Matsuura, K., Schuldt, J.C.N.: Efficient Constructions of Signcryption Schemes and Signcryption Composability. In: Roy, B., Sendrier, N. (eds.) INDOCRYPT 2009. LNCS, vol. 5922, pp. 321–342. Springer, Heidelberg (2009)
14. Tan, C.H.: Insider-secure hybrid signcryption scheme without random oracles. In: Availability, Reliability and Security – ARES 2007, pp. 1148–1154 (2007)
15. Tan, C.H.: Insider-secure signcryption KEM/Tag-KEM schemes without random oracles. In: Availability, Reliability and Security – ARES 2008, pp. 1275–1281 (2008)
16. Tan, C.H.: Signcryption Scheme in Multi-user Setting without Random Oracles. In: Matsuura, K., Fujisaki, E. (eds.) IWSEC 2008. LNCS, vol. 5312, pp. 64–82. Springer, Heidelberg (2008)
17. Zheng, Y.: Digital Signcryption or How to Achieve Cost (Signature & Encryption) $<<$ Cost(Signature) + Cost(Encryption). In: Kaliski Jr., B.S. (ed.) CRYPTO 1997. LNCS, vol. 1294, pp. 165–179. Springer, Heidelberg (1997)

Secure Accumulators from Euclidean Rings without Trusted Setup

Helger Lipmaa

Institute of Computer Science, University of Tartu, Estonia

Abstract. Cryptographic accumulators are well-known to be useful in many situations. However, the most efficient accumulator (the RSA accumulator) it is not secure against a certificate authority who has herself selected the RSA modulus n. We generalize previous work and define the root accumulator in modules over Euclidean rings. We prove that the root accumulator is secure under two different pairs of assumptions on the module family and on the used hash function. Finally, we propose a new instantiation of the root accumulator, based on class groups of imaginary quadratic order, that combines the best properties of previous solutions. It has short (non)membership proofs like the RSA accumulator, and at the same time it is secure against a malicious certificate authority. Up to this point, this seems to be the only unique application of class groups of imaginary quadratic orders, and we hope that this paper will motivate more research on cryptography in the said groups.

Keywords: Class groups of imaginary quadratic order, cryptographic accumulators, Euclidean rings.

1 Introduction

Cryptographic accumulators have been proven to be extremely useful in the public-key infrastructure, anonymous credential systems and many other applications. Briefly, in a cryptographic accumulator, for any document set S, a server can compute a short digest $\mathsf{Dig}(S)$, such that for any candidate document m one can find a succinct (non)membership proof $\mathsf{Proof}(m, S)$ of m (not) belonging to S. The digest $\mathsf{Dig}(S)$ is published, and everybody can obtain it in an authenticated manner. Finally, different clients use the verification algorithm Ver. It is required that $\mathsf{Ver}(m, \mathsf{Dig}(S), \mathsf{Proof}(m, S)) = \mathsf{Member}$ if $m \in S$, and (in some of the papers like [5,6,23]) $\mathsf{Ver}(m, \mathsf{Dig}(S), \mathsf{Proof}(m, S)) = \mathsf{NotMember}$ if $m \notin S$. Accumulators are required to be collision-resistant, that is, it should be difficult to construct a triple (m, S, p), such that $m \notin S$ but $\mathsf{Ver}(m, \mathsf{Dig}(S), p) = \mathsf{Member}$ [1].

One can construct collision-resistant accumulators (with nonmembership proofs) based on hash-trees, see [5,6]. However, hash-tree based solutions have relatively long — logarithmic in $|S|$ — (non)membership proofs. The more succinct RSA accumulator was introduced in [2], further studied in [1,25,26,11], and proven to be collision-resistant in [1]. Further accumulators have been proposed in say [24,10].

F. Bao, P. Samarati, and J. Zhou (Eds.): ACNS 2012, LNCS 7341, pp. 224–240, 2012.

Unfortunately, one cannot rely on the accumulating party (say, the certificate authority) to honestly generate the value $\mathsf{Dig}(S)$.[1] In particular, she could publish d (not necessarily knowing the corresponding S) such that she can later generate both membership and nonmembership proofs for some selected elements m. To tackle this situation, Buldas, Laud and Lipmaa [5,6] required accumulators to be *undeniable* in the next sense: it should be infeasible to generate a tuple $(m, d, p, \overline{p})$, such that $\mathsf{Ver}(m, d, p) = \mathsf{Member}$ but $\mathsf{Ver}(m, d, \overline{p}) = \mathsf{NotMember}$. (The same security requirement — under different names — has been independently reinvented in say [9].) Thus, in the case of certificate management, when a client sees a certificate m, digest d and (say) a proof p that m was revoked, she can be certain that there does not exist a contradictory proof $\overline{p}$ that m was not revoked. Buldas, Laud and Lipmaa also constructed a concrete undeniable accumulator based on hashed search trees. (They called it an undeniable attester since it is not based on the RSA accumulator.) Because their solution is based on hashed search trees, it is trapdoorless and thus secure against a malicious server. Unfortunately, there the (non)membership proofs p have length that is logarithmic in the size of S.

For a long time, it was not known how to construct short nonmembership proofs for the RSA accumulators. Only in 2007, Li, Li and Xue [23] showed how to do that. In their modification to the RSA accumulator, a membership proof consists of one group element and a nonmembership proof consists of one group element and one exponent. Unfortunately, in the case of the RSA accumulator, the server can generate the RSA modulus n herself, and thus knowing the factorization of n she can efficiently break the accumulator. That means that the Li-Li-Xue construction is only secure in the trusted setup model where n is generated by a trusted third party who does not disclose its factorization to the server. Sander [25] tried to eliminate the trapdoor in the RSA accumulator but his construction, while trapdoorless, is very inefficient. Moreover, from the perspective of a client who just started to use the accumulator, it still does not guarantee that the server does not know the trapdoor. Our goal is to get rid of the trusted setup assumption, and to achieve efficiency that is comparable to that of the RSA accumulator.

Our Contributions. We first substantially generalize the RSA accumulator as modified by Li, Li and Xue. The generalized *root accumulator* works in $\mathcal{R}_{\mathcal{D}}$, which is a family of modules D over Euclidean rings R, and uses a hash function (family) H. This generalization serves two different purposes. First, by generalizing the algebraic setting to the widest one, it may become possible in the future to find other more efficient instantiations of the primitive. (Even if at this

[1] The original motivation of this line of research is digital time stamping, where the digest over answers to time-stamping queries is computed by the time-stamping authority [17,7,8]. Cryptographic methods are precisely in place to counter the case where the authority may be malicious. In particular, a malicious time-stamping authority can clearly compute a spurious value of $\mathsf{Dig}(S)$. See [5] for more discussion and motivation.

moment, the only known instantiations consist of Abelian groups D and $R = \mathbb{Z}$, with the module operation $\circ : R \times D \to D$ defined as $\alpha \circ x := x^\alpha$.) Second, the construction of the root accumulator depends crucially on the existence of the Extended Euclidean Algorithm in the underlying ring. In addition, most of the security reductions of this paper make an explicit use of the Extended Euclidean Algorithm. Thus, we think it is methodologically useful to explicitly point out that the Extend Euclidean Algorithm algorithm must exist in the underlying algebraic structure, and must be efficient. While modules over rings have been used in cryptography before, see [16], we are unaware of any previous use of modules over Euclidean rings in cryptography. Thus, this generalization may be a contribution by itself.

Before proving the security of the root accumulator, we must define the corresponding security notions and underlying security assumptions. The first technical difficulty (and novelty) there is that because we want the accumulator to be secure without trusted setup, the security definitions will become more involved. In particular, an accumulator must have a public key divided into two parts, one of which (say, the RSA modulus n) is generated by using a public randomness known by the adversary, and another one (say, a generator of a large subgroup in $\mathbb{Z}_n^*$) can be chosen by using a non-public randomness. Because it was the trapdoor in n that we were worried about, this division is fine for our purposes. (We leave it as an interesting open question to solve the second part in an accountable way.) Similarly, when defining the security assumptions, we must consider the case where the adversary knows the randomness that is used when choosing the module (again, in the case of the RSA accumulator this corresponds to the adversary knowing the factorization of n) where the root accumulator will be run.

Then, we show that the root accumulator is both collision-resistant and undeniable if either (a) $\mathcal{R}_\mathcal{D}$ is a *strong prime root module family* and H is a *prime-valued injective function [1]*, or (b) $\mathcal{R}_\mathcal{D}$ is a *strong divisible root module family [13]* and H is a *division-intractable function family [15]*. (Corresponding security definitions are given later in the paper.)

Based on those results, we show that if factorization is hard in the Euclidean ring, then the security of the root accumulator—given that H is prime-valued injective—is based on a presumably weaker assumption than the strong root assumption (e.g., the security of the RSA accumulator is based on a presumably weaker assumption than the strong RSA assumption). We also show that the strong divisible root assumption is equivalent to the strong root assumption (which is known to be secure in the generic group model [14]), given that the module satisfies another seemingly unrelated *small root assumption*. (The latter is related but generalizes significantly the small root assumption of [13].)

As a concrete instantiation, we propose to use class groups of imaginary quadratic orders with a large discriminant Δ where $-\Delta$ is a prime [4]. Many previous cryptographic schemes are based on the strong root assumption in such groups. Importantly, Δ can be chosen by a malicious adversary with only negligibly changing her probability of breaking the root accumulator. While the applicability of class groups of imaginary quadratic order has been studied quite extensively in the cryptographic literature (see [3,18] for an overview), one has been mostly interested in such groups because they are one of the very few group families known (in addition to say multiplicative groups of residue rings and (hyper)elliptic curve groups) that are suitable for cryptographic use. We show that there is a natural cryptographic problem—construction of secure accumulators without trusted setup—for which class groups of imaginary quadratic order are the *only* known suitable group family. We hope that this will generate additional interest in cryptography based on such groups.

Basic Notation. We assume that Member, $\mathsf{NotMember}$ and Error are special symbols. k denotes the security parameter. The working time of all algorithms and the security of all primitives is measured as a function of the security parameter k. $\text{negl}(k)$ denotes an arbitrary negligible function in k, $\text{poly}(k)$ denotes an arbitrary polynomial function in k. PPT means probabilistic polynomial time. We note that in the context of this paper, the adversary can always be non-uniform; however, our reductions themselves are all uniform. If S is a set, then $x \leftarrow S$ denotes random sampling, and $x \leftarrow S(\omega)$ denotes random sampling while using ω as the random tape. If A is an algorithm, then $x \leftarrow A(y)$ denotes random sampling of the output of A, given input y.

2 Collision-Resistant and Undeniable Accumulators

First, we will state the syntax of accumulators that allow nonmembership proofs as in [5,6,23]. (In [5,6], an accumulator with nonmembership proofs was called an *attester*.) Informally, an accumulator is a mechanism that for each candidate element m and a set S produces a succinct (non)membership proof that attests to the fact that $m \in S$ or $m \notin S$. Based on m, the short digest of S and the corresponding proof (and without access to any other information), one can later verify whether $m \in S$ or not.

Definition 1 (Accumulator). *Let M, D and P be three sets (the message set, the digest set and the proof set correspondingly). A quadruple* $\mathsf{Acc} = (\mathsf{Gen}, \mathsf{Proof}, \mathsf{Dig}, \mathsf{Ver})$ *of PPT algorithms is a* (strong) accumulator*, if it satisfies the next conditions:*

Generating algorithm $\mathsf{Gen}(1^k)$ *outputs a public key* pk.
Membership algorithm $\mathsf{Proof}_{\mathsf{pk}}(m, S)$: *If $m \in M$ and $S \subseteq M$, then it outputs a* membership proof $p \in P$, *otherwise it outputs* Error.
Digest algorithm $\mathsf{Dig}_{\mathsf{pk}}(S)$: *If $S \subseteq M$ then it outputs a* digest $d \in D$, *otherwise it outputs* Error.
Verification algorithm $\mathsf{Ver}_{\mathsf{pk}}(m, d, p)$: *If $m \in M$, $d \in D$ and $p \in P$ then it outputs either* Member *or* $\mathsf{NotMember}$, *otherwise it outputs* Error.

An accumulator must satisfy the next correctness property: for valid $\mathsf{pk} \in \mathsf{Gen}(1^k)$, *$m \in M$ and $S \subseteq M$,* $\mathsf{Ver}_{\mathsf{pk}}(m, \mathsf{Dig}_{\mathsf{pk}}(S), \mathsf{Proof}_{\mathsf{pk}}(m, S))$ *outputs* Member *if $m \in S$, and* $\mathsf{NotMember}$ *if $m \notin S$.* □

Note that because all algorithms work in probabilistic polynomial time, it is always implicitly required that $|S| = \text{poly}(k)$.

Definition 2 (Security in Trusted Setup Model). *Let* $\mathsf{Acc} = (\mathsf{Gen}, \mathsf{Proof}, \mathsf{Dig}, \mathsf{Ver})$ *be an accumulator.* Acc *is* collision-resistant [1] *(in the trusted setup model) if*

$$\Pr\left[\begin{matrix}\mathsf{pk} \leftarrow \mathsf{Gen}(1^k), (m, S, p) \leftarrow A(\mathsf{pk}) : \\ m \notin S \wedge \mathsf{Ver}_{\mathsf{pk}}(m, \mathsf{Dig}_{\mathsf{pk}}(S), p) = \mathsf{Member}\end{matrix}\right] = \text{negl}(k)$$

for any PPT adversary A. Acc *is* undeniable [5,6,23] *(in the trusted setup model) if*

$$\Pr\left[\begin{matrix}\mathsf{pk} \leftarrow \mathsf{Gen}(1^k), (m, d, p, \overline{p}) \leftarrow A(\mathsf{pk}) : \\ \mathsf{Ver}_{\mathsf{pk}}(m, d, p) = \mathsf{Member} \wedge \mathsf{Ver}_{\mathsf{pk}}(m, d, \overline{p}) = \mathsf{NotMember}\end{matrix}\right] = \text{negl}(k)$$

for any PPT adversary A.

Proofs p and $\overline{p}$ are *contradictory* if for some m and d, $\mathsf{Ver}_{\mathsf{pk}}(m, d, p) = \mathsf{Member}$ and $\mathsf{Ver}_{\mathsf{pk}}(m, d, \overline{p}) = \mathsf{NotMember}$.

It was proven in [1] that the RSA accumulator [2] is collision-resistant (in the trusted setup model). However, in several potential usage scenarios of accumulators, the trusted setup assumption is really inappropriate. For example, imagine the setting (similar to digital time stamping [17,7], where cryptographic methods are introduced precisely to obtain security against a corrupt authority) where a certificate authority periodically revokes certificates. Instead of periodically publishing certificate revocation lists, she publishes their short digests. To every client who wants to check whether or not some particular certificate was revoked during that period, she also sends a succinct (non)membership proof with respect to this revocation list. If the accumulator is undeniable (without any trusted setup), the client can be certain that nobody else has a contradictory proof.

In the case of the RSA accumulator, the certificate authority may know the factorization $n = PQ$ of the RSA modulus n. (The same attack is also valid in other scenarios, obviously.) Even when using threshold methods to generate n, there is always a coalition of parties who know P and Q. In striking contrast with many other cryptographic applications, here we cannot assume that the client herself participated in the generation of n, since she might have not been using the services of this certificate authority at the that time. See [5,6] for more motivation and [25] for an early paper on trapdoorless RSA accumulator.

We will now define security in the case without trusted setup. The RSA accumulator is not secure without trusted setup, because even a semi-honest party who generates n can later cheat (e.g., by revealing its prime factors after being adaptively corrupted). We tackle this problem by introducing a new Setup algorithm (that generates the algebraic structure we are working in), and requiring that the adversary must have access to the random tape ω of Setup. On the other hand, Gen's random tape must remain hidden. (In the case of RSA accumulator, the latter corresponds to the part that is used while generating a generator of some large subgroup of $\mathbb{Z}_n^*$.) We thus naturally augment the definition of accumulators with the Setup algorithm, and assume that all other algorithms get the output of Setup as one of the inputs.

Definition 3 (Security without Trusted Setup). *Let* $\mathsf{Acc} = (\mathsf{Setup}, \mathsf{Gen}, \mathsf{Proof}, \mathsf{Dig}, \mathsf{Ver})$ *be an accumulator.* Acc *is* collision-resistant *(without trusted setup) if*

$$\Pr_{\omega}\left[\begin{array}{l}\mathsf{parm} \leftarrow \mathsf{Setup}(1^k, \omega), \mathsf{pk} \leftarrow \mathsf{Gen}(1^k, \mathsf{parm}),\\ (m, S, p) \leftarrow A(\omega, \mathsf{parm}, \mathsf{pk}) :\\ m \notin S \wedge \mathsf{Ver}_{\mathsf{pk},\mathsf{pk}}(m, \mathsf{Dig}_{\mathsf{parm},\mathsf{pk}}(S), p) = \mathsf{Member}\end{array}\right] = \mathrm{negl}(k)$$

for any PPT adversary A. Acc *is an* undeniable accumulator *(without trusted setup) if*

$$\Pr_{\omega}\left[\begin{array}{l}\mathsf{parm} \leftarrow \mathsf{Setup}(1^k, \omega), \mathsf{pk} \leftarrow \mathsf{Gen}(1^k, \mathsf{parm}),\\ (m, d, p, \overline{p}) \leftarrow A(\omega, \mathsf{parm}, \mathsf{pk}) :\\ (\mathsf{Ver}_{\mathsf{parm},\mathsf{pk}}(m, d, p) = \mathsf{Member}) \wedge\\ (\mathsf{Ver}_{\mathsf{parm},\mathsf{pk}}(m, d, \overline{p}) = \mathsf{NotMember})\end{array}\right] = \mathrm{negl}(k)$$

for any PPT adversary A.

Note that this is somewhat similar to security definitions in the common reference string model, where parm is honestly chosen, and the adversary (usually a simulator) can choose parm herself together with a corresponding trapdoor ω. In fact, one can consider a stronger requirement, where ω is not only known to the adversary but actually chosen by her. However, in this case in all subsequent security assumptions one would have to assume that the assumptions hold even if the adversary can choose the underlying module. Unfortunately, no module families are known where such security assumptions would hold. On the positive side, checking that pk is generated correctly is more plausible than checking that the trapdoor information ω is not known to the adversary. For example, pk can be generated by using verifiable randomness published in newspapers or the NIST beacon (`http://www.nist.gov/itl/csd/ct/nist_beacon.cfm`).

3 Module-Based Cryptography

Many public-key primitives are based on groups. We generalize the group-based setting to the module-based one. For this we generalize several well-known notions and introduce a few new ones. In the next section, we propose an accumulator that is based on a module over a Euclidean ring. Within this paper, all rings are commutative.

Algebraic Background and Definitions. A (left) R-module over the ring R consists of an Abelian group $(D, +)$ and an operation $R \times D \to D$ (that we denote by $\alpha \circ g$), such that for all $\alpha, \beta \in R$ and $x, y \in D$, we have (a) $\alpha \circ (x + y) = \alpha \circ x + \alpha \circ y$, (b) $(\alpha + \beta) \circ x = \alpha \circ x + \beta \circ x$, (c) $(\alpha \cdot \beta) \circ x = \alpha \circ (\beta \circ x)$, and (d) $1 \circ x = x$.

A commutative ring R with identity is called an *integral domain* if for all $\alpha, \beta \in R$, $\alpha\beta = 0$ implies $\alpha = 0$ or $\beta = 0$. A ring R is *Euclidean* if it is an integral domain and there exists a function $\deg : R \to \mathbb{Z}^+$, called the Euclidean degree, such that (a) if $\alpha, \beta \in R$ with $\alpha\beta \neq 0$ and $\alpha \neq 0$, then $\deg(\alpha) \leq \deg(\alpha\beta)$ and (b) if $\alpha, \beta \in R$ then there exist $\gamma, \delta \in R$ such that $\alpha = \gamma\beta + \delta$ with either $\delta = 0$, or $\delta \neq 0$ and $\deg(\delta) < \deg(\beta)$. Every Euclidean ring possesses a multiplicative identity. An element α of R which is neither 0 nor 1 is called *irreducible* if there are no non-1 elements β and γ with $\alpha = \beta \cdot \gamma$. Define $\textsc{Irr}(R)$ to be the set of irreducible elements of R.

Some examples of Euclidean rings R are $\mathbb{Z}$ with $\deg(\alpha) := |\alpha|$, $\mathbb{Z}[i]$ (the ring of Gaussian integers) with $\deg(\alpha) := |\alpha|^2$, $K[X]$ for arbitrary field K with $\deg(\alpha)$ being the degree of polynomial α when $\alpha \neq 0$, the ideals of polynomial ring $\mathbb{Z}_p[X]$ (that are modules over $\mathbb{Z}_p[X]$), and arbitrary field K where $\deg(\alpha) := 1$ when $\alpha \neq 0$. An example of a non-commutative Euclidean ring is the polynomial ring $P[x]$ over a skew field (division ring) P. In all such cases one can talk about the irreducible elements of R.

Intractable Problems in Modules. Because we want the accumulator to be secure without trusted setup, it must also be the case that in the underlying security assumptions the adversary can see the coins used while selecting the concrete module.

Definition 4 (Security Assumptions without Trusted Setup). *Let $\mathcal{R}_\mathcal{D} = ((R_i)_{D_i})$ be a family of modules with $i \in I$ and an efficient deterministic algorithm* $\mathsf{Setup}(1^k, \omega)$ *that picks some $i \in I$. We assume that A is a stateful algorithm.*

1. *$\mathcal{R}_\mathcal{D}$ is a* discrete logarithm module family *if for every PPT adversary A,*
$$\Pr_\omega[R_D \leftarrow \mathsf{Setup}(1^k, \omega), (x, y) \leftarrow D, \alpha \leftarrow A(x, y, \omega) : \alpha \circ y = x] = \mathrm{negl}(k) \enspace .$$
2. *$\mathcal{R}_\mathcal{D}$ is an* order module family *if for every PPT adversary A,*
$$\Pr_\omega[R_D \leftarrow \mathsf{Setup}(1^k, \omega), x \leftarrow D, y \leftarrow A(x, \omega) : \mathrm{ord}(x) = y] = \mathrm{negl}(k) \enspace .$$
3. *$\mathcal{R}_\mathcal{D}$ is a* root module *if for every PPT adversary A,*
$$\Pr_\omega[R_D \leftarrow \mathsf{Setup}(1^k, \omega), x \leftarrow D, \alpha \leftarrow R, y \leftarrow A(x, \alpha, \omega) : \alpha \circ y = x] = \mathrm{negl}(k) \enspace .$$
4. *$\mathcal{R}_\mathcal{D}$ is a* strong prime root module *if for every PPT adversary A,*
$$\Pr_\omega\left[\begin{matrix} R_D \leftarrow \mathsf{Setup}(1^k, \omega), x \leftarrow D, (y, \alpha) \leftarrow A(x, \omega) : \\ (\alpha \circ y = x) \wedge (\alpha \in \textsc{Irr}(R)) \end{matrix}\right] = \mathrm{negl}(k) \enspace .$$

Setup Algorithm $\mathsf{Setup}(1^k, \omega)$:
Generate random $i \leftarrow I$ according to ω. Let $\mathsf{parm} \leftarrow i$.
Generating Algorithm $\mathsf{Gen}(1^k, \mathsf{parm})$:
Generate random $g \leftarrow D_i$. Publish $\mathsf{pk} \leftarrow g$.
Digest Algorithm $\mathsf{Dig}_{\mathsf{parm},\mathsf{pk}}(S)$:
1. If $i \notin I$ or $g \notin D_i$, then return Error.
2. If $S \not\subseteq M_i$, then output Error.
3. Otherwise, output $(\prod_{s \in S} H(s)) \circ g$.

(Non)Membership Proof Algorithm $\mathsf{Proof}_{\mathsf{parm},\mathsf{pk}}(m, S)$:
1. If $i \notin I$ or $g \notin D_i$, then return Error.
2. If $m \notin M_i$ or $S \not\subseteq M_i$, then return Error.
3. If $m \in S$ then define $\mathsf{Proof}_{\mathsf{pk}}(m, S) := (\prod_{s \in S \setminus \{m\}} H(s)) \circ g$.
4. Otherwise, let $\delta \leftarrow \prod_{s \in S} H(s) \in R_i$. Because R_i is Euclidean and $\gcd(H(m), \delta) = 1$, there exist $\alpha, \beta \in R_i$, such that $\alpha \cdot H(m) + \beta \cdot \delta = 1$. Let $\mathsf{Proof}_{\mathsf{pk}}(m, S) := (\alpha \circ g, \beta)$.

Verification $\mathsf{Ver}_{\mathsf{parm},\mathsf{pk}}(m, d, p)$:
1. If $i \notin I$ or $g \notin D_i$, then return Error.
2. If $m \notin M_i$ or $d \notin D_i$, then return Error.
3. If $p \in D_i$, then check whether $H(m) \circ p = d$. If it is, then return Member, else return Error.
4. Otherwise, if $p = (q, \beta) \in D_i \times R_i$, then check whether $H(m) \circ q + \beta \circ d = g$. If it is, then return NotMember, else return Error.
5. Otherwise, return Error.

Fig. 1. Root accumulator for $(\mathcal{R}_\mathcal{D}, H)$

5. *$\mathcal{R}_\mathcal{D}$ is a* strong root module *if*

$$\Pr_\omega \left[\begin{array}{l} R_D \leftarrow \mathsf{Setup}(1^k, \omega), x \leftarrow D, (y, \alpha) \leftarrow A(x, \omega) : \\ (\alpha \circ y = x) \wedge (\alpha \neq 1) \end{array} \right] = \mathrm{negl}(k)$$

for every PPT adversary A.

6. *$\mathcal{R}_\mathcal{D}$ is a* strong divisible root module *if*

$$\Pr_\omega \left[\begin{array}{l} R_D \leftarrow \mathsf{Setup}(1^k, \omega), x \leftarrow D, (y, \alpha, \beta) \leftarrow A(x, \omega) : \\ ((\alpha\beta) \circ y = \beta \circ x) \wedge (\alpha \neq 1) \end{array} \right] = \mathrm{negl}(k)$$

for every PPT adversary A.

In the trusted setup model, one does not require security in the case A knows ω, and thus A may be able to break the assumption by obtaining access to it.

The strong prime root and (to certain extent) the strong divisible root assumption are novel, while others are generalizations of well-known assumptions. The assumptions are ordered starting from the "weakest" one, see Sect. 6. For example, if one can solve the discrete logarithm problem in R_D then one also clearly solve the order problem.

The cryptographically familiar example of modules is R_D with $\alpha \circ x := x^\alpha$ for $x \in D$ and $\alpha \in R = \mathbb{Z}$. The order assumption for groups is well known—for the RSA group, it is also called the RSA assumption. That RSA groups are strong root groups was postulated in [1] (the corresponding assumption being called the strong RSA assumption). Damgård and Fujisaki [13] enlisted some candidate strong divisible root groups. Note that in the case of RSA groups the assumptions can only hold in the trusted setup model.

4 Accumulator with Prime-Valued Injective Functions

In this section, we propose the root accumulator for R-modules that generalizes previous work of [23] that considered only the setting of RSA groups, and prime inputs m.

Setting. For some set M and Euclidean ring R, function $f : D \to R$ is a *prime-valued injective function* if it is an injective function $D \to \textsc{Irr}(R)$. We will not propose new prime-valued injective functions, see [1,15] for some existing designs. Let $\mathcal{R}_\mathcal{D} = (R_i)_{D_i}$ for $i \in I$, where D_i is an Abelian group and R_i is a Euclidean ring. Let H be a prime-valued injective function $H : M_i \to R_i$. Here, I depends on k and H depends on i. The root accumulator is depicted by Fig. 1.

Security Proofs.

Theorem 1. *The root accumulator satisfies the correctness property.*

Proof. Assume all participants are honest. Thus $i \in I$, $g \in D_i$, $m \in M_i$ and $d = \mathsf{Dig}_{\mathsf{pk}}(S) \in D_i$. We need to show that if $m \in S$ then $\mathsf{Ver}_{\mathsf{pk}}(m, d, p) = \mathsf{Member}$, and if $m \notin S$ then $\mathsf{Ver}_{\mathsf{pk}}(m, d, p) = \mathsf{NotMember}$. First, if $m \in S$ then $p = (\prod_{s \in S \setminus \{m\}} H(s)) \circ g$. Thus,

$$H(m) \circ p = H(m) \circ ((\prod_{s \in S \setminus \{m\}} H(s)) \circ g) = (\prod_{s \in S} H(s)) \circ g = d \ .$$

Second, if $m \notin S$ then $p = (q, \beta) \in D_i \times R_i$, with $q = \alpha \circ g$ and $\alpha \cdot H(m) + \beta \cdot \delta = 1$ for some α. But then

$$H(m) \circ q + \beta \circ d = (\alpha \cdot H(m)) \circ g + (\beta \cdot \delta) \circ g = 1 \circ g = g \ ,$$

since $\delta = \prod_{s \in S} H(s)$ and $d = \delta \circ g$. □

The next two proofs show that in some sense, collision-resistancy and undeniability of the root accumulator are equivalent, though their reductions to the same underlying problem have different costs. In general, it seems to be difficult to prove that every undeniable accumulator is collision-resistant, because in the case of undeniability the adversary has to return a (possibly fake) digest, while in the case of the collision-resistancy, the adversary has to return a set which may not be easily computable from the fake digest. Moreover, clearly not every collision-resistant accumulator is undeniable; see [5,6] for discussions.

Theorem 2 (Sufficient Conditions (with Prime-Valued Injective Functions)). *Let H be a prime-valued injective function. (1) If $\mathcal{R}_\mathcal{D}$ is a strong prime root module family, then the root accumulator is collision-resistant. (2) If $\mathcal{R}_\mathcal{D}$ is a strong prime root module family, then the strong root accumulator is undeniable.*

Proof. (1) Construct a machine B to break the strong prime root assumption using as the oracle an adversary A that breaks the collision-resistancy of the root accumulator.

1. B obtains $i \leftarrow \mathsf{Setup}(1^k, \omega)$, and random ω.
2. B obtains his challenge $x \leftarrow D_i$.
3. B queries $(m, S, p) \leftarrow A(x, \omega)$.
4. B sets $\delta \leftarrow \prod_{s \in S} H(s)$, $d \leftarrow \delta \circ x$. B finds a pair (α, β), such that $\alpha \cdot H(m) + \beta \cdot \delta = 1$.
5. B returns $(\alpha \circ x + \beta \circ p, H(m))$.

Since $m \notin S$ and H is prime-valued injective, thus $\gcd(H(m), \delta) = 1$. Because R_i is a Euclidean ring, (α, β) can be found efficiently by using the Extended Euclidean Algorithm. Then

$$\begin{aligned} H(m) \circ (\alpha \circ x + \beta \circ p) =& (\alpha \cdot H(m)) \circ x + (\beta \cdot H(m)) \circ p \\ =& (\alpha \cdot H(m)) \circ x + (\beta \cdot \delta) \circ x \\ =& (\alpha \cdot H(m) + \beta \cdot \delta) \circ x = x \ . \end{aligned}$$

Clearly if A is successful then B is successful. B's running time is dominated by the running time of A and by the time it takes to execute the Extended Euclidean algorithm (and thus, R_i has to be Euclidean).

(2) Construct a machine B to break the strong root assumption using as the oracle an adversary A that breaks the undeniability of the root accumulator.

1. B obtains $i \leftarrow \mathsf{Setup}(1^k, \omega)$, and random ω.
2. B obtains his challenge $x \leftarrow D_i$. B sets $\mathsf{pk} = (i, x)$.
3. B queries $(m, d, p, \overline{p}) \leftarrow A(\mathsf{pk}, \omega)$, where $\overline{p} = (q, \alpha)$.
4. B returns $(q + \alpha \circ p, H(m))$.

If A is successful, then $H(m) \circ p = d$ and $H(m) \circ q + \alpha \circ d = x$. Thus,

$$H(m) \circ (q + \alpha \circ p) = H(m) \circ q + (\alpha \cdot H(m)) \circ p = x \ .$$

Therefore, B breaks the strong root problem with the same probability that A breaks the undeniability of the root accumulator, in time that is dominated by A's running time. (For this reduction to go through, R_i does not have to be Euclidean.) □

5 Accumulator with Division-Intractable Function Family

One of the drawbacks of the root accumulator (as described in the previous section) is that prime-valued injective functions (see [1] for some examples) may be inefficient. In this section, we consider a variation of the root accumulator that works with potentially more efficient division-intractable functions [15]. However, due to that, it is based on a (probably) stricter assumption on $\mathcal{R}_\mathcal{D}$.

Division-intractable functions. Let I be an index set. As always, let R_D be an R-module over Euclidean ring R. A hash function family $\mathcal{H} = \{\mathcal{H}_i\}$ with $H : M_i \to R_i$ for every $H_i \in \mathcal{H}_i$ is a *division-intractable function family* [15] if

$$\Pr\left[\begin{array}{l} H_i \leftarrow \mathcal{H}_i, (m, S) \leftarrow A(H_i) : \\ (S \subseteq D_i) \wedge (m \in D_i \setminus S) \wedge (H(m) \mid \prod_{s \in S} H(s)) \end{array}\right] = \text{negl}(k)$$

for any PPT adversary A. Clearly, every prime-valued collision-resistant function is a division-intractable function family by itself. Division-intractable function families can be more efficient than prime-valued injective functions, see [15] for some constructions.

One can instantiate the root accumulator with a division-intractable function family, by letting the hash function $H \leftarrow \mathcal{H}$ to be a part of the public key pk. One also has to modify the definition of the non-membership proof. Namely, if $m \notin S$, Proof works as follows:

- Let $\delta \leftarrow \prod_{s \in S} H(s)$. For $\gamma \leftarrow \gcd(H(m), \delta)$, find $\alpha, \beta \in R_i$, such that $\alpha \cdot H(m) + \beta \cdot \delta = \gamma$. Let $\mathsf{Proof}_{\mathsf{pk}}(m, S) := (\alpha \circ g, \beta, \gamma)$.

Analogously, verification of non-membership is modified as follows:

- If $p = (q, \beta, \gamma) \in D_i \times R_i \times R_i$, then check whether $H(m) \circ q + \beta \circ d = \gamma \circ g$, $\gamma \mid H(m)$, and $\gamma \neq H(m)$. If it is, then return $\mathsf{NotMember}$, else return Error.

(Note that $\gamma \neq H(m)$ because $\mathcal{H}$ is division-intractable.)

Theorem 3 (Sufficient Conditions (with Division-Intractable Function Families).). *Let $\mathcal{H}$ be a division-intractable function family. Let $\mathcal{R}_\mathcal{D}$ be a family of modules over Euclidean rings. (1) If $\mathcal{R}_\mathcal{D}$ is a strong divisible root module family, then the root accumulator is collision-resistant. (2) If $\mathcal{R}_\mathcal{D}$ is a strong divisible root module family, then the root accumulator is undeniable.*

Proof. (1) Construct a machine B that breaks either the strong divisible root assumption of R_D or the division-intractability of $\mathcal{H}$ using as an oracle an adversary A who can break the collision-resistancy of the root accumulator.

1. B obtains $i \leftarrow \mathsf{Setup}(1^k, \omega)$, and random ω.
2. B obtains his challenge $x \leftarrow D_i$, and $H \leftarrow \mathcal{H}$.
3. B sets $\mathsf{pk} \leftarrow (i, x, H)$.
4. B queries $(m, S, p) \leftarrow A(\mathsf{pk}, \omega)$.
5. B sets $\delta \leftarrow \prod_{s \in S} H(s)$.
6. B sets $\beta^* \leftarrow \gcd(H(m), \delta)$, $\alpha^* \leftarrow H(m)/\beta^*$. If $\alpha^* = 1$, then B aborts.
7. By using the Extended Euclidean Algorithm, B computes γ and γ', such that $\gamma \cdot H(m) + \gamma' \cdot \delta = \beta^*$.
8. B returns $(\gamma \circ x + \gamma' \circ p, \alpha^*, \beta^*)$.

Clearly,

$$\begin{aligned}(\alpha^* \cdot \beta^*) \circ (\gamma \circ x + \gamma' \circ p) =& H(m) \circ (\gamma \circ x + \gamma' \circ p) \\ =& (\gamma \cdot H(m)) \circ x + (\gamma' \cdot H(m)) \circ p \\ =& (\gamma \cdot H(m)) \circ x + (\gamma' \cdot \delta) \circ x = \beta^* \circ x \ .\end{aligned}$$

Thus, if A is successful and B does not abort, then B is successful. But if B aborts, then $\beta^* = H(m)$ and thus B has broken the division-intractability of $\mathcal{H}$.

(2) Construct a machine B that breaks either the strong divisible root assumption of R_D or the division-intractability of $\mathcal{H}$ using as oracle an adversary A who can break the undeniability of the root accumulator.

1. B obtains $i \leftarrow \mathsf{Setup}(1^k, \omega)$, and random ω.
2. B obtains his challenge $x \leftarrow D_i$, and $H \leftarrow \mathcal{H}$.
3. B sets $\mathsf{pk} \leftarrow (i, x, H)$.
4. B queries $(m, d, p, \overline{p}) \leftarrow A(\mathsf{pk}, \omega)$, where $\overline{p} = (q, \beta, \gamma)$.
5. B sets $\alpha^* \leftarrow H(m)/\gamma$. If $\alpha^* = 1$, then B aborts.
6. B returns $(q + \beta \circ p, \alpha^*, \gamma)$.

If A is successful, then $H(m) \circ p = d$ and $H(m) \circ q + \beta \circ d = \gamma \circ g$ for $\gamma \mid H(m)$ and $\gamma \neq H(m)$. Thus,

$$(\alpha^* \gamma) \circ (q + \beta \circ p) = H(m) \circ q + (\beta H(m)) \circ p = H(m) \circ q + \beta \circ d = \gamma \circ g \ .$$

Thus, if A is successful and B does not abort, then B is successful in breaking the strong divisible root assumption. But if B aborts, then B has broken the division-intractability of $\mathcal{H}$. □

Now, we show that independently of the properties of R_D, the family $\mathcal{H}$ must be division-intractable.

Lemma 1. *If the root accumulator is collision-resistant, then $\mathcal{H}$ is division-intractable.*

Proof (Sketch.). By contradiction: assume an adversary finds a pair (m, S), $m \notin S$, such that $\prod_{s \in S} H(s) = \alpha \cdot H(m)$, for some $\alpha \in R_i$. Now,

$$d = (\prod_{s \in S)} H(s) \circ g = (\alpha \cdot H(m)) \circ g = H(m) \circ (\alpha \circ g) = H(m) \circ p$$

with $p = \alpha \circ g$, and therefore the adversary has broken the accumulator. □

6 Relations between New Assumptions

Clearly, if $\mathcal{R_D}$ is a strong root module family, then it is also a strong prime root module family. (If it is difficult to find (y, α) such that $\alpha \circ y = x$, then it is also difficult to find (y, α) with α irreducible, such that $\alpha \circ y = x$.) The opposite holds only if factorization is easy in the Euclidean ring. Moreover, if $\mathcal{R}_\mathcal{D}$ is a strong prime root module family, then it is clearly a root module family. Thus, the strong prime root assumption is in its strength somewhere between the root assumption and the strong root assumption. Because we showed that root accumulator with prime-valued injective function H is secure if and only if the underlying module module is strong root module family, we get

Theorem 4. *Let H be a prime-valued injective function. If factorization is difficult in the underlying Euclidean ring, then the security of root accumulator is based on a security assumption that is weaker than the strong root assumption.*

(In particular, the security of RSA accumulator can be based on an assumption that is weaker than the strong RSA assumption.)

Clearly, if $\mathcal{R}_\mathcal{D}$ is a strong divisible root family, then $\mathcal{R}_\mathcal{D}$ is also a strong root module family. (To break the strong divisible root assumption, just return $(g, \alpha, 1)$, where (g, α) was returned by an adversary who breaks the strong root assumption.) To show that the proposed strong divisible root assumption in this special case is not too strong, we reduce its security to the strong root assumption conditionally to the small root assumption that generalizes an earlier assumption of the same name by Damgård and Fujisaki [13]. (In their paper, it was assumed that $\beta = 2$.) See [13] for discussion.

Theorem 5. *Let $\mathcal{R}_\mathcal{D}$ be defined as always. Let the next two assumptions hold: (a) $\mathcal{R}_\mathcal{D}$ is a strong root module family, (b) For any $i \in I$, it is intractable to find elements $g \in D_i$ such that $\alpha \circ g = 0$ for some α with non-minimal non-zero degree* $\deg(\alpha)$*, but $\beta \circ g \neq 0$ for some* $\deg(\beta) < \deg(\alpha)$ *(we call this a* small root assumption*). Then $\mathcal{R}_\mathcal{D}$ is a strong divisible root module family.*

Proof. Assume that adversary A breaks the strong divisible root assumption. Construct a machine B that breaks one of the two premises as follows.

1. B obtains $i \leftarrow \mathsf{Setup}(1^k, \omega)$ and ω.
2. B gets his challenge $x \leftarrow D$ of the strong root problem game.
3. B obtains $(y, \alpha, \beta) \leftarrow A(x, \omega)$.
4. B returns (y, α).

Assume that A is successful. Then $(\alpha \cdot \beta) \circ y = \beta \circ x$ and $\alpha \neq 1$. If $\beta = 1$ then we are done. Otherwise, denote $w \leftarrow \alpha \circ y - x$. If $w = 0$ then we are done. Otherwise,

$$\beta \circ w = \beta \circ (\alpha \circ y) - \beta \circ x = (\alpha \cdot \beta) \circ y - \beta \circ x = \beta \circ x - \beta \circ x = 0 \ .$$

Choose some $\beta' \in R$ with $\deg(\beta') < \deg(\beta)$. By the small root assumption, also $\beta' \circ w = 0$. Thus, by application of the Euclidean algorithm, $\alpha = x \cdot \beta' + q$ such that $\deg(q) < \deg(\beta')$. By the choice of β',

$$\alpha \circ w = x \circ (\beta' \circ w) + q \circ w = q \circ w \ .$$

By the small root assumption, $w = 0$ and thus $\alpha \circ y = x$. □

7 Example Instantiations

7.1 RSA Accumulator

In the RSA accumulator, as modified by [1,26], the public parameters contain $n = PQ$ that is a product of two safe primes, and $D_n = \mathbb{Z}_n^*$, R_n with $\alpha \circ g := g^\alpha \pmod n$. If the factorization of n is known to a collusion of parties (say, generated by a malicious server or in a threshold manner by several parties who are all later corrupted), they can jointly compute membership proofs of any element m by defining $p \leftarrow \mathsf{Dig}_{n,g}(S)^{H(m)^{-1} \pmod{\phi(n)}} \pmod n$. Therefore, the RSA accumulator is not collision-resistant without trusted setup.

7.2 Root Accumulator in Class Groups of IQ Order

Class Group Preliminaries. Let Δ be a negative integer such that $\Delta \equiv 0, 1 \pmod 4$. The ring $\mathcal{O}_\Delta = \mathbb{Z} + \frac{\Delta + \sqrt{\Delta}}{2} \cdot \mathbb{Z}$ is an *imaginary quadratic order* of *discriminant* Δ. Its field of fractions is $\mathbb{Q}(\sqrt{\Delta})$. The discriminant Δ is *fundamental* if Δ is square-free if $\Delta \equiv 1 \pmod 4$ or $\Delta/4$ is square-free if $\Delta \equiv 0 \pmod 4$. The ring $\mathcal{O}_\Delta$ is a *maximal order* if Δ is fundamental. The *fractional ideals* of $\mathcal{O}_\Delta$ are of form $q(a\mathbb{Z} + (b + \sqrt{\Delta})/2\mathbb{Z})$ with $q \in \mathbb{Q}$, $a \in \mathbb{Z}^+$, $b \in \mathbb{Z}$ and $4a \mid (b^2 - \Delta)$. Therefore, a fractional ideal can be represented by a triple (q, a, b). An ideal (q, a, b) is *integral* if $q = 1$; an integral ideal can be represented by a pair (a, b). Two fractional ideals $\mathfrak{a}, \mathfrak{b} \subseteq \mathcal{O}_\Delta$ are *equivalent* if for some nonzero $\alpha \in \mathbb{Q}(\sqrt{\Delta})$, $\mathfrak{a} = \alpha\mathfrak{b}$. The set of equivalence classes forms an Abelian group under ideal multiplication; this group is called the *class group* and denoted by $\mathrm{Cl}(\Delta)$. The class group is always finite, its order is called the *class number* and denoted by $h(\Delta) := |\mathrm{Cl}(\Delta)|$.

For an integral ideal there exists a $c \in \mathbb{Z}^+$, such that $\Delta = b^2 - 4ac$. An ideal is called *reduced* if (a) $\gcd(a, b, c) = 1$, (b) $-a < b \le a \le c$ and (c) $b \ge 0$ if $a = c$. Every equivalence class contains exactly one reduced ideal. Thus, every element of $\mathrm{Cl}(\Delta)$ can be represented by a reduced ideal of $\mathcal{O}_\Delta$, and checking equality of two ideal classes means comparing the representatives. The neutral element of $\mathrm{Cl}(\Delta)$ is represented by $(1, \Delta \pmod 2)$. The inverse of the ideal class represented by (a, b) is the ideal class represented by $(a, -b)$. The group operation in $\mathrm{Cl}(\Delta)$ is ideal multiplication followed by reduction; a group operation requires $O(\log^2 |\Delta|)$ bit-operations. If (a, b) is reduced then $a \le \sqrt{|\Delta|/3}$. For more information on computations in class groups see [12, Chapter 5] or [3]; for algorithms see [21,3].

Class Groups and Cryptography. Class groups were first proposed for use in cryptography by Buchmann and Williams [4]. The class number is not efficiently computable if Δ is fundamental, but the even part of $h(\Delta)$ can be efficiently computed if the prime factorization of Δ is known. All problems of Def. 4 can be instantiated to the class groups, and the known efficient algorithms for tackling the discrete logarithm, order, and root problem are tightly connected [3].

General number fields sieve [22], the best currently known factorization algorithm, runs in time $L_n[1/3, \sqrt[3]{\frac{64}{9}}]$. The best currently known algorithm (MPQS) for the root problem in maximal orders runs in time $L_\Delta[1/2, 1 + o(1)]$ [21]. Even this time is only empirically suggested, the best rigorous algorithm for computing the discrete logarithm runs in time $L_\Delta[\frac{1}{2}, \frac{3}{4}\sqrt{2} + o(1)]$ assuming the Extended Riemann Hypothesis [27]. On the other hand, if $\mathcal{O}_\Delta$ is non-maximal then the discrete logarithm problem in $\mathrm{Cl}(\Delta)$ can be reduced to the discrete logarithm problem in multiplicative groups of finite fields [20]. On the other hand, a $(p-1)$-like algorithm can compute the class number efficiently, given that $h(\Delta)$ is smooth [19]. Hamdy and Möller estimate the probability that a $h(\Delta)$ is B-smooth for randomly chosen k-bit Δ, and conclude that if k-bit discriminants are large enough to guarantee security against the MPQS algorithm, then the probability to find a sufficiently smooth $h(\Delta)$ by choosing k-bit Δ's randomly and applying the $(p-1)$-like algorithm to them, is negligible. Their heuristic, that we also follow, is that the same holds true even when $|\Delta|$ is chosen to be a k-bit random prime.

In general, we rely on the next properties of the class groups:

- If $-\Delta$ is a random k-bit prime, for large k, then computing the roots in the class group $\mathrm{Cl}(\Delta)$ as well as the order of a random element from $\mathrm{Cl}(\Delta)$ is assumed to be intractable. In particular, the length of the discriminant Δ is reasonable: to achieve the same security as with $k = 1536$ in the RSA case, it seems to be sufficient to take $k \approx 1000$ [19].
- if $-\Delta$ is prime and Δ is fundamental, then with high probability 0.9775 the class group $\mathrm{Cl}(\Delta)$ of imaginary quadratic order is cyclic. Moreover, then $h(\Delta)$ is odd.

Root Accumulator in Class Groups of Imaginary Quadratic Order. Let us now concentrate on the case where $R_i = \mathbb{Z}$, and D_i is a class group of imaginary quadratic order, with $\alpha \circ x := x^\alpha$ in D_i.

In the setup phase, for class groups, choose a random negative k-bit prime fundamental discriminant Δ, that is, let $-\Delta$ be a random k-bit prime with $\Delta \equiv 1 \mod 4$. For a k-bit prime i, let $D_i := \mathrm{Cl}(-i)$. Note that for a random k-bit negative Δ, $h(\Delta)$ is not smooth [19], and it was also conjectured in [19] that this also holds when Δ is a random negative k-bit prime.

Another assumption that we make (but that is common to all previous papers on class group-based cryptography) is that with probability $1-\mathrm{negl}(k)$, a random element of $\mathrm{Cl}(\Delta)$ is a generator of some sufficiently large subgroup of $\mathrm{Cl}(\Delta)$.

Word of Caution. While the (weaker) root assumption is a well known assumption in class groups, we are not aware of any a priori use of the strong root assumption in class groups except [13]. Since also [13] did not analyze the strong root assumption but only used it, we must warn that this assumption is yet almost unstudied in the class groups. However, we hope that the current paper provides new incentive to study strong root assumption (and related assumptions) in class groups. A disproof of such assumptions would constitute a major result by itself.

Acknowledgments. The author was supported by Estonian Science Foundation, grant #9303 and the European Union through the European Regional Development Fund. We thank Valdis Laan, Safuat Hamdy and Lauri Tarkkala for useful comments.

References

1. Barić, N., Pfitzmann, B.: Collision-Free Accumulators and Fail-Stop Signature Schemes without Trees. In: Fumy, W. (ed.) EUROCRYPT 1997. LNCS, vol. 1233, pp. 480–494. Springer, Heidelberg (1997)
2. Benaloh, J.C., de Mare, M.: One-Way Accumulators: A Decentralized Alternative to Digital Signatures. In: Helleseth, T. (ed.) EUROCRYPT 1993. LNCS, vol. 765, pp. 274–285. Springer, Heidelberg (1994)
3. Buchmann, J., Hamdy, S.: A Survey on IQ Cryptography. Technical Report TI-4/01, TU Darmstadt, Fachbereich Informatik (March 21, 2001)
4. Buchmann, J.A., Williams, H.C.: A Key-exchange System Based on Imaginary Quadratic Fields. Journal of Cryptology 1(2), 107–118 (1988)
5. Buldas, A., Laud, P., Lipmaa, H.: Accountable Certificate Management Using Undeniable Attestations. In: Jajodia, S., Samarati, P. (eds.) ACM CCS 2000, Athens, Greece, November 2-4, pp. 9–18. ACM Press (2000)
6. Buldas, A., Laud, P., Lipmaa, H.: Eliminating Counterevidence with Applications to Accountable Certificate Management. Journal of Computer Security 10(3), 273–296 (2002)
7. Buldas, A., Laud, P., Lipmaa, H., Villemson, J.: Time-Stamping with Binary Linking Schemes. In: Krawczyk, H. (ed.) CRYPTO 1998. LNCS, vol. 1462, pp. 486–501. Springer, Heidelberg (1998)
8. Buldas, A., Lipmaa, H., Schoenmakers, B.: Optimally Efficient Accountable Time-Stamping. In: Imai, H., Zheng, Y. (eds.) PKC 2000. LNCS, vol. 1751, pp. 293–305. Springer, Heidelberg (2000)
9. Camacho, P., Hevia, A., Kiwi, M., Opazo, R.: Strong Accumulators from Collision-Resistant Hashing. In: Wu, T.-C., Lei, C.-L., Rijmen, V., Lee, D.-T. (eds.) ISC 2008. LNCS, vol. 5222, pp. 471–486. Springer, Heidelberg (2008)
10. Camenisch, J., Kohlweiss, M., Soriente, C.: An Accumulator Based on Bilinear Maps and Efficient Revocation for Anonymous Credentials. In: Jarecki, S., Tsudik, G. (eds.) PKC 2009. LNCS, vol. 5443, pp. 481–500. Springer, Heidelberg (2009)
11. Camenisch, J., Lysyanskaya, A.: Dynamic Accumulators and Application to Efficient Revocation of Anonymous Credentials. In: Yung, M. (ed.) CRYPTO 2002. LNCS, vol. 2442, pp. 61–76. Springer, Heidelberg (2002)
12. Cohen, H.: A Course in Computational Algebraic Number Theory. Graduate Texts in Mathematics. Springer (1995)
13. Damgård, I., Fujisaki, E.: A Statistically-Hiding Integer Commitment Scheme Based on Groups with Hidden Order. In: Zheng, Y. (ed.) ASIACRYPT 2002. LNCS, vol. 2501, pp. 125–142. Springer, Heidelberg (2002)
14. Damgård, I., Koprowski, M.: Generic Lower Bounds for Root Extraction and Signature Schemes in General Groups. In: Knudsen, L.R. (ed.) EUROCRYPT 2002. LNCS, vol. 2332, pp. 256–271. Springer, Heidelberg (2002)
15. Gennaro, R., Halevi, S., Rabin, T.: Secure Hash-and-Sign Signatures without the Random Oracle. In: Stern, J. (ed.) EUROCRYPT 1999. LNCS, vol. 1592, pp. 123–139. Springer, Heidelberg (1999)

16. Groth, J., Sahai, A.: Efficient Non-interactive Proof Systems for Bilinear Groups. In: Smart, N.P. (ed.) EUROCRYPT 2008. LNCS, vol. 4965, pp. 415–432. Springer, Heidelberg (2008)
17. Haber, S., Stornetta, W.S.: How to Time-Stamp a Digital Document. In: Menezes, A., Vanstone, S.A. (eds.) CRYPTO 1990. LNCS, vol. 537, pp. 437–455. Springer, Heidelberg (1991)
18. Hamdy, S.: Computations in Class Groups of Imaginary Quadratic Number Fields. In: Innovations in Information Technology, Dubai, UAE, November 19-21, pp. 1–5 (2006)
19. Hamdy, S., Möller, B.: Security of Cryptosystems Based on Class Groups of Imaginary Quadratic Orders. In: Okamoto, T. (ed.) ASIACRYPT 2000. LNCS, vol. 1976, pp. 234–247. Springer, Heidelberg (2000)
20. Hühnlein, D., Takagi, T.: Reducing Logarithms in Totally Non-maximal Imaginary Quadratic Orders to Logarithms in Finite Fields. In: Lam, K.-Y., Okamoto, E., Xing, C. (eds.) ASIACRYPT 1999. LNCS, vol. 1716, pp. 219–231. Springer, Heidelberg (1999)
21. Jacobson Jr., M.J.: Subexponential Class Group Computation in Quadratic Orders. PhD thesis, Technische Universität Darmstadt, Fachbereich Informatik, Darmstadt, Germany (1999)
22. Lenstra, A.K., Lenstra, J. H.W. (eds.): The Development of the Number Field Sieve. Lecture Notes in Mathematics, vol. 1554. Springer, Heidelberg (1993)
23. Li, J., Li, N., Xue, R.: Universal Accumulators with Efficient Nonmembership Proofs. In: Katz, J., Yung, M. (eds.) ACNS 2007. LNCS, vol. 4521, pp. 253–269. Springer, Heidelberg (2007)
24. Nguyen, L.: Accumulators from Bilinear Pairings and Applications. In: Menezes, A. (ed.) CT-RSA 2005. LNCS, vol. 3376, pp. 275–292. Springer, Heidelberg (2005)
25. Sander, T.: Efficient Accumulators without Trapdoor Extended Abstract. In: Varadharajan, V., Mu, Y. (eds.) ICICS 1999. LNCS, vol. 1726, pp. 252–262. Springer, Heidelberg (1999)
26. Sander, T., Ta-Shma, A., Yung, M.: Blind, Auditable Membership Proofs. In: Frankel, Y. (ed.) FC 2000. LNCS, vol. 1962, pp. 53–71. Springer, Heidelberg (2001)
27. Vollmer, U.: Asymptotically Fast Discrete Logarithms in Quadratic Number Fields. In: Bosma, W. (ed.) ANTS 2000. LNCS, vol. 1838, pp. 581–594. Springer, Heidelberg (2000)

Linear Fault Analysis of Block Ciphers

Zhiqiang Liu[1,*], Dawu Gu[1], Ya Liu[1], and Wei Li[2]

[1] Department of Computer Science and Engineering,
Shanghai Jiao Tong University, Shanghai 200240, P.R. China
{ilu_zq,dwgu,liuya0611}@sjtu.edu.cn
[2] School of Computer Science and Technology,
Donghua University, Shanghai 201620, P.R. China
liwei.cs.cn@gmail.com

Abstract. Differential fault analysis (DFA) has already been applied to attack many block ciphers with the help of inducing some faults at the last few rounds of block ciphers. Currently, a general countermeasure against DFA is to protect the last few rounds of block ciphers by means of redundancy. In this paper, we present a new fault attack on block ciphers called linear fault analysis (LFA), in which linear characteristics for some consecutive rounds of a block cipher will be utilized instead of exploiting differential distributions of S-Boxes within the block cipher in DFA. Basically, the new approach can handle the case that faults are induced several rounds earlier compared to DFA, thus leading to a threat to the protected implementations (against DFA) of block ciphers. For the purpose of illustration, we mount an effective attack on SERPENT by adopting LFA and achieve a good cryptanalytic result on SERPENT. We hope that our work enriches the picture on the applicability of fault attacks to block ciphers and could be beneficial to the security evaluation of block ciphers.

Keywords: Differential Fault Analysis, Linear Fault Analysis, Block Ciphers, SERPENT.

1 Introduction

In recent years, side channel attacks [1] have become important and efficient cryptanalytic tools in analyzing various cryptographic devices. These attacks exploit easily accessible information such as power consumption, running time, input-output behavior under malfunctions, and so on, and then evaluate such leaked information with the help of statistical methods. To some extent, side channel attacks are often more powerful than classical approaches such as differential cryptanalysis [2], linear cryptanalysis [3], related-key cryptanalysis [4], integral cryptanalysis [5], algebraic attack [6], and so on.

---

* This work has been supported by National Natural Science Foundation of China (No. 61073150 and No. 61003278), the Opening Project of Shanghai Key Laboratory of Integrate Administration Technologies for Information Security, and the Fundamental Research Funds for the Central Universities.

F. Bao, P. Samarati, and J. Zhou (Eds.): ACNS 2012, LNCS 7341, pp. 241–256, 2012.

As one of side channel attacks, fault analysis is a class of implementation attacks that disturb cryptographic computations so as to recover secret keys. To speak specifically, when an encryption is executed under faulty condition, an error occurs at some intermediate state, which results in a faulty output. The faulty output is then used as leaked information to help retrieve secret key. Since the fault analysis was first introduced in 1997 by Boneh *et al* [7], various methods of fault analysis have been proposed and studied. Among them, differential fault analysis(DFA) [8] can be regarded as the most effective cryptanalytic method against block ciphers. In fact, DFA derives information about the secret key of a block cipher by using differences between correct and faulty ciphertexts. Generally, an attacker gets faulty ciphertexts by giving external impact on a device with voltage variation, glitch, laser, etc [9]. To date, much research work has been devoted to DFA on a variety of block ciphers such as DES [8,10], AES [11,12,13,14,15,16,17,18], IDEA [19], CLEFIA [20], SMS4 and MacGuffin [21], ARIA [22] and Camellia [23]. Such work demonstrates the vulnerability of block ciphers towards DFA and the subsequent need of including countermeasures in embedded implementations of block ciphers. Moreover, some extensions to DFA have been presented in [24,25,26] in order to make fault attack more efficient.

As a matter of fact, most of DFA techniques target the last few rounds of a block cipher, i.e., faults will be triggered at the last few rounds of the cipher so as to induce information leakage. Consequently, the general countermeasure against DFA is to protect the last few rounds of the cipher by means of redundancy. However, the implementation of the countermeasure against DFA is more costly and less efficient along with the number of protected rounds increasing, thus for a block cipher, the practical implementations used to thwart DFA will cover as less protected rounds as possible.

In this paper, we propose a new fault attack on block ciphers called linear fault analysis (LFA), in which linear characteristics for some consecutive rounds of a block cipher will be utilized instead of exploiting differential distributions of S-Boxes within the block cipher in DFA. Generally, the new approach can deal with the case that faults are injected several rounds earlier compared to DFA as long as suitable linear approximations exist. Thus based on our new method, one may mount an effective attack on a block cipher even if the cipher has already been protected against DFA. Furthermore, in order to demonstrate the validity of LFA, we apply it to analyze the block cipher SERPENT which is a candidate of Advanced Encryption Standard and rated just behind the AES Rijndael. To the best of our knowledge, there isn't any known DFA attack on SERPENT which can be done by inducing faults at the round earlier than the penultimate round of the cipher, as a result, the countermeasure against DFA on SERPENT can be implemented by protecting the last two rounds of the cipher. However, we present an effective attack on the protected implementation of SERPENT by means of LFA. As a new extension of fault attack, LFA may be beneficial to the security evaluation of block ciphers.

The rest of the paper is organized as follows. Section 2 introduces the notations used throughout this paper and shows the possibility and rationality of fault injection briefly. Section 3 presents our new fault attack on block ciphers, that is, linear fault analysis. Section 4 applies our novel approach to mount an effective attack on the DFA-proof implementation of SERPENT. Finally, Section 5 summarizes the paper.

2 Preliminaries

The following notations are used throughout the paper.

- $\oplus$ denotes bitwise exclusive OR (XOR).
- $\cdot$ denotes bitwise inner product.
- $|x|$ denotes the absolute value of a real number x.
- $\circ$ denotes the composition operation.
- $\#S$ denotes the cardinality of a set S.
- 0x denotes the hexadecimal notation.

2.1 Fault Injection

In practice, many block ciphers have been implemented in cryptographic devices such as smart cards and RFID tags. Generally, we can assume that a cryptographic device with fixed secret key is under the control of the attacker who could use the device to encrypt (or decrypt) arbitrary plaintexts (or ciphertexts). Accordingly, the attacker is able to deliberately interfere the normal operation of the device with electromagnetic perturbations, voltage variations, clock glitches and lasers so as to induce faults [9].

With the development of fault injection techniques, many fault models have been established, among which single bit error model and single byte error model are the most well-studied and applicable. Actually, in order to trigger single bit error or single byte error at certain intermediate state within the encryption/decryption process of a block cipher, laser technique could be adopted since a laser with certain energy and wavelength could interfere fixed parts of the memory/registers without damaging them, resulting in single bit error or single byte error at some internal state accurately [27].

3 Linear Fault Analysis

We now present a new fault attack on block ciphers called linear fault analysis (LFA), in which linear characteristics for some consecutive rounds of a block cipher will be utilized instead of exploiting differential distributions of S-Boxes within the block cipher in DFA.

3.1 Fault Model and Assumption

Our attack is applicable in the single bit error model as well as single byte error model. To speak specifically, for a given block cipher, the attacker has the capability to choose plaintexts to encrypt, and during the encryption process, he can repeatedly induce single bit error or single byte error at the input of some certain round of the block cipher so as to obtain the corresponding right and faulty ciphertexts. Note that the values and positions (within the impacted round) of the faults injected by the attacker are unknown and randomly distributed.

3.2 Principle of Linear Fault Analysis

In the following we will demonstrate the principle of linear fault analysis (LFA) under the condition of single bit error model. Firstly, we will introduce the definition of linearly active input set with respect to a linear approximation and give a claim related to linear fault analysis.

Definition 1. Let E_1 be a block cipher and $\Gamma_P \cdot P \oplus \Gamma_C \cdot C = \Gamma_K \cdot K$ (also denoted as $\Gamma_P \rightarrow \Gamma_C$) be a linear approximation for E_1. Let $S_{\Gamma_P \rightarrow \Gamma_C}$ be a set consisting of all bits of P involving in the item $\Gamma_P \cdot P$. Then $S_{\Gamma_P \rightarrow \Gamma_C}$ is denoted as the linearly active input set with respect to the linear characteristic $\Gamma_P \rightarrow \Gamma_C$.

Claim 1. Let E be a block cipher and decompose the cipher into $E = E_1 \circ E_0$, where E_0 represents the first part of the cipher and E_1 represents the last part. Let $\Gamma_P \cdot P \oplus \Gamma_C \cdot C = \Gamma_K \cdot K$ be a linear approximation for E_1 with probability $1/2 + \varepsilon$ and $S_{\Gamma_P \rightarrow \Gamma_C}$ be the linearly active input set with respect to the linear characteristic $\Gamma_P \rightarrow \Gamma_C$. Suppose that an attacker has the ability to induce single bit error at the input of E_1 repeatedly and the error bits don't belong to the set $S_{\Gamma_P \rightarrow \Gamma_C}$, then an effective distinguisher $\Gamma_C \cdot C_1 \oplus \Gamma_C \cdot C_2 = 0$ for the cipher E with probability $1/2 + 2\varepsilon^2$ can be derived by the attacker.

Next we will show the reasonability of Claim 1 in detail. First of all, we find that the effect of injecting single bit error at the input of E_1 repeatedly with error bits not in the set $S_{\Gamma_P \rightarrow \Gamma_C}$, is somewhat like constructing a particular differential-linear distinguisher that is composed of a special truncated differential characteristic *unknown input difference* $\rightarrow \nabla$ for E_0 with probability 1 and a linear characteristic $\Gamma_P \rightarrow \Gamma_C$ for E_1 with probability $1/2 + \varepsilon$ such that $\nabla \cdot \Gamma_P = 0$. For any pair consisting of a right ciphertext C_1 under E and the corresponding faulty ciphertext C_2 derived under the above condition, we have that both

$$\Gamma_P \cdot E_1^{-1}(C_1) \oplus \Gamma_C \cdot C_1 = \Gamma_K \cdot K$$

and

$$\Gamma_P \cdot E_1^{-1}(C_2) \oplus \Gamma_C \cdot C_2 = \Gamma_K \cdot K$$

hold with probability $1/2+\varepsilon$. Assume that these two equations are uncorrelated, and take into account the condition that the error bit induced at the input of E_1 is not in the set $S_{\Gamma_P \rightarrow \Gamma_C}$, we immediately obtain that

$$\Gamma_C \cdot C_1 \oplus \Gamma_C \cdot C_2 = 0$$

holds with probability

$$(1/2 + \varepsilon)^2 + (1/2 - \varepsilon)^2 = 1/2 + 2\varepsilon^2.$$

Thus the distinguisher by checking the parity of $\Gamma_C \cdot C_1 \oplus \Gamma_C \cdot C_2$ can be utilized to distinguish the cipher E from a random permutation since for such a ciphertext pair (C_1, C_2), the equation $\Gamma_C \cdot C_1 \oplus \Gamma_C \cdot C_2 = 0$ holds with probability 1/2 for a random permutation. □

Based on the result given in the above Claim as well as the Algorithm 2 proposed in [3], we can mount a key recovery attack on $E' = E_2 \circ E = E_2 \circ E_1 \circ E_0$ (where E_2 represents the last round of the cipher E') by guessing part of the last round subkey of E'. The general attack procedure can be described as follows:

Step 1. Given the linear characteristic $\Gamma_P \to \Gamma_C$ for E_1, collect N pairs of ciphertexts, each pair consisting of a right ciphertext C_1^i under E' and the corresponding faulty ciphertext C_2^i derived by injecting single bit error at any position of the input of E_1, where $1 \le i \le N$.

Step 2. Let K_g denote the bits of the last round subkey which are related to the item $\Gamma_C \cdot E_2^{-1}(C_j^i)$, i.e., $\Gamma_C \cdot E_2^{-1}(C_j^i)$ can be obtained by performing a partial decryption of the ciphertext C_j^i with the guessed value of K_g, where $1 \le i \le N, 1 \le j \le 2$. Then for each possible value of K_g, do the following:
(1). Initialize a counter T_{K_g} firstly.
(2). For each ciphertext pair (C_1^i, C_2^i), implement the partial decryptions of C_1^i and C_2^i respectively and compute the parity of $\Gamma_C \cdot E_2^{-1}(C_1^i) \oplus \Gamma_C \cdot E_2^{-1}(C_2^i)$. If the parity is 0, increase the relevant counter T_{K_g} by 1, and decrease by 1 otherwise.
(3). Store the value of K_g as well as the value of the corresponding $|T_{K_g}|$.

Step 3. For all possible values of K_g, compare the stored values and take the value of K_g as the correct key information if the value of the corresponding $|T_{K_g}|$ is maximal.
Note that in the above key recovery attack, the single bit error can be triggered **at any position of the input of** E_1. Here we want to discuss the rationality of the key recovery attack. On the one hand, if the guessed value of K_g is correct, then for any ciphertext pair (C_1^i, C_2^i) in which C_2^i is derived by inducing single bit error at the input of E_1 such that the error bit is not in the set $S_{\Gamma_P \to \Gamma_C}$, the equation $\Gamma_C \cdot E_2^{-1}(C_1^i) \oplus \Gamma_C \cdot E_2^{-1}(C_2^i) = 0$ holds with probability $1/2 + 2\varepsilon^2$, and for any ciphertext pair (C_1^i, C_2^i) where C_2^i is obtained by injecting single bit error at the input of E_1 such that the error bit belongs to the set $S_{\Gamma_P \to \Gamma_C}$, the equation $\Gamma_C \cdot E_2^{-1}(C_1^i) \oplus \Gamma_C \cdot E_2^{-1}(C_2^i) = 1$ holds with probability $1/2 + 2\varepsilon^2$. Thus in the case that the guessed value of K_g is correct, we can estimate $|T_{K_g}|$ by

$$|T_{K_g}| \approx N \times |1 - 2 \times \#S_{\Gamma_P \to \Gamma_C}/n| \times 4\varepsilon^2,$$

where n is the block size of the cipher E', and $\#S_{\Gamma_P \to \Gamma_C}$ is not equal to $n/2$. On the other hand, if the guessed value of K_g is wrong, according to the Wrong-Key Randomization Hypothesis given in [28], it's assumed that the wrong guess of K_g results in a random-looking parity of $\Gamma_C \cdot E_2^{-1}(C_1^i) \oplus \Gamma_C \cdot E_2^{-1}(C_2^i)$. Consequently, the value of $|T_{K_g}|$ approximates to 0 in this case. So it is feasible to distinguish the correct value of K_g from all wrong guesses of K_g by applying the above key recovery attack if given sufficient ciphertext pairs (C_1^i, C_2^i), where C_2^i is gained by triggering single bit error at any position of the input of E_1. Following the technique introduced in [3], the number of ciphertext pairs required in our key recovery attack can be estimated as

$$c_N \times \frac{1}{|1 - 2 \times \#S_{\Gamma_P \to \Gamma_C}/n|} \times \frac{1}{4\varepsilon^4},$$

where the coefficient c_N, which is closely related to the number of guessed subkey bits and the desired success rate of our attack, can be measured by using the approach given in [29].

Furthermore, if an adversary has the capability to induce single bit error at the input of E_1 repeatedly as well as the ability to get several linear characteristics $\Gamma_P^i \to \Gamma_C^i$ $(1 \le i \le m)$ for E_1, where the calculations of $\Gamma_C^i \cdot E_2^{-1}$ and $\Gamma_C^j \cdot E_2^{-1}$ $(i \ne j)$ influence different bits of the last round subkey of E', then the adversary can mount the above key recovery attack m times so as to recover more bits of the last round subkey (note that ciphertext pairs could be multiplexed partially or entirely among these attacks). After the adversary derives enough bits of the last round subkey, he can guess the left unknown bits of the subkey by means of exhaustive search if needed, and then the last round of E' can be stripped. Repeat the above procedure until the adversary can recover the secret key of the cipher E'.

Regarding the linear fault analysis under the condition of single byte error model, similar result can be derived by the same means as above.

4 A Key Recovery Attack on SERPENT by Using LFA

In order to illustrate the effectiveness of LFA, we mount a key recovery attack on the block cipher SERPENT by using LFA in this section. Since there isn't any known DFA attack on SERPENT which can be done by inducing faults at the round earlier than the penultimate round of the cipher so far, the general countermeasure against DFA on SERPENT could be implemented by protecting the last two rounds of the cipher if taking into account the cost and efficiency of the implementation. However, our effective attack shows that LFA could be a threat to the protected implementation of SERPENT.

4.1 A Brief Description of SERPENT

The SERPENT block cipher was proposed by Anderson *et al* in 1998 [30]. As a candidate of Advanced Encryption Standard, it was rated just behind the AES Rijndael. SERPENT has a classical SPN structure with 32 rounds and 128-bit block size. It accepts keys of 128, 192 and 256 bits and consists of the following operations:

- an initial permutation IP;
- 32 rounds, each consisting of a key mixing operation, a passage through 32 S-boxes and a linear transformation (except the last round, where the linear transformation is replaced by an additional key mixing operation);
- a final permutation FP.

In our description we adopt the notations of [30] in the bitsliced version. The intermediate value just before the round i (i.e., the $(i+1)$-th round) is denoted by B_i (a 128-bit value), where $0 \leq i \leq 31$. Each B_i is composed of four 32-bit words X_0, X_1, X_2 and X_3, where bit j of X_k is the bit $4j+k$ of the 128-bit value B_i $(0 \leq j \leq 31,\ 0 \leq k \leq 3)$. The four bits, bit j of X_3, X_2, X_1 and X_0, consist of the nibble j (i.e., the $(j+1)$-th nibble), with the bit from X_3 as the most significant bit.

SERPENT uses 8 distinct 4-bit to 4-bit S-boxes S_i $(0 \leq i \leq 7)$ successively along the rounds and consequently, each S-box is used in exactly four different rounds (i.e., S_0 is used in round 0, S_1 is used in round 1, ..., after S_7 is used in round 7, S_0 will be adopted again in round 8, then S_1 in round 9, and so on).

As for each round function R_i $(0 \leq i \leq 31)$, a single S-box will be used 32 times in parallel. For instance, R_0 uses 32 copies of S_0, and the $(j+1)$-th copy of S_0 takes the nibble j as the input and then outputs the value according to the S-box, where $0 \leq j \leq 31$.

The cipher can be formally described as follows:

- $B_0 \leftarrow P$,
- $B_{i+1} \leftarrow R_i(B_i) \quad 0 \leq i \leq 31$,
- $C \leftarrow B_{32}$,

where P, C denote plaintext and ciphertext respectively, and round function R_i can be expressed as below:

$$R_i(X) = LT(\hat{S}_i(X \oplus K_i)) \quad i = 0, \ldots, 30,$$
$$R_i(X) = \hat{S}_i(X \oplus K_i) \oplus K_{32} \quad i = 31,$$

where $\hat{S}_i$ denotes the application of the S-box $S_{(i \bmod 8)}$ 32 times in parallel, LT denotes the linear transformation, and K_i denotes the subkey of round i (note that both K_{31} and K_{32} are the subkeys of round 31). Please refer to [30] for detailed information about the 8 S-boxes, the linear transformation and the key schedule algorithm.

4.2 Attacking SERPENT

We now present a key recovery attack on SERPENT under the condition of single bit error model, and a similar attack can be mounted on SERPENT for the case of single byte error model. Firstly, we construct twelve 2-round linear characteristics $\Gamma_P^i \rightarrow \Gamma_C^i$ $(1 \leq i \leq 12)$ for the rounds from round 29 to round 30 of SERPENT, and assume that single bit error can be injected at the input of the round 29 repeatedly and randomly, then by applying the method given in Section 3 twelve times, 128 bits of K_{32} can be retrieved from the attack.

In order to describe the linear characteristics adopted in our attack, 32 hexadecimal digits will be used to denote the masks corresponding to 32 nibbles respectively, where the $(j+1)$-th hexadecimal digit (numbered from right to left) corresponds to the nibble j, $0 \leq j \leq 31$. Please refer to Appendix for the depiction of the twelve 2-round linear characteristics used in our attack.

According to the approach given in Section 3, we could derive twelve distinguishers for the 31 rounds from round 0 to round 30 of SERPENT as below:

$$\Gamma_C^i \cdot SERPENT_{lr}^{-1}(C_1) \oplus \Gamma_C^i \cdot SERPENT_{lr}^{-1}(C_2) = 0, \quad 1 \leq i \leq 12, \qquad (1)$$

where $SERPENT_{lr}^{-1}$ means the inverse of the last round of SERPENT, C_1 is a right ciphertext under SERPENT and C_2 is the corresponding faulty ciphertext obtained by inducing single bit error at the input of the round 29 of SERPENT. Moreover, for the case $1 \leq i \leq 9$, the i-th equation in (1) holds with probability $1/2+2^{-9}$, and for the case $10 \leq i \leq 12$, the i-th equation in (1) holds with probability $1/2 + 2^{-7}$. Thus a key recovery attack can be mounted on SERPENT based on the above twelve distinguishers. Following gives the detailed description of the attack in three phases.

Phase 1. For the i-th ($1 \leq i \leq 3$) distinguisher given in equations (1), do the following:

Step 1. Collect N_i pairs of ciphertexts, each pair consisting of the right ciphertext C_1 under SERPENT and the corresponding faulty ciphertext C_2^j ($1 \leq j \leq N_i$) derived by randomly injecting single bit error at the input of the round 29 of SERPENT.

Step 2. Let K_g denote the 8 bits of K_{32} which are relevant to the two active nibbles (i.e., S-boxes) influenced by the distinguisher. Initialize 2^8 counters $\{T_l\}_{0\leq l\leq 2^8-1}$ (the size of each counter could be set to $\lceil \log_2^{N_i} \rceil$ bits), where T_l corresponds to l which represents the possible value of the 8 bits of C_2^j entering the above two active nibbles. For each faulty ciphertext C_2^j, increase the counter T_l by 1 if the corresponding 8-bit value of the C_2^j is equal to l. Then for each possible value of K_g, do the following:

(a). Initialize a counter T_{K_g} with the counter size being $\lceil \log_2^{N_i} \rceil$ bits and implement the partial decryption of C_1.

(b). For each value of l, decrypt the above two active nibbles and then calculate the parity in terms of the distinguisher. If the parity is 0, increase the counter T_{K_g} by the value of T_l, and decrease by the value of T_l otherwise.

(c). Store the value of K_g as well as the value of the corresponding $|T_{K_g}|$.

Step 3. For all possible values of K_g, compare the stored values and take the value of K_g as the correct key information if the value of the corresponding $|T_{K_g}|$ is maximal.

Thus the 24 bits of K_{32} related to the nibbles 20, 21, 22, 25, 26 and 27 could be recovered in this phase.

Phase 2. For the i-th ($4 \leq i \leq 6$) distinguisher given in equations (1), the attack steps are the same as those in Phase 1 except the step 2 which is described as below.

Step 2. Let K_g denote the 12 bits of K_{32} which are relevant to the three active nibbles (i.e., S-boxes) impacted by the distinguisher. Initialize 2^{12} counters $\{T_l\}_{0 \le l \le 2^{12}-1}$ (the size of each counter could be set to $\lceil \log_2^{N_i} \rceil$ bits), where T_l corresponds to l which represents the possible value of the 12 bits of C_2^j entering the above three active nibbles. For each faulty ciphertext C_2^j, increase the counter T_l by 1 if the corresponding 12-bit value of the C_2^j is equal to l. Then for each possible value of K_g, do the following:

(a). Initialize a counter T_{K_g} with the counter size being $\lceil \log_2^{N_i} \rceil$ bits and implement the partial decryption of C_1.

(b). For each value of l, decrypt the above three active nibbles and then calculate the parity in terms of the distinguisher. If the parity is 0, increase the counter T_{K_g} by the value of T_l, and decrease by the value of T_l otherwise.

(c). Store the value of K_g as well as the value of the corresponding $|T_{K_g}|$.

Accordingly, the 36 bits of K_{32} corresponding to the nibbles 0, 1, 2, 3, 4, 5, 10, 11 and 12 could be obtained in this phase.

Phase 3. After the above 60 bits of K_{32} have been retrieved, we can mount attacks on SERPENT sequentially in terms of the 7th, 8th, 9th, 10th, 11th and 12th distinguishers given in equations (1), and these attacks are similar to that in Phase 1. Finally we can get all the 128 bits of K_{32}.

After that, we rewrite the encryption algorithm of SERPENT in an equivalent way by swapping the order of the linear transformation in round 30 and the key mixing operation (with K_{31}) in round 31, then for the modified cipher, the part after the XOR operation with $LT^{-1}(K_{31})$ can be stripped. Furthermore, we mount an attack on the reduced-round cipher similarly to the above and recover the 128 bits of K_{31}. Thus following the key schedule algorithm of SERPENT, we can derive the secret key from K_{31} and K_{32}.

For the attack in terms of the i-th ($1 \le i \le 12$) distinguisher given in equations (1), the necessary number of ciphertext pairs, each pair consisting of a right ciphertext under SERPENT and the corresponding faulty ciphertext derived by triggering single bit error at the input of the round 29 of SERPENT randomly, could be estimated as

$$c_{N_i} \times \frac{1}{1 - 2 \times 6/128} \times \frac{1}{(2^{-9})^2}, \quad 1 \le i \le 9,$$

or

$$c_{N_i} \times \frac{1}{1 - 2 \times 3/128} \times \frac{1}{(2^{-7})^2}, \quad 10 \le i \le 12,$$

where the coefficient c_{N_i} is closely related to the number of guessed subkey bits and the desired success rate of the attack. Thus for the attack corresponding to the i-th distinguisher, according to the Theorem 2 proposed in [29], $2^4 \times \frac{32}{29} \times 2^{18} \approx 2^{22.14}$ and $2^{3.8} \times \frac{64}{61} \times 2^{14} \approx 2^{17.87}$ ciphertext pairs are needed in the cases $1 \le i \le 9$ and $10 \le i \le 12$ respectively so as to achieve a high success probability of 1 approximately. Note that the ciphertext pairs could be multiplexed in our key recovery attack on SERPENT, consequently, the data complexity of our attack with success probability of about 1 can be estimated as $2 \times 2^{22.14} = 2^{23.14}$ ciphertext pairs or $2^{23.14}$ faulty ciphertexts (taking the attack for recovering K_{31} into account as well).

The time complexity of our attack is dominated mainly by the decryptions of the active nibbles in the attack based on the 7th distinguisher of equations (1). As a result, the time complexity of our attack is around $2 \times 2^{22.14} \times 2^{16} \times \frac{6}{32 \times 32} \approx 2^{31.73}$ SERPENT encryptions (taking the attack for retrieving K_{31} into account as well).

The memory complexity of our attack is primarily owing to storing the value of $|T_{K_g}|$ in the attack based on the 7th distinguisher of equations (1) as well as keeping the required faulty ciphertexts. Considering the fact that the attacks for obtaining K_{32} and K_{31} are implemented sequentially, the memory complexity of our attack can be approximated as $(23 \times 2^{16} + 128 \times 2^{22.14})/8 \approx 2^{26.14}$ bytes.

4.3 Experiments and Results

We use a PC with i3 M380 processor(2.53 GHz) and 4G DDR memory to do the experiments of our key recovery attack on SERPENT. The software platform of the experiments is Visual C++, and fault inductions are simulated in this platform. Under this condition we implement 100 experiments of our attack on SERPENT with randomly generated secret keys.

The main procedure of each experiment is as follows. At first, a correct ciphertext is obtained by encrypting a given plaintext under SERPENT with a secret key. Secondly, we trigger single bit error at the input of round 29 of SERPENT randomly and repeatedly to derive $2^{22.14}$ faulty ciphertexts and then retrieve the 128 bits of K_{32} by the means presented in Section 4.2. After that, we inject single bit error at the input of round 28 of SERPENT randomly and repeatedly to generate $2^{22.14}$ faulty ciphertexts and then obtain the 128 bits of K_{31} by the means similar to the above. Finally, the secret key is recovered from K_{32} and K_{31} with the help of the key schedule algorithm of SERPENT.

Among all the 100 experiments, there are 92 experiments such that the recovered secret keys are equal to the corresponding correct ones. Consequently, our experimental results match the theoretical analysis given in Section 4.2 well.

5 Conclusion

In this paper, we have proposed a new fault attack on block ciphers called linear fault analysis (LFA), in which linear characteristics for some consecutive rounds of a block cipher will be utilized instead of exploiting differential distributions of S-Boxes within the block cipher in DFA. Generally, our new approach can deal with the case that faults are triggered several rounds earlier compared to DFA as long as suitable linear approximations exist, as a result, one may mount an effective attack on a block cipher by applying LFA even if the cipher has already been protected against DFA.

In order to demonstrate the validity of LFA, we have applied it to analyze the block cipher SERPENT. Basically, the countermeasure against DFA on SERPENT can be implemented by protecting the last two rounds of the cipher since there isn't any known DFA attack on SERPENT so far which can be done by inducing faults at the round earlier than the penultimate round of the cipher. However, with the help of LFA, we have presented an effective attack on the protected implementation of SERPENT. Although the attack has a data complexity which seems impractical for real cryptographic devices, it does show that LFA could be a potential threat to the protected implementations (against DFA) of block ciphers. Moreover, it is expected that further results could be derived by applying linear hulls and non-linear approximations in LFA.

Finally, the implementation of redundancy (a simple and widely used countermeasure against fault attack) is more costly and less efficient along with the number of protected rounds increasing, thus for a block cipher, the number of protected rounds must be chosen very carefully in order to prevent security flaws as well as keep the corresponding implementation economical and efficient. We hope that our work could be helpful in determining this number.

References

1. Kelsey, J., Schneier, B., Wagner, D., Hall, C.: Side Channel Cryptanalysis of Product Ciphers. In: Quisquater, J.-J., Deswarte, Y., Meadows, C., Gollmann, D. (eds.) ESORICS 1998. LNCS, vol. 1485, pp. 97–110. Springer, Heidelberg (1998)
2. Biham, E., Shamir, A.: Differential Cryptanalysis of DES-like Cryptosystems. In: Menezes, A., Vanstone, S.A. (eds.) CRYPTO 1990. LNCS, vol. 537, pp. 2–21. Springer, Heidelberg (1991)
3. Matsui, M.: Linear Cryptanalysis Method for DES Cipher. In: Helleseth, T. (ed.) EUROCRYPT 1993. LNCS, vol. 765, pp. 386–397. Springer, Heidelberg (1994)
4. Biham, E.: New Types of Cryptanalytic Attacks Using Related Keys. In: Helleseth, T. (ed.) EUROCRYPT 1993. LNCS, vol. 765, pp. 398–409. Springer, Heidelberg (1994)
5. Knudsen, L., Wagner, D.: Integral Cryptanalysis (Extended Abstract). In: Daemen, J., Rijmen, V. (eds.) FSE 2002. LNCS, vol. 2365, pp. 112–127. Springer, Heidelberg (2002)
6. Courtois, N.T., Pieprzyk, J.: Cryptanalysis of Block Ciphers with Overdefined Systems of Equations. In: Zheng, Y. (ed.) ASIACRYPT 2002. LNCS, vol. 2501, pp. 267–287. Springer, Heidelberg (2002)
7. Boneh, D., DeMillo, R.A., Lipton, R.J.: On the Importance of Checking Cryptographic Protocols for Faults. In: Fumy, W. (ed.) EUROCRYPT 1997. LNCS, vol. 1233, pp. 37–51. Springer, Heidelberg (1997)
8. Biham, E., Shamir, A.: Differential Fault Analysis of Secret Key Cryptosystems. In: Kaliski Jr., B.S. (ed.) CRYPTO 1997. LNCS, vol. 1294, pp. 513–525. Springer, Heidelberg (1997)
9. Bar-El, H., Choukri, H., Naccache, D., Tunstall, M., Whelan, C.: The Sorcerer's Apprentice Guide to Fault Attacks. In: FDTC 2004 in Association with DSN 2004, pp. 330–342 (2004)
10. Rivain, M.: Differential Fault Analysis on DES Middle Rounds. In: Clavier, C., Gaj, K. (eds.) CHES 2009. LNCS, vol. 5747, pp. 457–469. Springer, Heidelberg (2009)

11. Blömer, J., Seifert, J.-P.: Fault Based Cryptanalysis of the Advanced Encryption Standard (AES). In: Wright, R.N. (ed.) FC 2003. LNCS, vol. 2742, pp. 162–181. Springer, Heidelberg (2003)
12. Chen, C.-N., Yen, S.-M.: Differential Fault Analysis on AES Key Schedule and Some Countermeasures. In: Safavi-Naini, R., Seberry, J. (eds.) ACISP 2003. LNCS, vol. 2727, pp. 118–129. Springer, Heidelberg (2003)
13. Dusart, P., Letourneux, G., Vivolo, O.: Differential Fault Analysis on A.E.S. In: Zhou, J., Yung, M., Han, Y. (eds.) ACNS 2003. LNCS, vol. 2846, pp. 293–306. Springer, Heidelberg (2003)
14. Giraud, C.: DFA on AES. In: Dobbertin, H., Rijmen, V., Sowa, A. (eds.) AES 2005. LNCS, vol. 3373, pp. 27–41. Springer, Heidelberg (2005)
15. Kim, C.H., Quisquater, J.-J.: New Differential Fault Analysis on AES Key Schedule: Two Faults Are Enough. In: Grimaud, G., Standaert, F.-X. (eds.) CARDIS 2008. LNCS, vol. 5189, pp. 48–60. Springer, Heidelberg (2008)
16. Piret, G., Quisquater, J.-J.: A Differential Fault Attack Technique against SPN Structures, with Application to the AES and KHAZAD. In: Walter, C.D., Koç, Ç.K., Paar, C. (eds.) CHES 2003. LNCS, vol. 2779, pp. 77–88. Springer, Heidelberg (2003)
17. Takahashi, J., Fukunaga, T., Yamakoshi, K.: DFA Mechanism on the AES Key Schedule. In: FDTC 2007, pp. 62–74 (2007)
18. Kim, C.H.: Improved Differential Fault Analysis on AES Key Schedule. IEEE Transactions on Information Forensics and Security 7(1), 41–50 (2012)
19. Clavier, C., Gierlichs, B., Verbauwhede, I.: Fault Analysis Study of IDEA. In: Malkin, T. (ed.) CT-RSA 2008. LNCS, vol. 4964, pp. 274–287. Springer, Heidelberg (2008)
20. Chen, H., Wu, W., Feng, D.: Differential Fault Analysis on CLEFIA. In: Qing, S., Imai, H., Wang, G. (eds.) ICICS 2007. LNCS, vol. 4861, pp. 284–295. Springer, Heidelberg (2007)
21. Li, W., Gu, D., Wang, Y.: Differential Fault Analysis on the Contracting UFN Structure, with Application to SMS4 and MacGuffin. Journal of Systems and Software 82(2), 346–354 (2009)
22. Li, W., Gu, D., Li, J.: Differential Fault Analysis on the ARIA Algorithm. Information Sciences 10(178), 3727–3737 (2008)
23. Zhou, Y., Wu, W., Xu, N., Feng, D.: Differential Fault Attack on Camellia. Chinese Journal of Electronics 18(1), 13–19 (2009)
24. Phan, R.C.-W., Yen, S.-M.: Amplifying Side-Channel Attacks with Techniques from Block Cipher Cryptanalysis. In: Domingo-Ferrer, J., Posegga, J., Schreckling, D. (eds.) CARDIS 2006. LNCS, vol. 3928, pp. 135–150. Springer, Heidelberg (2006)
25. Kim, C.H.: Efficient Methods for Exploiting Faults Induced at AES Middle Rounds, `http://eprint.iacr.org/2011/349`
26. Derbez, P., Fouque, P.-A., Leresteux, D.: Meet-in-the-Middle and Impossible Differential Fault Analysis on AES. In: Preneel, B., Takagi, T. (eds.) CHES 2011. LNCS, vol. 6917, pp. 274–291. Springer, Heidelberg (2011)
27. Dutertre, J.M., Mirbaha, A.P., Naccache, D., Ribotta, A.L., Tria, A.: Reproducible Single-Byte Laser Fault Injection. In: PASTIS 2010 (2010)
28. Harpes, C., Kramer, G.G., Massey, J.L.: A Generalization of Linear Cryptanalysis and the Applicability of Matsui's Piling-Up Lemma. In: Guillou, L.C., Quisquater, J.-J. (eds.) EUROCRYPT 1995. LNCS, vol. 921, pp. 24–38. Springer, Heidelberg (1995)

29. Selçuk, A.A.: On Probability of Success in Linear and Differential Cryptanalysis. Journal of Cryptology 21(1), 131–147 (2008)
30. Anderson, R., Biham, E., Knudsen, L.R.: SERPENT: A Proposal for the Advanced Encryption Standard, NIST AES Proposal (1998)

Appendix: The Linear Characteristics Used in Section 4.2

The 2-round linear characteristic $(\Gamma_P^1 \rightarrow \Gamma_C^1)$ with $p = 1/2 + 2^{-5}$

$$\begin{array}{rcl}
\Gamma_P^1 = & \mathrm{0x00E0000000000000000000000000000E} & \\
& \downarrow S_5 & \mathrm{Pr} = 1/2 + 2^{-3} \\
& \mathrm{0x00400000000000000000000000000008} & \\
& \downarrow LT & \\
& \mathrm{0x00000000000000000000000080000000} & \\
& \downarrow S_6 & \mathrm{Pr} = 1/2 + 2^{-3} \\
& \mathrm{0x00000000000000000000000010000000} & \\
& \downarrow LT & \\
& \mathrm{0x00000010000A00000000000000000000} & = \Gamma_C^1,
\end{array}$$

The 2-round linear characteristic $(\Gamma_P^2 \rightarrow \Gamma_C^2)$ with $p = 1/2 + 2^{-5}$

$$\begin{array}{rcl}
\Gamma_P^2 = & \mathrm{0x0E0000000000000000000000000000E0} & \\
& \downarrow S_5 & \mathrm{Pr} = 1/2 + 2^{-3} \\
& \mathrm{0x04000000000000000000000000000080} & \\
& \downarrow LT & \\
& \mathrm{0x00000000000000000000000800000000} & \\
& \downarrow S_6 & \mathrm{Pr} = 1/2 + 2^{-3} \\
& \mathrm{0x00000000000000000000000100000000} & \\
& \downarrow LT & \\
& \mathrm{0x0000010000A000000000000000000000} & = \Gamma_C^2,
\end{array}$$

The 2-round linear characteristic $(\Gamma_P^3 \rightarrow \Gamma_C^3)$ with $p = 1/2 + 2^{-5}$

$$\begin{array}{rcl}
\Gamma_P^3 = & \mathrm{0xE0000000000000000000000000000E00} & \\
& \downarrow S_5 & \mathrm{Pr} = 1/2 + 2^{-3} \\
& \mathrm{0x40000000000000000000000000000800} & \\
& \downarrow LT & \\
& \mathrm{0x00000000000000000000008000000000} & \\
& \downarrow S_6 & \mathrm{Pr} = 1/2 + 2^{-3} \\
& \mathrm{0x00000000000000000000001000000000} & \\
& \downarrow LT & \\
& \mathrm{0x000010000A0000000000000000000000} & = \Gamma_C^3,
\end{array}$$

The 2-round linear characteristic ($\Gamma_P^4 \rightarrow \Gamma_C^4$) with $p = 1/2 - 2^{-5}$

$$
\begin{array}{rcl}
\Gamma_P^4 = & \texttt{0x00E0000000000000000000000000000E} & \\
 & \downarrow S_5 & \Pr = 1/2 + 2^{-3} \\
 & \texttt{0x00400000000000000000000000000008} & \\
 & \downarrow LT & \\
 & \texttt{0x00000000000000000000000080000000} & \\
 & \downarrow S_6 & \Pr = 1/2 - 2^{-3} \\
 & \texttt{0x00000000000000000000000040000000} & \\
 & \downarrow LT & \\
 & \texttt{0x00000000000000000000080000002004} & = \Gamma_C^4,
\end{array}
$$

The 2-round linear characteristic ($\Gamma_P^5 \rightarrow \Gamma_C^5$) with $p = 1/2 - 2^{-5}$

$$
\begin{array}{rcl}
\Gamma_P^5 = & \texttt{0x0E0000000000000000000000000000E0} & \\
 & \downarrow S_5 & \Pr = 1/2 + 2^{-3} \\
 & \texttt{0x04000000000000000000000000000080} & \\
 & \downarrow LT & \\
 & \texttt{0x00000000000000000000000800000000} & \\
 & \downarrow S_6 & \Pr = 1/2 - 2^{-3} \\
 & \texttt{0x00000000000000000000000400000000} & \\
 & \downarrow LT & \\
 & \texttt{0x00000000000000000000800000020040} & = \Gamma_C^5,
\end{array}
$$

The 2-round linear characteristic ($\Gamma_P^6 \rightarrow \Gamma_C^6$) with $p = 1/2 - 2^{-5}$

$$
\begin{array}{rcl}
\Gamma_P^6 = & \texttt{0xE0000000000000000000000000000E00} & \\
 & \downarrow S_5 & \Pr = 1/2 + 2^{-3} \\
 & \texttt{0x40000000000000000000000000000800} & \\
 & \downarrow LT & \\
 & \texttt{0x00000000000000000000008000000000} & \\
 & \downarrow S_6 & \Pr = 1/2 - 2^{-3} \\
 & \texttt{0x00000000000000000000004000000000} & \\
 & \downarrow LT & \\
 & \texttt{0x00000000000000000008000000200400} & = \Gamma_C^6,
\end{array}
$$

The 2-round linear characteristic $(\Gamma_P^7 \rightarrow \Gamma_C^7)$ with $p = 1/2 + 2^{-5}$

$$\begin{array}{rcl}
\Gamma_P^7 = & \texttt{0x00E0000000000000000000000000000E} & \\
& \downarrow S_5 & \Pr = 1/2 + 2^{-3} \\
& \texttt{0x00400000000000000000000000000008} & \\
& \downarrow LT & \\
& \texttt{0x00000000000000000000000080000000} & \\
& \downarrow S_6 & \Pr = 1/2 + 2^{-3} \\
& \texttt{0x00000000000000000000000080000000} & \\
& \downarrow LT & \\
& \texttt{0x004000000000000008000010800A0002} & = \Gamma_C^7,
\end{array}$$

The 2-round linear characteristic $(\Gamma_P^8 \rightarrow \Gamma_C^8)$ with $p = 1/2 + 2^{-5}$

$$\begin{array}{rcl}
\Gamma_P^8 = & \texttt{0x0E0000000000000000000000000000E0} & \\
& \downarrow S_5 & \Pr = 1/2 + 2^{-3} \\
& \texttt{0x04000000000000000000000000000080} & \\
& \downarrow LT & \\
& \texttt{0x00000000000000000000000800000000} & \\
& \downarrow S_6 & \Pr = 1/2 + 2^{-3} \\
& \texttt{0x00000000000000000000000800000000} & \\
& \downarrow LT & \\
& \texttt{0x04000000000000008000010800A00020} & = \Gamma_C^8,
\end{array}$$

The 2-round linear characteristic $(\Gamma_P^9 \rightarrow \Gamma_C^9)$ with $p = 1/2 + 2^{-5}$

$$\begin{array}{rcl}
\Gamma_P^9 = & \texttt{0xE0000000000000000000000000000E00} & \\
& \downarrow S_5 & \Pr = 1/2 + 2^{-3} \\
& \texttt{0x40000000000000000000000000000800} & \\
& \downarrow LT & \\
& \texttt{0x00000000000000000000008000000000} & \\
& \downarrow S_6 & \Pr = 1/2 + 2^{-3} \\
& \texttt{0x00000000000000000000008000000000} & \\
& \downarrow LT & \\
& \texttt{0x4000000000000008000010800A000200} & = \Gamma_C^9,
\end{array}$$

The 2-round linear characteristic ($\Gamma_P^{10} \to \Gamma_C^{10}$) with $p = 1/2 + 2^{-4}$

$$
\begin{array}{rcll}
\Gamma_P^{10} = & & \text{0x0000000000000000000000000000000E} & \\
& & \downarrow S_5 & \Pr = 1/2 - 2^{-2} \\
& & \text{0x00000000000000000000000000000004} & \\
& & \downarrow LT & \\
& & \text{0x00000400000000000000000000008000} & \\
& & \downarrow S_6 & \Pr = 1/2 - 2^{-3} \\
& & \text{0x00000D0000000000000000000000B000} & \\
& & \downarrow LT & \\
& & \text{0x000810800A30060A400018010A12A002} & = \Gamma_C^{10},
\end{array}
$$

The 2-round linear characteristic ($\Gamma_P^{11} \to \Gamma_C^{11}$) with $p = 1/2 + 2^{-4}$

$$
\begin{array}{rcll}
\Gamma_P^{11} = & & \text{0x000000000000000000000000000000E0} & \\
& & \downarrow S_5 & \Pr = 1/2 - 2^{-2} \\
& & \text{0x00000000000000000000000000000040} & \\
& & \downarrow LT & \\
& & \text{0x00004000000000000000000000080000} & \\
& & \downarrow S_6 & \Pr = 1/2 - 2^{-3} \\
& & \text{0x0000D0000000000000000000000B0000} & \\
& & \downarrow LT & \\
& & \text{0x00810800A30060A400018010A12A0020} & = \Gamma_C^{11},
\end{array}
$$

The 2-round linear characteristic ($\Gamma_P^{12} \to \Gamma_C^{12}$) with $p = 1/2 + 2^{-4}$

$$
\begin{array}{rcll}
\Gamma_P^{12} = & & \text{0x00000000000000000000000000000E00} & \\
& & \downarrow S_5 & \Pr = 1/2 - 2^{-2} \\
& & \text{0x00000000000000000000000000000400} & \\
& & \downarrow LT & \\
& & \text{0x00040000000000000000000000800000} & \\
& & \downarrow S_6 & \Pr = 1/2 - 2^{-3} \\
& & \text{0x000D0000000000000000000000B00000} & \\
& & \downarrow LT & \\
& & \text{0x0810800A30060A400018010A12A00200} & = \Gamma_C^{12}.
\end{array}
$$

Cryptanalysis of 256-Bit Key HyRAL via Equivalent Keys

Yuki Asano, Shingo Yanagihara, and Tetsu Iwata

Nagoya University, Japan
{y_asano,s_yanagi}@echo.nuee.nagoya-u.ac.jp,
iwata@cse.nagoya-u.ac.jp

Abstract. HyRAL is a blockcipher whose block size is 128 bits, and it supports the key lengths of $128, 129, \ldots, 256$ bits. The cipher was proposed for the CRYPTREC project, and previous analyses did not identify any security weaknesses. In this paper, we consider the longest key version, 256-bit key HyRAL, and present the analysis in terms of equivalent keys. First, we show that there are $2^{51.0}$ equivalent keys (or $2^{50.0}$ pairs of equivalent keys). Next, we propose an algorithm that derives an instance of equivalent keys with the expected time complexity of $2^{48.8}$ encryptions and a limited amount of memory. Finally, we implement the proposed algorithm and fully verify its correctness by showing several instances of equivalent keys.

Keywords: Cryptanalysis, blockcipher, HyRAL, equivalent key.

1 Introduction

HyRAL is a blockcipher whose block size is 128 bits, and it supports the key lengths of $128, 129, \ldots, 256$ bits [6,7,8]. The overall structure of HyRAL is the Generalized Feistel Structure with four branches, and 128-bit key HyRAL consists of 24 rounds, and 129-, 130-, ..., 256-bit key HyRAL consist of 32 rounds. The CRYPTREC project, running in Japan, is maintaining the e-Government recommended ciphers list, which was first established in 2003, and the list is planned to be revised in 2013 [3]. A call for algorithms was announced in 2009, and HyRAL is one of the proposed algorithms for the call [6]. The security of HyRAL against various attacks has been evaluated. The security against differential attacks [2] and linear attacks [15,16] is analyzed in [6,9,24], impossible differential attacks [1] is analyzed in [17,18], saturation attacks [5] and higher order differential attacks [13,14] is analyzed in [19,20,21,22,23,25,26], and boomerang attacks is analyzed in [10]. [27] also presents security analyses against various attacks, and so far, no security weaknesses have been identified. Therefore, identifying a security weakness of this cipher is of interest from a cryptanalytic view point.

For a blockcipher $E : \{0,1\}^k \times \{0,1\}^n \to \{0,1\}^n$ of k-bit keys and an n-bit block, two distinct keys $K, K' \in \{0,1\}^k$ that satisfy $E_K(M) = E_{K'}(M)$ for all $M \in \{0,1\}^n$ are called *equivalent keys* [12]. In this paper, out of the 129 key lengths supported in the specification of HyRAL, we consider the longest key

F. Bao, P. Samarati, and J. Zhou (Eds.): ACNS 2012, LNCS 7341, pp. 257–274, 2012.

version, 256-bit key HyRAL, and present the analysis in terms of equivalent keys. We show the following three results.

- First, we show that there are $2^{51.0}$ equivalent keys (or $2^{50.0}$ pairs of equivalent keys).
- Next, we propose an algorithm that derives an instance of equivalent keys with the expected time complexity of $2^{48.8}$ encryptions and a limited amount of memory.
- Finally, we implement the proposed algorithm and fully verify its correctness by showing several instances of equivalent keys.

The first result is obtained by an analysis of the differential characteristic of the particular component in the cipher called the Key Generation Algorithm, which we write KGA. KGA is used twice in the cipher, and their outputs are xor'ed to generate round keys. We show that, for KGA, there exist differential characteristics of probability higher than 2^{-128}, and hence the output differences collide with probability higher than 2^{-256}. Equivalent keys are obtained as the result of this internal collision. In general, the existence of equivalent keys directly implies the theoretical cryptanalysis of the cipher, as the time complexity of the brute-force attack becomes less than the time complexity implied by its key length.

The second result is the main technical contribution of this paper. We develop an algorithm to generate the input of KGA that follows the differential characteristic. The core of the algorithm is to derive the input of KGA that satisfies conditions on the 64-bit intermediate variables. More precisely, it inverts the 5 round Generalized Feistel Structure that has a feed forward at the input of the 5th round, where there are conditions that the xor of three 32-bit input variables of the 5th round is fixed to some constant, and that the xor of three 32-bit output variables of the 5th round is also fixed to some constant.

We obtain the third result by making use of a supercomputer system. It is well known that obtaining concrete instances of equivalent keys implies that we obtain collisions on the Davies-Meyer compression function based on the blockcipher. It is also easy to obtain collisions on the Merkle-Damgård hash function based on the compression function. Therefore, the existence of equivalent keys limits the use of the cipher in the widely deployed hash function mode.

With respect to the status of HyRAL in the CRYPTREC project, we note that the results of this paper were reported to the project [11], and it was announced in June 2011 that, based on the results, HyRAL did not proceed to the second round evaluation process [4].

2 Specification of HyRAL

We first define notation used throughout this paper. For an integer $n \geq 0$, $\{0,1\}^n$ is the set of n-bit strings. For two bit strings X and Y of the same length, $X \oplus Y$ is their xor. For integers $n, m \geq 0$ and a bit string $X \in \{0,1\}^{nm}$, $(X[1], \ldots, X[m]) \stackrel{n}{\leftarrow} X$ is the partitioning operation into n-bit strings, i.e., $X[1], \ldots, X[m]$ are unique n-bit strings such that $(X[1], \ldots, X[m]) = X$.

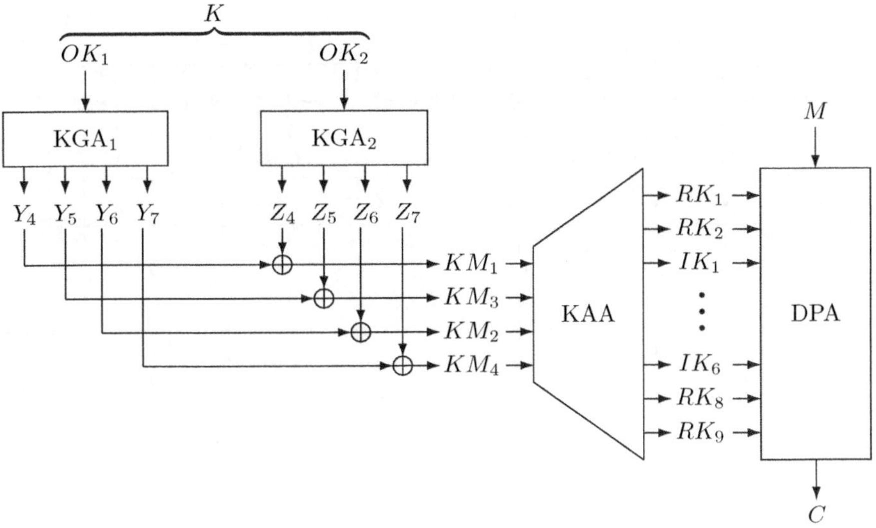

Fig. 1. The overall structure of 256-bit key HyRAL

Outline of 256-Bit Key HyRAL. HyRAL is a blockcipher whose block size is 128 bits, and it supports the key lengths of $128, 129, \ldots, 256$ bits. This paper deals with the 256-bit key version only, and the specifications of other key lengths are in [6,7,8].

The overall structure of 256-bit key HyRAL is shown in Fig. 1. The inputs are a key $K \in \{0,1\}^{256}$ and a plaintext $M \in \{0,1\}^{128}$, and the output is a ciphertext $C \in \{0,1\}^{128}$. 256-bit key HyRAL consists of the Key Generation Algorithm (KGA), the Key Assignment Algorithm (KAA), and the Data Processing Algorithm (DPA). KGA is used twice by changing the internal constants, and they are respectively denoted KGA_1 and KGA_2. For given $K \in \{0,1\}^{256}$ and $M \in \{0,1\}^{128}$, the encryption proceeds as follows.

1. Let $(OK_1, OK_2) \stackrel{128}{\leftarrow} K$. That is, let OK_1 be the most significant 128 bits of K, and OK_2 be the least significant 128 bits.
2. We then run KGA_1 and KGA_2 with OK_1 and OK_2, respectively, to generate $(Y_4, Y_5, Y_6, Y_7) \leftarrow \text{KGA}_1(OK_1)$ and $(Z_4, Z_5, Z_6, Z_7) \leftarrow \text{KGA}_2(OK_2)$, where $Y_i, Z_i \in \{0,1\}^{128}$.
3. Let $(KM_1, KM_3, KM_2, KM_4) \leftarrow (Y_4 \oplus Z_4, Y_5 \oplus Z_5, Y_6 \oplus Z_6, Y_7 \oplus Z_7)$, where $KM_i \in \{0,1\}^{128}$. We write $(KM_1, KM_3, KM_2, KM_4) = KM$.
4. Next, we run KAA with KM to generate $(RK_1, \ldots, RK_9, IK_1, \ldots, IK_6) \leftarrow \text{KAA}(KM)$, where $RK_i, IK_i \in \{0,1\}^{128}$.
5. Finally, we run DPA with $(RK_1, \ldots, RK_9, IK_1, \ldots, IK_6)$ and the plaintext M to generate the ciphertext $C \leftarrow \text{DPA}(RK_1, \ldots, RK_9, IK_1, \ldots, IK_6, M)$, and then C is returned.

In KAA, (KM_1, KM_3, KM_2, KM_4) are first parsed into 32-bit strings, and then $(RK_1, \ldots, RK_9, IK_1, \ldots, IK_6)$ are generated by taking their linear

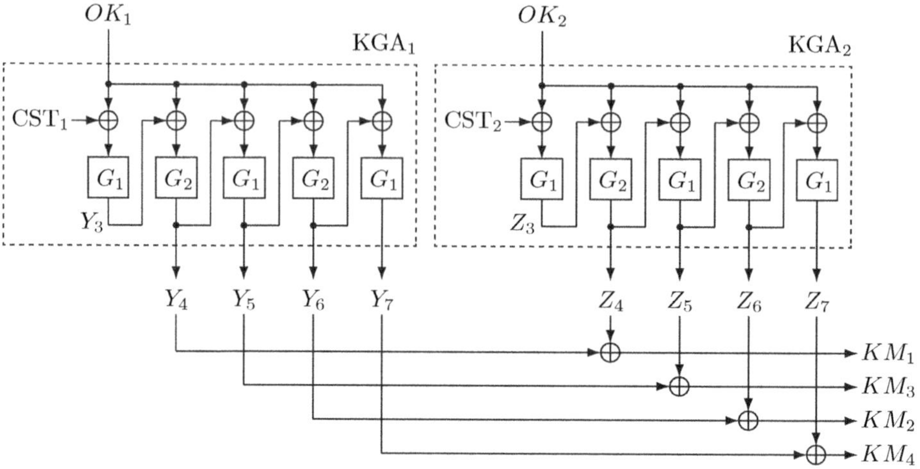

Fig. 2. KGA_1 and KGA_2

combinations. The overall structure of DPA is the 32 round Generalized Feistel Structure with four branches. Further details of KAA and DPA are not necessary in order to present the results of this paper, and their specifications can be found in [6,7,8].

The Key Generation Algorithms KGA_1 *and* KGA_2. KGA_1 and KGA_2 are shown in Fig. 2. For the input $OK_1 \in \{0,1\}^{128}$, KGA_1 outputs $(Y_4, Y_5, Y_6, Y_7) \in \{0,1\}^{512}$. Similarly, KGA_2 takes $OK_2 \in \{0,1\}^{128}$ and outputs $(Z_4, Z_5, Z_6, Z_7) \in \{0,1\}^{512}$. KGA_1 and KGA_2 internally use G_1 and G_2 functions, which are keyless permutations over $\{0,1\}^{128}$. KGA_1 and KGA_2 differ only in the internally used constants. The following 128-bit constants, CST_1 and CST_2, are used in KGA_1 and KGA_2, respectively, where the prefix `0x` indicates that the value is in the hexadecimal form.

$$\begin{cases} \mathrm{CST}_1 = \texttt{0xf9251a2365cd3c2e8066cbbbfe316b7b} \\ \mathrm{CST}_2 = \texttt{0x5de28625656b71ff9ffb1e12eef127f5} \end{cases}$$

G_1 *and* G_2 *Functions.* G_1 and G_2 functions are shown in Fig. 3.

G_1 and G_2 functions take $(X_1^{(1)}, X_2^{(1)}, X_3^{(1)}, X_4^{(1)}) \in \{0,1\}^{128}$ as the input, and output $(X_1^{(5)}, X_2^{(5)}, X_3^{(5)}, X_4^{(5)}) \in \{0,1\}^{128}$. They consist of four rounds of the Generalized Feistel Structure with four branches. G_1 function internally uses f_1, f_2, f_3, and f_4 functions, and G_2 function internally uses f_5, f_6, f_7, and f_8 functions.

$f_1, \ldots, f_8$ *Functions.* $f_1, \ldots, f_8$ functions are keyless permutations over $\{0,1\}^{32}$. They take a 32-bit string as the input and generate a 32-bit string as the output. The structure of f_i function is the SP-network, and a detailed specification is presented in Appendix A.

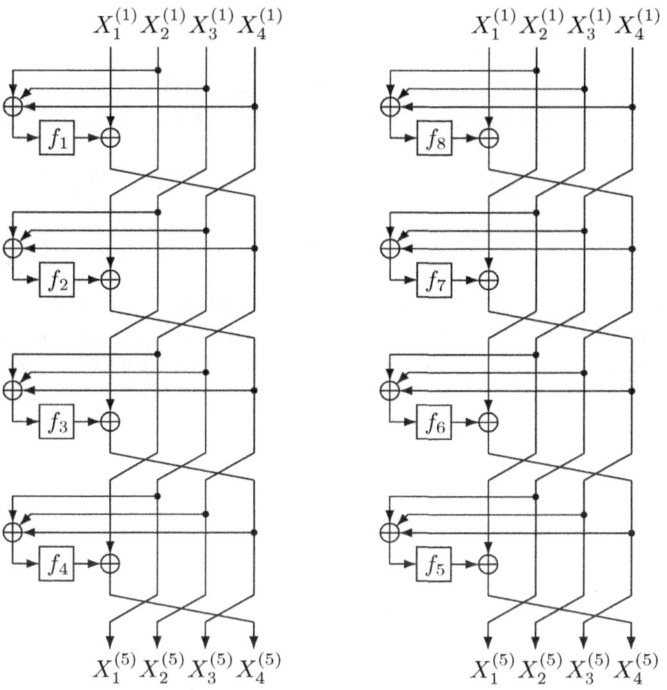

Fig. 3. G_1 function (left) and G_2 function (right)

3 Existence of Equivalent Keys

Overall Strategy. We make use of the differential cryptanalysis of Biham and Shamir [2] to show the existence of equivalent keys.

Let $(OK_1, OK_2) \in \{0,1\}^{256}$ be the key. Let ΔOK_1 be the input difference for KGA_1 and $(\Delta Y_4, \Delta Y_5, \Delta Y_6, \Delta Y_7)$ be the corresponding output difference. Similarly, let ΔOK_2 and $(\Delta Z_4, \Delta Z_5, \Delta Z_6, \Delta Z_7)$ be the input and output differences of KGA_2, respectively. We have

$$(\Delta Y_4, \Delta Y_5, \Delta Y_6, \Delta Y_7) = \text{KGA}_1(OK_1) \oplus \text{KGA}_1(OK_1 \oplus \Delta OK_1) \quad (1)$$

and

$$(\Delta Z_4, \Delta Z_5, \Delta Z_6, \Delta Z_7) = \text{KGA}_2(OK_2) \oplus \text{KGA}_2(OK_2 \oplus \Delta OK_2). \quad (2)$$

If the two output differences collide and

$$(\Delta Y_4, \Delta Y_5, \Delta Y_6, \Delta Y_7) = (\Delta Z_4, \Delta Z_5, \Delta Z_6, \Delta Z_7) \quad (3)$$

holds, we see that the differences are canceled by the xor operation and the input difference $(\Delta KM_1, \Delta KM_3, \Delta KM_2, \Delta KM_4)$ of KAA becomes null. Therefore, if

(3) holds, we have the following equivalent keys.

$$(K, K') = \begin{cases} ((OK_1, OK_2), (OK_1 \oplus \Delta OK_1, OK_2 \oplus \Delta OK_2)) \\ ((OK_1 \oplus \Delta OK_1, OK_2 \oplus \Delta OK_2), (OK_1, OK_2)) \\ ((OK_1 \oplus \Delta OK_1, OK_2), (OK_1, OK_2 \oplus \Delta OK_2)) \\ ((OK_1, OK_2 \oplus \Delta OK_2), (OK_1 \oplus \Delta OK_1, OK_2)) \end{cases} \quad (4)$$

In this paper, these are counted as four equivalent keys (or two pairs of equivalent keys).

Since KGA_1 and KGA_2 are the same algorithms except for the internally used constants, we may regard them identically as long as we consider their differential characteristics. In what follows, let $\mathrm{KGA} \in \{\mathrm{KGA}_1, \mathrm{KGA}_2\}$ be the Key Generation Algorithm. We next analyze the differential characteristic of KGA.

Differential Characteristic of KGA. We regard one round of G_1 and G_2 functions as one round of KGA. Then KGA is a function that consists of 20 rounds in total. For $r = 1, 2, \ldots, 20$, we write $f_i^{(r)}$ for f_i function used in the r-th round.

Let $\Delta OK \in \{0,1\}^{128}$ be the input difference of KGA and $(\Delta Y_4, \Delta Y_5, \Delta Y_6, \Delta Y_7) \in \{0,1\}^{512}$ be the corresponding output difference.

For $r = 1, 2, \ldots, 20$, let $\Delta X^{(r)} = (\Delta X_1^{(r)}, \Delta X_2^{(r)}, \Delta X_3^{(r)}, \Delta X_4^{(r)}) \in \{0,1\}^{128}$ be the input difference of the r-th round, and $\Delta Z^{(r)} = (\Delta Z_1^{(r)}, \Delta Z_2^{(r)}, \Delta Z_3^{(r)}, \Delta Z_4^{(r)}) \in \{0,1\}^{128}$ be its output difference. A *differential characteristic* is a tuple

$$((\Delta X^{(1)}, \Delta Z^{(1)}), \ldots, (\Delta X^{(20)}, \Delta Z^{(20)}))$$

of the input and output differences of each round, that satisfies the following conditions: First, it corresponds with the input and output differences of KGA, and hence we have $\Delta X^{(1)} = \Delta OK$, $\Delta Z^{(8)} = \Delta Y_4$, $\Delta Z^{(12)} = \Delta Y_5$, $\Delta Z^{(16)} = \Delta Y_6$, and $\Delta Z^{(20)} = \Delta Y_7$. Second, for $r = 1, 2, \ldots, 20$, we have $(\Delta X_2^{(r)}, \Delta X_3^{(r)}, \Delta X_4^{(r)}) = (\Delta Z_1^{(r)}, \Delta Z_2^{(r)}, \Delta Z_3^{(r)})$. Third, for $r \in \{4, 8, 12, 16\}$, we have $\Delta X^{(r+1)} = \Delta Z^{(r)} \oplus \Delta OK$, and for $r \in \{1, 2, \ldots, 19\} \setminus \{4, 8, 12, 16\}$, we have $\Delta X^{(r+1)} = \Delta Z^{(r)}$.

For a differential characteristic $((\Delta X^{(1)}, \Delta Z^{(1)}), \ldots, (\Delta X^{(20)}, \Delta Z^{(20)}))$, its probability is defined as

$$\mathrm{DCP}^{\mathrm{KGA}}((\Delta X^{(1)}, \Delta Z^{(1)}), \ldots, (\Delta X^{(20)} \Delta Z^{(20)})) = \prod_{1 \le r \le 20} \mathrm{DP}^{f_i^{(r)}}(\Delta I_i^{(r)}, \Delta O_i^{(r)}),$$

where $\Delta I_i^{(r)} = \Delta X_2^{(r)} \oplus \Delta X_3^{(r)} \oplus \Delta X_4^{(r)}$ is the input difference of $f_i^{(r)}$, $\Delta O_i^{(r)} = \Delta X_1^{(r)} \oplus \Delta Z_4^{(r)}$ is the corresponding output difference, and the differential probability $\mathrm{DP}^{f_i}(\Delta I_i, \Delta O_i)$ of f_i function for the input difference ΔI_i and the output difference ΔO_i is defined as

$$\mathrm{DP}^{f_i}(\Delta I_i, \Delta O_i) = \frac{\#\{I \mid f_i(I) \oplus f_i(I \oplus \Delta I_i) = \Delta O_i\}}{2^{32}}.$$

For a given differential characteristic, we say that f_i function is *active* if its input difference is non-zero. In KGA, there are 20 f_i functions and hence the maximum number of active f_i functions is 20. In the following lemma, we show that there exists a differential characteristic with only four active f_i functions.

Lemma 1. *For* KGA, *there exists a differential characteristic with four active f_i functions.*

Proof. Let $\delta \in \{0,1\}^{32}$ be any non-zero bit string. Let $\Delta OK = (\delta,\delta,\delta,\delta)$ be the input difference of KGA, and $(\Delta Y_4, \Delta Y_5, \Delta Y_6, \Delta Y_7)$ be the output difference, where $\Delta Y_4 = (\delta,\delta,0,0)$, $\Delta Y_5 = (0,0,0,\delta)$, $\Delta Y_6 = (\delta,\delta,\delta,\delta)$, and $\Delta Y_7 = (0,0,0,0)$. Consider the differential characteristic given in Table 1, which is also shown in Fig. 4. Then one can verify that there are four active f_i functions, which are $f_1^{(1)}$, $f_7^{(6)}$, $f_3^{(11)}$, and $f_5^{(16)}$. □

We see that the input and output differences of active f_i functions in the differential characteristic of Lemma 1 are both δ. Under the condition that both the input and output differences are the same, we have counted the number of active f_i functions for the 15 non-zero input differences, which are $(0,0,0,\delta)$, $(0,0,\delta,0)$, $(0,0,\delta,\delta), \ldots, (\delta,\delta,\delta,\delta)$. The results are summarized in Table 2.

From the table, we see that the number of active f_i functions of Lemma 1 is the minimum among the 15 differential characteristics.

Differential Probability of f_i Function. For f_i function, let $\mathrm{DP}^{f_i}(\delta)$ be the probability that both the input and output differences of f_i function are δ, i.e.,

$$\mathrm{DP}^{f_i}(\delta) = \mathrm{DP}^{f_i}(\delta,\delta) = \frac{\#\{I \mid f_i(I) \oplus f_i(I \oplus \delta) = \delta\}}{2^{32}}.$$

The probability of the differential characteristic in Lemma 1 depends only on δ, and we write the probability as $\mathrm{DCP}^{\mathrm{KGA}}(\delta)$, which is given as

$$\mathrm{DCP}^{\mathrm{KGA}}(\delta) = \mathrm{DP}^{f_1}(\delta) \times \mathrm{DP}^{f_3}(\delta) \times \mathrm{DP}^{f_5}(\delta) \times \mathrm{DP}^{f_7}(\delta).$$

We present the following lemma with respect to $\mathrm{DCP}^{\mathrm{KGA}}(\delta)$.

Lemma 2. *There exists non-zero $\delta \in \{0,1\}^{32}$ such that* $\mathrm{DCP}^{\mathrm{KGA}}(\delta) > 2^{-128}$.

Proof. For all the $(2^{32}-1)$ possible values `0x00000001`, ..., `0xffffffff` of non-zero $\delta \in \{0,1\}^{32}$, we computed the value of $\mathrm{DCP}^{\mathrm{KGA}}(\delta)$. The results are summarized in Table 3. From the table, we see that there exist 89938 values of δ such that $\mathrm{DCP}^{\mathrm{KGA}}(\delta) > 2^{-128}$. □

We note that, for $\delta =$ `0xd7d7d0d7`, we have

$$\mathrm{DP}^{f_1}(\delta) = 2^{-25}, \mathrm{DP}^{f_3}(\delta) = 2^{-26}, \mathrm{DP}^{f_5}(\delta) = 2^{-26}, \text{ and } \mathrm{DP}^{f_7}(\delta) = 2^{-26}. \quad (5)$$

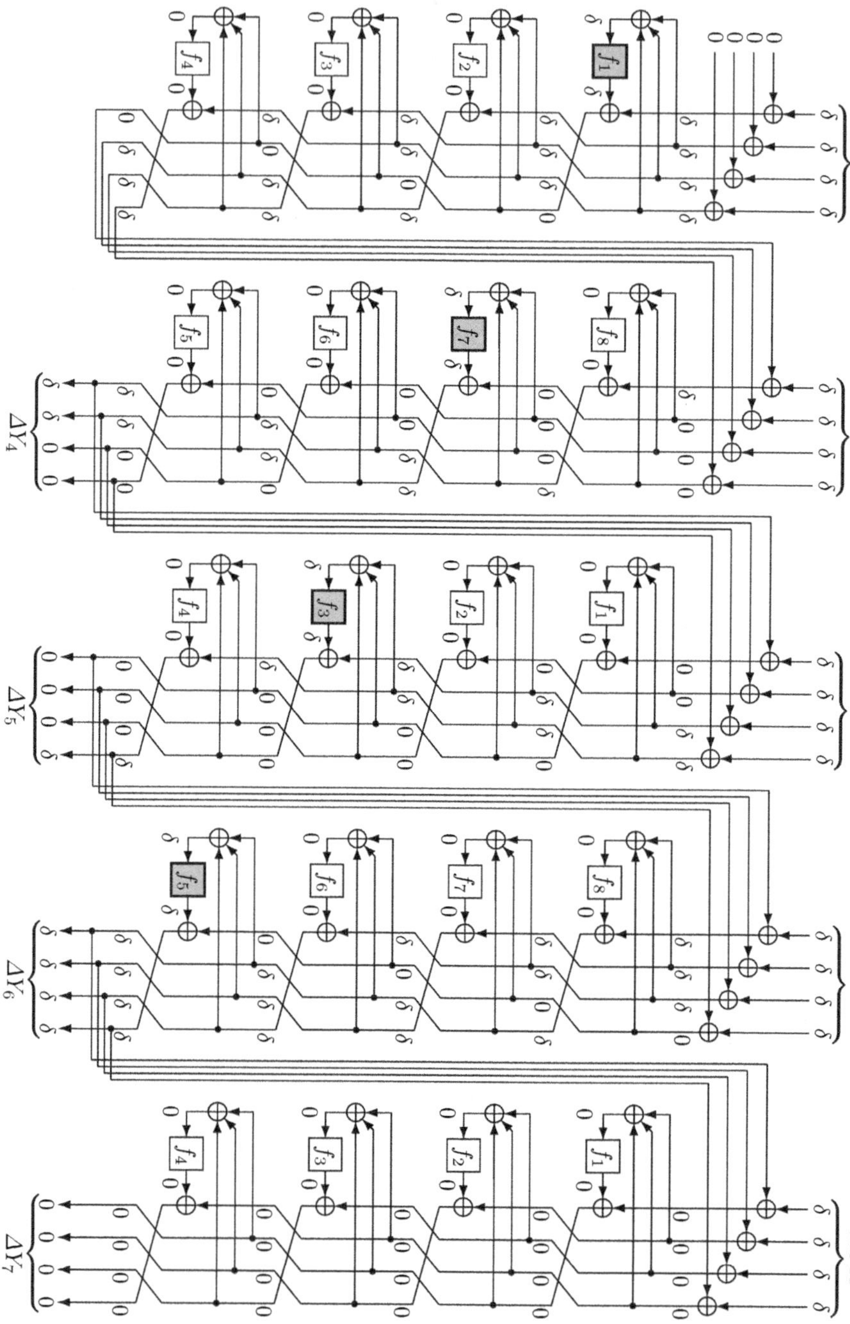

Fig. 4. The differential characteristic and active f_i functions (shown in gray) of Lemma 1

Table 1. The differential characteristic and active f_i functions of Lemma 1

r	Input diff. $\Delta X^{(r)}$	Output diff. $\Delta Z^{(r)}$	Active f_i function
1	$(\delta,\delta,\delta,\delta)$	$(\delta,\delta,\delta,0)$	f_1
2	$(\delta,\delta,\delta,0)$	$(\delta,\delta,0,\delta)$	
3	$(\delta,\delta,0,\delta)$	$(\delta,0,\delta,\delta)$	
4	$(\delta,0,\delta,\delta)$	$(0,\delta,\delta,\delta)$	
5	$(\delta,0,0,0)$	$(0,0,0,\delta)$	
6	$(0,0,0,\delta)$	$(0,0,\delta,\delta)$	f_7
7	$(0,0,\delta,\delta)$	$(0,\delta,\delta,0)$	
8	$(0,\delta,\delta,0)$	$(\delta,\delta,0,0)$	
9	$(0,0,\delta,\delta)$	$(0,\delta,\delta,0)$	
10	$(0,\delta,\delta,0)$	$(\delta,\delta,0,0)$	
11	$(\delta,\delta,0,0)$	$(\delta,0,0,0)$	f_3
12	$(\delta,0,0,0)$	$(0,0,0,\delta)$	
13	$(\delta,\delta,\delta,0)$	$(\delta,\delta,0,\delta)$	
14	$(\delta,\delta,0,\delta)$	$(\delta,0,\delta,\delta)$	
15	$(\delta,0,\delta,\delta)$	$(0,\delta,\delta,\delta)$	
16	$(0,\delta,\delta,\delta)$	$(\delta,\delta,\delta,\delta)$	f_5
17	$(0,0,0,0)$	$(0,0,0,0)$	
18	$(0,0,0,0)$	$(0,0,0,0)$	
19	$(0,0,0,0)$	$(0,0,0,0)$	
20	$(0,0,0,0)$	$(0,0,0,0)$	

Table 2. The number of active f_i functions for a given input difference

Input diff. ΔOK	Number	Input diff. ΔOK	Number
$(0,0,0,\delta)$	9	$(\delta,0,0,\delta)$	10
$(0,0,\delta,0)$	9	$(\delta,0,\delta,0)$	10
$(0,0,\delta,\delta)$	10	$(\delta,0,\delta,\delta)$	7
$(0,\delta,0,0)$	9	$(\delta,\delta,0,0)$	10
$(0,\delta,0,\delta)$	10	$(\delta,\delta,0,\delta)$	7
$(0,\delta,\delta,0)$	10	$(\delta,\delta,\delta,0)$	7
$(0,\delta,\delta,\delta)$	7	$(\delta,\delta,\delta,\delta)$	4
$(\delta,0,0,0)$	9		

Table 3. Examples of δ that satisfies $\mathrm{DCP}^{\mathrm{KGA}}(\delta) > 2^{-128}$ and the number of such δ

$\mathrm{DCP}^{\mathrm{KGA}}(\delta)$	Example of δ	Number
2^{-103}	`0xd7d7d0d7`	1
2^{-104}	`0xc5c5d254`	1
2^{-105}	`0x4e4ec554`	1
2^{-106}	`0x3c3cf4ff`	8
2^{-107}	`0x6161f9d9`	1
2^{-108}	`0x054d9797`	34
2^{-109}	`0x0101019a`	157
2^{-110}	`0x0159591a`	1579
2^{-111}	`0x0101e818`	7685
2^{-112}	`0x01010520`	80471

Existence of Equivalent Keys. We are now ready to present our main result of this section. Fix any δ such that $\mathrm{DCP}^{\mathrm{KGA}}(\delta) > 2^{-128}$. For randomly chosen $OK_1 \in \{0,1\}^{128}$, (1) is satisfied for

$$\begin{cases} \Delta OK_1 = (\delta,\delta,\delta,\delta), \\ \Delta Y_4 = (\delta,\delta,0,0), \Delta Y_5 = (0,0,0,\delta), \Delta Y_6 = (\delta,\delta,\delta,\delta), \Delta Y_7 = (0,0,0,0) \end{cases} \quad (6)$$

with at least a probability of $\mathrm{DCP}^{\mathrm{KGA}}(\delta)$. This implies that at least $2^{128} \times \mathrm{DCP}^{\mathrm{KGA}}(\delta)$ values of $OK_1 \in \{0,1\}^{128}$ satisfy (1). Similarly, at least $2^{128} \times \mathrm{DCP}^{\mathrm{KGA}}(\delta)$ values of $OK_2 \in \{0,1\}^{128}$ satisfy (2) for

$$\begin{cases} \Delta OK_2 = (\delta,\delta,\delta,\delta), \\ \Delta Z_4 = (\delta,\delta,0,0), \Delta Z_5 = (0,0,0,\delta), \Delta Z_6 = (\delta,\delta,\delta,\delta), \Delta Z_7 = (0,0,0,0). \end{cases} \quad (7)$$

If we fix a value of $(OK_1, OK_2) \in \{0,1\}^{256}$ that satisfies (1), (6), (2), and (7), we see that (3) is also satisfied, and hence we obtain four equivalent keys (or two pairs of equivalent keys) of (4). From Table 3 and by eliminating the duplications, the number of equivalent keys can be derived as

$$\frac{4 \times (2^{50} \times 1 + 2^{48} \times 1 + 2^{46} \times 1 + 2^{44} \times 8 + \cdots + 2^{32} \times 80471)}{4} \geq 2^{51.0},$$

and the number of pairs is the half of $2^{51.0}$, which is $2^{50.0}$.

From the discussions above, we obtain the following theorem.

Theorem 1. *In 256-bit key HyRAL, there exist* $2^{51.0}$ *equivalent keys (or* $2^{50.0}$ *pairs of equivalent keys).*

4 Derivation of Equivalent Keys

From the result of the previous section, we know that there are $2^{51.0}$ equivalent keys in 256-bit key HyRAL. In this section, we consider the problem of deriving a concrete instance of equivalent keys.

4.1 Equivalent Key Derivation Algorithm

As in the previous section, let $\mathrm{KGA} \in \{\mathrm{KGA}_1, \mathrm{KGA}_2\}$. Recall that one round of G_1 and G_2 functions are regarded as one round of KGA, and hence KGA is a function that consists of 20 rounds in total. Let $OK \in \{OK_1, OK_2\}$ be the input of KGA, and let $(K_1, K_2, K_3, K_4) \stackrel{32}{\leftarrow} OK \in \{0,1\}^{128}$ be its partition into 32-bit strings. Similarly, let $\mathrm{CST} \in \{\mathrm{CST}_1, \mathrm{CST}_2\}$ be the constant used in KGA, and let $(C_1, C_2, C_3, C_4) \stackrel{32}{\leftarrow} \mathrm{CST} \in \{0,1\}^{128}$ be its partition into 32-bit strings. KGA is the function that consists of 20 rounds in total, and we write the input and output strings of $f_i^{(r)}$ as $I_i^{(r)} \in \{0,1\}^{32}$ and $O_i^{(r)} \in \{0,1\}^{32}$, respectively, where $r = 1, 2, \ldots, 20$ and $f_i^{(r)}$ is f_i function used in the r-th round. Figure 5 shows the first 8 rounds of KGA.

We consider the case of $\delta = \texttt{0xd7d7d0d7}$. For $i \in \{1,3,5,7\}$, let $\mathcal{I}_i$ be a list of $I_i \in \{0,1\}^{32}$ that satisfies $f_i(I_i) \oplus f_i(I_i \oplus \delta) = \delta$. From (5), $\mathcal{I}_1$ consists of 128 elements, and each of $\mathcal{I}_3$, $\mathcal{I}_5$, and $\mathcal{I}_7$ consists of 64 elements, and we may thus write down the lists as $\mathcal{I}_1 = \{I_1[0], \ldots, I_1[127]\}$, $\mathcal{I}_3 = \{I_3[0], \ldots, I_3[63]\}$, $\mathcal{I}_5 = \{I_5[0], \ldots, I_5[63]\}$, and $\mathcal{I}_7 = \{I_7[0], \ldots, I_7[63]\}$.

Now if we can derive (K_1, K_2, K_3, K_4) that satisfies

$$I_1^{(1)} \in \mathcal{I}_1, I_7^{(6)} \in \mathcal{I}_7, I_3^{(11)} \in \mathcal{I}_3, \text{ and } I_5^{(16)} \in \mathcal{I}_5,$$

then this implies that we have derived OK that satisfies (1) and (6), or (2) and (7).

It is easy to obtain (K_1, K_2, K_3, K_4) that satisfies one of the four conditions, $I_1^{(1)} \in \mathcal{I}_1$, since this is simply (K_1, K_2, K_3, K_4) such that $K_2 \oplus C_2 \oplus K_3 \oplus C_3 \oplus K_4 \oplus C_4 \in \mathcal{I}_1$. In the following lemma, we show that one can derive (K_1, K_2, K_3, K_4) that satisfies two of the four conditions, namely, one can derive (K_1, K_2, K_3, K_4) such that both $I_1^{(1)} \in \mathcal{I}_1$ and $I_7^{(6)} \in \mathcal{I}_7$ are satisfied.

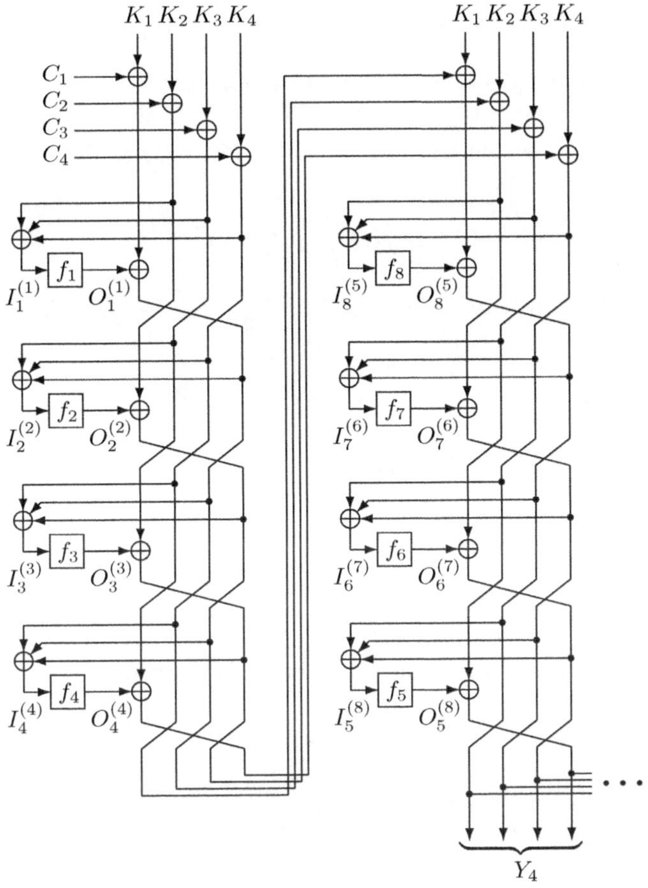

Fig. 5. The first 8 rounds of KGA

Lemma 3. *For arbitrarily fixed* $\tilde{K}_1$, $I_1^{(1)}$, $I_8^{(5)}$, *and* $I_7^{(6)}$, *where* $\tilde{K}_1 = K_1 \oplus K_3$, *the corresponding value of* (K_1, K_2, K_3, K_4) *can be derived.*

Proof. Since $I_1^{(1)}$ and $I_8^{(5)}$ are fixed, $O_1^{(1)} = f_1(I_1^{(1)})$ and $O_8^{(5)} = f_8(I_8^{(5)})$ are also fixed. To simplify the notation, let $\tilde{C}_1, \ldots, \tilde{C}_5$ be the fixed constants defined as $\tilde{C}_1 = C_1 \oplus C_3 \oplus C_4 \oplus O_1^{(1)}$, $\tilde{C}_2 = C_1 \oplus C_3 \oplus I_1^{(1)} \oplus O_1^{(1)}$, $\tilde{C}_3 = C_1 \oplus C_4 \oplus I_1^{(1)} \oplus O_1^{(1)}$, $\tilde{C}_4 = C_2 \oplus C_3 \oplus C_4$, and $\tilde{C}_5 = C_1 \oplus C_2 \oplus O_1^{(1)} \oplus I_7^{(6)}$. We also let $\tilde{K}_2 = K_1 \oplus K_3 \oplus K_4$ and $\tilde{K}_3 = K_1 \oplus K_4$.

First, $I_1^{(1)}$ has to satisfy $I_1^{(1)} = K_2 \oplus C_2 \oplus K_3 \oplus C_3 \oplus K_4 \oplus C_4$, which is equivalent to

$$K_2 = I_1^{(1)} \oplus C_2 \oplus K_3 \oplus C_3 \oplus K_4 \oplus C_4. \tag{8}$$

Next, since $I_2^{(2)} = \tilde{K}_2 \oplus \tilde{C}_1$, we have

$$O_2^{(2)} = f_2(\tilde{K}_2 \oplus \tilde{C}_1). \tag{9}$$

Similarly, since $I_3^{(3)}$ can be written as $I_3^{(3)} = \tilde{K}_1 \oplus \tilde{C}_2 \oplus O_2^{(2)}$ by using (8), we obtain

$$O_3^{(3)} = f_3(\tilde{K}_1 \oplus \tilde{C}_2 \oplus O_2^{(2)}). \tag{10}$$

Besides, since $I_4^{(4)}$ can be written as $I_4^{(4)} = \tilde{K}_3 \oplus \tilde{C}_3 \oplus O_2^{(2)} \oplus O_3^{(3)}$ by using (8), we obtain

$$O_4^{(4)} = f_4(\tilde{K}_3 \oplus \tilde{C}_3 \oplus O_2^{(2)} \oplus O_3^{(3)}). \tag{11}$$

Now since the input of the 5th round is $(C_1 \oplus O_1^{(1)}, C_2 \oplus O_2^{(2)}, C_3 \oplus O_3^{(3)}, C_4 \oplus O_4^{(4)})$ and $I_8^{(5)}$ is fixed,

$$I_8^{(5)} = \tilde{C}_4 \oplus O_2^{(2)} \oplus O_3^{(3)} \oplus O_4^{(4)} \tag{12}$$

has to be satisfied. Furthermore, since $I_7^{(6)}$ is fixed, $I_7^{(6)} = C_1 \oplus C_3 \oplus C_4 \oplus O_1^{(1)} \oplus O_3^{(3)} \oplus O_4^{(4)} \oplus O_8^{(5)}$ needs to be satisfied, which is equivalent to

$$\tilde{C}_5 \oplus O_2^{(2)} \oplus I_8^{(5)} = O_8^{(5)} \tag{13}$$

by using (12).

At this point, since $\tilde{C}_5$, $I_8^{(5)}$, and $O_8^{(5)}$ are all fixed, $O_2^{(2)}$ that satisfies (13) is uniquely determined. As we have now fixed $O_2^{(2)}$, $\tilde{K}_2$ that satisfies (9) is also uniquely determined, which is $\tilde{K}_2 = f_2^{-1}(O_2^{(2)}) \oplus \tilde{C}_1$. We also see that since $O_2^{(2)}$ is now fixed and $\tilde{K}_1$ is a fixed constant, $O_3^{(3)}$ that satisfies (10) is now uniquely fixed. Upon fixing both $O_2^{(2)}$ and $O_3^{(3)}$, we obtain unique $O_4^{(4)}$ that satisfies (12), and for these fixed $O_2^{(2)}$, $O_3^{(3)}$, and $O_4^{(4)}$, we obtain the corresponding $\tilde{K}_3$, which is $\tilde{K}_3 = f_4^{-1}(O_4^{(4)}) \oplus \tilde{C}_3 \oplus O_2^{(2)} \oplus O_3^{(3)}$.

Finally, we obtain (K_1, K_2, K_3, K_4) as $(K_1, K_2, K_3, K_4) \leftarrow (\tilde{K}_1 \oplus \tilde{K}_2 \oplus \tilde{K}_3, \tilde{K}_1 \oplus \tilde{K}_3 \oplus I_1^{(1)} \oplus \tilde{C}_4, \tilde{K}_2 \oplus \tilde{K}_3, \tilde{K}_1 \oplus \tilde{K}_2)$. □

We are now ready to present the basic version of our equivalent key derivation algorithm based on Lemma 3.

1. Fix arbitrarily $I_1^{(1)}$ and $I_7^{(6)}$ that satisfy $I_1^{(1)} \in \mathcal{I}_1$ and $I_7^{(6)} \in \mathcal{I}_7$.
2. Fix arbitrarily $I_8^{(5)}$ and $\tilde{K}_1$.
3. Derive (K_1, K_2, K_3, K_4) by using Lemma 3.
4. Compute $I_3^{(11)}$ from (K_1, K_2, K_3, K_4) by following the specification of 256-bit key HyRAL, and proceed to Step 5 if $I_3^{(11)} \in \mathcal{I}_3$ is satisfied. Otherwise return to Step 2.

5. Compute $I_5^{(16)}$ from (K_1, K_2, K_3, K_4) by following the specification of 256-bit key HyRAL, and output (K_1, K_2, K_3, K_4) and halt if $I_5^{(16)} \in \mathcal{I}_5$ is satisfied. Otherwise return to Step 2.

If we assume that $I_3^{(11)}$ and $I_5^{(16)}$ are independently and uniformly distributed random strings over $\{0,1\}^{32}$, then the probability that both $I_3^{(11)} \in \mathcal{I}_3$ and $I_5^{(16)} \in \mathcal{I}_5$ are satisfied is $(64/2^{32})^2 = 2^{-52}$, since there are 64 elements in each of $\mathcal{I}_3$ and $\mathcal{I}_5$. Therefore, we may expect that the algorithm returns (K_1, K_2, K_3, K_4) after trying 2^{52} values of $(I_8^{(5)}, \tilde{K}_1)$.

4.2 Time Complexity of the Algorithm

In the basic algorithm presented in Sect. 4.1, the test $I_3^{(11)} \in \mathcal{I}_3$ is executed for 2^{52} different values of $(I_8^{(5)}, \tilde{K}_1)$. This test of $I_3^{(11)} \in \mathcal{I}_3$ is the main cost in the time complexity of the algorithm, and the following lemma can be used in the actual implementation.

Lemma 4. *For arbitrarily fixed $\tilde{K}_1$, $I_1^{(1)}$, $O_1^{(1)}$, $I_8^{(5)}$, $I_7^{(6)}$, and $O_7^{(6)}$, the corresponding value of $I_3^{(11)}$ can be derived by seven computations of f_i functions.*

Proof. $I_3^{(11)}$ can be derived by the following steps.

1. $O_8^{(5)} \leftarrow f_8(I_8^{(5)})$
2. $O_2^{(2)} \leftarrow \tilde{C}_5 \oplus I_8^{(5)} \oplus O_8^{(5)}$
3. $\tilde{K}_2 \leftarrow f_2^{-1}(O_2^{(2)}) \oplus \tilde{C}_1$
4. $O_3^{(3)} \leftarrow f_3(\tilde{K}_1 \oplus \tilde{C}_2 \oplus O_2^{(2)})$
5. $O_4^{(4)} \leftarrow \tilde{C}_4 \oplus I_8^{(5)} \oplus O_2^{(2)} \oplus O_3^{(3)}$
6. $O_6^{(7)} \leftarrow f_6(C_1 \oplus C_2 \oplus C_4 \oplus O_1^{(1)} \oplus O_2^{(2)} \oplus O_4^{(4)} \oplus O_8^{(5)} \oplus O_7^{(6)})$
7. $O_5^{(8)} \leftarrow f_5(C_1 \oplus C_2 \oplus C_3 \oplus O_1^{(1)} \oplus O_2^{(2)} \oplus O_3^{(3)} \oplus O_8^{(5)} \oplus O_7^{(6)} \oplus O_6^{(7)})$
8. $O_1^{(9)} \leftarrow f_1(I_1^{(1)} \oplus O_2^{(2)} \oplus O_3^{(3)} \oplus O_4^{(4)} \oplus O_7^{(6)} \oplus O_6^{(7)} \oplus O_5^{(8)})$
9. $O_2^{(10)} \leftarrow f_2(\tilde{K}_2 \oplus \tilde{C}_1 \oplus O_3^{(3)} \oplus O_4^{(4)} \oplus O_8^{(5)} \oplus O_6^{(7)} \oplus O_5^{(8)} \oplus O_1^{(9)})$
10. $I_3^{(11)} \leftarrow \tilde{K}_1 \oplus C_1 \oplus C_3 \oplus I_1^{(1)} \oplus O_1^{(1)} \oplus O_2^{(2)} \oplus O_4^{(4)} \oplus O_8^{(5)} \oplus O_7^{(6)} \oplus O_5^{(8)} \oplus O_1^{(9)} \oplus O_2^{(10)}$

We see that the above steps run with seven computations of f_i functions. □

In the proof of Lemma 4, one can run Steps 1, 2, and 3 without using $\tilde{K}_1$. Therefore, one possible implementation is to search 2^{52} values of $(I_8^{(5)}, \tilde{K}_1)$ by searching 2^{20} values of $I_8^{(5)}$, and for each value of $I_8^{(5)}$, we first run Steps 1, 2, and 3 and then search all the 2^{32} possible values of $\tilde{K}_1$. Then the main cost of running the algorithm becomes 5×2^{52} computations of f_i functions assuming that 2^{26} computations of f_i functions can be ignored.

In order to derive both OK_1 and OK_2, we need to run the algorithm twice by changing the constant (C_1, C_2, C_3, C_4), and hence the time complexity of the algorithm is 10×2^{52} computations of f_i functions, which amount to running $2^{48.8}$ encryption functions as there are 96 f_i functions in the encryption function of 256-bit key HyRAL. We note that the memory requirement of the algorithm is small.

Table 4. Summary of the implementation. The "Cores" column indicates the number of cores used in running the program

	System	Queue name	Cores	Search range of $I_8^{(5)}$	Number of $(I_8^{(5)}, \tilde{K}_1)$	Running time
OK_1	HX600	h1024	1024	0x00000000, ..., 0x0000ffff	2^{48}	8h 48min 56s
		h1024	1024	0x00010000, ..., 0x0001ffff	2^{48}	8h 28min 4s
OK_2	FX1	f1024	1024	0x00000000, ..., 0x0003ffff	2^{50}	50h 36min 2s
		f512	512	0x00040000, ..., 0x0007ffff	2^{50}	92h 24min 15s
	HX600	h256	256	0x00080000, ..., 0x0009ffff	2^{49}	67h 42min 47s
		h256	256	0x000a0000, ..., 0x000bffff	2^{49}	67h 29min 1s
		h256	256	0x000c0000, ..., 0x000dffff	2^{49}	67h 34min 55s
		h256	256	0x000e0000, ..., 0x000fffff	2^{49}	67h 29min 57s

5 Deriving Equivalent Keys

We have implemented our algorithm in Sect. 4.2 on a supercomputer system. The systems we used are the server systems called HX600 and FX1. HX600 has 96 nodes, which are equivalent to 384 CPUs or 1536 cores, it has a total of 6TB of memory, and the CPU is AMD Opteron 8380 (4 cores, 2.5GHz). FX1 has 768 nodes, which are equivalent to 768 CPUs or 3072 cores, it has 24TB of memory, and CPU is SPARC64 VII (4 cores, 2.52GHz). We used C language for the implementation of the algorithm, and MPI library for the message passing library for the parallel process execution.

The values of δ, $I_1^{(1)}$, and $I_7^{(6)}$ that were used in the implementation are δ = 0xd7d7d0d7, $I_1^{(1)}$ = 0x17170c17, and $I_7^{(6)}$ = 0x1717292b. For deriving OK_1, we searched 2^{17} values of $I_8^{(5)}$, and for each value of $I_8^{(5)}$, we searched all the 2^{32} possible values of $\tilde{K}_1$. The program was divided into two programs by halving the search range of $I_8^{(5)}$, and a total of 2^{49} values of $(I_8^{(5)}, \tilde{K}_1)$ were tested. For deriving OK_2, we searched 2^{20} values of $I_8^{(5)}$, and for each value of $I_8^{(5)}$, we searched all the 2^{32} possible values of $\tilde{K}_1$. The program was divided into six programs depending on the range of $I_8^{(5)}$, and a total of 2^{52} values of $(I_8^{(5)}, \tilde{K}_1)$ were tested. The summary of the implementation is in Table 4.

As a result, we have successfully derived one value of OK_1 and three values of OK_2. The values, together with the corresponding values of $I_8^{(5)}$ and $\tilde{K}_1$, are in Table 5.

Table 5. Results of running the algorithm in Sect. 4.2

OK_1	0x2fd918837136d461f4bc99938907dd0b ($I_8^{(5)}$ = 0x00014b73, $\tilde{K}_1$ = 0xdb658110)
OK_2	0xa20ed0f467141b2a3b038abb5f61d59e ($I_8^{(5)}$ = 0x0005b394, $\tilde{K}_1$ = 0x990d5a4f)
	0xe3a1902aa60b6c3582a9131527d43b2f ($I_8^{(5)}$ = 0x000f8a7f, $\tilde{K}_1$ = 0x6108833f)
	0x3218a5b25828a0b7d2122283894cc63b ($I_8^{(5)}$ = 0x000f9953, $\tilde{K}_1$ = 0xe00a8731)

For δ = `0xd7d7d0d7`, $\Delta OK_1 = \Delta OK_2 = (\delta, \delta, \delta, \delta)$, and OK_1 and OK_2 in Table 5, (K, K') in (4) are all equivalent keys, which can be verified by the reference code available in [6].

6 Discussions

The existence of equivalent keys generally implies that the cipher is theoretically cryptanalyzed, as the time complexity of the brute-force attack becomes less than the time complexity implied by its key length. As there are $2^{50.0}$ pairs of equivalent keys in 256-bit key HyRAL, the search space of the brute-force attack is reduced from 2^{256} to $2^{256} - 2^{50.0}$. Although the fraction of equivalent keys, $2^{51.0}$, compared to 2^{256} is small, as a practical implication of identifying equivalent keys, in the rest of this section, we discuss well known observations that one can obtain collisions on the Davies-Meyer compression function based on 256-bit key HyRAL, and on the Merkle-Damgård hash function based on the compression function.

The Davies-Meyer Compression Function. Let $E : \{0,1\}^k \times \{0,1\}^n \to \{0,1\}^n$ be a blockcipher with k-bit keys and an n-bit block. The Davies-Meyer compression function $h : \{0,1\}^{n+k} \to \{0,1\}^n$, one of the standard constructions of a compression function, is defined as $h(H, M) = E_M(H) \oplus H$.

Let E be 256-bit key HyRAL. If we let (M, M') be one of the equivalent keys (K, K') in (4), then for any $H \in \{0,1\}^{128}$, we have $h(H, M) = h(H, M')$. Therefore, for each equivalent keys (K, K') in (4), one can generate 2^{128} different collisions $((H, M), (H, M'))$ on h.

The Merkle-Damgård Hash Function. Let $h : \{0,1\}^{n+k} \to \{0,1\}^n$ be a compression function. The Merkle-Damgård hash function $\mathcal{H} : \{0,1\}^* \to \{0,1\}^n$ is the construction of a hash function from h, and is defined as follows. Let $H_0 \in \{0,1\}^n$ be a fixed initial value. For an input string $M \in \{0,1\}^*$, let $\tilde{M} \in \{0,1\}^{mn}$ be the padded string in a standard and appropriate way, and let $(M_1, M_2, \ldots, M_m) \stackrel{n}{\leftarrow} \tilde{M}$ be its partition into n-bit strings. The hash value $\mathcal{H}(M)$ is $H_m \in \{0,1\}^n$, where $H_i \leftarrow h(H_{i-1}, M_i)$ for $i = 1, 2, \ldots, m$.

Let E be 256-bit key HyRAL, h be the Davies-Meyer compression function based on E, and $\mathcal{H}$ be the Merkle-Damgård hash function based on h. Let $M, M' \in \{K, K'\}^m$ be bit strings such that $M \neq M'$, where (K, K') is any equivalent keys in (4). Assume that the standard padding is used, e.g., appending a bit "1" and then bits "0" followed by the encoding of the length of the input, then we have $\mathcal{H}(M) = \mathcal{H}(M')$ and hence we obtain a collision on $\mathcal{H}$.

For example, for $m = 3$, $(M, M') = ((K, K, K), (K', K', K'))$, $((K, K, K'), (K', K', K))$, $((K, K', K), (K', K, K'))$, and $((K, K', K'), (K', K, K))$ all satisfy $\mathcal{H}(M) = \mathcal{H}(M')$. Similarly, when M and M' are m blocks in length, we obtain 2^{m-1} different collisions.

7 Summary

We presented the analysis of 256-bit key HyRAL in terms of equivalent keys. We showed that there are $2^{50.0}$ pairs of equivalent keys, leading to the theoretical cryptanalysis of the cipher as a blockcipher with 256-bit keys. We also developed the algorithm to derive an instance of equivalent keys, and demonstrated that we were able to derive concrete instances with the current computing environment.

Acknowledgments. The experiment in Sect. 5 was conducted using a supercomputer system at Information Technology Center, Nagoya University. The authors would like to thank Hideki Ando for advice on the experiment. This work was supported in part by CRYPTREC and in part by MEXT KAKENHI, Grant-in-Aid for Young Scientists (A), 22680001.

References

1. Biham, E., Biryukov, A., Shamir, A.: Cryptanalysis of Skipjack Reduced to 31 Rounds Using Impossible Differentials. In: Stern, J. (ed.) EUROCRYPT 1999. LNCS, vol. 1592, pp. 12–23. Springer, Heidelberg (1999)
2. Biham, E., Shamir, A.: Differential Cryptanalysis of DES-like Cryptosystems. J. Cryptology 4(1), 3–72 (1991)
3. Cryptography Research and Evaluation Committees (CRYPTREC), `http://www.cryptrec.go.jp/english/index.html`
4. Cryptography Research and Evaluation Committees (CRYPTREC): CRYPTREC Report, Report of the Scheme Committee (2010) (in Japanese), `http://www.cryptrec.go.jp/english/report.html`
5. Daemen, J., Knudsen, L.R., Rijmen, V.: The Block Cipher SQUARE. In: Biham, E. (ed.) FSE 1997. LNCS, vol. 1267, pp. 149–165. Springer, Heidelberg (1997)
6. Hirata, K.: Submission Documents of HyRAL to the CRYPTREC Project (2010), `http://www.cryptrec.go.jp/english/topics/cryptrec_20101001_callforattack.html`
7. Hirata, K.: The 128bit Block Cipher HyRAL (Hybrid Randomization Algorithm): Common Key Block Cipher. In: Proceedings of the 2010 International Symposium on Intelligence Information Processing and Trusted Computing, IPTC 2010, pp. 9–14. IEEE Computer Society, Washington, DC (2010), `http://dx.doi.org/10.1109/IPTC.2010.179`
8. Hirata, K.: The 128bit Blockcipher HyRAL. In: The 2010 Symposium on Cryptography and Information Security, 1D1-1, SCIS 2010 (2010) (in Japanese)
9. Igarashi, Y., Takagi, Y., Kaneko, T.: Security Evaluation of HyRAL against Linear Cryptanalysis. In: The 2010 Symposium on Cryptography and Information Security, 1D1-3, SCIS 2010 (2010) (in Japanese)
10. Inoue, T., Kaneko, T.: Security Evaluation of HyRAL against Boomerang Attack. IEICE Tech. Rep. 111(142), 1–6 (2011) (in Japanese); IT 2011-07-14
11. Iwata, T.: Security Evaluation Report of HyRAL. In: Technical Report of CRYPTREC, Investigation Reports Related to Cryptographic Techniques in FY 2010 (2011) (in Japanese)
12. Knudsen, L.R.: Cryptanalysis of LOKI. In: Matsumoto, T., Imai, H., Rivest, R.L. (eds.) ASIACRYPT 1991. LNCS, vol. 739, pp. 22–35. Springer, Heidelberg (1993)

13. Knudsen, L.R.: Truncated and Higher Order Differentials. In: Preneel, B. (ed.) FSE 1994. LNCS, vol. 1008, pp. 196–211. Springer, Heidelberg (1995)
14. Lai, X.: Higher Order Derivatives and Differential Cryptanalysis. In: Blahut, R.E., Massey, J.L. (eds.) Communications and Cryptography: Two Sides of One Tapestry, pp. 227–233. Kluwer Academic Publishers, Norwell (1994)
15. Matsui, M.: Linear Cryptanalysis Method for DES Cipher. In: Helleseth, T. (ed.) EUROCRYPT 1993. LNCS, vol. 765, pp. 386–397. Springer, Heidelberg (1994)
16. Matsui, M.: The First Experimental Cryptanalysis of the Data Encryption Standard. In: Desmedt, Y.G. (ed.) CRYPTO 1994. LNCS, vol. 839, pp. 1–11. Springer, Heidelberg (1994)
17. Shibayama, N., Igarashi, Y., Kaneko, T., Hangai, S.: Impossible Differential Attack on HyRAL. In: Forum on Information Technology, L-022 (2010) (in Japanese)
18. Shibayama, N., Igarashi, Y., Kaneko, T., Hangai, S.: On Impossible Differential of HyRAL Using MDS Characteristic. In: The 2010 IEICE Engineering Sciences Society Conference, A-7-8 (2010) (in Japanese)
19. Shibayama, N., Igarashi, Y., Kaneko, T., Hangai, S.: Security Evaluation of HyRAL against Saturation Cryptanalysis. In: The 33rd Symposium on Information Theory and its Applications, SITA 2010, 10.1 (2010) (in Japanese)
20. Shibayama, N., Igarashi, Y., Kaneko, T., Hangai, S.: Higher Order Differential Attack on HyRAL. IEICE Tech. Rep. 110(443), 341–347 (2011) (in Japanese); ISEC 2010-123
21. Shibayama, N., Igarashi, Y., Kaneko, T., Hangai, S.: Security Evaluation of HyRAL against Saturation Cryptanalysis (II). IEICE Tech. Rep. 111(123), 103–109 (2011) (in Japanese); ISEC 2011-19
22. Shibayama, N., Kaneko, T., Hangai, S.: New Saturation Characteristics of HyRAL. IEICE Tech. Rep. 111(455), 53–60 (2012) (in Japanese); ISEC 2011-81
23. Taga, B., Tanaka, H.: Higher Order Differential Characteristics of HyRAL. In: The 2011 Symposium on Cryptography and Information Security, 2B2-2, SCIS 2011 (2011) (in Japanese)
24. Takagi, Y., Igarashi, Y., Kaneko, T.: Security Evaluation of HyRAL against Differential Attack. In: The 2010 Symposium on Cryptography and Information Security, 1D1-2, SCIS 2010 (2010) (in Japanese)
25. Yamaguchi, Y., Shibayama, N., Kaneko, T.: Higher Order Differential Property of HyRAL (II). In: The 2012 Symposium on Cryptography and Information Security, 1C3-4, SCIS 2012 (2012) (in Japanese)
26. Yamaguchi, Y., Igarashi, Y., Kaneko, T.: Higher Order Differential Property of HyRAL. In: The 63rd Joint Conference of Electrical and Electronics Engineers in Kyushu, 02-1A-06 (2010) (in Japanese)
27. Youm, H.Y., Song, J.H., Lee, S.Y.: Security Analysis of HyRAL. In: Technical Report of CRYPTREC, Investigation Reports Related to Cryptographic Techniques in FY 2010 (2011)

A Details of $f_1, \ldots, f_8$ Functions

We present the details of the specification of $f_1, \ldots, f_8$ functions that were omitted from Sect. 2.

$f_1, \ldots, f_8$ functions are permutations over $\{0,1\}^{32}$. For a given input $I = (x_1, x_2, x_3, x_4) \in \{0,1\}^{32}$, f_i function generates the output as follows.

1. Let $(x_1, x_2, x_3, x_4) \leftarrow T_i(x_1, x_2, x_3, x_4)$.
2. Let $(x_1, x_2, x_3, x_4) \leftarrow (S(x_1), S(x_2), S(x_3), S(x_4))$.
3. Compute (o_1, o_2, o_3, o_4) by

$$\begin{pmatrix} o_1 \\ o_2 \\ o_3 \\ o_4 \end{pmatrix} \leftarrow \begin{pmatrix} \texttt{0x03} & \texttt{0x03} & \texttt{0x02} & \texttt{0x01} \\ \texttt{0x01} & \texttt{0x02} & \texttt{0x02} & \texttt{0x02} \\ \texttt{0x07} & \texttt{0x03} & \texttt{0x01} & \texttt{0x02} \\ \texttt{0x07} & \texttt{0x04} & \texttt{0x05} & \texttt{0x03} \end{pmatrix} \cdot \begin{pmatrix} x_1 \\ x_2 \\ x_3 \\ x_4 \end{pmatrix} \oplus \begin{pmatrix} \texttt{0x11} \\ \texttt{0x22} \\ \texttt{0x44} \\ \texttt{0x88} \end{pmatrix},$$

   where the arithmetic is over GF(2^8) defined by the irreducible polynomial $p(x) = x^8 + x^4 + x^3 + x + 1$.
4. The output is $O = (o_1, o_2, o_3, o_4)$.

T_i function is defined in Table 6. The S-box S is the composition of an affine mapping over GF(2) and the inversion over GF(2^8). Table 7 shows the input and output of the S-box. The values are in hexadecimal form. The input x is regarded as two hexadecimal digits, and if the first digit is i and the last is j, then the output is a value written in the i-th row and j-th column. For example, $S(\texttt{0x12}) = \texttt{0x06}$.

Table 6. T_i function

i	$T_i(x_1, x_2, x_3, x_4)$
1	(x_1, x_2, x_3, x_4)
2	(x_2, x_3, x_4, x_1)
3	(x_3, x_4, x_1, x_2)
4	(x_4, x_1, x_2, x_3)
5	(x_4, x_3, x_2, x_1)
6	(x_3, x_2, x_1, x_4)
7	(x_2, x_1, x_4, x_3)
8	(x_1, x_4, x_3, x_2)

Table 7. S-box S

	0	1	2	3	4	5	6	7	8	9	a	b	c	d	e	f
0	16	5e	d3	af	36	43	a6	49	33	93	3b	21	91	df	47	f4
1	b6	70	06	d0	81	82	fa	a1	10	b5	3c	ba	97	85	b7	79
2	ed	5c	ca	05	87	bf	24	4c	51	ec	17	61	22	f0	3e	18
3	a7	64	13	ab	e9	09	25	54	2d	31	69	f5	37	67	fe	1d
4	0b	28	a3	2f	e4	0f	d4	da	1b	fc	e6	ac	53	04	27	a9
5	94	8b	d5	c4	90	6b	f8	9d	c5	db	ea	e2	ae	63	07	7a
6	5b	23	34	38	03	8c	46	68	cd	1a	1c	41	7d	a0	9c	dd
7	08	4e	e3	d7	1e	b3	50	5d	c6	0e	ad	cf	d6	eb	0d	b1
8	fb	7c	c3	2e	65	48	b8	8f	ce	e7	62	d2	12	4a	c8	26
9	a5	8e	3d	76	86	57	bc	bd	11	75	71	78	1f	ef	e0	0c
a	de	6a	6d	32	84	72	8a	d8	f9	dc	9a	89	9f	88	14	2a
b	9b	9e	d9	95	b9	a4	02	f7	96	73	56	be	7f	80	7e	83
c	00	01	f6	8d	7b	d1	52	cb	b0	e1	c7	e5	29	c0	4f	e8
d	58	3f	cc	fd	ee	b2	40	ff	99	2b	5f	60	aa	4b	b4	74
e	2c	45	6c	92	66	42	39	f3	77	bb	19	59	20	6f	35	f2
f	c1	0a	15	98	a2	c2	44	30	55	4d	c9	a8	5a	f1	6e	3a

Distinguishers beyond Three Rounds of the RIPEMD-128/-160 Compression Functions

Yu Sasaki[1] and Lei Wang[2]

[1] NTT Secure Platform Laboratories, NTT Corporation
3-9-11 Midori-cho, Musashino-shi, Tokyo 180-8585 Japan
sasaki.yu@lab.ntt.co.jp
[2] The University of Electro-Communications
1-5-1 Choufugaoka, Choufu-shi, Tokyo, 182-8585 Japan
lei.wang@uec.ac.jp

Abstract. This paper presents differential-based distinguishers against ISO standard hash functions RIPEMD-128 and RIPEMD-160. Second-order differential paths are constructed on reduced steps of their compression functions. These lead to 4-sum attacks on 47 steps (out of 64 steps) of RIPEMD-128 and 40 steps (out of 80 steps) of RIPEMD-160. Then new properties called a *(partial) 2-dimension sum* and *q-multi-second-order collision* are considered. The partial 2-dimension sum is generated on 48 steps of RIPEMD-128 and 42 steps of RIPEMD-160, with a complexity of 2^{35} and 2^{36}, respectively. Theoretically, 2-dimension sums are generated faster than the brute force attack up to 52 steps of RIPEMD-128 and 51 steps of RIPEMD-160, with a complexity of 2^{101} and 2^{158}, respectively. The attacks on RIPEMD-128 can also be regarded as q-multi-second-order collision attacks. The practical attacks are implemented and generated examples are presented. We stress that our results do not impact to the security of full RIPEMD-128 and RIPEMD-160 hash functions.

Keywords: RIPEMD-128, RIPEMD-160, double-branch structure, 2-dimension sum, q-multi-second-order collision.

1 Introduction

Hash functions are taking important roles in various aspects of the cryptography. Since the collision resistance of MD5 and SHA-1 were broken by Wang *et al.* [1,2], cryptographers have looked for stronger hash function designs. While various new designs are discussed in the SHA-3 competition [3], some of existing hash functions seem to have much higher security than the MD4-family. Evaluating such hash functions are useful especially if they are standardized internationally.

RIPEMD-128 and RIPEMD-160 [4] are hash functions standardized by ISO [5]. RIPEMD-160 is also included in the recommended ciphers list of the Cryptography Research and Evaluation Committees (CRYPTREC) set up by the

F. Bao, P. Samarati, and J. Zhou (Eds.): ACNS 2012, LNCS 7341, pp. 275–292, 2012.

Japanese Government [6]. RIPEMD-160 is standardized in the SSL protocol [7], and is included in the OpenSSL Cryptography and SSL/TLS Toolkit [8]. RIPEMD-128 and RIPEMD-160 are implemented in various cryptographic libraries. For example, BouncyCastle [9], FlexiProvider [10], and GNU Crypto [11] for JAVA, and Crypto++ [12] for C++. The use of RIPEMD-128 and RIPEMD-160 in HMAC is explained in RFC [13,14].

RIPEMD-128 and -160 adopt the narrow-pipe Merkle-Damgård structure. Their compression functions adopt the double-branch structure, which takes a previous chaining variable H_{i-1} and a message block M_{i-1} as input and computes two compression functions $\mathrm{CF}^{\mathrm{L}}(H_{i-1}, M_{i-1})$ and $\mathrm{CF}^{\mathrm{R}}(H_{i-1}, M_{i-1})$. The output H_i is computed by merging H_{i-1}, $\mathrm{CF}^{\mathrm{L}}(H_{i-1}, M_{i-1})$, and $\mathrm{CF}^{\mathrm{R}}(H_{i-1}, M_{i-1})$. Due to the double size of the internal state and the difficulties of controlling the two functions simultaneously, only a few results were published before. Note that, in order to prevent the recent meet-in-the-middle preimage attacks [15,16,17], some hash functions adopt a structure, which applies the feed-forward function several times, e.g. ARIRANG [18]. Ohtahara *et al.* pointed out that such a structure can be viewed as the double-branch structure [19]. Hence, analyzing the double-branch structure is useful even for the future hash function design.

RIPEMD-128 produces 128-bit digests and its compression function consists of 4 rounds, 64 steps. RIPEMD-160 produces 160-bit digests and its compression function consists of 5 rounds, 80 steps. Mendel *et al.* investigated the differential property of the compression functions of RIPEMD-128 and -160 [20]. They applied the linear approximation and showed low Hamming weight differential paths up to 3 rounds (48 steps) for RIPEMD-128 and up to some intermediate step in the third round (steps 33 – 48) for RIPEMD-160. Although [20] is useful to obtain some intuition for collision attacks, a lot of work is necessary to complete the attacks. The complexity and even the possibility of the attacks are unclear. Other previous results are the ones by Ohtahara *et al.* [21] and Wang *et al.* [22], which investigated preimage attacks. [21] showed that the first 33 steps of RIPEMD-128 and the first 31 steps of RIPEMD-160 can be attacked while [22] showed that intermediate 36 steps of RIPEMD-128 can be attacked.

In this paper, boomerang type differential properties are discussed. The boomerang attack was first proposed by Wagner for analyzing block-ciphers [23]. It divides the function $E(\cdot)$ into two subparts E_0 and E_1 such that $E(\cdot) = E_1 \circ E_0(\cdot)$. Let the probabilities of differential paths for E_0 and E_1 be p and q, respectively. The boomerang attack exploits the fact that a second order differential property with a probability p^2q^2 exists for the entire function E. Aumasson *et al.* [24] applied the boomerang attack to the internal cipher of the hash function Skein. However, the goal of the attack is still recovering the secret key. After that Birukov *et al.* [25] and Lamberger and Mendel [26] independently applied this property on the compression function so as to mount distinguishers. Then, Sasaki [27] showed a straight-forward application of the framework of [25,26] to the MD4-family (using the single-branch structure) consisting of up to 5 rounds. With the straight-forward technique by [27], it is unclear how to attack the single-branch structure consisting of more than 5 rounds.

Our Contributions

This paper presents differential distinguishers against the compression functions of RIPEMD-128 and RIPEMD-160. The results are summarized in Table 1.

Our first target is the 4-sum property. Then, new differential property called a *2-dimension sum* and a *q-multi-second-order collision* are considered. Note that the partial 4-sum and partial 2-dimension sum can be introduced naturally.

Then, differential paths are constructed on reduced RIPEMD-128 and -160 compression functions. The differential path construction is based on the framework of the boomerang distinguisher by [25,26]. Our strategy is regarding CF^{L} as the first part of the boomerang attack (E_0) and CF^{R} as the second part (E_1), hence the entire function (E) is viewed as the single-branch structure with 8 and 10 rounds for RIPEMD-128 and -160, respectively. This simplifies the differential path construction because the differential paths for CF^{L} and CF^{R} can be analyzed almost independently. However, because the straight-forward application of the framework to the MD4-family [27] can only work up to 5 rounds, several improvements are necessary to extend the number of attacked rounds.

On RIPEMD-128, we use the local collision to construct differential paths. This leads to a long differential path satisfied with a high probability. As a result, 4-sums and partial 2-dimension sums are generated with a practical complexity up to 47 and 48 steps, respectively. In addition, 2-dimension sums are theoretically generated faster than the brute force attack up to 52 steps.

On RIPEMD-160, the local collision involves more message words than RIPEMD-128, and thus using the local collision is inefficient. Instead, we show an interesting non-linear differential property of RIPEMD-160 which can avoid the quick propagation of the difference. As a result, 4-sums and partial 2-dimension sums are generated with a practical complexity up to 38 and 40 steps, respectively. In addition, 2-dimension sums are theoretically generated faster than the brute force attack up to 43 steps. If the attack target starts from the second round, the numbers of attacked steps become 40, 42, and 51 for 4-sums, partial 2-dimension sums, and theoretical 2-dimension sums.

Paper Outline. In Sect. 2, differential properties are discussed. In Sect. 3, the specification of RIPEMD-128 and -160 are explained. In Sect. 4, attacks on RIPEMD-128 are explained. In Sect. 5, attacks on RIPEMD-160 are explained. Finally, the paper is concluded in Sect. 6.

2 Differential Properties to Be Distinguished

We summarize differential properties discussed in previous papers, which are 4-sums and second-order collisions, and introduce new properties called *2-dimension sums* and *q-multi-second-order collision.*

Table 1. Attacks on the compression functions. "2D-sum" and "q-multi-2nd" denote 2-dimension sum and q-multi-second-order collision respectively.

Target	#Steps	Property	Information Theoretic Bound	Generic Attack	Complexity	Reference
RIPEMD-128	33	preimage	2^{128}	2^{128}	2^{119}	[21]
(total 64 steps)	36†	preimage	2^{128}	2^{128}	2^{123}	[22]
	45	4-sum	2^{32}	2^{42}	2^{27}	Ours
	47	4-sum	2^{32}	2^{42}	2^{39}	Ours
	48	q-multi-2nd	2^{55} for $q = 17$	-	2^{53}	Ours
	52	q-multi-2nd	2^{108} for $q = 53$	-	2^{107}	Ours
RIPEMD-160	31	preimage	2^{160}	2^{160}	2^{148}	[21]
(total 80 steps)	38	4-sum	2^{40}	2^{53}	2^{42}	Ours
	40	partial 2D sum	2^{64}	2^{64}	2^{42}	Ours
	43	2D sum	2^{160}	2^{160}	2^{151}	Ours
	40‡	4-sum	2^{40}	2^{53}	2^{36}	Ours
	42‡	partial 2D sum	2^{64}	2^{64}	2^{36}	Ours
	51‡	2D sum	2^{160}	2^{160}	2^{158}	Ours

†: The attacked steps start from an intermediate step.
‡: The attacked steps start from the second round.

2.1 Previously Discussed Properties

4-sum is a set of 4 different inputs (I_0, I_1, I_2, I_3) where the sum of the corresponding outputs is 0, namely $\mathrm{CF}(I_0) \oplus \mathrm{CF}(I_1) \oplus \mathrm{CF}(I_2) \oplus \mathrm{CF}(I_3) = 0$. If the function is ideal, finding 4-sums requires at least $2^{n/4}$ queries for n-bit output. Therefore, if the 4-sum is obtained faster than $2^{n/4}$ computations, CF is regarded as non-ideal. Apart from the information theoretic bound ($2^{n/4}$), the current best generic attack to find 4-sums is a generalized birthday attack [28], which requires $2^{n/3}$ computations and $2^{n/3}$ memory. Hence, if 4-sums are generated with a complexity between $2^{n/4}$ and $2^{n/3}$, CF is said to be weak because the same property cannot be detected on other functions with the current knowledge.

The second-order collision [29,26] is a special form, in other words, a subset of the 4-sum. It can be viewed as limiting the form of input values on the 4-sum property. [29,26] defined the derivative at a point α for a function f as

$$\Delta_{(\alpha)} f(y) = f(y + \alpha) - f(y).$$

Then, the second-order derivative[1] at (α, β) is defined as

$$\begin{aligned}\Delta_{(\alpha,\beta)} f(y) &= \Delta_{(\beta)}(\Delta_{(\alpha)} f(y)) \\ &= f(y + \alpha + \beta) - f(y + \beta) - f(y + \alpha) + f(y).\end{aligned}$$

The second-order collision attack is finding (α, β, y) such that $\Delta_{(\alpha,\beta)} f(y) = 0$. Previous work [29,26] showed that the information theoretic bound is $3 \cdot 2^{n/3}$

[1] In [29,26], the n-th order derivative is discussed rather than the specific case $n = 2$.

because the problem is essentially finding three parameters α, β, y with an n-bit relation. On the other hand, the current best generic attack requires $2^{n/2}$.

2.2 2-Dimension Sums and Suitability for Double-Branch Structure

Similarly to the framework of the rebound attack [30], which limits the input and output differences before making any query, we introduce a new differential property, which we call *2-dimension sums.* The 2-dimension sum is a special form, in other words, a subset of the second-order collision. We further introduce limitations to the form of input values on the second-order collision. The problem is changed to find a value y such that $\Delta_{(\alpha,\beta)} f(y) = 0$ for two pre-specified values α and β. The 2-dimension sum is different from the second-order collision only in the sense that α and β are pre-specified. The information theoretic bound for the 2-dimension sum is 2^n because the problem is essentially finding an n-bit value satisfying an n-bit condition. A generic attack for this problem is also 2^n computations; choose the value of y and check that the corresponding $\mathrm{CF}(y) \oplus \mathrm{CF}(y \oplus \alpha) \oplus \mathrm{CF}(y \oplus \beta) \oplus \mathrm{CF}(y \oplus \alpha \oplus \beta)$ is 0.

The 2-dimension sum is particularly useful to attack the double-branch structure. The attacker can construct a pseudo-near-collision path for $\mathrm{CF^L}$ with setting input chaining variable difference to α. Then, a pseudo-near-collision path for $\mathrm{CF^R}$ is independently constructed with setting other difference β. If the product of the probability of each path (after the message modification) is higher than $2^{-n/2}$, the 2-dimension sum can be generated faster than 2^n by using the boomerang attack approach [25,26]. Different from the original RIPEMD, RIPEMD-128 and -160 adopt very different functions as $\mathrm{CF^L}$ and $\mathrm{CF^R}$. Therefore, the independence of the path construction for $\mathrm{CF^L}$ and $\mathrm{CF^R}$ greatly helps the attacker. More detailed discussion is given in Sect. 4, and 5.

Note that the *partial 2-dimension sum* is naturally introduced, where $\mathrm{CF}(y) \oplus \mathrm{CF}(y \oplus \alpha) \oplus \mathrm{CF}(y \oplus \beta) \oplus \mathrm{CF}(y \oplus \alpha \oplus \beta)$ becomes 0 only for the specified partial bits, say d bits. In this case, the complexity of the generic attack is 2^d and thus a valid distinguisher must find it faster than 2^d computations.

2.3 q-Multi-second-order Collision

By following the framework of the q-multicollision [31], we introduce a notion of a q-multi-second-order collision on a function $f : \{0,1\}^X \to \{0,1\}^Y$, which is a set of two non-zero differences and q distinct inputs $\{\Delta, \nabla, x_1, x_2, \cdots, x_q\}$ satisfying

$$f(x_1) - f(x_1 + \Delta) + f(x_1 + \Delta + \nabla) - f(x_1 + \nabla) = 0,$$
$$\cdots$$
$$f(x_q) - f(x_q + \Delta) + f(x_q + \Delta + \nabla) - f(x_q + \nabla) = 0.$$

Information-Theoretic Bound. Let an adversary make k distinct queries $x_1, \ldots x_k$ to a random function. For any specified Δ and ∇, k distinct queries at most contribute to k quartets as below; $(x_1, x_1+\Delta, x_1+\nabla, x_1+\Delta+\nabla), \ldots, (x_k, x_k+\Delta, x_k+\nabla, x_k+\Delta+\nabla)$. So the number of q-tuple quartets is $\binom{k}{q}$ for any specified Δ and ∇. There are less than 2^{2X} values for (Δ, ∇). Thus in total the number of q-tuple quartets satisfying the form is at most $2^{2X} \cdot \binom{k}{q}$. One such q-tuple has a probability $2^{-q\times Y}$. We get the following inequality.

$$2^{2X} \cdot \binom{k}{q} \geq 2^{qY},$$

$$\frac{k(k-1)\cdots(k-(q-1))}{q!} \geq 2^{qY-2X},$$

$$k > \sqrt[q]{q!} \cdot 2^{Y-\frac{2X}{q}}. \tag{1}$$

2.4 Remarks for the Motivation of Studying Weak Properties

Some may say that studying weak non-ideal properties such as the partial 2-dimension sum is meaningless. In fact, compared to the collision, the impact of finding weak differential properties are very limited. However, the security of symmetric primitives is usually evaluated and get trusted by demonstrating many cryptanalytic attempts. Therefore, we believe that not only investigating the standard properties but also extending the number of steps as much as possible with any non-ideal property is useful to understand the state-of-the-art about the security. Especially, such an activity is important for RIPEMD-128 and RIPEMD-160 because they are standardized and implemented in various environments but only a few cryptanalyses were presented so far. This paper is not claiming that weak distinguishers working for more steps are better than standard attack scenarios with a smaller number of attacked steps. Considering various approaches leads to better understanding and may be useful in future.

3 Specifications

RIPEMD-128/-160 were proposed by Dobbertin *et al.* [4] as stronger hash functions than RIPEMD [32]. They take a message of arbitrary length as input and produce 128-bit and 160-bit hash digests, respectively. Because our attack target is the compression function, we omit the description of the domain extension.

3.1 RIPEMD-128

The compression function of RIPEMD-128 takes a 128-bit chaining variable H_{i-1} and a 512-bit message block M_{i-1} as input and outputs a 128-bit chaining variable H_i. M_i is divided into sixteen 32-bit message words $m_0, m_1, \ldots, m_{15}$. Let p_j^L be a 128-bit chaining variable and a_j^L, b_j^L, c_j^L and d_j^L be 32-bit variables satisfying $p_j^L = a_j^L \| b_j^L \| c_j^L \| d_j^L$, where $0 \leq j \leq 64$. Similarly, p_j^R and $a_j^R, b_j^R, c_j^R, d_j^R$ are defined. The computation for $\mathrm{CF^L}$ is as follows.

$$p_0^L \leftarrow H_{i-1}, \qquad p_{j+1}^L \leftarrow \mathrm{SF}_j^L(p_j^L, m_{\pi^L(j)}) \text{ for } j = 0, 1, \ldots, 63,$$

Table 2. Boolean functions, message expansions, and rotation numbers

	$f_x(X, Y, Z)$
$x = 0, \ldots, 15$	$X \oplus Y \oplus Z$
$x = 16, \ldots, 31$	$(X \wedge Y) \vee (\neg X \wedge Z)$
$x = 32, \ldots, 47$	$(X \vee \neg Y) \oplus Z$
$x = 48, \ldots, 63$	$(X \wedge Z) \vee (Y \wedge \neg Z)$
$x = 64, \ldots, 79$	$X \oplus (Y \vee \neg Z)$

	$\pi^L(j)$															
$j = 0, \ldots, 15$	0	1	2	3	4	5	6	7	8	9	10	11	12	13	14	15
$j = 16, \ldots, 31$	7	4	13	1	10	6	15	3	12	0	9	5	2	14	11	8
$j = 32, \ldots, 47$	3	10	14	4	9	15	8	1	2	7	0	6	13	11	5	12
$j = 48, \ldots, 63$	1	9	11	10	0	8	12	4	13	3	7	15	14	5	6	2
$j = 64, \ldots, 79$	4	0	5	9	7	12	2	10	14	1	3	8	11	6	15	13

	$\pi^R(j)$															
$j = 0, \ldots, 15$	5	14	7	0	9	2	11	4	13	6	15	8	1	10	3	12
$j = 16, \ldots, 31$	6	11	3	7	0	13	5	10	14	15	8	12	4	9	1	2
$j = 32, \ldots, 47$	15	5	1	3	7	14	6	9	11	8	12	2	10	0	4	13
$j = 48, \ldots, 63$	8	6	4	1	3	11	15	0	5	12	2	13	9	7	10	14
$j = 64, \ldots, 79$	12	15	10	4	1	5	8	7	6	2	13	14	0	3	9	11

Table 3. Computations for the Output of the Compression Function

	RIPEMD-128	RIPEMD-160
$H_i^{(a)}$	$H_{i-1}^{(b)} + c_{64}^L + d_{64}^R$	$H_{i-1}^{(b)} + c_{80}^L + d_{80}^R$
$H_i^{(b)}$	$H_{i-1}^{(c)} + d_{64}^L + a_{64}^R$	$H_{i-1}^{(c)} + d_{80}^L + e_{80}^R$
$H_i^{(c)}$	$H_{i-1}^{(d)} + a_{64}^L + b_{64}^R$	$H_{i-1}^{(d)} + e_{80}^L + a_{80}^R$
$H_i^{(d)}$	$H_{i-1}^{(a)} + b_{64}^L + c_{64}^R$	$H_{i-1}^{(e)} + a_{80}^L + b_{80}^R$
$H_i^{(e)}$	–	$H_{i-1}^{(a)} + b_{80}^L + c_{80}^R$

where SF_j^L is a step function for $\mathrm{CF^L}$ and performs the following computation.

$$a_{j+1}^L \leftarrow d_j^L, \qquad b_{j+1}^L \leftarrow (a_j^L + f_j(b_j^L, c_j^L, d_j^L) + m_{\pi^L(j)} + k_j^L) \lll s_j^L,$$
$$c_{j+1}^L \leftarrow b_j^L, \qquad d_{j+1}^L \leftarrow c_j^L,$$

where '+' represents the addition on modulo 2^{32}, '$\lll s$' represents left cyclic shift by s bits, f_x is a Boolean function, $\pi^L(j)$ is the message expansion, and k_j^L is a constant. $\mathrm{CF^R}$ is similarly described. The values of $s_j^R, \pi^R(j), k_j^R$ are different and f_{63-j} is used in step j. Details of $f_x, \pi^L(j), \pi^R(j)$ are in Table 2. Finally, the output $H_i = H_i^{(a)} \| H_i^{(b)} \| H_i^{(c)} \| H_i^{(d)}$ is computed as shown in Table 3.

3.2 RIPEMD-160

The compression function of RIPEMD-160 is almost the same as the one for RIPEMD-128. The chaining variable size is 160 bits, and thus 160-bit intermediate states are represented by five 32-bit variables, e.g. $p_j^L = a_j^L \| b_j^L \| c_j^L \| d_j^L \| e_j^L$. The step functions SF^L and SF^R are iteratively computed 80 times ($0 \leq j \leq 79$). The details of the computation of SF^L are as follows.

$$a_{j+1}^L \leftarrow e_j^L, \quad b_{j+1}^L \leftarrow ((a_j^L + f_j(b_j^L, c_j^L, d_j^L) + m_{\pi^L(j)} + k_j^L) \lll s_j^L) + e_j^L,$$
$$c_{j+1}^L \leftarrow b_j^L, \quad d_{j+1}^L \leftarrow c_j^L \lll 10, \quad e_{j+1}^L \leftarrow d_j^L.$$

Most of the parameters are shared with RIPEMD-128. The details are described in Table 2. In the computations of SF^R, the Boolean function in step j is $f_{79-j}(b_j^R, c_j^R, d_j^R)$. The other computations are similarly specified as SF^L. Finally, the output chaining variable H_i is computed as shown in Table 3.

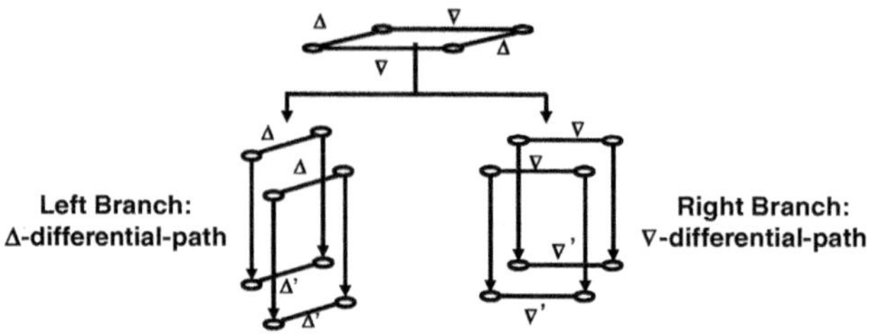

Fig. 1. Our Attack Strategy on RIPEMD-128 and -160

4 Attacks on RIPEMD-128

We construct differential distinguishers against compression function of RIPEMD-128. Hereafter, we compute the difference in modular subtraction because the main operation of RIPEMD-128/-160 is the modular addition. The message differences, differential path, and sufficient conditions against RIPEMD-128 are shown in Tables 4, 5 and 6 respectively.

4.1 Overall Strategy

A graphical description of our strategy is given in Fig. 1. First, we construct a Δ-differential-path $\Delta \xrightarrow{\Delta^M} \Delta'$ in the left branch, and a ∇-differential-path $\nabla \xrightarrow{\nabla^M} \nabla'$ in the right branch. Then we try to search for an input of the compression function (H, M) such that (H, M), $(H + \Delta, M + \Delta^M)$, $(H + \nabla, M + \nabla^M)$, and $(M + \Delta + \nabla, H + \Delta^M + \nabla^M)$ satisfy the following conditions.

- The difference propagations between (H, M) and $(H + \Delta, M + \Delta^M)$ and between $(H + \nabla, M + \nabla^M)$ and $(H + \nabla + \Delta, M + \nabla^M + \Delta^M)$ follow Δ-differential-path in the left branch.
- The difference propagations between (H, M) and $(H + \nabla, M + \nabla^M)$ and between $(H + \Delta, M + \Delta^M)$ and $(H + \Delta + \nabla, M + \Delta^M + \nabla^M)$ follow ∇-differential-path in the right branch.

For such a (H, M), we obtain the relationship; $\mathrm{CF}(H, M) + \mathrm{CF}(H + \Delta + \nabla, M + \Delta^M + \nabla^M) - \mathrm{CF}(H + \Delta, M + \Delta^M) - \mathrm{CF}(H + \nabla, M + \nabla^M) = 0$, where CF is the compression function of RIPEMD-128.

Note that the terminology "4-sum" is somehow strange to discuss the above modular subtraction. However, the terminology "second-order collision" usually considers the limitation of the inputs, and discusses essentially different properties. Hence, to avoid the confusion, we use "4-sum" for the above property.

4.2 Constructing Δ-Differential-Path

We should keep the differential path as simple as possible in order to maximize its probability. One natural approach is to restrict the difference propagations. Particularly, we expect that f functions do not produce new differences.

Table 4. Differential path construction for 3-round RIPEMD-128

round	$\pi^L(j)$	$\pi^R(j)$
1	⓪ 1 2 3 4 5 ⑥ 7 8 9 10 11 12 13 14 15	5 14 7 0 9 2 11 4 13 ⑥ 15 8 1 10 3 12
	Δ MM ← Δ | constant	MM ←
2	7 4 13 1 10 ⑥ 15 3 12 ⓪ 9 5 2 14 11 8	⑥ 11 3 7 0 13 5 10 14 15 8 12 4 9 1 2
	constant | ← LC → | constant	∇ | constant
3	3 10 14 4 9 15 8 1 2 7 ⓪ ⑥ 13 11 5 12	15 5 1 3 7 14 ⑥ 9 11 8 12 2 10 0 4 13
	constant | Δ Δ →	constant | ∇ →

The f functions of the first and the third rounds in the left branch have weak absorption property. By weak absorption property, we mean that a bit difference is produced by f with the probability 1 if a (particular) input variable has a difference at that bit. So we must make the Δ-differential-path short in these two rounds. In order to achieve it, we generate a *local collision* in the second round of the left branch, and pick a message difference Δ^M which appears at a very beginning step in the first round and at a very late step in the third round. Such a strategy maximizes the probability of the whole differential path. Finally we choose the message difference Δ^M as below

$$\Delta m_0 = -2^{10}; \text{ and } \Delta m_6 = 2^{31}.$$

Δ is determined backwards according to the differential path in the first round.

$$\Delta a_0 = 2^8; \ \Delta b_0 = 0; \ \Delta c_0 = 2^{31} + 2^{16}; \text{ and } \Delta d_0 = 2^{31} + 2^{16}.$$

Δ' changes with the number of the attacked steps. Moreover, we do not specify Δ' according to *amplified probability* using multiple outside differential paths.

Remarks on Multiple Differential Path. At step 44 of Δ-differential-path, a difference of $*2^5$ is produced by the f function. For this difference, we do not limit the sign for each pair, but we need the condition that the signs are identical between two pairs. Hence, the probability to satisfy this condition is 2^{-1} rather than 2^{-2}. We also set 2 similar conditions at steps 46 and 47.

4.3 Constructing ∇-Differential-Path

The f function of the second round in the right branch has weak absorption property. So we must make ∇-differential-path short in the second round of the right branch. At the same time, the f function in the first round of the right branch has strong absorption property. By strong absorption property, we mean that no bit difference is produced by f by setting conditions if only one input variant has a difference on that bit. So we decide to generate a relatively long but simple sub-path in the first round, which should be ended by the message difference in the second round. We should pick a message difference which appears at a very beginning step in the second round and at a very late step in the third round. The whole differential path consists of 2 sub-paths: a long path going through the whole first round and ending at a beginning step in the second round; and a short path at the late steps in the third round. Finally we choose the message difference ∇^M and corresponding ∇ as below

$$\nabla m_6 = -2^{30};\ \nabla a_0 = 2^{20};\ \nabla b_0 = 0;\ \nabla c_0 = 0;\ \text{and}\ \nabla d_0 = 2^6.$$

4.4 Searching for (H, M)

Firstly, we search for (H, M)s which satisfy the differential path for the first 17 steps in both branches. The complexity of finding such (H, M)s, i.e. the complexity for satisfying the first 17 steps can be ignored by applying the message modification technique (e.g. [1,2]) and optimizing the computation order. More precisely, the message modification is a technique to efficiently satisfy all conditions in the first round. It exploits the property that each step in the first round is computed with a message word which is not fixed yet. For example, to satisfy the conditions of the variable b_{j+1} in the first round, you can iterate the computation in step j many times by only changing the value of $m_{\pi(j)}$ without influencing the previous steps. Hence, by satisfying the conditions step by step, the complexity is greatly reduced. Moreover, modifying the message word m_{12} never impacts to the sufficient conditions in the first round. This is because m_{12} is used in late steps of the first round in both branches (See Table. 4). Thus, once we obtain an (H, M) satisfying the differential path up to step 17, we can generate up to 2^{32} valid (H, M) by modifying m_{12}. As a summary, the complexity for satisfying the differential path for the first 17 steps is (approximately) 2^{-32} times of the complexity of satisfying the whole differential path, and thus can be ignored.

For the remaining steps (after step 17), we simply satisfy the path in the brute-force manner. Hence, the entire attack complexity only depends on the number of conditions after step 17.

4.5 Complexity Evaluation and Experiments

Besides counting the number of conditions in Δ- and ∇-differential-paths after step 17, we also experimentally verify the amplified probability for outside paths.

Attack on 45 steps. There are 9 and 7 conditions in the Δ- and ∇-differential-paths, respectively. Each condition must be satisfied in two pairs and thus its probability is 2^{-2}. However, 1 condition in the Δ-path is the one discussed in the remarks in Sect. 4.2, which satisfied with probability 2^{-1}. Overall, the complexity is $2^{31(=2(8+7)+1)}$. We then experimentally check the amplified probability, and the final complexity is 2^{27}. This implies that 45 steps of the compression function is non-ideal with respect to the 4-sum property.

Attack on 46 and 47 steps. We experimentally checked the amplified probability, and the final complexities on 46 and 47 steps are 2^{34} and 2^{39} respectively. With respect to the 4-sum property, our attack on 47 steps is faster than the current best generic attack [28].

Table 5. Differential Paths for 2-Dimension Sums on 3-Round RIPEMD-128

$*$ means that the $+/-$sign of difference is not specified. Δb_j^L is 0 for all index j which are not listed below. Similarly Δb_j^R is 0 for all index j which are not listed below.

Path for $\mathrm{CF^L}$		Path for $\mathrm{CF^R}$	
$\Delta a_0^L = 2^8$; $\Delta b_0^L = 0$; $\Delta c_0^L = *2^{31} + 2^{16}$; $\Delta d_0^L = *2^{31} + 2^{16}$;		$\Delta a_0^R = 2^{20}$; $\Delta b_0^R = 0$; $\Delta c_0^R = 0$; $\Delta d_0^R = 2^6$;	
Step j	Δb_j^L	Step j	Δb_j^R
2	$*2^{31}$	1	2^{28}
3	$*2^{31}$	2	2^{15}
22	2^8	5	2^9
43	-2^{21}	6	2^{30}
44	$*2^5$	9	2^{16}
46	$*2^{26}$	13	2^{30}
47	$-2^{28} * 2^{12}$	39	-2^4
48	$*2^{10}$	43	-2^9
		47	-2^{16}

Attack on 48 steps. The complexity to satisfy the path is 2^{48}. We then introduce q-multi-second-order collisions. Generating them by following the path requires $q \cdot 2^{48}$. When $q = 17$, the complexity is less than 2^{53}. On the other hand, from Eq.(1), the generic case requires more than $\sqrt[17]{17!} \cdot 2^{128-\frac{2(512+128)}{17}} \approx 2^{55.55}$. Hence, our attack is faster than the generic case.

Attack on 52 steps. ∇-differential-path in the fourth round of the right branch becomes very complicated because the f function does not have the absorption property. Thus we only verified the amplified probability in the fourth round. As a result, the complexity to obtain a 4-sum is 2^{101}. We then consider the q-multi-second-order collision for $q = 53$. Our attack requires $53 \cdot 2^{101} < 2^{107}$, while the generic case in Eq.(1) requires $2^{108.21}$.

The attack was implemented on single PC. The generated 46-step 2-dimension sum, which is 3-round (48-step) partial 2-dimension sum, is shown in Table. 8.

Table 6. Sufficient Conditions of Attacks on 3-Round RIPEMD-128

Left Branch	Right Branch
$c_{0,16}^L = 0$; $d_{0,16}^L = 0$; $b_{1,16}^L = b_{0,16}^L$;	$b_{0,6}^R = c_{0,6}^R$; $b_{0,28}^R = 1$; $b_{0,15}^R = 0$; $c_{0,28}^R = 0$;
no carry in b_2^L; no carry in b_3^L;	$d_{0,6}^R = 0$; $b_{1,28}^R = 0$, $b_{1,15}^R = 1$; $b_{2,15}^R = 0$
$b_{22,8}^L = 0$; $c_{22,8}^L = d_{22,8}^L$; $b_{24,8}^L = 0$; $b_{25,8}^L = 1$	$b_{3,9}^R = 0$; $b_{4,15}^R = b_{3,15}^R$; $b_{4,30}^R = 0$; $b_{4,9}^R = 1$;
$b_{43,21}^L = 1$; $b_{43,5}^L = 0$; $d_{43,21}^L = 0$; $b_{44,21}^L = 1$;	$b_{5,9}^R = 0$; $b_{5,30}^R = 1$; $b_{6,30} = 0$; $b_{7,9}^R = b_{6,9}^R$;
$b_{45,26}^L = 0$; $b_{45,5}^L = 1$; $b_{46,28}^L = 0$, $b_{46,12}^L = 0$;	$b_{7,16}^R = 0$; $b_{8,30}^R = b_{7,30}^R$; $b_{8,16}^R = 1$; $b_{9,16}^R = 0$;
no carry in b_{46}^L; $b_{47,28}^L = 1$; $b_{47,26}^L = 1$;	$b_{11,16}^R = b_{10,16}^R$; $b_{12,30}^R = 0$; $b_{13,30}^R = 0$;
no carry in b_{47}^L; no carry in b_{48}^L;	$b_{15,30}^R = b_{14,30}^R$; $b_{39,4}^R = 1$; $b_{40,4}^R = 0$;
(H, M) and $(H + \nabla, M + \nabla^M)$:	$c_{39,4}^R = d_{39,4}^R$; $b_{41,4}^R = 1$; $b_{42,9}^R = b_{41,9}^R$; $b_{43,9}^R = 1$;
share same bit values on $b_{44,5}^L$, $b_{46,26}^L$, $b_{47,12}^L$;	$b_{44,9}^R = 0$; $b_{45,9}^R = 0$; $b_{46,16}^R = b_{45,16}^R$; $b_{47,16}^R = 1$;

5 Attacks on RIPEMD-160

Two attacks are presented on RIPEMD-160; in the first scenario, the attack target is starting from the first round and in the second scenario, the attack target is starting from the second round.

In the first scenario, the f functions of both branches do not have the absorption property in the first and third rounds. This makes the efficient differential path construction hard. On the other hand, in the second scenario, the absorption property is available in both branches in the first and third rounds. The differential path is more efficient than the first scenario, and the number of attacked steps is beyond 3 rounds (up to 52 steps).

5.1 Overall Strategy and Relatively Slow Differential Propagation

Different from RIPEMD-128, using the local-collision to construct the path is not efficient in RIPEMD-160. For RIPEMD-128, the local-collision is formed only with differences in 2 message words. However, in RIPEMD-160, we need the difference in 3 message-words due to the direct addition from chaining variable e_j. Hence, we stop using the local-collision. Instead, we insert the difference only into 1 message word that appears in a late step of the second round, and just propagate it to the third round as much as possible. The problem is that the differential propagation in RIPEMD-160 seems much quicker than RIPEMD-128 due to the direct addition from chaining variable e_j, and thus not so many steps can be attacked. However, we explain an useful property of RIPEMD-160 where we can limit the impact of the differential propagation. In fact, this is the main reason why we can attack more than 3 rounds in the second scenario.

Cancelling Differences between e_j and f_{j+1}. Assume that, in some step, there is no difference in chaining variables and a message difference is inserted. If the difference is not propagated through f in the following 3 steps, only chaining variable e has the difference. This situation is illustrated in Fig. 2. Let j be the step index of this chaining variable and $(\Delta a_j, \Delta b_j, \Delta c_j, \Delta d_j, \Delta e_j) = (0, 0, 0, 0, +2^n)$. In step j, e_j is directly added to compute b_{j+1}, thus the difference $+2^n$ is always propagated to b_{j+1}. As shown in Fig. 2, Δe_j and Δb_{j+1} can cancel each other in step $j+1$ through f_{j+1}.

Assume that the difference 2^n in b_{j+1} does not cause the carry, and thus only n-th bit has the difference. In step $j+1$, if the difference in the n-th bit of b_{j+1} is output from f_{j+1}, moreover if its sign is opposite (-2^n), the cancellation occurs.

In the attack starting from the first round, we utilize this property in the third round where $f_x(X, Y, Z)$ is $(X \vee \neg Y) \oplus Z$ in both branches. $Y = 1$ and $Z = 1$ are the conditions for this event. On the other hand, in the attack starting from the second round, $f_x(X, Y, Z)$ in the left branch is $(X \wedge Z) \vee (Y \wedge \neg Z)$. Then, the sign of Δf_{j+1} cannot be opposite of Δe_j, and we need a different strategy. In step j, we make a carry in b_{j+1} as shown in Fig. 3. Therefore, n-th bit position changes in the opposite direction as Δe_j. Finally, in step $j+1$, by propagating the difference in the n-th bit and by absorbing the difference in the $(n+1)$-th bit, the cancellation occurs.

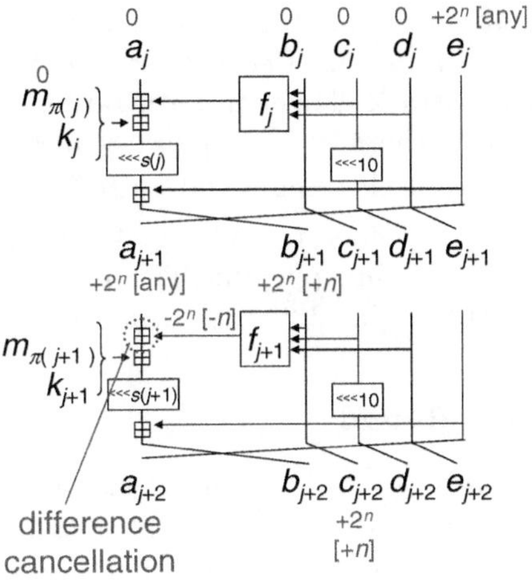

Fig. 2. Difference cancellation between e_j and f_{j+1}. The sign of Δf_{j+1} must be opposite of Δe_j. Information in '[ ]' is the bitwise difference. '[any]' represents that the bitwise difference is irrelevant.

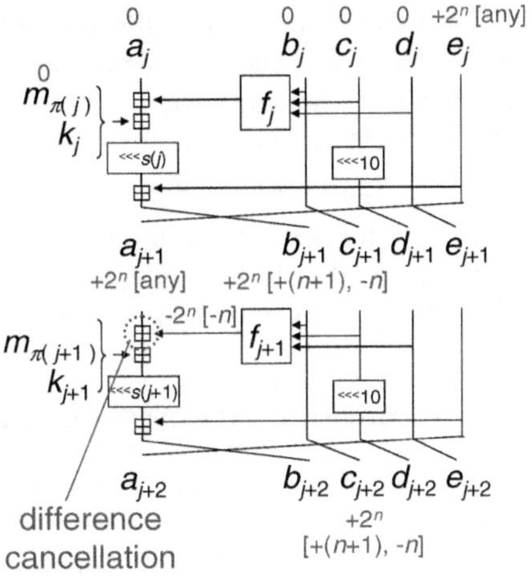

Fig. 3. Difference cancellation with considering the carry effect for the case that the sign of Δf_{j+1} is always the same as Δe_j

Table 7. Differential path construction for the first 3-rounds of RIPEMD-160

round	$\pi^L(j)$	$\pi^R(j)$
1	0 1 ② 3 4 5 6 7 8 9 10 11 12 13 14 15	5 14 7 0 9 ② 11 4 13 6 15 8 1 10 3 12
	← Δ| constant	MM ← ∇| constant
2	7 4 13 1 10 6 15 3 12 0 9 5 ② 14 11 8	6 11 3 7 0 13 5 10 14 15 8 12 4 9 1 ②
	constant | Δ →	constant | ∇
3	3 10 14 4 9 15 8 1 ② 7 0 6 13 11 5 12	15 5 1 3 7 14 6 9 11 8 12 ② 10 0 4 13
	→ Δ	→ ∇

5.2 Scenario 1: Attack from the First Round

The message differences for the attack from the first round is shown in Table 7. We need to insert both of the Δ-difference and ∇-difference in the 10th bit of m_2. To avoid the contradiction of two paths, the differences and the values of m_2 must be carefully chosen. We choose the following message differences;

$$\Delta m_2 = m_2^2 - m_2^1 = m_2^4 - m_2^3 = +2^{10}, \nabla m_2 = m_2^3 - m_2^1 = m_2^4 - m_2^2 = -2^{10}. \quad (2)$$

To achieve this, we first choose m_2^1 and then compute $m_2^2 \leftarrow m_2^1 + 2^{10}$ and $m_2^3 \leftarrow m_2^1 - 2^{10}$. m_2^4 should be $m_2^1 + 2^{10} - 2^{10}$ and thus identical with m_2^1. Hence, the attack only requires 3 messages m_2^1, m_2^2, m_2^3 rather than the standard message quartet. The differential paths and sufficient conditions are shown in Tables 10 and 11.

For the differential path in Table 10, the first round of $\mathrm{CF^L}$ and $\mathrm{CF^R}$ can be guaranteed with the message modification in negligible time. Hence, the attack cost only depends on the differential path from the second round. As a result, we can generate 4-sums up to 38 steps with 2^{42} computations. Because $2^{42} < 2^{160/3}$, the attack runs faster than the generic 4-sum attack. This can be regarded as partial 2-dimension sums up to 40 steps because the newly computed values in the last 2 steps are not used to compute two output chaining variables $H_{i-1}^{(b)}$ and $H_{i-1}^{(c)}$. The attack was implemented on a PC. The generated 38-step 4-sum (or 40-step partial 2-dimension sum) is shown in Table 9 in Appendix. Theoretically, 2-dimension sums can be generated up to 43 steps with 2^{151} computations.

5.3 Scenario 2: Attack from the Second Round

The overall strategy is the same as the first scenario. The details of the attack such as the message difference, differential path, and sufficient conditions are optimized for this scenario. Different from the first scenario, the f functions in the third round (round 4) have the absorption property. Hence, the differential propagation can be controlled more efficiently and this enables us to attack more rounds. Due to the limited space, we only show the message differences and how to propagate them in Table 12. The complexity to generate 4-sums up to 40 steps is 2^{36} computations, which is faster than the generic 4-sum attack using the generalized birthday attack. A 40-step 2-dimension sum, which can also be regarded as 42-step partial 2-dimension sum, was generated in our experiment. Due to the limited space, the generated data is omitted.

6 Concluding Remarks

We presented distinguishers against compression functions of RIPEMD-128 and -160. Differential paths were constructed by regarding CF^{L} as the first part and CF^{R} as the second part. This enabled us to analyze CF^{L} and CF^{R} independently. On RIPEMD-128, the local collision was applied to construct differential paths. Partial 2-dimension sums were generated for 48 steps. Theoretically, the attack works up to 52 steps. On RIPEMD-160, the difference cancelation between Δe_j and Δf_{j+1} was exploited. Partial 2-dimension sums were generated up to 40 steps. Theoretically, the attack works up to 43 steps. If the attack starts from the second round, more rounds can be attacked. We stress that our results do not impact to the security of full RIPEMD-128 and RIPEMD-160 hash functions.

Acknowledgements. Lei Wang was supported by Grant-in-Aid for JSPS Fellows (23001043).

References

1. Wang, X., Yu, H.: How to Break MD5 and Other Hash Functions. In: Cramer, R. (ed.) EUROCRYPT 2005. LNCS, vol. 3494, pp. 19–35. Springer, Heidelberg (2005)

2. Wang, X., Yin, Y.L., Yu, H.: Finding Collisions in the Full SHA-1. In: Shoup, V. (ed.) CRYPTO 2005. LNCS, vol. 3621, pp. 17–36. Springer, Heidelberg (2005)
3. U.S. Department of Commerce, National Institute of Standards and Technology: Federal Register /Vol. 72, No. 212/Friday, November 2, 2007/Notices (2007), http://csrc.nist.gov/groups/ST/hash/documents/FR_Notice_Nov07.pdf.
4. Dobbertin, H., Bosselaers, A., Preneel, B.: RIPEMD-160: A Strengthened Version of RIPEMD. In: Gollmann, D. (ed.) FSE 1996. LNCS, vol. 1039, pp. 71–82. Springer, Heidelberg (1996)
5. International Organization for Standardization: ISO/IEC 10118-3:2004, Information technology – Security techniques – Hash-functions – Part 3: Dedicated hash-functions (2004)
6. Cryptography Research and Evaluation Committees (CRYPTREC): e-Government recommended ciphers list (2003), http://www.cryptrec.go.jp/english/images/cryptrec_01en.pdf
7. Freier, A., Karlton, P., Kocher, P.: The Secure Sockets Layer (SSL) Protocol Version 3.0. Internet Engineering Task Force (IETF), RFC 6101 (2001), http://www.ietf.org/rfc/rfc6101.txt
8. Project, T.O. (crypto - OpenSSL cryptographic library), http://www.openssl.org/docs/crypto/ripemd.html
9. The Legion of the Bouncy Castle (Bouncy Castle Crypto APIs), http://www.bouncycastle.org/
10. Technische Universität Darmstadt (FlexiProvider), http://www.flexiprovider.de/
11. The GNU Crypto project: (GNU Crypto), http://www.gnu.org/software/gnu-crypto/
12. Crypto++: (Crypto++ Library 5.6.1 API Reference), http://www.cryptopp.com/
13. Kap, J.: Test Cases for HMAC-RIPEMD160 and HMAC-RIPEMD128. Internet Engineering Task Force (IETF), RFC 2286 (1998), http://www.ietf.org/rfc/rfc2286.txt
14. Keromyti, A., Provos, N.: The Use of HMAC-RIPEMD-160-96 within ESP and AH. Internet Engineering Task Force (IETF), RFC 2857 (2001), http://www.ietf.org/rfc/rfc2857.txt
15. Guo, J., Ling, S., Rechberger, C., Wang, H.: Advanced Meet-in-the-Middle Preimage Attacks: First Results on Full Tiger, and Improved Results on MD4 and SHA-2. In: Abe, M. (ed.) ASIACRYPT 2010. LNCS, vol. 6477, pp. 56–75. Springer, Heidelberg (2010)
16. Sasaki, Y., Aoki, K.: Finding Preimages in Full MD5 Faster Than Exhaustive Search. In: Joux, A. (ed.) EUROCRYPT 2009. LNCS, vol. 5479, pp. 134–152. Springer, Heidelberg (2009)
17. Sasaki, Y., Aoki, K.: Meet-in-the-Middle Preimage Attacks on Double-Branch Hash Functions: Application to RIPEMD and Others. In: Boyd, C., González Nieto, J. (eds.) ACISP 2009. LNCS, vol. 5594, pp. 214–231. Springer, Heidelberg (2009)
18. Chang, D., Hong, S., Kang, C., Kang, J., Kim, J., Lee, C., Lee, J., Lee, J., Lee, S., Lee, Y., Lim, J., Sung, J. (ARIRANG), Available at NIST home page: http://csrc.nist.gov/groups/ST/hash/sha-3/Round1/submissions_rnd1.html
19. Ohtahara, C., Okada, K., Sasaki, Y., Shimoyama, T.: Preimage Attacks on Full-ARIRANG: Analysis of DM-Mode with Middle Feed-Forward. In: Jung, S., Yung, M. (eds.) WISA 2011. LNCS, vol. 7115, pp. 40–54. Springer, Heidelberg (2012)

20. Mendel, F., Pramstaller, N., Rechberger, C., Rijmen, V.: On the Collision Resistance of RIPEMD-160. In: Katsikas, S.K., López, J., Backes, M., Gritzalis, S., Preneel, B. (eds.) ISC 2006. LNCS, vol. 4176, pp. 101–116. Springer, Heidelberg (2006)
21. Ohtahara, C., Sasaki, Y., Shimoyama, T.: Preimage Attacks on Step-Reduced RIPEMD-128 and RIPEMD-160. In: Lai, X., Yung, M., Lin, D. (eds.) Inscrypt 2010. LNCS, vol. 6584, pp. 169–186. Springer, Heidelberg (2011)
22. Wang, L., Sasaki, Y., Komatsubara, W., Ohta, K., Sakiyama, K. (Second) Preimage Attacks on Step-Reduced RIPEMD/RIPEMD-128 with a New Local-Collision Approach. In: Kiayias, A. (ed.) CT-RSA 2011. LNCS, vol. 6558, pp. 197–212. Springer, Heidelberg (2011)
23. Wagner, D.: The Boomerang Attack. In: Knudsen, L.R. (ed.) FSE 1999. LNCS, vol. 1636, pp. 156–170. Springer, Heidelberg (1999)
24. Aumasson, J.-P., Çalık, Ç., Meier, W., Özen, O., Phan, R.C.-W., Varıcı, K.: Improved Cryptanalysis of Skein. In: Matsui, M. (ed.) ASIACRYPT 2009. LNCS, vol. 5912, pp. 542–559. Springer, Heidelberg (2009)
25. Biryukov, A., Nikolić, I., Roy, A.: Boomerang Attacks on BLAKE-32. In: Joux, A. (ed.) FSE 2011. LNCS, vol. 6733, pp. 218–237. Springer, Heidelberg (2011)
26. Lamberger, M., Mendel, F.: Higher-order differential attack on reduced SHA-256. Cryptology ePrint Archive, Report 2011/037 (2011), `http://eprint.iacr.org/2011/037`
27. Sasaki, Y.: Boomerang Distinguishers on MD4-Family: First Practical Results on Full 5-Pass HAVAL. In: Miri, A., Vaudenay, S. (eds.) SAC 2011. LNCS, vol. 7118, pp. 1–18. Springer, Heidelberg (2012)
28. Wagner, D.: A Generalized Birthday Problem. In: Yung, M. (ed.) CRYPTO 2002. LNCS, vol. 2442, pp. 288–303. Springer, Heidelberg (2002)
29. Biryukov, A., Lamberger, M., Mendel, F., Nikolić, I.: Second-Order Differential Collisions for Reduced SHA-256. In: Lee, D.H. (ed.) ASIACRYPT 2011. LNCS, vol. 7073, pp. 270–287. Springer, Heidelberg (2011)
30. Mendel, F., Rechberger, C., Schläffer, M., Thomsen, S.S.: The Rebound Attack: Cryptanalysis of Reduced Whirlpool and Grøstl. In: Dunkelman, O. (ed.) FSE 2009. LNCS, vol. 5665, pp. 260–276. Springer, Heidelberg (2009)
31. Biryukov, A., Khovratovich, D., Nikolić, I.: Distinguisher and Related-Key Attack on the Full AES-256. In: Halevi, S. (ed.) CRYPTO 2009. LNCS, vol. 5677, pp. 231–249. Springer, Heidelberg (2009)
32. RIPE Integrity Primitives Berlin, Heidelberg, New York: Integrity Primitives for Secure Information Systems, Final RIPE Report of RACE Integrity Primitives Evaluation, RIPE-RACE 1040 (1995)

A Data for Experiments and Attacks on RIPEMD-160

Table 8. 4-sum on 46-steps and partial 2-dimension sum on 48-steps of RIPEMD-128

H_i^1	0x400268ec; 0x159b2e00 0x 6a66026; 0x268c3594;
M_i^1	0x7b69e00f; 0x7a1da89e; 0xb6760e61; 0x222860d3; 0x20d0343c; 0x36333ccb; 0x44a67388; 0x9034d3f9; 0x19810c83; 0x9cba7f1e; 0x43aa3459; 0xe8987de1; 0xf48cfeb0; 0x36b56404; 0xca71b589; 0x4e4956b3;
46-Step H_{i+1}^1	0x8e492929; 0x34c37860; 0x085981da; 0x3a28780d;
3-Round H_{i+1}^1	0xf57d1452; 0x00cc6f47; 0x9c7fe2e0; 0x5cdae22d;
H_i^2	0x400269ec; 0x159b2e00; 0x86a76026; 0xa68d3594;
M_i^2	0x7b69df0f; 0x7a1da89e; 0xb6760e61; 0x222860d3; 0x20d0343c; 0x36333ccb; 0xc4a67388; 0x9034d3f9; 0x19810c83; 0x9cba7f1e; 0x43aa3459; 0xe8987de1; 0xf48cfeb0; 0x36b56404; 0xca71b589; 0x4e4956b3;
46-Step H_{i+1}^2	0x4b37a7fb; 0xe0a9ebf0; 0x09e98a18; 0x17b730cd;
3-Round H_{i+1}^2	0x672c3692; 0x5e5c2707; 0xe9e5bbda; 0xc6e4b82a;
H_i^3	0x401268ec; 0x159b2e00; 0x06a66026; 0x268c35d4;
M_i^3	0x7b69e00f; 0x7a1da89e; 0xb6760e61; 0x222860d3; 0x20d0343c; 0x36333ccb; 0x04a67388; 0x9034d3f9; 0x19810c83; 0x9cba7f1e; 0x43aa3459; 0xe8987de1; 0xf48cfeb0; 0x36b56404; 0xca71b589; 0x4e4956b3;
46-Step H_{i+1}^3	0xd2ae4d5e; 0x5b495822; 0x13ac118a; 0x9c22aa6a;
3-Round H_{i+1}^3	0x5955db12; 0x62b6a1a4; 0xe0a50755; 0xa0eb49c4;
H_i^4	0x401269ec; 0x159b2e00; 0x86a76026; 0xa68d35d4;
M_i^4	0x7b69df0f; 0x7a1da89e; 0xb6760e61; 0x222860d3; 0x20d0343c; 0x36333ccb; 0x84a67388; 0x9034d3f9; 0x19810c83; 0x9cba7f1e; 0x43aa3459; 0xe8987de1; 0xf48cfeb0; 0x36b56404; 0xca71b589; 0x4e4956b3;
46-Step H_{i+1}^3	0x8f9ccc30; 0x072fcbb2; 0x153c19c8; 0x79b1632a;
3-Round H_{i+1}^3	0xcb04f456; 0xc0465964; 0x2e0ae04f; 0x0afb77c2;
46-Step 4-sum	0x00000000; 0x00000000; 0x00000000; 0x00000000;
3-Round 4-Sum	0xfffff704; 0x00000000; 0x00000000; 0x00065801;

Table 9. 38-step 4-sum and 40-step partial 2D sum on RIPEMD-160 from 1st round

H_i^1	0x4144c3a7; 0x8a965cea; 0x647e4d03; 0x04e7a03c; 0x18814c3e;
M_i^1	0x15f04e2b; 0xb2c328cd; 0x8eea7e12; 0xa1d55a25; 0xbff66c59; 0x399570f1; 0x997d1fd4; 0xea8f403e; 0xab607923; 0x712bc2a1; 0x32c11766; 0xafaf4bfe; 0x7297f9d4; 0xe2c4573f; 0x96dc27dc; 0x566d9d73;
38 Steps H_{i+1}^1	0x33d3fb92; 0x753e7a17; 0x50fb34b1; 0x1874be98; 0x48de951c;
H_i^2	0x4146bfa7; 0x8a965cea; 0x647e4d03; 0x04e7a43c; 0x08814c3e;
M_i^2	0x15f04e2b; 0xb2c328cd; 0x8eea7a12; 0xa1d55a25; 0xbff66c59; 0x399570f1; 0x997d1fd4; 0xea8f403e; 0xab607923; 0x712bc2a1; 0x32c11766; 0xafaf4bfe; 0x7297f9d4; 0xe2c4573f; 0x96dc27dc; 0x566d9d73;
38 Steps H_{i+1}^2	0x07e43a48; 0x1482c47c; 0x473df79e; 0xefed372c; 0x55037e85;
H_i^3	0x404cc5a7; 0x8a9e5cea; 0x647e4c03; 0x04e7a23c; 0x18814c3e;
M_i^3	0x15f04e2b; 0xb2c328cd; 0x8eea8212; 0xa1d55a25; 0xbff66c59; 0x399570f1; 0x997d1fd4; 0xea8f403e; 0xab607923; 0x712bc2a1; 0x32c11766; 0xafaf4bfe; 0x7297f9d4; 0xe2c4573f; 0x96dc27dc; 0x566d9d73;
38 Steps H_{i+1}^3	0xad046562; 0x61c2166a; 0x9d7dc2b3; 0x901e2f34; 0x12d8aa5c;
H_i^4	0x404ec1a7; 0x8a9e5cea; 0x647e4c03; 0x04e7a63c; 0x08814c3e;
M_i^4	0x15f04e2b; 0xb2c328cd; 0x8eea7e12; 0xa1d55a25; 0xbff66c59; 0x399570f1; 0x997d1fd4; 0xea8f403e; 0xab607923; 0x712bc2a1; 0x32c11766; 0xafaf4bfe; 0x7297f9d4; 0xe2c4573f; 0x96dc27dc; 0x566d9d73;
38 Steps H_{i+1}^3	0x8114a418; 0x010660cf; 0x93c085a0; 0x6796a7c8; 0x1efd93c5;
38 Steps 4-sum	0x00000000; 0x00000000; 0x00000000; 0x00000000; 0x00000000;
40 Steps 4-Sum	noisy data noisy data 0x00000000; 0x00000000; noisy data

Table 10. Differential paths for 2D sums on RIPEMD-160 in the first scenario

Path for CF^{L}		Path for CF^{R}	
$\Delta a_0^L = 2^{17} - 2^{10}$; $\Delta b_0^L = 0$; $\Delta c_0^L = 0$; $\Delta d_0^L = 2^{10}$; $\Delta e_0^L = -2^{28}$;		$\Delta a_0^R = -2^{24} + 2^{19} + 2^9$; $\Delta b_0^R = 2^{19}$; $\Delta c_0^R = -2^8$; $\Delta d_0^R = 2^9$; $\Delta e_0^R = 0$;	
Step j	Δb_j^L	Step j	Δb_j^R
29	-2^{21}	1	-2^0
33	-2^{31}	32	2^{21}
36	-2^{16}	35	-2^{14}
39	2^7	36	$+2^{31}$
40	$-2^{26} - 2^2$	38	2^{30}
41	$-2^{26} + 2^{19}$	39	$-2^{24} - 2^{15}$
42	-2^{25}	40	2^9
43	2^{25}	41	-2^{20}
		42	$2^{15} + 2^8 + 2^6$
		43	$-2^{25} - 2^{24} - 2^2$

Table 11. Sufficient conditions on RIPEMD-160 in the first scenario

Left Branch	Right Branch
$b_{0,10}^L = c_{0,10}^L$; no carry in d_0^L and e_0^L;	$b_{0,9}^R = 0$; $c_{0,19}^R = 1$; $d_{0,8}^R = 0$;
no carry in b_{29}^L; $b_{28,21}^L = b_{27,21}^L$; $b_{30,21}^L = 0$;	no carry in b_0^R, c_0^R and d_0^R;
$b_{31,31}^L = 1$; no carry in b_{33}^L; $b_{32,31}^L = 1$;	$b_{0,0}^R = 0$; $b_{0,18}^R = 1$; $c_{0,9}^R = 1$; $c_{0,22}^R = 1$;
$b_{34,31}^L = 1$; $b_{35,31}^L \vee \neg b_{34,31}^L = 1$; $b_{35,16}^L = 0$;	$b_{0,22}^R = 0$; $b_{1,29}^R = 1$; $b_{2,10}^R = 1$;
no carry in b_{36}^L and b_{37}^L;	no carry in b_{32}^R; $b_{31,21}^R = 0$; $b_{33,21}^R = 1$;
$b_{37,16}^L = 1$; $b_{36,9}^L = 1$; $b_{35,9}^L = 1$;	$b_{34,31}^R \vee \neg b_{33,31}^R = 1$; no carry in b_{35}^R;
$b_{38,9}^L = 1$; $b_{38,26}^L \vee \neg b_{37,26}^L = 1$; $b_{38,7}^L = 0$;	$b_{34,14}^R = 0$; no carry in b_{36}^R; $b_{34,14}^R = 0$;
no carry in b_{39}^L; $b_{39,19}^L \vee \neg b_{38,19}^L = 1$; $b_{40,7}^L = 1$;	no carry in b_{36}^R; $b_{36,14}^R = 1$; $b_{35,31}^R = 0$;
no carry in b_{40}^L; $b_{39,26}^L = 1$; $b_{39,2}^L = 0$;	$b_{37,31}^R = 1$; $b_{37,24}^R \vee \neg b_{36,24}^R = 1$; no carry in b_{38}^R;
$b_{38,26}^L = 1$; no carry in b_{41}^L; $b_{41,26}^L = 1$;	$b_{37,30}^R = 0, b_{38,9}^R \vee \neg b_{37,9}^R = 1$; no carry in b_{39}^R; $b_{39,30}^R = 1$;
$b_{41,2}^L = 1$; $b_{40,26}^L = 0$; $b_{40,24}^L = 1$;	$b_{38,24}^R = 1$; $b_{38,15}^R = 0$; $b_{37,24}^R = 1$; $b_{40,24}^R = 1$
$b_{39,24}^L = 1$; $b_{41,17}^L \vee \neg b_{40,17}^L = 1$; no carry in b_{42}^L;	no carry in b_{40}^R; $b_{40,15}^R = 1$; $b_{39,9}^R = 1$; $b_{38,9}^R = 1$;
$b_{42,24}^L = 1$; $b_{42,19}^L = 1$; $b_{42,4}^L = 0$;	$b_{40,8}^R \vee \neg b_{39,8}^R = 1$; no carry in b_{41}^R; $b_{41,9}^R = 1$;
$b_{41,25}^L = 0$; $b_{41,4}^L = 1$; $b_{42,12}^L \vee \neg b_{41,12}^L = 1$;	$b_{40,20}^R = 0$; $b_{41,25}^R \vee \neg b_{40,25}^R = 1$; $b_{41,2}^R \vee \neg b_{40,2}^R = 1$;
	no carry in b_{42}^R; $b_{42,20}^R = 1$; $b_{41,15}^R = 0$; $b_{41,8}^R = 1$;
	$b_{42,19}^R \vee \neg b_{41,19}^R = 1$;

Table 12. Differential path construction for the intermediate 3-rounds of RIPEMD-160

round	$\pi^L(j)$	$\pi^R(j)$
2	7 4 13 1 10 6 15 3 ⑫ 0 9 5 2 14 11 8	6 11 3 7 0 ⑬ 5 10 14 15 8 12 4 9 1 2
	MM ← Δ| constant	MM ← ∇| constant
3	3 10 14 4 9 15 8 1 2 7 0 6 13 11 5 ⑫	15 5 1 3 7 14 6 9 11 8 12 2 10 0 4 ⑬
	constant |Δ	constant |∇
4	1 9 11 10 0 8 ⑫ 4 13 3 7 15 14 5 6 2	8 6 4 1 3 11 15 0 5 12 2 ⑬ 9 7 10 14
	→ Δ	→ ∇

Zero-Value Point Attacks on Kummer-Based Cryptosystem*

Fangguo Zhang[1], Qiping Lin[1], and Shengli Liu[2]

[1] School of Information Science and Technology
Sun Yat-sen University, Guangzhou 510006, China
[2] Dept. of Computer Science and Engineering,
Shanghai Jiao Tong University, Shanghai 200240, China
isszhfg@mail.sysu.edu.cn

Abstract. The Zero-Value Point (ZVP) attack, one of side channel attacks, is very powerful to recover the secret information of elliptic curve cyrptosystem (ECC) on memory constraint devices by monitoring their power consumptions. In the ZVP attack, the zero-value registers are used in point addition and doubling formula of ECC to resist randomizations. Hence, the countermeasures against the differential power analysis (DPA), like Coron's and Joye-Tymen's randomization, do not work for the ZVP attack. The Kummer surface is a variety associated to the Jacobian of a genus 2 curve with a map. The pseudo-group structure on the Kummer surface defines a scalar multiplication, which is more efficient than that in HECC and comparable to ECC, especially in constraint environments. We inspect the pseudo-addition and doubling formula of the Kummer surface and show how to find zero-value registers. Our analysis shows that the scalar multiplication on the Kummer surface suffers from the ZVP attack, hence all Kummer-based cryptosystems are inevitable to the ZVP attack.

Keywords: Zero-value point attack, Kummer-based cryptosystem, Differential power analysis, Scalar multiplication.

1 Introduction

Elliptic Curve cryptosystem involves only a short key, and is more preferable to other public key variants for cryptographic applications in memory-constraint device. The concept of Hyperelliptic Curve Cryptosystem (HECC) was proposed by Koblitz [18] as a generalization of ECC. In a HECC, the jacobian of a hyperelliptic curve defined over a finite field is used to fulfil the discrete-logarithm-based cryptographic algorithms and protocols. Hyperelliptic curves have richer algebraic structures and are based on a smaller field than elliptic curves to achieve the same level of security. In cryptographic applications, scalar multiplications

* This work is supported by the the National Natural Science Foundation of China (No. 61070168, 61170229 and U1135001) and Scientific innovation projects of Shanghai Education Committee (Grant 12ZZ021).

F. Bao, P. Samarati, and J. Zhou (Eds.): ACNS 2012, LNCS 7341, pp. 293–310, 2012.

are essential and their computations decide the efficiency of the ECC or HECC cryptosystem.

Cantor's pioneering work [4] gave an algorithm to do scalar multiplications for hyperelliptic curves. In the early 1990s, Flynn [11] gave an explicit description of the Jacobian of a genus 2 hyperelliptic curve and suggested a more efficient arithmetic on scalar multiplications.

As for any hyperelliptic curves C of genus 2, there exists a map $J(C) \to K$ sending two opposite points of the Jacobian (except for the 2-torsion points) to a point of the Kummer surface K. The map is not a group-homomorphism, hence the Kummer surface is not structured as a group. However, the map does transform the group structure of the Jocobian to a pseudo-group structure on the Kummer surface. Fast formula for the arithmetic on the Kummer surface was developed with the theory of Theta functions and optimized in [12,14]. Using Montgomery ladder, it is enough for the pseudo-group structure to define scalar multiplications on the Kummer surface. This makes possible the Kummer-based cryptosystems. For example, Smart et al. [28] proposed a Diffie-Hellman protocol implemented on Kummer surface in 1999. It was shown in [12] that the discrete logarithm problem on Kummer sufaces is polynomial time equivalent to the discrete logarithm in the Jocobian of the hyperelliptic curve of genus 2. Fast arithmetic on the Kummer surface was also developed, which is much more efficient than that in ECC or HECC, especially in constraint environment. Therefore, many works have been devoted to fast arithmetic on the Kummer surface [7,8,12,14].

1.1 Side Channel Attacks and Zero-Value Point (ZVP) Attack on Scalar Multiplication

Due to the short key size and fast arithmetics, ECC, HECC and Kummer-Based Cryptosystems are especially adaptable to low-cost and memory-limited cryptographic devices like smartcards. However, cryptographic hardware operations can be an easy target to power analysis by Side Channel Attacks [19,20]. The Side Channel Attacks analyze the instantaneous power consumption of a device to derive the secret information stored in it. Over the years, power analysis evolves from the timing attack, the Simple Power Analysis (SPA), to the Differential Power Analysis (DPA) and Zero-Value Point attack (ZVP), etc.

ECC, HECC and Kummer-Based Cryptosystems generally base their security on the Discrete-Logarithm related assumptions on some additive (pseudo) group, and scalar multiplications are the main computation involved.

In [5], Coron showed how SPA and DPA work on scalar multiplications in ECC. Let d be the secret, and P be a point on ECC, the scalar multiplication dP needs to compute the addition of two different points and the doubling of a point. In the "double and addition" algorithm for dP, each bit of d determines whether both of doubling and addition or only doubling involved, and this causes different power consumption. Simple Power Analysis (SPA) just uses this difference to determine every bit of d.

There are many countermeasures to avoid SPA, like Coron's "double-and-add-always" method [5] and "Montgomery" method [24], "Hesse" type [27,17] or "Jacobi" type [21] of computing doubling and addition in a unified formula. But all the anti-SPA methods do not prevent the Differential Power Analysis (DPA), a much more powerful analysis.

To avoid the DPA, the proposals are to use randomized projective homogeneous coordinates [5] or Jacobian coordinates [21], to work in a random isomorphism of the elliptic curve, or to work in random field isomorphism [16].

In 2003, Goubin observed that randomization does not work very well for some special points[13]. Goubin [13] analyzed two kinds of special points, namely $(x, 0)$ and $(0, y)$, on elliptic curves. Points $(x, 0)$ and $(0, y)$ are expressed as $(X : 0 : Z)$ and $(0 : Y : Z)$ in Jacobian coordinates. Randomization of the Jocobian coordinates results in points $(r^2X : 0 : rZ)$ and $(0 : r^3Y : rZ)$ for some random integer $r \neq 0$. One of the coordinate remains to be zero after the randomization. Power analysis then takes advantage of those special points to derive the secret scalar d. However, Smart showed that Goubin's power-analysis attack for elliptic curves can be easily avoided [26].

Akishita et al. [3] extended Goubin's attack with the so-called "Zero-Value Point" (ZVP) attacks. The ZVP attack is not limited to zero-coordinate points like $(x, 0)$ or $(0, y)$. It collects those points Q such that the computation of the scalar multiplication dQ leads to 0 in the intermediate computation of the doubling or addition formulas. Lots of elliptic curves, including the SECG random curves over prime fields, suffer from the ZVP attack. It seems that the ZVP attack is one of the most powerful attack up to now. In [3], Akishita et al. showed the conditions that the zero-value points exist in elliptic curve, and the ZVP attack suggests a new security criteria for secure implementation of ECC. [10] gave a survey on known side-channel attacks and countermeasures for ECC.

As for HECC, the power analysis works in the same principle, and the only difference is that scalar multiplication dP works in divisor class groups of hyperelliptic curves, instead of the additive group of points on ECC. Avanzi [1] generalized Goubin's attacks to divisor class groups of hyperelliptic curves, and provided a generalization of the countermeasures.

As far as we know, there is no research work on power analysis of Kummer-based cryptosystems up to now. We will fulfill the analysis of Kummer-based cryptosystems in this paper. Although the Kummer Surface is constructed from hyperelliptic curves with $g = 2$, the ZVP attack on Kummer-based cryptosystems is much different from the ZVP attack on HECC. As we will show in this paper, there are more special points which can be taken advantage of by ZVP attacks on a Kummer surface, compared to its corresponding hyperelliptic curve, and it is also much more difficult to resist ZVP attacks on a Kummer surface.

1.2 Our Contributions

In this paper, we study how to find special points on a Kummer surface, i.e., those points result in zero-value register during the computation of scalar multiplications. We also provide how to use those zero-value points to implement

ZVP attacks on Kummer-based cryptosystem. Finally we show that there are no efficient ways to avoid the ZVP attacks on Kummer surfaces, and propose countermeasures against the ZVP attacks.

Organization. The rest of this paper is organized as follows. In section 2, we review scalar multiplication on a Kummer surface. In section 3, we focus on analyzing what are special points of a Kummer surface, and estimate the number of special points. In section 4, we describe how to carry out the zero-value point attacks on Kummer surface. In section 5, we give the countermeasures against this attacks. Section 6 concludes the paper.

2 Scalar Multiplication on the Kummer Surface

In this section, we will recall the pseudo-addition and doubling algorithm and the Montgomery scalar multiplication on the Kummer surface. The Kummer surface is defined as follows. Let

$$C : y^2 + h(x)y = f(x) \tag{1}$$

be a curve of genus 2 defined over a field F, where $\deg(f) \leq 6$ and $\deg(h) \leq 3$. Let J be the Jacobian variety of C. Then a hypersurface K in $\mathbb{P}^3$ can be associated to the Jacobian variety J. The associated hypersurface K is called the Kummer surface. If the characteristic of F is not 2, i.e., $\mathrm{char}(F) \neq 2$, the curve C of genus 2 has a reduced form

$$C : y^2 = f(x). \tag{2}$$

Cassels and Flynn [6] constructed the Kummer surface associated to the Jacobian variety J of curve $C : y^2 = f(x)$. If $\mathrm{char}(F) = 2$, the general form Eq.(1) can be reduced to the case $\deg(h) = 2$. Duquesne [9] considered how to construct the Kummer surface for this case. Gaudry [12] and Lubicz [14] proposed formulas for the arithmetic of Kummer surfaces based on the theory of algebraic theta functions. Müller [25] developed the general expressions of the Kummer surface for the more general form Eq.(1), which applies to any $\mathrm{char}(F)$.

Given a genus 2 curve C, the Kummer surface is obtained by a map $J(C) \to K$, which maps two opposite points of the Jacobian of a genus 2 curve into one point in K. The quotient of J by the negation map results in the variety K, i.e., the Kummer surface. However, the map is not a group homomorphism, and the Kummer surface does not have a group structure. But the map endows a pseudo-group structure on the Kummer surface, over which a scalar multiplication can be defined with the help of a so-called Montgomery ladder. The scalar multiplication on the Kummer surface associated to a genus 2 curve can be used to design genus 2 cryptosystems. Due to the map $J(C) \to K$, the discrete logarithm problem on the Kummer surfaces can be proved to be polynomial time equivalent to the discrete logarithm problem in the corresponding Jacobian [28].

If $\mathrm{char}(F) \neq 2$, there exists a fast formulae for the scalar multiplication on the Kummer surface associated to a genus 2 curve C using a Montgomery ladder. This make Kummer-based cryptosystems more efficient than hyperelliptic curve cryptosystems, especially in some hardware configurations, like smart cards. Let a genus 2 curve C be given by

$$C : y^2 = x^5 + f_4x^4 + f_3x^3 + f_2x^2 + f_1x + f_0.$$

Suppose $P_1 = (x, y)$ and $P_2 = (u, v)$ are affine points on curve C. According to [6], a projective embedding of the Kummer surface is given by

$$k_1 = 1, k_2 = x + u, k_3 = xu, k_4 = \frac{F_0(x, u) - 2yv}{(x - u)^2},$$

where

$$F_0(x, u) = (x + u)xu + 2f_4(xu)^2 + f_3(x + u)xu + 2f_2(xu) + f_1(x + u) + 2f_0.$$

The functions k_1, k_2, k_3, k_4 satisfy the quartic equation

$$K(k_1, k_2, k_3, k_4) = K_2(k_1, k_2, k_3)k_4^2 + K_1(k_1, k_2, k_3)k_4 + K_0(k_1, k_2, k_3) = 0,$$

where

$$\begin{aligned}
K_2(k_1, k_2, k_3) &= k_2^2 - 4k_1k_3,\\
K_1(k_1, k_2, k_3) &= -2k_2k_3^2 - 4k_1k_3^2f_4 - 2k_1k_2k_3f_3 - 4k_1^2k_3f_2 - 2k_1^2k_2f_1 - 4k_1^3f_0,\\
K_0(k_1, k_2, k_3) &= k_3^4 - 2k_1k_3^3f_3 - 4k_1k_2k_3^2f_2 - 4k_1k_2^2k_3f_1 - 4k_1k_2^3f_0\\
&\quad +k_1^2k_3^2f_3^2 - 4k_1^2k_3^2f_2f_4 + 2k_1^2k_3^2f_1 - 4k_1^2k_2k_3f_1f_4 + 4k_1^2k_2k_3f_0\\
&\quad -4k_1^2k_2^2f_0f_4 - 2k_1^3k_3f_1f_3 - 4k_1^3k_2f_0f_3 + k_1^4f_1^2 - 4k_1^4f_0f_2.
\end{aligned}$$

The points of the Kummer surface are of form $(k_1 : k_2 : k_3 : k_4)$. Given two points A and B on the Kummer surface, we can compute $A + B$ if and only if we know $A - B$ firstly. Smart et al. [28] was the first one who proposed to compute scalar multiplication using Montgomery ladder on the Kummer surface. Montgomery ladder can be used to compute the scalar multiplication, because $A - B$ is unchanged at every step of the Montgomery ladder. Duquesne proposed the pseudo-addition and doubling algorithms for $\mathrm{char}(F) \neq 2$ on the Kummer surface in [7] and [2, chapter 14]. Recently, Lin et al. [22] pointed out some minor errors in the algorithms and revised them. Below we recall the revised algorithms in [22].

Let F_q be a field of characteristic $p \neq 2, 3$ and let C/F_q be a curve of genus 2. Let K_C denote the Kummer surface of C. Assume that the difference $A - B$ is known and the third coordinate of $A - B$ on the Kummer surface is 1 (remember we are in $\mathbb{P}^3(F_q)$) where $A = (k_1 : k_2 : k_3 : k_4)$ and $B = (l_1 : l_2 : l_3 : l_4)$ are two points on the Kummer surface K_C. Then the Kummer surface coordinates for $A + B$ are as follows:

$$\begin{aligned}
k_1(A + B) &= \varphi_{11}(A, B),\\
k_2(A + B) &= 2(\varphi_{12}(A, B) - k_1(A + B)k_2(A - B)),\\
k_3(A + B) &= k_1(A - B)\varphi_{33}(A, B),\\
k_4(A + B) &= 2(\varphi_{14}(A, B) - k_1(A + B)k_4(A - B)),
\end{aligned}$$

where

$$\varphi_{11}(A,B) = ((k_4l_1 - k_1l_4) + (k_2l_3 - k_3l_2))^2;$$

$$\begin{aligned}\varphi_{12}(A,B) = &((k_2l_3 - k_3l_2) + (k_1l_4 - k_4l_1))(f_3(k_1l_3 - k_3l_1) + (k_2l_4 - k_4l_2))\\ &+2(k_1l_3 - k_3l_1)(f_2(k_1l_3 - k_3l_1) + (k_1l_2 - k_2l_1) - (k_3l_4 - k_4l_3))\\ &+2f_4(k_1l_4 - k_4l_1)(k_2l_3 - k_3l_2),\end{aligned}$$

$$\varphi_{33}(A,B) = ((k_3l_4 - k_4l_3) + (k_1l_2 - k_2l_1))^2,$$

$$\begin{aligned}\varphi_{14}(A,B) = &((k_1 + k_3)(l_1 - l_3) + k_1l_3 - k_3l_1)(f_3((k_1l_4 + k_4l_1) - (k_2l_3 + k_3l_2))\\ &+2((k_1l_2 + k_2l_1) - (k_3l_4 + k_4l_3)) + 2f_4(k_1l_1 - k_3l_3))\\ &+2f_2(k_1l_4 - k_2l_3)(k_4l_1 - k_3l_2)\\ &+((k_2 + k_4)(l_2 - l_4) + k_2l_4 - k_4l_2)((k_2l_3 + k_3l_2) - (k_1l_4 + k_4l_1)).\end{aligned}$$

Let $A = (k_1 : k_2 : k_3 : k_4)$, then the Kummer coordinates for $2A = (\delta_1 : \delta_2 : \delta_3 : \delta_4)$ are given by

$$\delta_1 = 2\varphi_{14}(A,A), \delta_2 = 2\varphi_{24}(A,A), \delta_3 = 2\varphi_{34}(A,A), \delta_4 = \varphi_{44}(A,A),$$

where

$$\begin{aligned}\varphi_{14}(A,A) = &2(k_1^2 - k_3^2)(f_3(k_1k_4 - k_2k_3) + 2(k_1k_2 - k_3k_4)\\ &+f_4(k_1^2 - k_3^2)) + 2(k_1k_4 - k_2k_3)(k_4^2 - k_2^2 + f_2(k_1k_4 - k_2k_3)),\end{aligned}$$

$$\begin{aligned}\varphi_{24}(A,A) = &(k_1^2 + k_3^2)(8k_1k_3 - f_3(k_1^2 + k_3^2)) + 2(k_2^2 + k_4^2)(k_3^2 + k_1^2 + k_2k_4 + f_3k_1k_3)\\ &+k_1k_3((f_3^3 - 8f_3 + 8f_2^2 + 8f_4^2 - 4f_2f_3f_4)k_1k_3\\ &+(7 - f_3^2 + 8f_2f_4)k_2k_4 + 8f_2(k_3k_4 + k_1k_2) + 8f_4(k_2k_3 + k_1k_4))\\ &+k_2k_4(4f_2(k_1k_4 + k_2k_3) + 4f_4(k_1k_2 + k_3k_4) + f_3(k_2k_4 + 4(k_1^2 + k_3^2))),\end{aligned}$$

$$\begin{aligned}\varphi_{34}(A,A) = &2(k_1^2 - k_3^2)(f_3(k_1k_2 - k_3k_4) + f_2(k_1^2 - k_3^2) + 2(k_1k_4 - k_2k_3))\\ &+2(k_1k_2 - k_3k_4)(k_2^2 - k_4^2 + f_4(k_1k_2 - k_3k_4)) + k_1k_2(k_1^2 - k_2^2),\end{aligned}$$

$$\begin{aligned}\varphi_{44}(A,A) = &(k_1^2 + k_3^2)((f_3^2 - 4f_2f_4 - 2)(k_1^2 + k_3^2) - 8f_2(k_3k_4 + k_1k_2)\\ &-8f_4(k_1k_4 + k_2k_3) - 4f_3k_1k_3 - 12k_2k_4 + (k_2^2 + k_4^2)(k_2^2 + k_4^2 - 2f_3(k_1^2 + k_3^2))\\ &+k_1k_3(8f_2(k_2k_3 + k_1k_4) + 8f_4(k_1k_2 + k_3k_4) + (16 + 8f_2f_4 - 2f_3^2)k_1k_3)\\ &+2K_2k_4(2f_3k_1k_3 - k_2k_4).\end{aligned}$$

In Algorithm 1, we show how to use Montgomery ladder to compute scalar multiplication on the Kummer surface. At each step, the algorithm performs one addition and one doubling, which makes this method resistant to Simple Power Attacks.

Algorithm 1. Montgomery scalar multiplication algorithm for Kummer surface

---

Input: A point D on the Kummer surface and $d = (d_{n-1}, \cdots, d_1, d_0)_2$.
Output: dD.

1. $(A, B) \leftarrow ((0, 0, 0, 1), D)$;
2. for $i = n - 1$ down to 0 do
3. If $d_i = 0$, $(A, B) \leftarrow (2A, A + B)$;
4. If $d_i = 1$, $(A, B) \leftarrow (A + B, 2B)$;
5. end for
6. return A.

---

3 Special Points on Kummer Surface

In this section, we will find some special points of Kummer surface with respect to different pseudo-addition and doubling algorithms. We will also estimate the number of those special points. Special points will serve the zero-value attack in the next section.

In [3], the proposed ZVP attack utilizes the auxiliary register to take zero-values and reduces the computation overhead in ECC. Those points, which results in zeros in the auxiliary registers, are called Zero-Value points of ECC. Similarly, we can also define those special points on Kummer surface, which results in a reduction of computation overhead, as zero-value points.

As pointed earlier, scalar multiplications on Kummer surface involve the Montgomery adder, and different implementations of pseudo-addition and doubling cooperating with the Montgomery adder result in different Montgomery scalar multiplication algorithms. We will first analyze Duquesne's pseudo-addition and doubling formula [7,22] for the characteristic $p > 3$, and show how to find special points and evaluate the number of special points. Then we will apply our analysis on Gaudry's algorithm of pseudo-addition and doubling formula [12] which use Theta function, and other algorithms of pseudo-addition and doubling formula [8,14] for Characteristic 2.

3.1 The Possible Special Points for Duquesne's Pseudo-addition and Doubling Formula

The special points can occur on the computation of the pseudo-addition or the doubling. Now we analyze Duquesne's pseudo-addition and doubling formula [7,22] for the characteristic $p > 3$ as follows.

Theorem 1. *Let F_q be a field of characteristic $p \neq 2, 3$ and let C/F_q be a curve of genus 2. Let K be a Kummer surface of C and $(k_1 : k_2 : k_3 : k_4)$ a point on K. The point $(k_1 : k_2 : k_3 : k_4)$ is a special point for Montgomery scalar multiplication algorithm, which is instantiated with Duquesne's pseudo-addition and doubling formula [7,22], on the Kummer surface, if either of the following conditions are satisfied ($ord(K)$ denotes the order of K)*

1. $k_i = 0, i \in \{1, 2, 3, 4\}$;
2. $k_1 = \pm k_3 \mod ord(K)$.

Proof.
Kummer-based cryptosystems make use of Montgomery ladder (see algorithm 1), so both of pseudo-addition and doubling are needed per bit of the exponentiation. The cost of scalar multiplication algorithm is $59M + 12S$ according to [22], where M denotes field multiplication and S squaring. Here we assume $M = S$, then the total cost is $71M$ (We assume that $f_2^2, f_3^2, f_4^2, f_2 f_4, f_2 f_3 f_4$ were precomputed. We also assume that before computing pseudo-addition and double, we first precompute $\{k_i l_j\}_{i,j=1,\ldots,4}$ and $\{k_i k_j\}_{i,j=1,\ldots,4}$). We examine the conditions listed in the theorem.

Case $k_1 = 0$. The intermediate values related to k_1 in the doubling formulas are as follows:

$$\begin{aligned}
&k_1^2, k_1k_2, k_1k_3, k_1k_4, f_3k_1k_3,\ k_1k_2(k_1^2 - k_2^2),\\
&k_1k_3((f_3^2 - 8f_3 + 8f_2^2 + 8f_4^2 - 4f_2f_3f_4)k_1k_3 + (7 - f_3^2 + 8f_2f_4)k_2k_4\\
&+8f_2(k_3k_4 + k_1k_2) + 8f_4(k_2k_3 + k_1k_4)),\\
&k_1k_3(8f_2(k_2k_3 + k_1k_4) + 8f_4(k_1k_2 + k_3k_4) + (16 + 8f_2f_4 - 2f_3^2)k_1k_3).
\end{aligned}$$

The intermediate values related to k_1 in the pseudo-addition formulas are $k_1l_1, k_1l_2, k_1l_3, k_1l_4$.
When $k_1 = 0$, the cost of doubling formulas can save $9M + 1S$, whereas the cost of addition formulas save $4M$. Therefore, the total cost reduces to $57M$ (let $M = S$) per bit of the Montgomery ladder.

Case $k_3 = 0$. It is the same as $k_1 = 0$, for k_1 and k_3 are symmetric in the Montgomery ladder.

Case $k_2 = 0$. The intermediate values related to k_2 in the doubling formulas are:

$$\begin{aligned}
&k_2^2, k_1k_2, k_2k_3, k_2k_4, (7 - f_3^2 + 8f_2f_4)k_2k_4,\\
&k_2k_4(4f_2(k_1k_4 + k_2k_3) + 4f_4(k_1k_2 + k_3k_4) + f_3(k_2k_4 + 4(k_1^2 + k_3^2))),\\
&k_1k_2(k_1^2 - k_2^2),\ 2k_2k_4(2f_3k_1k_3 - k_2k_4).
\end{aligned}$$

The intermediate values related to k_2 in the pseudo-addition formulas are k_2l_1, k_2l_3, k_2l_4.
It saves $7M + 1S$ in the doubling formulas in case of $k_2 = 0$, whereas it saves $3M$ in the addition formulas.

Case $k_4 = 0$. It is the same as $k_2 = 0$, for k_2 and k_4 are also symmetric in the formulas.

Case $k_1 = \pm k_3 \mod ord(K)$**.** In this case, $k_1^2 - k_3^2 = 0$. The intermediate values relating to $k_1^2 - k_3^2$ are as following.

$$2(k_1^2 - k_3^2)(f_3(k_1k_4 - k_2k_3) + 2(k_1k_2 - k_3k_4) + f_4(k_1^2 - k_3^2)),$$
$$2(k_1^2 - k_3^2)(f_3(k_1k_2 - k_3k_4) + f_2(k_1^2 - k_3^2) + 2(k_1k_4 - k_2k_3)).$$

Therefore, it saves $4M$ in the doubling formulas while it remains the same in the addition formulas.

We summarize the results in the following table.

Table 1. The cost of pseudo-addition and doubling in 4 cases

	standard	$k_1 = 0$ or $k_3 = 0$	$k_2 = 0$ or $k_4 = 0$	$k_1 = \pm k_3$
pseudo-addition	$31M$	$27M$	$28M$	$31M$
doubling	$40M$	$30M$	$32M$	$36M$
total	$71M$	$57M$	$60M$	$67M$

3.2 The Number of Special Points for Duquesne's Formula

Now we classify zero-value points on Kummer surface into two types, and estimate the number of points of each type.

type 1 special points: those points satisfying $k_i = 0, i \in \{1, 2, 3, 4\}$;
type 2 special points: those points satisfying $k_1 = \pm k_3 \mod ord(K)$.

Theorem 2. *Let* $C : y^2 = f(x)$ *be a genus 2 curve over a finite field* F_q *and* κ *be the map from the Jacobian of* C *into Kummer surface* K*. Then the number of* ***type 1*** *special points on this Kummer surface is about* $4q$*.*

Proof. We only consider the imaginary hyperelliptic curves of form

$$f(x) = x^5 + f_4x^4 + f_3x^3 + f_2x^2 + f_1x + f_0.$$

Then the points $(k_1 : k_2 : k_3 : k_4)$ on the Kummer surface satisfy the following function

$$K(k_1, k_2, k_3, k_4) = K_2(k_1, k_2, k_3)k_4^2 + K_1(k_1, k_2, k_3)k_4 + K_0(k_1, k_2, k_3) = 0.$$

Now we estimate the number of special points. In the case of $k_1 = 0$, the number of special points are equal to the number of solutions of equation $K(0, k_2, k_3, k_4) = 0$ in affine case, where

$$K_2(0, k_2, k_3) = k_2^2,\ K_1(0, k_2, k_3) = -2k_2k_3^2,\ K_0(0, k_2, k_3) = k_3^4.$$

That is

$$k_3^4 - 2k_2k_4k_3^2 + k_2^2k_4^2 = 0,\ \Rightarrow k_3^2 = k_2k_4.$$

Because $k_2, k_3, k_4 \in F_q$, it is easy to see the number of solutions to the equation $k_3^2 = k_2k_4$ is about q^2. Therefore, there are about q special points for the case of $k_1 = 0$ (Note that the points of Kummer surface are in the projective space $\mathbb{P}^3(F_q)$).

We can estimate the number of special points when $k_i = 0, i \in \{2, 3, 4\}$ in the same way. Therefore, the total number of special points on Kummer surface is about $4q$. □

Theorem 3. *Let $C : y^2 = f(x)$ be a genus 2 curve over a finite field F_q and κ be the map from the Jacobian of C into Kummer surface K. Then the number of* ***type 2*** *special points on this Kummer surface is about $2q$.*

Proof. Let $(k_1 : k_2 : k_3 : k_4)$ be a point on the Kummer surface. Now we consider the **type 2 special point** , i.e., $k_1 = \pm k_3$. Since we consider the point in $\mathbb{P}^3(F_q)$, we assume $k_1 = 1$, then $k_3 = 1$ or $k_3 = -1$ as well.

The point $(1, k_2, 1, k_4)$ satisfy the quartic equation

$$K(1, k_2, 1, k_4) = K_2(1, k_2, 1)k_4^2 + K_1(1, k_2, 1)k_4 + K_0(1, k_2, 1) = 0,$$

where

$$\begin{aligned} K_2(1, k_2, 1) &= k_2^2 - 4, \\ K_1(1, k_2, 1) &= -2k_2(1 + f_1 + f_3) - 4(f_0 + f_2 + f_4), \\ K_0(1, k_2, 1) &= (1 - 2f_3 + f_3^2 + 2f_1 + f_1^2 - 4f_2f_4 - 2f_1f_3 - 4f_0f_2) \\ &\quad +k_2(-4f_2 - 4f_1f_4 + 4f_0 - 4f_0f_3) - 4k_2^2(f_0f_4 + f_1) - 4k_2^3f_0. \end{aligned}$$

Let $A_1 = 1 + f_1 + f_3$, $A_2 = f_0 + f_2 + f_4$, $A_3 = 1 - 2f_3 + f_3^2 + 2f_1 + f_1^2 - 4f_2f_4 - 2f_1f_3 - 4f_0f_2$, $A_4 = -4f_2 - 4f_1f_4 + 4f_0 - 4f_0f_3$, $A_5 = f_0f_4 + f_1$, $A_6 = f_0$, then k_2 and k_4 satisfy the following equation

$$(k_2^2 - 4)k_4^2 - (2A_1k_2 + 4A_2)k_4 - 4A_6k_2^3 - 4A_5k_2^2 + A_4k_2 + A_3 = 0.$$

The number of solutions to the above equation is about q. In this proof, we have assumed $k_1 = 1$ which is the first coordinate of the point. So the point denotes its equivalence class. The case of $k_3 = -1$ can be estimated in the same way. Therefore, there are about $2q$ special points for **type 2**. □

In total, there are about $6q$ special points for ZVP, whereas the number of points on a Kummer surface are about q^2 in a finite field F_q. Thus the special points account for $6/q$ of the total points.

In the following, we will give a simple example to show that there are about $6q$ points that have zero values during the computation of scalar multiplication on the Kummer surface. The data in Table 2 was computed via the Magma computer algebra system [23].

Example 1. Consider a hyperelliptic curve $y^2 = x^5 + x^3 + x + 4$ over F_{11}. The order of the Jacobian is 107. The corresponding Kummer surface is:

$$\begin{aligned} &x_1^4 + 6x_1^3x_2 + 9x_1^3x_3 + 6x_1^3x_4 + 5x_1^2x_2x_3 + 9x_1^2x_2x_4 + 3x_1^2x_3^2 + 6x_1x_2^3 \\ &+7x_1x_2^2x_3 + 9x_1x_2x_3x_4 + 9x_1x_3^3 + 7x_1x_3x_4^2 + x_2^2x_4^2 + 9x_2x_3^2x_4 + x_3^4. \end{aligned}$$

Choose a random point $P = (1 : 1 : 8 : 7)$ on Kummer surface with order 107. The result of the scalar multiplication iP are listed in Table 2.

Table 2. iP, $i = 1, \cdots, 107$

2P, 105P	(1 : 1 : 2 : 9)	28P, 79P	(0 : 1 : 7 : 5)
3P, 104P	(1 : 2 : 9 : 6)	29P, 78P	(1 : 7 : 8 : 3)
4P, 103P	(1 : 8 : 8 : 7)	30P, 77P	(1 : 7 : 0 : 6)
5P, 102P	(1 : 0 : 8 : 1)	31P, 76P	(0 : 1 : 6 : 3)
6P, 101P	(1 : 4 : 7 : 2)	32P, 75P	(1 : 5 : 0 : 0)
7P, 100P	(1 : 9 : 10 : 4)	33P, 74P	(1 : 8 : 10 : 3)
8P, 99P	(1 : 3 : 6 : 2)	34P, 73P	(1 : 1 : 1 : 10)
9P, 98P	(1 : 0 : 0 : 9)	35P, 72P	(1 : 0 : 3 : 9)
10P, 97P	(1 : 3 : 10 : 4)	36P, 71P	(1 : 6 : 3 : 8)
11P, 96P	(0 : 1 : 10 : 1)	37P, 70P	(1 : 4 : 6 : 9)
12P, 95P	(1 : 5 : 7 : 7)	38P, 69P	(1 : 10 : 0 : 3)
13P, 94P	(1 : 3 : 3 : 6)	39P, 68P	(1 : 6 : 4 : 0)
14P, 93P	(1 : 4 : 8 : 8)	40P, 67P	(1 : 3 : 9 : 9)
15P, 92P	(1 : 4 : 6 : 1)	41P, 66P	(1 : 6 : 1 : 2)
16P, 91P	(1 : 4 : 9 : 2)	42P, 65P	(1 : 5 : 5 : 10)
17P, 90P	(1 : 6 : 4 : 9)	43P, 64P	(1 : 1 : 6 : 8)
18P, 89P	(1 : 6 : 0 : 10)	44P, 63P	(1 : 5 : 1 : 5)
19P, 88P	(1 : 9 : 2 : 9)	45P, 62P	(1 : 1 : 3 : 1)
20P, 87P	(1 : 5 : 5 : 9)	46P, 61P	(1 : 10 : 4 : 1)
21P, 86P	(1 : 7 : 0 : 0)	47P, 60P	(1 : 10 : 0 : 0)
22P, 85P	(1 : 9 : 1 : 9)	48P, 59P	(1 : 2 : 9 : 0)
23P, 84P	(1 : 5 : 0 : 5)	49P, 58P	(0 : 1 : 0 : 0)
24P, 83P	(1 : 10 : 1 : 1)	50P, 57P	(1 : 0 : 8 : 4)
25P, 82P	(1 : 6 : 7 : 10)	51P, 56P	(1 : 3 : 5 : 8)
26P, 81P	(0 : 1 : 5 : 3)	52P, 55P	(1 : 10 : 3 : 10)
27P, 80P	(1 : 6 : 0 : 3)	53P, 54P	(1 : 1 : 2 : 4)
P, 106P	(1 : 1 : 8 : 7)	107P	(0 : 0 : 0 : 1)

We can see there are 27 special points except $(0 : 0 : 0 : 1)$ in Table 2, which constitute for 51% of the total (107-1)/2=53 Kummer points.

3.3 Special Points due to other Pseudo-addition and Doubling Formula

The special points vary with different implementations of pseudo-addition and doubling. Other variant algorithms for pseudo-addition and doubling on Kummer surface are Gaudry's algorithm [12] using Theta function, Duquesne's algorithm [8] and GL's algorithm [14] which works for characteristic 2. Similarly, we can exploit special points for these algorithms. We summarize the results in the following theorems, and we omit the proofs since they are similar to that of Theorem 1.

Possible special points with Gaudry's algorithm [12]

Theorem 4. *Let K be a Kummer surface and $(k_1 : k_2 : k_3 : k_4)$ be a point over K. The point $(k_1 : k_2 : k_3 : k_4)$ is a special point for Montgomery scalar multiplication algorithm, which is instantiated with Gaudry's pseudo-addition and doubling formula [12], on the Kummer surface, if either of the following conditions is satisfied*

1. $k_1 = k_2$ *and* $k_3 = k_4$;
2. $k_1 = k_3$ *and* $k_2 = k_4$;
3. $k_2 = k_3 = k_4$.

Possible special points with Duquesne's Algorithm [8]

Theorem 5. *Let F_q be a field of characteristic $p = 2$ and let C/F_q be a curve of genus 2. Let K be a Kummer surface of C and $(k_1 : k_2 : k_3 : k_4)$ be a point over K. The point $(k_1 : k_2 : k_3 : k_4)$ is a special point for Montgomery scalar multiplication algorithm, which is instantiated with Duquesne's pseudo-addition and doubling formula [8] for characteristic $p = 2$, if there exists an $i \in \{1, 2, 3, 4\}$ such that $k_i = 0$.*

Possible special points with GL's Algorithm [14]

Theorem 6. *Let F_q be a field of characteristic $p = 2$ and K be a Kummer surface. Let $(k_1 : k_2 : k_3 : k_4)$ be a point over K. The point $(k_1 : k_2 : k_3 : k_4)$ is a special point for Montgomery scalar multiplication algorithm, which is instantiated with GL's pseudo-addition and doubling formula [14] for characteristic $p = 2$, if either one of the following conditions is satisfied*

1. $k_i = 0$, $i \in \{1, 2, 3, 4\}$;
2. $k_1 = k_2$ *and* $k_3 = k_4$;
3. $k_1 = k_4$ *and* $k_2 = k_3$;
4. $k_1 = k_3$ *and* $k_2 = k_4$.

4 Zero-Value Point Attacks on Kummer Surface

According to Table 1, the special points of type I and II lead to some loss of computational overhead, compared with common points on the Kummer surface. How to make use of this property to exploit the secret scalar is just a kind of Zero-Value-Point (ZVP) attack. The special points on the Kummer space may not all have zeros in some coordinates, but the special points play the same role as the zero-value points in the ZVP attack in [3]. Hence those special points can be regarded as the generalized zero-value points. Here we show some general idea how the ZVP attack exploits the computational difference of the special points to reveal the secret scalar d..

Let $d = (d_{n-1}, d_{n-2}, \ldots, d_0)_2$ be the secret scalar. Suppose that it is free to compute dQ for any point Q on the Kummer surface. The ZVP attack will reveal the secret scalar bit by bit by adaptively choosing the base point Q. Suppose that the $n-j-1$ most significant bits $(d_{n-1}, d_{n-2}, \cdots, d_{j+1})_2$ are known, the ZVP attack will guess the $n-j$-th bit d_j due to the fact that the zero-value point will result in different computational overhead of pseudo-addition and doubling depending on $d_j = 0$ or $d_j = 1$.

The previous zero-value point attacks(e.g., [2, Chapter 29].) applied to Montgomery ladder scalar multiplication over ECC or HECC. However, things are different on the Kummer surface. As will show in the next subsection, when a special point happens (**type 1** or **type 2**), the doubling formulas will results in a heavier computational loss than the pseudo-addition formulas.

4.1 General Zero-Value Point Attacks on Kummer Surface

In the Montgomery scalar multiplication algorithm of computing dP in section 3, both a pseudo-addition and a doubling are involved in each loop. If the current bit $d_i = 0$, (A, B) is updated by $(2A, A+B)$; if $d_i = 1$, (A, B) is updated by $(A+B, 2B)$. Note that $B = A + P$ always holds in each loop.

If both A and B are common (not special) points then the computation in each loop will be $71M$. However, things are different when A or B is a special point, as shown in Table 3 and 4.

Table 3. Computational loss when $A = (k_1 : k_2 : k_3 : k_4)$ is a special point and $B = (l_1 : l_2 : l_3 : l_4)$ is a normal point

	when $k_1 = 0$ or $k_3 = 0$	when $k_2 = 0$ or $k_4 = 0$	when $k_1 = \pm k_3$
$d_i = 0$ Computational loss of $(2A, A+B)$	$14M$	$10M$	$4M$
Ratio of loss	$14/71 = 19.7\%$	$10/71 = 14.1\%$	$4/71 = 5.6\%$
$d_i = 1$ Computational loss of $(A+B, 2B)$	$4M$	$3M$	0
Ratio of loss	$4/71 = 5.6\%$	$3/71 = 4.2\%$	0%

When A is a special point, it may lead to a significant loss of computational overhead for $d_i = 0$, but slight loss for $d_i = 1$. On the other hand, when B is a special point, it may lead to a significant loss of computational overhead for $d_i = 1$, but slight loss for $d_i = 0$. This phenomenon helps us to design a ZVP attack to guess the value of bit d_i.

Assume that the $n-j-1$ most significant bits $(d_{n-1}, d_{n-2}, \cdots, d_{j+1})_2$ of scalar d are known. Now we want to determine the value of the j-th bit $d_j \in \{0, 1\}$. If we have a correct guess of d_j, then $A = \left(\sum_{i=j}^{n-1} d_i 2^{i-j}\right) P$ and $B = \left(\sum_{i=j}^{n-1} d_i 2^{i-j} + 1\right) P$ after $n-j$ loops in the Montgomery scalar multiplication algorithm. If we choose the base point P such that $A = \left(\sum_{i=j}^{n-1} d_i 2^{i-j}\right) P$ or $B =$

Table 4. Computational loss when $A = (k_1 : k_2 : k_3 : k_4)$ is a normal point and $B = (l_1 : l_2 : l_3 : l_4)$ is a special point

	when $l_1 = 0$ or $l_3 = 0$	when $l_2 = 0$ or $l_4 = 0$	when $l_1 = \pm l_3$
$d_i = 0$ Computational loss of $(2A, A+B)$	$4M$	$3M$	0
Ratio of loss	$4/71 = 5.6\%$	$3/71 = 4.2\%$	0%
$d_i = 1$ Computational loss of $(A+B, 2B)$	$14M$	$10M$	$4M$
Ratio of loss	$14/71 = 19.7\%$	$10/71 = 14.1\%$	$4/71 = 5.6\%$

$\left(\sum_{i=j}^{n-1} d_i 2^{i-j} + 1\right) P$ is a special value, the power consumption will be different from the normal consumption during the $n - j + 1$-*th* loop.

On the other hand, if the guess of d_j is not correct, then the power consumption will be as normal as usual during the $n - j + 1$-*th* loop.

Let H be a set of elements including the **type 1** and **type 2** special points. The next algorithm shows how to implement the attack in detail.

Algorithm 2. Zero-value point attack on Kummer surface

Input: H, $(d_{n-1}, d_{n-2}, \cdots, d_{j+1})_2$;
Output: d_j;

0) Guess $d_j = 0$;
1) for $l = 1$ to m
2) Choose an element $P_l \in H$. Compute $k = \sum_{i=j}^{n-1} d_i 2^{i-j}$ and compute point $P'_l = k^{-1} P_l$. Check that
$$(k - d_j + \overline{d}_j) P'_l \text{ not in } H,$$
where $\overline{d}_j = (d_j + 1) \mod 2$, otherwise choose another value for $P_l \in H$.
3) Compute $C_l = dP'_l$ and record the power consumption of C_l as T_l;
4) end for;
5) Compute the average power consumption $T = \frac{1}{m} \sum_{l=1}^{m} T_l$;

1′) for $l = 1$ to m
2′) Choose an element $Q_l \in H$. Compute $k = \sum_{i=j}^{n-1} d_i 2^{i-j} + 1$ and compute point $Q'_l = k^{-1} Q_l$. Check that
$$(k - d_j + \overline{d}_j) Q'_l \text{ not in } H,$$
where $\overline{d}_j = (d_j + 1) \mod 2$, otherwise choose another value for $Q_l \in H$.
3′) Compute $C'_l = dQ'_l$ and record the power consumption of C'_l as T'_l;

4′) end for;

5′) Compute the average power consumption $T' = \frac{1}{m}\sum_{l=1}^{m} T'_l$;

6) If either T or T' is smaller than the normal average power consumption, output $d_j = 0$; otherwise output $d_j = 1$.

Note: When the guess of d_j is correct, Algorithm 2 also suggests a guess of d_{j-1} according to Table 3 and 4. If the power consumption of T is much less than that of T', then $d_{j-1} = 0$; if the power consumption of T' is much less than that of T, then $d_{j-1} = 1$.

4.2 A Variant of Zero-Value Point Attack on Kummer Surface

We observed that there is a big different power consumption between the pseudo-addition and doubling formulas when a special point happens. We can exploit this difference to get a simplified attack.

Given the $n-j-1$ most significant bits $(d_{n-1}, d_{n-2}, \cdots, d_{j+1})_2$ of scalar d, we have $A = \left(\sum_{i=j+1}^{n-1} d_i 2^{i-j-1}\right) P$ and $B = \left(\sum_{i=j+1}^{n-1} d_i 2^{i-j-1} + 1\right) P$ after $n-j-1$ loops in the Montgomery scalar multiplication algorithm. If we choose the base point P such that $A = \left(\sum_{i=j}^{n-1} d_i 2^{i-j}\right) P$, the power consumption of $d_j = 0$ will be much less than that of $d_j = 1$ in the $(n-j)$-th loop.

Let H be a set of elements including the **type 1** and **type 2** special points and $d = (d_{n-1}, \cdots, d_1, d_0)$. Suppose that the most significant bits $d_{n-1}, d_{n-2}, \cdots, d_{j+1}$ of the secret scalar d are known and now we want to discover the next bit d_j.

Algorithm 3. New ZVP attack on Kummer surface

Input: H, $(d_{n-1}, d_{n-2}, \cdots, d_{j+1})_2$;
Output: d_j;

1) for $l = 1$ to m

2) Choose an element $P_l \in H$ and compute $k = \sum_{i=j+1}^{n-1} d_i 2^{i-j-1}$ and $P'_l = k^{-1} P_l$. Check that
$$(k+1)P'_l \text{ not in } H,$$

3) Compute $C_l = dP'_l$ and record the power analysis of C_l as T_l;

4) end for;

5) Compute the average power analysis $T = \frac{1}{m}\sum_{l=1}^{m} T_l$;

6) If T is much less than normal, then output $d_j = 0$, otherwise output $d_j = 1$.

---

5 Countermeasures Against Zero-Value Point Attacks

Smart showed the method to avoid the ZVP attack on ECC in [26]. In 2004, Avanzi [1] generalized the ZVP attack to HECC and also provided the countermeasures, consisting of scalar randomization and message blinding. Below we will show the principles of those countermeasures and why scalar randomization does not work for the ZVP attack on Kummer-based cryptosystems.

Denote D as a point on a Kummer surface and d a secret scalar. In the following, the order of the Kummer surface is assumed to be known.

Message Blinding Method: To compute dD, we compute an additional scalar multiplication $S = dB$ first, where B is a random point of large order. Then, the scalar multiplication dD is computed as

$$dD = d(D + B) - S.$$

The above blinding process works well for ECC and HECC, it doesn't work on the Kummer surface. The reason is that the computation of $d(D+B)$ needs the information of $D - B$ and the computation of $d(D + B) - S$ needs the value of $d(D + B) + S$ beforehand. Generally it is impossible to get the value of $D - B$ or $d(D + B) + S$. Therefore, the message blinding method won't work on the Kummer surface.

Scalar Blinding Method: The idea of scalar blinding method is to change the representation of the scalar. Given the order of the Kummer surface K, denoted by $ord(K)$, it is easy to see that $dD = (d + i * ord(K))D$ holds for any integer i. Set $d' = d + r \cdot ord(K)$ with r a random integer, then $dD = d'D$.

The additional computation caused by the scalar blinding depends on the bit length of r. To resist the zero-value point attacks, r should be big enough, yet not too big (for example, $r \approx 2^{20}$).

It seems that the scalar blinding method is the only one to resist the zero-value point attack on the Kummer surface.

6 Conclusion

In this paper, we proposed zero-value point attacks on the Kummer-based cryptosystem. We found some special points on Kummer surface and estimated the number of those special points on Kummer surface. We showed that there are as many as $6q$ special points on a Kummer surface associated with a hyperelliptic curve of genus 2, compared to about q ZVPs on the corresponding hyperelliptic curve over $\mathbb{F}_q$. On the other hand, most of the countermeasures against ZVP attacks on HECC do not work for Kummer surfaces, which makes ZVP attacks a more powerful side-channel attacks for Kummer-based cryptosystems. We pointed out that the only possible countermeasure to avoid ZVP attacks is to blind random scalars to scalar multiplications.

References

1. Avanzi, R.M.: Aspects of Hyperelliptic Curves over Large Prime Fields in Software Implementations. In: Joye, M., Quisquater, J.-J. (eds.) CHES 2004. LNCS, vol. 3156, pp. 148–162. Springer, Heidelberg (2004)
2. Avanzi, R., Cohen, H., Doche, C., Frey, G., Lange, T., Nguyen, K., Vercauteren, F.: Handbook of elliptic and hyperelliptic curve cryptography. CRC Press, Boca Raton (2005)
3. Akishita, T., Takagi, T.: Zero-Value Point Attacks on Elliptic Curve Cryptosystem. In: Boyd, C., Mao, W. (eds.) ISC 2003. LNCS, vol. 2851, pp. 218–233. Springer, Heidelberg (2003)
4. Cantor, D.G.: Computing on the Jacobin of a hyperelliptic curve. Math. Comp. 48, 95–101 (1987)
5. Coron, J.-S.: Resistance against Differential Power Analysis for Elliptic Curve Cryptosystems. In: Koç, Ç.K., Paar, C. (eds.) CHES 1999. LNCS, vol. 1717, pp. 292–302. Springer, Heidelberg (1999)
6. Cassels, J.W.S., Flynn, E.V.: Prolegomena to a middlebrow arithmetic of curves of genus 2. Cambridge University Press, Cambridge (1996)
7. Duquesne, S.: Montgomery Scalar Multiplication for Genus 2 Curves. In: Buell, D.A. (ed.) ANTS 2004. LNCS, vol. 3076, pp. 153–168. Springer, Heidelberg (2004)
8. Duquesne, S.: Montgomery Ladder for All Genus 2 Curves in Characteristic 2. In: von zur Gathen, J., Imaña, J.L., Koç, Ç.K. (eds.) WAIFI 2008. LNCS, vol. 5130, pp. 174–188. Springer, Heidelberg (2008)
9. Duquesne, S.: Traces of the group law on the Kummer surface of a curve of genus 2 in characteristic 2. Math. Comput. Sci. 3, 173–183 (2010)
10. Fan, J., Guo, X., De Mulder, E., Schaumont, P., Preneel, B., Verbauwhede, I.: State-of-the-art of secure ECC implementations: a survey on known side-channel attacks and countermeasures. In: Hardware-Oriented Security and Trust (HOST 2010), pp. 76–87. IEEE (2010)
11. Flynn, E.V.: The group law on the jacobian of a curve of genus 2. J. Reine. Angew. Math. 439, 45–69 (1995)
12. Gaudry, P.: Fast genus 2 arithmetic based on Theta functions. Journal of Mathematical Cryptology 1, 243–265 (2007)
13. Goubin, L.: A Refined Power-Analysis Attack on Elliptic Curve Cryptosystems. In: Desmedt, Y.G. (ed.) PKC 2003. LNCS, vol. 2567, pp. 199–210. Springer, Heidelberg (2002)
14. Gaudry, P., Lubicz, D.: The arithmetic of characteristic 2 Kummer surfaces and of elliptic Kummer lines. Finite Fields and Their Applications 15(2), 246–260 (2009)
15. Harley, R.: Fast arithmetic on genus 2 curves (2000), `http://cristal.inria/fr/~harley/hyper`
16. Joye, M., Tymen, C.: Protections against Differential Analysis for Elliptic Curve Cryptography. In: Koç, Ç.K., Naccache, D., Paar, C. (eds.) CHES 2001. LNCS, vol. 2162, pp. 377–390. Springer, Heidelberg (2001)
17. Joye, M., Quisquater, J.-J.: Hessian Elliptic Curves and Side-Channel Attacks. In: Koç, Ç.K., Naccache, D., Paar, C. (eds.) CHES 2001. LNCS, vol. 2162, pp. 402–410. Springer, Heidelberg (2001)
18. Koblitz, N.: Hyperelliptic cryptosystems. J. Cryptology 1, 139–150 (1989)
19. Kocher, P., Jaffe, J., Jun, B.: Introduction to Differential Power Analysis and Related Attacks. Technical Report, Cryptography Research Inc. (1998), `http://www.cryptography.com/dpa/technical/index.html`

20. Kocher, P.C., Jaffe, J., Jun, B.: Differential Power Analysis. In: Wiener, M. (ed.) CRYPTO 1999. LNCS, vol. 1666, p. 388. Springer, Heidelberg (1999)
21. Liardet, P.-Y., Smart, N.P.: Preventing SPA/DPA in ECC Systems Using the Jacobi Form. In: Koç, Ç.K., Naccache, D., Paar, C. (eds.) CHES 2001. LNCS, vol. 2162, pp. 391–411. Springer, Heidelberg (2001)
22. Lin, Q., Zhang, F.: Scalar Multiplication on Kummer Surface Revisited. IEICE Trans. Fundamentals E95-A(1), 410–413 (2012)
23. MAGMA Computational Algebra System, `http://magma.maths.usyd.edu.au/magma/`
24. Montgomery, P.L.: Speeding up the Pollard and elliptic curve methods of factorization. Mathematics of Computation 48(177), 243–264 (1987)
25. Müller, J.S.: Explicit Kummer surface formulas for arbitrary characteristic. LMS J. Comput. Math. 13, 47–64 (2010)
26. Smart, N.P.: An Analysis of Goubin's Refined Power Analysis Attack. In: Walter, C.D., Koç, Ç.K., Paar, C. (eds.) CHES 2003. LNCS, vol. 2779, pp. 281–290. Springer, Heidelberg (2003)
27. Smart, N.P.: The Hessian Form of an Elliptic Curve. In: Koç, Ç.K., Naccache, D., Paar, C. (eds.) CHES 2001. LNCS, vol. 2162, pp. 118–125. Springer, Heidelberg (2001)
28. Smart, N., Siksek, S.: A fast Diffie-Hellman protocol in genus 2. J. of Cryptology 12, 67–73 (1999)

PICARO – A Block Cipher Allowing Efficient Higher-Order Side-Channel Resistance

Gilles Piret[1], Thomas Roche[2], and Claude Carlet[3]

[1] Oberthur Technologies, 71-73, rue des Hautes Pâtures, 92726 Nanterre, France
g.piret@oberthur.com
[2] ANSSI, 51, Bd de la Tour-Maubourg, 75700 Paris 07 SP, France
thomas.roche@ssi.gouv.fr
[3] LAGA, Universities of Paris 8 and Paris 13, CNRS
2, Rue de la Liberté, 93526 Saint-Denis Cedex, France
claude.carlet@univ-paris8.fr

Abstract. Many papers deal with the problem of constructing an efficient masking scheme for existing block ciphers. We take the reverse approach: that is, given a proven masking scheme (Rivain and Prouff, CHES 2010) we design a block cipher that fits well the masking constraints. The difficulty of implementing efficient masking for a block cipher comes mainly from the S-boxes. Therefore the choice of an adequate S-box is the first and most critical step of our work. The S-box we selected is non-bijective; we discuss the resulting design and security problems. A complete design of the cipher is given, as well as some implementation results.

1 Introduction

In a *side-channel attack* (SCA for short), the attacker observes — at runtime — the execution environment (timing, power, electromagnetic radiation, etc.) of a secret-dependent operation. From this observation, the attacker might either be able to identify (part of) the secret (the attack is then called *Simple Side-Channel Analysis*, SSCA) or get noisy information about internal states of the cryptographic operation. In the latter case, when the accessed internal values are a simple combination of a known variable and (part of) the secret, the attacker will recover the secret from a statistical treatment of multiple cryptographic operation execution, the attack is then called *Differential Side-Channel Analysis* (DSCA for short). DSCA fits particularly well block ciphers which build their security from piling-up simple and cryptographically weak operations. By accessing internal states, the attack bypasses the cipher strength. Since the seminal work of Kocher *et al.* [27], DSCA (and its numerous variants and extensions) are a constant threat against embedded devices that implement cryptographic primitives. The development of DSCA countermeasures is a dynamic and challenging research domain where the ultimate goal is to find the good trade-off between security and performances. Many countermeasures focus on noise addition techniques, which should increase the attack complexity. For instance, inserting random delays during the cipher execution is a common practice in order to render

F. Bao, P. Samarati, and J. Zhou (Eds.): ACNS 2012, LNCS 7341, pp. 311–328, 2012.

difficult finding the manipulation of secret-dependent variables inside multiple side-channel traces. This kind of countermeasures indeed increases the classical DSCA attack complexity and it can be done with relatively small overhead on the algorithm complexity (see for instance [14]). However these countermeasures cannot be proven robust, *i.e.* an optimal attacker would be able to recognize and suppress the random delays. This is actually what claim Durvaux *et al.* in a very recent paper [17], where Hidden Markov Chain inference techniques are used to point out dummy operations from real ones, discarding almost perfectly the countermeasure proposed in [14]. In fact, the only known countermeasure that possesses security proofs is the so-called *Masking Schemes* where the cipher's secret-dependent internal values are randomized from one execution to another (*e.g.* [12,21]). However, such countermeasures usually induce a high performance overhead, making their implementation difficult if not impracticable in small embedded devices. Many works have been dedicated to building a masking scheme with low cost that fits the existing block ciphers (mainly DES and AES). In the present paper we will take the problem the other way around: we will study a proven masking scheme and propose a new construction of block cipher that fits well the masking constraints. Hence, we come up with the design of a cipher that ensures resistance to conventional cryptanalysis methods, with special care for the S-boxes (that are used to introduce non-linearity in the cipher design, and are usually the most challenging part to implement when a masking scheme is used) in order to lower the performance overhead of masking.

The paper is organised as follows: In the next section, the basics about masking techniques are recalled and Rivain and Prouff's Boolean masking scheme [43] is described; our block cipher construction will follow the design criteria derived from this scheme. In Section 3, we propose a new S-box having a good trade-off between efficiency, conventional security and masking efficiency. The main limitation of the new S-box is its non-bijectiveness but the use of a Feistel network allows us to build a full block cipher from the S-box. We exhibit in Section 4 a devastating attack on Feistel schemes if no special care is taken on the diffusion layer of the round function. In connection with this attack, various specific cryptanalysis techniques of Feistel networks are recalled in Section 5. The full round function is described in Section 6. Finally, in Section 7, a complete design specification of a full block cipher is proposed as well as a performance analysis compared to the AES block cipher.

2 Preliminaries on Higher-Order Masking Schemes

As recalled in introduction, a Differential Side-Channel Attack compiles leaked information from side-channel observations of internal states of a block cipher in order to recover some knowledge about the secret key. The strength of DSCA comes from the statistical treatments of the leaked information that makes the attack particularly robust to noise (from measurement setup, concurrent operations, etc...). Many improvements have been proposed on the original *Differential Power Analysis* introduced by Kocher *et al.* in 1999 [27]. Among them the *Correlation Power Analysis* [9] and the *Mutual Information Analysis* [20] propose different statistical treatments to enhance the attack complexity with respect to the noise and leakage model. Another notable extension of DSCA is the so-called *Higher-Order DSCA* (HO-DSCA for short), already mentioned in [27], that upgrades the attacker model: in a dth-order DSCA attack, it is able to observe d different internal variables in a single cipher execution.

Countermeasures by masking are certainly the most studied countermeasures against (HO-) DSCA because of their security proofs. However, the performance overhead due to a masking scheme that thwarts HO-DSCA is such that they are hardly used in practice. Our goal here is to point out the operations that make a masking scheme costly and propose a block cipher that avoids as much as possible such operations. To this end we will focus on a recent (HO-)Masking scheme introduced by Rivain and Prouff [43].

2.1 Masking Schemes

A dth*-order masking scheme* is a countermeasure at the algorithmic level that thwarts d^{th}-order DSCA; the idea is to randomize the data processed by the symmetric cipher such that there exists no set of d processed data that together depend on the secret. A proof of security on a masking scheme ensures that this property holds, in addition, the data complexity of a HO-DSCA attack increases exponentially with its order, assuming the presence of noise (as showed by Chary *et al.* [12]), which always exists in practice. All together, with a high enough order with respect to the noise level, these properties make the masking approach the only sound countermeasure against DSCA.

Several 1st-order masking schemes have been proposed (*e.g.* [8, 34, 37]) and some specific 2nd-order masking schemes [42, 45] but until recently very few schemes could be extended to any order d. The first provable dth-order masking scheme was proposed by Ishai *et al.* in 2003 [23]; the construction has been extended in 2010 by Rivain and Prouff [43]. This work was followed by two new propositions [25, 39] for which our own work would apply just the same as for Rivain and Prouff's construction. A third publication by Genelle *et al.* [19] was also proposed in 2011; it is dedicated to very specific non-linear functions (power functions) which would not leave enough room for us in the research of new S-boxes.

2.2 Rivain-Prouff's Scheme

Let us consider an intermediate variable $V \in GF(2^n)$ of the targeted block cipher, the variable V is called *sensitive* if its value depends on a secret key K and on a known variable (*e.g.* the plaintext P), for instance $V = K \oplus P$. The manipulation of a sensitive variable should be avoided due to DSCA attacks, therefore, in a *dth-order Boolean Masking Scheme* (as Rivain and Prouff's scheme), its manipulation is replaced by the manipulation of $d+1$ shares $(V_0, V_1, \cdots, V_d)$ such that

$$V = V_0 \oplus V_1 \oplus \cdots \oplus V_d \ . \quad (1)$$

A dth-order Masking scheme is an algorithm that modifies the cipher sub-functions in order to only manipulate such sharing of sensitive variables (ideally without ever re-constructing the sensitive variables or decreasing the sharing order).

In [43], the authors propose such an algorithm for each atomic operation: affine functions ($v \mapsto A_f(v)$), addition ($(v, w) \mapsto v \oplus w$) and multiplication ($(v, w) \mapsto v \times w$).

Remark 1. Any function can be decomposed in a sequence of such atomic operations, which gives a great genericity to the masking scheme (this approach is classic in Secure Multi-Party Computations, a research area that is very close to our problem and on which most of the dth-order masking scheme are based [23,39]). The drawback is that those atomic functions shall be executed explicitly (pre-computed tables, commonly used to evaluate S-boxes are not an option).

Affine functions and additions over shared variables can be applied straightforwardly, the masking overhead will solely correspond to d times the original operation complexity. In the case of multiplication, when it is not linear over $GF(2^n)$, the masking scheme is more expensive: it costs $(d+1)^2$ field multiplications, $2d(d+1)$ XORs and the generation of $d(d+1)/2$ random n-bit values. In a block cipher like the AES, each of the 160 S-box computations needs at the least 4 such multiplications in $GF(2^8)$, making the cost of the masking scheme mostly carried by the non-linear multiplications.

This study leads naturally to the following constraints that an S-box should satisfy in order to be efficiently masked: the S-box should have a simple expression as a polynomial and minimum number of non-linear field multiplications in this form.

3 Research of a "Good" S-Box

S-boxes are non-linear functions from $\mathrm{GF}(2)^n$ to $\mathrm{GF}(2)^m$ where n and m are positive integers. We also use the terminology (n, m)*-functions*. The vector spaces $\mathrm{GF}(2)^n$ and $\mathrm{GF}(2)^m$ can be endowed with the structure of field. This gives, when m divides n (and in particular when $m = n$), the possibility of designing S-boxes as polynomial functions over finite fields.

3.1 Design Constraints

S-boxes must allow resistance to several logical attacks. The three main attacks to be withstood are the linear attack [32], the differential attack [3] and the higher order differential attack [26]. An attack which is not yet efficient but represents some threat for the design of future block ciphers is the algebraic attack [13]. Designing an S-box, which is fastly implementable, allows high resistance to the first three attacks and would not be potentially weak against a future efficient version of the fourth one is a difficult challenge. Historically, the S-boxes of the DES have been found by clever random computer search. This was possible thanks to the relatively small size of these (6, 4)-functions. The S-box of the AES was too big for that; it has been the result of a theoretical work by K. Nyberg [36] on the so-called inverse function $GF(2^n) \to GF(2^n) : x \mapsto x^{-1}$. This function has very good properties: it is a permutation (which is necessary for using it as an S-box in an SPN and is quite useful for a Feistel cipher as we shall see below), it achieves the highest known nonlinearity when n is even (in the case of AES $n = 8$; it is common to choose n as a power of 2 because it makes software implementation easier), and has very high resistance to the differential attack and to the higher order differential attack. It happens that, since 1993, no other function in a number of variables equal to a power of 2 and gathering these properties could be found (see a survey in [10]). Note however that the inverse function is almost the worst possible against the algebraic attack.

The criteria listed above are those that an S-box should satisfy in black box cryptography. We need to add the requirements derived for side-channel resistance (see Section 2) and the practical design constraints:

1. Higher-Order Masking against HO-SCA attacks implementable without slowing down the cryptosystem too much (reducing the overhead leads to minimizing the number of non-linear multiplications, see Section 2, and is also related to Constraint 2).
2. Efficiency (*i.e.* reduce the number of instructions and allow the operations to be performed in small fields).
3. Function in 8 variables (the number of variables must be large enough for allowing good resistance to the three main known logical attacks; the choice of 8 helps satisfying Constraint 2 and allows compatibility with standard block size).

We describe now in more details the *function's criteria*. Given an (n, n)-function in the form:

$$f : X \to \sum_{i=0}^{2^n-1} a_i X^i,\ a_i \in GF(2^n) \tag{2}$$

its important parameters are:

- Non-linearity: $nl(f) = 2^{n-1} - \frac{1}{2} \max_{a,b \neq 0} \left| \sum_X (-1)^{b \cdot f(X) + a \cdot X} \right|$, where $a \cdot X$ is an inner product in $GF(2^n)$; in practice, $a \cdot X = tr(aX)$ where tr is the trace function $tr(a) = a + a^2 + a^{2^2} + \ldots + a^{2^{n-1}}$.
  To thwart linear cryptanalysis [32], the nonlinearity must be close to the best known nonlinearity of vectorial functions in even numbers of variables: $2^{n-1} - 2^{n/2}$ (that is 112 for $n = 8$).
- Differentiality: $\delta = \max_{a \neq 0, b} (\#\{X \mid f(a + X) + f(X) = b\})$.
  Because of the differential cryptanalysis [4], it should be 2 (then the function is called Almost Perfect Nonlinear APN [35]) or 4 (then the function is called differentially 4-uniform [35]), or at most 6.

- Algebraic degree: $d = \max_i (\omega_2(i) \mid a_i \neq 0)$, where $\omega_2(i)$ is the Hamming weight of the binary expansion of i.
  Because of the higher differential attack [26], it should be at least 3 and preferably at least 4.
- Graph algebraic Immunity: equals the minimal algebraic degree of a nonzero Boolean function vanishing on the graph $G_f = \{(X, f(X));\ X \in GF(2^n)\}$ of the function (that is, the minimal algebraic degree of an annihilator of the graph); this parameter is not related to an efficient attack yet, but 1 is definitely too small and 2, as in the case of the inverse function, is risky.
- Minimum number of non-linear multiplications.
  The evaluation of $f : X \to \sum_{i=0}^{2^n-1} a_i X^i$ involves a number, say k, of non-linear multiplications (by opposition to linear transformations such as, in characteristic 2, an exponentiation by a power of 2, i.e. a monomial of degree a power of 2). Higher Order masking schemes like the one proposed by Rivain and Prouff [43] slow down significantly the S-box implementation, the overhead being directly related to the number of non-linear multiplications.

3.2 Bijective vs Non-Bijective S-Box

Considering two comparable functions (with respect to their execution efficiency as well as the above mentioned criteria), a bijective S-box is much more interesting than a non bijective one, because in the latter case we have to solve the problem of making the cipher invertible anyway. Moreover we will see in Section 4 that a non-bijective S-box induces security flaws.

However, it is a matter of fact that the research of good (meaning efficient and cryptographically strong) non-linear functions is much harder when only considering bijective functions, especially in an number of variables that is a power of two, where the inverse function is considered the only good candidate.

The selected function is a non-bijective function, specially efficient in the number of operations necessary for evaluating it and involving only operations in a small Galois Field (of 16 elements), operations that can then be tabulated on standard platforms.

3.3 S-Box Description

A possible S-box candidate is proposed in [11]. It is not expressed as a polynomial of the form (2), but as the concatenation of two bivariate polynomials whose variables live in $GF(2^{n/2})$:

$$f : \mathrm{GF}(2^{n/2})^2 \to \mathrm{GF}(2^{n/2})^2 : (x, y) \mapsto (xy, (x^3 + \omega)(y^3 + \omega')), \tag{3}$$

where xy is the product of x and y in the field $GF(2^{n/2})$. This S-box has the desired properties when $n/2$ is even and ω, ω' and $\frac{\omega}{\omega'}$ belong to $GF(2^{n/2}) \setminus \{x^3, x \in GF(2^{n/2})\}$. In particular, for $n = 8$, we have:

- $\delta = 4$.
- $nl = 94$.
- algebraic degree: 4.
- number of non-linear multiplications: 4 in $GF(2^4)$.

S-box Instantiation. To represent elements in GF(2^4) we chose to work in field representation GF(2)[x]/$P(x)$ with $P(X) = X^4 + X^3 + 1$. Moreover we need to make a choice in the family of S-boxes described above, that is to choose ω and ω'. We took $\omega = 02_x$, $\omega' = 04_x$ (in hexadecimal notation).

3.4 Masked S-Box Cost Evaluation

As explained in Section 2, Rivain and Prouff's higher order masking scheme [43] uses a sequence of field multiplications and additions to compute the masked S-box. It can be easily checked that one needs at most 2 additions, 2 square operations and 4 multiplications in $GF(2^4)$ to evaluate our S-box. By comparison, the AES S-box is computed with 3 raisings to some power 2^i (i.e. X^{2^i}) and 4 multiplications in $GF(2^8)$ (see [43]).

The cost of the higher order masking scheme is linear in the masking order for additions and linear operations (like "X^{2^i}" operations in fields of characteristic 2) whereas it is quadratic for non-linear multiplications. Hence the overhead in the number of operations to evaluate a masked AES S-box and our S-box seems at first glance quite the same. Table 1 details the number of operations in $GF(2^4)$ that are needed to evaluate our S-box.

Table 1. Number of Operations

	# additions	# squarings	# multiplications	# random values (4-bit)
Unmasked	2	2	4	0
dth-order masked	$(8d+2)(d+1)$	$2(d+1)$	$4(d+1)^2$	$2d(d+1)$

In practice, the field size will play an important role in the runtime as a field operation cost is directly dependent on the field size. Decreasing the field size to 2^4 allows to tabulate the field multiplication in a lookup table with much less memory, which makes it possible even when it is very constrained (contrary to the case of $GF(2^8)$). As a matter of fact, using tower field methods, the evaluation of higher-order masked AES's S-box in $GF(2^4)$ by Kim *et al.* [25] has been shown to be faster (for masking order 2 and 3) than the original evaluation in $GF(2^8)$ (from [43]), even though the number of non-linear multiplications turned to 5 in $GF(2^4)$.

4 From the S-Box to the Cipher

4.1 Using a Feistel Network with SP-Type Round Function

The non-bijectivity of the S-box we selected requires us to use an adequate structure, in order to make the cipher invertible. A well-known way to use non-bijective round functions to build a block cipher is to use a Feistel network. Therefore we could think of embedding our S-box in an SP-Type F-function as considered in many papers ([46–48] and several others), and using this F-function as the round function of a Feistel network.

An *SP-type F-function* $F : \mathrm{GF}(2)^n \times \mathrm{GF}(2)^n \to \mathrm{GF}(2)^n$ is defined as follows.

Definition 1. *Let m the number of S-boxes in a round, and t the size of the S-box, with $mt = n$. Consider $\gamma, \theta : \mathrm{GF}(2)^n \to \mathrm{GF}(2)^n$ with*

- *γ the function generated by concatenating m S-boxes.*
- *θ a linear diffusion layer.*

Then an SP-type F-function *F is defined as $F(x, k) = \theta(\gamma(x \oplus k))$.*

One round of a Feistel network with round function $F : \mathrm{GF}(2)^n \to \mathrm{GF}(2)^n$ is defined as

$$\Psi(F) : \mathrm{GF}(2)^{2n} \to \mathrm{GF}(2)^{2n} : \langle L, R\rangle \to \langle R, L \oplus F(R)\rangle \tag{4}$$

Then

Definition 2. *An* SP-type R-round Feistel Network *is the composition*

$$\bigcirc_{i=1}^{R} \Psi(F(., k^i)) \tag{5}$$

where F is an SP-type F-function and the k^i's are round keys derived from the master key K by a key schedule algorithm.

4.2 Why It Is Not a Good Idea

This approach is actually not applicable as such with our S-box S, as it lends itself to a little-known but devastating attack.

Consider $a, b \in \mathrm{GF}(2^8)$ such that $S(a) = S(b)$ (two such inputs always exist as S is not injective). Let us denote $\Delta = a \oplus b$. Consider two plaintexts $P = \langle L, R\rangle = \langle(l_1, \ldots, l_m), (r_1, \ldots, r_m)\rangle$ and $P' = \langle L, R'\rangle = \langle(l_1, \ldots, l_m), (r'_1, \ldots, r'_m)\rangle$ ($l_i, r_i, r'_i \in \mathrm{GF}(2^8)$) such that

$$P \oplus P' = \langle(0, \ldots, 0), (\Delta, 0, \ldots, 0)\rangle. \tag{6}$$

Assuming the first round key $k^1 = (k^1_1, ..., k^1_m)$ is uniformly distributed, with probability at least[1] $2/2^8$ we have

$$F(R, k^1) = F(R', k^1) \tag{7}$$

[1] It is greater than that if $\exists c (a \neq c \neq b)$ such that $S(c) = S(c \oplus \Delta)$.

As a matter of fact, the inputs $R \oplus k^1$ and $R' \oplus k^1$ to the S-box layer differ in their first byte only. Thus (7) is satisfied when $S(r_1 \oplus k_1^1) = S(r_1' \oplus k_1^1)$. This equality is satisfied if $r_1 \oplus k_1^1 = a$ or $r_1 \oplus k_1^1 = b$.

Attack Complexity. Let us consider a SP-type Feistel Network with R rounds and a block size of n bits. The round function's S-Box (S) is non-injective and we denote by DP_0 its maximum 0-*output differential probability*: $DP_0 = \max_{a \neq 0}(\#\{x \text{ s.t. } S(x \oplus a) \oplus S(x) = 0\})/2^m$, where m is the size of the S-box input (the best case scenario from the security point of view is when $DP_0 = 2/2^m$, this is the case for our S-box). The differential attack we study here assumes that the attacker chooses pairs of plaintexts P, P' such that the input difference at the beginning of the first round function is null everywhere except on the input of a single S-box where the difference is equal to $\Delta = \texttt{argmax}_{a \neq 0}(\#\{x \text{ s.t. } S(x \oplus a) \oplus S(x) = 0\})$.

The differential characteristic over R rounds considered in this attack is such that the round function's input differential for even rounds (*resp.* odd rounds) is equal to $\langle (0, \ldots, 0), (\Delta, 0, \ldots, 0) \rangle$ (*resp.* null). Assuming that the round keys are independent and uniformly distributed (*i.e.* the classical Markov Cipher assumption), it is easy to evaluate the probability of such a differential characteristic Ω:

$$\Pr[\Omega] = (DP_0)^{R/2} . \tag{8}$$

Given the differential characteristic probability it is well known (see for instance [29]) that the differential cryptanalysis data complexity can be approximated by

$$\mathcal{C} = \frac{2}{\Pr[\Omega]} . \tag{9}$$

Hence, in order to get an attack complexity higher than exhaustive search, we would need to assure that $\Pr[\Omega] \leq \frac{1}{2^{n-1}}$. Considering the best non-injective S-box for $m = 8$ ($DP_0 = 2^{-7}$) and $n = 128$ this means that $(2^{-7})^{R/2} \leq 2^{-127}$, therefore the number of rounds R should be greater than 36.

4.3 Linear Counterpart to the Previous Attack

Several results show that some duality exists between linear and differential attacks [33]. Therefore it is not surprising that a linear attack exists that is as powerful as the differential attack we exposed in the previous section.

This attack is based on the following theorem (see for instance [30]).

Theorem 1. *A Boolean transformation F is invertible if and only if every output parity (i.e. every component function $\lambda \cdot F$) is a balanced binary Boolean function of input bits.*

Applying this theorem to our S-box S we obtain

Corollary 1. *There exists a linear mask $\lambda \in \mathrm{GF}(2)^8$ such that, for x random and uniformly distributed, $\lambda \cdot S(x) = 0$ is satisfied with probability $p = 1/2 + \epsilon$ where the bias, denoted by ϵ, is not null.*

Let us denote by $L^i = (l^i_1, \ldots, l^i_m)$ the left part and by $R^i = (r^i_1, \ldots, r^i_m)$ the right part of the input to round $i+1$, and by $Y^i = (y^i_1, \ldots, y^i_m)$ the output of the ith F-function. Then if we use mask λ to approximate the first S-box (only) of first round, we have a linear characteristic on 2 rounds of the form

$$\lambda \cdot l^0_1 = \lambda \cdot l^2_1 \qquad \text{with probability } 1/2 + \epsilon \tag{10}$$

As a matter of fact, we have: $\lambda \cdot y^1_1 = 0 \Leftrightarrow \lambda \cdot l^0_1 \oplus \lambda \cdot r^1_1 = 0 \Leftrightarrow \lambda \cdot l^0_1 \oplus \lambda \cdot l^2_1 = 0$.

This characteristic is iterative. Therefore R rounds can be approximated with probability $1/2 + 2^{R/2-1}\epsilon^{R/2}$. It is well known (see for instance [32]) that the linear cryptanalysis data complexity, given a characteristic of bias ϵ, is given by $\mathcal{C} = \frac{1}{2(\epsilon)^2}$. Assuming that the bias associated to λ is $2^{-8/2}$ (the smallest known non-zero bias for an 8-bit S-box), 42 rounds are required to have an attack complexity above 2^{127} for the whole cipher. Moreover, in the case of our S-box, the maximal bias ϵ (over all output linear maps λ) is equal to $\frac{22}{2^8}$ which leads to $R \geq \frac{126}{7-\log(22)} > 49$.

5 Comparison of Specific Attacks on Feistel Ciphers with Non-bijective Round Function

We have seen that the use of non-bijective functions introduces vulnerabilities in the design of a Feistel cipher. Some of these vulnerabilities can be found in the literature. We divide them in three categories: The differential cryptanalysis [3] and its linear counterpart (this corresponds to the attack described in Section 4), Rijmen *et al.* 's non-surjective attack [41] and the Davies and Murphy attack [2, 16, 28].

5.1 Non-injective Round Functions

The first attack described in Section 4 can be seen as a particular case of differential cryptanalysis. We already mentioned that the initial paper of Biham and Shamir on differential cryptanalysis [3] already proposed a similar differential characteristic construction for the DES cipher, but is not the best for the DES because of its expansion function. Another example is the McGuffin cipher proposed by Schneier and Blaze [7] that did not have a similar expansion transformation and was then completely exposed against such strong differential cryptanalysis. Rijmen *et al.* described the attack in [40].

5.2 Non-surjective and Unbalanced Round Function

Rijmen *et al.* propose in [41] an attack on Feistel schemes assuming that the round function is non-surjective and extend their attack when the round function is simply unbalanced. The linear attack presented in Section 4 is based on the fact that the S-box is unbalanced (Theorem 1 and then Corollary 1). This attack corresponds exactly to the Rijmen and Preneel attack where only linear masks of the round output bits are considered. Moreover, the SP-Network structure of the round function allows us (similarly to the differential version) to consider a single S-box instead of the full round function (as in [41]). This property conceals the strength of the attack; thus an expansion transformation is needed to ensure that more than a unique S-box per 2 successive rounds is involved in the linear characteristic. This will lower the linear (*resp.* differential) characteristic bias and then increase the attack's data complexity (see Section 7.1).

Before that, it is important to note that introducing an expansion step before the round key mixing step and the S-box evaluation may open a new vulnerability in the scheme: the Davies and Murphy attack.

5.3 Unbalanced Round Functions with Key Dependent Output Distribution

Davies and Murphy proposed in [16] an attack on Feistel schemes assuming that the round functions are unbalanced and the output distribution is dependent on some key bits. This seminal paper was followed by many others, among them a first improvement by Biham and Biryukov [2] and then a second improvement proposed by Kunz-Jacques and Muller [28]. In the latter article, a parallel is drawn between Davis and Murphy attack and the linear cryptanalysis; moreover, the initial attack is optimized by the use of a distinguisher that evaluates divergence between univariate distributions (through linear projections) instead of divergence between multivariate distributions.

The use of an expansion step induces the S-box output distribution to be somewhat dependent on the secret key. In [28], Kunz-Jacques and Muller exhibit the tight relation between Davies and Murphy attack and linear cryptanalysis. As a matter of fact, in the case of DES, they could show that the classical Davies and Murphy attack would not be more efficient than a restriction of it where only linear combinations of the round outputs are considered. This restricted Davies and Murphy attack falls into linear cryptanalysis and then is naturally bounded by the linear cryptanalysis complexity bound found for the DES.

6 Expansion and Compression Function

The attacks we described in Section 4 exploit the fact that it is possible to choose (pairs of) plaintexts such that one S-box only is active in the first round (and that this property can be propagated to the following odd rounds). If we deny this possibility to the attacker, we thwart these attacks.

Using linear codes to ensure good diffusion in block ciphers is a well-known idea (see [15], and many other works). We show how to use them slightly differently from what is usually done in order to render impossible the attacks discussed in Section 4 and Section 5.

- Let $(a_1, \ldots, a_8) \in \mathrm{GF}(2^8)^8$ be the input of the round function. We encode it with a linear code of dimension 8 and length $8 + \ell$ over $\mathrm{GF}(2^8)$, before performing the key addition and the S-boxes layer. That is, if G is the generator matrix of such $[8+\ell, 8]$ code, we compute

$$(b_1, \ldots, b_{8+\ell}) = (a_1, \ldots, a_8) \cdot G \quad (11)$$

We call this computation the *expansion layer* E. If d is the minimal distance of the code, it is trivial that the minimal number of active S-boxes is d. In order to maximize it, we use a MDS code $[8+\ell, 8, \ell+1]$. An easy way to construct such code is to use a shortened Reed-Solomon code.

- After E comes a key addition layer, which will use $8(8+\ell)$ key bits, and the non-linear layer that consists in $8+\ell$ S-boxes in parallel.
- Finally the state must be compressed from $8+\ell$ to 8 bytes. Note that the expansion layer only defeats the differential attack we exposed in Section 4.2, not the linear one described in Section 4.3. Therefore the linear *compression layer* C must ensure that every non-zero linear mask approximating the output of the S-box layer has as many active S-boxes as possible. If we denote by H the compression matrix and we write $(d_1, \ldots, d_8) = (c_1, \ldots, c_{8+\ell}) \cdot H$, a linear approximation of the output of the round can be written as $(\beta_1, \ldots, \beta_8) \cdot (d_1, \ldots, d_8)^T = (\beta_1, \ldots, \beta_8) \cdot H^T \cdot (c_1, \ldots, c_{8+\ell})^T$. We define $(\beta'_1, \ldots, \beta'_{8+\ell})$ as the linear mask at the output of the S-boxes corresponding to $(\beta_1, \ldots, \beta_8)$. Thus we have $(\beta'_1, \ldots, \beta'_{8+\ell}) = (\beta_1, \ldots, \beta_8) \cdot H^T$. In order to maximize the number of active S-boxes, we must choose H^T such as to lower bound the byte Hamming weight of $(\beta'_1, \ldots, \beta'_{8+\ell})$. The best choice H^T for this purpose is again to choose H^T as the generator matrix of an MDS code. We decide to take $H^T = G$.

We choose $\ell = 6$, which offers a good compromise between the number of rounds and the computational cost of one round. We built the matrix G such as to make its implementation efficient. More precisely, we tried to minimize the number of non-zero coefficients, to use a small number of different coefficients, and to use coefficients with a small Hamming weight

The resulting matrix G is as follows. Its elements belong to the Galois field $GF(2^8)$ defined as $GF(2)[X]/(1+X^2+X^3+X^4+X^8)$.

$$G = \begin{pmatrix} 01 & 00 & 00 & 00 & 00 & 00 & 00 & 00 & 01 & 01 & 0A & 01 & 09 & 0C \\ 00 & 01 & 00 & 00 & 00 & 00 & 00 & 00 & 05 & 01 & 01 & 0A & 01 & 09 \\ 00 & 00 & 01 & 00 & 00 & 00 & 00 & 00 & 06 & 05 & 01 & 01 & 0A & 01 \\ 00 & 00 & 00 & 01 & 00 & 00 & 00 & 00 & 0C & 06 & 05 & 01 & 01 & 0A \\ 00 & 00 & 00 & 00 & 01 & 00 & 00 & 00 & 09 & 0C & 06 & 05 & 01 & 01 \\ 00 & 00 & 00 & 00 & 00 & 01 & 00 & 00 & 01 & 09 & 0C & 06 & 05 & 01 \\ 00 & 00 & 00 & 00 & 00 & 00 & 01 & 00 & 0A & 01 & 09 & 0C & 06 & 05 \\ 00 & 00 & 00 & 00 & 00 & 00 & 00 & 01 & 01 & 0A & 01 & 09 & 0C & 06 \end{pmatrix} \quad (12)$$

7 Full Description of the Block Cipher

One round of the block cipher is pictured in Figure 1. In the next section we analyze the number of rounds required to achieve good security.

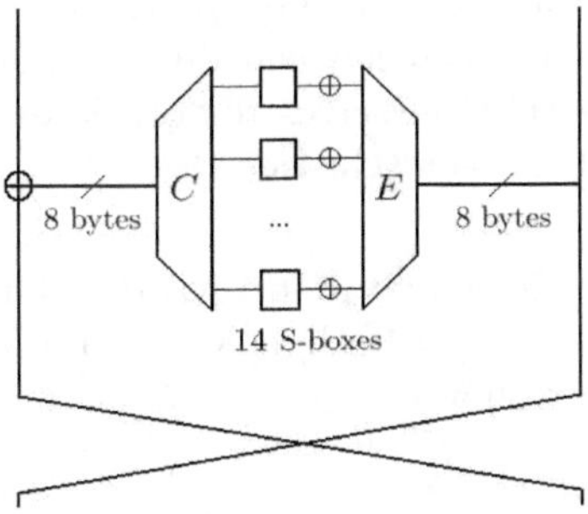

Fig. 1. One round of the cipher

7.1 Evaluation of the Number of Rounds

Differential Cryptanalysis As our S-box is differentially 4-uniform, the probability of any non-trivial 1-round characteristic is at most $(4/2^8)^7$. Therefore a differential characteristic over $2t$ rounds has probability at most $(4/2^8)^{7t}$. In order to upper bound the probability of any differential characteristic by 2^{-127}, at least $2t = \frac{127}{3\cdot 7} \simeq 6$ rounds are necessary.

Linear Cryptanalysis Our S-box has non-linearity $nl = 94$; hence, the bias of its best linear approximation is $\frac{128-94}{256}$. Over one round the bias of any non-trivial linear characteristic is at most $1/2 \cdot (34/128)^7$, and over $2t$ rounds it is $1/2 \cdot (34/128)^{7t}$. As the data complexity of linear cryptanalysis is $\mathcal{C} = \frac{1}{2\epsilon^2}$, we must have

$$\frac{1}{2 \cdot (1/2 \cdot (\frac{34}{128})^{7t})^2} \geq 2^{128} \tag{13}$$

which gives a lower bound of $2t = \frac{127}{7(7-\log_2(34))} = 9.5$ rounds in order to ensure an attack complexity at least 2^{128}.

A security margin must be added to take linear hull effects into account, and to deal with nR- (i.e. key guess) attacks. It is why we decided to use 12 rounds.

7.2 The Key Schedule

We have to derive 12 round keys $k^1, \ldots, k^{12}$ of 112 bits each[2] from one 128-bit master key K. We want our scheme to resist known attacks on a key schedule algorithm, in particular related-key attacks [1, 24] and slide attacks [5, 6]. A detailed security analysis of the key schedule will be published in an extended version of this paper, available on the ePrint (`http://eprint.iacr.org/`).

The key schedule must also be easy to implement; one very desirable property is the ability to derive round keys *on-the-fly* in both encryption and decryption

[2] 112 bits are required because of the use of the expansion layer.

mode (which is possible for DES, but not for AES). It is our belief that designing highly complicated non-linear key schedules is not mandatory to have good security. It is why we restrict ourselves to rotations, bitwise additions and bit selection in the design of the key schedule.

Round Key Derivation in Encryption Mode. The round keys are extracted from an *extended key* $(\kappa^1, \kappa^2, ..., \kappa^{12})$ by a simple bit selection. The κ^i's are 128-bit long and computed as follows:

$$\begin{cases} \kappa^1 = K \\ \kappa^i = T(K) \ggg \Theta(i) & \text{for } i = 2, 4, 6, 8, 10, 12 \\ \kappa^i = K \ggg \Theta(i) & \text{for } i = 3, 5, 7, 9, 11 \end{cases} \quad (14)$$

where $\ggg j$ is the right-rotation of j bits, and Θ is given by the following table:

i	2	3	4	5	6	7	8	9	10	11	12
$\Theta(i)$	1	16	17	32	33	85	86	101	102	117	118

Regarding T, it is defined as follows. Let us write $K = (K^{(1)}, K^{(2)}, K^{(3)}, K^{(4)})$, where $K^{(i)} \in \mathrm{GF}(2)^{32}$. Then

$$\begin{pmatrix} T(K)^{(1)} \\ T(K)^{(2)} \\ T(K)^{(3)} \\ T(K)^{(4)} \end{pmatrix} = \begin{pmatrix} 0\,1\,1\,1 \\ 1\,0\,1\,1 \\ 1\,1\,0\,1 \\ 1\,1\,1\,0 \end{pmatrix} \cdot \begin{pmatrix} K^{(1)} \\ K^{(2)} \\ K^{(3)} \\ K^{(4)} \end{pmatrix} \quad (15)$$

We note that T is involutive. Therefore it is easy to derive κ^{i+1} from κ^i by applying T followed by a rotation of a given number of bits. We describe such iterative computation as

$$\begin{cases} \kappa^1 = K \\ \kappa^i = T(\kappa^{i-1}) \ggg \theta(i) & \text{pour } i = 2...12 \end{cases} \quad (16)$$

where θ is

i	2	3	4	5	6	7	8	9	10	11	12
$\theta(i)$	1	15	1	15	1	52	1	15	1	15	1

The round key k^i is obtained from κ^i by extracting the 112 leftmost bits of κ^i: if $\kappa^i = (\kappa^i_1, \ldots, \kappa^i_{16})$ ($\kappa^i_j \in \mathrm{GF}(2^8)$), then $k^i = (\kappa^i_1, \ldots, \kappa^i_{14})$.

Round Key Derivation in Decryption Mode. Given K, the extended key *for decryption* $\kappa'^1 \ldots \kappa'^{12}$ is computed as

$$\begin{cases} \kappa'^1 = T(K) \lll 10 \\ \kappa'^i = T(\kappa'^{i-1}) \lll \theta'(i) & \text{for } i = 2...12 \end{cases} \quad (17)$$

where $\lll j$ is the left-rotation of j bits, and θ' is given by the following table:

i	2	3	4	5	6	7	8	9	10	11	12
$\theta'(i)$	1	15	1	15	1	52	1	15	1	15	1

We remark that once $k'^1 = T(K) \lll 10$ is computed, the sequences of round keys in encryption and decryption mode, respectively, only differ by the direction of the rotations (right for encryption, left for decryption). Again, k'^i ($i \in \{1, \ldots, 12\}$) is computed from κ'^i ($i \in \{1, \ldots, 12\}$) by considering the 112 leftmost bits.

7.3 Performance Analysis

Our block cipher[3] has been implemented (not by the authors, see acknowledgments) on a smart card based on an 8-bit micro-controller, with 4 different masking levels: without masking, and with maskings of order 1, 2, and 3. This implementation has been compared with state-to-the-art implementations of (masked) AES on the same platform. The results are given in Table 2.

Table 2. Implementation results of AES and our algorithm using different masking orders

Number of Kcycles: ciphering		
Version	**AES**	**Our algorithm**
Unprotected	**2**	**26**
Masked order 1	**129**	**94**
Masked order 2	**271**	**160**
Masked order 3	**470**	**253**

We remark that AES in its non-masked version is definitely much faster than the non-masked version of our algorithm. However once we consider masked versions, our algorithm takes the lead, and the difference between both algorithms increases with the order of the masking.

8 Conclusion

This article illustrates how pertinent it is to have side-channel resistance in mind when building a block cipher. To thwart higher order side-channel attacks we focus on the use of masking schemes, of which the complexity is mainly impacted by the cost of S-box implementation. We emphasize on a new criteria for the design of S-boxes and present a construction that shows a good trade-off

[3] To be precise, we need to mention that the block cipher implemented is a preliminary version, which differs from the cipher we described in the compression layer (which was also a [14, 8]-MDS code but different from the one used in the expansion step). We believe that this difference in the block cipher design will not significantly change the performance results given here.

between efficiency and security. The non-bijectivity of the S-box requires us to use a Feistel Network. We point out a weakness in the straightforward use of Feistel Networks when the S-boxes are non-bijective. We propose to circumvent it by the use of MDS codes to build optimal expansion and compression layers. As an achievement of our work, a new block cipher is fully specified.

Acknowledgments. This paper has been initiated during the project *Secure Algorithm*, sponsored by the French Ministry of Finances through the *Systematic Pole*. We would like to thank all members of the project (from Nagra, Telecom ParisTech, Thales, UVSQ) for the many interesting interactions we had.

More specifically, we would like to thank Emmanuel Prouff for his careful proofreading and comments on preliminary versions of this paper. We also thank Reda Najibi, Laurent Albanèse and Jean-Bernard Fisher: Reda implemented the algorithm during his internship at Nagra, under direction of Laurent and Jean-Bernard. They provided us with the performance figures of Section 7.3.

References

1. Biham, E.: New Types of Cryptanalytic Attacks Using Related Keys. In: Helleseth, T. (ed.) EUROCRYPT 1993. LNCS, vol. 765, pp. 398–409. Springer, Heidelberg (1994)
2. Biham, E., Biryukov, A.: An Improvement of Davies' Attack on DES. In: De Santis, A. (ed.) EUROCRYPT 1994. LNCS, vol. 950, pp. 461–467. Springer, Heidelberg (1995)
3. Biham, E., Shamir, A.: Differential Cryptanalysis of DES-like Cryptosystems. In: Menezes, A., Vanstone, S.A. (eds.) CRYPTO 1990. LNCS, vol. 537, pp. 2–21. Springer, Heidelberg (1991)
4. Biham, E., Shamir, A.: Differential Cryptanalysis of DES-like Cryptosystems. Journal of Cryptology 4(1), 3–72 (1991)
5. Biryukov, A., Wagner, D.: Slide Attacks. In: Knudsen, L.R. (ed.) FSE 1999. LNCS, vol. 1636, pp. 245–259. Springer, Heidelberg (1999)
6. Biryukov, A., Wagner, D.: Advanced Slide Attacks. In: Preneel, B. (ed.) EUROCRYPT 2000. LNCS, vol. 1807, pp. 589–606. Springer, Heidelberg (2000)
7. Blaze, M., Schneier, B.: The MacGuffin Block Cipher Algorithm. In: Preneel, B. (ed.) FSE 1994. LNCS, vol. 1008, pp. 97–110. Springer, Heidelberg (1995)
8. Blömer, J., Guajardo, J., Krummel, V.: Provably Secure Masking of AES. In: Handschuh, H., Hasan, M.A. (eds.) SAC 2004. LNCS, vol. 3357, pp. 69–83. Springer, Heidelberg (2004)
9. Brier, E., Clavier, C., Olivier, F.: Correlation Power Analysis with a Leakage Model. In: Joye, M., Quisquater, J.-J. (eds.) CHES 2004. LNCS, vol. 3156, pp. 16–29. Springer, Heidelberg (2004)
10. Carlet, C.: Vectorial Boolean Functions for Cryptography (Chapter 9). In: Crama, Y., Hammer, P.L. (eds.) Boolean Models and Methods in Mathematics, Computer Science, and Engineering, pp. 398–469. Cambridge University Press (2010), Prel. version: `http://www.math.univ-paris13.fr/~carlet/pubs.html`
11. Carlet, C.: Relating three nonlinearity parameters of vectorial functions and building APN functions from bent functions. Des. Codes Cryptogr. 59(1-3), 89–109 (2011)

12. Chari, S., Jutla, C., Rao, J., Rohatgi, P.: Towards Sound Approaches to Counteract Power-Analysis Attacks. In: Wiener (ed.) [49], pp. 398–412
13. Charpin, P., Pasalic, E.: On Propagation Characteristics of Resilient Functions. In: Nyberg, K., Heys, H.M. (eds.) SAC 2002. LNCS, vol. 2595, pp. 175–195. Springer, Heidelberg (2003)
14. Coron, J.-S., Kizhvatov, I.: Analysis and improvement of the random delay countermeasure of ches 2009. In: Mangard, Standaert (eds.) [31], pp. 95–109
15. Daemen, J., Rijmen, V.: The Design of Rijndael. Springer (2002)
16. Davies, D.W., Murphy, S.: Pairs and triplets of DES s-boxes. J. Cryptology 8(1), 1–25 (1995)
17. Durvaux, F., Renauld, M., Standaert, F.-X., van Oldeneel tot Oldenzeel, L., Veyrat-Charvillon, N.: Cryptanalysis of the ches 2009/2010 random delay countermeasure. Cryptology ePrint Archive, Report 2012/038 (2012), `http://eprint.iacr.org/`
18. Feigenbaum, J.: EUROCRYPT 1991. LNCS, vol. 547. Springer, Heidelberg (1991)
19. Genelle, L., Prouff, E., Quisquater, M.: Thwarting higher-order side channel analysis with additive and multiplicative maskings. In: Preneel, Takagi (eds.) [38], pp. 240–255
20. Gierlichs, B., Batina, L., Tuyls, P., Preneel, B.: Mutual Information Analysis. In: Oswald, E., Rohatgi, P. (eds.) CHES 2008. LNCS, vol. 5154, pp. 426–442. Springer, Heidelberg (2008)
21. Goubin, L., Patarin, J.: DES and Differential Power Analysis. In: Koç, Ç.K., Paar, C. (eds.) CHES 1999. LNCS, vol. 1717, pp. 158–172. Springer, Heidelberg (1999)
22. Helleseth, T. (ed.): EUROCRYPT 1993. LNCS, vol. 765. Springer, Heidelberg (1994)
23. Ishai, Y., Sahai, A., Wagner, D.: Private Circuits: Securing Hardware against Probing Attacks. In: Boneh, D. (ed.) CRYPTO 2003. LNCS, vol. 2729, pp. 463–481. Springer, Heidelberg (2003)
24. Kelsey, J., Schneier, B., Wagner, D.: Key-Schedule Cryptanalysis of IDEA, G-DES, GOST, SAFER, and Triple-DES. In: Koblitz, N. (ed.) CRYPTO 1996. LNCS, vol. 1109, pp. 237–251. Springer, Heidelberg (1996)
25. Kim, H., Hong, S., Lim, J.: A fast and provably secure higher-order masking of AES s-box. In: Preneel, Takagi (eds.) [38], pp. 95–107
26. Knudsen, L.R.: Truncated and Higher Order Differentials. In: Preneel, B. (ed.) FSE 1994. LNCS, vol. 1008, pp. 196–211. Springer, Heidelberg (1995)
27. Kocher, P., Jaffe, J., Jun, B.: Differential Power Analysis. In: Wiener (ed.) [49], pp. 388–397
28. Kunz-Jacques, S., Muller, F.: New Improvements of Davies-Murphy Cryptanalysis. In: Roy, B. (ed.) ASIACRYPT 2005. LNCS, vol. 3788, pp. 425–442. Springer, Heidelberg (2005)
29. Lai, X., Massey, J.L., Murphy, S.: Markov ciphers and differential cryptanalysis. In: Feigenbaum (ed.) [18], pp. 17–38
30. Lidl, R., Niederreiter, H.: On orthogonal systems and permutation polynomials in several variables. Acta Arith. 22, 257–265 (1973)
31. Mangard, S., Standaert, F.-X. (eds.): CHES 2010. LNCS, vol. 6225. Springer, Heidelberg (2010)
32. Matsui, M.: Linear cryptanalysis method for DES cipher. In: Helleseth (ed.) [22], pp. 386–397
33. Matsui, M.: On correlation between the order of S-boxes and the strength of DES. In: Santis (ed.) [44], pp. 366–375
34. Nikova, S., Rijmen, V., Schläffer, M.: Secure hardware implementation of nonlinear functions in the presence of glitches. J. Cryptology 24(2), 292–321 (2011)

35. Nyberg, K.: Perfect nonlinear S-boxes. In: Feigenbaum (ed.) [18], pp. 378–386
36. Nyberg, K.: Differentially Uniform Mappings for Cryptography. In: Helleseth, T. (ed.) EUROCRYPT 1993. LNCS, vol. 765, pp. 55–64. Springer, Heidelberg (1994)
37. Oswald, E., Mangard, S., Pramstaller, N., Rijmen, V.: A Side-Channel Analysis Resistant Description of the AES S-Box. In: Gilbert, H., Handschuh, H. (eds.) FSE 2005. LNCS, vol. 3557, pp. 413–423. Springer, Heidelberg (2005)
38. Preneel, B., Takagi, T. (eds.): CHES 2011. LNCS, vol. 6917. Springer, Heidelberg (2011)
39. Prouff, E., Roche, T.: Higher-order glitches free implementation of the AES using secure multi-party computation protocols. In: Preneel, Takagi (eds.) [38], pp. 63–78
40. Rijmen, V., Preneel, B.: Cryptanalysis of McGuffin. In: Preneel, B. (ed.) FSE 1994. LNCS, vol. 1008, pp. 353–358. Springer, Heidelberg (1995)
41. Rijmen, V., Preneel, B., Win, E.D.: On weaknesses of non-surjective round functions. Des. Codes Cryptography 12(3), 253–266 (1997)
42. Rivain, M., Dottax, E., Prouff, E.: Block Ciphers Implementations Provably Secure Against Second Order Side Channel Analysis. Cryptology ePrint Archive, Report 2008/021 (2008), `http://eprint.iacr.org/`
43. Rivain, M., Prouff, E.: Provably secure higher-order masking of aes. In: Mangard, Standaert (eds.) [31], pp. 413–427
44. De Santis, A. (ed.): EUROCRYPT 1994. LNCS, vol. 950. Springer, Heidelberg (1995)
45. Schramm, K., Paar, C.: Higher Order Masking of the AES. In: Pointcheval, D. (ed.) CT-RSA 2006. LNCS, vol. 3860, pp. 208–225. Springer, Heidelberg (2006)
46. Shirai, T., Preneel, B.: On Feistel Ciphers Using Optimal Diffusion Mappings Across Multiple Rounds. In: Lee, P.J. (ed.) ASIACRYPT 2004. LNCS, vol. 3329, pp. 1–15. Springer, Heidelberg (2004)
47. Shirai, T., Shibutani, K.: Improving Immunity of Feistel Ciphers against Differential Cryptanalysis by Using Multiple MDS Matrices. In: Roy, B., Meier, W. (eds.) FSE 2004. LNCS, vol. 3017, pp. 260–278. Springer, Heidelberg (2004)
48. Shirai, T., Shibutani, K.: On Feistel Structures Using a Diffusion Switching Mechanism. In: Robshaw, M. (ed.) FSE 2006. LNCS, vol. 4047, pp. 41–56. Springer, Heidelberg (2006)
49. Wiener, M. (ed.): CRYPTO 1999. LNCS, vol. 1666. Springer, Heidelberg (1999)

Wide Collisions in Practice

Xin Ye and Thomas Eisenbarth

Department of Mathematical Sciences
Florida Atlantic University, Boca Raton, FL 33431, USA
{xye2,teisenba}@fau.edu

Abstract. Detecting collisions in the power consumption of a modern cryptographic engine is usually difficult due to small leakages and possible countermeasures. Wide collisions offer a much stronger leakage that significantly facilitates their detection. This is the first time wide collisions are exploited in a power analysis attack. In this work we introduce a collision detection method based on the detection of characteristic outliers. Detection results are compared to optimized subspace-based templates. We show that the outlier detection method, while not requiring a template building phase, is almost as effective in detecting collisions as the template-based approach.

1 Motivation

After several years of their exploration, side channel attacks remain a threat to embedded cryptographic implementations. Especially various power analysis attacks have been developed [8,9,10] and improved [1,7,12], as have countermeasures against them [11]. Power based collision attacks [13] are mostly disregarded because detection of collisions is usually more sensitive to noise and countermeasures than most other DPA attacks. However, the AES-specific wide collisions as described in [4] offer a huge advantage. Instead of trying to detect a single collision of a byte during an SubBytes operation of the AES algorithm, wide collisions result in a colliding column for a whole round in addition to two SubBytes byte collisions in the prior and anterior rounds. Hence, such collisions are much easier to detect due to increased leakage. This work is the first one to present results of applying wide collisions using power analysis.

Related Work. In 2004, Schramm *et al.* mounted the first collision attack on AES, pointing out the important fact that the collision attack significantly reduces the attack complexity by combining the analytical and side channel approach [13]. Bogdanov in 2007 [2] generalized the concept of internal collisions and improved the attack under the assumption that byte collisions are detectable. In the following year two multiple-differential methods – binary and ternary voting— were raised for collision detection in [6] and multiple-differential collision attacks MDCA were proposed in [3]. A differential cache-collision timing attack on AES was mounted on an ARM microprocessor in 2010 in [4]. This

F. Bao, P. Samarati, and J. Zhou (Eds.): ACNS 2012, LNCS 7341, pp. 329–343, 2012.

is a chosen plaintext attack based on the occurrence of wide collisions and its algebraic properties.

Another important approach, the template based attack, was firstly introduced in [8] in 2002. Archambeau *et al.* in 2006 [1] suggested a principal subspace-based template attack which overcomes the problem of selection of important points and specification of the minimal distance between points. Recently Bogdanov *et al.* [5] combined collision attack with divide-and-conquer attacks as DPA and template attack to further reduce the computational complexity.

Our Contribution. In this paper we propose a new practical method – the *outlier method* – for detecting wide collisions in AES with a high success probability. It is for the situation where the creation of templates is impossible. It has no requirement on any prior knowledge concerning the leakage model, but it requires knowledge of leaking time instances in the power traces. The attack is based on the assumption that two traces are more likely to form a wide collision pair if they are far away from the mean trace of all the measurements and, at the same time close to each other. The results of the outlier method attack are compared to the detection rates of the PCA based template attack proposed in [1]. The strength of the template-based approach lies in that the principal components are not computed from all the traces as a whole, but instead from the 256 bin average traces — each of which is the mean trace of a bin that has been trained. The idea is to magnify of Inter-Bins Variation through PCA. We extend this idea to an iterative PCA algorithm to make further separation amongst those bins that are still close to each other in a previous iteration.

The organization of this paper is as follows. In Section 2 we review the collision attacks, wide collisions and template attack. In Section 3 we describe the performed attacks. Section 3.1 gives a detailed description of the outlier method. In Section 3.2 the concepts of Inter-Bins Variation and Inner-Bin Variation are introduced. Sections 3.3 and 3.4 introduce the PCA based template attack followed by further improvements through repeated applications of PCA. In Section 4 we analyze the influencing factors for the outlier method and compare the results for the template based collision detection in conditions of reduced templates in the time domain, full templates in the time domain and the principal subspace.

2 Background

The following gives an overview on collision attacks in general and describes the properties of wide collisions. Furthermore, the template attack is introduced.

2.1 Collision Attack

In general, an internal collision in a cryptographic primitive occurs if some specific target function ϕ produces the same output value y for two different inputs x_1, x_2, that is, $\phi(x_1) = y = \phi(x_2)$. Internal collisions in AES were defined by [13] and generalized by [2]. Collisions occur in the output of the MixColumns transformation in each round function of AES, which takes two different plaintexts (or internal states) as input and outputs the same byte value. For example, an internal collision at byte 0 in round 1 occurs for two plaintexts P, Q, where $P = (p_{ij})$, $Q = (q_{ij})$, $i, j = \{0, 1, 2, 3\}$ if

$$02 \cdot p'_{00} \oplus 03 \cdot p'_{10} \oplus 01 \cdot p'_{20} \oplus 01 \cdot p'_{30} = 02 \cdot q'_{00} \oplus 03 \cdot q'_{10} \oplus 01 \cdot q'_{20} \oplus 01 \cdot q'_{30}$$

where all the p'_{ij}, q'_{ij} refer to the byte values of the internal state before the MixColumns operation. Since each $p'_{i0} = S(k_{ii} \oplus p_{ii})$ and $q'_{i0} = S(k_{ii} \oplus q_{ii})$, the above equation contains information about a part of the key. The idea of side channel collision attack on AES is therefore to detect internal collisions from the side channel leakages and then to reduce the number of possible subkey candidates by making use of the above equations, as described in [13] and [2]. One of the main challenges is to get reliable detection of such collisions.

2.2 Wide Collisions

Wide collisions for AES are defined in [4]. They are a special case of internal collisions in the following way. Specific plaintexts are chosen to satisfy the condition that the bytes off the diagonal are pairwisely equal i.e. $p_{ij} = q_{ij}$ for all $i \neq j$. After such chosen plaintexts entering the encryption engine, one can track the byte of an internal collision occurring at the first round MixColumns. After the ShiftRows operation of the next round, the entire column where this collision byte was shifted to will collide. Therefore 4 more internal byte collisions can be observed after the second round MixColumns.

Consequently, this gives rise to one byte collision in Round 2 SubBytes and four additional byte collisions in Round 3 SubBytes, resulting in a total of five byte collisions. This phenomenon is referred to as *wide collision*. For example, if two plaintexts collide at byte 0 after Round 1 MixColumns ($p'_{00} = q'_{00}$), they will continue to collide in Round 2 SubBytes ($S(k'_{00} \oplus p'_{00}) = S(k'_{00} \oplus q'_{00})$). After ShiftRows, all the four bytes in the Column 0 are pairwisely equal and thus we get four additional collisions after Round 2 MixColumns ($p''_{i0} = q''_{i0}$, $i = 0, 1, 2, 3$), which will remain colliding in Round 3 SubBytes ($S(k''_{i0} \oplus p''_{i0}) = S(k''_{i0} \oplus q''_{i0})$).

The wide collision attack is described as a three stage algorithm [4]: an online stage where side channel leakage is measured for the chosen plaintexts, a collision detection stage which returns several pairs of plaintexts which most likely give wide collisions, and finally a key recovery stage. It is mentioned that every 4 wide collisions for each of the diagonally chosen plaintexts sufficiently reduce the subkey space with a remaining uncertainty of 2^8. Hence, a total of 16 correctly detected collisions reduces the number of remaining key candidates to 2^{32}, which can be exhaustively searched on an average PC within minutes.

2.3 Template Attack

Template attack is a powerful side channel attack.In [8] power traces are characterized with a multivariate Gaussian distribution model. The attack assumes that each individual trace follows a multivariate normal distribution with the center and covariance that are identical to the ones of the correct bin. In other words, traces of the same bin form a Gaussian distribution. Therefore each bin can be characterized by a template $(\mathbf{m}, C)$, where $\mathbf{m}$ refers to the mean trace and the C is the covariance matrix.

A template attack consists of template building phase and template matching phase. In the building phase, one characterizes the device with a set of templates $(\mathbf{m}_i, C_i)$, each of which is created for one bin of traces. In the matching phase each analyzed trace $\mathbf{x}$ is matched to the template that it most likely belongs to. That is, it outputs the template which gives the highest probability density computed by

$$Pr\left(\mathbf{x}; (\mathbf{m}_i, C_i)\right) = \frac{\exp\left(-\frac{1}{2}\left(\mathbf{x} - \mathbf{m}_i\right)^T C_i^{-1}\left(\mathbf{x} - \mathbf{m}_i\right)\right)}{\sqrt{(2\pi)^N \det\left(C_i\right)}}$$

where N is the length of $\mathbf{m}_i$, the number of points in the mean trace. This is called the full template matching.

A simplified approach is referred as the reduced template matching in which the dependency amongst different data points are disregarded. That is, for each template the covariance matrix is replaced with the identity matrix so that the computation of the probability density is simplified as

$$Pr\left(\mathbf{x}; \mathbf{m}_i\right) = \frac{\exp\left(-\frac{1}{2}\left(\mathbf{x} - \mathbf{m}_i\right)^T\left(\mathbf{x} - \mathbf{m}_i\right)\right)}{\sqrt{(2\pi)^N}}$$

Such reduced model is equivalent to computing the Euclidean distance between each analyzed trace $\mathbf{x}$ and the bin average traces $\mathbf{m}_i$ of the templates. It determines as the correct matching the template $\mathbf{m}_j$ that is nearest to the analyzed trace $\mathbf{x}$, i.e. $\| \mathbf{x} - \mathbf{m}_j \| \leq \| \mathbf{x} - \mathbf{m}_i \|$, for all i.

3 Practical Collision Attacks

In classical collision attacks, bins are formed in the way that certain bytes of the intermediate state of the cipher collide to the same value. When the same value is processed by certain operations (e.g. MixColumns, SubBytes in AES) of the cryptographic primitive, the pattern of the power consumption of these operations should be highly similar. Consequently, all possible 256 values of a given byte give rise to 256 bins, hence 256 different patterns for each colliding byte position. In a wide collision scenario, traces of each bin provide at least 5 internal byte collisions, spanning from MixColumns in round one up to SubBytes in round three of an AES execution. This results in a highly similar power consumption over a relatively long time period, especially for serial implementations. Hence, detecting wide collisions should be much easier than detecting simple collisions.

Conventionally, leaking points of the power traces refer to a subset of samplings of measurement or its transformation which represents the characteristics of the pattern of the power traces. Locating these points usually requires knowledge of the implementation and leakage properties of the platform or profiling. As an example, the two plots in Figure 1 show a single trace (the upper plot) over the region from round one MixColumns to round three SubBytes and a differential trace (the lower plot) calculated as the absolute value of difference between an average of traces of bin 7 and the average of all traces obtained. It can be seen from this figure that peaks, which indicate the location of promising leaking points, are spread out all over this region.

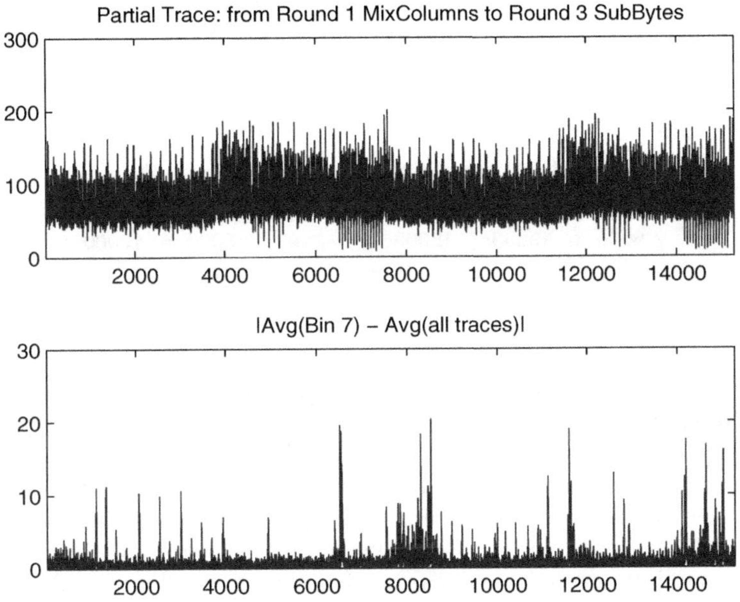

Fig. 1. Important Point Distributions

Another observation of leakages in the wide collision attack is that the positions of peaks are not invariant with respect to all wide collision bins. Some bins share one or more positions of leakages, while no pair of two different bins follows an identical pattern. Figure 2 gives an intuitive idea of the distribution of leakages for some bins.

Hence, picking only a single point from all important points for the purpose of collision detection is rather risky because only few bins leak at this point. In other words, if the closeness of two traces at one fixed point is the only criterion for wide collision detection, the detection can only succeed in rare cases because power traces of other bins that do not leak locally at this point are dominantly influenced by noise. Such traces make it difficult for a correct

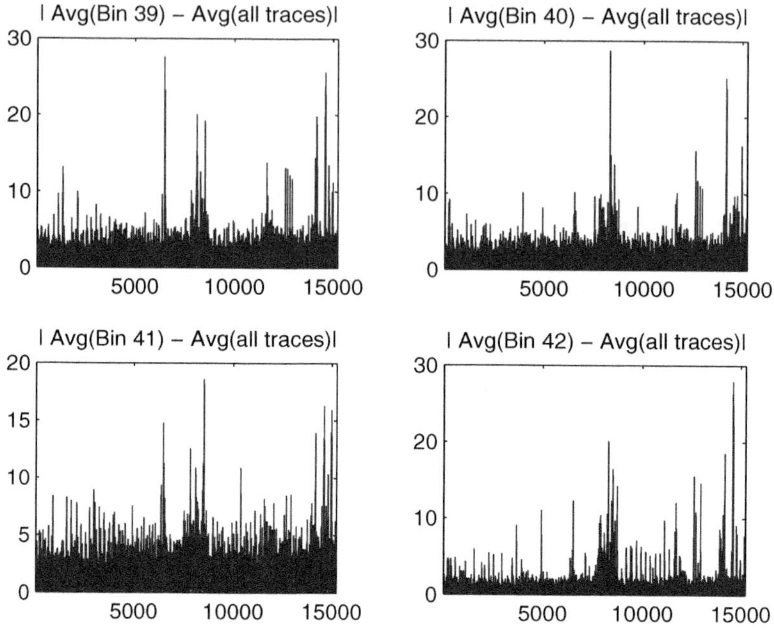

Fig. 2. Important Points for different Bins

collision detection, since, while randomly scattered, they are far more numerous than colliding traces.

We distinguish two different kinds of multiple-point-based approaches: One builds on a template-based detection, the other does not require templates. When generating templates is not possible, we propose an outlier method which assumes that a pair of traces forms a collision in the case they are close to each other and simultaneously far away from the average of all traces. As a comparison, we show that wide collisions can easily be detected using templates. The challenge in this case is to discover the characteristics of each pattern and to correctly recognize each individual power trace from all the patterns with high probability.

Furthermore, we introduce the concepts of Inner-Bin-Variation and Inter-Bins-Variation as two parameters determining the effect of collision detection. We propose methods with the application of Principal Component Analysis (PCA) and iterative PCA so that the idea of maximizing Inter-Bins-Variation is realized.

3.1 Outlier Method

Generally, the outlier method assumes that two traces in the outlier region – with distance sufficiently far away from the average of all traces – are more likely to form a collision pair if additionally they are sufficiently close to each other. It includes one distance function $dist(\mathbf{x}, \mathbf{y})$ that gives a distance metric between two trace representatives $\mathbf{x}$ and $\mathbf{y}$. It also includes two distance parameters R and r, where R is the outlier lower bound ratio determining if one trace is inside the outlier region, and r is the mutual distance upper bound ratio determining if two traces are close by enough. Both R and r should be a number between 0 and 1. The procedure of the outlier method as follows:

Step 1: Use $\left(\mathbf{x}^{(1)}, ..., \mathbf{x}^{(D)}\right)$ as the representation of D traces collected in the online stage and compute the average $\bar{\mathbf{x}}$ of all trace representation $\mathbf{x}^{(i)}$ at this point by $\bar{\mathbf{x}} = \sum \mathbf{x}^{(i)}/D$

Step 2: Compute the distances vector $\mathbf{d} = \left(d^{(1)}, ..., d^{(D)}\right)$ where each entry $d^{(i)} = dist\left(\bar{\mathbf{x}}, \mathbf{x}^{(i)}\right)$ is the distance between the trace $\mathbf{x}^{(i)}$ and the average trace $\bar{\mathbf{x}}$. Find the maximum element of the vector $\mathbf{d}$ by $maxd = \max(\mathbf{d})$.

Step 3: Find the set A of outliers by

$$A = \left\{\mathbf{x}^{(i)} \mid d^{(i)} \geq R \cdot maxd\right\}$$

It is a collection of trace representations with distance of no less than $R{\cdot}maxd$ from the average trace $\bar{\mathbf{x}}$. Figure 3 gives an example of the location of the outlier region.

Step 4: Find the list of pairwise distance

$$B = \left\{d_{ij} = dist\left(\mathbf{x}^{(i)}, \mathbf{x}^{(j)}\right) \text{ where } \mathbf{x}^{(i)}, \mathbf{x}^{(j)} \in A, i \neq j\right\}$$

Note that if there are n outliers in set A, then the set B contains distances of $n(n-1)/2$ pairs of traces.

Step 5: Find the set C by

$$C = \left\{\left(\mathbf{x}^{(i)}, \mathbf{x}^{(j)}\right) \mid d_{ij} \leq r \cdot maxd \text{ , } \mathbf{x}^{(i)}, \mathbf{x}^{(j)} \in A\right\}$$

This is a filteration from the set B of those pairs with mutual distance greater than $r \cdot maxd$. The set C is the output of the outlier method containing pairs of traces that are promising candidates for collisions.

Please note that in general the distance function $dist(\cdot)$ is a combination of all components. In practice, one could use a Euclidean distance for the *dist* function, viewing the trace representatives as elements in a finite dimensional vector space and assuming that components are independent from one another and each component contributes equally to the resulting distance.

Pros and Cons. The existence of leaking points is a necessary prerequisite for the wide collision detection. The points depend on the target device and implementation and should be chosen wisely by the attacker. Usually, either prior knowledge about the implementation or a profiling phase is needed. For detecting significant leakage points, SPA or DPA methods can be applied.

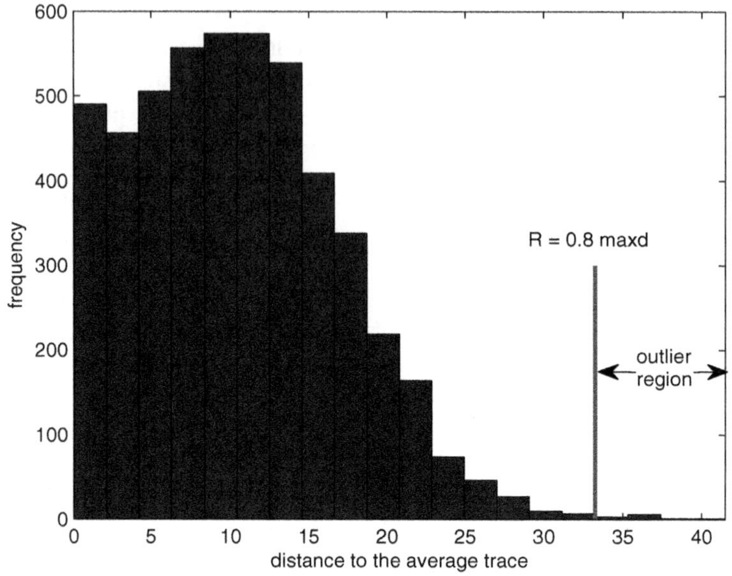

Fig. 3. outlier detection step 3

Choice of the parameters R and r is subjective. In practice, decreasing R while increasing r will eventually result in more non-colliding pairs that are detected as collisions. On the other side, decreasing r and increasing R will eventually increase to the set of detected collisions that does not contain enough pairs for key recovery.

The Euclidean distance is a convenient metric, since it is a straightforward combination of the influence of each leaking point. However, it is often weaker than a template attack on the same points. Using Euclidean distance makes two additional non-justified assumptions: that the points leak independently of one another and that they contribute equally to the output distance. For improved detection results, it can be replaced by some function $g(\cdot)$ such that the influence of different points can be more accurately reflected.

3.2 Inter-Bins Variation and Inner-Bin Variation

One way of selecting significant leakage points for collision detection is described in [5] and [6]. Both publications describe the *maxmin* function for selecting the most informative point of power traces and computing characteristics from traces at this point. Specific speaking, they first find for each fixed time point the lowest signal difference between all pairs of traces. They then find the time point that gives rise to the biggest one among all the lowest signal differences and consider that point to be the best choice. The logic of this method is that the lowest signal difference determines the level of difficulty of trace separtion at each time point. The larger such lowest signal difference, the easier the separation of traces. However, this method makes use of a single point of the power trace, while ignoring all other remaining points. It yields for the attack a strong reliance on the single point that has been selected, which still suffers the risk of being influenced by the signal noise for each individual collision detection.

In contrast to single point selection, we propose an improved method. First, one should only include the leaking part of the power traces, i.e. the targeted round. For example, in wide collision attack, we only analyze the region starting from round 1 MixColumns to round 3 SubBytes. In this situation, two parameters – *inter-bins variation* (ITV) and *inner-bin variation* (INV) – determine the ease and probability of correct collision detection.

ITV describes the variation of the characteristics of the averaged traces $\mathbf{m}_i$ of each bin value. It is computed as a Euclidean distance as

$$\mathrm{ITV} = \sqrt{\sum_i (\mathbf{m}_i - \bar{\mathbf{m}})^2}$$

where $\bar{\mathbf{m}} = \sum \mathbf{m}_i/256$ is the mean trace of all the bin average traces. An increased ITV indicates an easier separation of the bins and more accurate pattern matching of each individual trace. Notice that the computation of ITV can also be applied to any representation of the traces. We refer the notion of maximizing ITV as computing the maximal ITV amongst all representations of the traces. Maximizing the ITV is therefore desired for the successful collision detection.

INV describes the variation of the characteristics of each individual trace $\mathbf{x}_j^{(i)}$ from that of the averaged trace $\mathbf{m}_i = \mathrm{avg}_j \left(\mathbf{x}_j^{(i)}\right)$ of a particular bin B_i. That is

$$\mathrm{INV}\,(i) = \sqrt{\sum_j \left(\mathbf{x}_j^{(i)} - \mathbf{m}_i\right)^2}$$

INV can similarly be applied to any representation of the traces and we refer minimizing INV as computing the minimal of INV amongst all representations of the traces. Note that INV is a tuple of 256 entries, corresponding to the inner-bin variation of 256 bins, and minimizing one entry does not imply small values for the rest of entries. Hence, although minimizing INV is desired, it might not be practically feasible.

One should note that the existence of the bin average traces does not necessarily guarantee the feasibility of its computation. In fact, only if one can build up templates for the 256 bins, one can also obtain a raw representation of the average traces of bins. Since each representation gives a computational result for the ITV, the adversary would profit from a representation that maximizes the ITV.

In our experience, we realize the magnification of ITV through finding the representation of the average traces in the principal subspace, which are detailed in 3.3 and 3.4.

3.3 Template-Based Collision Detection

In cases where creating templates is possible, the attacker can build a template for each bin of collision in the time domain, as detailed in [11] and described in Section 2.3. Point selection can be automated by using *principal component analysis* (PCA) [1]. Template-based collision detection can achieve good detection rates, as shown in Section 4.2. PCA is a three step algorithm:

1. Finding the mean vector $\bar{\mathbf{x}}$ and the centered data matrix $\mathbf{X}' = (\mathbf{x}'_1, ..., \mathbf{x}'_D)^T$ of all the raw data record $\mathbf{X} = (\mathbf{x}_1, ..., \mathbf{x}_D)^T$, where $\mathbf{x}'_i = \mathbf{x}_i - \bar{\mathbf{x}}$;
2. Computing the covariance matrix $\mathbf{S} = \frac{1}{D}(\mathbf{X}')^T(\mathbf{X}')$ and its d eigenvectors $(\mathbf{v}_1, ..., \mathbf{v}_d)$ corresponding to the largest d eigenvalues $(\lambda_1, ..., \lambda_d)$ of $\mathbf{S}$;
3. Projecting $\mathbf{X}$ into the subspace spanned by the d eigenvectors (also called components) $\mathbf{Y} = \mathbf{X}(\mathbf{v}_1, ..., \mathbf{v}_d)$.

PCA performs an orthogonal projection into a subspace called principal subspace. The projection maximizes the variance of the data. Hence, a point selection with minimal information loss becomes possible.

Constructing templates in principal subspace is only one additional step to the build-up of the templates in the time domain. That is, the raw traces need to be projected into the principal subspace before the construction of templates. For this step, the principal components could be obtained in two ways: from all the raw traces $\mathbf{X}$, or from the bin average matrix $\mathbf{M} = (\mathbf{m}_0, ..., \mathbf{m}_{255})^{\mathbf{T}}$, consisting bin average traces $\mathbf{m}_i$ of bin B_i where $\mathbf{m}_i = \text{avg}\{\mathbf{x}_j \mid \mathbf{x}_j \in B_i\}$. The consequence of applying PCA for the first choice is the maximization of the variance amongst individual traces and for the second approach amongst bin average traces. It is clear that the second method —computing principal components from bin averages— is desired because it achieves the goal of maximizing the ITV. After getting the projected traces in the principal subspace, the regular template building — the computation of bin averages and bin covariance — is performed as discussed in Section 2.3.

Finally, in the template matching phase, each analyzed trace is firstly projected onto principal components, then matched to the closest template through the evaluation of the probability densities. Every two different analyzed traces that are matched to the same template form a collision pair.

3.4 Template-Based Collision Detection Using Iterative PCA

A further improvement can be achieved by repeatedly conducting PCA. This gives rise to an iterative algorithm using projection. That is, in the template building phase, if two bins are too close or overlapping after the first PCA projection, one can repeat PCA projection (for which the computation of components only involves the average traces of that two bins) to further separate those two bins. The algorithm is given as follows:

Step 1: Use $\mathbf{M}^{(1)} = \{\overline{\mathbf{m}}_0, ..., \overline{\mathbf{m}}_{255}\}$ to compute the first set of principal components $\mathbf{V}^{(1)} = (\mathbf{v}_1, ..., \mathbf{v}_r)$ and the projection of each trace $\mathbf{P}^{(1)} = \mathbf{X} \cdot \mathbf{V}^{(1)}$.

Step 2: Partition the 256 bins into $C_\alpha^{(1)}$ and $C_\beta^{(1)}$ where bins from $C_\alpha^{(1)}$ can be clearly separated from other bins, while bins from $C_\beta^{(1)}$ are still clustered with some other bins. That is, if bin $B_i \in C_\beta^{(1)}$, then there exists bin $B_j \in C_\beta^{(1)}$ such that traces of bin B_i are not separable from traces in B_j.

Step 3: If $C_\beta^{(1)}$ is not empty, compute the set $\mathbf{M}^{(2)}$ which consists of the averages of projected traces of non-separable bins in $C_\beta^{(1)}$

$$\mathbf{M}^{(2)} = \left\{ \overline{\mathbf{m}}_i = \mathrm{avg}_j \left\{ \mathbf{p}_j \mid \mathbf{p}_j \in B_i \right\} \mid B_i \in C_\beta^{(1)} \right\}$$

Then based on $\mathbf{M}^{(2)}$, compute a second set of principal components $\mathbf{V}^{(2)}$ and obtain another set of projected traces $\mathbf{P}^{(2)} = \mathbf{P}^{(1)} \cdot \mathbf{V}^{(2)}$ from the previously projected ones.

Step 4: Repeat steps 2 and 3 until after k iterations $C_\beta^{(k)}$ is empty so that all bins are sufficiently separated.

4 Experimental Results

All experiments have been performed on a smart card featuring an 8-bit microcontroller based software implementation of AES. The measurements have been performed using a Tektronix digital sampling oscilloscope with an 8 bit A/D converter. The sampling rate of 50MS/s provides about 12 sampling points per clock cycle.

We first evaluate the detection rate of the outlier method. We explore the impact on the detection rate of several parameters: the number of promising leaking points, the choices of the outlier lower bound ratio R and the mutual distance upper bound ratio r, as well as the number of traces being investigated. Detection results are shown in Tables 1 through 3. The first column contains the analyzed influencing factor. The second column is the average size of set A, i.e. the average number of the outliers, as described in Section 3.1. Again, if n traces were in the outlier region, $n(n-1)/2$ pairs are further analyzed by computing pairwise closeness. The third column shows the size of the set C, i.e. the average number of output pairs of promising collisions. The next column counts the number of correctly detected collisions, that is the detected pairs which actually form wide collisions. The last column is the ratio between the third and the fourth column, i.e. the ratio of correctly detected collisions.

For comparison we apply template-based collision detection in three different scenarios, (1) reduced templates in the time domain, (2) full template in time domain and (3) full template in the principal subspaces. Figure 4 shows how many traces per bin are necessary for training good templates and further indicates the asymptotic recognition rate for the three cases for our platform.

4.1 Results for the Outlier Method

We apply the outlier method as detailed in Section 3.1 to detect wide collision from the power traces. In our experiment, we define the distance function with the norm $\|\cdot\|_1$. That is,

$$dist\,(\mathbf{x},\mathbf{y}) = \parallel \mathbf{x}-\mathbf{y} \parallel_1 = \sum_i |x_i - y_i|$$

gives the distance from trace $\mathbf{x} = (x_1, ..., x_t)$ to $\mathbf{y} = (y_1, ..., y_t)$. In our experiments we use 3000 traces and fix the parameter R to 0.9 and r to 0.3. We locate between 1 to 8 promising leaking points to analyze the influence of the combination of leaking points. Table 1 confirms that using multiple points results in a better collision detection rate comparing to the case of locating only one important point per power trace.

Table 1. Collision Detection: Single point vs. multiple points

number of leaking points	number of outliers	number of detected pairs	number of correct detection	average rate of correct detection
1	19.6	127.7	21.9	23.0%
4	30.6	46.3	33.4	71.1%
6	110.7	126.3	105.4	86.2%
8	81.7	88.1	82.3	93.7%

Next, we explore different choices of the parameter pair (R, r) to analyze the effect on the collision detection rate. As explained in the algorithm in Section 3.1, R is the parameter determining which traces are in the region of "outliers" (the set A), sufficiently far away from the center of all traces. The larger R is, the fewer traces are considered as outliers. On the other hand r is the parameter that determines if two outliers are close enough to each other. The smaller r is, the fewer pairs of traces are detected as collisions, namely the smaller the cardinality of the set C . Our experiments use 3000 traces (the same as above) and fix 6 locations of promising leaking points. They confirm that the stricter the choice of (R, r), i.e. the larger choice of R and the smaller choice of r, the more accurate the detection is, as shown in Table 2. As a last analysis of the

Table 2. Collision Detection: Choices of R and r

choice of (R, r)	number of outliers	number of detected pairs	number of correct detection	average rate of correct detection
$(0.7, 0.2)$	382.1	807	551	68.4%
$(0.8, 0.2)$	110.7	126.3	105.4	86.2%
$(0.9, 0.2)$	19.9	8.3	7.7	89.6%
$(0.9, 0.3)$	19.9	16.1	12.9	81.3%
$(0.9, 0.4)$	19.9	22.9	13.9	60.8%

outlier method, we explore the relationship between the successful detection and the number of traces being used in the experiment. In this experiment, 6 leaking points are fixed, the parameter R is set to 0.8 and r is 0.2. It is found that increasing the number of traces yields the increase in the number of outlier traces and the number of pairs being detected, meanwhile the detection rate does not significantly increase, as shown in Table 3.

Table 3. Collision Detection: Impact of number of traces used

number of traces	number of outliers	number of detected pairs	number of correct detection	average rate of correct detection
1000	37.1	13.4	12.1	93.6%
3000	81.7	88.1	82.3	93.7%
5000	118.7	217.1	200.1	93.7%
7000	127.3	277	256.9	94.3%

4.2 Results for Template-Based Detection

In the preceding outlier method, it is assumed that one can locate several leaking points and these points are independent of each other and contribute equally to the computation of distance function. The same assumptions hold if a reduced template attack is mounted in the time domain. But if a full template attack is applied, these assumptions do not need to be fulfilled. If the templates are built in the time domain, one only needs to locate good leaking points. While templates built in principal subspaces, as described in Section 3.3, even locating leaking points is no longer necessary. This is because the attacker can make use of all the region of power traces that corresponds to wide collisions operations. Our experiment compares the method using reduced template and the full template to see how the dependency amongst leaking points helps with assigning an analyzed trace to its collision bin. We also compare the recognition rate between templates in the time domain and in the principal subspace through which we can verify that magnifying ITV enhances the recognition rate.

Our experiments use 8 leaking points in the time domain. The counterpart in PCA is 8 principal components that correspond to the 8 largest eigenvalues computed as specified in Section 3.4. We use 2560 to 5632 traces to build up templates so that each template makes use of 10 to 22 traces. We test 1000 traces to match to the templates and interpret #(*correct* recognition)/1000 as the recognition rate. From Figure 4 we can draw the following conclusions:

1. Using reduced templates in the time domain gives very stable recognition rate for a wide range of the number of used training traces for the templates. The asymptotic rate is at approx. 0.8. This is lower than the result for the two full models when sufficient training traces are provided. Therefore, the assumption of independence of leaking points cannot provide a strong collision distinguisher.
2. Full templates with PCA gain better recognition results comparing to the full templates in the time domain. Full templates in principal subspace gain close to 0.95 recognition rate given more than 20 training traces per bin. While

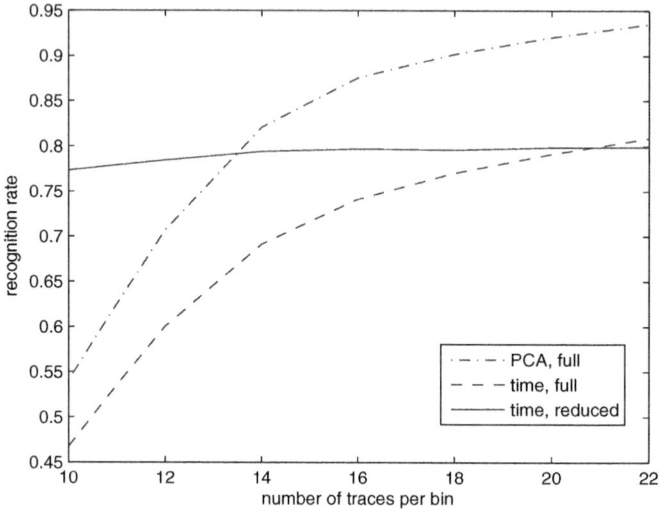

Fig. 4. Templates in the time domain and in the principal subspace

full model in the time domain achieves only around 0.8 to 0.85. This confirms the contribution of PCA for maximizing ITV in terms of recognition.

3. Full models have stricter requirements on the number of training traces. In particular, if the attacker chooses n leaking points in the time domain or n principal components in the principal subspace, then the number of training traces per bin cannot be less than n, otherwise a singular covariance matrix C is an unavoidable result and this makes the computation of probability density infeasible. Even when the least number of training traces is satisfied, the computation of the covariance matrix can still be remarkably impacted by the noise in the side channel. That is why the recognition rate for both of the full templates is low when fewer than 14 traces per bin are used.

5 Conclusion

This paper presents the first wide collision based power analysis attack. One major finding is the outlier collision detection method. It shows that power traces that are close to each other in the outlier region have a highly increased chance of forming a collision pair. It is a strong method for detecting wide collisions. Unlike template attacks, it does not require extensive profiling. However, it is shown that template-based approaches are a great method for detecting collisions if template building is possible. Different ways of building collision-detecting templates are compared. Using PCA for finding independent strong leaking points seems to be better than hand-picking points in the time domain. However, sufficiently many measurements for each bin must be available during template building.

Acknowledgment. The work described in this paper has been supported in part by the National Science Foundation under Grant No. 1054776.

References

1. Archambeau, C., Peeters, E., Standaert, F.-X., Quisquater, J.-J.: Template Attacks in Principal Subspaces. In: Goubin, L., Matsui, M. (eds.) CHES 2006. LNCS, vol. 4249, pp. 1–14. Springer, Heidelberg (2006)
2. Bogdanov, A.: Improved Side-Channel Collision Attacks on AES. In: Adams, C., Miri, A., Wiener, M. (eds.) SAC 2007. LNCS, vol. 4876, pp. 84–95. Springer, Heidelberg (2007)
3. Bogdanov, A.: Multiple-Differential Side-Channel Collision Attacks on AES. In: Oswald, E., Rohatgi, P. (eds.) CHES 2008. LNCS, vol. 5154, pp. 30–44. Springer, Heidelberg (2008)
4. Bogdanov, A., Eisenbarth, T., Paar, C., Wienecke, M.: Differential Cache-Collision Timing Attacks on AES with Applications to Embedded CPUs. In: Pieprzyk, J. (ed.) CT-RSA 2010. LNCS, vol. 5985, pp. 235–251. Springer, Heidelberg (2010)
5. Bogdanov, A., Kizhvatov, I.: Beyond the Limits of DPA: Combined Side-Channel Collision Attacks. IEEE Transactions on Computers PP(99), 1 (2011)
6. Bogdanov, A., Kizhvatov, I., Pyshkin, A.: Algebraic Methods in Side-Channel Collision Attacks and Practical Collision Detection. In: Chowdhury, D.R., Rijmen, V., Das, A. (eds.) INDOCRYPT 2008. LNCS, vol. 5365, pp. 251–265. Springer, Heidelberg (2008)
7. Brier, E., Clavier, C., Olivier, F.: Correlation Power Analysis with a Leakage Model. In: Joye, M., Quisquater, J.-J. (eds.) CHES 2004. LNCS, vol. 3156, pp. 16–29. Springer, Heidelberg (2004)
8. Chari, S., Rao, J.R., Rohatgi, P.: Template Attacks. In: Kaliski Jr., B.S., Koç, Ç.K., Paar, C. (eds.) CHES 2002. LNCS, vol. 2523, pp. 13–28. Springer, Heidelberg (2003)
9. Gierlichs, B., Batina, L., Tuyls, P., Preneel, B.: Mutual Information Analysis. In: Oswald, E., Rohatgi, P. (eds.) CHES 2008. LNCS, vol. 5154, pp. 426–442. Springer, Heidelberg (2008)
10. Kocher, P.C., Jaffe, J., Jun, B.: Differential Power Analysis. In: Wiener, M. (ed.) CRYPTO 1999. LNCS, vol. 1666, pp. 388–397. Springer, Heidelberg (1999)
11. Mangard, S., Oswald, E., Popp, T.: Power Analysis Attacks: Revealing the Secrets of Smartcards. Springer (2007)
12. Rechberger, C., Oswald, E.: Practical Template Attacks. In: Lim, C.H., Yung, M. (eds.) WISA 2004. LNCS, vol. 3325, pp. 440–456. Springer, Heidelberg (2005)
13. Schramm, K., Leander, G., Felke, P., Paar, C.: A Collision-Attack on AES. In: Joye, M., Quisquater, J.-J. (eds.) CHES 2004. LNCS, vol. 3156, pp. 163–175. Springer, Heidelberg (2004)

A General Construction for 1-Round δ-RMT and $(0, \delta)$-SMT

Reihaneh Safavi-Naini, Mohammed Ashraful Alam Tuhin, and Pengwei Wang

Department of Computer Science, University of Calgary
{rei,maatuhin,pengwwan}@ucalgary.ca

Abstract. In Secure Message Transmission (SMT) problem, a sender $\mathcal{S}$ is connected to a receiver $\mathcal{R}$ through N node disjoint bidirectional paths in the network, t of which are controlled by an adversary with *unlimited computational power*. $\mathcal{S}$ wants to send a message m to $\mathcal{R}$ in a *reliable* and *private* way. It is proved that SMT is possible if and only if $N \geq 2t+1$. In Reliable Message Transmission (RMT) problem, the network setting is the same and the goal is to provide reliability for communication, only. In this paper we focus on 1-round δ-RMT and $(0, \delta)$-SMT where the chance of protocol failure (receiver cannot decode the sent message) is at most δ, and in the case of SMT, privacy is perfect.

We propose a new approach to the construction of 1-round δ-RMT and $(0, \delta)$-SMT for all connectivities $N \geq 2t + 1$, using list decodable codes and message authentication codes. Our concrete constructions use folded Reed-Solomon codes and multireceiver message authentication codes. The protocols have optimal transmission rates and provide the highest reliability among all known comparable protocols. Important advantages of these constructions are, (i) they can be adapted to all connectivities, and (ii) have simple and direct security (privacy and reliability) proofs using properties of the underlying codes, and δ can be calculated from parameters of the underlying codes.

We discuss our results in relation to previous work in this area and propose directions for future research.

1 Introduction

In a Secure Message Transmission (SMT) system a sender is connected to a receiver through N wires, t of which are controlled by the adversary. Wires are abstractions of bidirectional node disjoint paths in a network. The adversary's control of a wire is by taking complete control of a node or a link on the path, allowing them to stop, inject or change arbitrarily, the messages that are sent on the path. The goal of the system is to provide *reliability* and *privacy* for the transmitted messages against an adversary with *unlimited computational* power without assuming any prior shared key between the sender and the receiver. In *Perfectly Secure Message Transmission* (PSMT) systems, the adversary will not learn anything about the message, and the receiver always correctly receives the sent message. SMT protocols can have one or more *rounds* and their communication efficiency for a given number of rounds is measured by the *transmission*

F. Bao, P. Samarati, and J. Zhou (Eds.): ACNS 2012, LNCS 7341, pp. 344–362, 2012.

rate which is the total number of communicated bits per one message bit. Protocols with the lowest rate for a given number of rounds, are called *optimal.* The initial motivation for this model was to simulate secure links between nodes in a distributed setting (e.g., multi-party computation [1,2]), where there are no direct secure links between nodes but there are multiple paths that connects the two nodes. In recent years, however, the protocols, and in particular 1-round SMT protocols, have found other applications including key agreement and key strengthening in wireless sensor networks (e.g., [22]).

It has been shown [3] that 1-round PSMT is possible if and only if $N = 3t+1$. To increase the number of corrupted wires that can be tolerated by the protocol without increasing the number of rounds, one may sacrifice some reliability. A 1-round $(0, \delta)$-SMT protocol provides perfect privacy and bounds probability of error in receiving the message by δ. These protocols can be constructed for $N \geq 2t + 1$.

A related scenario, known as *Reliable Message Transmission (RMT),* is when the only requirement is the reliability of communication assuming the same network and adversary model (N wires, t of which are controlled by the adversary) as the SMT problem. A trivial protocol for reliable transmission when $N \geq 2t+1$ is by sending the message on all the wires and using majority voting at the receiver to recover the correct message. This will correctly recover the message as only $t \leq \frac{N-1}{2}$ wires are corrupted. However the transmission rate of this protocol is N, which grows linearly with N (similar to repetition codes) and so the goal of δ-RMT protocols is to achieve optimal transmission rate (which is *constant* for 1-round when $N = 2t + K$, where $K \geq 1$ is a constant).

In this paper we consider 1-round δ-RMT and 1-round $(0, \delta)$-SMT protocols.

Towards a systematic construction of 1-round δ-RMT and $(0, \delta)$-SMT. All existing optimal 1-round (0, δ)-SMT protocols use complex combinations of secret sharing and authentication systems for message encoding, together with elaborate secret reconstruction and verification algorithms to construct a decoding algorithm for the SMT. A disadvantage of these complex and clever constructions is the difficulty of verifying their properties. It was shown [12] that the proofs of security of the 1-round (0, δ)-SMT protocol in [18] were not correct.

A limitation of these constructions in practice is that a protocol is designed for a specific type of connectivity (for example, $N = 2t + 1$, or $N = 2t + K, K > 1$, or $N = (2 + c)t, c > \frac{1}{t}$), and when used in a setting with different types of connectivity, the optimality of the protocol cannot be guaranteed. This means that for optimality guarantee, one may need to implement multiple protocols in cases that the connectivity is not known beforehand. Similar observations can be made for optimal 1-round δ-RMT with $N \geq 2t + 1$. The only construction with optimal rate is given in [16]. The construction uses a complex combination of secret sharing for encoding and an elaborate verification for the decoding.

In contrast to the above constructions, there is a simple and elegant construction of an optimal 1-round PSMT [7] for $N = 3t + 1$, that uses Reed-Solomon (RS) code. The encoding and decoding in this construction are encoding and decoding of RS-codes. The construction also works for higher connectivity of the form, $N = 3t + K, K > 1$, and sends K messages instead of one. For this construction the receiver only has to implement the decoder of an RS-code which has many well-known implementations. For less connectivity, $2t + 1 \leq N \leq 3t$, 1-round PSMT is not possible. However, it is an open question if it is possible to have simple and modular constructions for δ-RMT and $(0, \delta)$-SMT using known primitives such as RS-codes.

Our Contributions

A general construction of a 1-round δ-RMT. We give a general construction of a 1-round δ-RMT for $N \geq 2t + 1$ from two components: a list decodable code and a Message Authentication Code (MAC). In a (ρ, L)-list decodable code (LD code) of length n, the number of codewords within distance ρn of any received word is at most L.

The basic idea of the construction is as follows. An information block $I_{\mathcal{S}}$ is first appended with a tag generated by a symmetric key authentication mechanism, to form a message $m_{\mathcal{S}}$, which is then encoded using the LD code, and each component is sent over a wire. LD code allows correction of up to t adversarial errors and so the decoder will obtain a list of L closest codewords to the received word. The key for the authentication mechanism will be generated by the sender and sent along the wires to the receiver and so parts that are sent over the corrupted wires, will be corrupted. The authentication information together with the key information will allow the receiver to use the corresponding verification mechanism to recognize the correct codeword in the decoded list. The authentication mechanism must ensure that despite partly corrupted keys, the receiver will output the correct codeword. We describe our approach and prove a general theorem that proves reliability of the construction, with a value of δ that can be calculated from the parameters of the underlying components.

We then give a concrete construction when $N = 2t + 1$ using a Folded RS-code and a new multireceiver MAC that we propose. The result is an optimal δ-RMT with *the smallest* δ (highest reliability) among all known optimal δ-RMT protocols.

The drawback of this construction is that the receiver algorithm in RMT is exponential. For higher connectivities of the form $N = (2+c)t, c > \frac{1}{t}$ however, we will have an optimal δ-RMT with efficient (polynomial) receiver algorithm. The main challenge in this construction is choosing the authentication mechanism and its parameters, as well as parameters of FRS-code to achieve the required performance. We give details of these selections for $N = 2t + 1$, and for higher connectivities we omit the details because of space. (We will provide details for SMT when $N = (2+c)t, c > \frac{1}{t}$, which gives a good idea of challenges of designing δ-RMT for these connectivities.)

Constructing 1-round $(0, \delta)$-SMT from FRS-codes. Although it is possible to give a general construction of $(0, \delta)$-SMT using an approach similar to δ-RMT, for clarity of results and because of space limitations we limit ourselves to concrete constructions. We first describe a construction for a 1-round $(0, \delta)$-SMT for $N =$

$2t+1$ using FRS-codes and a multireceiver MAC, and then extend the result to the case where $N = (2+c)t, c > \frac{1}{t}$. For $N = 2t + K, K > 1$, a similar approach can be used resulting in an optimal 1-round $(0, \delta)$-SMT.

The construction of 1-round PSMT for $N = 3t + 1$ [7] uses an RS-code to encode the message, and the receiver algorithm uses unique decoding algorithm of RS-codes. For connectivity $2t + 1 \leq N \leq 3t$, the minimum distance of the RS-code will be $t + 1 \leq d \leq 2t$, and so unique decoding for t adversarial errors is not possible.

We use *list decoding* to correct errors beyond unique error correcting radius of the code, and use an authentication mechanism to recognize the sent codeword. There are however two major challenges:

(i) In SMT the sender and the receiver *do not share a secret key* and so the key for the authentication mechanism (based on MACs) must be delivered to the receiver over the wires, some of which are corrupted.
(ii) For $N = 2t+1$, and code dimension $k = t+1$ which is dictated by the perfect privacy requirement, the code rate is $R = k/N = \frac{t+1}{2t+1}$ and so the percentage of errors that needs to be corrected is $\rho = \frac{t}{2t+1} = 1 - R$, which is the information theoretic *list decoding capacity* of the code. Codes that can achieve this capacity and have efficient decoders need special construction.

An important property of the resulting $(0, \delta)$-SMT protocols is that *they have the lowest δ* and so the highest reliability among all known optimal 1-round (0, δ)-SMT protocols with comparable connectivity (for protocols that output correct messages, or **Fail**.) Our proposed constructions of multireceiver MACs provide optimal and near optimal (different by a factor of 2) forgery probabilities and are of independent interest.

Advantages of the Approach

A general construction for 1-round δ-RMT. LD codes and multireceiver MACs are both well-established primitives with numerous efficient constructions. The advantage of a general construction using these two primitives as building blocks, is that one can choose appropriate constructions for a given setting. Moreover, advances in LD codes and MACs can result in better construction for δ-RMT systems. The instantiation of the LD code with FRS-code is directly adaptable (with revised message structure and code parameters) to $(0, \delta)$-SMT. This means that the main building block of the receiver will stay the same in both cases and development of more efficient decoding for FRS-codes will translate into more efficient receiver algorithms for SMT.

1-round $(0, \delta)$-SMT and RS-codes. FRS-codes are in fact RS-codes with blocks of symbols interpreted as elements of a larger field. The construction of $(0, \delta)$-SMT from FRS-codes provides an elegant and systematic construction for 1-round $(0, \delta)$-SMT *for all connectivities* ($N \geq 2t + 1$) using RS-codes.

The decoding algorithm is the same for RMT and SMT and in all cases consists of a two step algorithm: list decoding of the FRS-code, and a message verification algorithm based on the MAC for every element of the list. The proof of privacy is based on the properties of the FRS-codes and are intuitive. The proof of reliability (calculation of δ) is also intuitive with concrete values depending on the parameters of the FRS-codes and the MAC.

A unified approach to 1-round δ-RMT and $(0, \delta)$-SMT. The above shows a unified approach for the construction of these two primitives, reliable communication without or with privacy, for all connectivities $N \geq 2t + 1$. This means that the sender and receiver can use a modular construction in which the module with the higher complexity, which is the list decoding module, is implemented once and its parameters are adjusted depending on the required properties (reliability only, or both reliability and privacy) and the choice of the MAC determined by the given N and t.

δ-RMTs are similar to error correcting codes with protection against adversarial errors with the difference that adversary's view is limited to the t corrupted wires. In error correcting codes with protection against adversarial channels, the adversary can see the whole codeword before choosing the error pattern. In δ-RMT the adversary only sees the positions that it corrupts.

Related Work

Srinathan et al. designed an efficient and optimal 1-round δ-RMT protocol [16]. This protocol uses a complex combination of secret sharing by the sender and elaborate verification of the received information by the receiver to determine the correct message block. There are two optimal and efficient 1-round $(0, \delta)$-SMT protocols for $N = 2t + 1$ [13,21]. The protocol in [13] is based on the 1-round δ-RMT protocol of [16] and has similar complexity. Moreover, to achieve the optimal transmission rate for higher connectivity, this protocol can not be directly used as the encrypted message blocks are broadcasted. The protocol of [21] suffers from the same problem as that of [13].

On the other hand, for $N = (2 + c)t, c > \frac{1}{t}$, there are two optimal and efficient 1-round $(0, \delta)$-SMT protocols [21,19]. The protocol of [19] uses two SMT protocols in two levels. The first SMT protocol for the lowest connectivity is used many times on different subsets of the wires. The second protocol which works for higher connectivity is applied to virtual wires obtained from the actual physical wires. The shortcoming of this protocol is that to achieve optimal rate, the sender needs to send a very large information block (of size at least N^3 field elements). The protocol in [21] also uses multiple secret sharings in a clever way which results in a lower δ than that of [13].

2 Background and Primitives

In SMT problem, there is an incomplete network, that connects a sender $\mathcal{S}$ to a receiver $\mathcal{R}$. The sender and the receiver are connected by N vertex-disjoint paths, also known as *wires* or channels. The network is undirected and communication on the wires is synchronous and bidirectional. Both $\mathcal{S}$ and $\mathcal{R}$ are honest. The goal is to enable $\mathcal{S}$ to send a message m, drawn from a message space $\mathcal{M}$ with a probability distribution $\Pr(m)$, to $\mathcal{R}$ such that $\mathcal{R}$ receives the message *correctly* and *privately*.

In a *message transmission protocol*, the sender $\mathcal{S}$ chooses m from a message space $\mathcal{M}$ with a probability distribution $\Pr(m)$, and uses a protocol with one or more rounds, to send the message to the receiver. In each protocol round, $\mathcal{S}$ or $\mathcal{R}$, constructs a protocol message that is sent over the wires to the other party. A protocol message is received by the recipient of the round, possibly in a corrupted form, before the next round starts. At the end of the protocol, the receiver outputs a message m', or outputs a **Fail**.

We consider only 1-round protocols. The adversary $\mathcal{A}$ has unlimited computational power and can corrupt and control a subset of wires: the adversary can eavesdrop, block or modify the communication over the corrupted wires. $\mathcal{A}$ can corrupt at most t out of the N wires and the corrupted wires are unknown to $\mathcal{S}$ and $\mathcal{R}$.

Denote by $V_{\mathcal{A}}(M_{\mathcal{S}}, r_{\mathcal{A}})$ the random variable that denotes the view of the adversary $\mathcal{A}$ when attacking the protocol assuming the sender has chosen $M_{\mathcal{S}}$ and $r_{\mathcal{A}}$ is the random coins of the adversary. Let the statistical distance of two random variables X, Y defined over a set $\mathcal{U}$ be defined as, $\Delta(X,Y) = \frac{1}{2}\sum_{u\in\mathcal{U}} |\Pr[X=u] - \Pr[Y=u]|$.

Definition 1. *A message transmission protocol between $\mathcal{S}$ and $\mathcal{R}$ is an (ε,δ)-Secure Message Transmission ((ε,δ)-SMT) protocol if the following two conditions are satisfied:*

- *Privacy: For every two messages $m_0, m_1 \in \mathcal{M}$ and every $r \in \{0,1\}^*$ used by the adversary,*
  $\Delta(V_{\mathcal{A}}(m_0,r), V_{\mathcal{A}}(m_1,r)) \le \varepsilon$*, where the probability is over the randomness of $\mathcal{S}$ and $\mathcal{R}$.*
- *Reliability: $\mathcal{R}$ outputs the message m with probability $\ge 1-\delta$, and **Fail** with probability $\le \delta$. That is, the receiver never outputs an incorrect message and,* $\Pr[$*Receiver outputs* **Fail**$] \le \delta$.

This is the definition of reliability used by Kurosawa et. al. [12]. The original definition of reliability in [8] however assumes that the receiver always outputs a message m' and δ-reliability is, $\Pr[m' \neq m] \le \delta$. Kurosawa et al. require that the receiver be sure that the received message is correct. When $\varepsilon = 0$, the protocol is said to achieve *perfect privacy*, and when $\delta = 0$, the protocol is said to achieve *perfect reliability*. A *δ-Reliable Message Transmission (δ-RMT)* protocol is a protocol between $\mathcal{S}$ and $\mathcal{R}$ in the same network setting, and only requiring the reliability of transmission.

It was shown [8] that (0, δ)-SMT is possible if and only if $N \ge 2t+1$ and 1-round PSMT is possible if and only if $N \ge 3t+1$ [3]. 1-round δ-RMT protocols exist if and only if $N \ge 2t+1$ [8].

Communication efficiency of RMT and SMT protocols is in terms of the *number of rounds*, and *transmission rate*. The *number of rounds of an SMT protocol* is the number of interactions between $\mathcal{S}$ and $\mathcal{R}$. *Transmission rate of RMT and SMT protocols* is the ratio of the total communication to the length of the message: that is the communication cost of sending one bit.

Lower bounds on transmission rates of 1-round δ-RMT and 1-round $(0, \delta)$-SMT protocols are $\Omega(\frac{N}{N-t})$ [14] and $\Omega(\frac{N}{N-2t})$ [13], respectively. Protocols whose transmission rate asymptotically match the associated lower bounds are called *optimal.* For 1-round δ-RMT with $N = 2t + 1$, the lower bound on the transmission rate is $\Omega(1)$. For 1-round $(0, \delta)$-SMT protocols with $N = 2t + 1$ and $N = (2 + c)t, c > \frac{1}{t}$, optimal protocols must have transmission rates $O(N)$ and $O(1)$, respectively.

Computation efficiency of RMT and SMT protocol is the amount of computation performed by $\mathcal{S}$ and $\mathcal{R}$ throughout the protocol. A protocol that needs exponential (in N) computation for $\mathcal{S}$ and $\mathcal{R}$, is called *inefficient.* Efficient protocols need polynomial (in N) computation.

2.1 Folded Reed-Solomon Codes

A (k, n) linear error correcting code over $\mathbf{F_q}$ is a subspace of dimension k of the n dimensional vector space over $\mathbf{F_q}$. The *information rate* of a linear error correcting code is, $R = \frac{k}{n}$. A decoder takes a corrupted word and determines the most likely codeword that was sent. In unique decoding, the closest (Hamming distance) codeword to the received one is found. In list decoding, a list of codewords within a radius from the received word is found. For a constant ρ, let ρn denotes the number of errors that can be corrected by the decoder.

Definition 2. *A code C with the encoding function $LD : \mathbf{F_q}^k \to \mathbf{F_q}^n$ is (ρ, L)-list decodable (LD) if the number of codewords within distance ρn of any received word is at most L. That is for every word $y \in \mathbf{F_q}^n$, there are at most L codewords at distance ρn (where ρ is the relative distance) or less from y.*

The *list decoding capacity* $\rho_{cap}(R)$ of a code with rate R is the information theoretic limit of list decodability and is given by $\rho_{cap}(R) = 1 - R = 2\rho_U(R)$, where $\rho_U(R) = (1 - R)/2$, is the unique decoding radius of the code. It is shown [6] that for sufficiently large alphabet size $\rho_{cap}(R)$ can reach $1 - R - \epsilon$. So for any code rate, *list decoding can potentially correct twice as many errors as unique decoding [10].*

To achieve this potential however, one needs special constructions that guarantee that the list size is bounded by L. Efficient list decoding algorithms are polynomial time. *Folded Reed-Solomon codes (FRS-codes)*, proposed by Guruswami et al. [10], is a special type of RS-codes that corrects up to a fraction $\rho = 1 - R - \epsilon$ of errors, for any rate R and arbitrary $\epsilon > 0$ using a polynomial time list decoding algorithm. The list size, however for some parameter choices of the code, becomes exponential.

Description of Folded Reed-Solomon Codes. FRS encoding and decoding are defined using an RS-code of length n and dimension k over a finite field $\mathbf{F_q}$, using two parameters, u, the *folding parameter* and s, which determines the $(s + 1)$-variate interpolation used in the decoding. Let u be an integer, called the *folding parameter*, such that n in divisible by u. n is chosen as the largest integer that is less than $q = |\mathbf{F_q}|$ and is divisible by u.

Let γ be a generator of $\mathbf{F_q}^*$, the multiplicative group of the field $\mathbf{F_q}$. A codeword $(f(1), f(\gamma), f(\gamma^2), \cdots, f(\gamma^{n-1}))$ of $RS[n,k]$ is the evaluation of a polynomial $f(x)$, of degree at most $k-1$ over $\mathbf{F_q}$, at the (ordered) points $1, \gamma, \gamma^2, \cdots, \gamma^{n-1}$.

Definition 3. *The u-folded FRS-code is a code with block length $N = n/u$ over $\mathbf{F_q}^u$. The encoding of a message, represented by a polynomial $f(x)$ of degree at most $k-1$ over $\mathbf{F_q}$, is obtained as u-tuples, $(f(\gamma^{ju}), f(\gamma^{ju+1}), \cdots, f(\gamma^{ju+u-1}))$, for $0 \le j < N$. In other words, a codeword of the u-folded RS-code is in one-to-one correspondence with codewords of the RS-code C, and is obtained by grouping consecutive u−tuples of components of C.*

$$\begin{bmatrix} f(1) & f(\gamma^u) & \cdots & f(\gamma^{u(N-1)}) \\ f(\gamma) & f(\gamma^{u+1}) & \cdots & f(\gamma^{u(N-1)+1}) \\ \vdots & \vdots & \ddots & \vdots \\ f(\gamma^{u-1}) & f(\gamma^{2u-1}) & \cdots & f(\gamma^{uN-1}) \end{bmatrix}$$

Decoding u-folded RS-code uses $(s+1)$-variate interpolation followed by a list pruning step. In the full version of the paper [20], we give an outline of the original decoding algorithm.

Linear-algebraic list decoding. In [9] a variant decoding for FRS-code is given in which the interpolation uses polynomial $Q(X, Y_1, \cdots, Y_s)$ where degree of Y_i is 1. This variant has a simpler exposition and allows a simpler way of choosing the code parameter, s and u. However it can correct less errors. By appropriate choice of s and u, the code can reach $1 - R - \epsilon$ radius and so asymptotically is optimal. Lemma 1 below, given in [9], gives the condition that needs to be satisfied by the two parameters and the number of errors to be corrected.

Lemma 1. *In linear-algebraic list decoding, for every integer u and s, the linear interpolation FRS decoding algorithm successfully list decodes to a radius $N - T$ as long as the agreement parameter T satisfies:*

$$T \ge N(\frac{1}{s+1} + \frac{s}{s+1}\frac{uR}{u-s+1}).$$

T is the number of correct positions. The algorithm outputs a list of size at most $|\mathbf{F_q}|^{s-1} = q^{s-1}$ codewords.

2.2 Multireceiver Message Authentication Codes

A one-time MAC in information-theoretic setting, is a shared key cryptographic primitive, defined by two functions: a MAC function that takes a message $m \in \mathcal{M}$ and the shared key $k_{MAC} \in \mathcal{K}$ and outputs a tag $MAC(m, k_{MAC})$, which is appended to the message, and a verification function, $V((m', x'), k_{MAC})$, which outputs 1 if (m', x') is a valid pair for the key k_{MAC}, and 0, otherwise. The following definition is for security of one-time MAC.

Definition 4. *A one-time MAC,* $MAC : q^{l_{msg}} \times q^{l_{key}} \rightarrow q^{l_{tag}}$ *has forgery probability* γ *if the best success chance of a computationally unbounded adversary with access to a message and tag pair* $(m, x), x = MAC(m, k_{MAC})$*, to construct a different pair* (m', x') *where* $m \neq m'$*, and* $V((m', x'), k_{MAC}) = 1$ *is at most* γ*, where the probability is taken over all keys.*

Multireceiver authentication codes [5] allow a sender to efficiently send a message to a group of N receivers such that each receiver can individually verify the message, using his individual shared key k_i with the sender. The sender is honest but upto t receivers can be corrupted and attempt to forge a message to be acceptable by an uncorrupted receiver. In a $(t+1, N)$-multireceiver message authentication system, there are N receivers and at most t receivers can be corrupted.

Definition 5. *A one-time* $(t+1, N)$*-multireceiver authentication code (multireceiver MAC) with* N *receivers and* $k_{MAC} = (k_S, k_1, ..., k_N)$*, is* γ*-secure if the best success chance of any colluding set of receivers (size at most* t*) with access to a message and tag pair,* $(m, x, x = MAC(m, k_S))$ *in forging a different message, tag pair* (m', x')*, where* $m \neq m'$*, and* $V_i((m', x'), k_i) = 1$*, is at most* γ*, and the probability is over all unknown keys.*

2.3 New Constructions for Multireceiver MAC

The basic construction of $(t+1, N)$-multireceiver MACs is for authenticating a single message. To authenticate a block of messages one can use a one-time multireceiver MAC multiple times, or for more efficiency, use separately designed multireceiver MAC for message blocks. In the following, we give two new constructions for $(t+1, N)$-multireceiver MACs for message blocks that are used in the RMT and SMT constructions of this work. Construction 1 is a generalization of [5] for $d > 1$ messages. Construction 2 is built over a brand new MAC.

Construction 1.

Let $m = (m_1, \cdots, m_d)$, where $m_i \in \mathbf{F_q}^{s'}, i = 1, ..., d$, be the message block.

- *Key distribution:* A Trusted Initializer does the following: (i) randomly generates $d+1$ polynomials $P_1(z), P_2(z), \cdots, P_{d+1}(z)$, each of degree at most t, over $\mathbf{F_q}^{s'}$; chooses N random distinct elements $z_1, z_2, \cdots, z_N$, where $z_i \in \mathbf{F_q}^{s'}, i = 1, ..., N$; makes $z_1, z_2, \cdots, z_N$ public, assigns z_i to receiver i and privately sends $k_i = (P_1(z_i), P_2(z_i), \cdots, P_{d+1}(z_i))$ to receiver i, for $1 \leq i \leq N$ and to the sender.
- *Constructing authenticated messages:* The sender computes the authentication tag as:
$$A(z) = P_1(z)m_1 + P_2(z)m_2 + \cdots + P_d(z)m_d + P_{d+1}(z).$$
The authenticated messages consist of the message block and the tag polynomial, $((m_1, m_2, \cdots, m_d), A(z))$.
- *Verification:* Receiver i accepts $(m_1, m_2, \cdots, m_d, A(z))$ if and only if $A(z_i) = P_1(z_i)m_1 + P_2(z_i)m_2 + \cdots + P_d(z_i)m_d + P_{d+1}(z_i) \mod q^{s'}$.

The above construction is a $(t+1, N)$-multireceiver MAC for authentication of a block of size d. The size of the tag is $t+1$ elements of $\mathbf{F_q}^{s'}$ and so only depends on the collusion size (rather than the total number of receivers).

The following Theorem is proved in the full version of the paper [20].

Theorem 1. *For construction 1, the forgery probability is bounded as* $\gamma \leq q^{-s'}$.

Construction 2.
This multireceiver MAC is built on a new one-time MAC which has message block size $\binom{t+2}{2} - 1$ (each block element from $\mathbf{F_q}^{s'}$) and has forgery probability bounded by $\frac{2}{q^{s'}}$.

- *Key distribution:* Same as Construction 1, with $d = t$.
- *Constructing authenticated messages:* For $m = (m_1, m_2, ..., m_{\binom{t+2}{2}-1})$, the sender computes,

$$A(z) = m_1 P_1(z) + \cdots + m_t P_t(z) + m_{t+1} P_1(z)^2 + \cdots + m_{2t} P_t(z)^2$$
$$+ m_{2t+1} P_1(z) P_2(z) + \cdots + m_{\binom{t+2}{2}-1} P_{t-1}(z) P_t(z) + P_{t+1}(z).$$

- *Verification:* Receiver i accepts $(m_1, m_2, \cdots, m_{\binom{t+2}{2}-1}, A(z))$ if and only if

$$A(z_i) = m_1 P_1(z_i) + \cdots + m_t P_t(z_i) + m_{t+1} P_1(z_i)^2 + \cdots$$
$$+ m_{\binom{t+2}{2}-1} P_{t-1}(z_i) P_t(z_i) + P_{t+1}(z_i).$$

Here $m_i \in \mathbf{F_q}^{s'}$, and $P_1(z), P_2(z), \cdots, P_{t+1}(z)$ are polynomials of degree at most t over $\mathbf{F_q}^{s'}$. The MAC function is a linear sum (coefficients being the message block) of all products of at most two polynomials from the set, $\{P_1(z), P_2(z), \cdots, P_t(z)\}$. Finally $P_{t+1}(z)$ is used to mask the result. The final MAC value is a polynomial over $\mathbf{F_q}^{s'}$. The size of the message block (over $\mathbf{F_q}^{s'}$) that is authenticated by the MAC, is $\binom{t+2}{2} - 1$.

Theorem 2. *For construction 2, the forgery probability is bounded as* $\gamma \leq 2q^{-s'}$.

The proof outline is provided in the full version of the paper [20].

3 Construction of 1-Round δ-RMT for $N \geq 2t+1$

Construction 3: A general construction of 1-round δ-RMT. The protocol requires a (ρ, L) LD code of dimension k and length N over $\mathbf{F_q}$, with $\rho = \frac{t}{N}$ and list size L, and a $(t+1, N)$-multireceiver MAC with message space $\mathbf{F_q}^{k'}$, $k' = k - l_{tag} < k$, and forgery probability ϵ. Here l_{tag} is the length of tag in terms of $\mathbf{F_q}^{s'}$ elements.

- **Sender Algorithm**:
  1) Securely generates keys $(k_S, k_1, ..., k_N)$ for a multireceiver MAC and assigns the key k_i to the ith wire, W_i.
  2) Constructs the message block $m_S, m_S \in \mathbf{F_q}^k$ to be sent to the receiver as,

$$m_S = (I_S, MAC(I_S, k_S)). \quad (1)$$

The sender constructs the codeword c_S of the LD code by encoding m_S as $c_S = LD(m_S)$. The sender sends the ith component of the codeword c_S, and k_i through wire W_i. k_S is kept by the sender.

- **Receiver Algorithm**:
  1) Parses the received (corrupted) N-vector; separates the key k_i from the ith component (possibly corrupted) of the received word for $i = 1, \cdots, N$; constructs the corrupted codeword and uses LD decoding algorithm to obtain a list (of size L) of codewords that are at distances at most $\rho = \frac{t}{N}$ from the received word. *The list will always include the correct codeword.*
  2) To identify the sent codeword, the receiver (i) parses each codeword in the list into a message, tag pair $(\widehat{m_i}, \widehat{t_i}), i = 1, \cdots, L$; (ii) for each message $\widehat{m_i}$, uses all keys $k_j, j = 1, ..., N$, and checks the verification equations $V_j(\widehat{m_i}, k_j) = \widehat{t_i}$. The message is accepted if at least $t+1$ verification equations are passed; otherwise the codeword is rejected.
  The decoding algorithm of the SMT succeeds if there is a unique codeword that is accepted by the verification algorithm above. Otherwise, the receiver outputs a **Fail**.

Theorem 3. *The above construction is a 1-round δ-RMT protocol for $N = 2t+1$, with $\delta = 1 - (L-1)\epsilon$ and transmission rate $\frac{N(1+|k_i|)}{k - l_{tag}}$.*

The proof is omitted due to lack of space.

3.1 An Optimal δ-RMT

In the following we give an instantiation of the general construction above using (i) an FRS-code for the LD code, and (ii) Construction 2 for multireceiver MAC. The multireceiver MAC allows authentication of a message block of size $\sim t^2$ field elements requiring $\sim t$ field element for encoding of m_S as, to be sent on each wire, resulting in optimal transmission rate.

Selecting Parameters of the FRS-code. Let $N = 2t + 1$ for the RMT. We consider a u-folded RS-code of length N with length $n = Nu$ for the underlying RS-code. Using the message format in (1) and using a block of size $(\binom{t+2}{2} - 1)$ of elements of $\mathbf{F_q}^{s'}$ for information block we will have the required dimension for the code as,

$$k = |m_S| = (\binom{t+2}{2} - 1)s' + ts' + s' = \frac{s'(t^2 + 5t + 2)}{2}. \quad (2)$$

For a code of length $N = 2t + 1$ and dimension k as (2), we must choose folding parameter u, number of decoding variable s, and the finite field sizes s' and q, to ensure that decoding succeeds for linear-algebraic decoding (outlined in Section 2.1) for radius $\rho = t/N$ (t errors in the FRS-code). That is, the inequality 3 below, is satisfied.

$$t + 1 \geq N(\frac{1}{s+1} + \frac{s}{s+1}\frac{uR}{u-s+1}). \quad (3)$$

Let $0 < \sigma < 1$ and set $s = \frac{N}{\sigma} - 1$. Furthermore, choose $s' = \left\lfloor \frac{2u(t+1-2\sigma)}{t^2+5t+2} \right\rfloor$, and $u > s^2 - 1$. the following shows that with these choices of parameters inequality (3) is satisfied. We have,

$$t+1 > t+1-\sigma = \sigma + (t+1-2\sigma) \overset{(a)}{=} \sigma + \frac{s'(t^2+5t+2)}{2u} \overset{(b)}{=} \sigma + \frac{k}{u} \quad (4)$$

where (a) is because of the choice of s' and (b) is because of of the value of k in (2). Note that because $s \geq 1$, we have $\frac{1}{u} > \frac{s}{s+1}\frac{1}{u-s+1}$ and so, (4) gives the following:

$$t+1 > \sigma + \frac{k}{u} > \sigma + \frac{s}{s+1}\frac{k}{u-s+1} = N(\frac{1}{s+1} + \frac{s}{s+1}\frac{uR}{u-s+1})$$

Finally we can choose q to be the smallest prime that is bigger than the codeword length $n = Nu$.

The Protocol. The construction uses (i) an FRS-code with parameters u and s obtained above and (ii) Construction 2 of the multireceiver MAC in Section 2.3. The final protocol is as follows.

- **Sender Algorithm**:
  1. Uses the key generation algorithm of Construction 2 and obtains for wire $W_i, i = 1, ..., N$, the associated key,
  $$k_i = (P_1(z_i), P_2(z_i), \cdots, P_{t+1}(z_i)).$$
  The tag for the information part $I_S = (m_0, m_1, \cdots, m_{(\binom{t+2}{2}-1)}), m_i \in \mathbf{F_q}^{s'}$:
  $$A(z) = m_1 P_1(z) + \cdots + m_t P_t(z) + m_{t+1} P_1(z)^2 + \cdots + m_{2t} P_t(z)^2$$
  $$+ m_{2t+1} P_1(z) P_2(z) + \cdots + m_{\binom{t+2}{2}-1} P_{t-1}(z) P_t(z) + P_{t+1}(z).$$
  2. The message m_S is of the form (1), $m_S = (m_0, m_1, \cdots, m_{(\binom{t+2}{2}-1)}, A(z))$. The dimension of the FRS-code is $(\binom{t+2}{2} - 1)s' + ts' + s'$ over $\mathbf{F_q}$, where $s' = \left\lfloor \frac{2u(t+1-2\sigma)}{t^2+5t+2} \right\rfloor$.
  3. The sender encodes the message to a codeword c_S using the FRS encoding algorithm. Wire j, $1 \leq j \leq 2t+1$, transmits the jth component of c_S and k_j.
- **Receiver Algorithm**: Uses the two step decoding of Construction 3 for FRS-code as the LD code, and Construction 2 as the multireceiver MAC. The algorithm outputs the correct message or **Fail**.

Theorem 4. *The above construction is a δ-RMT with* $\delta = \frac{2(t+1)}{q^{s'-s+1}}$, *which is equal to* $\frac{N+1}{q^{s'-s}}$, *when* $N = 2t+1$. *The transmission rate is constant.*

The proof is given in the full version of the paper [20].

Comparison with Related Work

For $N = 2t+1$, this protocol has $\delta = \frac{N+1}{q}$. The value of δ for the only other known optimal 1-round δ-RMT protocol [13], is $\frac{N^2(N-1)}{q}$ $(q \geq \frac{N^2(N-1)}{\delta})$. The field size required in our construction is Nu.

Table 1 compares our protocol with the protocol in [13]. For simplicity of comparison we have used $s' = s$, resulting in $\delta = \frac{t+1}{q}$. The comparison shows that for all connectivities the proposed protocol has a much higher reliability. The field size although asymptotically is larger (N^4 and N^3, respectively) for concrete values (for example $N < 1000$, and $\delta < 10^{-3}$) could be smaller or comparable. Decoder efficiency for higher connectivities is the same. For $N = 2t+1$, however, our protocol has exponential cost.

Table 1. Comparison of 1-round δ-RMT protocols for different connectivities which never outputs an incorrect message; here Comp. refers to computation complexity, *Poly.* refers to polynomial (in N), *Exp.* refers to exponential (in N), and $\mathbf{F_q}$ is the field

Author	Comp. $N = 2t+1$	Comp. $N = (2+c)t$	q	δ	Optimality
[13]	*Poly.*	*Poly.*	$\geq\approx \frac{N^3}{\delta}$	$\leq\approx \frac{N^3}{q}$	Yes
This Work	*Exp.*	*Poly.*	$\geq Nu$	$\leq \frac{2(t+1)}{q} \approx \frac{N+1}{q}$	Yes

4 1-Round $(0, \delta)$-SMT

To use the approach of Construction 3 for $(0, \delta)$-SMT, one needs to ensure that the view of the adversary does not leak any information about the information block I_S. Using FRS-code for LD code allows us to achieve this goal by choosing the dimension of the code to be at least $t+1$. Code parameters need to be chosen such that decoding up to $\rho = \frac{t}{N}$ is achievable. For $N = 2t+1$ given in Section 4.1, this requires the FRS-code to achieve the list decoding capacity. The construction for $N = (2+c)t, c > \frac{1}{t}$, given in Section 4.2, uses Construction 2 for multireceiver MAC to allow easier calculation of code parameters while maintaining optimal asymptotic performance. Same approach can also be used for connectivity $N = 2t + K$, where $K > 1$ is a constant. Details are omitted because of space.

In all cases, decoding is the two step Receiver Algorithm of Construction 3.

4.1 A Construction for 1-Round $(0, \delta)$-SMT for $N = 2t+1$

The construction uses the approach of Construction 3, but with a different message structure to guarantee perfect privacy.

Message Structure. The message m_S consists of three parts: (i) information part $I_S = (m_0, m_1, \cdots, m_{\sigma u-1}), m_i \in \mathbf{F_q}^{s'}$; (ii) ut random elements $(a_1, a_2, \cdots, a_{ut})$, $a_i \in \mathbf{F_q}$ that are used to ensure privacy that the ut captured components do not reveal anything about I_S; (iii) $MAC(X, k_S)$ where $X = (m_0, m_1, \cdots, m_{\sigma u-1}, a_1, ..., a_{s'd-\sigma u})$.

That is,

$$m_S = (m_0, m_1, \cdots, m_{\sigma u-1}, a_1, a_2, \cdots, a_{ut}, MAC(X, k_S)).$$

Here σ is a positive constant. To have optimal rate, the information block size must be a constant fraction of u. Using Construction 1, this size is $s'd$, where d is message length for the MAC. To find s', the information block (I_S) and the first $s'd - \sigma u$ elements a_i are used to form d blocks of size s', where $d = \lceil \frac{\sigma u}{s'} \rceil$. Each s' block is interpreted as an element of ${\mathbf{F_q}}^{s'}$ and MAC calculations are performed over ${\mathbf{F_q}}^{s'}$. The ut random elements appended to I_S will ensure perfect privacy. The MAC is only computed on X. The total length of m_S to be encoded by the FRS-code is $ut + \sigma u + s'(t+1)$ were $\sigma < 1$ is a constant. The key k_i for wire i consists of $d+1$ elements $P_1(z_i), P_2(z_i), \cdots, P_{d+1}(z_i)$ over ${\mathbf{F_q}}^{s'}$. The multireceiver MAC value for X is:

$$MAC(X, k_S) = A(z) = P_1(z)x_1 + P_2(z)x_2 + \cdots + P_d(z)x_d + P_{d+1}(z).$$

The codeword of the FRS-code that is constructed for m_S, will have N components, each an element of ${\mathbf{F_q}}^u$. The adversary's view will contain only t elements of ${\mathbf{F_q}}^u$ and will be independent from I_S.

Parameters of the FRS-code. The u-folded RS-code will have $N = 2t+1, n = Nu$, and $k = ut + \sigma u + ts' + s'$. We must choose u, s, and the field size q, to ensure that decoding succeeds for linear-algebraic decoding (outlined in Section 2.1) up to radius $\rho = t/N$.

According to Lemma 1,

$$t + 1 \geq N(\frac{1}{s+1} + \frac{s}{s+1}\frac{uR}{u-s+1}).$$

We set the parameter $s = (N/\sigma) - 1$, $u > s^2 - 1$, and $s' = \left\lfloor \frac{u(1-3\sigma)}{t+1} \right\rfloor$, where $0 < \sigma < \frac{1}{3}$ is a positive constant. By using these values of σ, u, s', one can verify that the inequality is satisfied. Finally we can choose q to be the smallest prime that is bigger than the codeword length $n = Nu$.

Construction 4: (0, δ)-SMT Protocol for $N = 2t + 1$.

- **Sender Algorithm**:
  1. Uses the key generation of Construction 1 and obtains for wire $W_i, i = 1, ..., N$, the associated key,
  $$k_i = (P_1(z_i), P_2(z_i), \cdots, P_{d+1}(z_i)).$$
  2. Constructs m_S: Forms $I_S = (m_0, m_1, \cdots, m_{\sigma u-1}), m_i \in {\mathbf{F_q}}^{s'}$ and calculates the tag,
  $$A(z) = x_1P_1(z) + x_2P_2(z) + \cdots + x_dP_d(z) + P_{d+1}(z),$$
  for d and s' chosen as above. Here $X = (m_0, ..., m_{\sigma u-1}, a_1, ..., a_{s'd-\sigma u})$. m_S is of the form (1) given by, $(m_0, m_1, \cdots, m_{\sigma u-1}, a_1, a_2, \cdots, a_{ut}, A(z))$.
  3. Constructs c_S and message transcript: The sender encodes the message to a codeword c_S using the FRS encoding algorithm. Wire j, $1 \leq j \leq 2t+1$, transmits the jth component of c_S and k_j.
- **Receiver Algorithm**: Uses the decoding algorithm of Construction 3.

Theorem 5. *The SMT protocol described above is a* $(0, \delta)$*-SMT for* $N = 2t+1$*, with* $\delta = \frac{t+1}{q^{s'-s+1}}$.

The proof is omitted due to lack of space, but is given in the full version of the paper [20].

Transmission rate: The transmission rate is $\frac{uN+(s'd+s')N}{\sigma u} = \mathcal{O}(N)$ and it is optimal for 1-round (0, δ)-SMT for $N = 2t + 1$.

Computation complexity: The list size is at most q^{s-1}. Each element of the list must be verified and so the complexity of SMT decoding algorithm is $\mathcal{O}(q^N)$ (as $s = O(N)$).

Comparison with Related Work. Table 2 compares the protocol with 1-round (0, δ)-SMT protocols that have the property that the output is either the correct message or **Fail**. For simplicity of comparison we have used $s' = s$, resulting in $\delta = \frac{t+1}{q} = \frac{N+1}{2q}$, when $N = 2t + 1$. The table shows that δ for this construction is the lowest. The minimum field size however is larger and decoding is computationally inefficient. In Section 4.2 we show that both these shortcomings can be removed for higher connectivities.

Table 2. Comparison of 1-round $(0, \delta)$-SMT protocols for $N = 2t + 1$; here Comp. refers to computation complexity, Poly. refers to polynomial (in N), Exp. refers to exponential (in N) and $\mathbf{F_q}$ is the field. The protocols never output incorrect message. Here b is a constant and λ is the probability that the cheater wins in a secret sharing scheme with a cheater.

Author	Comp.	q	δ	Optimality
[12]	*Exp.*	$> N$	$\leq (\binom{N}{t+1} - 1)\lambda \approx N^{(N+1)/2}\lambda$	Yes
[17]	*Poly.*	$\geq \frac{2N^3}{\delta}$	$\leq \frac{N^3}{q}$	Yes
[4]	*Poly.*	$\geq bt(t+1) \approx N^2$	$\leq \frac{t(t+1)}{q} \approx \frac{N^2}{q}$	No
[21]	*Poly.*	$\geq bt(t+1) \approx N^2$	$\leq \frac{t(t+1)}{q} \approx \frac{N^2}{q}$	Yes
This Work	*Exp.*	$\geq Nu \approx N^4$	$\leq \frac{t+1}{q} \approx \frac{N+1}{2q}$	Yes

4.2 1-Round $(0, \delta)$-SMT for $N = 2t + ct, c > 1/t$

Let $N = 2t + ct, c > \frac{1}{t}$. We use the same approach as Construction 4, using Construction 2 for multireceiver MAC and choose parameters of the FRS-code and multireceiver MAC such that the SMT construction has optimal rate and efficient computation. The message m_S has the format of (1) and can be written as $(m_0, m_1, \cdots, m_{(\binom{t+2}{2}-1)}, a_1, a_2, \cdots, a_{ut}, MAC(X, k_S))$, where $X = I_S = (m_0, m_1, \cdots, m_{(\binom{t+2}{2}-1)})$ is the information block. The MAC function is over $\mathbf{F_q}^s$, where s is the parameter of FRS decoding (instead of $\mathbf{F_q}^{s'}$ for $N = 2t+1$). The tag value is:

$$A(z) = m_1 P_1(z) + \cdots + m_t P_t(z) + m_{t+1} P_1(z)^2 + \cdots + m_{2t} P_t(z)^2$$
$$+ m_{2t+1} P_1(z) P_2(z) + \cdots + m_{\binom{t+2}{2}-1} P_{t-1}(z) P_t(z) + P_{t+1}(z),$$

where $m_i \in \mathbf{F_q}^s$ and polynomials $P_1(z), P_2(z), \cdots, P_{t+1}(z)$ are over $\mathbf{F_q}^s$ and have degree t.

Parameters $s, u,$ and q are chosen to allow the receiver to decode up to t errors. The number of correct wires $t + ct$ must satisfy,

$$t + ct \geq N(\frac{1}{s+1} + \frac{s}{s+1} \frac{uR}{u-s+1}).$$

For a constant c_0, let $u = c_0 t$. We show that list decoding upto $\rho = \frac{t}{N}$ is possible for a constant value of $s = s_0$ when $t > 7c_0 + 2s_0 + 1$, and the value of c satisfies $c > \frac{s_0}{c_0} + \frac{1}{s_0}$. The details are in the full version of the paper [20]. Table 4.2 below gives example values for s_0 and c, and the size of the resulting list. The complete protocol is given below.

Table 3. Values of c and the list size for different values of s_0

s_0	list size	c
$s_0 = 1$	q^0	$c \approx \frac{1}{c_0} + 1$
$s_0 = 2$	q	$c \approx \frac{2}{c_0} + 1/2$
$s_0 = 3$	q^2	$c \approx \frac{3}{c_0} + 1/3$

SMT Protocol for $N = (2 + c)t, c > \frac{1}{t}$.

- **Sender Algorithm**:
  1. Uses the key generation of Construction 2 and obtains for wire $W_i, i = 1, ..., N$, the associated key,
  $$k_i = (P_1(z_i), P_2(z_i), \cdots, P_{t+1}(z_i)).$$
  2. The message m_S is,
  $$m_S = (m_0, m_1, \cdots, m_{(\binom{t+2}{2}-1)}, a_1, a_2, ..., a_{ut}, A(z)),$$
  where the tag for the information block $I_S = (m_0, m_1, \cdots, m_{(\binom{t+2}{2}-1)}), m_i \in \mathbf{F_q}^{s_0}$ is,
  $$A(z) = m_1 P_1(z) + \cdots + m_t P_t(z) + m_{t+1} P_1(z)^2 + \cdots + m_{2t} P_t(z)^2$$
  $$+ m_{2t+1} P_1(z) P_2(z) + \cdots + m_{\binom{t+2}{2}-1} P_{t-1}(z) P_t(z) + P_{t+1}(z).$$
  3. The FRS-code is over $\mathbf{F_q}^u$ and has dimension $k = ut + (\binom{t+2}{2} - 1)s_0 + ts_0 + s_0$. The sender encodes the message to a codeword c_S using the FRS encoding algorithm. Wire j, $1 \leq j \leq 2t + ct$, transmits the jth component of c_S and k_j.
- **Receiver Algorithm**:
  Uses the SMT decoding algorithm of Construction 3.

Theorem 6. *The protocol above is a 1-round (0, δ)-SMT for* $N = (2+c)t, c > \frac{1}{t}$ *with optimal transmission rate, and has efficient (polynomial time) decoding. The value of* δ *is given by* $\frac{2(t+1)}{q}$ *and is the smallest among all known protocols with the same connectivity.*

The proof outline is given in the full version of the paper [20].

Comparison with Related Work

There has been two other optimal (transmission rate) and efficient (computation) 1-round (0, δ)-SMT protocols for higher connectivity ($N = (2 + c)t, c > \frac{1}{t}$) [19,21]. The protocol presented in Section 4.2 has the least δ. The comparison of these protocols is outlined in Table 4.

Table 4. Comparison of 1-round $(0, \delta)$-SMT protocols for $N = (2 + c)t, c > \frac{1}{t}$; here Comp. refers to computation complexity, Poly. refers to polynomial (in N), Exp. refers to exponential (in N) and $\mathbf{F_q}$ is the field. Here ν is a parameter used in wire-virtualization which refers to the number of *physical* wires in each *virtual* wire.

References	Comp.	δ	Optimality	Outputs Incorrect Message
[19]	*Poly.*	$\leq \frac{N^{\nu}t(t+1)}{q} \approx \frac{N^{\nu+2}}{q}$	Yes	Yes [1]
[21]	*Poly.*	$\leq \frac{t(t+1)}{q} \approx \frac{N^2}{q}$	Yes	No
This Work	*Poly.*	$\leq \frac{2(t+1)}{q} \approx \frac{N}{q}$	Yes	No

5 Concluding Remarks

We showed a novel general approach to the construction of 1-round δ-RMT and (0, δ)-SMT protocols using LD codes and MACs. The approach has a number of advantages, (i) it is general, unifies construction of 1-round δ-RMT and (0, δ)-SMT protocols, and is applicable to all connectivities including $N = 2t+K, K \geq 1$, where K is a constant, (ii) relies on well-studied mathematical objects (list decodable codes and MACs) and so allow a wide range of instantiations; this also allows direct translation of advances in those areas into better constructions for 1-round δ-RMT and (0, δ)-SMT, and finally (iii) resulting in proofs of security (privacy and reliability) to be intuitive and easily verifiable. Instantiation of this general approach, using FRS-codes and our proposed multireceiver MACs result in constructions that have optimal transmission rates and the smallest δ, when $N = 2t + 1$ and $N = (2 + c)t, c > \frac{1}{t}$. For $N = 2t + 1$ the protocol is not computationally efficient for our instantiations. It is an interesting open problem if this general construction can be instantiated to achieve efficient and optimal construction for $N = 2t + 1$. Another interesting open question is whether δ can be further lowered while maintaining optimality. Another open problem is establishing lower bound on δ and constructing protocols that can achieve the

[1] The authors in [19] mention that their protocol can be modified to output only correct message block by using a different sub-protocol.

bound. We note that 1-round δ-RMT, can be seen as error correcting code in the traditional setting where channel corruption is adversarial and adversary has a limited view of the codeword (only t component).

References

1. Ben-Or, M., Goldwasser, S., Wigderson, A.: Completeness Theorems for Non-cryptographic Fault-tolerant Distributed Computation (extended abstract). In: Proc. of STOC, pp. 1–10 (1988)
2. Chaum, D., Crepeau, C., Damgard, I.: Multiparty Unconditionally Secure Protocols (Extended Abstract). In: Proc. of FOCS, pp. 11–19 (1988)
3. Dolev, D., Dwork, C., Waarts, O., Yung, M.: Perfectly Secure Message Transmission. Journal of the ACM 40(1), 17–47 (1993)
4. Desmedt, Y., Erotokritou, S., Safavi-Naini, R.: Simple and Communication Complexity Efficient Almost Secure and Perfectly Secure Message Transmission Schemes. In: Bernstein, D.J., Lange, T. (eds.) AFRICACRYPT 2010. LNCS, vol. 6055, pp. 166–183. Springer, Heidelberg (2010)
5. Desmedt, Y., Frankle, Y., Yung, M.: Multi-receiver/Multi-sender network security: efficient authenticated multicast/feedback. In: IEEE Infocom, pp. 2045–2054 (1992)
6. Elias, P.: List Decoding for Noisy Channels. Technical Report 335, MIT Research Lab of Electronics (1957)
7. Fitzi, M., Franklin, M., Garay, J., Vardhan, S.H.: Towards Optimal and Efficient Perfectly Secure Message Transmission. In: Vadhan, S.P. (ed.) TCC 2007. LNCS, vol. 4392, pp. 311–322. Springer, Heidelberg (2007)
8. Franklin, M.K., Wright, R.N.: Secure Communication in Minimal Connectivity Models. Journal of Cryptology 13(1), 9–30 (2000)
9. Guruswami, V.: Linear-algebraic List Decoding of Folded Reed-Solomon Codes. CoRR abs/1106.0436 (2011)
10. Guruswami, V., Rudra, A.: Explicit Codes Achieving List Decoding Capability: Error-Correction With Optimal Redundancy. IEEE Transactions on Information Theory 54(1) (2008)
11. Guruswami, V., Sudan, M.: Improved Decoding of Reed-Solomon and Algebraic-Geometric Codes. In: FOCS, pp. 28–39 (1998)
12. Kurosawa, K., Suzuki, K.: Almost Secure (1-Round, n-Channel) Message Transmission Scheme. In: Desmedt, Y. (ed.) ICITS 2007. LNCS, vol. 4883, pp. 99–112. Springer, Heidelberg (2009)
13. Patra, A., Choudhary, A., Srinathan, K., Rangan, C.: Unconditionally Reliable and Secure Message Transmission in Undirected Synchronous Networks: Possibility, Feasibility and Optimality. IJACT 2(2), 159–197 (2010)
14. Srinathan, K.: Secure Distributed Communication. PhD Thesis, IIT Madras (2006)
15. Sudan, M.: Decoding of Reed Solomon Codes beyond the Error-Correction Bound. J. Complexity 13(1), 180–193 (1997)
16. Srinathan, K., Patra, A., Choudhary, A., Rangan, C.P.: Probabilistic Perfectly Reliable and Secure Message Transmission – Possibility, Feasibility and Optimality. In: Srinathan, K., Rangan, C.P., Yung, M. (eds.) INDOCRYPT 2007. LNCS, vol. 4859, pp. 101–122. Springer, Heidelberg (2007)
17. Srinathann, K., Choudhary, A., Patra, A., Rangann, C.: Efficient Single Phase Unconditionally Secure Message Transmission with Optimum Communication Complexity. In: PODC, p. 457 (2008)

18. Srinathan, K., Narayanan, A., Pandu Rangan, C.: Optimal Perfectly Secure Message Transmission. In: Franklin, M. (ed.) CRYPTO 2004. LNCS, vol. 3152, pp. 545–561. Springer, Heidelberg (2004)
19. Safavi-Naini, R., Tuhin, M.A.A., Shi, H.: Optimal Message Transmission Protocols with Flexible Parameters. In: Proceedings of the 6th ACM Symposium on Information, Computer and Communications Security (ASIACCS 2011), pp. 453–458 (2011)
20. Safavi-Naini, R., Tuhin, M.A.A., Wang, P.: A General Construction for 1-round δ-RMT and $(0, \delta)$-SMT. Cryptology ePrint Archive (2012)
21. Tuhin, M.A.A., Safavi-Naini, R.: Optimal One Round Almost Perfectly Secure Message Transmission (Short Paper). In: Danezis, G. (ed.) FC 2011. LNCS, vol. 7035, pp. 173–181. Springer, Heidelberg (2012)
22. Wu, J., Stinson, D.R.: Three Improved Algorithms for Multi-path Key Establishment in Sensor Networks Using Protocols for Secure Message Transmission. In: IEEE Transactions on Dependable and Secure Computing, pp. 929–937 (2011)

A Prefiltering Approach to Regular Expression Matching for Network Security Systems

Tingwen Liu[1,2], Yong Sun[3], Alex X. Liu[4], Li Guo[3], and Binxing Fang[1,3]

[1] Institute of Computing Technology, Chinese Academy of Sciences, Beijing, China
[2] Graduate University of Chinese Academy of Sciences, Beijing, China
[3] National Engineering Laboratory for Information Security Technologies, Beijing
[4] Dept. of Computer Science and Engineering, Michigan State University
{liutingwen,suny}@software.ict.ac.cn, alexliu@cse.msu.edu

Abstract. Regular expression (RegEx) matching has been widely used in various networking and security applications. Despite much effort on this important problem, it remains a fundamentally difficult problem. DFA-based solutions can achieve high throughput, but require too much memory to be executed in high speed SRAM. NFA-based solutions require small memory, but are too slow. In this paper, we propose RegexFilter, a prefiltering approach. The basic idea is to generate the RegEx print of RegEx set and use it to prefilter out most unmatched items. There are two key technical challenges: the generation of RegEx print and the matching process of RegEx print. The generation of RegEx is tricky as we need to tradeoff between two conflicting goals: filtering effectiveness, which means that we want the RegEx print to filter out as many unmatched items as possible, and matching speed, which means that we want the matching speed of the RegEx print as high as possible. To address the first challenge, we propose some measurement tools for RegEx complexity and filtering effectiveness, and use it to guide the generation of RegEx print. To address the second challenge, we propose a fast RegEx print matching solution using Ternary Content Addressable Memory. We implemented our approach and conducted experiments on real world data sets. Our experimental results show that RegexFilter can speedup the potential throughput of RegEx matching by 21.5 times and 20.3 times for RegEx sets of Snort and L7-Filter systems, at the cost of less than 0.2 Mb TCAM chip.

Keywords: regular expression, prefilter, RegEx print, TCAM.

1 Introduction

Regular expressions (RegExes) have been widely used in a variety of network and security applications, such as anti-virus scanners [1], network intrusion detection and prevention systems [2], firewalls, traffic classification and monitoring [3]. In intrusion detection and prevention systems, RegExes are used to specify attack signatures. In traffic classification and monitoring, RegExes are used to specify the signature of application protocols, thus allowing the classification and monitoring of network traffic based on application protocols. The widespread

F. Bao, P. Samarati, and J. Zhou (Eds.): ACNS 2012, LNCS 7341, pp. 363–380, 2012.

usage is because of the expressive power, simplicity and flexibility of RegExes in specifying signatures.

The RegEx matching problem can be defined as follows: given a set R of RegExes, at run time, for each incoming item i (*e.g.*, packets), we want to get RegEx set $O(R, i)$ whose members are all matched by the item. RegEx matching, as the core operation of many applications, needs to be done at high speed with small memory. However, despite much work that has been done, this remains a fundamentally difficult problem. As a set of RegExes can be formally represented as a Deterministic Finite Automata (DFA) or Nondeterministic Finite Automata (NFA), prior RegEx matching solutions often fall into two categories: DFA-based and NFA-based. First, DFA-based solutions may achieve high speed because at any time there is only one active state, but may require too much memory. For applications running on networking devices such as intrusion detection and prevention systems and application firewalls, RegEx matching needs to be done in high speed SRAM, which has small capacity in terms of a few megabytes. Second, NFA-based solutions require small memory, but cannot achieve high speed because at any time there may be many active states [4].

In this paper, we propose RegexFilter, a prefiltering approach to RegEx matching for network security systems. Given a RegEx set R, we want to construct another RegEx set R' so that any unmatched item of R' is also an unmatched item of R. An *unmatched item of a RegEx set* is an item that does not match any RegEx in the set. Moreover, we want the matching efficiency of R' to be much higher than that of R; thus, we can use R' as a prefilter procedure of R: Given an item i, we first match it against R' and get set $O(R', i)$, if $O(R', i)$ is empty, then it for sure does not match any member in R and therefore we can skip this item safely; otherwise $O(R', i)$ is not empty, then we continue to match it against $T(R, O(R', i))$, where $O(R, i) \subseteq T(R, O(R', i)) \subseteq R$, and $T(R, O(R', i))$ can be obtained reversely from $O(R', i)$. Because most items are unmatched items for network security systems [5] and the matching cost of R' is much less than that of R, the overall throughput of this prefiltering approach is much higher than directly matching against R. We call R' the RegEx print of R. According to the main idea, RegexFilter divides the RegEx matching process into two stages: filtering stage and verifying stage. The filtering stage performs high-speed RegEx print matching on each arriving item. If one RegEx print is matched, the corresponding RegEx will be checked in the verifying stage.

There are two main technical challenges to implementing RegexFilter. The first challenge is the construction of the RegEx print for a given RegEx set. On one hand, we want the RegEx print to filter out as many unmatched items as possible. On the other hand, we want the matching efficiency of the RegEx print to be as high as possible. These two goals are unfortunately conflicting. With the RegEx print being the original RegEx set, the RegEx print can filter out all unmatched items, but the matching efficiency is the lowest. With the RegEx print being empty, the matching efficiency of zero cost is the highest, but it cannot filter out any unmatched item. We need to carefully tradeoff between these two conflicting goals. The second challenge is the matching of items against RegEx prints. We want this process to be as fast as possible to achieve high overall RegEx matching throughput.

To address the first challenge, we first propose a method to generate all possible RegEx prints for each RegEx in the given RegEx set. Second, we propose an estimation method to quantitatively measure the filtering effectiveness and complexity of RegEx prints. Third, we reduce the problem of selecting the RegEx prints with high filtering effectiveness and low complexity among all candidate RegEx prints of a RegEx to the classical 0-1 knapsack problem. In this paper, we use dynamic programming to choose RegEx prints among all candidates with the goal of maximizing filtering effectiveness while keep the complexity of a RegEx print less than a predefined threshold.

To address the second challenge, we propose a Ternary Content Addressable Memory (TCAM) based solution for RegEx print matching. As larger TCAMs have lower lookup frequency, require more power, generate more heat, and also have high hardware cost, we want to minimize the TCAM space required to encode the RegEx print DFA. Unfortunately, as minimizing TCAM space is NP-hard, we propose a heuristic method that uses the Quine-McCluskey algorithm to reduce TCAM space.

We make three key contributions in this paper. First, we propose an efficient method to generate RegEx prints. In particular, we propose some measurement tools for RegEx complexity and filtering effectiveness, and then use the tools to guide the generation of RegEx print. Second, we propose an efficient method of implementing RegEx print matching based on TCAMs. Third, we implemented our approach and conducted experiments on real-world RegEx sets and traffic traces. Our experimental results show that RegexFilter can speedup the throughput of RegEx matching by 21.5 times and 20.3 times for RegEx sets of Snort and L7-Filter systems, at the cost of less than 0.2 Mb TCAM chip.

The rest of the paper is organized as follows. We review related work in Section 2. In Sections 3 and 4 we explain the generation of RegEx prints and the implementation of high-speed RegEx print matching in TCAM for RegexFilter respectively. In Section 5, we present experimental results. Finally, We give conclusions and future work in Section 6.

2 Related Work

As DFA is the preferred representation of RegEx matching, recent work has focused on reducing the huge memory usage of DFA-based RegEx matching [4,6,7,8,9,10,11]. However, they achieve memory reduction only for signature sets of simple or specific RegExes. None of them can achieve high-speed RegEx matching for real-world signature sets that contain thousands of complex RegExes. However, these solutions are orthogonal to our work as they focus on improving RegEx matching in our verifying stage. Meiners et al. [12] propose a well-designed TCAM-based RegEx matching solution that introduces three novel techniques to reduce TCAM space and improve matching speed. The solution cannot work on real-world signature sets directly as the composite DFAs are too big to be encoded in a TCAM chip. However, it can be used in our RegEx print matching as the RegEx print DFA is small enough even for real-world RegEx sets. In fact, we design a more effective solution, that goes a step further and tackles the prefix problem in [12].

Several string-based prefiltering techniques have been developed to improve the performance of RegEx matching [5,13,14,15,16]. To the best of our knowledge, the most outstanding one is sigMatch [5]. The sigMatch technique organizes a signature set into a (processor) cache-efficient q-gram index structure, called the sigTree. For a given signature set, sigMatch requires that each signature has at least one string of length b to construct its sigTree. For these signatures that do not satisfy the requirement, sigMatch rewrites them into multiple signatures that have at least one string of length b in enumerating idea. Then, for each signature, sigMatch picks exactly one discriminative substring (without any meta-characters of RegExes) of length $b + \beta$ as its fingerprint. The first b bytes of the substring map the signature to a sigTree node, and the next β bytes following the b bytes in the substring are used to hash into the Bloom Filter at that node using a "set" of hash functions. Linked lists are used in sigTree nodes for short signatures that do not have a substring of length $b + \beta$.

These string-based prefiltering techniques have three major drawbacks. First, they suffer from the problem of member set explosion when they are applied to RegExes with character subclasses. An example is RegEx `^[a-z][a-z0-9]{5,15}` that matches user ID starting with an English alphabet followed by some alphanumeric characters.

These techniques have to enumerate all possible strings represented by the RegEx, the size of which is more than 26^{16}. Obviously this step is time-consuming and impracticable. Second, the fingerprints generated by these techniques are strings, which do not include the positioning of RegExes. For example anchor ^ in the above RegEx, which indicates that successful matchings must start from the beginning position of items. This inability leads to the problem that they may have poor filtering effectiveness. Third, there needs to be one Bloom Filter for each possible length of fingerprints [17]. The hardware cost can be prohibitive if fingerprints have a large number of distinct lengths. RegexFilter addresses these problems by generating short RegExes as fingerprints for each original RegEx without enumerating its all possible strings, and performing RegEx print matching with DFA representation in TCAM.

Ficara et al. [18] propose the first RegEx-based prefiltering solution that uses sampling technique to accelerate RegEx matching. The main idea is to sample a byte every θ bytes over packet payloads (θ is the sampling period). The sampled payloads are then used to match with a proper sampled DFA constructed from sampled RegExes. The method can process normal packets θ times faster at the cost of false-positive alarms. However, sampling RegExes correctly sometimes is very hard. Moreover, the sampled DFA may still experience state explosion . For example, RegEx |ab.1024cd— is sampled into two RegExes |a.512c— and |b.512d— given sampling period $\theta = 2$, the sampled DFA constructed from the two sampled RegExes has billions of states.

3 RegEx Print Generation

For a given RegEx, we first present a method to find all of its possible RegEx prints in this section, and then we introduce an algorithm to selectively generate good RegEx prints that satisfy our goal.

3.1 RegEx Print

Before presenting our work, we give some definitions to be used first. A RegEx r is a string over symbol set $\Sigma \cup \{\epsilon, |, \cdot, {}^*, (,)\}$, which is recursively defined as the empty character ϵ; a character $\alpha \in \Sigma$; and (r_1), $r_1 \cdot r_2$, $r_1|r_2$, and $r_1{}^*$, where r_1 and r_2 are RegExes. It represents a set of strings without enumerating them explicitly over alphabet Σ, which is defined recursively on the structure of r as follows:

- if $r{=}\epsilon$, $S(r){=}\{\epsilon\}$, the empty string
- if $r{=}\alpha$ $(\alpha \in \Sigma)$, $S(r){=}\{\alpha\}$, a single string of one character
- if $r{=}(r_1)$, $S(r){=}S(r_1)$
- if $r{=}r_1 \cdot r_2$, $S(r){=}S(r_1) \cdot S(r_2)$, where $S(r_1) \cdot S(r_2)$ is the set of strings w such that $w{=}w_1w_2$, with $w_1 \in S(r_1)$ and $w_2 \in S(r_2)$. The operator '$\cdot$' represents the classical concatenation of strings
- if $r{=}r_1|r_2$, $S(r){=}S(r_1) \cup S(r_2)$, the union of the two sets. The operator '$|$' is called union operator.
- if $r{=}r_1{}^*$, $S(r){=}S(r_1)^*{=}\bigcup_{i=0}^{\infty} S(r_1)^i$, where $S^0{=}\{\epsilon\}$ and $S^i{=}S \cdot S^{i-1}$ for any string set S. That is, the result is the set of strings formed by a concatenation of zero or more strings represented by r_1. The operator '*' is called star operator.

In order to construct an automata (NFA or DFA) for RegExes, most of the constructions use a binary tree representation as an intermediate form. The leaves of the tree are labeled with the characters of alphabet Σ or the symbol ϵ, and the internal nodes are labeled with the operators. The nodes that are labeled with '$|$' or '$\cdot$' have two children, while nodes labeled with '*' have only one child. Prior work describe how to parse a RegEx to obtain its parse tree recursively, in fact this conversion is reversible. Given a parse tree, its original RegEx can be obtained recursively just like the parsing process. In our work, we perform the generation of RegEx prints over the parse-tree representation for a given RegEx.

Definition 1. *Given a RegEx r, its Expression Size, denoted by $\mathbb{ES}(r)$, is the number of strings represented by r, namely $\mathbb{ES}(r) = |S(r)|$.*

For the given RegEx r, how to calculate its $\mathbb{ES}$ value is an open question. One simple method is to enumerate all the strings represented by r according to the definition, and then count the size. However, it is very inefficient as mentioned above. In this paper, we propose a novel method that can calculate $\mathbb{ES}(r)$ approximately, as shown in the following recursive way:

- if r=ϵ, $\mathbb{ES}(r) = 1$
- if r=α ($\alpha \in \Sigma$), $\mathbb{ES}(r) = 1$
- if r=(r_1), $\mathbb{ES}(r) = \mathbb{ES}(r_1)$
- if r=$r_1 \cdot r_2$, $\mathbb{ES}(r) = \mathbb{ES}(r_1) \times \mathbb{ES}(r_2)$ (We can infer that $\mathbb{ES}(r) = \mathbb{ES}(r_1)^n$ if r=$r_1\{n\}$)
- if r=$r_1|r_2$, $\mathbb{ES}(r) = \mathbb{ES}(r_1) + \mathbb{ES}(r_2)$
- if r=$r_1{}^*$, $\mathbb{ES}(r) = \sum_{t=0}^{\infty} \mathbb{ES}(r_1)^t = \infty$ ($\mathbb{ES}(r_1)$ is an integer no less than 1)

The easiest cases are single characters and ϵ. For operators '·', '|' and '*', the equations do not hold strictly, because there may be the same strings among all the combinations. For example, RegEx `a(b|`ϵ`)(b|`ϵ`)c` represents a set of strings {`ac, abc, abbc`} (because of $\epsilon \cdot r = r \cdot \epsilon = r$), its real expression size should be 3. According to our calculating method, we get $\mathbb{ES}$(`a(b|`ϵ`)(b|`ϵ`)c`) = $\mathbb{ES}$(`a`) × $\mathbb{ES}$(`b|`ϵ) × $\mathbb{ES}$(`b|`ϵ) × $\mathbb{ES}$(`c`) $= 1 \times 2 \times 2 \times 1 = 4$. Fortunately, our method produces approximate $\mathbb{ES}$ values that are very close to the real values.

Similarly, we define the Expression Size of a RegEx set $R = \{r_1, \cdots, r_n\}$, denoted by $\mathbb{ES}(R)$, as the number of strings represented by $r_1|\cdots|r_n$. We can infer that $\mathbb{ES}(R) = \sum_{i=1}^{n} \mathbb{ES}(r_i)$ according to our calculating method of $\mathbb{ES}$.

Obviously, a RegEx set has larger or equal $\mathbb{ES}$ value than any of its RegEx prints. Meanwhile, a RegEx set has higher complexity than any of its RegEx prints, where the complexity is regarded as state size for DFA-based matching solutions in this paper. Inspired by the insight, we argue that $\mathbb{ES}$ can be used as a measurement tool to compare the complexity even for two totally different RegEx sets. This speculation is reasonable intuitively: a RegEx set with larger $\mathbb{ES}$ value represents more strings, and more strings consume more states when constructing Aho-Corasick complete automata (a special DFA for string matching). A persuasive example is a string signature, such as r_1=`ACNS`, and a RegEx signature that contains constrained repetitions of wildcards, such as r_2=|AC.10NS—. After calculating we know $\mathbb{ES}(r_1) = 1$ and $\mathbb{ES}(r_2) = 256^{10}$, meanwhile the DFA of r_1 has 5 states and the DFA of r_2 has more than one thousand states. An important application of our $\mathbb{ES}$ tool is that it can be used to deal with state explosion pertinently by combining with previous work [11,10,19,9]. Because we can locate the accurate positions that will lead to state explosion quantitatively by calculating $\mathbb{ES}$, while previous work solve the problem qualitatively and empirically.

A signature with better filtering effectiveness is matched with a lower probability. A string, which is a RegEx too, only represents itself. Moreover, the length of a string is fixed. It is easy to prove that the matching probability of a string is inverse to the size of alphabet Σ to the power of its length over random inputs of infinite length. Some prefiltering work tend to generate longer strings as fingerprints because they will be matched much less frequently than shorter ones. However, it is hard to measure the matching probability of a RegEx, because a RegEx usually represents many strings of different lengths. Motivated by the insight that shorter strings have much higher matching probability, we speculate that the matching probability of a RegEx mainly depends on its Minimum Expression Length and its Shortest Expression Size, which are defined as follows.

Definition 2. *Given a RegEx r, the Minimum Expression Length of r, denoted by $\mathbb{L}(r)$, is the number of characters in s, where s is the shortest string in set $S(r)$.*

Unlike in the calculation of $\mathbb{ES}$ value, the same strings in $S(r)$ do not change $\mathbb{L}$ value for the given RegEx r. Thus, we can calculate $\mathbb{L}(r)$ accurately in the following recursive way:

- if $r{=}\epsilon$, $\mathbb{L}(r) = 0$
- if $r{=}\alpha$ $(\alpha \in \Sigma)$, $\mathbb{L}(r) = 1$
- if $r{=}(r_1)$, $\mathbb{L}(r) = \mathbb{L}(r_1)$
- if $r{=}r_1 \cdot r_2$, $\mathbb{L}(r) = \mathbb{L}(r_1) + \mathbb{L}(r_2)$
- if $r{=}r_1|r_2$, $\mathbb{L}(r) = min(\mathbb{L}(r_1), \mathbb{L}(r_2))$
- if $r{=}r_1{}^*$, $\mathbb{L}(r) = 0$

Definition 3. *Given a RegEx r, the Shortest Expression Size of r, denoted by $\mathbb{SES}(r)$, is the number of strings of length $\mathbb{L}(r)$ in set $S(r)$.*

Similar to the calculation of $\mathbb{ES}$, we can calculate $\mathbb{SES}$ approximately with the following method:

- if $r{=}\epsilon$, $\mathbb{SES}(r) = 1$
- if $r{=}\alpha$ $(\alpha \in \Sigma)$, $\mathbb{SES}(r) = 1$
- if $r{=}(r_1)$, $\mathbb{SES}(r) = \mathbb{SES}(r_1)$
- if $r{=}r_1 \cdot r_2$, $\mathbb{SES}(r) = \mathbb{SES}(r_1) \times \mathbb{SES}(r_2)$
- if $r{=}r_1|r_2$:
  - if $\mathbb{L}(r_1)$ is bigger than $\mathbb{L}(r_2)$, $\mathbb{SES}(r) = \mathbb{SES}(r_2)$
  - if $\mathbb{L}(r_1)$ is smaller than $\mathbb{L}(r_2)$, $\mathbb{SES}(r) = \mathbb{SES}(r_1)$
  - if $\mathbb{L}(r_1)$ equals to $\mathbb{L}(r_2)$, $\mathbb{SES}(r) = \mathbb{SES}(r_1) + \mathbb{SES}(r_2)$
- if $r{=}r_1{}^*$, $\mathbb{SES}(r) = 1$

Definition 4. *The matching probability of a RegEx r, denoted by $\mathbb{MP}(r)$, is defined as the ratio of $\mathbb{SES}(r)$ to $\mathbb{SCS}(r)$, where $\mathbb{SCS}(r)$ represents the total number of strings of length $\mathbb{L}(r)$ over alphabet Σ.*

Definition 5. *A RegEx r is dividable if and only if it can be rewritten into the form of $r_1 \cdot r_2$, where r_1 and r_2 are RegExes, and $S(r_1) \neq \{\epsilon\}$, $S(r_2) \neq \{\epsilon\}$. RegEx r is atomic if it is not dividable.*

Lemma 1. *Given a RegEx $r = r_1 \cdots r_n$, where r_t is atomic $(1 \leq t \leq n)$. We abbreviate RegEx r of this type as $r = r_{[1,n]}$ later. Then 1) $r_{[i,j]}$ $(1 \leq i \leq j \leq n)$ is a RegEx print of r; 2) RegEx r has at most $\frac{n(n+1)}{2}$ RegEx prints.*

Proof. 1) First, $r_{[i,j]}$ is obviously a RegEx. Second, it is a fingerprint, because any input T matched by r will be matched by $r_{[i,j]}$: T must contains one string s in set $S(r)$, while s is the concatenation of a string in set $S(r_{[1,i-1]})$, a string in set $S(r_{[i,j]})$ and a string in set $S(r_{[j+1,n]})$.

2) Since 1) is right, the proof of 2) is simple and trivial.

3.2 RegEx Print Generation Algorithm

For a RegEx set R, we want to generate a set of RegEx prints that produces a DFA with as few states as possible and prefilters as many items as possible. However, comparing among all possible RegEx print sets is high-cost. In this section, we present a novel algorithm that can achieve the goal with low cost by pruning uncompetitive RegEx print sets. Our algorithm involves three stages: selecting stage, refining stage and deciding stage, which are described below.

Selecting Stage. As compiling the RegEx print set into a composite DFA within limited memory is a hard requirement, we try to limit the DFA state size, at the time we want to bypass the time-consuming process of DFA construction. The main idea is to select those RegEx prints whose $\mathbb{ES}$ values are no more than a predefined expression size threshold β for each RegEx. Detailedly speaking, for a given dividable RegEx $r_{[1,n]}$, RegEx print $r_{[i,j]}$ is selected in this stage if it satisfies the following conditions: 1) $\mathbb{ES}(r_{[i,j]}) \leq \beta$; 2) $\mathbb{ES}(r_{[s,t]}) > \beta$ for $s \leq i$ and $t > j$, or $s < i$ and $t \geq j$. We can easily prove that $r_{[i,j]}$ is also a RegEx print of $r_{[s,t]}$.

Before describing our algorithm, we introduce a theorem first.

Theorem 1. *For any dividable RegEx $r = r_1 \cdot r_2$, where r_1, r_2 are RegExes, the following two conditions hold: 1) $\mathbb{ES}(r) \geq \mathbb{ES}(r_1)$, meanwhile $\mathbb{MP}(r) \leq \mathbb{MP}(r_1)$.*

Proof. According to the above calculation methods, we know that $\mathbb{ES}$ value is not less than one meanwhile $\mathbb{MP}$ value is not more than one for any RegEx. Thus $\mathbb{ES}(r_2) \geq 1$, $\mathbb{MP}(r_2) \leq 1$. Then we can infer that: 1) $\mathbb{ES}(r) = \mathbb{ES}(r_1) \times \mathbb{ES}(r_2) \geq \mathbb{ES}(r_1)$, moreover $\mathbb{ES}(r) = \mathbb{ES}(r_1)$ only when r_2 represents the string set $\{\epsilon\}$ or $\{\alpha\}$ ($\{\alpha \in \Sigma\}$); 2) $\mathbb{MP}(r) = \frac{\mathbb{SES}(r)}{|\Sigma|^{\mathbb{L}(r)}} = \frac{\mathbb{SES}(r_1) \times \mathbb{SES}(r_2)}{|\Sigma|^{\mathbb{L}(r_1)+\mathbb{L}(r_2)}} = \mathbb{MP}(r_1) \times \mathbb{MP}(r_2) \leq \mathbb{MP}(r_1)$, moreover $\mathbb{MP}(r) = \mathbb{MP}(r_1)$ only when the shortest expressing string set of r_2 is its shortest complete string set.

Theorem 1 accords with our intuition: a RegEx requires more resource but has better filtering effectiveness than any of its RegEx print. Our algorithm works recursively in post-order traversal from the root node of the parse tree of $r_{[1,n]}$ to select RegEx prints. Figure 1 shows the selecting process of RegEx "`a[bc]d.[bc]`" that has five atoms. Given $\beta = 256$, we begin the selecting stage from the first atom in step 1, *curr* pointer keeps moving to the next atom if $\mathbb{ES}$ value of the RegEx print between *begin* pointer and *curr* pointer is less than or equal to β. When *curr* pointer arrives at the fourth atom ".", condition 1 does not hold, thus RegEx print `a[bc]d` is selected. Because we can infer that $\mathbb{ES}$ value of any RegEx print that contains "`a[bc]d.`" is longer than 256 according to Theorem 1, we begin step 2 from the second atom. Although RegEx print `[bc]d` satisfies the two conditions at the same time, it is included in the already selected RegEx print "`a[bc]d`". According to Theorem 1 we know that `a[bc]d` has higher $\mathbb{MP}$ value than `[bc]d`, thus `[bc]d` is not selected. Step 3, 4 and 5 follow the same idea to select RegEx prints.

For RegEx $r_{[1,n]}$, the $\mathbb{ES}$ value of a atom, supposing r_i, may be larger than β. Obviously, any RegEx print containing r_i will not be selected. However, these

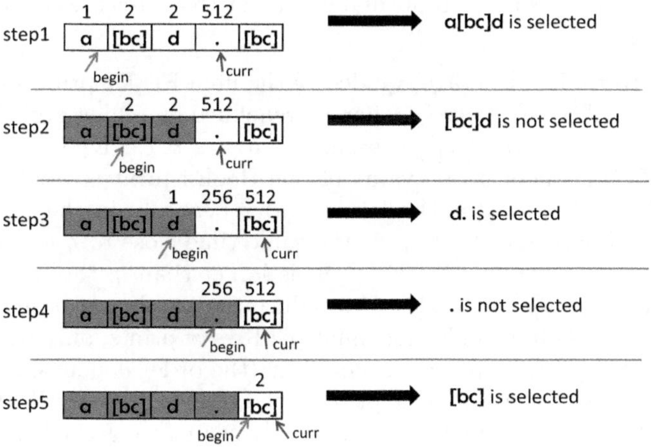

Fig. 1. Selecting RegEx prints for RegEx `a[bc]d.[bc]` with $\beta = 256$. In each step atoms in blue indicate that they are covered by already selected RegEx prints, the next selected RegEx print must contain at least one atom that is not in blue.

RegEx prints may have low $\mathbb{MP}$ value while all the remaining RegEx prints have high $\mathbb{MP}$ value. To address this problem, we rewrite r into multiple RegExes until no atom has $\mathbb{ES}$ value larger than β or no atom is in the form of the union of RegExes. For instance, given $\beta = 256$, RegEx $r =$`(tele|phone|AC.*NS)[^a]` has only one RegEx print `[^a]` whose $\mathbb{MP}$ value is close to 1. We can rewrite it into two RegExes: $r_1 =$`(tele|phone)[^a]`, $r_2 =$`AC.*NS[^a]`, then we can select RegEx prints with the same $\mathbb{ES}$ value and lower $\mathbb{MP}$ value for rewritten RegExes.

Refining Stage. In this stage, we refine the RegEx prints selected in the last stage in the following steps. Step 1, for each selected RegEx prints, the first atom (the last atom) whose $\mathbb{MP}$ value equals 1 should be removed repeatedly. For the example in Figure 1, the wildcard in "`d.`" introduces very limited filtering effectiveness by requiring one random symbol after "`d`". However it will introduce much state in the RegEx print DFA. Thus, we need to remove the wildcard. RegEx prints of the example become `a[bc]d`, `d` and `[bc]` now. Step 2, these RegEx prints, whose atoms are included by other RegEx prints, should be deleted according to Theorem 1. For the above example, RegEx print `d` has only one atom that is included by RegEx print `a[bc]d`. We should delete RegEx print `d` because it is meaningless: at any time RegEx `a[bc]d` is matched means `d` must be matched. One thing to notice that, RegEx print `[bc]` should not be deleted if allowing one RegEx has more than one RegEx print (see in deciding stage). Because its atoms are not included by `a[bc]d`. Step 3, RegEx prints with positioning (or anchors) should be kept as many as possible. Because positioning add the limitation that these RegEx prints must be matched at the special positions of items. As a result, they will not be matched frequently even for these RegEx prints with high $\mathbb{MP}$ value. In our implementation, we regard the filtering effectiveness of ^ as a normal character to decrease $\mathbb{MP}$ value while not increase $\mathbb{ES}$ value.

Deciding Stage. In this stage, we decide the final RegEx print for each RegEx by limiting its $\mathbb{MP}$ value no larger than a predefined matching probability threshold η. Given a RegEx $r = r_{[1,n]}$, assuming it has k RegEx prints $(p_1, \cdots, p_k)$ left after refining stage: if $\mathbb{MP}$ value of one RegEx print is smaller than η, we removing the first atom (or the last atom) repeatedly in the RegEx print to make its $\mathbb{MP}$ value larger than η, at the same time close to η as much as possible; if $\mathbb{MP}$ value of any one RegEx print is larger than η, then we use multiple RegEx prints together for RegEx r to reduce the number of items that need to be verified. Because if a RegEx has multiple RegEx prints, only when an item is matched by these RegEx prints sequentially (the order depends on the position of the first atom of these RegEx prints), the item will be verified. For RegEx prints `a[bc]d` and `[bc]` in Figure 1, the former one matched does not imply that the latter one must be matched in sequential matching order. One vivid data item is `abdea`. How to choose the final RegEx prints is an open problem at present. We want the sum of $\mathbb{ES}$ value of these final RegEx prints is smaller than or equal to β and the product of $\mathbb{MP}$ value of these final RegEx prints is as large as possible. Obviously this is a typical 0-1 knapsack problem. In this paper, we achieve the goal using the classical dynamic programming solution for 0-1 knapsack problem.

4 Regex Print Matching

As the RegEx print set in RegexFilter can be compiled into a composite DFA within limited memory, prior TCAM-based DFA matching solutions can be used here directly for high-speed RegEx print matching. TCAM is a special type of memory which takes input of data as key to look-up address. It has the following three capacities: i) ternary states encoding: 0's, 1's, and *'s where *'s stand for either 0 or 1, enabling one TCAM entry to encode multiple DFA transitions; ii) parallel content lookup, enabling TCAM to complete lookups in a single operation no matter the number of occupied TCAM entries; iii) first-match semantic, making TCAM to return the index of the first address for the content that the key matches.

Meiners et al. [12] propose a well-designed TCAM-based RegEx matching solution, which uses three novel techniques to reduce TCAM space and improve RegEx matching speed: transition sharing, table consolidation, and variable striding. The main idea is to encode multiple DFA transitions into a TCAM entry by the help of TCAM capacities. However, the character bundling algorithm used to encode transitions inside each state in the work is designed for TCAM-based packet classification applications, which produce prefix TCAM entries: the predicate of each entry is a prefix bit string (*e.g.*, 01**) where no 0 and 1 behind *. In fact, a ternary TCAM entry allows * to appear at any positions (*e.g.*, 0**1), which means it misses the opportunity of encoding transitions created by non-prefix entries.

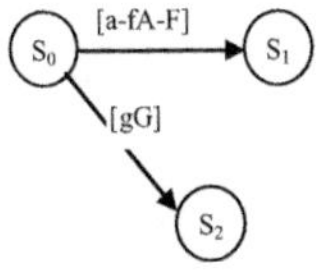

(a) Transitions of State S_0

TCAM		SRAM
Curr State	Input Sym	Dest State
S_0	01000111	S_2
	01100111	S_2
	01000000	S_0
	01100000	S_0
	01000000	S_1
	01100000	S_1
	********	S_0

(b) Character Bundling

TCAM		SRAM
Curr State	Input Sym	Dest State
S_0	01*00111	S_2
	01*00**1	S_1
	01*00*1*	S_1
	01*001**	S_1
	********	S_0

(c) Transition Logic Simplification

TCAM		SRAM
Curr State	Input Sym	Dest State
S_0	01*00111	S_2
	*****000	S_0
	01*00***	S_1
	********	S_0

(d) Optimal Encoding

Fig. 2. Outgoing transitions of state S_0 and their different encodings in TCAM

In this section we go a step further and tackle the prefix problem. For any DFA state S_i, it has 256 transitions (assuming alphabet Σ is ASCII), which have the same current state and completely different 8-bit input symbols. Obviously encoding all the 256 transitions into TCAM entries can be regard as the simplification of a logic function with 8-input 1-output, hereinafter referred to as a transition logic function. However. the outputs of the transition logic function have $|S|$ possible values, which is different from the classical logic function that always have a logic value of either "0" or "1". Using exhaustive searching method can get the optimal solution, but its cost is higher than that of the classical Quine-McCluskey algorithm.

To reduce the complexity we add a limitation of encoding the transitions with the same output together: treating a destination state as value "1" of classical logic functions, and simplifying the transitions to the state with Quine-McCluskey algorithm; then setting their destination states to irrelevant term (any state is allowed), and encoding the remaining transitions. We give a detailed description with the example of encoding the transitions in Figure 2 (a). Current state S_0 moves to state S_1 if the input symbol is in character range `[a-fA-F]`, moves to state S_2 along g and G, and moves to itself for the remaining input symbols. As long as the occupied TCAM entries are arranged according to the encoding order of destination states, the capacity of first-match semantic ensures the correctness of lookup results. Figure 2 (c) shows the result with the encoding order $S_2 \rightarrow S_1 \rightarrow S_0$, which occupies 5 TCAM entries, while character bundling algorithm occupies 7 entries, as shown in Figure 2 (b). One thing to notice is that the occupied TCAM entries of our encoding is relevant to the encoding order.

In this paper, we do not address the encoding order problem by testing all possible orders. We propose a near-optimal solution to the problem based on the distribution of transitions with the same destination state: for each DFA state, its 256 outgoing transitions moves to few destination states, furthermore the distribution of these transitions is very uneven. This observation can be verified by the statistical results over the whole state set averagely. Given a state S_i in state set S, assuming its 256 outgoing transitions move to M_i different destination states, thereinto N_{ij} transitions move to the j-th destination state ($1 \leq j \leq M_i$, and N_{i*} is in descending order. If not, sorting them). Obviously,

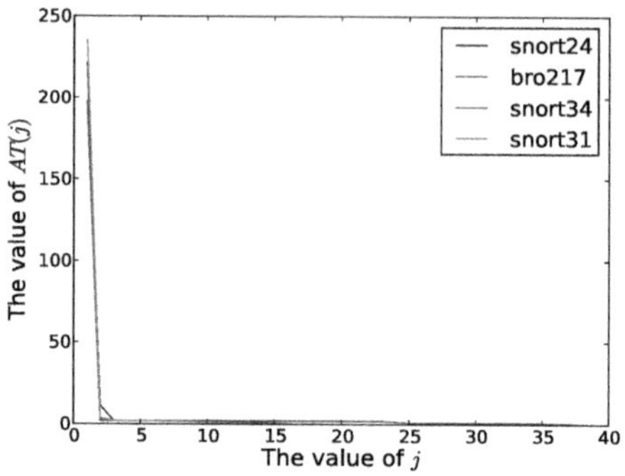

Fig. 3. The distribution of $AT(j)$ for Bro217, Snort24, Snort31 and Snort34

$\sum_{j=1}^{j\leq M_i} N_{ij} = 256$. For state set S, the number of different destination states per state ADS is defined as $\frac{\sum_{i=0}^{i<|S|} M_i}{|S|}$, its average number of j-th most transitions $AT(j)$ is defined as $\frac{\sum_{i=0}^{i<|S|} N_{ij}}{|S|}$.

Taking RegEx sets bro217, snort24, snort31 and snort34 (widely used in prior experiments) as examples, the corresponding DFAs has 36.8, 14.4, 11.6 and 12.3 different destination states in average. The distribution of $AT(j)$ for the DFAs is shown in Figure 3: the transitions to the most destination state ($AT(0)$) account for 90% proportions; the transitions to other destination states roughly equal, and the value is no more than 2 in most case. Therefore, a preferable encoding order of is determined heuristically by the number of the transitions to the same destination state. Because when the number is one or two, encoding them will consume the same TCAM entries regardless of the order. Nevertheless, encoding them first can improve subsequent results. In Figure 2 (c), encoding the transitions to state S_0 only consumes one TCAM entries.

An important thing to note is that the heuristic order does not guarantee to minimize the transition logic function. The optimal encoding is shown in Figure 2 (d), which encodes partial transitions to state S_0 first. This change makes one TCAM entry enough to encode all the transitions to state S_1.

5 Experimental Results

In this section, we first give a brief description of our experimental setup. Then we evaluate RegexFilter on the metrics of memory consumption and matching performance. At last we show how RegexFilter changes as the change of expression size threshold β and matching probability threshold η.

5.1 Experimental Setup

In this paper, we evaluate RegexFilter on RegEx sets extracted from two real-world systems, namely L7-Filter [3] and Snort [2]. L7-Filter is a popular open-source application layer traffic classifier for Linux. It re-assembles the payload

Table 1. Comparison of state size among SinstrFilter, MulstrFilter and RegexFilter

RegEx sets	# of RegExes		# of DFA states / # of RegExes unable to handle		
	original	rewritten	SinstrFilter	MulstrFilter	RegexFilter
backdoor	158	161	1452 / 0	2128 / 0	2302 / 0
l7filter	107	166	1317 / 8	1541 / 8	2847 / 0

content of a flow and identifies its application level protocol through RegEx matching. The latest version of L7-Filter has 112 RegExes for traffic classification. In this paper, we remove five RegExes that are overmatched, and select the remaining ones to constitute our experimental RegEx set. Snort is a famous open-source intrusion detection system, which can be configured to perform protocol analysis, content inspecting over online traffic to detect a variety of worms, attacks and probes. We consider all the RegExes in `backdoor.rules` file of Snort systems. In Both L7-Filter and Snort systems, each RegEx is compiled into one automaton; at run time all automatons are used to match each incoming item sequentially.

In this paper, we compare RegexFilter with SinstrFilter and MulstrFilter. SinstrFilter chooses a single string that is longest as the fingerprint for each RegEx while MulstrFilter uses all the strings as fingerprints. All the three filters can work in two modes. One is item-filter mode, which matches an item with all RegExes of R in verifying stage, if the item passes through filtering stage. The other is pair-filter mode that matches an item with RegEx set $T(R, O(R', i))$ in verifying stage, $T(R, O(R', i))$ contains all the RegExes that match the item successfully as described in section 1.

5.2 Experimental Evaluation

In our evaluation, we use threshold $\log_2(\beta) = 16$ and $-\log_{256}(\eta) = 6$ to generate RegEx print for RegexFilter. As shown in Table 1, RegexFilter can construct a small Regex print DFA with less than three thousand states for each RegEx set. On the contrary, both backdoor set and l7filter set produce a composite DFA with more than one million states. One thing to notice is that it is impossible to extract any strings for 8 RegExes of l7filter set. One example is RegEx `^[a-z][a-z0-9 -_]+`, which is used to classify Finger traffic. Both SinstrFilter and MulstrFilter have to experience the step of enumerating all possible strings represented by these RegExes. In our experiment, SinstrFilter and MulstrFilter do not generate fingerprints for these eight RegExes because of the high cost of enumeration.

In this paper, we estimate the throughput of TCAM-based fingerprint matching using Agrawal and Sherwood's TCAM model, which makes the assumption that each TCAM chip is manufactured with a 0.18 μm process. Table 2 shows the results of TCAM-based fingerprint matching of SinstrFilter, MulstrFilter

Table 2. TCAM size and throughput for RegEx print DFAs

Filters	backdoor set			l7filter set		
	TCAM size	TPS	Throughput	TCAM size	TPS	Throughput
SinstrFilter	0.050 Mb	1.007	7.27 Gbps	0.046 Mb	1.018	7.27 Gbps
MulstrFilter	0.073 Mb	1.003	7.27 Gbps	0.054 Mb	1.015	7.27 Gbps
RegexFilter	0.15 Mb	1.03	5.44 Gbps	0.17 Mb	1.61	5.44 Gbps

and RegexFilter on TCAM size and throughput for backdoor and l7filter sets [1]. TCAM size is the TCAM memory required to encode the corresponding DFAs. We calculate its value by multiplying the number of entries by the TCAM width. For all the fingerprint DFAs, we need at most 15 state ID bits, thus TCAM width 36 is enough to store the lookup key. TPS means TCAM entries Per State, which is calculated by dividing the number of TCAM entries required by the number of states. DFA engine takes fixed stride over inputs with one character each transition, the throughput is estimated by the number of TCAM lookups that can be performed in a second for a given number of TCAM entries by 8 bits.

We can draw the following conclusions from Table 2. First, our encoding can reduce TCAM memory required sharply, whose maximum value is less than 0.2 Mb for all the fingerprint DFAs. Second, the TPS value is far less than 256, precisely close to 1, which means that our encoding is possible to encode a DFA with more than one millon states into a 72 Mb TCAM. Consequently, it makes Regex-Filter to work on a set of thousands of RegExes. Third, TCAM-based matching can achieve high throughput, the value is over 5 Gbps for all the fingerprint DFAs. Fourth, SinstrFilter and MulstrFilter are superior to RegexFilter on the performance of fingerprint matching. The primary reason is that their fingerprint DFAs have less states and occupies less TCAM entries than that of RegexFilter.

A key criterion to measure filtering effectiveness is verifying rate, which is defined as the percentage of matches performed in verifying stage when with filtering stage among the total number of matches performed when without filtering stage. We make a comparison over a real traffic trace captured in 2010 from a backbone network. The result is shown in Figure 4. The percentages of items (each item is a flow here) that are matched by backdoor set and l7filter set are 10.4% and 91.4%. The malicious ratio on the normal traffic for backdoor set is higher than that in real Snort system, the primary reason may be we skip the packet classification step before RegEx matching in Snort. Although more than 90% items pass through filtering stage for l7filter, verifying rate is still very small in pair-filter mode. The result confirms our assumption that an item is usually matched by limited RegExes. Our experiment presents RegexFilter shows the minimum verifying rate, and recalls all matched pairs of items and RegExes.

[1] Our implementation first removes transitions redundancy among state using shadow encoding technology in [12], and then encoding each state's remaining labeled transitions with transition logic simplification method instead of character bundling.

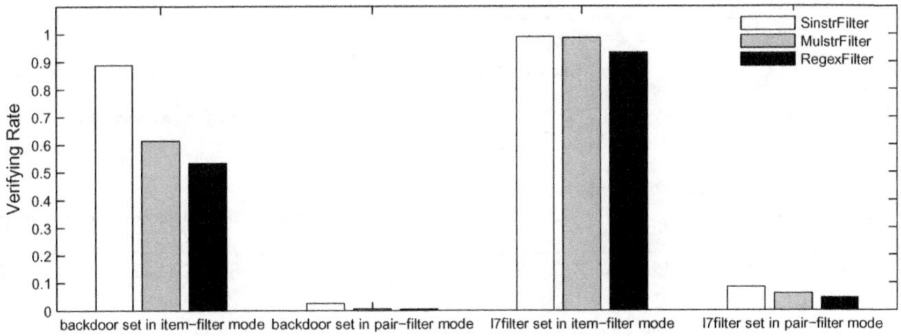

Fig. 4. Comparison among SinstrFilter, MulstrFilter and RegexFilter on verifying rate over a real traffic

After knowing the performance of TCAM-based fingerprint matching and verifying rate, we can estimate the potential throughput by determining the time required to process a byte as the sum of the time required by TCAM-based fingerprint matching and the expected time required by verifying stage to process a byte under the verifying rate. In pair-filter mode, RegexFilter can improve the throughput by 21.5 times for backdoor set and by 20.3 times for l7filter set. Our RegexFilter achieves the potential throughput as 1.2 times and 1.7 times high on backdoor and l7filter set comparing with TCAM-based SinstrFilter, and achieves the potential throughput as 0.79 times and 1.3 times high comparing with TCAM-based MulstrFilter. One thing to notice is that MulstrFilter introduces additional time to validate whether all fingerprints of the same RegEx are matched in sequence, which is not included in the above estimation.

5.3 Effect of Expression Size Threshold

In this section we evaluate the effect of threshold β for RegexFilter. As β is used to bound the number of strings represented by a RegEx print, state size of RegEx print DFA is expected to grow along with the increase of β.

Figure 5 shows the change of RegEx print DFA state size for different β on our experimental RegEx sets. All RegEx prints are generated under threshold $-\log_{256}(\eta) = 6$. From Figure 5 we can find that backdoor set experiences almost the same state size of for different β. Because each RegEx in backdoor set has string fingerprints. As for l7filter set, state size of RegEx print DFA initially decreases rapidly as the increase of β, and then increases after a certain limit. This behavior is because some RegExes do not have any RegEx print for small β, *i.e.* RegEx `^[\x14\x1c\$].{6,15}[\xc6-\xff]` used to classify "network time protocol" traffic in L7-Filter. We add these RegExes into the set of RegEx prints to ensure that no false-negative matches occur, as a result the fingerprint DFAs experience state explosion for small β. As β increases further, RegexFilter can generate RegEx prints for these RegExes. Therefore RegEx print DFA becomes compact suddenly, and then increases in the number of states stably.

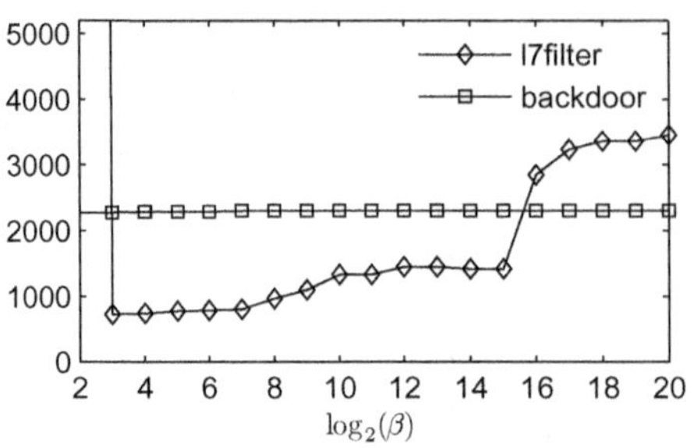

Fig. 5. State size of fingerprint DFAs as a function of $\log_2(\beta)$

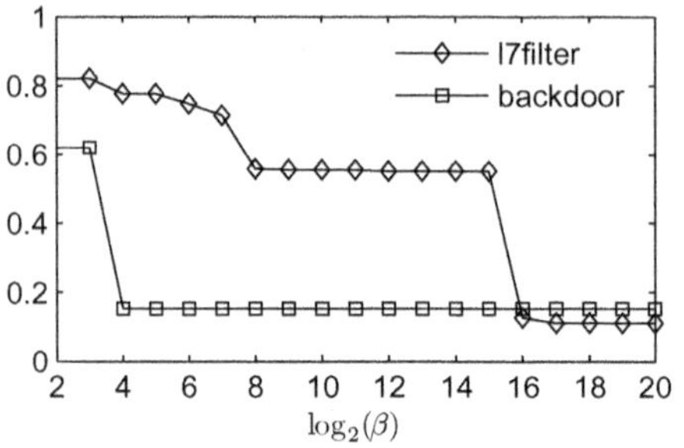

Fig. 6. Verifying rate in item-filter mode as a function of $\log_2(\beta)$

Figure 6 shows that how verifying rate of RegexFilter in item-filter mode varies along with the increase of β. The input trace is a synthetic file generated by regex-tool [20] with $p_m = 0.15$. Owing to the same reason as described in the last paragraph, RegexFilter presents fluctuant curve on verifying rate for l7filter set. In one word, verifying rate decreases as the increase of β in the whole.

5.4 Effect of Matching Probability Threshold

Threshold η, which is used to bound the maximum matching probability of each RegEx print, is the other parameter that can impact the generation of RegEx prints in RegexFilter. In our evaluation, we keep $\log_2(\beta)$ fixed at 16 because backdoor set and l7filter set presents normal behavior on state size and filtering rate for $\log_2(\beta) \geq 14$. It gives a good trade-off when $\log_2(\beta)$ is 16.

Figure 7 and Figure 8 show the effect of threshold η on state size of RegEx print DFA and verifying rate respectively. As can be seen in Figure 7, decreasing η, namely increasing $-\log_{256}(\eta)$, will increase the state size of RegEx print DFAs, as more symbols will be included for higher η. The state size of RegEx print DFA experience sublinear growth for backdoor set, while it grows slowly

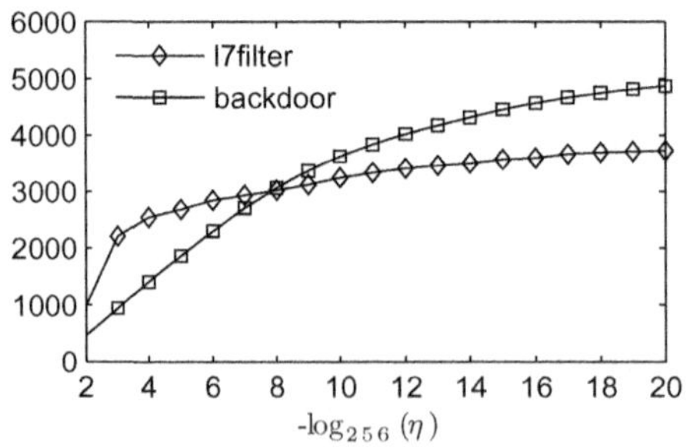

Fig. 7. State size of fingerprint DFAs as a function of $-\log_{256}(\eta)$

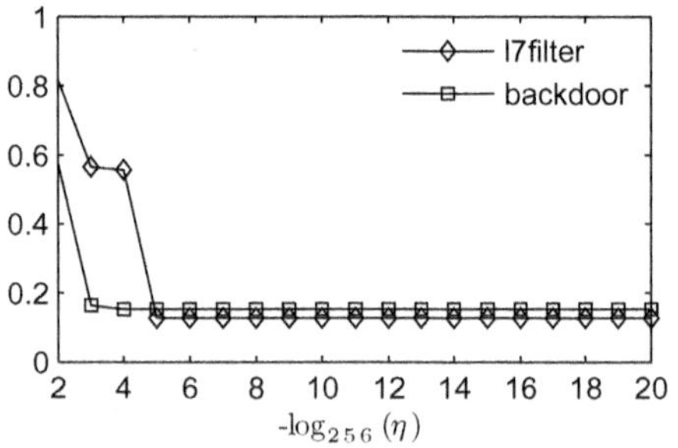

Fig. 8. Verifying rate in item-filter mode as a function of $-\log_{256}(\eta)$

and flatly for l7filter set. The primary reason is that little RegExes in l7filter107 set have the RegEx print whose $\mathbb{MP}$ value is lower than 256^{-8}, on the contrary some RegExes of backdoor set have long string fingerprints. We expect that the growth rate of RegEx print DFA state size will slow down gradually and become zero after a certain value.

From Figure 8, we find that increasing $-\log_{256}(\eta)$ does not improve verifying rate any more after a certain value, which are 5 and 3 for the two RegEx sets respectively. This indicates that RegExes in real-world systems are distinct enough with $-\log_{256}(\eta) = 6$.

6 Conclusions

In this paper, we present RegexFilter, a high-speed and memory-efficient technique to improve the throughput of RegEx matching for network and security applications. Our solution leverage the insights that an item is usually matched by limited RegExes, and most items do not match any member in real-word RegEx sets. Thus we try to speedup RegEx matching by quickly finding these RegExes that may match each arriving item as little as possible. First, we develop a novel method that generate RegEx prints to filter a large number of unmatched items with little memory requirement. The method utilizes some new tools to guide the generation of RegEx prints without constructing DFAs. Second, we propose an non-prefix encoding algorithm to minimize the TCAM entries required for TCAM-based RegEx matching. As a result, RegexFilter can perform RegEx print matching quickly.

We evaluate our work on some reasonable metrics and compare it with other two solutions. The preliminary experimental results show that our TCAM-based RegexFilter is suitable to accomplish the filtering task in high-speed for sets of large-scale and complex RegExes. As part of future work, we will beef our work and explore its extension for multi-cores.

Acknowledgments. This work was supported by the National High-Tech Research and Development Plan of China under Grant No. 2011AA010703; the National Natural Science Foundation of China under Grant No. 61070026 and No. 61003295.

References

1. Kojm, T.: Clam Anti-virus Signature Database, `http://www.clamav.net`
2. Roesch, M.: Snort - Lightweight Intrusion Detection for Networks. In: Proc. USENIX LISA, pp. 229–238 (1999)
3. Levandoski, J., Sommer, E., Strait, M.: Application Layer Packet Classifier for Linux, `http://l7-filter.sourceforge.net/`
4. Yu, F., Chen, Z., Diao, Y., Lakshman, T.V., Katz, R.H.: Fast and Memory-Efficient Regular Expression Matching for Deep Packet Inspection. In: Proc. ACM/IEEE ANCS, pp. 93–102 (2006)

5. Kandhan, R., Teletia, N., Patel, J.M.: SigMatch: Fast and Scalable Multi-Pattern Matching. Proceedings of the VLDB Endowment 3(1-12), 1173–1184 (2010)
6. Kumar, S., Dharmapurikar, S., Yu, F., Crowley, P., Turner, J.: Algorithms to Accelerate Multiple Regular Expressions Matching for Deep Packet Inspection. In: Proc. ACM SIGCOMM, pp. 339–350 (2006)
7. Ficara, D., Giordano, S., Procissi, G., Vitucci, F., Antichi, G., Di Pietro, A.: An Improved DFA for Fast Regular Expression Matching. ACM SIGCOMM Computer Communication Review 38(5), 29–40 (2008)
8. Liu, T., Yang, Y., Liu, Y., Sun, Y., Guo, L.: An Efficient Regular Expressions Compression Algorithm From A New Perspective. In: Proc. IEEE INFOCOM, pp. 2129–2137 (2011)
9. Becchi, M., Crowley, P.: A Hybrid Finite Automaton for Practical Deep Packet Inspection. In: Proc. ACM CoNEXT Conference, pp. 1–12 (2007)
10. Smith, R., Estan, C., Jha, S., Kong, S.: Deflating the Big Bang: Fast and Scalable Deep Packet Inspection with Extended Finite Automata. ACM SIGCOMM Computer Communication Review 38(4), 207–218 (2008)
11. Kumar, S., Chandrasekaran, B., Turner, J., Varghese, G.: Curing Regular Expressions Matching Algorithms from Insomnia, Amnesia, and Acalculia. In: Proc. ACM/IEEE ANCS, pp. 155–164 (2007)
12. Meiners, C.R., Patel, J., Norige, E., Torng, E., Liu, A.X.: Fast Regular Expression Matching using Small TCAMs for Network Intrusion Detection and Prevention Systems. In: Proc. USENIX Security Symposium, p. 8 (2010)
13. Watson, B.W.: A New Regular Grammar Pattern Matching Algorithm. In: Díaz, J. (ed.) ESA 1996. LNCS, vol. 1136, pp. 364–377. Springer, Heidelberg (1996)
14. Cho, J., Rajagopalan, S.: A Fast Regular Expression Indexing Engine. In: Proceedings of the 18th International Conference on Data Engineering, pp. 419–430 (2002)
15. Yang, C.C., CHENG, C.M., WANG, S.D.: Two-phase Pattern Matching for Regular Expressions in Intrusion Detection Systems. Journal of Information Science and Engineering 26, 1563–1582 (2010)
16. Ramaswamy, R., Kencl, L., Iannaccone, G.: Approximate Fingerprinting to Accelerate Pattern Matching. In: Proceedings of the 6th ACM SIGCOMM Conference on Internet Measurement, pp. 301–306 (2006)
17. Broder, A., Mitzenmacher, M., Mitzenmacher, A.: Network Applications of Bloom Filters: A Survey. In: Internet Mathematics. Citeseer (2002)
18. Ficara, D., Antichi, G., Pietro, A.D., Giordano, S., Procissi, G., Vitucci, F.: Sampling Techniques to Accelerate Pattern Matching in Network Intrusion Detection Systems. In: Proc. IEEE ICC, pp. 1–5. IEEE (2010)
19. Tang, Y., Jiang, J., Hu, C., Liu, B.: Managing DFA History with Queue for Deflation DFA. Journal of Network and Systems Management, 1–26 (2011)
20. Becchi, M.: Regular Expression Processor, `http://regex.wustl.edu/`

iHTTP: Efficient Authentication of Non-confidential HTTP Traffic*

Jason Gionta[1], Peng Ning[1], and Xiaolan Zhang[2]

[1] North Carolina State University, Raleigh, NC, USA
{jjgionta,pning}@ncsu.edu
[2] IBM T.J. Watson Research Center, Howthorne, NY, USA
cxzhang@us.ibm.com

Abstract. HTTPS is the standard protocol for protecting information sent over the World Wide Web. However, HTTPS adds substantial overhead to servers, clients, and networks [1, 2]. As a result, website owners often pass on HTTPS and resort to only HTTP for hosting websites, leaving clients and servers vulnerable to attacks [3, 4]. Techniques have been proposed to only enable authentication and integrity of HTTP (response) data [2, 5–7]. However, they all suffer from vulnerabilities and poor performance. In this paper, we propose iHTTP, a new approach for enabling lightweight, efficient authentication and verification of HTTP (response) data. We adaptively handle different data encodings to allow for better performance without effecting user experience. We introduce a novel technique, Sliding-Timestamps, to allow iHTTP clients to authenticate the freshness of response data to prevent replay attacks and amortize signing costs. We also introduce Opportunistic Hash Verification to reduce client public key operations required to authenticate full web pages. We show in our experimental evaluation that iHTTP provides similar performance to HTTP, and higher throughput and lower maximum response time than HTTPS and HTTPi, the most recent HTTP authentication approach [7], for Client-Static data.

1 Introduction

HTTP [8] is the most popular protocol used to construct the World Wide Web because it is lightweight, flexible, and scalable. However, HTTP provides no security protection and as a result technologists have accepted HTTP over TLS/SSL (i.e., HTTPS) [9] as the standard for providing authentication, integrity, and confidentiality while sacrificing being lightweight, flexible, and scalable. Other security protocols such as SHTTP [10] and HTTPA [11] have also been proposed; however, they suffer from similar flexibility and scalability problems. As a result,

* This work is supported by U.S. National Science Foundation (NSF) under grant 0910767, the U.S. Army Research Office (ARO) under grant W911NF-08-1-0105 managed by NCSU Secure Open Systems Initiative (SOSI)), and an IBM Open Collaboration Faculty Award.

F. Bao, P. Samarati, and J. Zhou (Eds.): ACNS 2012, LNCS 7341, pp. 381–399, 2012.

data owners often choose HTTPS to provide data confidentiality and HTTP to provide efficient data delivery.

While HTTP data is non-confidential, the integrity of HTTP data is still very important. Recent research has shown that the lack of data integrity can have averse effects on website owners and clients. For example, Vratonjic et al. provided a case study of ad frauds in which HTTP web content was modified on the fly to include or rewrite advertisements. The study estimated that a WiFi hot spot with 100 users could raise $49,400 in annual revenue via HTTP ad content rewriting [3]. Research by Stamm et al. demonstrated an attack on HTTP called "Drive-By Pharming", where a malicious javascript reconfigures a victim router's DNS to redirect clients to fake pages. As a result, attackers are able to launch phishing attacks or present bogus data to clients [4].

The discussion above highlights the distinct need to protect even non-sensitive HTTP content from malicious modification. A simple solution is to enable HTTPS for all websites. However, HTTPS adds substantial overhead to both the server and the clients [1]. Furthermore, HTTPS does not support network caching, which is known to significantly improve performance [12] and reduce upstream bandwidth usage [2]. There clearly exists a need to provide data authentication and integrity of HTTP data without sacrificing flexibility and scalability as with HTTPS.

1.1 Previous Work

A viable solution to the above problems must have a minimal impact on performance, flexibility, and scalability, which existing security techniques such as HTTPS clearly lack. Towards this end, recent research has attempted to provide lightweight integrity verification and authentication mechanisms for HTTP.

Web Tripwires was proposed to verify if web content has changed between the server response and client rendering by embedding a javascript measurement agent in HTML pages [13]. However, Web Tripwires provides no security protection for the measurement agent itself and can be easily bypassed.

HTTPI provides authentication and data integrity of HTTP data by using a pre-established session key for keyed hashing of client requests and server responses [6]. HTTPI sessions are described as long-lived with no indication of key management. This of course poses significant threats to the security of the system [14]. Furthermore, the evaluation of HTTPI does not consider the cost of TLS/SSL handshakes. Indeed, HTTPI in some cases can show performance as poorly as HTTPS.

Lesniewski-Laas et al. used HTTPS to only provide authentication and integrity while enabling caching [5], which leads to a significant reduction in the origin server loads. However, this technique requires the modification of and the trust in network caches, making it difficult to adopt.

To decouple authentication and integrity from key management, SINE seeks to provide data integrity and origin authentication by digitally signing HTTP (response) data [2], which is signed once and used to serve many client requests to enable network caching. However, SINE is vulnerable to replay attacks of stale authenticated data, and does not support chunked transfer encoding as introduced in HTTP 1.1. HTTPi further enables support of chunked transfer encoding, and subsumes SINE by providing network caching and progressive rendering of both chunked and non-chunked data [7].

SINE and HTTPi are promising in enabling data origin authentication and integrity without key management. However, they suffer from issues that seriously undermine their usability. SINE enables the caching of authenticators for long periods of time by setting a static expiration [2]. This feature allows attackers to launch replay attacks using non-expired data. In other words, clients cannot verify the freshness of data. HTTPi attempts to address the freshness issue by requiring that each response to a client request, or a chunk of a response, be digitally signed, thus severely impacting the server performance [7]. In both SINE and HTTPi, clients are required to perform at least one public key operation per server response. As a result, clients potentially are required to make hundreds of public key operations to render a single HTML page.

1.2 Our Contributions

Among the previous research, SINE and HTTPi are the most promising candidates for lightweight HTTP integrity protection. However, both fail to address the issues outlined previously. In this paper, we propose a new approach called iHTTP to address these problems. In particular, we seek an authentication scheme with performance similar to HTTP. Our approach is inspired by SINE and HTTPi. However, we propose new techniques to achieve stronger security features without sacrificing performance.

We observe that HTTP response data can be categorized into two types: *Client-Static data*, and *Client-Unique data*. Client-Static data refers to site resources which are not specific to a client. For example, many clients requesting a single image via a common URL will receive the same HTTP content. Client-Unique data refers to response data that is directed to a specific client. For example, a client requesting a common URL which returns the client's WAN IP address will return unique HTTP content. In this case each response is unique to the client who requested it. In this paper, we focus on techniques to authenticate and verify Client-Static data.

iHTTP adopts three techniques to achieve lightweight authentication of HTTP response data. The first is to handle encoding data adaptively. HTTP 1.1 data can be encoded in two ways: Content and Transfer [8]. We observe that servers and clients handle encoded data in different manners. For example, compressed data is buffered while non-compressed data can be processed as a stream. We introduce a rule to adaptively apply existing integrity techniques based on encoding format. By processing unique encodings differently, we can reduce the server response size and the number of cryptographic operations without sacrificing flexibility or user experience.

The second technique is the decoupling of freshness verification and signature generation. Previous signature-based HTTP integrity techniques [2, 7] tightly couple data freshness and signature generation. As a result, these techniques either suffer from performance issues or are susceptible to replay attacks. To address this issue, we present a technique called *Sliding-Timestamps*, which decouples signature generation from data freshness authentication by using authenticated hash chain values to calculate an extended timestamp. Servers simply need to release specific hash values to extend the freshness of signatures, which forgoes signing data to update freshness.

The third technique is aimed at reducing the cost of client verification. Signature-based HTTP integrity techniques require clients to verify at least one signature per response. As a result, clients may be required to verify hundreds of responses to render a single web page, which hinders user experience and requires unnecessary computation. Towards this end, we present a technique called *Opportunistic Hash Verification* to provide clients an opportunity to verify responses without signature verification. Using the descriptive nature of HTML, servers can provide contextual authentication information about anticipated future responses. This will reduce client overhead for rendering complex web pages.

We validate iHTTP through a prototype implementation and experimental evaluation. In our experiments, we compare iHTTP with existing standard protocols, HTTP and HTTPS, and the most recent signature-based HTTP integrity technique HTTPi [7]. We show that iHTTP outperforms HTTPS and HTTPi significantly. Furthermore, our results show iHTTP achieves similar throughput to native HTTP for Client-Static data. We also provide an evaluation of the impact of Client-Unique data on signature-based HTTP techniques.

The rest of the paper is organized as follows. Section 2 discusses our design goals and assumptions. Section 3 reviews some signature-based HTTP integrity techniques, on which iHTTP is based. Section 4 presents the iHTTP protocol in detail. Section 5 provides security and performance analysis of iHTTP. Section 6 reports the implementation and evaluation, and Section 7 concludes this paper.

2 Design Goals, Assumptions, and Threat Model

Design Goals: The high-level design goals for iHTTP are given below:

- Data Origin Authentication: Clients should be able to verify whether the received iHTTP data was generated by a trusted identifiable source.
- Data Integrity: Clients should be able to verify whether the received iHTTP response data has been modified by intermediate parties.
- Content Freshness: Client should be able to verify whether the received iHTTP response data is "out of date".
- Low Performance Impact: iHTTP will have minimal impact on servers and clients, allowing for high throughput and low response time.
- Flexibility: iHTTP should allow caching of iHTTP data without modifying network caches or proxies.
- Standards Compatible: iHTTP should be HTTP 1.1 [8] compatible as it is currently the latest HTTP specification and widely adopted on the Internet.

Assumptions: We assume that our client and server machines are trusted and server private keys are protected. We also assume that the clocks on client and server machines are loosely synchronized. We assume data sent over iHTTP is non-confidential as our goal is to provide efficient authentication and integrity protection of HTTP data. Finally, we assume that iHTTP is serving Client-Static data as opposed to Client-Unique data.

Threat Model: We assume that attackers have the ability to intercept, add, modify, delete, reorder, and store all data sent between the client and the server. Attackers can sign data with authentic certificates not associated with the origin server's domain and/or IP address. Finally, attackers can slow down the delivery of data for limited periods of time. The client assumes a reasonable response time from the server.

We consider the following attacks out of the scope of this paper: Attacks aimed at disrupting network availability or undermining cryptographic primitives. Such attacks constitute general attacks on network implementations and protocol weaknesses. We also do not defend against vulnerabilities targeting web applications or scripting software.

3 Preliminaries

In this section we discuss two signature-based HTTP integrity techniques, Naive and Progressive Authentication, presented in [2, 7] for the authentication of HTTP content. These techniques form the foundation of the new techniques we develop in this paper. The notation used in this paper is included in Figure 1. Note that authenticator refers to the collective group of information sent to clients to authenticate and verify HTTP responses.

K – Private Key provided by Certificate
$Sign_K$\{X\} – Sign X with private key K
H(X) – Hash of X
A.t – Timestamp
A.e – Expiration
A.u – Object's URL
A.l – Message length
HTTP – HTTP Response
HTTP.Headers – HTTP Dependent HTTP headers
HTTP.Body – HTTP response body
HTTP.Content – Response Specific Authentication Data
SH – $Sign_K$\{H(A.t|A.e|A.u|A.l|HTTP.Headers|HTTP.Content)\}
Authenticator – SH, A.t, A.e, A.u, A.l, HTTP.Content

Fig. 1. Notation

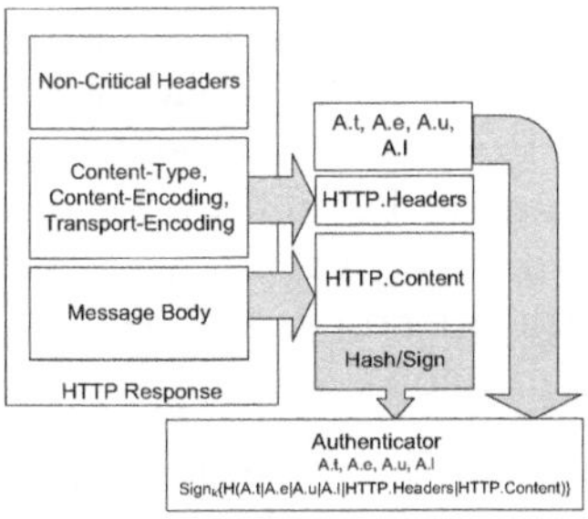

Fig. 2. Naive Authentication

Naive Technique: With the Naive Technique, a server first buffers the HTTP response prior to sending data. The server then generates the authenticator specific to the HTTP response by signing the hash of a server timestamp (A.t), expiration (A.e), requested URL (A.u), and HTTP response body ($Sign_k\{H(\texttt{A.t}|\texttt{A.e}|\texttt{A.u}|HTTP.Headers|HTTPBody)\}$). The authenticator

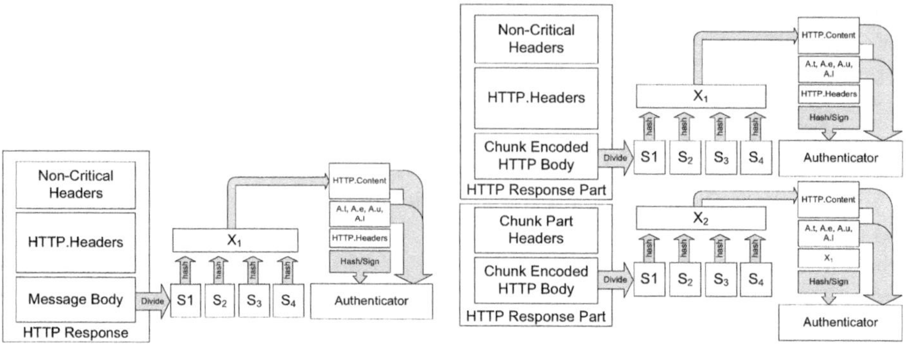

(a) Progressive Authentication (b) Chunked-Transfer Authentication

Fig. 3. Progressive Authentication Techniques

is then sent as an HTTP header along with the associated timestamp, expiration, and URL. Clients can then verify the integrity of the HTTP response by first buffering the HTTP message, performing one hash operation over the data, and one (public key) signature verification operation. Figure 2 depicts this approach.

The Naive Technique lacks progressive processing and rendering support, which hinders user experience. Thus, Progressive Authentication was proposed to overcome this shortfall [7].

Progressive Authentication of Data Streams: Progressive Authentication enables the authentication of data streams by providing hashes of data blocks as part of the authenticator [7]. To enable progressive rendering, the server first divides the entire HTTP response body into equally sized segments $(S_1, ..., S_n)$. Each segment is then hashed to get $H(S_i)$ and the hashes are concatenated into a list $X_1 = H(S_1)|...|H(S_n)$. The signature for the response is generated as $Sign_k(H(\texttt{A.t}|\texttt{A.e}|\texttt{A.u}|\texttt{A.l}|HTTP.Headers|X_1)) = SH_1$ [7]. The authenticator for Progressive Authentication includes $(X_1, \texttt{A.t}, \texttt{A.e}, \texttt{A.l}, SH_1)$ and is sent along as an HTTP header. Upon receiving the authenticator, the client immediately verifies $X_1, \texttt{A.t}, \texttt{A.e}, \texttt{A.l}$, and $HTTP.Headers$ via the signature SH_1. Once the authenticator is verified, X_1 can be used to immediately authenticate any segment upon arrival [7].

Chunked-Transfer Support: Chunked Transfer-Coding was introduced in HTTP 1.1 to allow for servers to send partial information to clients without knowing the response size [8]. Chunked-Transfer Authentication enables support of chunk encoded data by generating an authenticator for each individual chunk. Essentially each chunk has a unique authenticator that protects the chunk data and chunk order. The authenticator of the first chunk is added as an HTTP header while the authenticators of subsequent chunks are embedded as part of data [7]. Figure 3(b) shows the process for creating the authenticator for chunk encoded data. Please refer to [7] for details.

4 Our Approach – iHTTP

iHTTP is an HTTP integrity approach for efficiently enabling HTTP authentication and integrity, preventing replay attacks, and reducing overhead for both clients and servers. iHTTP achieves these goals by adaptively handling data encoding, enabling Freshness Authentication with Sliding-Timestamps, and providing Opportunistic Hash Verification.

In the following, we first describe a generic authenticator generation process for handling different data-encodings, then expand the authenticator generation process to add Sliding-Timestamps to enable authenticator caching, and finally describe Opportunistic Hash Verification to reduce client verification cost.

4.1 Authenticator Generation

This subsection outlines the iHTTP authenticator content and generation process. Specifically, iHTTP adaptively uses the Naive and Progressive Authentication techniques to enable better performance by reducing cryptographic operations and payload size.

iHTTP uses cryptographic hash and digital signature to generate message authenticators. Several key pieces of information are required for providing authentication, including 1) a timestamp (`A.t`) used to verify the authenticator generation time, 2) an expiration timestamp (`A.e`) for preventing reuse of expired authenticators, 3) the requested URL (`A.u`) to link the response data to the requested URL, 4) a subset of HTTP headers ($HTTP.Headers$), 5) content length (`A.l`), and 6) a message content identifier ($HTTP.Content$) referring to a unique identifier that is generated based on the HTTP message body. After $HTTP.Content$ generation, the server hashes and signs the above items as $Sign_k\{H(\texttt{A.t}|\texttt{A.e}|\texttt{A.u}|\texttt{A.l}|HTTP.Headers|HTTP.Content)\}$. For simplicity we refer to this collective group of data as an authenticator.

$HTTP.Headers$ is included in the signature to ensure clients process response data in the correct manner by verifying the response format and properties. This prevents data misuse attacks in which the client is persuaded to handle data in a different manner than specified by the server. HTTP headers are categorized into two groups: End-to-end or hop-by-hop. End-to-end must be stored and forwarded by caches in the original form (with the exception of Content-Length, which may be modified by network caches), while hop-by-hop headers may be modified by caches [8]. Thus, all the non-modifiable headers in the response are included as part of $HTTP.Headers$. To handle the exception Content-Length, iHTTP uses `A.l` to verify the data length in the authenticator.

The Naive and Progressive Authentication techniques differ in the generation process for the message content identifier ($HTTP.Content$). We observe that no single previous technique provides the best performance for generating $HTTP.Content$ over all data types, encodings, and formats. Thus we adaptively apply the two techniques to achieve the best performance for generating $HTTP.Content$. Figure 1 contains the relevant notation and information with regards to the authenticator.

iHTTP determines the optimal authentication technique for *HTTP.Content* generation based on the response encoding. We observe that compressed responses require clients to buffer data prior to decompression. As a result, the limitation of Naive Technique pointed out by [2,7] does not apply to compressed data. By applying the Naive technique, iHTTP will reduce the number of hash operations from $O(n)$ to $O(1)$ and decrease the authenticator size by 1.4% [7]. iHTTP determines if responses are compressed by checking for compression tokens located in the Content-Encoding or Transport-Encoding headers. If a compression token is present, the Naive technique is used for the creation of *HTTP.Content*, which consists of a SHA-1 hash of the HTTP message body. Otherwise, Progressive Authentication is used for creating *HTTP.Content*, which is the concatenated list of SHA-1 hash segments representing the HTTP message body.

An iHTTP server generates a new authenticator when 1) the content being served for the requested URL has changed, or 2) when the authenticator expires.

Local Authenticator Caching: The techniques presented in this paper rely upon locally caching the server generated authenticators. The Local Authenticator Cache is only concerned with caching the latest generated authenticator per requested URL. Thus, once a new authenticator for a URL is generated, the previous URL specific authenticator can be discarded. Caching may be implemented using many different techniques and data structures. Specific caching implementations may provide better performance in different environments and platforms. Thus, determining the optimal caching mechanism for Local Authenticator Caching is orthogonal to this work.

4.2 Freshness Authentication

As discussed previously, authenticator caching is enabled using a static expiration time. This presents a dilemma to iHTTP. If the expiration time for an authenticator is set too long, an update to the content at the corresponding URL may have occurred before the expiration time. As a result, an attacker may replay the old data using the authenticator before its expiration time. This can certainly be mitigated by using a short expiration time. However, a short expiration time will result in frequent generation of the authenticators, which involve expensive public key operations.

In the following, we present a freshness authentication technique that can provide fresh authentication tokens with light overhead.

Sliding-Timestamp: The problem described above is a result of a tight coupling between authenticator generation and the client URL request. The tight coupling is due to the strong constraint of generating an authenticator to prove freshness. Thus, it is desirable to provide authenticator freshness verification without actually signing the authenticator data.

We propose to meet this goal by allowing the server to extend a given authenticator's timestamp using a server generated hash chain. The server generates the hash chain at the time of authenticator generation and signs the commitment of the hash chain to bind the hash chain and authenticator. Each intermediate hash

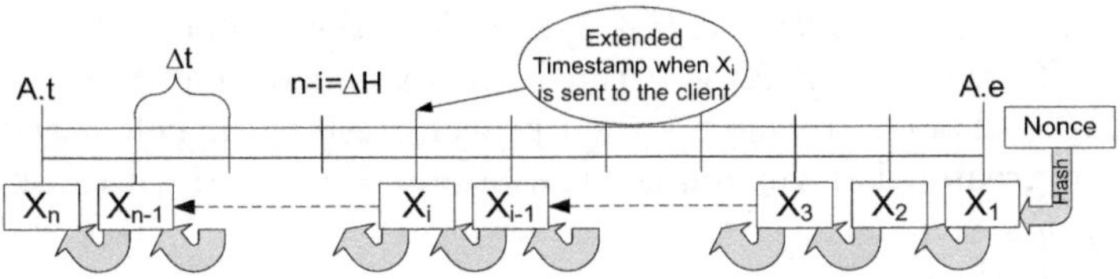

Fig. 4. Sliding-Timestamp Generated using a Hash Chain

in the chain represents a calculated amount of time to extend the authenticator's timestamp. The one-way nature of hash chains allows clients to authenticate the server's decision to extend the timestamp without requiring the server to sign the authenticator data again [15]. Figure 4 illustrates this approach.

The hash chain is created using a server generated random number N, the size of the chain n, and one-way cryptographic hash function H as $X_1 = H(N), ..., X_n = H(X_{n-1})$. The random number N is kept secret at the server; this value can be used to calculate intermediate hash values. X_i represents the ith hash in the chain. We consider the number of hash operations to generate X_n beginning with X_i as ΔH and we represent this process as $H^{\Delta H}(X_i) = X_n$. We introduce Δt as a short server-defined configurable time-increment used for extending the authenticator timestamp. Δt represents the time duration associated with each hash operation. Using the above properties, the extended timestamp is calculated by $\mathtt{A.t} + \Delta H * \Delta t$, where `A.t` is the authenticator timestamp.

The server generates a hash chain during each authenticator generation. The size of the hash chain, n, is determined by $\frac{\mathtt{A.e}-\mathtt{A.t}}{\Delta t}$. The server stores n, Δt, N, and X_n in the local server cache associated with each authenticator. The authenticator signature is modified to include Δt and X_n, i.e., the authenticator signature is generated as $Sign_k\{H(\mathtt{A.t}|\mathtt{A.e}|\mathtt{A.u}|\mathtt{A.l}|HTTP.Headers|HTTP.Content|\Delta t|X_n)\}$. Prior to sending an HTTP response, the server must generate the appropriate X_i to authenticate the freshness of the authenticator. X_i is calculated based on the current server timestamp (c), the authenticator's timestamp (`A.t`), Δt, and size of the hash chain (n). The server calculates $i = n - \lceil \frac{c-\mathtt{A.t}}{\Delta t} \rceil$ and then generates the ith hash of N as $H^i(N) = X_i$. (Alternatively, the server may pre-compute all hash values and use each appropriately.) The server sends X_i, ΔH, Δt, and X_n as part of the authenticator.

Figure 5 contains the updated final authenticator and signature. The server generates a new authenticator only if *HTTP.Content* for an HTTP response and locally cached authenticator do not match or if the authenticator expires. Otherwise, the server uses the locally cached authenticator for the response along with the intermediate hash chain value X_i to extend the timestamp.

Prior to verifying X_i and the extended timestamp, the client first verifies the authenticator contents via signature. To verify the extended timestamp, the client first verifies X_i is a member of the hash chain rooted at X_j, where X_j is a previously verified hash chain value for the same signature or $X_j = X_n$ when the authenticator signature has not been previously received. X_i is verified if

$X_j = H^{\Delta H\prime}(X_i)$ where $\Delta H\prime = \Delta H - \Delta j$ and Δj is the number of hash operations to generate X_n from X_j. If X_i is verified, the extended timestamp is calculated by `A.t` $+ \Delta H * \Delta t$. The request is then proven fresh if the calculated extended timestamp is greater than the request timestamp. If the authenticator is verified, the client stores that last verified authenticator where $\Delta j = \Delta H$ and $X_j = X_i$ for a unique URL.

Given X_i, the server provides X_{i-1} to extend the authenticator timestamp by Δt. Since it is infeasible for third parties to compute X_{i-1} from X_i given the one-way property of cryptographic hash functions, the clients can be sure only the server can provide X_{i-1}. As a result, this approach decouples the client request and authenticator generation and allows authenticator caching to prevent resigning when *HTTP.Content* matches the cache value for the requested URL.

K – Private Key provided by Certificate
$Sign_K\{X\}$ – Sign X with private key K
H(X) – Hash of X
A.t – Timestamp
A.e – Expiration
A.u – Object's URL
A.l – Message Length
N – Server nonce used to build the hash chain
n – Hash chain size
X_i – the ith hash generated in a hash chain
ΔH – Number of hash operations to convert X_i to X_n
Δt – Server defined time parameter to determine an extended timestamp
HTTP.Headers – iHTTP Dependent HTTP headers
iHTTP.Content – Data identifiers for authentication/integrity techniques
SH – $Sign_K$\{H(A.t|A.e|A.u|A.l|HTTP.Headers|iHTTP.Content|X_n|Δt)\}
Authenticator – SH, A.t, A.e, A.u, A.l, iHTTP.Content, Δt, X_i,ΔH

Fig. 5. iHTTP Authenticator

Network Caching of iHTTP Objects: Servers may direct network caches to store iHTTP data using standard HTTP 1.0/1.1 caching directives. In this manner, no changes are needed by network caches to enable clients and servers to use iHTTP. Thus, iHTTP can be adopted incrementally. However, iHTTP requires that the server use cache-directives to allow for effective caching while enabling iHTTP data freshness.

iHTTP uses the "must-revalidate" cache directive to force caches to revalidate every request with the origin server. This ensures the iHTTP server can respond with an updated fresh authenticator for each request. When a NOT-MODIFIED response is provided, the server will provide an updated authenticator and Sliding-Timestamp. The cache can overwrite any changed headers seamlessly and forward necessary data to the client [8].

Forcing network caches to query the origin server for each request does result in an unnecessary request when the cached authenticator is fresh. For example, two different clients may request the same file at the same time, resulting in the same authenticator. This limitation is due to the fact that iHTTP does not require cache modification. For caches wishing to handle iHTTP, caches can calculate if the cached authenticator is fresh based on the Sliding-Timestamp and bypass the origin request.

iHTTP also uses the "no-transform" directive to prevent caches from modifying response data and a set of end-to-end headers, which would cause authentication failure.

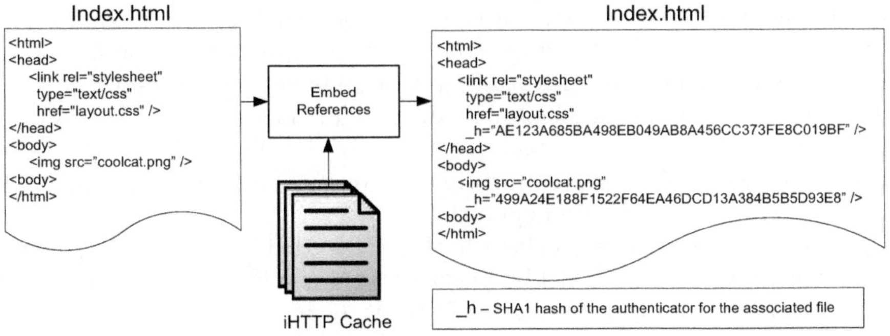

Fig. 6. Opportunistic Hash Verification

4.3 Opportunistic Hash Verification

In the above design, iHTTP clients have to authenticate each iHTTP response via at least one expensive signature verification operation. HTML templates often outline multiple HTTP objects required to render a full web page, which results in multiple responses per page.

In HTTP, clients make two types of requests: initial requests and supporting requests. Initial requests are associated with the top level document of any web page and will always require clients to verify the authenticator signature to ensure freshness. Client supporting requests, on the other hand, are made for supporting resources (e.g., iframes, images, javascript, css) required to render the HTML page.

Note that the number of supporting resources required to render modern websites is fairly large and continues to grow. For example, CNN.com contains 103 unique references to supporting files. Thus, enabling iHTTP for CNN would require that clients perform 104 expensive signature verification operations (including the initial request). This will pose a problem for resource constrained clients. In the following, we propose an Opportunistic Hash Verification technique to reduce the number of client signature verification operations when authenticating an entire web page.

Hash Embedding: The basic idea of Opportunistic Hash Verification is to amortize the expensive signature verification operations at the client by matching authenticated cryptographic hashes with client generated hashes based on the received iHTTP responses. This approach uses the descriptive properties of HTML to allow servers to provide contextual information about potential client requests. Authenticating an HTML document in turn verifies the contextual information and allows clients to bypass most signature verifications.

Specifically, the server will parse all responses containing HTML. HTML tags that may generate additional client requests (i.e., link, img, script, iframe, etc.) are located and each tag's source value is used to search the authenticator cache. The matched authenticator will be cryptographically hashed as $H(\mathtt{A.t}|\mathtt{A.e}|\mathtt{A.u}|\mathtt{A.l}|HTTP.Headers|HTTP.Content|\Delta t|X_n)$ and the hashes are embedded in the HTML as attributes of the associated tags. For example, Figure 6 depicts an HTML file which contains two supporting resources: a style sheet and image. The server uses the "src" attributes of these resources to look up the corresponding authenticator in the iHTTP local cache. The associated authenticator is hashed and embedded as part of the HTML. Using the embedded hash gives the client the opportunity to verify the expected authenticator without a signature verification operation.

Upon receiving a response, a client hashes the authenticator content and compares it with the verified embedded hash value (e.g., the "_h" attributes on the right side of Figure 6) for the requested URL. If the hash values are equal, the authenticator content is verified. To verify the freshness of the response, the client then uses the Sliding-Timestamp technique presented earlier. If the hash values match and the response is fresh, the *HTTP.Content* can be used immediately to verify the message body. In the case when the hash values of the authenticators do not match, the iHTTP client simply falls back to verifying the authenticator via public key operation as the content of the requested file may have changed in the short time between the server embedding the hash and the sending the HTTP response to the client supporting request.

5 Analysis

5.1 Security Analysis

Data Origin Authentication: An iHTTP server signs authenticators with a protected private key. The corresponding public key is certified by a trusted Certificate Authority and provided to iHTTP clients. Using the verified server certificate, clients can verify the authenticator signatures and thus verify that the data originated from the server that possesses the certificate's private key. Once the authenticator is verified, *HTTP.Content* authenticates the message body by matching one-way hashes of message content with *HTTP.Content*. When the Opportunistic Hash Verification is used, authenticating HTML responses in turn verifies the data origin of the supporting resources (e.g., iframes, images, javascript, css) through the embedded hash values.

Data Integrity: Clients verify the integrity of data through signature verifications and a series of matching of hash values. The response signature allows the client to verify the integrity of the authenticator data and HTTP headers. Once verified, the client can then use the authenticated *HTTP.Content* field to verify the integrity of the message body. When the Opportunistic Hash Verification technique is used, the authenticated HTML allows clients to use the embedded hash values to verify the integrity of the corresponding supporting resources. In other words, the embedded hashes verify the integrity of the authenticator. Part of the authenticator, *HTTP.Content*, can in turn verify the integrity of the message bodies of these supporting resources. Any modification of either the top level web page or a supporting resource will lead to a mismatch and can be detected.

Freshness: Data freshness authentication is provided by the authenticator timestamp (`A.t`) and the calculated Sliding-Timestamp ($\texttt{A.t} + \Delta H * \Delta t$). Without the knowledge of the private key of a server, an attacker will not be able to forge an invalid authenticator timestamp without being detected. Moreover, due to the one-way property of the cryptographic hash function H, the attacker cannot forge the hash value X_i used to extend the timestamp, either, unless X_i was released by the server. Assume the maximum clock difference between any client and a server is δ_{max}. The above analysis means that an attacker can hold an authenticated timestamp valid for at most $\Delta t + \delta_{max}$ long before a client identifies it as out of date. Since the clients and the server are loosely synchronized and Δt is chosen by the server, both Δt and δ_{max} can be kept pretty small. Finally, note that the attacker can generate negative influence only when the server modifies the data at the requested URL after the authenticated timestamp is released.

Mixed HTTP & iHTTP Content: A subtle security issue exists when an HTTP document requires an iHTTP resource. While clients can authenticate the received iHTTP data, clients cannot be sure that the received data was intended to be requested. Nevertheless, iHTTP provides the same guarentees as HTTPS since, in this case, the untrusted HTTP document can be misused to request any invalid resource.

5.2 Performance Analysis

Low Performance Impact: The performance of iHTTP is highly dependent on the authenticator generation process. iHTTP uses Sliding-Timestamps to assist in authenticator caching which amortizes the number of signature generation operations. As a result, iHTTP has both a throughput and response time comparable to HTTP.

However, iHTTP impacts the size of the response due to the addition of authenticators and embedded hashes. Thus, iHTTP requires more bandwidth to send a response which will effect overall throughput. The actual cost of iHTTP is dependent on the content encoding, hash size, and transfer coding. For non-compressed responses, the authenticator size is dependent on HTTP message content, or more precisely, $Hashsize * \frac{ContentSize}{BlockSize}$. Transfer coding applies one authenticator for each chunk of a response. Unfortunately, the number of chunks per response is dependent on the implementation of the server. Thus care should be taken when chunking data as the performance benefit of chunking can be overshadowed by the cost of enabling iHTTP. Finally, enabling Opportunistic Hash Verification adds a hash value for each HTTP object in a given HTML document. Hence, the size of the response is increased by $(NumberofUniqueReferences) * (HashSize)$. This overhead is also specific to the HTTP content.

Flexibility: iHTTP uses HTTP header directives to configure network caches to store authenticators and HTTP responses while enabling data freshness. Furthermore, iHTTP does not require changes to existing network infrastructure or software. Caches wishing to natively handle iHTTP can further improve performance, which will reduce server loads. Furthermore, no changes are needed for server generation software such as PHP or ASP.NET to enable iHTTP.

Standards Compatible: iHTTP uses standard based configurations to achieve compatibility (e.g., the use of standard HTTP headers to configure caches). iHTTP also supports any combination of chunked and compression encodings. In addition, iHTTP only requires minor modifications to clients and servers. iHTTP does not require changes to the network infrastructure or caches, and thus allows incremental deployment. As a result, iHTTP is backward compatiable with existing network infrastructures and non-iHTTP enabled clients.

5.3 Limitation

As discussed previously, HTTP responses can be classified as either Client-Unique or Client-Static. While iHTTP can correctly serve Client-Unique HTTP responses, iHTTP, as well as other signature-based HTTP integrity techniques [2, 7], are not suitable to handle Client-Unique data for two reasons:

First, Client-Unique responses can never be cached due to response data being unique per client, which requires data to be signed for each response. As a result, signature-based HTTP integrity techniques will perform at least as poorly as HTTPS. Moreover, existing signature-based HTTP integrity techniques do not provide mechanisms for allowing clients to authenticate response data as logically correct. As a result, attackers or intermediate parties can redirect authenticated fresh data to the wrong clients.

Similarly, iHTTP and signature-based HTTP integrity techniques do not authenticate or verify cookies associated with requests or responses. Thus, servers and clients cannot trust received cookies. However, cookies are widely used by servers to provide a unique client experience, and hence prominently used with Client-Unique data. When cookies are provided with Client-Static data, servers can exclude them from the authenticated data to still retain the benefits of iHTTP.

Servers wishing to authenticate Client-Unique data should resort to HTTPS or other integrity protocol designed to provide client specific authentication of HTTP data.

6 Implementation and Experimental Evaluation

6.1 Implementation

iHTTP requires modification of both client and server to enable its security features. On the server side, we implemented iHTTP as an Apache module for handling iHTTP requests and responses. The Apache module is responsible for authenticator generation, managing the authenticator cache, and embedding hash identifiers into HTML. We used Apache's Portable Runtime API to implement caching. To evaluate iHTTP against the latest proposed HTTP integrity technique, we also implemented HTTPi as an Apache module.

iHTTP is designed to handle both chunked and non-chunked data. For non-chunked data, the iHTTP authenticator is added as part of the HTTP headers.

Chunked data must be handled differently as chunks occurring after the first chunked do not contain HTTP headers. Since authenticators are associated with each chuck, iHTTP embeds the authenticator as chunked data.

On the client side, we developed a Firefox extension to enable iHTTP support. Our extension relies upon the Mozilla service interfaces for intercepting responses and rewriting data. The extension handles verification as well as support for Opportunistic Hash Verification.

6.2 Experimental Evaluation

Experimental Methodology: First, we provide a microbenchmark to investigate the costs of specific iHTTP operations (e.g., signing, hashing, caching, and hash embedding), which may impact the server performance. These operations also represent the operational costs for the HTTPi implementation. Next we give a macrobenchmark of the iHTTP server module to investigate the throughput and max response time for HTTP, HTTPS, iHTTP, and HTTPi. Finally, we benchmark iHTTP on a resource restricted client to measure the impact on the overall response time for rendering an entire page with Opportunistic Hash Verification enable and disabled.

Experimental Setup: Our hardware platform is an IBM HS22 X-Server with 16 cores and 32 GB of RAM. The iHTTP Apache module is installed with Apache 2.2 web server hosted on a virtual machine (VM) running CentOS 5. Apache is run with the standard configuration for server processes. The VM is configured with dedicated 4 CPU cores and 16 GB of RAM. To run our benchmarks, we created another VM running Ubuntu 11.10 with 2 dedicated CPU cores and 4 GB of RAM, also hosted on the IBM HS22 X-Server. Network communication is provided via ESXi virtual switch.

The iHTTP module is configured with a 2,048 bit SSL Certificate which represents the suggested key strength by NIST [16]. We note that larger keys will have a more severe impact on the performance on existing HTTP integrity protocols and thus iHTTP provides even more benefit as keys become larger.

To test resource constrained clients, we install our client module as a Firefox mobile plugin on a Motorola Droid 2 running Android based Cyanogenmod 7.

Server Microbenchmark: iHTTP has several operations for enabling the protocol that incur overhead. The main operations include hashing, signing, caching, and hash embedding. We instrument the Apache module to record the costs of these operations and display the results in Table 1. As expected, the time required to sign the authenticator is significantly more costly than the other operations involved with enabling iHTTP. Thus, we can assume a great savings by foregoing signing each chunk with hash based authentication and integrity verification.

Table 1. Server Microbench Results

Authenticator Creation	4.97771 ms
Signature Generation	4.3207 ms
Hash Embedding	0.13189 ms
Cache Search	0.08751 ms
SHA-1 Operation	0.00042 ms

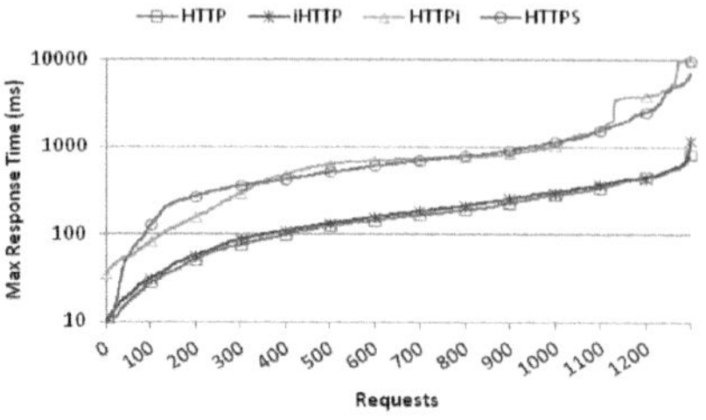

(a) Non-Chunked-Not-Compressed

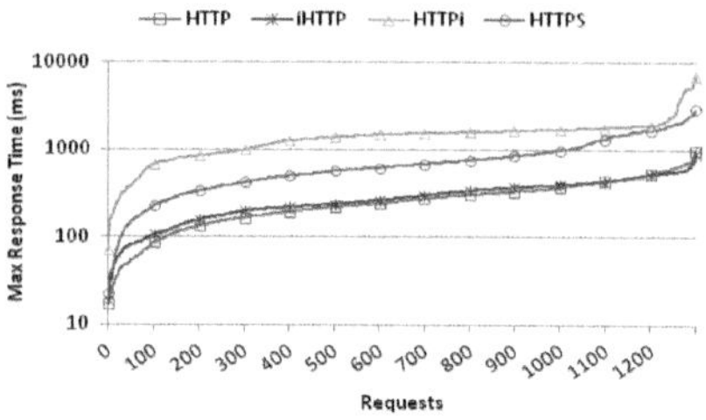

(b) Compressed

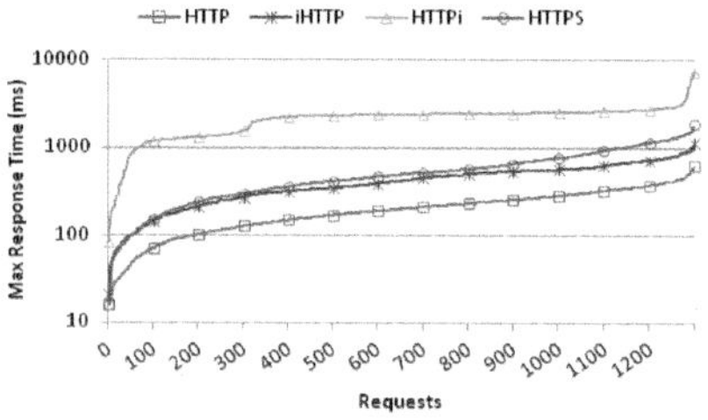

(c) Chunked

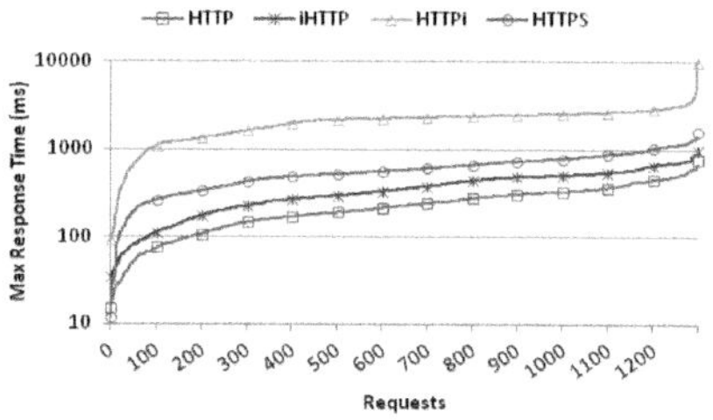

(d) Chunked-Compressed

Fig. 7. Distribution of Response Time

Server Macrobenchmarks: To measure the impact of iHTTP on overall performance, we run two different macrobenchmarks, JMeter Benchmark and SpecWeb2009 Benchmark, to investigate the impact given different website configurations. For these tests, our iHTTP module is configured with Δt=*1 second* and an authenticator expiration (`A.e`) of *30 seconds*(i.e., n=*30*). Δt and `A.e` impact the overall performance. However, due to the small overhead of one SHA-1 operation, the impact is limited when n is not significantly large(e.g., $n < 1000$).

JMeter Benchmark: We deployed a website representing a typical blog or personal website containing only Client-Static data. Four copies were deployed to represent each of the HTTP data formats, non-chunked-not-compressed, compressed, chunked, and chunked-compressed. The sites using chunked data are configured such that each HTML response contains five chunks. The landing page is 67.91 Kb in size and contains 16 HTTP objects. JMeter benchmark was configured to simulate 130 different simultaneous clients, each making 10 page requests for the site. Each page request resulted in 17 GET requests per page. The interactions were duplicated across the sites to ensure equality and each simulation was run separately. Figure 7 shows the response time with respect to the number of requests.

The figures show that iHTTP outperforms HTTPi in all cases. Furthermore, the response times of iHTTP are very close to HTTP for both Figures 7(a) and 7(b). iHTTP performs not as well for chunked and chunked-compressed data. This is expected since iHTTP must handle each chunk separately. In general, iHTTP performs better than HTTPS since authenticator caching helps amortize the number of signature operations.

Table 2 shows the response sizes and the throughputs of the HTML documents for Figure 7(a). Here we see the added cost of the authenticators and embedded hashes. The authenticator size is 1,139 bytes, and Opportunistic Hash Verification adds 640 bytes for the HTML page, which are 7% and 4%, respectively, for non-chunked-non-compressed responses. We note that both of these numbers are reliant on the size of the response data.

Table 2. JMeter Results for Figure 7(a)

(a) Response Size

HTTP	16087 bytes
iHTTP	17866 bytes
HTTPS	16087 bytes
HTTPi	17226 bytes

(b) Throughput

HTTP	267.2 req/sec
iHTTP	252.8 req/sec
HTTPS	114.6 req/sec
HTTPi	84.1 req/sec

SpecWeb2009 Benchmark: SpecWeb2009 allows us to investigate the impact of Client-Unique data on servers and protocols by simulating dynamic web applications. We deployed the SpecWeb2009 banking application, which consists of 15 pages and each page makes an average of 13.6 supporting requests with the minimum being 8 requests and maximum being 19 requests. We configured SpecWeb to simulate 150 simultaneous users for HTTP, HTTPS, iHTTP, and HTTPi configurations. Default configurations were used for KeepAlive and SSL sessions on the SpecWeb clients.

Table 3 shows that both iHTTP and HTTPi perform more poorly than HTTPS and HTTP. First, we observe that each of the 15 generated HTML pages generate Client-Unique content per URL request. Hence, the "account_summary.php" page will contain content specific to the user who requested it. In this case, iHTTP and HTTPi will be required to regenerate the authenticator for each client request.

Table 3. SpecWeb2009 Results

Protocol	Avg Resp	Bytes/Req
HTTP	544 ms	41,818
HTTPS	576 ms	41,828
iHTTP	647 ms	50,627
HTTPi	662 ms	52,147

Client Benchmark: This section compares iHTTP Opportunistic Hash Verification with plain signature-based HTTP integrity techniques. We do not provide comparison of iHTTP with HTTP and HTTPS, since previous research has already provided a thorough comparison of signature-based HTTP integrity techniques with HTTP and HTTPS [7].

We installed the iHTTP client on Firefox mobile version 8.0. We requested the landing page of our static website used in the JMeter benchmark and recorded the load time of 20 requests when both enabling and disabling Opportunistic Hash Verification from the server. The plain signature-based approach requires 16 additional public key operations by the client.

The average time per page load for the plain signature-based approach took 7.4321 second. Pages with Opportunistic Hash Verification enabled on average took 5.8291 seconds to load. This shows that Opportunistic Hash Verification reduces the computational overhead of the client by 21% compared to previous HTTP integrity techniques. Furthermore, the disparity of performance will increase with the number of HTTP objects outlined in the HTML page.

7 Conclusion

In this paper, we proposed a new protocol named iHTTP to enable lightweight authentication of Client-Static HTTP response data. The proposed iHTTP protocol adaptively handles different data encodings to allow for better performance without effecting user experience. It also uses a hash chain based Sliding-Timestamps to provide efficient freshness authentication without using public key operations, and exploits Opportunistic Hash Verification to reduce client public key operations. Our experimental evaluation demonstrated that iHTTP provides similar performance to HTTP, and higher throughput and lower maximum response time than HTTPS for Client-Static data.

References

1. Coarfa, C., Druschel, P., Wallach, D.S.: Performance analysis of tls web servers. ACM Trans. Comput. Syst. 24, 39–69 (2006)
2. Gaspard, C., Goldberg, S., Itani, W., Bertino, E., Nita-Rotaru, C.: Sine: cache-friendly integrity for the web. In: 5th IEEE Workshop on Secure Network Protocols, pp. 7–12 (2009)
3. Vratonjic, N., Freudiger, J., Hubaux, J.P.: Integrity of the web content: the case of online advertising. In: Proceedings of the 2010 International Conference on Collaborative Methods for Security and Privacy, CollSec 2010. USENIX Association, Berkeley (2010)
4. Stamm, S., Ramzan, Z., Jakobsson, M.: Drive-By Pharming. In: Qing, S., Imai, H., Wang, G. (eds.) ICICS 2007. LNCS, vol. 4861, pp. 495–506. Springer, Heidelberg (2007)
5. Lesniewski-Laas, C.: Ssl splitting and barnraising: Cooperative caching with authenticity guarantees. Master's thesis, Massachusetts Institute of Technology (2003)
6. Choi, T., Gouda, M.: Httpi: an http with integrity. In: Proceedings of 20th International Conference on Computer Communications and Networks (2011)
7. Singh, K., Wang, H., Moshchuk, A., Jackson, C., Lee, W.: Practical end-to-end web content integrity. In: Proceedings of the 21st International World Wide Web Conference, WWW 2012 (2012)
8. Fielding, R., Gettys, J., Mogul, J., Frystyk, H., Masinter, L., Leach, P., Berners-Lee, T.: Rfc 2616, hypertext transfer protocol – http/1.1 (1999)
9. Rescorla, E.: Http over tls. Internet RFC 2818 (2000)
10. Rescorla, E., Schiffman, A.: The secure hypertext transfer protocol – shttp (1999)
11. Torvinen, V., Arkko, J., Naeslund, M.: Hypertext transfer protocol (http) digest authentication using authentication and key agreement version-2. Internet RFC 4169 (2005)

12. Erman, J., Gerber, A., Hajiaghayi, M.T., Pei, D., Spatscheck, O.: Network-aware forward caching. In: Proceedings of the 18th International Conference on World Wide Web, WWW 2009, pp. 291–300. ACM, New York (2009)
13. Reis, C., Gribble, S.D., Kohno, T., Weaver, N.C.: Detecting in-flight page changes with web tripwires. In: Proceedings of the 5th USENIX Symposium on Networked Systems Design and Implementation. NSDI 2008, pp. 31–44. USENIX Association, Berkeley (2008)
14. Menezes, A.J., Vanstone, S.A., Oorschot, P.C.V.: Handbook of Applied Cryptography, 1st edn. CRC Press, Inc., Boca Raton (1996)
15. Perrig, A., Canetti, R., Tygar, J., Song, D.: Efficient authentication and signing of multicast streams over lossy channels. In: Proceedings of IEEE Symposium on Security and Privacy, pp. 56–73 (2000)
16. Barker, E., Roginsky, A.: Transitions: Recommendation for transitioning the use of cryptographic algorithms and key lengths. SP-800-131a, U.S. DoC/National Institute of Standards and Technology (2011)

ARC: Protecting against HTTP Parameter Pollution Attacks Using Application Request Caches

Elias Athanasopoulos[1], Vasileios P. Kemerlis[1], Michalis Polychronakis[1], and Evangelos P. Markatos[2]

[1] Department of Computer Science
Columbia University, New York, NY, USA
{elathan,vpk,mikepo}@cs.columbia.edu
[2] Institute of Computer Science
Foundation for Research and Technology – Hellas
markatos@ics.forth.gr

Abstract. HTTP Parameter Pollution (HPP) vulnerabilities allow attackers to exploit web applications by manipulating the query parameters of the requested URLs. In this paper, we present Application Request Cache (ARC), a framework for protecting web applications against HPP exploitation. ARC hosts all benign URL schemas, which act as generators of the complete functional set of URLs that compose the application's logic. For each incoming request, ARC exports the URL, extracts the associated schema, and searches for it in the set of already known benign schemas. In case the schema is not found, the request is rejected, and the event is recorded.

ARC can be transparently integrated with existing web applications without any modifications to the server and client code. It is implemented in Google's Go language and uses efficient data structures for storing the URL schemas, imposing negligible computational overhead on the web application server. When running on a 4-core Linux server, ARC can process hundreds of thousands of URL requests per second. A typical URL resolution is in the scale of microseconds.

Keywords: HPP, Web Security.

1 Introduction

Web applications are experiencing a variety of highly sophisticated attacks that stem from many different sources. Some of them exist due to fundamental design choices of the web platform [5], while others rise due to faulty browser implementations [4,20]. Some of them are based on deceiving users by creating specially crafted visual conditions [10,15], and others emanate from the complexity and the wide use of web applications in many different systems [27,7]. HTTP Parameter Pollution (HPP) is a recently discovered technique for exploiting web applications. HPP can be considered as an *injection* attack that targets URLs; one of the fundamental concepts of the web platform [33]. Web browsers communicate with web applications through HTTP requests and responses, which

F. Bao, P. Samarati, and J. Zhou (Eds.): ACNS 2012, LNCS 7341, pp. 400–417, 2012.

reference resources using URLs. This communication can be polluted by injecting parameters in the HTTP stream. These *injected* parameters form URLs, which if served, instruct the application to perform actions that were originally not part of the application's design. Thus, the control flow of the web application is altered according to an attacker's need.

To illustrate the attack in a short example (in Section 2 we give a detailed presentation of HPP and, more particularly, in Section 2.1 we discuss a formal threat model), consider an e-store application taking two arguments, namely a product identifier and an action, which affects the product state. A combination of a product identifier and the action *purchase* results in ordering a product. The product identifier and the action must be attached as parameters in a URL, which in turn is communicated to the application through the construction of an HTTP request. If the attacker manages to *pollute* the request with extra parameters, then the control flow of the application may change in numerous ways. The simplest manifestation of the vulnerability is for the attacker to inject a particular parameter multiple times. In case the parameter that carries out the product identifier is duplicated, then many different control flows can take place, depending on the parameter occurrence (first, last, or a combination of) that the application will give significance while the URL is parsed.

About 1,499 of 5,000 highly ranked in Alexa.com web sites are considered vulnerable to HPP exploitation according to the methodology outlined by Balduzzi *et al.* [3]. In this paper, we propose Application Request Cache (ARC), a framework that can *protect* web applications from HPP exploitation. ARC does not detect HPP vulnerabilities, although it can record HPP exploitation attempts. It is deployed at server side and works completely transparently. A web application can be protected, using ARC, from HPP exploitation by simply incorporating ARC in the application server. Note that clients need no further modifications. In contrast to PAPAS [3], which currently is the only available methodology for discovering HPP vulnerabilities, ARC aims at protecting the web application without auditing. ARC assumes that the web application *is* vulnerable and tries to protect it from being exploited. To this respect, ARC and PAPAS can be combined. The former as a protection layer and the latter as a periodic auditor.

ARC is based on the following fundamental concept. Each web application is characterized by a set of URL schemas, which act as generators of the complete functional set of URLs that compose the application's logic. A URL schema is extracted by a URL by masking out all variables that are assigned to the URL's parameters. Each control flow is triggered by having the application serving a URL, which stems from a particular URL schema. ARC collects all schemas taken from benign requests during a training phase.[1] At production time, for each incoming request, ARC extracts the URL and its schema, and searches for

[1] Notice that the term "cache" is frequently used to describe temporary storage that holds recently or frequently used elements for improving performance. In this work, we use the term "cache" to refer to storage that holds a set of benign URL schemas, which can generate all possible URLs that can be safely served by a web application.

it in a set of already known benign schemas. In case the schema is not found, the request is rejected and the event is recorded. An incoming *polluted* URL will have no schema stored in ARC and thus will be rejected. This methodology cannot only prevent HPP, but also certain types of XSS [11], where JavaScript is attached to HTTP parameters [37].

ARC is fast. Our prototype is developed using Google's Go, a very efficient programming language for constructing system tools. ARC stores all cached schemas in carefully selected data structures, which are implemented using `maps` and `slices`, as provided by Go. ARC also takes advantage of the multiprocessing features of Go, goroutines and `channels`. In a 4-core Linux server, ARC can process hundreds of thousands of URLs per second. A typical request resolution takes no more than a few microseconds.

Contributions. This paper contributes the following:

- We define a formal threat model for HPP; a new class of vulnerability targeting web applications.
- We design ARC, a framework that can efficiently protect web applications from HPP exploitation. The framework can be applied transparently in any application server. The web application and the available clients need no modifications.
- We implement and evaluate an ARC prototype. We implement ARC in Go, a fast strong typed C-like language by Google. ARC running on a 4-core Linux server, with 4 concurrently running goroutines, can process hundreds of thousands of URL requests per second. Memory requirements, in terms of RSS, from application to application increase linearly with the size of different URL schemas.

2 HTTP Parameter Pollution

Web sites have evolved from simple, mostly-static document repositories to complex, multi-tier applications. Although different organizational paradigms are possible (*e.g.,* 3-, 4-, and n-tier), modern web applications incorporate a mixture of technologies that are typically grouped into two parts: the *application* part and *presentation* part. The former runs on the server and consists of server-side code written in PHP, Perl, Java, ASP.NET, or even C/C++, whereas the latter is rendered by the client, *i.e.,* the web browser, and is made up of (D)HTML, JavaScript, Flash, *etc.* The two parts communicate over TCP using the HTTP protocol in a request-response manner. A typical form of communication involves a request issued from a web browser, for accessing a resource provided by the web application, using a request path defined very precisely in a URL [6]. The web browser issues an HTTP request, which embeds the URL describing the location of the resource, and if the web server can serve the request, it does so by returning the result in the form of an HTTP response. Otherwise, an error is returned, again as an HTTP response. The following simplified URL shows an example of an on-line purchase.

`http://www.e-store.com/purchase?item_id=42` (1)

The communication channel between the server and the client that is used for exchanging URLs can be attacked, affecting both the confidentiality and integrity of the application. An attacker can eavesdrop the communication and steal confidential information (*e.g.,* credit card numbers or account credentials), or modify a request issued by the client, before reaching the server, and hence break the integrity of the communication. Such attacks can be easily prevented, by forcing the web application to communicate with the client over HTTPS [17]. Nevertheless, carefully crafted injection attacks can still happen, even when HTTPS is in use. For instance, an attacker can lure an unsuspecting user to click on a hyperlink that targets a URL embedding some JavaScript code. Upon clicking the link, a request from the victim's browser is sent to a server. This request embeds JavaScript code, which, if not sanitized correctly by the web server, exists in the response and will be executed in the victim's browser. This is called Cross-Site Scripting (XSS) *reflection* attack. HPP is yet another injection technique for attacking web applications [21]. Instead of pushing JavaScript code in URLs, the attacker is polluting the URL by injecting her own combination of HTTP parameters. Consider the following URL that has the same HTTP parameter (*i.e.,* `item_id`) encoded twice.

`http://www.e-store.com/purchase?item_id=6&item_id=42` (2)

The result of processing this request depends on the web application's logic. There are three possible scenarios. If the application consumes the first (from left to right) occurrence of `item_id`, then the item with `id` 6 is purchased. On the other hand, if the application consumes the second occurrence of `item_id`, then the item with `id` 42 is purchased. Finally, it is possible that the application considers both values, or a concatenation of them, as a valid `id`. In that case, both items or item 642 (or 426) are purchased. This ambiguity in processing URL parameters is the core weakness behind HPP. The attacker is taking advantage that there is no standardized way of processing URL parameters, in order to exploit a web application by altering its the control flow.

To a large extent, HPP attacks are manifested by duplicating URL encoded parameters. However, it is also possible to launch an HPP attack without injecting the same parameter multiple times, but by constructing URLs that the web application does not handle correctly.

`http://www.e-store.com/purchase?item_id=42&action=empty_basket` (3)

Normally, the request shown in URL 3 results in purchasing item 42. However, due to the high complexity of modern web applications, each incoming request is processed by a series of scripts. Hence, the script chain of the imaginary web application may host a script for which the `action` parameter is significant. If such a script is executed, then the basket holding user products will be emptied.

Running Example. Suppose that Alice is the victim, e-store is an electronic commerce application, vulnerable to HPP, and Bob is the attacker, who runs his own web site. Bob's goal is to force Alice buying a different product than the one she originally intended to. Additionally, Bob has no access to the e-store and has not compromised Alice's host machine or her browser. However, Bob can lure Alice into visiting his site. Bob's site presents some offers that can be purchased from the e-store. The web application of e-store has an entrance page, which shows all items per category, in the following form:

```
<a href='http://www.e-store.com/show?category=1'>Show cat 1<a/>
<a href='http://www.e-store.com/show?category=2'>Show cat 2<a/>
...
<a href='http://www.e-store.com/show?category=9'>Show cat 9<a/>
```

Upon clicking one of the above links, the e-store application extracts the `category` parameter, and concatenates it with the purchase action and a list of available `ids` (`item_id`) for the selected category. Note that e-shop erroneously trusts `category` and does not verify it for validity before processing it.

```
<a href='http://www.e-store.com/purchase?category=7&item_id=1'>Buy item 1<a/>
<a href='http://www.e-store.com/purchase?category=7&item_id=2'>Buy item 2<a/>
...
<a href='http://www.e-store.com/purchase?category=7&item_id=99'>Buy item 99<a/>
```

Now, Bob is creating his own entrance page with offers that can be purchased from e-store and lures Alice to visiting his site. Bob's site has the following form:

```
<a href='http://www.e-store.com/show?category=1%26item_id=42'>Go to offer 1<a/>
<a href='http://www.e-store.com/show?category=2%26item_id=42'>Go to offer 2<a/>
...
<a href='http://www.e-store.com/show?category=9%26item_id=42'>Go to offer 9<a/>
```

Alice clicks one of the above hyperlinks and the e-store application extracts the `category` parameter, which in our case is `<number>%26item_id=42`, and performs the concatenation. The result is shown below (notice that `%26` has been compiled to `''&''`).

```
<a href='http://www.e-store.com/purchase?category=7&item_id=42&item_id=1'>Buy item 1<a/>
<a href='http://www.e-store.com/purchase?category=7&item_id=42&item_id=2'>Buy item 2<a/>
...
<a href='http://www.e-store.com/purchase?category=7&item_id=42&item_id=99'>Buy item 99<a/>
```

Assuming that the e-store application gives significance to the first parameter (from left to right) while parsing a URL, the product with identifier 42 will be purchased no matter which hyperlink Alice clicks.

2.1 Formal Threat Model

We now define a formal threat model for HPP vulnerabilities. A is a web application, and u_i is used for denoting any URL schema that has the following

```
              [URL]                                                    [URL schema]

http://www.e-store.com/purchase?item_id=42                      |   purchase?item_id=
http://www.e-store.com/purchase?item_id=30&discount=true        |   purchase?item_id=&discount=
```

Fig. 1. Examples of URLs (left) and their respective URL schemas (right). A URL schema expresses a family of HTTP requests that act as descriptors of valid control flows.

form: `action?`p_1`=&`p_2`=&...&`p_N`=`. The schema is composed of an action and a set of parameters that can take arbitrary values. *URL schemas express families of HTTP requests that are served by the web application and act as descriptors of valid control flows.* Figure 1, illustrates a set of URLs with their respective URL schemas.

$U_a = \{u_1, u_2, ..., u_n\}$ is a set that contains all benign URL schemas that A can handle. This means that for each incoming URL in U_a, a well defined control flow f takes place, according to the application's logic. More formally:

$$\forall u \in U_a \longrightarrow f \in F_L$$

$F_L = \{f_1, f_2, ..., f_N\}$ contains the control flows that can be handled safely by the web application. We denote as F_c the set of *all* possible control paths of A. Apparently, $F_L \subseteq F_c$ and $F_h = F_c - F_L$ is the set of all control flows that A can reach, but not initially programmed to execute.

We define the set $U_{hpp} = \{v_1, v_2, ..., v_N\}$ that contains all URL schemas that can initiate a control flow $f \in F_h$. Ideally, we want A to *reject* all incoming v for which the following relationship holds:

$$\forall v \in U_{hpp} \longrightarrow f \in F_h.$$

Notice that flows in F_h may have arbitrary consequences and force the web application to produce undesired results.

2.2 Extreme Cases

We have defined HPP as a technique that is based on the creation of URLs embedding a combination of legitimate, yet unexpected, HTTP parameters, which can drive a web application to an undesired state. So far, we have discussed only the case where a *combination* of parameters is not handled (sanitized) correctly. However, it is possible that HPP can be carried out using the following techniques, depending always on the complexity of the web application.

Parameter Sequence. An attacker may carefully construct URLs that contain valid HTTP parameters, but in a non-expected order. Depending on the complexity of a web application, it might be possible that the URLs trigger a series of server-side scripts, which if executed in a non-expected order, a surprising result occurs. Note that ARC can be configured so that it can protect against such attacks (see Section 3).

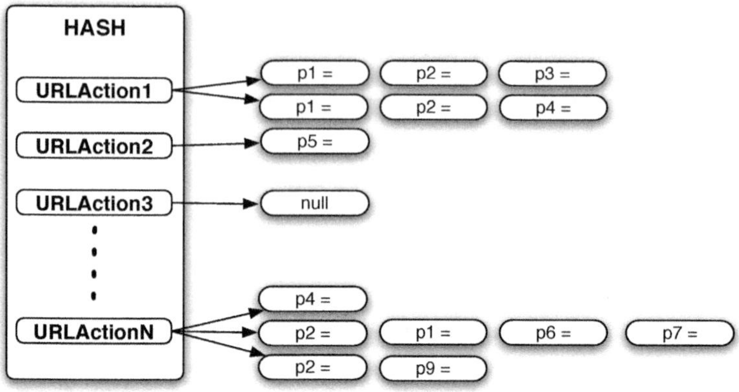

Fig. 2. The data structures used by ARC. A hash table, which holds references to linked lists hosting the set of the parameters of each schema. Each entry in the table has been produced by hashing the action part of a URL schema.

Parameter Values. An attacker may carefully construct URLs that have an expected sequence of HTTP parameters, but with erroneous values. This case is hard to prevent, since it stems from unsafe input sanitization. ARC is based on URL schemas, which have already masked out all values, and tries to prevent parameter *injection*. We believe that with minor modifications, ARC might be able to handle such scenarios, but it needs significant effort and knowledge of the web application's internals by the developer.

3 Application Request Cache

An ARC is a cache that stores all possible URL schemas supported by a web application's logic. Recall that a URL schema is characterized by an action and a set of parameters. Each parameter is not bounded by a specific range of values. URL schemas express generators of HTTP requests served by the web application and they act as descriptors of valid control flows. A URL schema describes a series of different control flows. For example, consider the following URL schema taken from the running example of this paper:

```
www.e-store.com/process-item?item_id=&action=
```

The schema is characterized by an action, in our example "`process-item`", and a set of two parameters: {`item_id, action`}. New control flows are created depending on the value each parameter of the set takes. If "`delete`" is assigned to "`action`" the product corresponding to a given "`item_id`" will be erased. If "`show`" is assigned to "`action`" the product corresponding to a given "`item_id`" will be rendered in the user's browser. ARC aims at collecting and maintaining all benign URL schemas supported by a web application. An ARC-enabled web application, checks every incoming HTTP request to verify if a benign URL schema for the particular request is already stored in the ARC. In the case there is no available schema, ARC does not forward the HTTP request to the application server, and the event is logged. There are two crucial things for the transparent and efficient operation of the system. First, the collection of URL schemas must take place in a controlled environment and in an automated fashion. Second, upon the schemas' cache has been built, ARC must resolve each incoming HTTP request as fast as possible.

Data Collection - Training. ARC needs to know in advance all valid URL schemas supported by the web application. Thus, ARC needs initially to be trained, while the web application is running in non-hostile environment and is receiving only legitimate traffic. It is common for many anomaly detection systems to require an initial training phase [30,25,29]. While in training phase, ARC passively monitors all web traffic received by the web application, filters out all URLs and extracts all URL schemas. These URL schemas are the generators of the complete set of legitimate HTTP requests the web application can serve without becoming HPP exploitable. Training is particularly easy for large companies, which perform extensive beta-testing prior publishing their applications in the wild. Passively monitoring a web application while it is being developed can produce the complete set of allowed schemas, since developers are used to test every new feature they implement. Training is also easy for applications that are based on frameworks for providing blog, forum, or other web services. This is because the application must be monitored *once* for extracting all URL schemas. The same cache can be used by all application instances.

Another option is to use a crawler or scanner for extracting all possible URLs the application provides. However, modern applications use dynamic interfaces implemented in AJAX [12], which many times perform requests towards the application server asynchronously using JavaScript. These requests cannot be easily captured by a crawler. However, today, there are efforts towards sophisticated crawlers that can handle the complexity and the dynamic nature of Web2.0 applications with rich interfaces. One such effort is Crawljax [22], which has been used by researchers for extracting the user interface of Web2.0 applications [8]. Finally, notice, that many frameworks assume that all URLs an application can handle is known [26,16,2] (see discussion in Section 6).

Data Structures. The data structures used by ARC is a hash table and a collection of linked lists. Each schema is stored in the cache in the following way. First, the *action* part of the schema is hashed. In the case there is no entry in the hash table with the same key, a new hash node is inserted at the index, which is equal to the key. Otherwise, a pointer of the currently occupied index of the hash table is fetched. This pointer holds references to linked lists, which host the set of the parameters of each schema. In the case that there is no list hosting the parameters of the new schema, a new one is created and a reference is assigned to the hash index. The data structures are schematically depicted in Figure 2. Observe the hash table that stores each action (from 1 to N), which is noted with `URLActionX`. Each hash entry stores pointers towards a series of linked lists. In the example ARC of Figure 2, `URLAction1` stores two pointers, meaning that this entry describes two different URL schemas, each of them described by three different parameters. In the same fashion, `URLAction2` stores one pointer towards a list that contains a single parameter, and `URLAction3` stores a pointer towards `null`, meaning the particular schema takes no parameters. Finally, `URLActionN` stores three pointers towards three lists, containing 1, 4, and 2 parameters, respectively.

Search Algorithm. It is trivial now to derive the search algorithm and its complexity, since we have analyzed the data structures employed by ARC in the previous part. For each incoming HTTP GET or POST request the URL schema is derived by parsing the request line. The action part (the part before the character "?") and the set of parameters (all left parts of expressions "`par=var`" delimited with each other by the character "`&`") are derived in this step. We assume that URLs follow the specification [6]. ARC can be extended to use a custom URL schema, for web applications that do not follow the specification, since ARC runs purely at the server side and, thus, can co-operate with the application server. We do not account for parsing operations in complexity, since all requests have to be parsed by the application server, no matter if ARC is enabled or not. When a URL schema is derived, the action part is hashed and is looked-up in the ARC table. The complexity of this operation is $O(1)$. Now, the set with the parameters of the schema has to be checked against all sets already stored with this action. We define as URL action density, ρ, the ratio of unique actions over all possible URL schemas. For example, a web application that supports 1,000 URL schemas and those include 100 unique actions, has $\rho = 0.1$. The density reversed approximates how many schemas are associated with a particular action, or how many lists are associated with each hash bucket. Assuming that an input schema has a number of parameters, N, then the complexity of the search is $O(\frac{N}{\rho})$. Thus, the complexity of the complete algorithm is $O(1) + O(\frac{N}{\rho}) \simeq O(N)$. Thus, the search algorithm has linear complexity with the number of parameters of each input schema.

Optimizations. We can substitute the linked lists with trees, in order to reduce the search time required for scanning the lists. The optimized version can reduce the search time and, thus, increase the URL throughput (see Section 4). However, security must be sacrificed, since cases described in Section 2.2 cannot be handled correctly. Thus, for the rest of this paper, we discuss and evaluate only the unoptimized ARC. A second approach is to use DFAs for searching the cache. Consider, for example, that each URL can be represented by a string, whose characters are selected from a space defined by all the different parameters, which can occur in all collected URL schemas. Although, a DFA has linear complexity in search, in practice, implementing regular expressions that can contain all the thousands of URL parameters used by a large web application is not considered trivial, due to intrinsic constraints of current off-the-shelf implementations. For example, PCRE [2] has a hard limit for the maximum size of a regular expression, and it needs special recompilation for changing this. Finally, we can use a single hash table for speeding up search. For each incoming schema, we can concatenate the action part and all parameters and feed the result to a hashing function. By definition, this will speed up the search to $O(1)$, no-matter the length of the URL parameters of each incoming schema. However, large web applications, using many URLs, will experience hash collisions, which can be resolved by incorporating linked lists. Thus for large web applications, this approach is, essentially, identical to the approach we follow for building ARC.

3.1 Implementation

We implemented an ARC prototype in Go [19]. Go is a programming language created by Google for fast system development. We created two versions, one single-threaded and one that utilizes 4 threads. In the world of Go, the term *goroutine* is used, instead of thread. Goroutines cannot be used standalone. There is no way for a goroutine to complete and communicate the result to the rest of the program, unless a channel is used. Thus, for the rest of this paper, we will refer to the single-threaded version as *single-channel ARC* and to the multi-threaded one as *4-channel ARC* . As far as the data structures are concerned, we use `maps` for implementing the hash and `slices` for implementing the linked lists. `Maps` and `slices` are standard data structures provided by Go. A `map` represents a relation of two data types, one serving as the key and one as the data holder. On the other hand, `slices` are similar to C arrays, but their size can be modified at run-time. The ARC implementation works as follows. First, it builds the cache by reading a collection of already stored URLs in the disk. It forms the cache (see Figure 2), which is maintained in memory (we evaluate the system's memory footprint in Section 4). For each incoming HTTP request the application server extracts the URL (and the POST parameters, if it is required) and forwards it to the ARC. The URL schema is extracted and the ARC looks up in the available cache for its existence. If the schema exists, the parsed form of the URL is forwarded to the application server, otherwise the incoming request is dropped and the event is logged. We implemented ARC and the application server in Go. However, with minimum changes, ARC can cooperate with any modular application server.

4 Evaluation

All experiments are carried out using artificially created traces. In this way, we are able to create large collections with thousands of URL schemas, in order to stress our implementation as much as possible. Initially, we create three different URL sets. The set is composed by URLs that are formed by a random action part and by a set of random strings representing URL parameters. Each parameter is a random string of size between 6 and 16 characters. Each set is characterized by

[2] `http://www.pcre.org`

Table 1. Properties of URL sets used in evaluation. Each set is characterized by 4 properties: (1) the amount of URLs the set includes, (2) the minimum number of parameters a random URL of the set may include, (3) the maximum number of parameters a random URL of the set may include, and (4) the density ρ of the set.

Web Application	URLs	Min Par.	Max Par.	ρ
Small	1,000	5	12	0.01
Medium	10,000	7	15	0.001
Heavy	100,000	12	20	0.001

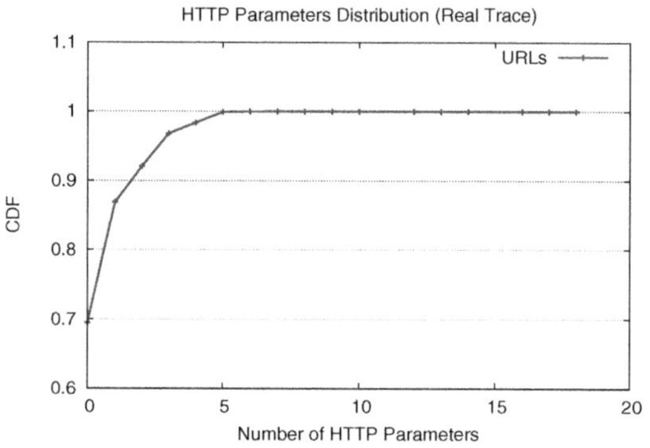

Fig. 3. Cumulative distribution function of HTTP parameters, as collected from a real-world trace, including HTTP/HTTPS traffic for the phpBB and phpMyAdmin applications. The plot depicts 1 million URLs sampled from a trace containing over 50 millions of captured URLs.

4 properties: (1) the amount of URLs the set includes, (2) the minimum number of parameters a random URL of the set may include, (3) the maximum number of parameters a random URL of the set may include, and (4) the density ρ of the set. Recall from Section 3, that ρ is defined as the ratio of unique actions over all possible URL schemas. Thus, we create three URL collections, each one representing a different web application. The first set contains 1,000 URLs, each one having 5 to 12 parameters, with $\rho = 0.01$. We will further refer to this set as *Small Application*. The second set contains 10,000 URLs, each one having 7 to 15 parameters, with $\rho = 0.001$. We will further refer to this set as *Medium Application*. Finally, the third set contains 100,000 URLs, each one having 12 to 20 parameters, with $\rho = 0.001$. We will further refer to this set as *Heavy Application*. We summarize all these details in Table 1.

The characteristics of the artificially created traces are based on real-world evidence. We monitored two well-known web applications, phpBB and phpMyAdmin, and managed to collect over 50 millions of URLs. We then analyzed a sample of 1 million URLs and measured the number of HTTP parameters per HTTP GET/POST request. We plot the CDF in Figure 3. Notice, that the majority of HTTP requests include less than 5 different parameters, and there

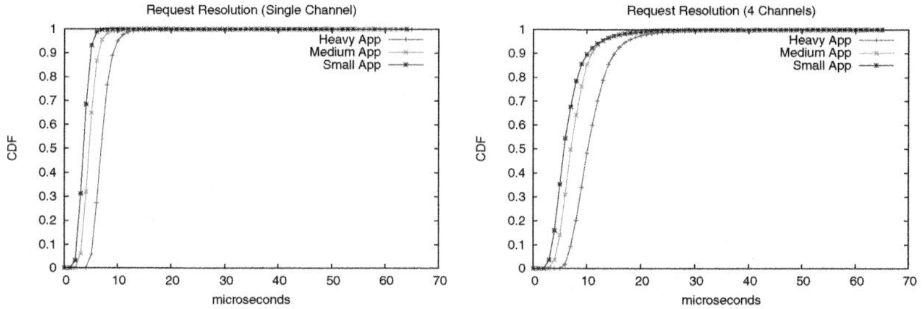

Fig. 4. Cumulative distribution function of all measured resolutions, for both the single-channel and the 4-channel version of the ARC, and for all different web applications. The majority of all request resolutions, about 98%, are completed in less than 10 microseconds.

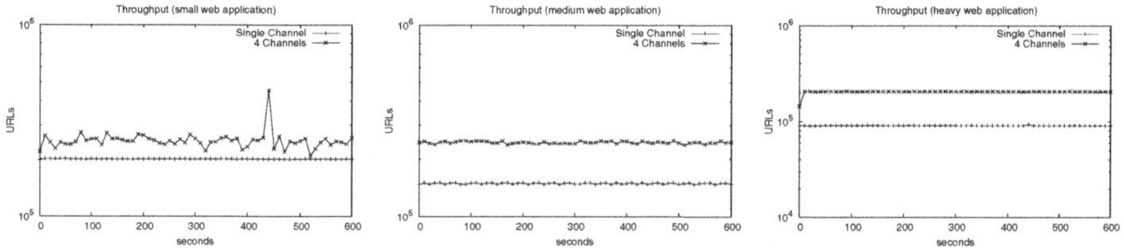

Fig. 5. Resolved requests per second for the small, medium, and heavy application, respectively. The 4-channel ARC significantly outperforms the single-channel one, serving hundreds of thousands requests per second, in all applications.

were not recorded HTTP requests containing more than 18 parameters. The tree different URL sets are precomputed and stored to files on disk. For each experiment, ARC loads the URLs, exports the schemas, and creates the caches as we described in Section 3 (see Figure 2). All information is maintained in memory. As far as the hardware setup is concerned, all experiments run in a Linux server, equipped with i7/2.93 GHz (4-cores) and 4 GB RAM.

4.1 Request Resolution

We are interested to identify the average time it takes for ARC to process one singe request. We run the ARC with one of the three URL sets, which correspond to a particular web application (small, medium, and heavy). We forward 1,000,000 URL requests towards ARC, after it has loaded all URLs and has built all data structures. All requests are taken randomly from the initial file that hosts the artificially created URLs. For each request we measure the time needed by ARC to find the URL schema that corresponds to the incoming URL request. The search time includes parsing the initial URL. We perform all measurements with the `Nanoseconds()` function, which is contained in the `time` package.

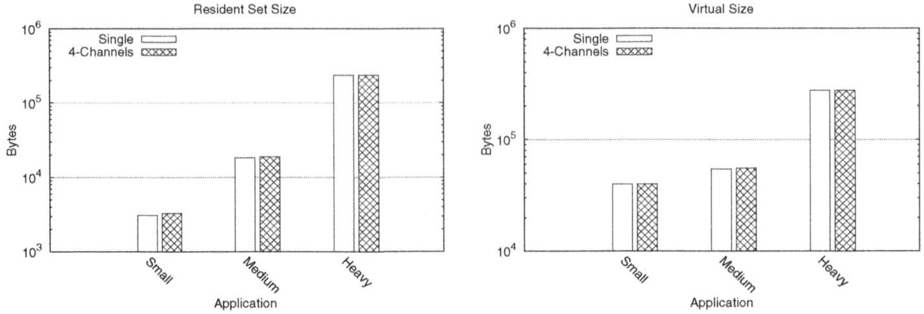

Fig. 6. Resident Set Size (RSS) and Virtual Size (VSIZE), both as reported by `ps(1)`, while running ARC for each one of the three applications. Notice, that both, 4-channel and single-channel, versions have similar memory requirements. Also, memory requirements, in terms of RSS, from application to application increase linearly.

In Figure 4 we plot the CDF of all measured resolutions, for both the single-channel and the 4-channel ARC, and for all different web applications. It is important to highlight the following. First, the majority of all request resolutions, about 98%, are completed in less than 10 microseconds. We consider that the performance is enough for not causing significant overhead to an application server, even in configurations that are based on commodity hardware. Second, the requests for the heavy application seem to be resolved a little bit slower than the medium one, and the requests of the medium one seem to be resolved a little bit slower than the small one. This is reasonable, since the heavy application is characterized by URLs that have more parameters that the ones of the medium and of the small ones. This has two consequences: (1) the parsing time is longer (recall, that we account for parsing in every search operation), and (2) the lists' size is larger or, more formally, N is larger (recall the complexity of the search algorithm, $O(N)$, presented in Section 3). Finally, notice that the 4-channel ARC behaves worse than the single-channel ARC (all CDFs are shifted to the right, in the right plot of Figure 4). Initially, this seems to be counterintuitive. However, it is not. The 4-channel version has the additional overhead of managing and context-switching the 4 goroutines. This affects slightly the performance of each request resolution. Nevertheless, the overall performance of the 4-channel version significantly outperforms the single-channel version, since the 1,000,000 requests are completed in shorter time. We quantify this in the following part.

4.2 Request Throughput

We configure ARC to run with each one of the three different applications for 600 seconds. We record how many requests ARC can resolve per second for the small, medium and heavy application, respectively. We run all experiments for both, 4-channel and the single-channel, ARC implementations. We present the results in Figure 5. Notice, that the 4-channel ARC significantly outperforms the single-channel one in all applications. Observe, that the 4-channel ARC can

```
package hello

import (
    "arc"
    ....
)

func init() {
    http.HandleFunc("/", handler)
    arc_stats = arc.Init()
}

func handler(w http.ResponseWriter, r *http.Request) {
    if (arc.FilterURL(r.URL.RawPath) == true) {
      fmt.Fprint(w, deliver_page(r))
    } else {
      fmt.Fprint(w, deliver_error("URL is not supported."))
    }
}
```

Fig. 7. An example web application written in Go, for running over Google's AppEngine, which incorporates ARC. Some functions are omitted for presentation purposes. Notice, that ARC integrates seamlessly with the rest of the code.

serve hundreds of thousands requests per second. This is to be expected, because the 4-channel ARC takes advantage of all 4 cores of the server. Thus, a typical request resolution maybe slightly faster for the single-channel ARC, but the overall throughput is much greater for the 4-channel ARC.

4.3 Memory Footprint

ARC stores all information (*i.e.,* all URL schemas) in memory for fastest access. The more the distinct URL schemas a web application has, the more the memory the ARC needs. In Figure 6 we plot the Resident Set Size (RSS) and the Virtual Size (VSIZE), both as reported by `ps(1)`, while running ARC for each one of the three applications. Notice that both versions (*i.e.,* 4-channel and single-channel) have similar memory requirements. This is to be expected, since both versions maintain memory in exactly the same way. Notice, also, that the memory requirements, in terms of RSS, from application to application increase linearly. Recall from Table 1, that the size of complexity, in terms of URL schemas, for each application increases by one order of magnitude.

5 Case Study

Google AppEngine [9] is a platform for deploying web applications. Recently, Google announced an SDK for building web applications in Go. Although, it is still experimental, it seems the ideal application server for incorporating ARC into. Notice, that ARC can be enabled in any application server as an external CGI script. A typical web application written in Go is composed as a package. There are many official Go packages for managing HTTP requests and URLs, which can be easily imported in the main package, which serves as the core of the application. Next, there is an initialization routine which assigns handlers for URLs matching a specific pattern, and, finally, there is a series of handlers that can serve incoming requests. Enabling ARC for an AppEngine application is trivial. In Figure 7 we present the skeleton of an example web application written in Go for running over Google's AppEngine. Some functions have been omitted for presentation reasons. There are three basic steps needed for enabling ARC. First, the `arc` package must be imported (line 4). Second, `arc.Init()` must be called for initializing the cache (line 10). This function reads all available URL schemas from a text file and organizes them to data structures in memory (see Section 3 for the description of the data structures used). Finally, a check is applied to the core request handler for filtering out all incoming URLs that are not compatible with any of the available stored schemas. This check is performed using the `arc.Filter()` function, which takes as a parameter the incoming URL in raw format (line 14) and returns a boolean value (`true` if the URL compiles to a valid schema, `false` otherwise).

6 Related Work

HPP is originally discovered by Luca Carettoni and Stefano di Paola in 2009 [21]. The most relevant research to ARC is PAPAS [3], which aims at detecting HPP vulnerabilities through a black-box scanning technique. In this respect, PAPAS and ARC are different, since ARC aims at preventing exploitation through HPP; ARC assumes that the application *is* vulnerable. Nevertheless, the two technologies can be combined. A web application, which rapidly changes, can use ARC for protection and occasionally scanned for new HPP vulnerabilities. HPP Finder [1] is a Chrome extension that scans web pages in real-time for detecting potential HPP exploits. Thus, the extension aims at protecting the end user from vulnerable (to HPP) web applications. However, HPP Finder has limited scope. It can identify only hyperlinks and forms that include a particular parameter multiple times. As we have already discussed in Section 2, HPP is a broader class of vulnerabilities that can be manifested when a parameter occurs multiple times in an HTTP request. Moreover, HPP Finder has many false positives, especially in pages with radio buttons. Therefore, HPP Finder is not considered a complete solution against HPP exploitation, but rather a precaution.

There are many frameworks for detecting and preventing XSS [18,23,14,28,26,32]. Robertson and Vigna [26] attempt to introduce structure in the web documents served by a web application, for taking advantage of it and detect potential injections. The framework needs a map of all URLs that the application supports in advance. In their context, this is called a `RouteMap` and it is similar to the `routes` package present in popular web development frameworks, such as Rails [16] and Pylons [2]. ARC needs also all URLs supported by a web application, in order to extract all possible URL schemas. However, ARC does not assume that this information is known (we have listed techniques in Section 3 for collecting URLs). Researchers have developed generic techniques for covering web exploitation [25,30]. These techniques share a common property with ARC; they are also based on a training phase for collecting features that characterize the benign behavior of the web application. These proposals are more generic, and, thus, suffer from false positives. ARC, on the other hand, is practical and focuses on HPP only. Web exploitation is not only XSS. HPP is among the recently discovered highly sophisticated techniques for attacking a web application [5,7,27,20,4,15,36]. To that end, many academic efforts aim at applying security concepts from operating systems to the web platform [34,35,13,24,31].

7 Conclusion

HTTP Parameter Pollution (HPP) is a recently discovered technique for exploiting web applications. Since web applications communicate with browsers using HTTP requests and responses, the communication can be polluted by injecting parameters that alter the control flow of the web application according to an attacker's need. In this paper, we constructed a formal threat model for HPP and we proposed Application Response Caches (ARC), a framework that can prevent HPP exploitation in web applications. ARC can be transparently enabled in an application server, without further modifications to the web application and to the clients. We implemented a single-channel and a 4-channel ARC prototype using Google's Go language. ARC running on a 4-core Linux server, with 4 concurrently running goroutines, can process hundreds of thousands of URL requests per second. A typical URL resolution is in the scale of microseconds. Memory requirements, in terms of RSS, from application to application increase linearly with the size of different URL schemas.

Acknowledgements. This work was supported in part by the FP7 project SysSec, the FP7-PEOPLE-2010-IOF project XHUNTER, and the FP7-PEOPLE-2009-IOF project MALCODE, funded by the European Commission under Grant Agreements No. 257007, No. 273765, and No. 254116, respectively, and by DARPA through Contract FA8650-11-C-7190. Any opinions, findings, conclusions, or recommendations expressed herein are those of the authors, and do not necessarily reflect those of the European Commission, US Government, or DARPA.

References

1. Athanasopoulos, E.: HPP Finder (2011), `http://www.ics.forth.gr/~elathan/extra/hpp/index.html`
2. Bangert, B., Gardner, J.: The Pylons Project, `http://pylonsproject.org` (last visited on July 2011)
3. Balduzzi, M., Gimenez, C., Balzarotti, D., Kirda, E.: Automated discovery of parameter pollution vulnerabilities in web applications. In: Proceedings of the 18th Network and Distributed System Security Symposium (2011)
4. Barth, A., Caballero, J., Song, D.: Secure Content Sniffing for Web Browsers or How to Stop Papers from Reviewing Themselves. In: Proceedings of the 30th IEEE Symposium on Security & Privacy, Oakland, CA (May 2009)

5. Barth, A., Jackson, C., Mitchell, J.C.: Robust Defenses for Cross-Site Request Forgery. In: Proceedings of the 15th ACM Conference on Computer and Communications Security, CCS (2008), http://crypto.stanford.edu/websec/csrf/csrf.pdf
6. Berners-Lee, T., Masinter, L., McCahill, M.: RFC 1738: Uniform Resource Locators (1994), http://www.ietf.org/rfc/rfc1738.txt
7. Bojinov, H., Bursztein, E., Boneh, D.: XCS: Cross Channel Scripting and Its Impact on Web Applications. In: CCS 2009: Proceedings of the 16th ACM Conference on Computer and Communications Security, pp. 420–431. ACM, New York (2009)
8. Chapman, P., Evans, D.: Automated Black-Box Detection of Side-Channel Vulnerabilities in Web Applications. In: Proceedings of the 18th ACM Conference on Computer and Communications Security, CCS 2011, pp. 263–274. ACM, New York (2011), http://doi.acm.org/10.1145/2046707.2046737
9. Ciurana, E.: Developing with Google AppEngine. Springer (2009)
10. Dhamija, R., Tygar, J., Hearst, M.: Why Phishing Works. In: Proceedings of the SIGCHI Conference on Human Factors in Computing Systems, pp. 581–590. ACM, New York (2006)
11. Fogie, S., Grossman, J., Hansen, R., Rager, A., Petkov, P.: XSS Attacks: Cross Site Scripting Exploits and Defense. Syngress Publishing (2007)
12. Garrett, J., et al.: Ajax: A New Approach to Web Applications. Adaptive Path 18 (2005)
13. Grier, C., Tang, S., King, S.: Secure Web Browsing with the OP Web Browser. In: Security and Privacy, pp. 402–416. IEEE (2008)
14. Gundy, M.V., Chen, H.: Noncespaces: Using Randomization to Enforce Information Flow Tracking and Thwart Cross-Site Scripting Attacks. In: Proceedings of the 16th Annual Network and Distributed System Security Symposium (NDSS), San Diego, CA, February 8-11 (2009)
15. Hansen, R., Grossman, J.: Clickjacking, technical Report, SecTheory (2008), http://www.sectheory.com/clickjacking.htm
16. Hansson, D.H., et al.: Ruby on Rails, http://www.rubyonrails.org (last visited on July 2011)
17. Jackson, C., Barth, A.: Forcehttps: Protecting High-security Web Sites from Network Attacks. In: Proceeding of the 17th International Conference on World Wide Web, WWW 2008, pp. 525–534. ACM, New York (2008), http://doi.acm.org/10.1145/1367497.1367569
18. Jim, T., Swamy, N., Hicks, M.: Defeating Script Injection Attacks with Browser-Enforced Embedded Policies. In: WWW 2007: Proceedings of the 16th International Conference on World Wide Web, pp. 601–610. ACM, New York (2007)
19. Baugh, J.P.: Go Programming (June 2010) ISBN: 1453636676
20. Lin-Shung, H., Zack, W., Chris, E., Collin, J.: Protecting Browsers from Cross-Origin CSS Attacks. In: CCS 2010: Proceedings of the 17th ACM Conference on Computer and Communications Security. ACM, New York (2010)
21. Carettoni, L., di Paola, S: HTTP Parameter Pollution (2009), https://www.owasp.org/images/b/ba/AppsecEU09_CarettoniDiPaola_v0.8.pdf
22. Mesbah, A., Bozdag, E., Deursen, A.: v.: Crawling AJAX by Inferring User Interface State Changes. In: Proceedings of the 2008 Eighth International Conference on Web Engineering, ICWE 2008, pp. 122–134. IEEE Computer Society, Washington, DC (2008), http://dx.doi.org/10.1109/ICWE.2008.24

23. Nadji, Y., Saxena, P., Song, D.: Document Structure Integrity: A Robust Basis for Cross-site Scripting Defense. In: Proceedings of the 16th Annual Network and Distributed System Security Symposium (NDSS), San Diego, CA, February 8-11 (2009)
24. Reis, C., Gribble, S.: Isolating web programs in modern browser architectures. In: Proceedings of the 4th ACM European Conference on Computer Systems (EuroSys), pp. 219–232. ACM (2009)
25. Robertson, W., Vigna, G., Kruegel, C., Kemmerer, R.: Using Generalization and Characterization Techniques in the Anomaly-based Detection of Web Attacks. In: Proceeding of the Network and Distributed System Security Symposium (NDSS), San Diego, CA (February 2006)
26. Robertson, W., Vigna, G.: Static Enforcement of Web Application Integrity Through Strong Typing. In: Proceedings of the 18th USENIX Security Symposium, Montreal, Quebec (August 2009)
27. Saxena, P., Hanna, S., Poosankam, P., Song, D.: FLAX: Systematic Discovery of Client-side Validation Vulnerabilities in Rich Web Applications. In: Proceedings of the 17th Annual Network and Distributed System Security Symposium (NDSS)
28. Sekar, R.: An Efficient Black-box Technique for Defeating Web Application Attacks. In: Proceedings of the 16th Annual Network and Distributed System Security Symposium (NDSS), San Diego, CA, February 8-11 (2009)
29. Sommer, R., Paxson, V.: Outside the closed world: On using machine learning for network intrusion detection. In: Proceedings of the 2010 IEEE Symposium on Security and Privacy, SP 2010, pp. 305–316. IEEE Computer Society, Washington, DC (2010), http://dx.doi.org/10.1109/SP.2010.25
30. Song, Y., Keromytis, A., Stolfo, S.: Spectrogram: A Mixture-of-Markov-Chains Model for Anomaly Detection in Web Traffic. In: Proceedings of the 16th Annual Network and Distributed System Security Symposium, NDSS (2009)
31. Tang, S., Mai, H., King, S.: Trust and Protection in the Illinois Browser Operating System. In: Proceedings of the 10th USENIX Conference on Operating Systems Design and Implementation (OSDI). USENIX (2010)
32. Ter Louw, M., Venkatakrishnan, V.: Blueprint: Precise Browser-neutral Prevention of Cross-site Scripting Attacks. In: Proceedings of the 30th IEEE Symposium on Security & Privacy, Oakland, CA (May 2009)
33. Berners-Lee, T.: Tim Berners-Lee on the WorldWideWeb project. USENET post (1991), http://groups.google.com/group/alt.hypertext/tree/browse_frm/thread/7824e490ea164c06/f61c1ef93d2a8398
34. Wang, H.J., Fan, X., Howell, J., Jackson, C.: Protection and Communication Abstractions for Web Browsers in MashupOS. In: Bressoud, T.C., Kaashoek, M.F. (eds.) SOSP, pp. 1–16. ACM (2007)
35. Wang, H.J., Grier, C., Moshchuk, A., King, S.T., Choudhury, P., Venter, H.: The Multi-Principal OS Construction of the Gazelle Web Browser. In: Proceedings of the 18th USENIX Security Symposium, Montreal, Canada (August 2009)
36. Weinberg, Z., Chen, E., Jayaraman, P., Jackson, C.: I Still Know What You Visited Last Summer. In: Proceedings of the 32th IEEE Symposium on Security & Privacy, Oakland, CA (May 2011)
37. XSSed.com: XSS exploit in key example, http://xssed.com/mirror/33541/

Tracking the Trackers: Fast and Scalable Dynamic Analysis of Web Content for Privacy Violations

Minh Tran[1], Xinshu Dong[2], Zhenkai Liang[2], and Xuxian Jiang[1]

[1] Department of Computer Science, North Carolina State University
{mqtran,xuxian_jiang}@ncsu.edu
[2] School of Computing, National University of Singapore
{xdong,liangzk}@comp.nus.edu.sg

Abstract. JavaScript-based applications are very popular on the web today. However, the lack of effective protection makes various kinds of privacy violation attack possible, including cookie stealing, history sniffing and behavior tracking. There have been studies of the prevalence of such attacks, but the dynamic nature of the JavaScript language makes reasoning about the information flows in a web application a challenging task. Previous small-scale studies do not present a complete picture of privacy violations of today's web, especially in the context of Internet advertisements and web analytics. In this paper we present a novel, fast and scalable architecture to address the shortcomings of previous work. Specifically, we have developed a novel technique called *principal-based tainting* that allows us to perform dynamic analysis of JavaScript execution with lowered performance overhead. We have crawled and measured more than one million websites. Our findings show that privacy attacks are more prevalent and serious than previously known.

Keywords: Privacy, Web security, Information flow, JavaScript, Dynamic analysis.

1 Introduction

Privacy violation is common in web applications today, especially with the excessive power of the JavaScript language. The Same-Origin Policy (SOP) [1] governs the accesses by JavaScript to web page and network resources, which ensures objects in a web application are only accessible by JavaScript from the same origin (defined as the tuple $< protocol, host, port >$). Unfortunately, a web application often needs to include third-party scripts in the same origin of the web application itself. If the scripts have privacy violations, they are free of restrictions from SOP. Besides, although XMLHttpRequest is restricted by SOP, its successor Cross-Origin Resource Sharing (CORS) has more flexibility in sending requests to different origins. Moreover, SOP does not prevent information leakage through requests for external resources, such as images, and CSS background.

This concern has motivated researchers to search for an answer. Work by Krishnamurthy and Wills [2] examined 75 mobile Online Social Networks (OSNs) and showed that all of these OSNs exhibit some leakage of private information to third parties. In a

F. Bao, P. Samarati, and J. Zhou (Eds.): ACNS 2012, LNCS 7341, pp. 418–435, 2012.

similar vein, Krishnamurthy et al. [3] surveyed 120 popular non-OSN websites. Using a proxy, they monitored the traffic and discovered 56% of the sites directly leak pieces of private information in clear text to third-party aggregators. From another perspective, Jang et al. [4] studied Alexa global top 50,000 websites and found multiple cases of privacy violation including history sniffing and behavior tracking. However, analysis on potential privacy-violation behaviors at the web page level is too coarse-grained, as it does not distinguish between different pieces of JavaScript on the page. On the other hand, although fine-grained taint tracking, such as [4], provides desirable accuracy in identifying information leakage behaviors, the performance overhead imposed by it has limited the scale of the study that can be performed on real-world web applications.

Our Approach. In this work, we propose an effective mechanism to perform a large-scale study on privacy-violating behaviors in real-world web applications. We observe that web applications often include many third-party libraries. Although these libraries share the same namespace, they are loosely coupled. The reason is threefold. First of all, by virtue of the JavaScript language, local objects declared inside a function are only accessible inside that scope. Furthermore, functions are typically wrapped inside a closure (e.g. jQuery), restricting accesses to their objects. And even if a function does declare objects in the global scope, other code will probably not access these objects because the code is oblivious to the existence of the objects. As a result, the objects are only accessed and modified by their creator. Based on this fact, we propose to track web application behaviors at the granularity of the JavaScript libraries, which greatly boosts the performance in tracking potential privacy violation behaviors in web applications. We thereafter refer to this novel technique as *principal-based tracking*.

In this paper, we present the design, implementation, and evaluation of *LeakTracker*, a fast and scalable tool to study the privacy violations in web contents. We perform principal-based tracking on web applications to identify the libraries that have suspicious behaviors in accessing user privacy information, where the principal is defined as the URL of the source of the JavaScript. For each piece of JavaScript code going to be compiled by the JavaScript runtime, we introduce a tag specifying which principal this piece of code belongs to. The resulting script, when executed, can introduce more scripts into the system and these new scripts will in turn be tagged, and this process continues. Our system then monitors the behaviors of scripts with different principals, and identifies suspicious behaviors in accessing users' private information.

To verify that such suspicious scripts do leak private information to external parties, we perform an additional variable-level dynamic taint analysis only on them, and for the rest of the web application, our system runs at full speed without tainting. By applying the principal-based tainting technique, we manage to reduce the significant performance overhead associated with application-wide taint analysis [5,6,4], while directing the power of taint analysis directly towards those suspicious script principals.

We have implemented a prototype of LeakTracker by extending the JavaScript engine of Mozilla Firefox. Our prototype tracks not only JavaScript code originated from web pages but also JavaScript from browser plugins (e.g. Flash/Silverlight) and extensions. Based on this prototype, we have developed a HoneyMonkey-style [7] crawler and deployed it on production systems. Our computer cluster consists of ten monkeys and one monkey controller to drive them. Compared to previous work, we substantially extend the scope of the study, by crawling the list of top one million global websites provided by Alexa. The cluster finished crawling these websites within one week, demonstrating that our design is fast and scalable. We have further evaluated the performance of LeakTracker with the SunSpider [8] benchmark suite, which indicates that our system incurs a reasonable performance impact.

In summary, this paper makes two primary contributions:

- **Principal-based tainting technique.** We design a less expensive dynamic tainting technique that tracks the taint flows in a web application. It tracks JavaScript originated from different sources, including that from NPAPI-based plugins and browser extensions. We show that our technique is robust in the presence of JavaScript obfuscation and redirection. Our implemented prototype confirms that the technique incurs reduced performance overhead.
- **Comprehensive evaluation through large-scale study.** We evaluate LeakTracker in terms of effectiveness and performance. With timeout parameter set at fifteen seconds, within a week of deployment, we finished crawling the global top one million websites. We show that even in such popular set of websites, privacy attacks are still prevalent. Specifically we found that 39.1% of websites exfiltrate certain private information about the user and her browser. Most alarmingly, 31,811 websites still spy on user behaviors and 7 websites still sniff on users' browsing history.

The rest of this paper is organized as follows. Section 2 provides an overview of the problem and existing works. Next, Section 3 and 4 detail the design and implementation of LeakTracker. After that, Section 5 presents our evaluation results. We cover the related work in Section 6 and finally Section 7 concludes this paper.

2 Background

In this section we review the economic and technical reasons why privacy violations come to exist, especially how the issue tends to relate to Internet advertisements (ads) and web analytics.

Internet advertising is an important business model to support free Internet content. To sustain the publishing effort and to earn a profit, web publishers display ads on their sites and get paid by the advertisers, or more often, the ad networks (who are in turn paid by the advertisers). The arrangement is somewhat similar to the situation in traditional media like television and newspapers. In other word, the user unwittingly gives up some of his time viewing ads in exchange for the content provided by the publishers. This win-win arrangement between the users, the publishers, and the advertisers is one of the reasons that contribute to the booming of web media since the late 1990s.

With the rising popularity of the web, advertisers get more and more sophisticated. Instead of serving the same ad to everyone visiting a particular website, advertisers or ad networks adopt a practice called *targeted advertising*. In this practice, they dynamically decide what kind of ads to display to the site's visitors. For example, the ads can be chosen based on the type of content users are viewing, thus making Internet advertising more relevant and presumably more helpful to visitors. The net result is generally more ad clicks, more sales, and ultimately more profit for the advertisers and the publishers.

Unfortunately, advertisers do not stop at this point. To further enhance the relevancy of their ads, they start to adopt more aggressively practices, namely, tracking and profiling visitors. According to a study by the Wall Street Journal (WSJ) [9] in August 2010, "the nation's 50 top websites on average installed 64 pieces of tracking technology onto the computers of visitors, usually with no warning." They also identified that between the web users and the advertisers, there are "more than 100 middlemen: tracking companies, data brokers and advertising networks". Data about the users' interests are collected, aggregated, packaged, and sold to the advertisers, who use analytics software to determine the most relevant ads to serve. For example, BlueKai, one of the major data brokers, sells 50 million pieces of personal information on a daily basis.

Besides the economic reasons for tracking, some technological advances make tracking more effective. For example, a study [10] by the Stanford Center for Internet and Society (CIS) found that Microsoft Advertising network places a syncing cookie on its **live.com** domain, effectively respawning the tracking cookie even though the user has cleared cookies in an attempt to preserve her privacy. This mechanism is referred to as "supercookies". Flash Local Shared Objects (LSO) can also be used to achieve the same purpose. The second tracking method, reported by the Panopticlick project at the Electronic Frontier Foundation [11], is called "fingerprinting". In this method, public configuration information provided by the web browser can be used to create a signature that uniquely identifies the browser and its user. Their study [12] shows that this method is highly effective: pick a browser at random, at best only one in 286,777 other browsers will share its signature, resulting in 18.1 bit of entropy.

In this paper, we focus on privacy attacks that happen when embedded JavaScript code abuses its privileges to track and leak confidential user information such as history, behaviors, interests and so forth. Out-of-band methods like traffic fingerprinting are out of the scope of this work. Our goal in this work is to conduct an extensive study and quantify the state of privacy violations in web contents.

3 System Design

We have developed a comprehensive and efficient system called LeakTracker to track privacy violations in web applications, whose overall architecture is shown in Figure 1. The parts in dark colors are major components of LeakTracker. In essence, we first identify JavaScript principals in a web application that have suspicious behaviors in accessing privacy information, and then apply the principles of dynamic taint analysis to track their executions. If the taint flows from critical sources to critical sinks our reference monitor logs an alert. In other words, by observing the flow of taint we can detect possible privacy information leakage in a web application.

Next, we will explain the principal-based tracking technique, followed by the elaboration on how we assign principals to JavaScript in different cases with principal tagging. Finally, we elaborate on the taint sources and sinks tracked by our system.

3.1 Principal-Based Tracking

Despite being a powerful technique, taint analysis incurs a significant performance overhead. This overhead is mainly associated with the introduction, propagation and

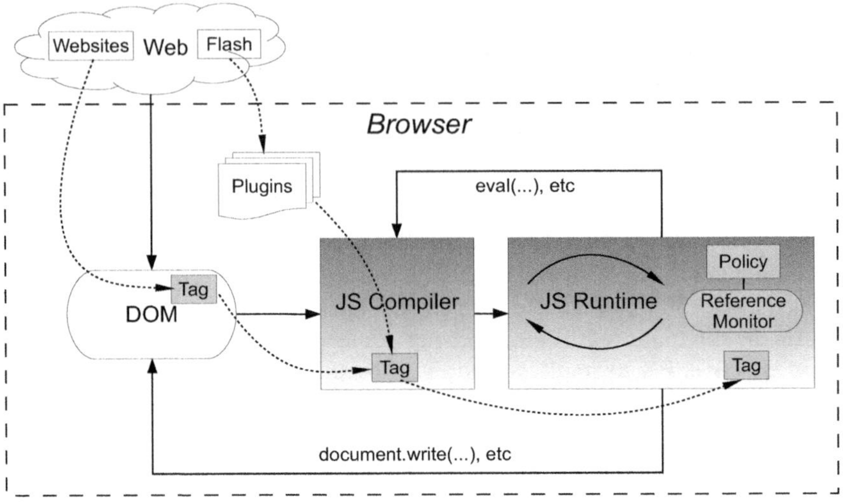

Fig. 1. LeakTracker architecture

checking of taint. One needs to perform these for every JavaScript bytecode instructions executed. As a result, the performance impact from taint analysis is prohibitive. To reduce this overhead, we introduce a novel technique called principal-based tainting. Our key insight is: even though an embedded JavaScript widget shares the global namespace with other JavaScript code, its objects are naturally isolated from other code. JavaScript libraries, such as jQuery, tend to wrap functions and variables inside closures, and even for global variables or functions, they are generally not accessed by other JavaScript since other scripts are oblivious to their existence. The net result is that the objects are only accessed and modified by the creator. As a result we only need to perform taint analysis on the relevant functions, instead of the entire web application as implemented in previous approaches [5,6,4]. This lessens the performance overhead introduced by taint analysis.

To differentiate relevant functions from irrelevant ones, we introduce the concept of principal. A principal is defined as the URL of the JavaScript code introduced to the page. We observe that modern web applications typically employ third-party tracking libraries to track user behaviors, and such JavaScript libraries put users' privacy information at risk. During our analysis on privacy violations, it is necessary to distinguish first-party and third-party logic mixed in a web application to further lower the need for full-blown taint analysis. For example, it is quite natural and even expected that a web page will set and later read the cookie object to restore view settings, user preferences, among other things. On the other hand, it would be suspicious for a third party code to read the cookie and then transmit its content to a third party server. In fact, this constitutes the cookie stealing attack. However, due to the complex interactions between principals, it is challenging to differentiate the owner of a piece of code and accurately assign a principal to each piece of JavaScript. We detail our principal assignment strategy next.

3.2 Principal Tagging

Although JavaScript libraries tend to be independent of each other during execution, they may modify or overwrite other pieces of scripts. For instance, a script principal can overwrite the handler on a DOM element created by another principal. In another case, a principal can introduce more code into the page by features like **document.write**, by modifying the DOM tree, and so forth. To address this challenge, we introduce a second technique termed principal tagging.

As can be seen in Figure 1, our system completely tracks the provenance of all JavaScript code and assigns appropriate principal tags. The flow of tags is as follows. We start with the most common source of JavaScript code: the DOM. A <**script**> tag requests the browser rendering engine to send the corresponding JavaScript code to the JavaScript compiler. Our instrumented JavaScript compiler, in addition to parsing and compiling JavaScript code into bytecode, determines the principal of the script from the DOM and attaches a tag to the resulting **JSScript** object. We note that a <**script**> tag is not the only way to introduce JavaScript code into a web application. Another way of doing so is to use unnamed JavaScript code blocks, e.g. event handler code. Yet JavaScript code can also be introduced into the system by browser plugins (e.g. Flash and Silverlight) or browser extensions. A list of all sources is shown in Table 1.

As Table 1 indicates, JavaScript code can also be generated dynamically by the running JavaScript code through the use of JavaScript language features like **eval()** or DOM APIs like **document.write()**. We interpose on these vectors and propagate the principal tag from the creator script to the created script accordingly. In this way, we always keep track of the true identity of every piece of code. We also note that by keeping the full URL of the script in the tag, our system offers much better precision and accuracy than the Same Origin Policy(SOP), which assigns the same origin to all scripts embedded in the same page.

3.3 Taint Sources and Sinks

The Reference Monitor (RM) component guards the taint sources and sinks according to our policy. In our system, we track all major taint sources and sinks. For example, to detect possible cookie stealing attacks, the RM keeps track of all principals that has accessed the **cookie** object. Similarly, we treat the **getComputedStyle** function as a taint source to detect history hijacking attacks. We also track the registration of event handlers and other sensitive sources, as further discussed in Section 4. After our

Table 1. Script Sources

Source	Type
DOM	Direct script
DOM	Unnamed script
Source of the plugin object	Plugins
Source of the extension	Extensions
JavaScript	JS features
JavaScript	DOM APIs

principal-based tracking technique identifies the script principals that exhibit suspicious behaviors in accessing user privacy information and sending data through the sinks (e.g. the network), we apply taint analysis on these principals to confirm whether they actually leak privacy information to the network. Upon detection of a policy violation (e.g. sending private information to a third party server) the RM will record the identity of the offender to the audit log, and let the operation succeed. We note again that the focus in this work is to detect policy violation and not to implement a protection system, so currently when any tainted data is flowed to the taint sink, we log such event with the principal of the JavaScript that generates such behaviors.

4 Implementation

We have implemented a prototype of LeakTracker by instrumenting Mozilla Firefox 3.6.13. The reference monitor is implemented with 700 source lines of code (SLOCs) in C++. About 1200 SLOCs is used to implement the principal tracker. We add about 2400 SLOCs into the SpiderMonkey JavaScript engine to track and propagate taint. Overall, we add 4300 SLOCs to Mozilla Firefox code base.

The crawler is implemented using AutoIT, a specialized scripting language for application automation. The monkey controller is implemented in 500 SLOCs. 300 SLOCs is needed to implement a monkey. The total number of SLOCs used to implement the crawling infrastructure is 800 SLOCs. In the following subsections, we examine each component of LeakTracker and present its implementation details.

4.1 Instrumented Browser

To implement the principal-based tainting technique, we first need to be able to track the principal of any piece of JavaScript code. We therefore instrument the browser to introduce and propagate the principal tag. The tag is stored as a property of the **JSScript** object. As shown in Table 1, there are six sources and each of them need to be handled a little differently. We provide details on each of them as follows. The first one is JavaScript code directly included on the page using the **script** tag. If the **src** property is specified, the tag will be set to the value of the **src** property. This case is relatively straight-forward. On the other hand, if script code is specified directly in the body of the **script** tag, then the principal should be carefully determined based on who created that **script** tag. If the creator was another JavaScript, LeakTracker will assign the principal tag of the created script to that of the creator. If, however, the **script** tag is part of the original HTML document then the newly created script's tag will have the principal

Table 2. DOM Functions and Properties

Family	Functions
Write	write, writeln
Timer	setTimeout, setInterval
DOM Tree Manipulation	createElement, insertBefore, appendChild, ...
DOM Node Property	innerHTML, text, textContent

of the hosting page. The second case is handled in a similar way. In both cases, we enhanced the Firefox's **JSTokenStream** class to achieve the goal.

In the third case, the principal of the script should be that of the embedded object and not of the hosting page. For example, for JavaScript code introduced by Flash plugin, the principal should be set to the URL of the Flash file. We modified the internal NPAPI implementation inside *nsNPAPIPlugin.cpp* for this purpose. The fourth case is handled similar to the first case, in that the newly created script will inherit the principal of the extension, which starts with "`chrome://`". The fifth case is the least complicated of all. In this one, we copy the tag directly from caller of **eval()** to the newly created script. The final case requires the most engineering effort, as we need to instrument every relevant DOM API functions. This careful implementation is needed to thwart obfuscation attempts. Table 2 lists the APIs that we instrument. With the above mechanisms in place, we can now realize the principal-based dynamic taint analysis technique.

Recall that we apply additional variable-level taint analysis to principals detected as suspicious by the principal-based tracking. Variable-level taint analysis helps confirm the information flow between the sources and the sinks. For example, if we introduce taint at any reads from the **cookie** object, and later find a tainted object being send to a remote server, we can conclude that the content of the cookie is being leaked to the server. Tracking taint propagation for the entire web application is very expensive. Therefore our instrumented JavaScript engine only performs taint propagation for the script originated from suspicious principals. For every piece of JavaScript code, upon determining that the principal is trusted, the engine switches to an alternate execution path that is free of taint propagation operations. From that moment on, the piece of code is executed at full speed to completion. This helps reduce the overall overhead incurred by taint analysis. On the other hand, if the principal is determined to be suspicious by the principal-based taint analysis, we propagate taint as the code executes. To achieve this goal, we instrumented the 235 opcodes of the JavaScript bytecode interpreter. These opcodes operate on the unified **jsval** type. We introduced a taint bit into this **jsval** structure. The instrumented opcodes then propagate this taint bit. In this way, taint analysis is achieved with minimum space overhead.

The second component of our instrumented browser is the reference monitor (RM). To implement the RM, we instrument the Firefox's JavaScript interpreter and introduce guards at the critical sinks. When our system tracks suspicious principals, the RM will inject taint into any objects derived from taint sources. There are four critical sinks that JavaScript code can use to send data to a remote server: XMLHttpRequest, form submission, CSS property misuse, and **src** property misuse. As further discussed in Section 5, our findings indicate that the last method is the most frequently used for exfiltrating data. Every time there is an access to a critical sink, the RM will check if the taint bit is present. If it is, the RM will raise an alert and record the principal of the executing code into the audit log.

4.2 Crawler

To explore and scan the web, we implement a HoneyMonkey-style[7] crawler. In this approach, a machine, which we refer to as the monkey controller, maps and dispatches tasks to individual monkeys. Note that a monkey can actually be a controller itself, and further map the tasks to its children. In this way, a tree is formed with monkeys as the leaves. Therefore the LeakTracker architecture supports scalability by allowing the flexible addition of computing power should the need arise.

We now discuss the monkey itself. Inside each monkey there is a software agent that acts as a liaison between the controller and the instrumented browser. The agent, upon receiving tasks from the controller, will drive the browser to visit websites. This approach is more advantageous than the use of an indexer in traditional crawler like Heritrix[13] in that a full-blown browser allows us to examine dynamically generated page while an indexer does not.

The monkey controller takes as input a list of websites, and divides it into tasks, each consists of one hundred websites. It then maps the tasks to idle monkeys, and instructs them to start crawling. The monkeys crawl the websites and when they have finished with the tasks, they compress the logs and send them to a log server. They then signal the controller for more tasks. The system continues to operate as described until there is no more tasks, at which point it remains idle. For each task, the software agent iterates over the list and drives our instrumented Firefox browser to visit each of the websites. It also sends keystrokes and mouse events to the browser to simulate keyboard and mouse activities. This is necessary to observe possible behavior tracking of websites. After letting the website run for a timeout period of fifteen seconds, the agent will close the browser, collect the logs and continue with the next website. Our empirical experience shows that the threshold of fifteen seconds is enough to let most websites finish loading.

5 Evaluation and Findings

We have deployed LeakTracker on production systems to survey the web. In our prototype, ten monkeys were used to run the crawling, each is a virtual machine running Microsoft Windows XP Service Pack 3 . The virtual machine was configured with one virtual CPU running at 3.0GHz and 320MB of memory. The entire computing cluster ran on an IBM eServer BladeCenter HS21 with VMWare ESX Server 3i as the hypervisor. Within one week our computing cluster finished crawling the Alexa global top one million websites, demonstrating that our design is fast and scalable. Overall, we found 817,831 instances of leakage in 391,837 out of one million websites (roughly 39.1%), resulting in an average of 2.08 instances per website. In the following sections, we provide detailed discussion of our findings.

5.1 General Findings

Examining the leakage cases identified in our experiment, we found that most of them are caused by web analytics software embedded in the websites. Manual examinations of a random sample of two hundred leakages reveal that most of them leak screen resolutions, color depth, and JavaScript version. More than a dozen of them transmit the full list of installed browser plugins to the tracking server. One typical case is shown below:

```
http://charter.122.2o7.net/b/ss/charternetprod/1/H.22.1/s436952264
8743?AQB=1&ndh=1&t=4%2F7%2F2011%209%3A49%3A29%204%20240&ce=UTF-8&
ns=charter&pageName=homepage&g=http%3A%2F%2Fcharter.net%2F&cc=USD&
ch=home&server=web12.charter.synacor.com&events=event1&c1=home&v1=
D%3Dc1&v7=D%3Dc7&v9=D%3Dc9&v17=D%3Dc17&c23=Repeat&v23=D%3Dc23&c
27=4&v27=D%3Dc27&c28=Less%20than%201%20day&v28=D%3Dc28&c29=10%
3A30am&v29=D%3Dc29&c30=thursday&v30=D%3Dc30&c51=non-customized&c52
=not%20a%20premium%20owner&c53=logged-out&s=1680x1050&c=24&j=1.7&v
=N&k=Y&bw=1680&bh=900&p=Test%20Plug-in%3BMozilla%20Default%20Plug-
in%3BGoogle%20Update%3BShockwave%20Flash%3BSilverlight%20Plug-In%3
BAdobe%20Acrobat%3BQuickTime%20Plug-in%207.6.8%3B&AQE=1
```

This is the actual query string captured by LeakTracker when it was being set to the **src** property of an image. The taint sources in this case are the **screen** object (`height`, `width` and `colorDepth` properties) and the **navigator** object (the `plugins` property). As pointed out in [12], the tracking server could then use the captured information and other public information such as IP addresses, User-Agent strings and so forth to compute a fingerprint of the user. We therefore consider this a potentially dangerous privacy attacks. To ease the examination effort, we implemented a classifier based on data provided by Ghostery (www.ghostery.com). This classifier allows us to identify the provider of each suspicious JavaScript principal found by our system. Running the classifier over our log database, we found 480 known providers and they are responsible for the majority of the leakages. Our system detects six new trackers that were not previously identified by Ghostery, one of which is ClickHeat, a behavior tracker (further discussed in section 5.2). We observe a long-tail effect in the distribution of leakages over trackers. In other words, some providers are more popular than others (for example Google Analytics is used in 339,147 sites, Quantcast - 34,351 sites) but the majority of leaks are in the long-tail.

Regarding the distribution of trackers per website (we count one distinct script file as one tracker, regardless of how many leaks resulted by that tracker), we found that the majority of them employ less than ten trackers. Specifically, 384,535 sites (98.1%) have less than ten trackers on their page. A minority of the sites, however, have a large amount of trackers, four of which even have more than 50. Overall, we found 7,302 sites that have more than ten trackers. The list of 20 websites with the most trackers is shown in Table 3. We also observe that less popular websites tend to have many more trackers than popular websites. As can be seen in Table 3, 19 out of 20 websites have ranking more than 200,000.

On the prevalence of the four popular kinds of attacks, we did not observe any case of cookie stealing or location hijacking. This finding is consistent with those reported by Jang et al. [4]. We, however, did find notable cases of history sniffing and behavior tracking, which we discuss in the following subsections.

5.2 Behavior Tracking Cases

We found multiple cases of behavior tracking in Alexa global top one million websites. In this kind of privacy attack, a website will stealthily record the behaviors of visiting users. It achieves this goal by installing event handlers on various relevant events like mouse movements, mouse hovering and so forth. The collected log is then silently transferred back to the server using XMLHttpRequest, form submission, CSS property

Table 3. 20 Sites Embedded the Most Trackers

Site	Rank	Trackers
pmcarpenter.blogs.com	430602	62
freestore.spb.ru	534501	58
minordetails.typepad.com	800405	55
thefraserdomain.typepad.com	307401	51
exopolitics.com	570650	46
travel.ru	10106	44
tuschistesmachistas.com	990528	44
meine-preis.de	686600	43
mylol.net	268601	42
sysopt.com	251904	40
thediscountdivaguide.com	733649	40
markwebberforum.com	857301	40
thelifeofluxury.com	299507	40
blackberrydownload.net	305801	39
bigwigbiz.com	505509	39
smallbizlabs.com	428284	38
democracyforums.com	762002	38
usaonlinemall.net	559693	38
community.web.id	286006	37
bestpriceproduct.com	838801	37

Table 4. Behavior Tracking Providers

Tracking Provider	Sites	Events Tracked
Tynt Insight	1602	copy, mouseover
ClickDensity	259	mousedown
ClickHeat	475	mousedown, focus
Omniture	6884	mousedown, keydown
Woopra	2263	mousedown, mousemove, keydown
Pagealizer	5	mousedown
Statcounter	15149	mousedown, load, unload
Visual Website Optimizer	651	blur, focus, focusin, focusout, load, resize, scroll, unload, click, dblclick, mousedown, mouseup, mousemove, mouseover, mouseout, mouseenter, mouseleave, change, select, submit, keydown, keypress, keyup
ClickTale	2187	load, unload, scroll, mousemove, mouseover, mouseout, mousedown, mouseup, click, contextmenu, resize, keydown, keyup, keypress, focus, blur, select, change
Etracker	1334	mousedown
Reinvigorate	722	mouseup
SeeVolution	203	onscroll, onpaste, keydown, on blur, change, focus, mousedown
Mouseflow	77	mousemove, mouseover, mousedown, mouseup

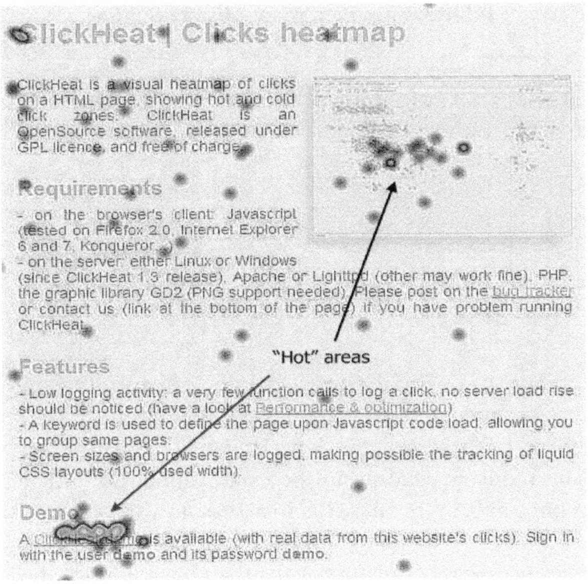

Fig. 2. ClickHeat

misuse, or **src** property misuse. The aggregated behavior data can then be used to generate a profile about the specific interests of that user. This proves to be very effective in tailoring highly relevant ads to that user. This is the main reason why user interest data command a premium in data exchange, as pointed out in [9] . As this level of surveillance is more than most people are comfortable with [14], we consider it a dangerous kind of privacy attack.

Table 4 summarizes our findings. The first column shows the name of the tracking provider we have identified. The number of sites that embeds a particular tracker is listed in the second column. The last column shows the event handlers identified by LeakTracker. Overall, we found that 31,811 websites spy on user behaviors, using trackers provided by 13 companies. All of them are interested in tracking mouse events, with most of them register event handlers for mousedown and mousemove. This preference is expected, as mouse cursor movement is shown to have strong correlation with eyes movement and thus user's attention [15]. Some of the trackers seem very aggressive in collecting data, as indicated by the number of event they track. For instance, Visual Website Optimizer registers handlers for 23 events, while ClickTale registers 18.

We identified one new behavior tracker not previously recognized by Ghostery, who refers to itself as ClickHeat. Figure 2 shows a demo by ClickHeat itself, where data collected from users visiting its websites are used to compute a heatmap. As can be seen in the heatmap, hot and bright colors like yellow and green represent most clicked areas while cold colors like blue represent less frequently clicked areas.

5.3 History Sniffing Cases

History sniffing attacks have received considerable attention recently. In this kind of attack, a website gains unauthorized accesses to its visitors' browsing history and

Table 5. Sites Performing History Sniffing

Site	Description	Inspected URLs
golfdigestschool.com	Sports	casinogolfschools.com, executivegolfschool.org, +18
webleads-tracker.fr	Business	quivisite.com, +2
gearyi.com	Business	businessol.com, +2
netfactor.com	Business	leadlander.com, +3
pagealizer.com	Business	yahoo.com, +13
beencounter.com	Business	sitebrand.com, +5
bitbang.it	Business	facebook.com, +2

exfiltrates the data. Due to its highly invasive nature, history sniffing is generally considered an unsavory practice and several high profile lawsuits have been made against the perpetrators [16,17]. These lawsuits have drawn considerable scrutiny over the behavioral advertising industry, leading to the establishment of AdChoices and opt-out programs. One of our goals in this paper is to assess how the current situation is.

As shown in Table 5, history sniffing attacks still exist in the wild and we observed several cases of history sniffing occurring in the long-tail of the web. Specifically, we found seven websites still sniff on users' browsing history, querying for a total of 53 websites. Portions of the attack code captured from `http://www.netfactor.com/` are shown below:

```
var b_c_urls = [ 'iuuq://xxx.fyb68768hkih878nqmf.dpn', 'iuuq://xxx
.mfbemboefs.dpn', 'iuuq://xxx.wjtjtubu.dpn', 'iuuq://xxx.efnboecbtf
.dpn'];
var b_c_urls_id = [ '-1', '7850', '12973', '12972'];
b_c["3"] = "8301265327807-1";
...
if (ciphertext.length>2)
{
ciphertext = ciphertext.replace("][", String.fromCharCode(92));
ciphertext = ciphertext.replace(")(", String.fromCharCode(39));
Decrypt();
}
...
style_to_add_to_page += ".beencounter-id-" + b_c_urls_id_ +
":visited {background:url("+String.fromCharCode(39)+"http://www.
beencounter.com/b.php?one=" + b_c_urls_id_ + b_c_url_ending
 +String.fromCharCode(39)+")}";

links_to_add_to_page += "<a href="+ String.fromCharCode(39) +
plaintext + String.fromCharCode(39) +" class="+String.fromCharCode(
39)+"beencounter-id-" + b_c_urls_id_ +String.fromCharCode(39) +
"> </a>";
...
document.write( "<style type=" + String.fromCharCode(39) + "text/css"
+String.fromCharCode(39)+ ">" );
document.write( style_to_add_to_page );
document.write( "</style>" );
document.write( "<div  style=" + String.fromCharCode(39) + "position:
absolute;left: -4000px ;text-indent:-999px"+
String.fromCharCode(39)+">" );
document.write("<img src="+String.fromCharCode(39)+"http://www.
```

```
beencounter.com/b.php?one=" + b_c_url_ending+
String.fromCharCode(39)+">");
document.write(links_to_add_to_page );
document.write( "</div>" );
```

We have reverse-engineered the attacks and found that all of them are caused by a tracker called Beencounter. In the following, we provide a quick description of the attack. The list of URLs to sniff for is pre-encrypted and stored in the **b_c_urls** array. This encryption apparently is to avoid detection. Each of the URL is associated with a predetermined unique ID stored in **b_c_urls_id**. At runtime, the URLs are decrypted and used to build a list of links, stored in **links_to_add_to_page**. The associated IDs are then used to build respective CSS styles and then stored in **style_to_add_to_page**. As can be seen in the attack code, the CSS **background** property is misused to trigger the browser into sending requests to the web server. Therefore when the links are inserted into the page, the web server can learn about which sites the user has visited, based on what is requested and what is not requested by the browser.

5.4 Performance

In order to assess the performance overhead of LeakTracker, we conducted experiments with SunSprider [8], an industry standard in browser benchmarking. The version of the benchmarking suite used was 0.9.1 and our experiments were done with a Dell Optiplex 760 workstation running Windows 7 x64 Service Pack 1. This workstation was equipped with an Intel Core 2 Quad Q9550 CPU (2.83Ghz 12 MB L2 Cache) and 4 GB of RAM. The baseline for evaluation was the original Mozilla Firefox 3.6.13. Each experiment was repeated ten times, and the average result was recorded. Overall, our instrumented browser completes the test in 1979ms versus 899.5ms of the original browser, resulting in a 2.2 times slowdown. We consider this satisfactory given the amount of additional computation needed to propagate and check taints, as a typical dynamic taint analysis system can incur up to 7.9x slowdown [4,5,6].

6 Related Work

Web-based attacks have received significant attention over the last few years. The threats from malicious web pages have motivated researcher into developing mechanisms to protect web surfers from being exploited. We first review related work on detection of malicious websites followed by a brief discussion on specific protection approaches.

6.1 Detection

Recent work by Curtsinger et al. [18] presents a mostly static approach to detect JavaScript malware. Their tool uses Bayesian classification of hierarchical features of the JavaScript AST to identify syntax elements that are highly predictive of malware. The results show that the system can achieve a very low false positive rate at negligible performance overhead. As a result, it can be deployed inside an end-user browser or as a first-level filter to reduce the workload of a dynamic, high-overhead but more effective approach like NOZZLE [19]. In a similar fashion to ZOZZLE, Prophiler by Canali et al. [20] combines JavaScript, HTML and URL features into one classifier that is able to quickly discard benign pages. According to their evaluation, the tool is able to reduce the load on a more costly dynamic analysis tools Wepawet [21] by 85%. In addition to using classifiers, it is also possible to detect attacks through the static analysis of other content retrieved from the web server, as shown in [22], [23], [24].

Work by Jang et al. [4] focuses on detecting privacy violating information flows in JavaScript web applications. Using dynamic taint analysis, they are able to detect the leaking of sensitive information through taint sinks and show four kinds of attack: cookie stealing, location hijacking, history sniffing, and behavior tracking. Their empirical study of Alexa global top 50,000 websites shows that many websites, including several in the top 100 sites, leak private information about users' browsing behavior. Their implementation uses a dynamic source-to-source rewriting approach where taints are injected, propagated and blocked within the rewritten JavaScript code. In contrast, LeakTracker implements taint tracking at one level below, inside the JavaScript engine itself. As a result, it incurs lower performance overhead and is more resilient to subversion. In the same vein as [4], using the Fiddler framework, Krishnamurthy et al. [3] surveyed 120 popular non-social-network websites and show that 56% of the sites directly leak pieces of private information in clear text to third-party aggregators.

6.2 Protection

Cross-site scripting (XSS) attacks can be prevented in several ways. BrowserShield [25], for instance, rewrites dynamic scripts into safe equivalents before sending them to the clients. BLUEPRINT [26], on the other hand, integrates with web applications to encode user-generated HTML content into a syntactically inert format and decodes it at the client. This allows them to bypass the anomalous parsing behaviors from client web browsers. ConScript [27] leverages aspect-oriented programming techniques to insert hooks into various interfaces, thus allowing the tool to restrict how JavaScript can interact with its environment. Document Structure Integrity (DSI) [28] takes a different approach and cast the XSS problem as a document structure problem where client and server have inconsistent views of the document structure. Their evaluation shows that the tool is effective and incurs low performance overhead.

The possibility of history sniffing attack was first considered nine years ago by Clover in a BUGTRAQ mailing list post in February of 2002 [29]. This, however, has not been considered seriously by the browser vendors until recently when Jang et al. [4] reported 46 popular websites did sniff on users' history. After the disclosure, a prevention mechanism was proposed by L. David Baron of Mozilla [30] and has been adopted in the latest version of major browsers. Right after that, in May 2011, Weinberg et al. [31] demonstrated a new kind of history sniffing attack that circumvents even Baron's defense and to date, no newer protection measure has been proposed.

Dynamic taint analysis has widespread application in the research community. Vogt et al. [5] instrumented the browser's JavaScript engine to track a taint bit that determines whether a piece of data is sensitive and report an XSS attack if this data is sent to a domain other than the page's domain. In the same vein, Dhawan et al. [6] used similar techniques to analyze confidentiality properties of JavaScript browser extensions for Firefox.

From another perspective, the JavaScript language can be constrained to allow sandboxing of untrusted widgets. AdSafety [32] uses a novel type system to analyze the robustness of web sandboxes. They found several new bugs in the implementation of AdSafe, a web sandbox by Yahoo. Concurrent with AdSafety, Taly et al. [33] also studies JavaScript reference monitors and devises a restricted version of the JavaScript language. They then develop a tool that can soundly prove that an API cannot be circumvented or subverted, hence ensuring the robustness of sandbox protection.

7 Conclusions

We have presented the design, implementation and evaluation of LeakTracker, a fast and scalable tool to study the privacy violations in web content. Particularly, we developed a novel technique called principal-based tainting that allowed us to perform dynamic analysis of JavaScript execution at reduced performance overhead. We have implemented a Firefox-based prototype of LeakTracker and deployed it on production systems. We have crawled and surveyed the Alexa global top one million websites. Our findings demonstrated that privacy attacks are more prevalent and serious than previously known.

Acknowledgments. We thank the anonymous reviewers for their insightful comments that helped improve this paper. This work is supported in part by the US National Science Foundation (NSF) under Grants 0855297, 0855036, 0910767, and 0952640, and the Singapore Ministry of Education under the grant R-252-000-460-112. Any opinions, findings, and conclusions or recommendations expressed in this material are those of the authors and do not necessarily reflect the views of the funding agencies.

References

1. Mozilla: Same origin policy for javascript, `https://developer.mozilla.org/En/Same_origin_policy_for_JavaScript`
2. Krishnamurthy, B., Wills, C.: Privacy Leakage in Mobile Online Social Networks. In: Workshop on Online Social Networks (2010)
3. Krishnamurthy, B., Naryshkin, K., Wills, C.: Privacy Leakage Vs. Protection Measures: The Growing Disconnect. In: Web 2.0 Security and Privacy (2011)
4. Jang, D., Jhala, R., Lerner, S., Shacham, H.: An Empirical Study of Privacy-Violating Information Flows in JavaScript Web Applications. In: 17th ACM Conference on Computer and Communications Security (2010)
5. Vogt, P., Nentwich, F., Jovanovic, N., Kirda, E., Kruegel, C., Vigna, G.: Cross-Site Scripting Prevention with Dynamic Data Tainting and Static Analysis. In: Network and Distributed System Security Symposium (2007)
6. Dhawan, M., Ganapathy, V.: Analyzing Information Flow in JavaScript-based Browser Extensions. In: ACSAC 2009: Proceedings of the 25th Annual Computer Security Applications Conference (2009)

7. Wang, Y.M., Beck, D., Jiang, X., Roussev, R., Verbowski, C., Chen, S., King, S.: Automated Web Patrol with Strider HoneyMonkeys: Finding Web Sites That Exploit Browser Vulnerabilities. In: Network and Distributed System Security Symposium (2006)
8. Webkit: SunSpider JavaScript Benchmark, http://www.webkit.org/perf/sunspider/sunspider.html
9. The Wall Street Journal: The Web's New Gold Mine: Your Secrets, http://online.wsj.com/article/SB10001424052748703940904575395073512989404.html
10. The Center for Internet and Society: Tracking the Trackers: Microsoft Advertising, http://cyberlaw.stanford.edu/node/6715
11. Electronic Frontier Foundation: Panopticlick, http://panopticlick.eff.org/
12. Eckersley, P.: How Unique Is Your Web Browser? In: Atallah, M.J., Hopper, N.J. (eds.) PETS 2010. LNCS, vol. 6205, pp. 1–18. Springer, Heidelberg (2010)
13. Heritrix: Heritrix, https://webarchive.jira.com/wiki/display/Heritrix/Heritrix
14. The Wall Street Journal: What They Know About You, http://online.wsj.com/article/SB10001424052748703999304575399041849931612.html
15. Cooke, L.: Is the Mouse a 'Poor Man's Eye Tracker?'. In: Society for Technical Communication Conference (2006)
16. Forbes: Class Action Lawsuit Filed Over YouPorn History Sniffing, http://www.forbes.com/sites/kashmirhill/2010/12/06/class-action-lawsuit-filed-over-youporn-history-sniffing/
17. Forbes: McDonald's, CBS, Mazda, and Microsoft Sued for 'History Sniffing', http://www.forbes.com/sites/kashmirhill/2011/01/03/mcdonalds-cbs-mazda-and-microsoft-sued-for-history-sniffing/
18. Curtsinger, C., Livshits, B., Zorn, B., Seifert, C.: Zozzle: Low-overhead Mostly Static JavaScript Malware Detection. In: 20th USENIX Security (2011)
19. Ratanaworabhan, P., Livshits, B., Zorn, B.: Nozzle: A Defense Against Heap-Spraying Code Injection Attacks. In: 18th USENIX Security (2009)
20. Canali, D., Cova, M., Kruegel, C., Vigna, G.: Prophiler: A Fast Filter for the Large-Scale Detection of Malicious Web Pages. In: Proceedings of the World Wide Web Conference, WWW (2011)
21. Cova, M., Kruegel, C., Vigna, G.: Detection and Analysis of Drive-by-Download Attacks and Malicious JavaScript Code. In: Proceedings of the World Wide Web Conference, WWW (2010)
22. Provos, N., Mavrommatis, P., Rajab, M.A., Monrose, F.: All Your iFRAMEs Point to Us. In: 17th USENIX Security (2008)
23. Seifert, C., Komisarczuk, P., Welch, I.: Identification Of Malicious Web Pages With Static Heuristics. In: Australasian Telecommunication Networks and Applications Conference (2008)
24. Spoor, R.J., Kijewski, P., Overes, C.: The HoneySpider Network: Fighting Client-side Threats. In: FIRST (2008)
25. Reis, C., Dunagan, J., Wang, H.J., Dubrovsky, O., Esmeir, S.: BrowserShield: Vulnerability-driven Filtering of Dynamic HTML. In: Proceedings of the Symposium on Operating Systems Design and Implementation (2006)
26. Louw, M., Venkatakrishnan, V.: Blueprint: Robust Prevention of Cross-site Scripting Attacks for Existing Browsers. In: Proceedings of the IEEE Symposium on Security and Privacy (2009)

27. Meyerovich, L., Livshits, B.: ConScript: Specifying and Enforcing Fine-Grained Security Policies for JavaScript in the Browser. In: Proceedings of the IEEE Symposium on Security and Privacy (2010)
28. Nadji, Y., Saxena, P., Song, D.: Document Structure Integrity: A Robust Basis for Cross-site Scripting Defense. In: Network and Distributed System Security Symposium (2009)
29. Clover, A.: CSS Visited Pages Disclosure. BUGTRAQ Mailing List Posting, http://seclists.org/bugtraq/2002/Feb/271
30. Baron, L.D.: Preventing Attacks on a User's History Through CSS: visited Selectors, http://dbaron.org/mozilla/visited-privacy
31. Weinberg, Z., Chen, E., Jayaraman, P.R., Jackson, C.: I Still Know What You Visited Last Summer: Leaking Browsing History Via User Interaction and Side Channel Attacks. In: 31st IEEE Symposium on Security and Privacy (May 2011)
32. Politz, J.G., Eliopoulos, S.A., Guha, A., Krishnamurthi, S.: ADsafety: Type-Based Verification of JavaScript Sandboxing. In: 20th USENIX Security (2011)
33. Taly, A., Erlingsson, U., Mitchell, J.C., Miller, M.S., Nagra, J.: Automated Analysis of Security-Critical JavaScript APIs. In: Proceedings of the IEEE Symposium on Security and Privacy (2011)

The Shy Mayor: Private Badges in GeoSocial Networks

Bogdan Carbunar[1], Radu Sion[2], Rahul Potharaju[3], and Moussa Ehsan[2]

[1] Florida International University Miami, FL
carbunar@cs.fiu.edu
[2] Stony Brook University, Stony Brook, NY
{sion,mehsan}@cs.stonybrook.edu
[3] Purdue University, West Lafayette, IN
rpothara@cs.purdue.edu

Abstract. Location based social or *g*eosocial networks (GSNs) have recently emerged as a natural combination of location based services with online social networks: users register their location and activities, share it with friends and achieve special status (e.g., "mayorship" badges) based on aggregate location predicates. Boasting millions of users and tens of millions of daily check-ins, such services pose significant privacy threats: user location information may be tracked and leaked to third parties. Conversely, a solution enabling location privacy may provide cheating capabilities to users wanting to claim special location status. In this paper we introduce new mechanisms that allow users to (inter)act privately in today's geosocial networks while simultaneously ensuring honest behavior. An Android implementation is provided. The Google Nexus One smartphone is shown to be able to perform tens of badge proofs per minute. Providers can support hundreds of million of check-ins and status verifications per day.

1 Introduction

Location based services offer information and entertainment services to mobile users, that rely on the geographical position of their mobile devices. A recently introduced but popular example, is the geosocial network (GSN) – a social network centered on the geographical position of its users. Services such as Foursquare [1], Yelp [2] or Gowalla [3] allow users to register or "check-in" their location, share it with their friends, leave recommendations and collect prize "badges". Badges are acquired by checking-in at certain locations, following a required pattern simultaneously with other users, i.e. multiplayer games, or obtaining the highest number of check-ins during a time window ("mayor" badge).

Besides keeping track of their friends' location, the user incentives for participation include receiving promotional deals, coupons and personalized recommendations. The main source of revenue for service providers lies in ad targeting. Boasting millions of users [4] and tens of millions of location check-ins per day [5], GSNs can provide personalized, location dependent ads. As such,

F. Bao, P. Samarati, and J. Zhou (Eds.): ACNS 2012, LNCS 7341, pp. 436–454, 2012.

the price of participation for users is steep: compromised location privacy. Service providers learn the places visited by each user, the times and the sequence of visits as well as user preferences (e.g., places visited more often) [6,7]. The implications are significant as service providers may use this information in ways that the users never intended when they signed-up (e.g., having their location information shared with third parties [8,9]).

While compromised privacy may seem a sufficient reason to avoid the use of such services, it may not be necessary. Instead, we propose here a framework where users themselves store and manage their location information. The provider's (oblivious) participation serves solely the goal of ensuring user correctness. This enables users to privately and securely check-in and acquire special location based status, e.g., in the form of badges. Badges are defined as aggregate predicates of locations. We then devise solutions to support a variety of such predicates, including (i) registering a pre-defined number of times at a location or set of locations, (ii) registering the most number of times (out of all the users) at a location and (iii) simultaneously registering with k other users at a location.

Given the recent surge of location privacy breaches and the ensuing liabilities issues [10], implementing privacy solutions may ultimately be in the service provider's best interest.

The problem is two-faceted. On one side, clients need strong privacy guarantees: The service provider should not learn user profile information, including (i) linking users to (location,time) pairs, (ii) linking users to any location, even if they achieve special status at that location and (iii) building user profiles – linking multiple locations where the same user has registered. On the other side, when awarding location-related badges, the service provider needs assurances of client correctness. Otherwise, since special status often comes with financial and social perks, clients have incentives to report fake locations [11], copy and share special status tokens, or check-in more frequently than allowed.

We note that, despite being seemingly attractive, the simple use of client pseudonyms as a means to provide client privacy during check-ins and special status requests is vulnerable to profile based de-anonymization attacks [12,13].

In this work we first define essential *privacy* and *correctness* properties for the aggregate location predicate problem. We then introduce SPOTR , a venue-oriented location verification protocol, that allows GSN providers to certify the locations claimed by users. SPOTR relies on single-use, 2 dimensional QR (Quick Response) codes, displayed on devices inside participating venues. Furthermore, we propose three privacy-preserving solutions, *GeoBadge*, *GeoM* and *MPBadge*, for the three aggregate location predicates described above. The solutions deploy cryptographic techniques such as zero-knowledge proofs, quadratic residuosity constructs, threshold secret sharing and blind signatures. Clients collect special, provider-issued tokens during check-ins, which they either aggregate to build generic, non-traceable badges, or use to build zero-knowledge proofs of ownership. Client correctness is partly ensured by the use of blind signatures of single-use tokens.

We have implemented and evaluated the performance of our solutions on a Revision C4 BeagleBoard, Google Nexus One smartphones and a 16 quad-core server. Experimental results are extremely positive. The GSN provider can support thousands of check-ins and special status verifications per second, while a smartphone can build strongly secure aggregate location and correctness proofs in just a few seconds.

2 Related Work

Location Cloaking: Anonymization, pseudonimization, location and temporal cloaking techniques (introducing errors in location reports to provide 1-out-of-k anonymity) have been initially proposed in [14], followed by a significant body of work [15,16,17,18]. These techniques are vulnerable to de-anonymization attacks [12,13]: the identity of a user frequently reporting a residential address may be revealed by computing intersection sets of of cloaked reports.

Location Verification: Saroiu and Wolman [19] introduced the location proof concept – a piece of data that certifies a receiver to a geographic location. The solution relies on special access points (APs), that are able to issue such signed proofs. Luo and Hengartner [20] extend this concept with client privacy, achieved with the price of requiring three independent trusted entities. Note that both solutions rely on the existence of specialized APs or cell-towers, that modify their beacons and are willing to participate and sign arbitrary information. To address the central management problems, Zhu and Cao [21] proposed the APPLAUS system, where co-located, Bluetooth enabled devices compute privacy preserving location proofs.

Proximity Alerts: Zhong et al. [22] have proposed three protocols that privately alert participants of nearby friends. Location privacy here means that users of the service can learn a friend's location only if the friend is nearby. Manweiler et al. [23] propose several cloaking techniques for private server-based location/time matching of peers. Narayanan et al. [24] proposed several other solutions for the same problem, introducing the use of location tags as a means to provide location verification. Our work is different, by enabling private and correct aggregate location predicates in GSNs.

This paper extends our previous work [25] with a location verification solution, SPOTR , detailed descriptions of the private aggregate location predicate protocols (*GeoBadge*, *GeoM* and *MPBadge*), proofs of correctness and privacy, details of Foursquare as well as implementation results of SPOTR , *GeoBadge* and *GeoM*.

Summary: Existing work has focused on (i) hiding user location from LBS providers and other parties and on (ii) enabling users to prove claimed locations. Besides proposing a novel, venue oriented approach for location verification, in this paper we focus on the next step, of anonymizing location aggregates defined by geosocial networks.

3 Model

3.1 The System

We consider a geosocial network provider, S. Each subscriber (or user) has an account with S. Subscribers are assumed to have mobile devices equipped with a GPS receiver and a Wi-Fi interface (present on most smartphones). To use the provider's services, a client application needs to be downloaded and installed. Subscribers can register and receive initial service credentials, including a unique user id; let Id_A denote the id of user A. In the following we use the terms *user* and *subscriber* to refer to users of the service and the term *client* to denote the software provided by the service and installed by users on their devices.

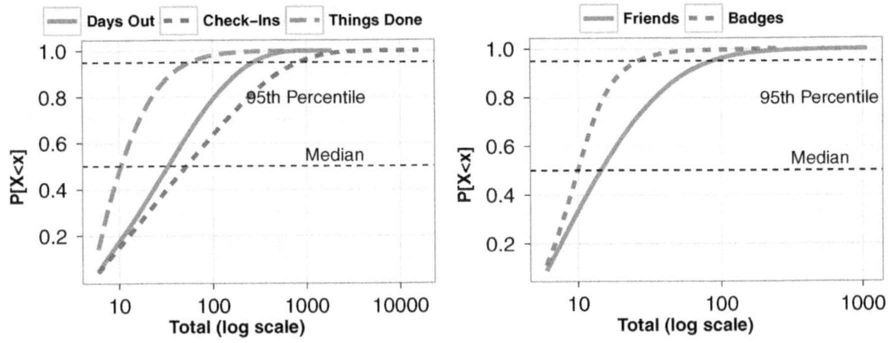

(a) (b)

Fig. 1. Foursquare stats: (a) CDF of days out, check-ins and things done by users. (b) Badge and friends evaluation.

Foursquare: In the following, we model the online geosocial network provider S after the most popular in existence to date, Foursquare [1]. In Foursquare, users report their location, through *check-ins* at venues of interest, share it with friends (e.g., imported from Facebook or discovered and invited on Foursquare) and are awarded points and "badges". A user with more check-in days at a venue than anyone else in the past 60 days becomes the "Mayor" of the venue. Foursquare has partnered with a long list of venues (bars, cafes, restaurants, etc) to reward the Mayor with freebies and specials. Foursquare imposes a discrete division of time, in terms of *epochs*. A user can check-in at one venue at most once per epoch. This strategy has made Foursquare quite popular, with a constantly growing user base, which we currently estimate at over 14 million users.

In order to understand the need for our solutions, we have collected profiles from 781,239 randomly selected Foursquare users. Our first question was how active are Foursquare users. Figure 1(a) shows the CDF of the number of check-ins, days out (days the user was actively performing check-ins) and things done (e.g., reviews left for a venue) by users. Note that 45% of the collected users have between 80 and 950 check-ins, for between 50 and 300 days of activity (at this time Foursquare is 2 years and a half old). This shows that many Foursquare users are very active. Our second question regards the popularity of badges

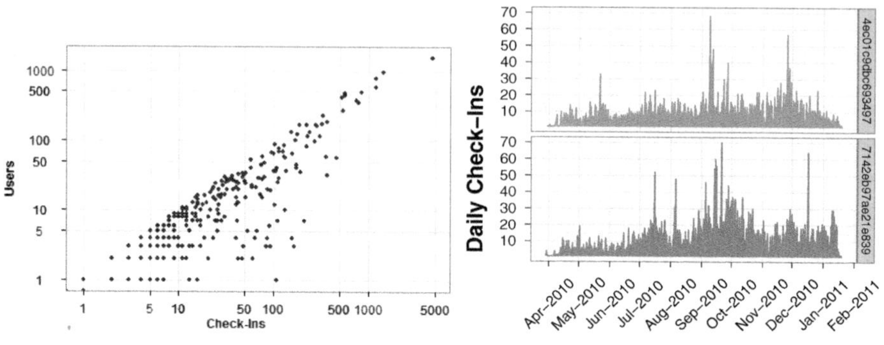

(a) (b)

Fig. 2. (a) Scatterplot check-ins vs. users in a small town. (b) Per-venue check-in distribution over time for two random venues.

in geosocial networks. Figure 1(b) shows the cumulative distribution function (CDF) of the number of badges earned by users as well as their friends. Note that 45% of the users (between the median and the 95th percentile) have between 10 and 50 badges and between 20 and 95 friends. This, coupled with the large numbers of check-ins reported strengthens our belief that private badge protocols are needed.

We corroborate the check-in data in a location-aware fashion: Figure 2(a) shows the scatter plot of check-ins vs. users in one of the most active locations in our dataset, the city of Babylon in Long Island, NY. Each point on the plot denotes a venue, the x axis shows the total number of check-ins recorded at the venue and the y axis shows the total number of users that have performed the check-ins. Note that a few venues record 1000-5000 check-ins, from more than 500 users. Most venues however range from a few tens to a few hundred check-ins and users. Finally, Figure 2(b) shows the evolution between August 2010 and February 2011 of the number of check-ins per day for two randomly selected venues. The number of check-ins range between 3 to almost 70 per day. Our conclusions are that Foursquare users are actively checking-in and venues record many daily check-ins. This data rich environment can be a goldmine for rogue GSN providers. Moreover, the number of recorded check-ins suggests that badges and mayorship are likely to become objects of contention. These points show that devising private and secure "badging" protocols is a problem of primary importance for GSNs.

Geo: A private GSN. A full-fledged privacy solution is composed of a set of protocols $Geo = \{Setup, RegisterVenue, Subscribe, CheckIn, StatVerify\}$. *Setup* is executed initially by the service provider to generate system-wide parameters and *RegisterVenue* is used to register a new venue with the provider S. *Subscribe* is initiated by a client when registering with the service. *CheckIn* is executed by a client to report its presence at a venue to S and *StatVerify* is executed when the client has accumulated sufficient check-ins and claims its special status. Each operation returns -1 to report failure or 0 for success.

We support three special status types. First, *location badges* (see Section 6), issued after the client runs $CheckIn$ during k different epochs at a venue V (e.g., "local" badge in Foursquare [1]) or after the client runs $CheckIn$ at k different, select, locations (e.g., "adventurer" badge). Second, *mayorships* (see Section 7), issued when the client has the largest number of $CheckIn$ runs, at most one per epoch, in the past m epochs at a given venue V. m is a system parameter. Third, *multi-player badges* (see Section 8), issued when the client runs $CheckIn$ simultaneously with s other users at the same location. s is a system parameter.

3.2 Privacy and Correctness Properties

Server Model. The provider S is honest, yet curious. S follows the protocol correctly, but is interested in collecting tuples of the format (Id, V, T), where Id is a user id, V is a venue and T is a time value. To this end, it may collude with existing clients and generate Sybil clients to track users of interest. The provider has no interest in colluding with users to issue badges without merit. To achieve privacy, intuitively, the provider should learn nothing about Geo clients. First, this includes the venues at which users run the $CheckIn$ function, how many times and when they run $CheckIn$ (in total and for any venue). We note that this necessarily includes also hiding correlations between venues where a given client has run $CheckIn$. We formalize this intuition using games run between an adversary $\mathcal{A}$ and a challenger $\mathcal{C}$. $\mathcal{A}$ controls the service provider and any number of clients, thus controls the initial parameter generation functionality (e.g., the $Setup$ function). $\mathcal{A}$ shares public parameters with $\mathcal{C}$. $\mathcal{C}$ controls two clients C_0 and C_1. $\mathcal{C}$ initially runs the $Subscribe$ function with $\mathcal{A}$ for the two clients and obtains their unique identifiers.

In a first CheckIn-Indistinguishability game, we model the adversary's inability to distinguish between clients during $CheckIn$ executions, even when the adversary controls an initial trace of $CheckIn$ executions. The game is defined for a given venue V.

CheckIn Indistinguishability (CI-IND). $\mathcal{A}$ generates l bits $c_1, .., c_l$ and sends them to $\mathcal{C}$. For each bit c_i, $\mathcal{C}$ executes $CheckIn(C_{c_i}(V), \mathcal{A})$. After processing all l bits, $\mathcal{C}$ flips a bit $b \in \{0,1\}$ and runs $CheckIn(C_b, \mathcal{A})$. $\mathcal{A}$ outputs a bit b'. A solution is said to be CI-IND if the advantage of $\mathcal{A}$ in the CI-IND game, $Adv(\mathcal{A}) = |Pr[b = b'] - 1/2$—, is negligible.

In a second, StatVerify-Indistinguishability game, the adversary (e.g., service provider) should be unable to distinguish between clients running $StatVerify$, *even if* the adversary is able to trace client $CheckIn$ executions.

StatVerify Indistinguishability (SV-IND). $\mathcal{C}$ performs l $CheckIn$ and m $StatVerify$ operations on behalf of C_0 and C_1, as requested by $\mathcal{C}$. A $StatVerify$ operation succeeds only if special status has been achieved by the corresponding client in the previous $CheckIn$ runs. $\mathcal{A}$ generates $k > 2s$ new bits $c_1, .., c_k$ such that at least s of them are 0 and at least s of them are 1. $\mathcal{A}$ sends $c_1, .., c_k$ to $\mathcal{C}$. For each bit c_i, $\mathcal{C}$ runs $CheckIn(C_{c_i}(V), \mathcal{A})$. Finally, $\mathcal{C}$ flips a coin $b \in \{0,1\}$

and runs $StatVerify(C_b(V,s),\mathcal{S})$. $\mathcal{A}$ outputs a bit b'. A solution is said to be SV-IND if the advantage of $\mathcal{A}$, $Adv(\mathcal{A}) = |Pr[b = b'] - 1/2|$, is negligible.

Note that even though the $CheckIn$ runs are executed for the same venue V, irrespective of the client, the SV-IND game is also suitable for the mayor badge – only one of the clients (C_b) will become mayor. However, for mayor badges, the value s needs to exceed the number of $CheckIn$ executions run on behalf of any client in the first step of the SV-IND game. Finally, we also need to allow the server to collect venue-based statistics:

Provider Usability. The service provider can count the number of $CheckIn$ executions for any venue as well as list the issued badges and mayorships.

Client Model. The client is assumed to be malicious. Malicious clients can be outsiders that are able to corrupt existing devices or may be insiders, i.e., subscribers, users that have installed the client. Malicious clients can try to cheat on their location (claim to be in a place where they are not [11]), attempt to prove a status they do not have, or disseminate credentials received from the server to other clients. The latter case includes any information received from the server, certifying presence at a specific location. Formally, we need a solution that has the following properties.

Status Safety. The challenger $\mathcal{C}$ controls the service provider and the adversary $\mathcal{A}$ controls any number of clients. The challenger runs first the *Setup* protocol and provides $\mathcal{A}$ with its public parameters. $\mathcal{A}$ runs *Subscribe* any number of times to generate clients. $\mathcal{A}$ then runs $CheckIn$ with $\mathcal{C}$ for any number of venues, but at most $k-1$ times for any venue. $\mathcal{A}$ runs $StatVerify$ with $\mathcal{C}$. The advantage of $\mathcal{A}$ is defined to be $Adv(\mathcal{A}) = Pr[StatVerify(C(params_C), S(priv_S)) = 1]$. We say that a solution is safe if $Adv(\mathcal{A})$ is negligible.

Note that a safe solution also prevents clients from running $CheckIn$ for venues where they are not located – otherwise $\mathcal{A}$ would succeed in $StatVerify$ with less than k $CheckIn$ runs at a site.

Token Non-distributability. No client or coalition thereof can use the same set of tokens more than once.

Token-Epoch Immutability. No client or coalition thereof can obtain more than one token per site per epoch.

4 Tools

Cryptographic Tools. We use a semantically secure cryptosystem, as well as unforgeable signature schemes. Let $S_X(M)$ denote the signature of a message M by participant X. Unforgeability is defined in terms of security "against one-more-forgery", where the user engaged in l runs of the signature algorithm with the signer cannot obtain more than l signatures. We also use blind signatures with the standard (i) blindness and (ii) unforgeability properties. Blindness means that the signer learns nothing about the signed messages. We use cryptographic hashes that are easy to compute and are (i) pre-image resistant, (ii) second pre-image resistant and (iii) collision resistant. We use $x \in_R S$ to denote the random choice of x from set S.

Anonymizers. We assume the existence of a network anonymizer, Mix, such as Tor [26]. Anonymizers or mix-nets [26,27] are tools that make communication untraceable and unlinkable. Untraceability implies the infeasibility of finding the identity of the issuer of a given set of messages. Unlinkability implies the infeasibility of discovering pairs of communicating entities.

Anonymous Authentication. We rely on anonymous authentication techniques with revocation and identity escrow, e.g., [28], performed over Mix, to enable users to prove they are service subscribers.

QR-Assumption. Given a large composite $n = pq$, where p and q are safe primes and given n but not p and q, it is computationally hard to decide if any value v, whose Jacobi symbol $(v|n)$ is 1, is a quadratic residue or not. v is a quadratic residue if there exists a value y such that $y^2 = v \ mod \ n$.

5 Spotr : Secure Location Verification

In this section, we propose SPOTR , a solution that allows the GSN provider to privately verify the claimed locations of clients. Since venues have the most incentives to correctly reward users, SPOTR relies on the co-operation of venue owners: owners need to install and operate a device inside their venues. We show that simple, off-the-shelf equipment is sufficient and no Internet connectivity is required – thus imposing solely a one time investment. SPOTR relies on Quick Response Codes (QR codes), 2D barcodes consisting of black modules arranged in a square pattern on a white background, that can store up to 2,953 bytes.

Let SPOTR_V denote the device installed at venue V. When registering SPOTR_V , the owner instructs SPOTR_V to generate a public/private key, store the private key, encode the public key in a QR code and display it on the screen. The owner takes a picture of the QR code, decodes the public key and reports it to S, the GSN provider. S associates with each venue, the owner's public key. At any time, SPOTR_V displays a QR code encoding $T, \Delta T, S_O(H(T, ctr))$, containing the time when the QR code was generated, an expiration increment ΔT and the owner O's signature on these values. During a check-in at V, the following takes place:

CheckIn$(C(Id, V, T, pub_S), S(priv_S, pub_O))$: The user approaches SPOTR_V , snaps a picture of the displayed QR code and sends it, along with the venue identity, over Mix, to S. With the public key pub_O of the owner O of venue V, S verifies the correctness of the received signature, and that the current time is between $[T, T + \Delta T]$. If the verifications succeed, S validates the check-in. Otherwise, it returns -1. SPOTR_V changes the QR code to encode a fresh timestamp when either (i) the current time approaches $T + \Delta T$ or (ii) it detects that the current QR code has been read (see Section 9 for implementation details).

We are exploring alternative, challenge-response based location verification protocols involving Wi-Fi/Bluetooth/NFC communication between the user smartphone and the venue's device. They are not included here due to space limitations. SPOTR will be used as a building block by all subsequent solutions, GeoBadge, GeoM and MPBadge. Its security is proved as part of *GeoBadge*.

6 Geo-Badge

GeoBadge is a private protocol that allows users to prove having visited the same location k times. At the end of the section we show how to adapt it to support private proofs of visiting k different places. *GeoBadge* works as follows: each subscribed client contacts the provider over the anonymizer Mix, authenticates anonymously, proves its current location and obtains a blindly signed, single use nonce and a share of a secret associated with the current venue. When k shares have been acquired (after k check-ins at the same venue) the client is able to reconstruct the secret - which is the proof required for the badge of the venue. The single use nonces prevent users from distributing received shares (or proofs).

GeoBadge extends *Geo* and provides the skeleton on which we build the subsequent solutions. Each client maintains a set Tk, storing all the tokens accumulated during $CheckIn$ runs. When the client accumulates enough tokens in Tk to achieve special status, it runs $StatVerify$, aggregating the tokens in Tk. In the following we instantiate each protocol, executed between a client C and the GSN provider S.

Setup: The server chooses a large prime p and generates a random key K. The server publishes p and keeps K secret.

RegisterVenue$(C(), S(priv_S))$: The client C that registers venue V, called the owner of the venue, sends to S its public key. For each new venue V, S generates a secret M_V randomly. S uses a threshold secret sharing solution to compute shares of M_V, by generating a polynomial Pol of degree $k-1$ whose free coefficient is M_V: $Pol(x) = M_V + c_1x + c_2x^2 + ... + c_{k-1}x^{k-1}$. S keeps Pol's coefficients secret but publishes the degree k and the verification value $Ver_V = H(M_V H_K(V) \ mod \ p)$. S stores Pol's coefficients for V, along with the public key of V's owner - to be used as part of SPOTR (see Section 5).

Subscribe$(C(), S(pub_S, priv_S))$: The communication in this step is performed over Mix, to hide C's location from S. C runs the setup stage of the Anonymous Authentication protocol of Boneh and Franklin [28] to obtain tokens that allow it later to authenticate anonymously with the server.

CheckIn$(C(Id, V, T, pub_S), S(priv_S))$: Let time T be during epoch e. The following actions are performed by a client C and the service provider S:

- **Anonymous Authentication:** C runs the anonymous authentication procedure of Boneh and Franklin [28] to prove to S that it is a subscriber. This step is performed over Mix.
- **Location Verification:** C runs SPOTR (Section 5) to prove presence at V.
- **Token Generation:** C generates a fresh random value R and sends the blinded R to S, as $O(R)$. S computes $x_e = H_K(e) \ mod \ p$ and $y_e = Pol(x_e) \ mod \ p$.

S sends to C (as a reply over the anonymizer) the tuple $(x_e, c_e, S_S(O(R)))$, where $c_e = H_K(V)y_e \ mod \ p$ and the last field denotes the blindly signed R. C "unblinds" the signed nonce, obtaining $s_e = S_S(R)$ and stores (x_e, c_e, s_e) into its token set Tk.

StatVerify$(C(Id, V, k, Tk, pub_S), S(priv_S))$: Let $Tk = \{(x_1, c_1, S_S(R_1)), .., (x_k, c_k, S_S(R_k))\}$. Let $l_j(x) = \Pi_{m=1..k, m \neq j} \frac{x - x_m}{x_j - x_m} \ mod \ p$ be the Lagrange coefficients. The following steps are executed, over Mix:

– C computes $SS = \Sigma_{j=1..k} c_j l_j(0)$. C verifies that $H(SS) = Ver_V$. If the verification fails, C outputs -1 and stops. Otherwise, it sends SS, along with the set of signed nonces, $(S_S(R_1), .., S_S(R_k))$ and the venue V to S.
– S verifies that (i) the k random values are indeed signed by it, (ii) that $R_1, .., R_k$ are unique and have not been used before and (iii) that $H(SS) = Ver_V$. If either verification fails, S outputs -1. Otherwise, S stores the values $R_1, .., R_k$, then issues a badge $S_S(\text{"}GeoBadge'', V, T_c)$ for the venue V, where T_c is the current issuance time. S sends this badge to C (as a reply over Mix).

6.1 Analysis

Correctness. The following holds due to Lagrange interpolation:

$$SS = \sum_{j=1}^{k} c_j l_j(0) = H_K(V) \sum_{j=1}^{k} Pol(x_j) l_j(0) = H_K(V) Pol(0) = H_K(V) M_V$$

Theorem 1. *GeoBadge is CI-IND.*

Proof. (Summary) Following the CI-IND game, $\mathcal{A}$'s view consists of the outcome of $l + 1$ anonymous authentication procedures, $l + 1$ venue signatures (from QR codes) and $l + 1$ blinded random values. The venue signatures carry no information identifying the client. The blinded random values are information theoretical secure. Then, if $\mathcal{A}$ can distinguish between C_0 and C_1 in the last step of the game, we can build an adversary that has a non-negligible advantage against either (i) the anonymous authentication solution of Boneh and Franklin [28] or (ii) the untraceability property of Mix.

Theorem 2. *GeoBadge is SV-IND.*

Proof. (Summary) At the completion of the SV-IND game $\mathcal{C}$ can reconstruct the SS values for both C_0 and C_1. $\mathcal{A}$ has published a pre-commitment for SS – Ver_V. Note that $\mathcal{C}$'s verification of $H(SS) = Ver_V$ prevents $\mathcal{A}$ from guessing b based on the value $\mathcal{C}$ to reconstruct during $StatVerify$. Thus, if the adversary has non-negligible advantage in the SV-IND game then we can also build an adversary that has non-negligible advantage against either (i) the untraceability property of Mix, (ii) the semantic security of the blinding algorithm E, or (iii) the information theoretic security of the threshold secret sharing mechanism.

Theorem 3. *GeoBadge provides Status Safety.*

Proof. (Summary) SPOTR efficiently prevents a single attacker from falsely claiming presence at V: without being present, the attacker is unable to predict or forge the signature displayed on SPOTR_V (see the security against one-more-forgery of the signature scheme from Section 4). Then, if there exists an adversary that has non-negligible advantage in the Badge-Safety game we can build an adversary that has a non-negligible advantage against (i) the pre-image resistance property of hashes (inverting $Ver_V = H(SS)$) or (ii) the information theoretic threshold secret sharing technique (including combining shares generated at multiple sites).

Note that trivially *GeoBadge* also provides the Token Non-Distributability property – the single use, server signed random nonces prevent more than one run of $StatVerify$ for a given set of tokens. The Token-Epoch Immutability property holds (no colluding clients can obtain more than one token for a venue during any epoch e), since the pair (x_e, c_e) is a deterministic function of e.

7 Geo-M

Using the Foursquare terminology, the user that has run $CheckIn$ the most number of times, at a venue S, within the past m epochs, becomes the mayor of the place. We now propose $GeoM$, a solution that allows users to achieve this status with privacy, while allowing anyone to verify correctness. *GeoM* extends *GeoBadge*: First, it allows clients to prove any number of check-ins, not just a pre-defined value k. Second, the check-ins are time constrained: clients have to prove that all check-ins have occurred in the past m epochs. Finally, client issued proofs can be published by the provider to be verified by any third party, without the risk of being copied and re-used by other clients.

GeoM achieves these features by requiring the service provider to issue only one token for each venue during any epoch. When a user has accumulated k tokens for a venue, it proves to the provider that it has k out of the m tokens given in the past m epochs for that venue. The proof is in zero knowledge (ZK) and if it verifies is published by the server.

Setup: The server generates two large safe primes p and q and the composite $n = pq$. Let N denote n's bit length. S publishes n and keeps p and q secret.

RegisterVenue$(C(), S(priv_S))$: For each newly registered venue V, S generates a new random seed r_V and uses it to initialize a pseudo-random number generator G_V. During every epoch e_i, for the venue V, S generates a fresh random token t_i, using G_V, and publishes $t_i^2 \ mod \ n$.

CheckIn$(C(Id, V, T, q, pub_S), S(priv_S))$: Inherits the Anonymous Authentication and Location Verification steps from *GeoBadge*. If they succeed, let time T be within epoch e_i, when the provider's published token value is $t_i^2 \ mod \ n$. C generates a random nonce R, engages in a blind signature protocol with S and obtains $S_S(R)$. S also sends to C the value t_i, the square root of the value

published for the epoch e_i. C stores t_i in the set Tk along with the signed nonce, $S_S(R)$. All communication takes place over Mix.
StatVerify$(C(Id, V, k, Tk, pub_S), S(priv_S))$: Without loss of generality, let $T = \{(t_1, S_S(R_1)), .., (t_k, S_S(R_k))\}$ be the set of all tokens issued by S for venue V in the past m epochs and let $\mathcal{T}^2 = \{t_1^2, t_2^2, .., t_m^2\}$ denote the corresponding published values. Note that the membership of $\mathcal{T}^2$ changes during every epoch. The client and the server run the following steps s times (ZK proof of the client knowing k square roots of values from $\mathcal{T}^2$). If successful, at the end of the s steps S will be convinced with probability $1 - 2^{-s}$.

- C generates $y_1, .., y_m \in_R \{0,1\}^N$ and a random permutation π_1. C computes the set $M = \pi_1\{t_1^2 y_1^2, .., t_m^2 y_m^2\}$ and sends it to S. Note that C does not need to know $t_1, .., t_m$ to compute M.
- C generates $z_1, .., z_k \in_R \{0,1\}^N$ and a random permutation π_2 and computes the set $Proof = \pi_2\{t_1 z_1, .., t_k z_k\}$, which it sends to S.
- S flips a coin b and sends it to C.
- If b=0, C sends $y_1, .., y_m$ to S, which then verifies that for every $t_i^2 \in \mathcal{T}^2$, $t_i^2(y_i)^2$ occurs once in M.
- If b=1, C generates and sends $A = \pi_2\{a_1 = z_1^{-1} y_1, .., a_k = z_k^{-1} y_k\}$. S verifies that for every $p_i \in Proof$ and corresponding a_i, $(p_i a_i)^2$ occurs in M once.

If any step fails, S outputs -1 and stops. Otherwise, it generates a signed "mayor" token $S_S(\text{“}Mayor'', V, T_c)$ for venue V issued at time T_c and sends it to C. All communication in this step is done over Mix. To reduce delays, the ZK proof can be non-interactive – in the standard way, by making the challenge bits depend in an unpredictable way on the values sent to the server. This allows C to send the entire proof at once. S publishes the ZK proof for the current "mayor", which can be downloaded and verified by any third party.

7.1 Analysis

Theorem 4. *The $StatVerify$ protocol of GeoM is a zero knowledge proof system of k square roots from $\mathcal{T}^2$.*

Proof. (Summary) To see that *GeoM* is a proof system, we need to prove completeness and soundness.

Completeness – an honest server will be convinced by an honest client of the correctness of the proof. If b=0, S is convinced that M is obtained from $\mathcal{T}^2$ by multiplication with quadratic residues, y_i^2. That is, for each $t_i \in \mathcal{T}^2$, $t_i^2 y_i^2 \in M$. If b=1, S is convinced that C knows the square roots of k elements in M. This is because C can provide a_i values that satisfy $(p_i a_i)^2 = (t_i z_i z_i^{-1} y_i)^2 = t_i^2 y_i^2 \in M$. In conjunction, these two cases prove to S that C knows the square roots of k elements from $\mathcal{T}^2$ with probability $1 - 2^{-s}$.

Soundness – if the statement is false, no cheating client can convince an honest server that the statement is true, except with small probability. Without loss of generality, let us assume that C knows only $k-1$ square roots of $\mathcal{T}^2$, $t_1, .., t_{k-1}$. If C expects the challenge to be $b = 0$, C generates $y_1, .., y_m$ as in the

protocol, builds M correctly but generates $Proof = \pi_2\{t_1z_1, .., t_{k-1}z_{k-1}, z_k\}$, where z_k is random. If the challenge ends up being $b = 1$, C has to produce one a_j value that is equal to $y_j z_j^{-1}(t_j^2)^{1/2}$, for one $j \in k..m$. Due to the QR-Assumption, C is unable even to tell whether any t_j^2 is a quadratic residue or not. If C expects the challenge to be 1, it builds $M\pi_1 = \{t_1^2 w_1^2, .., t_{k-1}^2 w_{k-1}^2, w_k^2, .., w_m^2\}$, where the w_i's are random. It then build Proof to be
$Proof = \pi_2\{t_1z_1, .., t_{k-1}z_{k-1}, z_k\}$. If $b = 1$, C can provide square roots for k values in M. If $b = 0$ however, C has to produce $m - k + 1$ values y_j such that $y_j = w_j(t_j^{-2})^{1/2}$, which contradicts again the QR-Assumption. The chance of a cheating client to succeed after s repetitions is 2^{-s}.

Zero Knowledge – if the statement is true, no cheating server learns anything except this fact. We prove this by following the approach from [29,30]. Specifically, let S^* be an arbitrary, fixed, expected polynomial time server Turing machine. We generate an expected polynomial time machine M^* that, without being given access to a client C (or the square roots of any elements from $\mathcal{T}^2$, produces an output whose probability distribution is identical to the probability distribution of the output of $< C, S^* >$.

While we skip details due to space limitations, we note that M^* is built by using S^* as a black box. For each of the s steps of the protocol, M^* flips a coin a and builds the sets M and $Proof$ anticipating that the challenge bit b will equal a. It then feeds these values to S^*, which then outputs b. If $b = a$, M^* outputs the transcript of the transaction and moves to the next step. Otherwise, it repeats the current step. M^* terminates in expected polynomial time (each of the s steps is executed on average twice). The probability distributions of the output of $< C, S^* >$ and of M^* are identical, which is proved by induction.

Theorem 5. *GeoM is CI-IND and SV-IND.*

Proof. (Intuition) The CI-IND proof is inherited from *GeoBadge*: *CheckIn* protocol differs solely in the provider's issuance of a square root value. For the SV-IND proof, we note that *StatVerify* is a ZK proof system. Then, an adversary with advantage in the SV-IND game can be used to build an adversary against *Mix*'s untraceability property.

Theorem 6. *GeoM provides Status Safety.*

Proof. Results directly from Theorem 4: *StatVerify* is a proof system of having k square roots from $\mathcal{T}^2$. A cheating client can succeed with probability 2^{-s}, where s is the number of proof iterations.

The single-use blindly signed nonces generated during *CheckIn* ensure the token non-distributability property of *GeoM*. *GeoM* trivially provides the token-epoch immutability property, as S issues a single token per venue per epoch.

8 Multi-player: MP-Badge

The multi-player badge is issued when a user presents proof of co-location and interaction with $k - 1$ other users at a venue V. k is a parameter that may depend on the venue V. We now present *MPBadge*, an extension of *GeoBadge* that provides this functionality with privacy. *MPBadge* relies on threshold signatures, where each client is able to provide a signature share and k unique signature shares generated at the same venue in the same epoch (see protocol $MP-CheckIn$). The shares can then be combined to produce a signed co-location proof. An additional difficulty here lies in the ability of an anonymous user to cheat: run *CheckIn* multiple times in the same epoch, obtain k signature shares and generate by itself the co-location proof. We solve this issue by allowing a user to run *CheckIn* only once per venue per epoch - using the blind signature generation, *BSGen*, protocol (see below).

Setup: The server S generates two large safe primes p and q and the composite $n = pq$. Let N denote n's bit length. S publishes n and keeps p and q secret.

RegisterVenue$(C(), S(priv_S))$: The following steps are executed:
– S stores a key table KT, indexed by venues and epochs. $KT[V, e]$ contains a unique key, used only for signing values for a venue V during epoch e. Let v denote the total number of venues supported.
– For each venue V and epoch e, S generates a value $M_{V,e} \in_R \{0,1\}^N$ and a random polynomial $Pol_{V,e}$ with degree $k-1$, whose free coefficient is $M_{V,e}$. $M_{V,e}$ and $Pol_{V,e}$ are secret.

BSGen$(C(Id, e, pub_S), S(priv_S))$: Executed once per epoch e by each client C (when active) with provider S, over an authenticated channel. C generates v random values, one for each venue in the system, $R_1, .., R_v$. C and S engage in a blind signature protocol, where each R_i is blindly signed by S with $KT[P_i, e]$. S records the epochs when C has executed this step and returns -1 if C attempts to run this step twice for the same epoch. Otherwise, the client obtains $BS_{KT[P_i,e]}(R)$, $\forall i = 1..v$.

CheckIn$(C(Id, V, T, n, pub_S), S(priv_S))$: C and S run the Anonymous Authentication and Location Verification steps of *GeoBadge*. If they succeed, C sends $R, BS_{KT[V,e]}(R)$ to S over Mix – the values correspond to the venue V and epoch e where C runs *CheckIn*. S verifies that (i) R has not been used before and (ii) the validity of its signature. If either step fails, S returns -1. Otherwise, S stores R and generates a share of $M_{V,e}$: (x_e, y_e), where x_e is random and $y_e = Pol_{V,e}(x_e)$. S sends (x_e, y_e) to C as a reply over Mix, and C stores them.

MP-CheckIn$(C_1(Id_1, V, T), C_2(Id_2, V, T, x_{e,2}, y_{e,2}))$: This step is executed when a client C_1 contacts a co-located client C_2 to build a co-location proof for V during epoch e (containing current time T). The communication is done over Mix. C_1 contacts C_2 with the message $M = (\text{``}MPBadge'', V, e)$. If C_2 has already executed *CheckIn* at venue V and epoch e, let $(x_{e,2}, y_{e,2})$ be its share of $M_{V,e}$. C_2 then generates $\sigma_{e,2} = M^{y_{e,2}}\ mod\ n$ and sends back to C_1 the tuple $(x_{e,2}, \sigma_{e,2}, R_2, BS_{V,e}(R_2)\ mod\ n)$. R_2 is the value that C_2 has had the server blindly sign: $BS_{V,e}(R_2)$. C_1 stores these values in the set Tk.

StatVerify$(C(Id, V, k, Tk, e, pub_S), S(priv_S))$: Without loss of generality, let $Tk = \{(x_{e,i}, \sigma_{e,i}, R_i, BS_{V,e}(R_i)\}$, $\forall i = 1..k$. C and S run the following steps:

– C computes $\sigma = \prod_{i=1}^{k} \sigma_i^{l_i(0)} = M^{\Sigma_i y_{e,i} l_i(0)} = M^{M_{V,e}}$. C sends σ, R_i, $BS_{V,e}(R_i)$, for all k R_i values received from co-located clients to S over Mix.

– S verifies that (i) the time when the communication of the previous step has been initiated is within epoch e, (ii) that $(\text{"}MPBadge\text{"}, V, e)^{M_{V,e}} = \sigma$ and (iii) that all $BS_{V,e}(R_i)$ signatures verify for venue V during epoch e. S checks that the exact set of k revealed blind signatures has not been used before more than k-1 times: S records the set of k blind signatures and allows it to be used only k times. Subsequent uses of the tokens are allowed, as long as the newly revealed set contains at least one fresh blind signature. If any verification fails, S outputs -1 and stops. Otherwise, S generates an MPBadge: $S_S(\text{"}MPBadge\text{"}, V, e, T_c)$, where T_c is the time of issue, and sends it over Mix to C.

While we omit the proofs due to space constraints, we note that $MPBadge$ is CI-IND and SV-IND.

9 Evaluation

Spotr Implementation: We have implemented SPOTR in Android and have tested it on a Revision C4 of the BeagleBoard [31] system, featuring an OMAP 3530 DCCB72 720 MHz and a Google Nexus One smartphone featuring a 1 GHz Scorpion processor, Adreno 200 GPU with 512 MB RAM. We use the ambient light sensor of the Nexus One to detect when anyone takes a picture of the displayed QR code (light level changes). Figure 3 shows a picture of the BeagleBoard displaying a generated QR code. The time to generate a QR code on the BeagleBoard is 50ms. The time to decode the QR code on the Nexus One is 190ms, at a distance of 20cm.

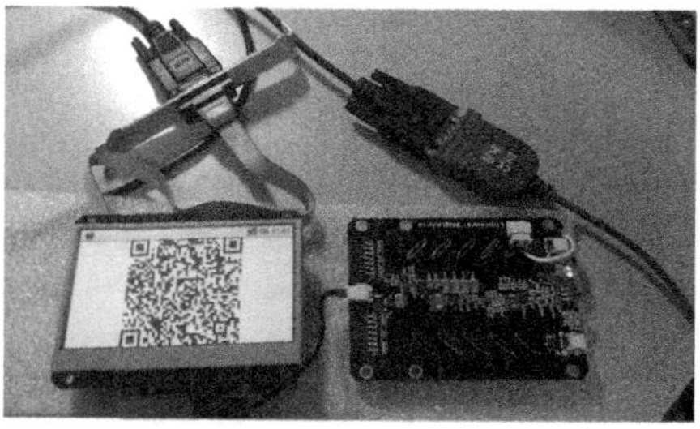

Fig. 3. SPOTR on BeagleBoard

We have implemented *GeoBadge* and *GeoM* in Android and Java and have tested the client side on the Nexus One smartphone and the server side on a 16 quadcore server featuring Intel(R) Xeon(R) CPU X7350 @ 2.93GHz and 128GB RAM. We have stress-tested the server side by sequentially sending multiple client requests. All the results shown in the following are computed as an average over at least 10 independent runs.

GeoBadge: We study the most compute-intensive functions of *GeoBadge*: *Setup*, the GSN provider side of *CheckIn*, the client and provider sides of *StatVerify*. We investigate first the dependence on the modulus bit size. The *Setup* cost, a one time cost for the GSN provider, ranges from 277ms for 512 bit keys to 16.49s for 2048 bit keys.

Figure 4(a) shows the performance of the remaining three components in milliseconds (ms) using a logarithmic y scale. The x axis is the modulus size, ranging from 512 to 2048 bits. The value of k, the number of *CheckIn* runs required to

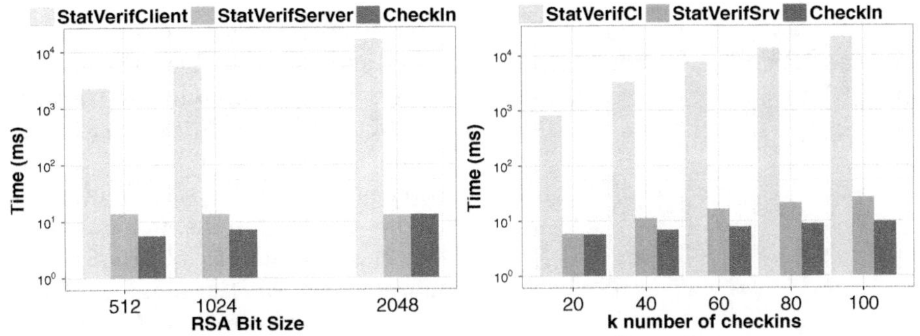

(a) (b)

Fig. 4. GeoBadge dependence on (a) modulus size, (b) k, the check-in count

acquire the badge is set to 50. On a single core, the $CheckIn$ cost, is 13ms even for a 2048 bit modulus size. The cost of the provider side of $StatVerify$ is almost constant for different key bit sizes, also around 13ms – on an OpenSSL sample, the cost of performing one signature verification for 2048 bit is 0.1ms, thus dwarfed by the cost of string operations. Thus, the provider can support more than 4800 $CheckIn$ or $StatVerify$ runs per second, or more than 412 million operations per day. The client side of $StatVerify$ requires 16.5s for 2048 bit keys, on the Nexus One.

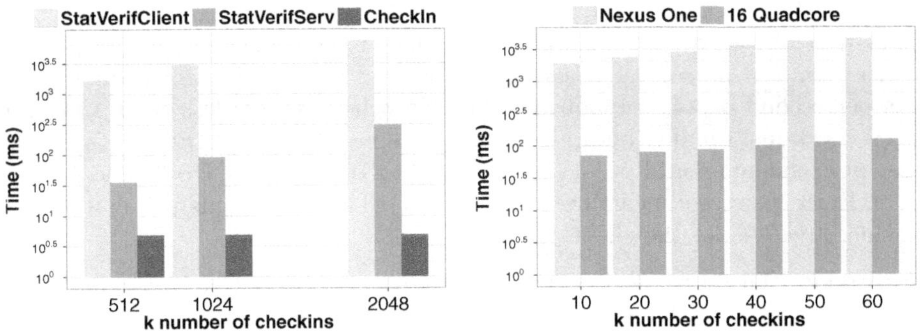

(a) (b)

Fig. 5. $GeoM$: (a) Dependence on N, the modulus size, (b) StatVerify client and server side, function of k, the number of check-ins

Figure 4(b) shows the performance dependency of the same protocols on k, the number of check-ins required, when the key size is set to 1024 bits. The client $StatVerify$ takes up to 21s when $k = 100$. The provider components are much faster: the $StatVerify$ takes less than 27ms, allowing the provider to support more than 2400 such operations per second (more than 207 million ops per day). The $CheckIn$ cost is even smaller, less than 10ms for k=100, allowing more than 6500 simultaneous check-ins, or more than 560 million check-ins per day.

GeoM: For the next experiment, we studied $GeoM$. We have first tested key bit sizes ranging from 512 to 2048. A one time occurrence for the GSN provider, the

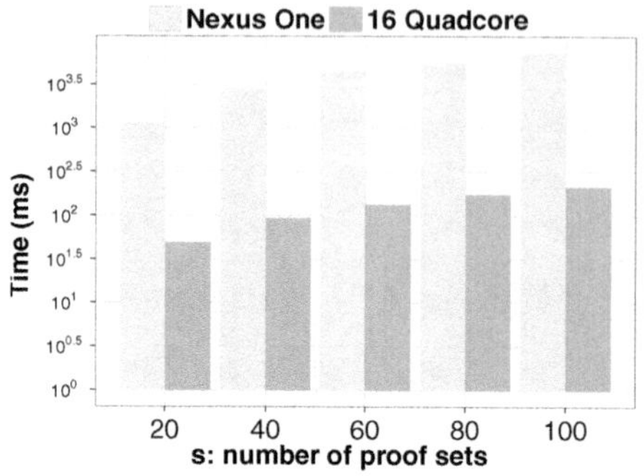

Fig. 6. StatVerify dependence on s, the number of proof iterations. y axis is time in milliseconds, in logarithmic scale

Setup cost ranges from 227ms to 1.5s and is negligible. Figure 5(a) shows the performance of $CheckIn$ (server side) and $StatVerify$ (client and server side) in ms, as a function of the key bit size. The y axis shows the time in ms, in logarithmic scale. s, the number of proof rounds is set to 40, m, the number of past epochs is set to 60 and k, the number of $CheckIn$ runs is set to 30. The client side $StatVerify$, executed on the Nexus One platform , requires between 1.7s to 7.5s. Since the provider is the bottleneck, the sensitive operations are $CheckIn$ and the provider side of $StatVerify$. These operations are fast: Requiring one table lookup and a signature generation, $CheckIn$ takes 4.8ms. On a 16 quadcore server, the provider can support more than 13,000 check-ins per second - more than 1.1 billion ops per day. The provider side of $StatVerify$ is less compute intensive than the client side: it ranges from 36ms to 309ms (form 2048 bit keys).

We further evaluate the dependency of $StatVerify$ (client and server side) on the value of k when the modulus size N is 1024, m=60 and s=40. Figure 5(b) shows that the server side exhibits small linear increases with k, but is only 124ms when $k = m = 60$. The server can support 512 simultaneous $StatVerify$ runs per second or 44+ million per day. The client side is less then 4.6s even for 60 check-ins. Finally, Figure 6 shows the dependency of $StatVerify$ on the value of s, the number of proof sets. N is set to 1024, m is set to 60 and k is set to 30. Both costs are linear: up to 211ms, or 18+ million runs per day for the provider and 7.2s for the client.

Summary. The server side overhead of *GeoBadge* and *GeoM* is small. The provider can support thousands of $CheckIn$s and $StatVerify$s per second. While on the order of a few seconds, the client side overhead of $StatVerify$ is not time sensitive and can be executed in the background.

10 Conclusions

We studied privacy issues related to aggregate location predicates in GSNs and proposed solutions that privately and securely enable aggregate location predicates. We showed that our solutions are efficient, as the provider can support between tens of millions to 1.1 billion operations per day. We leave for future work the issue of allowing the provider to privately collect aggregates over user information. This will address the provider's reluctance to offer services without being able to collect valuable user information.

References

1. Foursquare, https://foursquare.com/
2. Yelp, http://www.yelp.com
3. Gowalla, http://gowalla.com/
4. Indvik, L.: Foursquare Surpasses 3 Million User Registrations, http://mashable.com/2010/08/29/foursquare-3-million-users/
5. O'Dell, J.: Foursquare Day Sets Record with 3M+ Checkins, http://mashable.com/2011/04/20/foursquare-day-2/
6. Albanesius, C.: Apple location, privacy issue prompts house inquiry. PC Mag, http://www.pcmag.com/article2/0,2817,2365619,00.asp
7. Valentino-Devries, J.: Google defends way it gets phone data. Wall Street Journal (2011), http://online.wsj.com/article/SB10001424052748703387904576279451001593760.html
8. Krishnamurthy, B., Wills, C.E.: On the leakage of personally identifiable information via online social networks. In: WOSN, pp. 7–12 (2009)
9. Krishnamurthy, B., Wills, C.E.: On the leakage of personally identifiable information via online social networks. Computer Communication Review 40(1), 112–117 (2010)
10. Lowensohn, J.: Apple sued over location tracking in iOS. Cnet News (2011), http://news.cnet.com/8301-27076_3-20057245-248.html/
11. Gpscheat!, http://www.gpscheat.com/
12. Krumm, J.: Inference Attacks on Location Tracks. In: LaMarca, A., Langheinrich, M., Truong, K.N. (eds.) Pervasive 2007. LNCS, vol. 4480, pp. 127–143. Springer, Heidelberg (2007)
13. Golle, P., Partridge, K.: On the Anonymity of Home/Work Location Pairs. In: Tokuda, H., Beigl, M., Friday, A., Brush, A.J.B., Tobe, Y. (eds.) Pervasive 2009. LNCS, vol. 5538, pp. 390–397. Springer, Heidelberg (2009)
14. Gruteser, M., Grunwald, D.: Anonymous usage of location-based services through spatial and temporal cloaking. In: Proceedings of MobiSys (2003)
15. Hoh, B., Gruteser, M., Herring, R., Ban, J., Work, D., Herrera, J.-C., Bayen, R., Annavaram, M., Jacobson, Q.: Virtual Trip Lines for Distributed Privacy-Preserving Traffic Monitoring. In: Proceedings of ACM MobiSys (2008)
16. Olumofin, F., Tysowski, P.K., Goldberg, I., Hengartner, U.: Achieving Efficient Query Privacy for Location Based Services. In: Atallah, M.J., Hopper, N.J. (eds.) PETS 2010. LNCS, vol. 6205, pp. 93–110. Springer, Heidelberg (2010)
17. Pan, X., Meng, X., Xu, J.: Distortion-based anonymity for continuous queries in location-based mobile services. In: GIS, pp. 256–265 (2009)
18. Ghinita, G., Damiani, M.L., Silvestri, C., Bertino, E.: Preventing velocity-based linkage attacks in location-aware applications. In: GIS, pp. 246–255 (2009)
19. Saroiu, S., Wolman, A.: Enabling New Mobile Applications with Location Proofs. In: Proceedings of HotMobile (2009)
20. Luo, W., Hengartner, U.: VeriPlace: A Privacy-Aware Location Proof Architecture. In: Proceedings of ACM SIGSPATIAL GIS (2010)
21. Zhu, Z., Cao, G.: APPLAUS: A Privacy-Preserving Location Proof Updating System for Location-based Services. In: Proceedings of IEEE INFOCOM (2011)
22. Zhong, G., Goldberg, I., Hengartner, U.: Louis, Lester and Pierre: Three Protocols for Location Privacy. In: Proceedings of PETS (2007)

23. Manweiler, J., Scudellari, R., Cancio, Z., Cox, L.P.: We saw each other on the subway: secure, anonymous proximity-based missed connections. In: Proceedings of HotMobile (2009)
24. Narayanan, A., Thiagarajan, N., Lakhani, M., Hamburg, M., Boneh, D.: Location privacy via private proximity testing. In: Proceedings of NDSS (2011)
25. Carbunar, B., Sion, R.: Private geosocial networking. In: Proceedings of the ACM SIGSPATIAL GIS (2011)
26. Dingledine, R., Mathewson, N., Syverson, P.F.: Tor: The second-generation onion router. In: USENIX Security Symposium, pp. 303–320 (2004)
27. Reiter, M.K., Rubin, A.D.: Anonymous web transactions with crowds. Commun. ACM 42(2), 32–38 (1999)
28. Boneh, D., Franklin, M.: Anonymous Authentication With Subset Queries (Extended Abstract). In: Proceedings of CCS (1999)
29. Goldwasser, S., Micali, S., Rackoff, C.: The knowledge complexity of interactive proof systems. SIAM J. Comput. 18(1) (1989)
30. Goldreich, O., Micali, S., Wigderson, A.: Proofs that yield nothing but their validity or all languages in np have zero-knowledge proof systems. J. ACM 38(3) (1991)
31. Coley, G.: Beagleboard system reference manual. BeagleBoard.org (December 2009)

Detecting Social Spam Campaigns on Twitter

Zi Chu[1], Indra Widjaja[2], and Haining Wang[1]

[1] Department of Computer Science, The College of William and Mary, Williamsburg, VA 23187, USA
{zichu,hnw}@cs.wm.edu
[2] Bell Laboratories, Alcatel-Lucent, Murray Hill, NJ 07974, USA
iwidjaja@research.bell-labs.com

Abstract. The popularity of Twitter greatly depends on the quality and integrity of contents contributed by users. Unfortunately, Twitter has attracted spammers to post spam content which pollutes the community. Social spamming is more successful than traditional methods such as email spamming by using social relationship between users. Detecting spam is the first and very critical step in the battle of fighting spam. Conventional detection methods check individual messages or accounts for the existence of spam. Our work takes the collective perspective, and focuses on detecting spam campaigns that manipulate multiple accounts to spread spam on Twitter. Complementary to conventional detection methods, our work brings efficiency and robustness. More specifically, we design an automatic classification system based on machine learning, and apply multiple features for classifying spam campaigns. Our experimental evaluation demonstrates the efficacy of the proposed classification system.

Keywords: Spam Detection, Anomaly Detection, Machine Learning, Twitter

1 Introduction

With the tremendous popularity of online social networks (OSNs), spammers have exploited them for spreading spam messages. Social spamming is more successful than traditional methods such as email spamming by taking advantage of social relationship between users. One important reason is that OSNs help build intrinsic trust relationship between cyber friends even though they may not know each other in reality. This leads to users to feel more confident to read messages or even click links from their cyber friends. Facilitated by this fact, spammers have greatly abused OSNs and posted malicious or spam content, trying to reach more victims.

Detecting spam is the first and very critical step in the battle of fighting spam. Our work chooses Twitter as the battlefield. Currently, Twitter is the most popular micro-blogging site with 200 million users. Twitter has witnessed a variety of spam attacks. Conventional spam detection methods on Twitter mainly check

F. Bao, P. Samarati, and J. Zhou (Eds.): ACNS 2012, LNCS 7341, pp. 455–472, 2012.

individual tweets or accounts for the existence of spam [30, 16]. The tweet-level detection screens individual tweets to check whether they contain spam text content or URLs. As of August 2011, around 8.3 million tweets are generated per hour [9], and they demand near real-time delivery. Thus, the tweet-level detection would consume too much computing resources and can hardly meet time-stringent requirements. The account-level detection checks individual accounts for the evidence of posting spam tweets or aggressive automation behavior. Accounts violating the Twitter rules of spam and abuse [11] will get suspended. Suspending spam accounts is an endless cat and mouse game as it is easy for spammers to create new accounts to replace suspended ones.

Our work shifts the perspective from individual detection to collective detection and focuses on detecting spam campaigns. A spam campaign is defined as a collection of multiple accounts controlled and manipulated by a spammer to spread spam on Twitter for a specific purpose (e.g., advertising a spam site or selling counterfeit goods). Detecting spam campaigns is an important complement to conventional spam detection methods. Moreover, our work brings two additional benefits. (1) Efficiency. Our approach clusters related spam accounts into a campaign and generates a signature for the spammer behind the campaign. Thus, not only our work can detect multiple existing spam accounts at a given time, it can also capture future ones if the spammer maintains the same spamming strategies. (2) Robustness. There are some spamming methods which cannot be detected at individual level. For example, Twitter defines the behavior of "posting duplicate content over multiple accounts" as spamming. By grouping related accounts, our work can detect such a collective spamming behavior.

We have performed data collection for three months in 2011, and obtained a dataset with 50 million tweets posted by 22 million users. Using the dataset, we cluster tweets with the same final URL into a campaign, partitioning the dataset into numerous campaigns based on URLs. We perform a detailed analysis over the campaign data and generate a set of useful features to classify a campaign into two classes: spam or legitimate. Based on the measurement results, we present an automatic classification system using machine learning. We validate the efficacy of the classification system. The experimental results show high accuracy with low false positive rate.

The remainder of the paper is organized as follows. Section 2 presents a brief background of Twitter and covers related work of social spam detection. Section 3 details the data collection and measurements on Twitter. Section 4 describes our automatic classification system. Section 5 evaluates the system efficacy for detecting spam campaigns. Finally, Section 6 concludes the paper.

2 Related Work

As spammers often use Twitter-specific features to allure victims, we first briefly describe the background of Twitter and its working mechanism. Then, we survey related work in social spam detection and discuss the scope of our work.

2.1 Twitter and Related Social Spam Detection

Users post textual messages on Twitter, known as tweets. The tweet length is up to 140 characters, which limits the spam content the spammer can include in a tweet. Thus, embedding an external URL in a tweet becomes a routine for spammers to allure users to spam websites. A tweet may contain some textual features for better user interaction experience, which are also abused by spammers. A *hashtag*, namely a word or a phrase prefixed with the # symbol, is used to group tweets by their topic. For example, #Japan_Tsunami and #Egyptian_Revolution are two of the worldwide trending hashtags on Twitter in March 2011. Spammers may attach popular hashtags to unrelated spam tweets to increase the chance of being searched. This spamming trick is called hashtag hijacking. The *mention* feature, namely the @ symbol followed by a username in a tweet, enables the direct delivery of the tweet to the user. This feature facilitates spammers to directly send spam to targeted users.

Traditional spam methods include sending spam emails [31] and creating spam web content [27]. The past few years have witnessed the rapid rise of online social networks. One key feature of such systems is the reliance on content contributed by users. Unfortunately, the system openness coupled with the large user population has made OSNs an ideal target of social spammers. By exploiting the social trust among users, social spam may achieve a much higher success rate than traditional spam methods. For example, Grier *et al.* analyzed the click-through rate of spam on Twitter [21], and found out that around 0.13% of spam tweets generate a visit, orders of magnitude higher than click-through rate of 0.003% - 0.006% reported for spam email [24].

As a countermeasure, Twitter has released its rules against spam and abuse [11]. Accounts violating the rules will result in permanent suspension. The set of rules mainly define spam on Twitter in the following categories of content, behavior and social relationship. In the content category, it is forbidden to post content or URLs of any kinds of spam. Large numbers of unrelated @replies, mentions and #hashtags, or duplicate content are also disallowed. The behavior category covers both individual and collective behavioral codes. At the individual level, aggressive automation such as constantly running programs to post tweets without human participation is prohibited. At the collective level, using multiple accounts to post duplicate content is also considered as spamming. In terms of social relationship, one cannot follow a large number of users in a short amount of time, or have a small number of followers compared to the number of friends it is following, or create or purchase accounts in order to gain followers.

To avoid being detected by Twitter rules, social spammers have adopted a similar idea of email spam campaigns by coordinating multiple accounts to achieve a specific purpose. The spammer distributes the workload among spam accounts, thus individual accounts now may exhibit stealthy spam behavior and fly under the radar. Besides, multiple accounts also can spread spam to a wider audience. Some related studies have demonstrated the wide existence of spam campaigns on OSNs, such as Twitter and Facebook, respectively [21, 20]. The existing work mainly relies on the URL feature. More specifically, related messages with the shared final landing URL are clustered into a campaign. Then, the URL is looked up in URL blacklists. If the URL is blacklisted, the campaign is classified as a spam campaign; otherwise it is legitimate. Currently, the existing detection methods have some disadvantages listed as follows. First, URL blacklists have the lag effect, allowing more than 90% of visitors to click on a spam URL before it becomes blacklisted [21]. Furthermore, URL blacklists can only cover part of spam URLs, and thus some spam campaigns may escape detection. Second, some URL blacklists generate false positive errors as they only check the hostname component of a URL, instead of the whole URL. For example, the URL shortening service http://ow.ly is listed on the URIBL blacklist [13] because it is greatly abused by spammers. Although http://ow.ly/6eAci is a benign URL that redirects to a CNN's report of Hurricane Irene, it is blacklisted by URIBL based on the hostname. Third, the URL feature generates false negative errors. For instance, consider a campaign that advertises a benign website in an aggressive spamming way. The spammer manipulates multiple accounts to post duplicate tweets about the website. The URL feature cannot classify the tweets as a spam campaign since the website URL is benign and not blacklisted. The first two disadvantages may be overcome by improving blacklisting process, but the third cannot be fixed by merely using the URL feature. Thus, the other features, such as collective posting content and behavior, should also be included. This paper improves the existing work by introducing new features. The details of classification features are covered in Section 4.1.

2.2 Scope of This Paper

A variety of spam attacks exist on Twitter. This paper solely focuses on characterizing and detecting large-scale spam campaigns conducted on Twitter. The definition of spam in this paper is spreading malicious, phishing or scam[1] content in tweets. Spammers may carry different purposes, but spam campaigns exhibit a shared feature that, they either create or compromise a large number of Twitter accounts to spread spam to a wide range of audience. Our work does not screen individual tweets to detect spam, and may miss small spam campaigns[2]. As a complement to existing spam detection methods, the main contribution of this paper is detecting multiple related spam tweets and accounts in a robust and efficient way.

Note that after detecting a spam campaign, a site administrator may further classify the involved accounts into Sybil and compromised accounts, and process them accordingly. Here Sybil accounts refer to those created by spammers and exclusively used to post spam tweets. Compromised accounts refer to those used by legitimate users but hijacked by spammers to post spam without the permission of owners. Sybil accounts will be permanently suspended, while the owners

[1] We define a scam as any webpage that advertises a spectrum of solicitations, including but not limited to pornography, online gambling, fake pharmaceuticals.

[2] According to our clustering algorithm presented in Section 3.2, a single tweet may be clustered as a campaign if no other related tweets exist in the dataset.

of compromised accounts can be notified for spamming activities via their registration emails. The differentiation between these two types of accounts is out of the scope of this paper.

3 Characterization

3.1 Data Collection

To measure the pervasiveness of spam, we conduct the data collection on Twitter from February to April in 2011. Thanks to Twitter's courtesy of including our test accounts to its whitelist, our dataset accumulates more than 50 million tweets posted by around 22 million accounts. We develop a crawler in PHP which taps into Twitter's Streaming API [12] and Search API [14], respectively. The Streaming API outputs a small proportion of real-time global tweets via random sampling, and constitutes the majority of our dataset. The Search API enables the crawler running specific searches against the real-time index of recent tweets. Since this work studies spam campaigns, we exclude tweets without URLs, and focus on the remaining 8 million tweets with URLs in the dataset. Due to the limited length of tweets, most spam tweets contain URLs to allure users to visit external spam websites. Thus, we assume that tweets without URLs are not spam. As shown in Section 3.2, our clustering algorithm is based on shared URLs.

URL redirection is widely used on Twitter. Normal users apply URL shortening services, such as t.co and bit.ly, to convert arbitrarily long URLs to short ones to better fit in tweets. Spammers also use shortening and other redirection techniques to hide original spam URLs and to avoid blacklist detection. We develop a Firefox extension in JavaScript to automatically visit every URL in the dataset and convert to its final landing URL if redirection is used. Some spammers tend to use long redirection chains that involve multiple hops (such as original URL -> intermediate URL -> ... -> final URL) to hide their traces. The extension records the whole chain, and provides a classification feature.

3.2 Clustering

We develop a clustering algorithm that clusters tweets into campaigns based on shared final URLs[3]. The idea behind the algorithm is that those tweets that share the same final URL are considered related. A tweet is modeled as the <textual content, URL> pair. A given campaign, c_i, is denoted by a vector $c_i =< u_i, T_i, A_i >$, where u_i is the shared final URL i for the campaign, T_i is the set of tweets containing u_i, and A_i is the set of accounts that have posted tweets in T_i. Let C denote the current set of campaigns. The clustering procedure iteratively chooses without replacement an arbitrary tweet t in the dataset. If the tweet's URL is $u_{i'}$ and $c_{i'} \in C$, then the tweet is added in the campaign

[3] The subsequent campaign classification applies a variety of features, including both content and URL of tweets. More feature details are presented in Section 4.1.

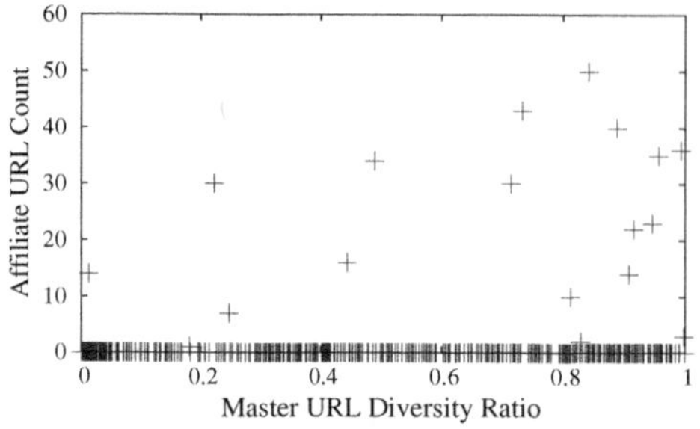

(a) Legitimate Campaigns

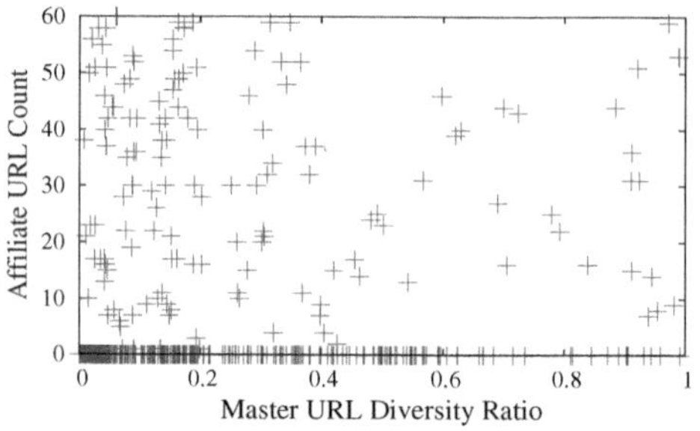

(b) Spam Campaigns

Fig. 1. URL Statistics of Campaigns

by updating $T_{i'} = T_{i'} \cup \{t\}$. If t's account, a, is also new, then an update $A_{i'} = A_{i'} \cup \{a\}$ is also performed. If $c_{i'} \notin C$, then a new campaign $c_{i'}$ is created and $C = C \cup \{c_{i'}\}$ is updated.

In our implementation, we store the dataset in MySQL database, and create a table for the clustering result. Every URL string is hashed, and the hash value is set as the table index. Two URL strings are compared by their indexed hash values to improve the clustering performance. Once complete, the dataset includes 5,183,656 campaigns. The largest contains 7,350 accounts with 9,761 tweets posted.

3.3 Ground Truth Creation

After campaigns have been clustered, we create a ground truth set containing samples labeled as spam and legitimate campaigns. We select some campaigns from our dataset, manually perform several heuristics tests, and use human expertise to label unknown campaigns. Due to the limited raw data returned by the Twitter API with low privilege, we favor the campaigns associated with a large number of accounts and tweets during the selection process as large campaigns carry abundant collective behavior characteristics. Small campaigns are excluded from our selection.

More specifically, we follow Twitter's spam rules during the manual inspection, and check both collective and individual features of an unknown campaign. First, we inspect the campaign's final URL. A batch script is performed to check the URL in five blacklists: Google Safe Browsing, PhishingTank, URIBL, SURBL and Spamhaus [1, 3, 13, 7, 6]. More details of the blacklist detection will be presented in Section 4.1. If the URL is captured by the first two blacklists, the related campaign is directly labeled as spam without further manual inspection required.

Second, we check the tweet content of the campaign. The human inspects the content to see if (1) it contains spam information, (2) it is unrelated with the URL's web content (namely, the URL is misleading), (3) duplicate or similar content is posted via single or multiple accounts. In addition, we also check content-related Twitter properties.

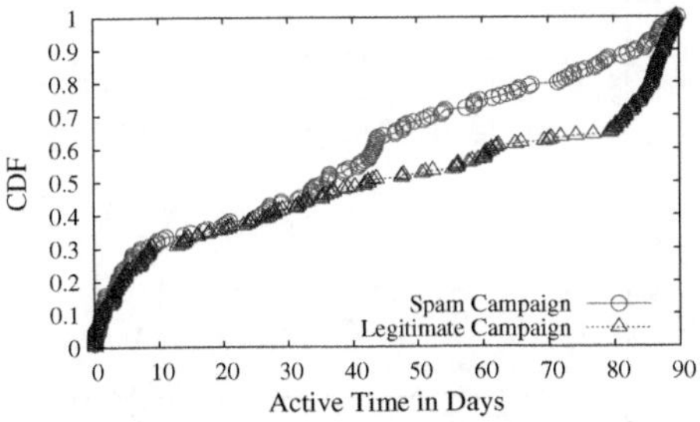

Fig. 2. CDF of Campaign Active Time

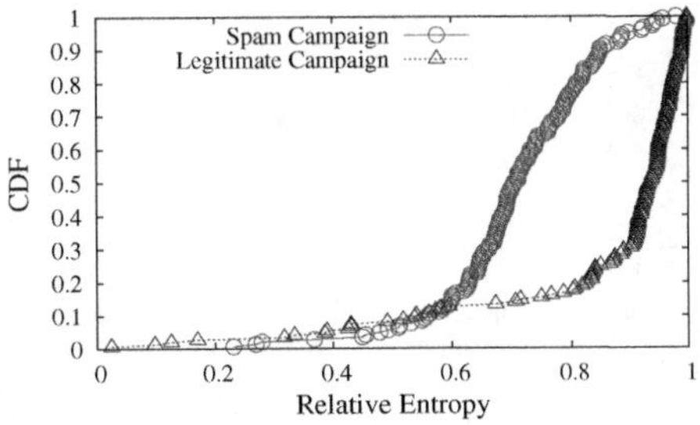

Fig. 3. CDF of Entropy of Posting Inter-arrivals

Third we check the automation degree exhibited in the campaign, as automation is a good indicator of spam. The script presents the human inspector with the posting device makeup, the median, and the entropy value of the posting inter-arrival timing sequence. The formal description of these features will be detailed in Section 4.1. Aggressive automation may raise the red flag, and influence the human's classification decision for the campaign.

By taking all of the above into consideration, the human inspector reaches the decision to label the campaign as spam or legitimate. In practice, we find out that most spam campaigns carry obvious characteristics of URL and content, making it easy to differentiate them from legitimate campaigns. We acknowledge that we may make mistakes in labeling campaigns, but believe that the error rate is very low. Finally, the ground truth set contains 744 spam campaigns and 580 legitimate ones.

3.4 Campaign Analysis

We now examine the characteristics of spam campaigns and compare with legitimate ones. The data analysis leads to the formal definition of classification features in Section 4.1.

We first discuss using URL statistics to reveal account connection in the campaign. We have observed that accounts in a legitimate campaign are usually run by independent users, while those involved in a spam campaign are often controlled by the same spammer. The URL statistics can provide hints of account connection. For clarity, we first define two terms: master URL and affiliate URL. For a normal URL such as http://biy.ly/5As4k3, affiliate URLs with it can be created by appending random strings as the query component to the URL, such as http://biy.ly/5As4k3?=xd56 and http://biy.ly/5As4k3?=7yfd. The original URL is denoted as master URL. Affiliate URLs help track the origin of click traffic. By assigning every account with a specific affiliate URL, the spammer can evaluate the spamming effect of individual accounts. This trick widely exists in online pyramid scams. Frequent appearance of affiliate URLs indicates strong connection among accounts. In contrast, different forms of master URLs indicate

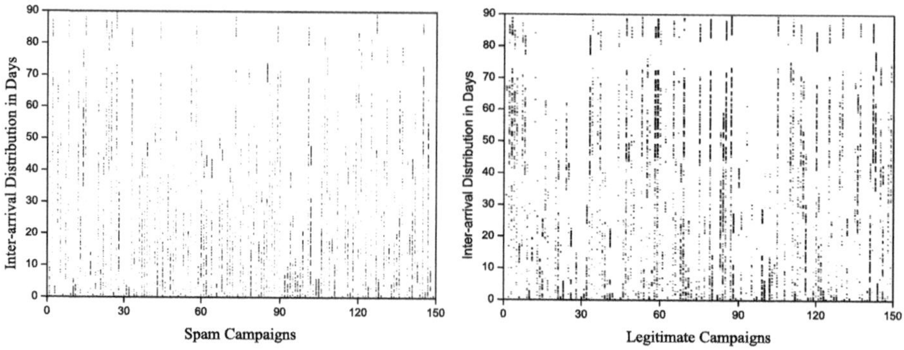

Fig. 4. Inter-arrival Timing Distribution of Campaigns

account independence. Although the tweets in a campaign share the same final URL, they may have different master URLs, such as http://bit.ly/1wgYxU and http://ow.ly/6jRqX[4]. We define the master URL diversity ratio as the number of unique master URLs over the number of tweets in a campaign. A low ratio indicates the wide usage of affiliate URLs and account dependence, whereas a high ratio indicates the account independence. Figure 1 shows that more than 50% of spam campaigns use affiliate URLs, while only 3.6% of legitimate campaigns contain affiliate URLs. The average master URL diversity ratio of spam campaigns is 0.225, much lower than that of legitimate campaigns, at 0.423.

Now we analyze the temporal properties of campaigns. We define the active time of a campaign as the time span between its first and last tweet in our dataset. We point out a limitation of our dataset as our collection runs for three months while a campaign may exist before and/or after the measured period. While the largest possible active time in our dataset is 90 days, the actual time may be greater. Figure 2 shows the cumulative distribution function (CDF) of active time (in days) of spam and legitimate campaigns. Around 40% of campaigns in both categories have active time less than 30 days. For those longer than 30 days, the average active time of legitimate campaigns is 72.0 days, greater than that of spam campaigns at 59.5 days. Thanks to the workload distribution among accounts, the spamming behavior of an account may be stealthy during its initial stage, and avoid Twitter's detection. It explains the equal proportions of both categories within the 30-day time window. The accumulation of spamming behavior and the increase of campaign size expose spam accounts, and many of them get suspended by Twitter. Beyond the 30-day window, the average active time of spam campaigns is clearly shorter than that of legitimate ones. However, more efforts need to be made to detect and eliminate spam campaigns in the initial stage for damage control.

The burstiness characterizes the overall workload distribution of spam campaigns. Figure 4 plots the inter-arrival timing pattern of two categories of campaigns. Due to space limit, each category contains 150 individual campaigns.

[4] All the URLs in this paragraph lead to http://twitter.com.

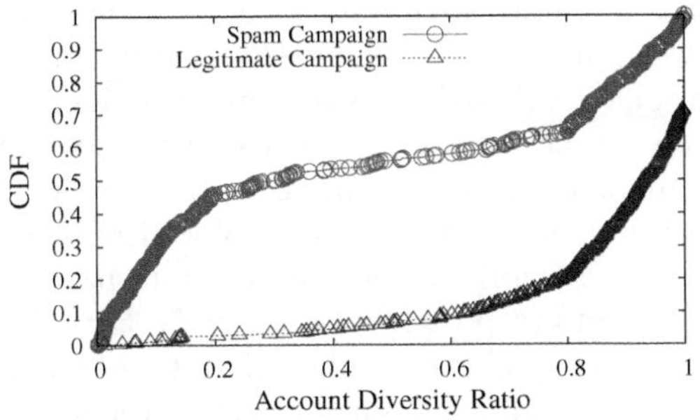

Fig. 5. CDF of Account Diversity Ratio of Campaigns

Each campaign is represented by a vertical strip. Each tweet corresponds to a tiny horizontal segment in the strip, and a block of intensive strips represent a burst of tweets in the campaign. A large number of spam campaigns show burstiness in the early stage. Some spammers aim to achieve the spamming goal in a quick way, and direct spam accounts to massively post tweets. Although the workload is distributed to multiple accounts, the collective inter-arrival pattern can reflect the overall campaign workload. The gradual suspension of spam accounts causes the stagnation in the late stage [5]. Many legitimate campaigns tend to take a while to grow up, and demonstrate burstiness in the late stage. A popular legitimate campaign generates the epidemic effect by making more users tweet about it, spreading to even the larger audience.

Entropy is another temporal property that detects periodic or regular timing of posting patterns in a campaign. In Information Theory, the entropy rate is a measure of the complexity of a random process [19]. A high entropy rate indicates a random process, whereas a low entropy rate indicates a regular one. More theoretical proofs can be found in our previous work [18]. To get relative entropy for every campaign, we normalize entropy values via dividing them by the maximum value of the campaign in the ground truth set. Figure 3 plots the CDF of relative entropy of posting inter-arrivals of both categories. The behavior of auto programs (namely Twitter bots) is often less complicated than that of humans, which can be measured by low entropy rate. In the range between [0.6, 1], the relative entropy of the legitimate category is clearly higher than that of the spam category. The majority of spam campaigns (and a large proportion of their accounts) run auto devices to post, driven by regular or pseudo-random timers. In contrast, tweets in legitimate campaigns are mostly posted by humans. The intrinsic irregularity and complexity of human behavior generates a higher entropy rate. We also find an interesting fact that, a small part of spam campaigns post their tweets manually, generating high entropy. We speculate it is either a form of click farm on Twitter, or some spammers are not professional, and do not run auto programs to tweet.

[5] We re-visit the accounts involved in a spam campaign, and observe that a high proportion of these accounts have been suspended by Twitter.

Finally we discuss a dilemma spammers often face, namely reusing spam accounts. If multiple tweets in the campaign are posted by an account, considering the tweets share the same final URL, the account exhibits the evidence of duplicated posting, which is an indicator of spam. We introduce the account diversity ratio feature. For normalization, this feature is defined as the number of accounts in the campaign over that of tweets. Figure 5 plots the CDF of this feature of both categories. Spammers want to operate accounts in a stealthy way, which requires individual accounts to post few tweets. In reality, it costs effort to get followers to a spam account, and the number of "influential" accounts owned by a spammer is limited. Thus, the spammer tends to repeatedly use accounts to post duplicate spam, causing the low ratio. The figure clearly demonstrates that, the account diversity ratio of legitimate campaigns is much higher than that of spam campaigns. In particular, about 28.8% legitimate campaigns have the ratio as 1, meaning every tweet in the campaign is posted by a unique account. The average ratio of legitimate campaigns is 86.4%, while that of spam campaigns is 45.0%. It further suggests that, legitimate campaigns have stronger account independence than spam campaigns.

4 Classification

In this section, we first present the design philosophy of the classification system. In particular, we formally describe classification features and introduce semantic similarity to detect duplicate content in a campaign. Then, we implement the classifier based on the Random Forest algorithm.

4.1 Classification Features

The classification involves a variety of features, ranging from individual tweet/account levels to a collective campaign level. No single feature is capable of discriminating effectively between spam and legitimate campaigns. Here we introduce these features used in our classification, and later the machine learning algorithm will decide the importance (namely weight) of the features during the training, which is shown in Section 5.1.

Tweet-level Features. We start with tweet-level features, as tweets are the atomic unit of Twitter. A tweet is modeled as the <textual content, original URL> pair.

Spam Content Proportion. Some spam tweets carry explicit spam information, such as "buy Viagra online without a prescription" and "get car loan with bad credit". We create a list of spam words with high frequency on Twitter to capture spam content based on our observation and some existing lists of spam trigger words [5, 2]. The tweet text is tokenized into words which are further checked in the spam word list. This feature is defined as the number of spam words over the total word number in a tweet .

URL Redirection. We develop a Firefox extension to check the original URL in the tweet. If URL redirection is used, it records the final landing URL. By recording the status change in the browser's address bar, the extension logs the whole redirection chain (such as original URL -> intermediate URL -> ... -> final URL). Besides the binary redirection flag, hop number also serves as a useful feature. Spammers tend to use multi-hop redirection to hide spam origins and avoid URL blacklists.

URL Blacklisting. We check the final URL in five blacklists including Google Safe Browsing, PhishingTank, URIBL, SURBL, and Spamhaus. Google Safe Browsing checks URLs against Google's constantly updated lists of suspected phishing and malware pages. PhishingTank focuses on phishing websites. The mechanisms of URIBL, SURBL and Spamhaus are similar. They contain suspicious websites that have appeared in spam emails. If the URL appears in any of the blacklists, the feature is set as true. As the tweets in a campaign share the same final URL, this operation only needs to be performed once.

Account-level Features. We also collect data of Twitter accounts involved in a campaign by calling Twitter's REST API [10], and present account-level features to characterize accounts.

Account Profile. An account has a self-introduction profile consisting of a short description text and homepage URL. We check whether the description contains spam or the URL is blacklisted.

Social Relationship. Tweets of an account can only be delivered to its followers. To achieve a wide influence, the spammer needs to accumulate a large number of followers. However, normal users are unlikely to follow spam accounts. A common trick shared by spammers is following a great number of users (either targeted or randomly selected), and expecting some of them to follow back. Many spam victims blindly follow back "spammer friends" without carefully checking those suspicious accounts. For an account, we calculate its friend count, follower count, and the ratio between them.

Account Reputation. Extended from the previous feature, we have observed that users are likely to follow "famous" accounts. This feature is calculated and normalized as $follower\ count/\ (follower\ count + friend\ count)$. A celebrity usually has many followers and few friends[6], and its reputation is close to 1. However, for a spammer with few followers and many friends, its reputation is close to 0.

Account Taste. Intuitively, the account chooses whom to follow (namely, friends), and this reflects its "taste". If it follows spammers, its "taste" is bad. By doing this, it helps spread spam to more users, making itself a "supporter" of spammers. This feature is defined as average *Account Reputation* of all the friends of the account.

Lifetime Tweet Number. Spam accounts may get suspended for aggressively posting spam. Due to the short lifetime, averagely spam accounts may post fewer

[6] For example, @Yankees, the official Twitter account of New York Yankees, has 400,000 followers and only 29 friends.

tweets. This feature shows the number of tweets an account has posted in lifetime when it is visited by our crawler.

Account Registration Date. Spammers may frequently create new accounts to replace suspended ones. Many spam accounts in our measurement have been created recently.

Account Verification. Twitter verifies accounts for celebrities and organizations. It is difficult for spammers to acquire verified accounts. This binary feature shows whether the account is verified or not.

Account Protection. For user privacy, an account that opts in the protection option makes its tweets invisible to general public, and only visible to approved followers. The option conflicts with the purpose of spreading spam to the wide audience, and may not be adopted by spam accounts.

Campaign-level Features. Collective features may reveal the characteristics of spam campaigns that cannot be observed through individual features. At last we present the campaign-level features as follows. The features of the account diversity ratio, the original URL diversity ratio, the affiliate link number and the entropy of inter-arrival timing have been explained in Section 3.4.

Hashtag Ratio. Spammers often hijack trending hashtags and append them to unrelated spam tweets to increase the chance of being searched and displayed. The feature is defined as the number of hashtags in the tweets over the number of tweets of the campaign.

Mention Ratio. Another trick spammers often play is using @mention to deliver spam to targeted users even without the existing social relationship. The feature is defined as the number of mentions in the tweets over the number of tweets of the campaign.

Content Self-similarity Score. A spam campaign may contain similar tweets created by spam content templates. Users in a legitimate campaign usually contribute content individually, and may not show a strong self-similarity. This feature measures the content self-similarity of the campaign. The details are presented in Section 4.2.

Posting Device Makeup. Twitter supports a variety of channels to post tweets, such as web, mobile devices, and 3rd-party tools. The 8 million tweets in our campaign dataset are posted by 44,545 distinct devices. In the perspective of behavior automation, they can be divided into two categories: manual and auto devices. Manual devices require direct human participation, such as tweeting via web browser or smart-phone. Auto devices are piloted programs that automatically perform tasks on Twitter, and require minimum human participation (such as importing Twitter account information). We manually label the top 100 devices as manual or auto, and use the tdash's API [8] to process the rest. In the campaign dataset, around 62.7% of tweets are posted by manual devices, and the rest 37.3% by auto devices. For every campaign, the script checks its posting devices against the labeled device list, and calculates the proportions of manual and auto devices as the value of posting device makeup.

4.2 Content Semantic Similarity

Spammers may use content templates to create similar spam tweets. Calculating semantic similarity can detect duplicate or similar content in multiple tweets in the campaign. The calculation is challenging as short messages like tweets do not carry as many semantic features as long texts (i.e. email bodies). Our work applies the Vector Space Model [29] that converts tweet texts into vectors, and then calculates the cosine similarity between them. Equation 1 denotes the cosine similarity between two n-dimensional vectors, A and B.

$$cos\_sim = \frac{A \bullet B}{\|A\|\|B\|} = \frac{\sum_{i=1}^{n} A_i \times B_i}{\sqrt{\sum_{i=1}^{n} A_i^2} \times \sqrt{\sum_{i=1}^{n} B_i^2}} \tag{1}$$

For implementation, we use SenseClusters, an open-source program [4], that clusters text messages based on contextual similarity. Given the set of tweets in the campaign, we treat it as a text corpus, and generate a vocabulary by extracting distinct words from the corpus. Then we generate an occurrence matrix with tweets as rows, and words in the vocabulary as columns. The value of $cell_{ij}$ is the TF-IDF (Term Frequency - Inverse Document Frequency) weight [15], which represents the occurrence frequency of $word_j$ in $tweet_i$. As the most intuitive approach, 1st-order similarity detects the number of exact words shared (or overlapped) between tweets. Because spam templates often adopt synonym interchanging for the purpose of obfuscation, our work applies 2nd-order similarity to measure similar tweets. Its general idea is to replace the context with something else that will still represent it, and yet likely provide more information from which similarity judgments can be made [28]. Given the tweet corpus, SenseClusters divides N tweets into K clusters based on the semantic sense on the fly.

We design Equation 2 to measure the self-similarity of the campaign's tweet content.

$$self\_sim\_score = \sum_{i=1}^{K} \frac{cluster_i\_size^{w1} * cluster_i\_sim^{w2}}{K^{w3}}, \tag{2}$$

where K is the number of semantic clusters in the campaign, and $w1$ to $w3$ are weight factors with their tuning procedure presented in Section 5.1.

4.3 Machine Learning Classifier

Our classification problem can be defined as follows. Given a campaign, $c =< u, T, A >$, the classifier determines c as a either spam or legitimate campaign. We choose Random Forest [17] as the machine learning algorithm[7], and train the classifier to make the binary decision. Random Forest serves as an ensemble classifier that includes multiple decision trees. The algorithm combines the bagging idea in [17] and random feature selection in [23] to construct a "forest" of decision trees with controlled variation. Suppose the training set contains M features, and each decision tree only uses $m(<< M)$ features to reach the decision. For classification, an unknown sample is pushed down the tree, and assigned with the class of the leaf node where the sample ends up. More details about decision tree can be found in [25]. Given a specific sample, every decision tree makes a classification decision (either spam or legitimate campaign in our case), and Random Forest applies the majority voting of all the trees to reach the final decision.

5 Evaluation

In this section, we first train the classifier. Then, we evaluate the accuracy of our classification system based on the ground truth set.

5.1 Training

As described in Section 3.3, our ground truth set consists of manually labeled campaigns. More specifically, 744 spam campaigns contain around 70,000 accounts and 131,000 tweets, whereas 580 legitimate campaigns contain around 150,000 accounts and 180,000 tweets.

Before training the classifier, we need to determine the content self-similarity feature by tuning the weight factors in Equation 2 with the following method. We choose Decision Tree as the tuner, and the feature represented by the self-similarity score as the only classification feature. We try different combinations of numeric values of $w1$ to $w3$. In every test round, a combination generates a different self-similarity score for a campaign in the ground truth set. The decision tree associates the self-similarity feature with the root as it is the only feature in the classification, and calculates the best split between spam and legitimate campaigns. The combination of ($w1 = 0.8, w2 = 0.5, w3 = 1$) generates the highest overall accuracy on the ground truth set, and is chosen for Equation 2. According to the algorithm of the SenseClusters, the cluster similarity score assigned is also no greater than 1. Note that $w1$ and $w2$ are decimal fractions, and they add more weights to the cluster size and cluster similarity. Furthermore, $w2$ makes cluster similarity more important than cluster size, as $w2$ is less than $w1$.

5.2 Cross Validation

By calculating the values of the features described in Section 4.1, a feature vector is generated for each campaign. Weka supports a collection of machine learning algorithms for classification, including mainstream categories of Bayes, trees and

[7] The reason is explained in Section 5.2

Table 1. Algorithm Performance Comparison

Feature	Accuracy (%)	FPR (%)	FNR (%)
Random Forest	94.5	4.1	6.6
Decision Table	92.1	6.7	8.8
Random Tree	91.4	9.1	8.2
KStar	90.2	7.9	11.3
Bayes Net	88.8	9.6	12.4
SMO	85.2	11.2	17.6
Simple Logistic	84.0	10.4	20.4
Decision Tree	82.8	15.2	18.8

so on [22]. We try multiple algorithms in each category, list and compare performance results for the top classifiers with accuracy greater than 80% in Table 1. For each classifier, we use Cross Validation with ten folds to train and test it over the ground truth set [26]. The dataset is randomly partitioned into ten complementary subsets with equal size. In each round, one out of ten subsets is retained as the test set to validate the classifier, while the remaining nine subsets are used as the training set to train the classifier. The individual results from ten rounds are averaged to generate the final estimation.

Table 1 lists three metrics for evaluating the classification performance sorted on accuracy. Considering the confusion matrix with spam campaigns as positive cases, *Accuracy* is the proportion of samples that are correctly identified, *False Positive Rate* (FPR) is the proportion of negatives cases that are incorrectly classified as positive, and *False Negative Rate* (FNR) is the proportion of positives cases that are incorrectly classified as negative. During evaluation, we expect to constrain the FPR low at the cost of accepting the medium FNR. Classifying benign campaigns as spam upsets legitimate users, while missing a small part of spam campaigns is tolerable. Random Forest achieves the highest accuracy, lowest FPR and FNR, and hence is selected as the final classifier for our dataset.

Some features play a more important role than others during the classification. Subsequently, we attempt to evaluate the discrimination weight each feature has. Similar to the tuning method for Equation 2, in each test, we use only one feature to independently cross validate the ground truth set with Decision Tree[8]. The one with the highest accuracy may be considered as the most important feature. Table 2 presents the performance results of the top 10 features, which are also sorted on accuracy. The Account Diversity Ratio feature has the highest accuracy at 85.6%. Technically this one is not difficult to bypass, because spammers could use a large amount of accounts to distribute the workload and lower the ratio. However, spam accounts with limited normal followers cannot generate the satisfying propaganda. We speculate that, in reality, spammers tend to repeatedly use "influential" accounts to deliver spam to a wide audience. The

[8] Random Forest transforms to Decision Tree in the case of single-feature classification. There is only one decision tree to build, and the single feature is associated with its root.

Table 2. Feature Performance Comparison

Feature	Accuracy (%)	FPR (%)	FNR (%)
Account Diversity Ratio	85.6	16.2	13.0
Timing Entropy	83.0	9.5	22.8
URL Blacklists	82.3	3.2	29.0
Avg Account Reputation	78.5	25.6	18.3
Active Time	77.0	16.2	28.3
Affiliate URL No	76.7	9.6	34.0
Manual Device %	74.8	10.3	36.8
Tweet No	75.4	28.6	21.5
Content Self Similarity	72.3	33.7	23.0
Spam Word Ratio	70.5	25.8	32.4

Timing Entropy feature captures the intrinsic complexity of human behavior, that is difficult for bot accounts to bypass. However, many spam campaigns involve manual accounts (probably in the form of click farm), that generate the high FNR at 22.8% for the feature.

We are particularly interested in the performance of the URL Blacklist feature, as it is used as the only feature for spam campaign detection in some existing work [21]. We present the performance comparison between our work based on Random-Forest-based classifier that applies multiple features and the previous work based on the single blacklist feature. Blacklists are haunted by the inevitable lag effect, and cannot include all spam sites "in-the-wild". Besides, blacklists cannot detect duplicate spamming over multiple accounts. These factors generate a high FNR at 29.0%. By using multi-dimensional features, our classifier manages to capture more spam campaigns that would have been missed by the blacklist feature, and lowers the FNR to 6.6%. The low FPR of the blacklist feature is caused by the fact that, some blacklists only check the hostname of URL, and mis-classify some benign web pages hosted by the blacklisted websites. The FPR of our approach (4.1%) is slightly higher than that of the blacklist feature (3.2%). Most importantly, our approach improves the accuracy from 82.3% to 94.5%.

6 Conclusion

Spam haunts social networks, as social relationship facilitates spam spreading. Conventional spam detection methods check individual accounts or messages for the existence of spam. In this paper, we exploit the collective detection approach to capturing spam campaigns with multiple accounts. Our work uses the features combining both content and behavior to distinguish spam campaigns from legitimate ones, and build an automatic classification framework. Our work can be applied to other social networks by integrating application-specific features. Spam detection is an endless cat-and-mouse game. As spamming methods may evolve in the future, some features may be added or replaced with new ones, and the classifier should also be re-trained with the up-to-date ground truth dataset.

References

[1] Google safe browsing api, http://code.google.com/apis/safebrowsing/ (accessed: August 27, 2011)

[2] The list of email spam trigger words, http://blog.hubspot.com/blog/tabid/6307/bid/30684/The-Ultimate-List-of-Email-SPAM-Trigger-Words.aspx (accessed: April 15, 2012)

[3] Phishtank, join the fight against phishing, http://www.phishtank.com/ (accessed: August 27, 2011)

[4] Senseclusters, http://senseclusters.sourceforge.net/ (accessed: September 2, 2011)

[5] Spam words by wordpress, http://codex.wordpress.org/Spam_Words (accessed: April 15, 2012)

[6] The spamhaus project, http://www.spamhaus.org/ (accessed: August 27, 2011)

[7] Surbl, http://www.surbl.org/lists (accessed: August 27, 2011)

[8] tdash's api of twitter applications statistics, http://tdash.org/stats/clients (accessed: September 6, 2011)

[9] Twitter blog: Your world, more connected, http://blog.twitter.com/2011/08/your-world-more-connected.html (accessed: August 17, 2011)

[10] Twitter rest api resources, https://dev.twitter.com/docs/api (accessed: August 30, 2011)

[11] The twitter rules, http://support.twitter.com/entries/18311-the-twitter-rules (accessed: August 17, 2011)

[12] Twitter's streaming api documentation, https://dev.twitter.com/docs/streaming-api (accessed: August 30, 2011)

[13] Uribl, realtime uri blacklist, http://www.uribl.com/about.shtml

[14] Using the twitter search api, https://dev.twitter.com/docs/using-search (accessed: August 30, 2011)

[15] Aizawa, A.: The feature quantity: an information theoretic perspective of tfidf-like measures. In: Proceedings of the 23rd Annual International ACM SIGIR Conference on Research and Development in Information Retrieval, pp. 104–111 (2000)

[16] Benevenuto, F., Magno, G., Rodrigues, T., Almeida, V.: Detecting spammers on twitter. In: Proceedings of the CEAS 2010 (2010)

[17] Breiman, L.: Random forests. Machine Learning 45, 5–32 (2001)

[18] Chu, Z., Gianvecchio, S., Wang, H., Jajodia, S.: Who is tweeting on twitter: human, bot or cyborg? In: Proceedings of the 2010 Annual Computer Security Applications Conference, Austin, TX, USA (2010)

[19] Cover, T.M., Thomas, J.A.: Elements of information theory. Wiley Interscience, New York (2006)

[20] Gao, H., Hu, J., Wilson, C., Li, Z., Chen, Y., Zhao, B.Y.: Detecting and characterizing social spam campaigns. In: Proceedings of the 10th Annual Conference on Internet Measurement, pp. 35–47 (2010)

[21] Grier, C., Thomas, K., Paxson, V., Zhang, M.: @spam: the underground on 140 characters or less. In: Proceedings of the 17th ACM Conference on Computer and Communications Security, pp. 27–37 (2010)

[22] Hall, M., Frank, E., Holmes, G., Pfahringer, B., Reutemann, P., Witten, I.H.: The weka data mining software: an update. SIGKDD Explor. Newsl. 11, 10–18 (2009)

[23] Ho, T.K.: The random subspace method for constructing decision forests. IEEE Transactions on Pattern Analysis and Machine Intelligence 20, 832–844 (1998)
[24] Kanich, C., Kreibich, C., Levchenko, K., Enright, B., Voelker, G.M., Paxson, V., Savage, S.: Spamalytics: an empirical analysis of spam marketing conversion. Commun. ACM 52, 99–107 (2009)
[25] Kohavi, R., Quinlan, R.: Decision tree discovery. In: Handbook of Data Mining and Knowledge Discovery, pp. 267–276. University Press (1999)
[26] McLachlan, G., Do, K., Ambroise, C.: Analyzing microarray gene expression data. Wiley (2004)
[27] Ntoulas, A., Najork, M., Manasse, M., Fetterly, D.: Detecting spam web pages through content analysis. In: Proceedings of the 15th International Conference on World Wide Web, pp. 83–92 (2006)
[28] Pedersen, T.: Computational approaches to measuring the similarity of short contexts: A review of applications and methods. CoRR, abs/0806.3787 (2008)
[29] Salton, G., Wong, A., Yang, C.S.: A vector space model for automatic indexing. Commun. ACM 18, 613–620 (1975)
[30] Stringhini, G., Kruegel, C., Vigna, G.: Detecting spammers on social networks. In: Proceedings of the 26th Annual Computer Security Applications Conference (2010)
[31] Xie, M., Yin, H., Wang, H.: An effective defense against email spam laundering. In: Proceedings of the 13th ACM Conference on Computer and Communications Security, pp. 179–190 (2006)

A New Framework for Privacy of RFID Path Authentication

Shaoying Cai[1], Robert H. Deng[1], Yingjiu Li[1], and Yunlei Zhao[2]

[1] Singapore Management University, 80 Stamford Road, Singapore 178902
{shaoyingcai.2009,robertdeng,yjli}@smu.edu.sg
[2] Fudan University, No. 825 Zhangheng Road, Shanghai, China 201203
ylzhao@fudan.edu.cn

Abstract. RFID-based path authentication enables supply chain managers to verify the exact path that a tag has taken. In this paper, we introduce a new oracle Move that models a tag's movement along a designed or an arbitrary path in a supply chain. With this oracle, we refine the existing security and privacy notions for RFID-based path authentication. In addition, we propose a new privacy notion, called path privacy, for RFID-based path authentication. Our privacy notion captures the privacy of both tag identity and path information in a single game. Compared to existing two-game based privacy notions, it is more rigorous, powerful, and concise. We also construct a new path authentication scheme. Our scheme does not require the entities in a supply chain to have any connection with each other except in the initial stage. It requires only 480 bits storage and no computational ability on each tag; thus it can be deployed on the standard EPCglobal Class 1 Generation 2 tags in the market.

1 Introduction

Supply chain is a network of multiple parties such as suppliers, transporters, storage facilities, distributors, and retailers that participate in the production, delivery, and sale of product [5]. It is difficult to monitor a supply chain since the involving parties are distributed at multiple locations or even across countries. So that supply chains are vulnerable to the counterfeiting problem, where an adversary injects fake goods into a supply chain. The counterfeiting problem has become a major threat to supply chains. According to the 2011 report of International Chamber of Commerce, it is estimated that the counterfeiting accounts for 5-7% of world trade, or about 600 billion U.S. dollars per year [6]. The ratio of counterfeiting is even higher in luxury market.

Radio Frequency IDentification (RFID) technology has been recently used to facilitate real-time monitoring of supply chains so as to thwart counterfeiting threats. In general RFID-enabled supply chains, each item is attached with a tag. The tag stores identity information of the item. A reader identifies an

F. Bao, P. Samarati, and J. Zhou (Eds.): ACNS 2012, LNCS 7341, pp. 473–488, 2012.

item through the interaction with the corresponding tag. Various tag authentication schemes (e.g.[2,5,8,9,10,11,12,13]) have been proposed to enable privacy-preserving identification of tags.[1] However, most of proposals require tags to have certain computational capability, which may incur unbearable cost in practice. Another common problem of deploying existing solutions in supply chain is that: to monitor a supply chain, the manager should have access to all the databases of the entities in the supply chain. This requires high-quality network connection and fine-grained access control, which may not be realistic in practice.

Recently, RFID-enabled path authentication was proposed by Blass, Elkhiyaoui and Molva [3,4], and extended later to be more practical [6,14], to tackle the counterfeiting problem in supply chains. In the proposal, which is named as TRACKER, the manager of a supply chain verifies the genuineness of tag by checking whether it has been processed by a series of reliable readers. Compared to the existing tag authentication schemes [2,5,8,9,10,11,12,13], the verification of a tag's genuineness is merely based on the credentials stored on the tag about the readers that have processed the tag along the path. TRACKER can be implemented with standard EPCglobal C1 G2 tags, which has several hundred bits storage and no computational ability. It does not require the entities in the supply chain to have any connection except in the initial stage. In this paper, we refine the privacy notions for path authentication and propose a more practical path authentication scheme. Our contributions include:

- We analyze the existing security and privacy notions for path authentication in RFID-enabled supply chain, including tag unlinkability and step unlinkability. We show that these two notions can be further refined to be more concise and formal.
- We propose a combined privacy notion that considers both tag unlinkability and step unlinkability for RFID-enabled supply chains. We analyze the relations among our new privacy notion, the tag unlinkability notion and the step unlinkability notion. We prove that our privacy notion implies tag unlinkability and step unlinkability.
- We propose a new path authentication solution using the standard EPC Class 1 Gen 2 tags without sharing the secret among supply chain parties. Compare to TRACKER, our solution is more efficient and requires less storage. We prove that our solution satisfies the security notion and the privacy notion.

2 Background

First, we model an RFID-enabled supply chain management system and the adversary in the system. Then we refine the security and privacy notions for RFID-enabled path authentication in supply chains.

[1] Most of the existing tag authentication schemes and their extensions are listed on http://www.avoine.net/rfid/

2.1 RFID-Enabled Supply Chain Management System

Supply chain is a network of multiple parties, which can be represented by a digraph $G = (V, E)$, where V is a set of vertices, E is a set of edges. Each vertex $v \in V$ represents one *step* in the supply chain. Note that each supply chain party may conduct several steps to process an item. Each directed edge $e \in E$, $e = \overrightarrow{v_i v_j}$, denotes that v_j is a possible next step to step v_i in the supply chain. A *path* is a finite sequence of steps $P = (v_0, \cdots, v_l)$, where $(v_i, v_{i+1}) \in E$, for $i \in \{0, l-1\}$. Every path shares the same source v_0. The last step v_l of a valid path $P_{valid_i} = (v_0, \cdots, v_l)$ represents a *check point*. Every item enters the supply chain from v_0, and goes through a path according to its own procedure. When it arrives at the check point, the manager will verify the item. Note that if a path consists of an empty set of steps (except v_0), we call it empty path, and denote it as "–".

An RFID-enabled supply chain system consists of an issuer I, a set of managers $\mathcal{M}$ and a set of normal readers $\mathcal{R}$. The issuer I is located at the source v_0 of the supply chain; a managers from $\mathcal{M}$ is placed at the end of each valid path and normal readers from $\mathcal{R}$ are placed at other places of a supply chain. The issuer I initializes a tag by storing certain information on the tag. While a tag goes through the supply chain, each reader in its path updates the content of the tag. Eventually, the tag arrives at a manager, the manager reads out the content of the tag and checks the validity of the tag. Formally, the system has the following functions:

- $\mathsf{Initialize}(\kappa)$: Given the security parameter κ, the system prepares a supply chain G, an issuer I and a set of l managers $\mathcal{M}$, a set of m readers $\mathcal{R}$ and a set of n tags $\mathcal{T}$, and a set of ν valid path $\mathcal{P}_{valid}$. We denote the content stored on any tag T_i as state S_{T_i}.
- $\mathsf{Read}(T_i)$: a function that returns back the current internal state S_{T_i} of T_i.
- $\mathsf{Write}(T_i)$: a function that writes a new state S'_{T_i} to T_i. Here we assume that the readers in each step are honest, that is, they update a tag only if the tag is authenticated.
- $\mathsf{PathCheck}(S^j_{T_i})$: a function that verifies whether tag T_i has gone through a valid path P_{valid}. If it is the case, it returns the valid path P_{valid}, else it returns $\emptyset$.

2.2 Adversary Model

We use the following the notations. If $A(\cdot, \cdot, \cdots)$ is a randomized algorithm, then $y \leftarrow A(x_1, x_2, \cdots; \rho)$ means that y is assigned with the unique output of algorithm A on inputs x_1, x_2, $\cdots$ and coins ρ, while $y \leftarrow A(x_1, x_2, \cdots)$ is a shorthand for first picking ρ at random and then setting $y \leftarrow A(x_1, x_2, \cdots)$. $y \leftarrow A^{O_1, \cdots, O_n}(x_1, x_2, ...)$ denotes that y is assigned with the output of algorithm A which takes $x_1, x_2, ...$ as inputs and has oracle accesses to $O_1, ..., O_n$. If S is a set, then $s \in_R S$ indicates that s is chosen uniformly at random from set S. Let $\Pr[E]$ denote the probability that an event E occurs. Let $\mathcal{N}$ denote the set of all integers. Let $\mathcal{R}$ denote the set of all real numbers. A function $f : \mathcal{N} \rightarrow \mathcal{R}$ is said to be *negligible* if for every $c > 0$ there exits a number $n_0 \in N$ such that $f(n) < \frac{1}{n^c}$ holds for all $n > n_0$.

An adversary $\mathcal{A}$, against RFID path authentication, is given accesses to four oracles $\mathcal{O} = \{O_1, O_2, O_3, O_4\}$. O_1, O_2, O_3 denote Read, Write, PathCheck functions, respectively. O_4 denotes a function Move($T_i, k, \mathcal{K}, b$), where $k \in \mathcal{N}$, $\mathcal{K} \in \{P, G\}$, $b \in \{0, 1\}$. Move ($T_i, k, \mathcal{K}, b$) is defined as follows:

- If $\mathcal{K} = G$, no matter whether $b = 0$ or $b = 1$, starting from the current step of T_i with internal state $S^j_{T_i}$, move the tag T_i forward $k \geq 1$ steps arbitrarily in the supply chain system G.
- If $\mathcal{K} = P$, works as follows: If $b = 1$, from the current step of T_i with internal state $S^j_{T_i}$, move the tag T_i forward $k \geq 1$ steps through the designated path $\mathcal{P}$ (the length of P is at least k steps). If $b = 0$, move tag T_i forward $k \geq 1$ steps according to any path that does not have a common step with P. The reader in each step updates the tag's state. Finally, Move(T_i, k, P, b) returns back the state transcript $\{S^{j+1}_{T_i}, \cdots, S^{j+k}_{T_i}\}$ of T_i from step $j + 1$ to $j + k$.

Note that oracle O_4 is a new oracle introduced in this paper. It is critical to precisely model various kinds of tag movement. In [4,3], the concept of path is not explicitly defined, and the operations on tag movement are specified through step-level oracles; thus, it is difficult to describe the tag movement at path level. While using O_4, any tag movement can be precisely represented by adjusting the parameters of **Move** function. The introducing of O_4 facilitates defining clear security and privacy notions.

The four oracles capture the adversary's ability to read from a tag, write into a tag, check the validity of a tag, and follow a tag through a designated path P (for the case of $\mathcal{K} = P$) or simply update the state of the tag by forwarding it arbitrarily in the system G (for the case of $\mathcal{K} = G$). We denote by $\mathcal{A}^{\mathcal{O}}(para)$ a probabilistic polynomial-time (PPT) algorithm $\mathcal{A}$ that, on input of some system public parameters $para$, runs a supply chain system via the four oracles in $\mathcal{O}$. An adversary is a (t, n_1, n_2, n_3, n_4)-adversary if it works in time t and makes oracle queries to O_μ without exceeding n_μ times, where $1 \leq \mu \leq 4$.

2.3 Existing Security and Privacy Notions

Security Notion. The security goal of our system is to prevent an adversary from inserting counterfeited goods to the supply chain. As the manager checks the authenticity of a tag merely based on the state stored on a tag, the system should prevent an adversary from forging a tag's internal state with a valid path that has not been actually taken by the tag in the supply chain. Since standard EPC C1 G2 tags have no computation capability, no reader authentication is performed. If a tag's state has been changed by an adversary, even if it has gone through a valid path, it is not considered as a valid tag by a manager.

The security for RFID path authentication means, it is infeasible for any probabilistic polynomial-time adversary $\mathcal{A}$ to create a state $S^l_{T_i}$ for a tag T_i

such that given $S^l_{T_i}$, a manager M outputs a valid path $P_{valid} = \{v_0, \cdots, v_l\}$ which T_i has not gone through. It is formalized by an experiment $\mathbf{Exp}_{\mathcal{A}}^{\mathbf{Security}}[\kappa]$ shown in Figure 1. The adversary $\mathcal{A}$ consists of two algorithms $\mathcal{A}_1$ and $\mathcal{A}_2$ which run in two phases, learning phase and challenge phase. Firstly, given parameter κ, the experiment setups the system through $\mathsf{Setup}(\kappa)$, and passes the public system parameters *para* to $\mathcal{A}_1$. In the learning phase, $\mathcal{A}_1$ is allowed to collect information by querying the four oracles without exceeding n_1, n_2, n_3, n_4 times, respectively. Then it generates a transcript *st* which contains the information about the system it gathered during the learning phase. In the challenge phase, $\mathcal{A}_2$ creates a tag T with state s^j_T using *st*. The tag T may have a new ID or an existing ID in the system. Then the game checks the validity of T_i through $\mathsf{Check}(s^j_T)$. $\mathbf{Exp}_{\mathcal{A}}^{\mathbf{Security}}[\kappa]$ outputs 1 if both of the following two conditions hold: $\mathsf{Check}(s^j_T)$ returns a valid path P_{valid}; and there exists $z \in \{1, \ldots, l\}$ such that the tag has not passed v_z in its z-th step, where l denotes the length of the path and v_z denotes the z-th step in P_{valid}. $\mathbf{Exp}_{\mathcal{A}}^{\mathbf{Security}}[\kappa]$ outputs 0, otherwise.

Experiment $\mathbf{Exp}_{\mathcal{A}}^{Security}[\kappa]$
1. run $\mathsf{Setup}(\kappa)$ to setup $I, \mathcal{R}, \mathcal{T}, \mathcal{M}$.
2. $\{st\} \leftarrow \mathcal{A}_1^{\mathcal{O}}(para)$. // the learning phase
3. $T \leftarrow \mathcal{A}_2(st)$. //the challenge phase
4. $s^j_T \leftarrow \mathsf{Read}(T)$.
5. output 1, if $\mathcal{P}_{valid} \longleftarrow \mathsf{PathCheck}(S^j_T)$,
   and there is a step $v_z \in P_{valid}$ which T has not gone through in its z-th step; output 0, otherwise.

Fig. 1. Security Experiment

Definition 21. *The advantage of $\mathcal{A}$, denoted $Adv_{\mathcal{A}}^{Security}(\kappa)$, in the security experiment is*

$$\left|\Pr[\mathbf{Exp}_{\mathcal{A}}^{\mathbf{Security}}[\kappa] = 1]\right|$$

Definition 22. *We say an RFID path authentication scheme is $(t, n_1, n_2, n_3, n_4, \epsilon)$-secure, if for any t-time adversary $\mathcal{A}$ who makes at most n_1, n_2, n_3, n_4 queries to O_1, O_2, O_3, O_4 respectively, $Adv_{\mathcal{A}}^{Security}(k) < \epsilon$ holds. The probability is taken over coins of $\mathcal{A}$ and the oracles.*

Privacy Notions. For an RFID-enabled supply chain system, Blass, Elkhiyaoui and Molva [3] considered two privacy notions: tag unlinkability and step unlinkability. Tag unlinkability corresponds to the privacy of a tag's identity. Step unlinkability corresponds to the privacy of a tag's path. Note that in the older version of TRACKER [4], there is another path privacy notion, namely path unlinkability, which is proven to be weaker than step unlinkability in [3].

Tag Unlinkability Briefly, tag unlinkability requires that no efficient adversary can link the state information stored in a tag to the tag's identity. In [4], the tag unlinkability is defined through a formal experiment. The experiment contains two phases: the learning phase and the challenge phase. An adversary $\mathcal{A}$ is provided with two tags T_0 and T_1. In the learning phase, the adversary can access the system and gather information without exceeding the constraints set by the game. In the challenge phase, the game updates the tags by moving them one more step further in the supply chain. The experiment then flips a coin $\delta \in_R \{0,1\}$, and provides the updated state of T_δ to the adversary. The adversary guesses the value of δ. The adversary wins the game if it can successfully guess δ with probability 1/2 plus a non-negligible quantity.

We slightly modify the experiment to $Exp_{\mathcal{A}}^{Tag-Unlinkability}[\kappa]$. In the learning phase, the adversary is allowed to access the oracles O_1, O_2, O_3, O_4 without exceeding n_1, n_2, n_3, n_4 times, respectively. Then, the adversary outputs two tags T_0 and T_1 together with a transcript st, where st is the information it has gathered. In the challenge phase, the experiment tosses a coin $\delta \in_R \{0,1\}$. The experiment moves the tag T_δ one step forward arbitrarily in the system G, and provides the updated state S_δ of tag T_δ to the adversary. With S_δ and the transcript st, the adversary guesses the value of δ, then outputs the guessed value δ'. If $\delta = \delta'$, the experiment outputs 1; else, the experiment outputs 0. The adversary wins the game if the experiment outputs 1 with probability 1/2 plus a non-negligible quantity.

A key difference between the original tag unlinkability notion [3] and our refined one is that, in the original notion, the challenge tags T_0 and T_1 are selected by the experiment, while in our notion, the challenge tags T_0 and T_1 are selected by the adversary; therefore, the adversary in our notion is stronger then the adversary in [3]. We depict $Exp_{\mathcal{A}}^{Tag-Unlinkability}[\kappa]$ in Figure 2.

Experiment $\mathbf{Exp}_{\mathcal{A}}^{Tag-Unlinkability}[\kappa]$
1. run $\mathsf{Setup}(\kappa)$ to setup $I, \mathcal{R}, \mathcal{T}, \mathcal{M}$.
   Denote by $para$ the public system parameter.
2. $\{T_0, T_1, st\} \leftarrow \mathcal{A}_1^{\mathcal{O}}(para)$.
3. $\delta \longleftarrow_R \{0,1\}$.
4. $S_\delta \leftarrow$Move$(T_\delta, 1, G, 1)$, i.e., move T_δ one step arbitrarily forward in the system G.
   Denote by $\mathcal{S}_\delta$ the updated state of T_δ.
5. $\delta' \leftarrow \mathcal{A}_2^{\mathcal{O}}(S_\delta, st)$.
6. output 1 if $\delta' = \delta$, 0 otherwise.

Fig. 2. Tag Unlinkability Experiment

Definition 23. *The advantage of $\mathcal{A}$, denoted $Adv_{\mathcal{A}}^{Tag-Unlinkability}(\kappa)$, in the tag unlinkability experiment is* $\left|\Pr[\mathbf{Exp}_{\mathcal{A}}^{\mathbf{Tag-Unlinkability}}[\kappa] = 1] - \frac{1}{2}\right|$

Definition 24. *An RFID path authentication scheme is $(t, n_1, n_2, n_3, n_4, \epsilon)$-tag-unlinkable, if for any t-time adversary $\mathcal{A}$ who makes at most n_1, n_2, n_3, n_4 queries to O_1, O_2, O_3, O_4, respectively, we have $Adv_{\mathcal{A}}^{Tag-Unlinkability}(\kappa) < \epsilon$. The probability is taken over the choice of δ, coins of $\mathcal{A}$ and the oracles.*

Step Unlinkability Step unlinkability requires that no efficient adversary is feasible to tell whether the two paths of any two different tags share a common step or not. In [3], the step unlinkability game is defined as follows. Firstly, the experiment randomly chooses a tag T for the adversary. In the learning phase, the adversary arbitrarily queries the oracle without exceeding the constraints. The adversary may gather information from the system. It may follow T, so that it knows the path of the targeted tag. In the challenge phase, the experiment provides the adversary with another tag T_c, the adversary lets T_c move forward along its path for several steps and then reads the state of T_c. Finally, the adversary is asked to guess whether T and T_c have a step in common besides v_0. The adversary breaks the path privacy if the probability of correct guessing is non-negligibly more than $\frac{1}{2}$.

The above experiment defined in [4] is based on the assumption that every tag passes through every step with the same probability. However, given a tag, in case that the probabilities of the tag to pass by different steps are not even, then, an adversary can trivially win the game. We give an example to illustrate the situation. Suppose there are four paths in the system, P_a, P_b, P_c and P_d and every tag will go through the fours paths with equal probability. P_a, P_b, P_c shares a common step v besides v_0, while P_d have no common step with the other three paths besides v_0. In case that the adversary learns that tag T has gone through path P_a, for any T_c the probability that it has a common step v with T is 75%. Thus the adversary will win the game with non-trivial advantage.

We modify the step unlinkability experiment to make it more rigorous. The new step unlinkability experiment $Exp_{\mathcal{A}}^{Step-Unlinkability}[\kappa]$ is shown in Figure 3. The experiment starts by setting the system $I, \mathcal{R}, \mathcal{T}, \mathcal{M}$ through $\mathsf{Setup}(\kappa)$. An adversary $\mathcal{A}$ runs two algorithms $\mathcal{A}_1$ and $\mathcal{A}_2$, respectively in the two phases. In the learning phase, $\mathcal{A}_1$ queries the oracle set $\mathcal{O}$ and outputs a tag T and transcript st. In the challenge phase, the experiment creates a new tag T_c, and then tosses a coin $\delta \in_R \{0,1\}$. The experiment sets a path P as follows: if $\delta = 0$, the path P does not have any common step with T's path; else the path P have certain common steps with T's path. After getting the path P, the experiment moves T_c along path P in k steps. $\mathcal{A}_2$ reads the state S_{T_c} of T_c, guesses the value of δ, and outputs the guessed value δ'. Note that S_{T_c} contains the states updated by the readers in path P. If the probability of $\delta' = \delta$ is non-negligibly more than $\frac{1}{2}$, the adversary wins the game.

Definition 25. *The advantage of $\mathcal{A}$, denoted $Adv_{\mathcal{A}}^{Step-Unlinkability}(k)$, in the step unlinkability experiment is* $\left|\Pr[\mathbf{Exp}_{\mathcal{A}}^{\mathbf{Step-Unlinkability}}[\kappa] = 1] - \frac{1}{2}\right|$

Definition 26. *An RFID path authentication scheme is $(t, n_1, n_2, n_3, n_4, \epsilon)$-step unlinkable, if for any t-time adversary $\mathcal{A}$ who makes at most n_1, n_2, n_3, n_4 queries to O_1, O_2, O_3, O_4, respectively, we have $Adv_{\mathcal{A}}^{Step-Unlinkability}(k) < \epsilon$. The probability is taken over the choice of δ, coins of $\mathcal{A}$ and oracles.*

Experiment $\mathbf{Exp}_{\mathcal{A}}^{Step-Privacy}[\kappa]$
1. run $\mathsf{Setup}(\kappa)$ to setup $T, \mathcal{R}, \mathcal{T}, \mathcal{M}$. Denote by *para* the public system parameter.
2. $\{T, k, st\} \leftarrow \mathcal{A}_1^{\mathcal{O}}(para)$.
3. create a new tag T_c.
4. randomly selects a bit $\delta \in \{0,1\}$.
5. if $\delta = 0$, selects a path P that dose not have any common step with T" path; else, select a path P that has one or more common steps with T's path. The length of the path is at least k.
6. $S_{T_c} \leftarrow$Move$(T_c, k, P, 1)$.
7. $\delta' \leftarrow \mathcal{A}_2(S_{T_c}, st)$.
8. output 1 if $\delta' = \delta$, 0 otherwise.

Fig. 3. Step Unlinkability Experiment

3 A New RFID Privacy Notion for Path Authentication

In this section, we propose a new privacy notion, named path privacy, for path authentication. This notion captures the privacy of tag identity and path information in a single game. We show that path privacy implies tag unlinkability and step unlinkability.

3.1 Path Privacy

In [3], two privacy notions, tag-unlinkability and step-unlinkability, should be used together to analyze the privacy of a path authentication scheme. These two notions are formulated separately (via four algorithms). We present a single game-based privacy notion, path-privacy, which implies tag unlinkability and step unlinkability.

The experiment $\mathbf{Exp}_{\mathcal{A}}^{\mathbf{Path-Privacy}}[\kappa]$ of path privacy is shown in Figure 4 and formalized as follows. The experiment consists of two phases: the learning phase and the challenge phase. An adversary $\mathcal{A}$ runs two algorithms $\mathcal{A}_1$ and $\mathcal{A}_2$, respectively in the two phases. The experiment sets up the system $I, \mathcal{R}, \mathcal{T}, \mathcal{M}$ through $\mathsf{Setup}(\kappa)$. In the learning phase, $\mathcal{A}_1$ queries the four oracles without exceeding n_1, n_2, n_3, n_4 times, respectively. $\mathcal{A}_1$ outputs two tags T_0, T_1, a path P that has at least k steps left for both tags, and state information st. In the challenge phase, the experiment firstly flips a coin δ. If $\delta = 1$, the experiment moves T_1 k steps along the path P, and T_1 is updated by k readers in the path. Let the state of T_1 be denoted as S_1. If $\delta = 0$, the experiment moves T_0 k steps without going through the path P (T_0 is updated by k readers that are not in the path). Let the state of T_0 be denoted as S_0. The Move operations are performed by the game challenger, and the adversary has no access to the readers and the tag during the Move operations. In the challenge phase, the experiment provides $\mathcal{A}_2$ with S_δ and st. $\mathcal{A}_2$ guesses the value of δ as δ'. If $\delta' = \delta$, the experiment outputs 1; else the experiment outputs 0. If the experiment outputs 1 with probability non-negligibly more than $\frac{1}{2}$, the adversary wins the game.

Experiment $\mathbf{Exp}_{\mathcal{A}}^{Path-Privacy}[\kappa]$ 1. run $\mathsf{Setup}(\kappa)$ to setup $I, \mathcal{R}, \mathcal{T}, \mathcal{M}$. Denote by $para$ the public system parameter. 2. $\{T_0, T_1, P, k, st\} \leftarrow \mathcal{A}_1^{\mathcal{O}}(para)$, where P is a path of length at least k, st is state information. 3. $\delta \leftarrow \{0, 1\}$. 4. $S_\delta \leftarrow$Move(T_δ, k, P, δ). Denote by $\mathcal{S}_\delta$ the state of $\mathcal{T}_\delta$. 5. $\delta' \leftarrow \mathcal{A}_2^{\mathcal{O}}(S_\delta, st)$. 6. output 1 if $\delta' = \delta$, 0 otherwise.

Fig. 4. Path Privacy Experiment

Definition 31. *The advantage of $\mathcal{A}$, denoted $Adv_{\mathcal{A}}^{Path-Privacy}(k)$, in the path privacy experiment is* $\left|\Pr[\mathbf{Exp}_{\mathcal{A}}^{\mathbf{Path-Privacy}}[\kappa] = 1] - \frac{1}{2}\right|$

Definition 32. *A RFID path authentication scheme is $(t, n_1, n_2, n_3, n_4, \epsilon)$-private, if for any t-time adversary $\mathcal{A}$ who makes at most n_1, n_2, n_3, n_4 queries to O_1, O_2, O_3, O_4, respectively, we have $Adv_{\mathcal{A}}^{Privacy}(k) < \epsilon$. The probability is taken over the choice of δ, coins of $\mathcal{A}$ and oracles.*

3.2 Relations among Privacy Notions

Now, we analyze the relations among our new privacy notion and the two existing privacy notions. We show that path-privacy is stronger than tag unlinkability and step unlinkability.

Theorem 1. *Path-privacy implies tag unlinkability.*

Proof. Path privacy implies that S_0 and S_1 in the path privacy experiment are computationally indistinguishable, even if the adversary $\mathcal{A}$ has full control over the supply chain system via the four oracle access *except that the random bit δ is blinded to $\mathcal{A}$*. Intuitively, tag unlinkability is implied by path privacy, as the ability of linking tag's state to tag's identity can be *directly* used to break path privacy.

In more details, we show that it is possible to construct an adversary $\mathcal{B}$ that $(t, n_1, n_2, n_3, n_4, \epsilon)$-breaks path privacy using $\mathcal{A}$ as a subroutine, where $\mathcal{A}$ is an adversary which can $(t, n_1, n_2, n_3, n_4, \epsilon)$-break tag unlinkability. Adversary $\mathcal{B}$ plays the path privacy game using adversary $\mathcal{A}$ as a subroutine; it is $\mathcal{A}$ who conducts the attacks to the system, while A aims to win the tag-unlinkability game. Firstly, $\mathbf{Exp}_{\mathcal{B}}^{\mathbf{Path-Privacy}}[\kappa]$ sets up the system $I, \mathcal{R}, \mathcal{T}, \mathcal{M}$ and publishes the public system parameter $para$. Then $\mathcal{B}$ passes $para$ to $\mathcal{A}$. $\mathcal{A}$ plays the tag-unlinkability game. In the learning phase, when $\mathcal{A}_1$ queries the oracles $\mathcal{O}$, the queries are transferred to $\mathcal{B}_1$, and $\mathcal{B}_1$ queries the oracles $\mathcal{O}$ for $\mathcal{A}_1$ in the path-privacy experiment. Then $\mathcal{A}_1$ outputs $\{T_0, T_1, st\}$. Upon receiving $\mathcal{A}_1$'s output, $\mathcal{B}_1$ chooses a path P and submits $\{T_0, T_1, P, 1, st\}$ to the path-privacy experiment. The experiment $\mathbf{Exp}_{\mathcal{B}}^{\mathbf{Path-Privacy}}[\kappa]$ chooses $\delta \in_R \{0, 1\}$, and returns

$S_\delta \leftarrow \text{Move}(\mathcal{T}_\delta, 1, P, 1)$ to $\mathcal{B}_2$. $\mathcal{B}_2$ transfers S_δ to $\mathcal{A}_2$. When $\mathcal{A}_2$ stops, $\mathcal{B}_2$ outputs whatever output by $\mathcal{A}_2$. It is clear that if $\mathcal{A}$ wins the tag unlinkability game, then $\mathcal{B}$ wins the path privacy game. We have:

$$\Pr[\mathbf{Exp}_{\mathcal{B}}^{\mathbf{Path-Privacy}}[\kappa] = \mathbf{1}] = \Pr[\mathbf{Exp}_{\mathcal{A}}^{\mathbf{Tag-Unlinkability}}[\kappa] = \mathbf{1}] \tag{1}$$

If $\mathcal{A}$ $(t, n_1, n_2, n_3, n_4, \epsilon)$-breaks tag-unlinkability, then $\mathcal{B}$ also $(t, n_1, n_2, n_3, n_4, \epsilon)$-breaks path privacy. □

Theorem 2. *Path privacy implies step unlinkability.*

Proof. Assuming that a system is not step-unlinkable, there exists an adversary $\mathcal{A}$ which can $(t, n_1, n_2, n_3, n_4, \epsilon)$-break its step unlinkability. We can construct an adversary $\mathcal{B}$ that breaks the path privacy using $\mathcal{A}$ as a subroutine.

$\mathbf{Exp}_{\mathcal{B}}^{\mathbf{Path-Privacy}}[\kappa]$ sets up the system $I, \mathcal{R}, \mathcal{T}, \mathcal{M}$ and publishes the public system parameter $para$. $\mathcal{B}$ passes $para$ to $\mathcal{A}$. If $\mathcal{A}$ can break the step unlinkability in $\mathbf{Exp}_{\mathcal{A}}^{\mathbf{Step-Unlinkability}}[\kappa]$. Then $\mathcal{B}$ can use $\mathcal{A}$ as a subroutine to break path-privacy. In the learning phase, $\mathcal{A}_1$ gathers the information of the system. In this process, $\mathcal{A}_1$ cannot query the oracles directly; instead, it submits the queries to $\mathcal{B}_1$ and then $\mathcal{B}_1$ queries the oracles $\mathcal{O}$ for $\mathcal{A}_1$. Then $\mathcal{A}_1$ outputs $\{T, st\}$. As $\mathcal{A}_1$ fully controls the system during the learning phase, then $\mathcal{A}_1$ knows the path of T. We denote the path by P, which is contained in st. Then $\mathcal{A}_1$ passes $\{T, k, st\}$ to $\mathcal{B}_1$. $\mathcal{B}_1$ creates two new tags T_0 and T_1 and outputs $\{T_0, T_1, P, k, st\}$. $\mathbf{Exp}_{\mathcal{B}}^{\mathbf{Path-Privacy}}[\kappa]$ tosses a coin δ. If $\delta = 0$, then the experiment moves T_0 without going through path P in k step, and the state of T_0 is denoted as S_0; else, the experiment moves T_1 through path P in k step, and the state of T_1 is denoted as S_1. The experiment returns S_δ to $\mathcal{B}_2$. $\mathcal{B}_2$ transfers S_δ to $\mathcal{A}_2$, and outputs whatever output by $\mathcal{A}_2$.

In the above path-privacy game, $\mathcal{B}_2$ is provided with the state S_δ. If $\delta = 0$, then the tag with state S_0 does not have any common step with T. If $\delta = 1$, then the tag with state S_1 has at least one common step with T. Given S_δ, $\mathcal{A}_2$ guesses whether the tag has a common step with T or not. $\mathcal{B}_2$ can directly use the result of $\mathcal{A}_2$. It is clear that:

$$\Pr[\mathbf{Exp}_{\mathcal{B}}^{\mathbf{Path-Privacy}}[\kappa] = \mathbf{1}] = \Pr[\mathbf{Exp}_{\mathcal{A}}^{\mathbf{Step-Unlinkability}}[\kappa] = \mathbf{1}] \tag{2}$$

Hence, if $\mathcal{A}$ $(t, n_1, n_2, n_3, n_4, \epsilon)$-breaks the step-unlinkability, then $\mathcal{B}$ also $(t, n_1, n_2, n_3, n_4, \epsilon)$-breaks the path privacy. □

4 A New RFID Path Authentication Protocol

We propose a new RFID-based path authentication scheme under the path privacy notion. Our path authentication scheme is suitable for a supply chain that where the paths of products are pre-determined. We use pseudorandom function and elliptic curve ElGamal encryption scheme as building blocks.

4.1 Building Blocks

Pseudorandom function Given a security parameter κ, let $m(\cdot)$ and $l(\cdot)$ be two positive polynomials in κ. We say that

$$\{F_k : \{0,1\}^{m(\kappa)} \longrightarrow \{0,1\}^{l(\kappa)}\}_{k \in_R \{0,1\}^\kappa} \tag{3}$$

is a PRF ensemble if the following two conditions hold:

1. Efficient evaluation: There exists a polynomial-time algorithm that on input k and $x \in \{0,1\}^{m(\kappa)}$ returns $F_k(x)$.
2. Pseudorandomness: A PPT oracle machine $\mathcal{A}(t, \varepsilon)$-breaks the PRF ensemble, if

$$|Pr[\mathcal{A}^{F_\kappa}(\kappa) = 1] - Pr[\mathcal{A}^{H_\kappa}(\kappa) = 1]| \geq \varepsilon \tag{4}$$

where F_κ is a random variable uniformly distributed over the multi-set $F_k, k \in_R \{0,1\}^\kappa$, H_κ is uniformly distributed among all functions mapping $m(\kappa)$-bit-long strings to $l(\kappa)$-bit-long strings, and the running time of $\mathcal{A}$ is at most t (here each oracle query accounts for one unit operation).

The PRF ensemble is pseudorandom, if for all sufficiently large κ, there exists no algorithm A that can (t, ϵ)-break the PRF ensemble, for any t that is polynomial in κ and any ϵ that is nonnegligible in κ [15].

Elliptic Curve ElGamal Cryptosystem An elliptic curve ElGamal cryptosystem provides the following usual set of operations:

- *Setup*: The system outputs an elliptic curve $\mathcal{E}$ over a finite field $\mathbb{F}_p$, where p is a large prime. Let P be a point on $\mathcal{E}(\mathbb{F}_p)$ of a large prime order q such that the discrete logarithm problem is intractable for $\mathcal{G} =< P >$.
- *Key generation*: The secret key is $sk \in \mathbb{F}_p$. The corresponding public key pk is the pair of points $(P, Y = sk \cdot P)$.
- *Encryption*: To encrypt a point $M \in \mathcal{E}$, one randomly selects $r \in \mathbb{F}_q$ and computes $E(M) = (U, V) = (r \cdot P, M + r \cdot Y)$. The ciphertext is $c = (U, V)$.
- *Decryption*: To decrypt a ciphertext $c = (U, V)$, one computes $D(c) = U - sk \cdot V = M$.

To encrypt message m, we need a point mapping algorithm to transform $m \in \mathbb{F}q$ to a point in the elliptic curve $\mathcal{E}$. $\mathcal{M}(m) = m \cdot P$ is a simple additively homomorphic and unreversed mapping $\mathcal{M} : \mathbb{F}_q \rightarrow \mathcal{E}$, where P is a point in $\mathcal{E}$ of large prime order q. This mapping is a one-to-one mapping from $\mathbb{F}_q$ to $G =< P >$: if $\exists m_1, m_2 \in \mathbb{F}_q$ such that $\mathcal{M}(m_1) = \mathcal{M}(m_2)$, then $m_1 = m_2 \bmod q$.

ElGamal system supports re-encryption operation denoted as ReE. Given a ciphertext $c = (U, V)$ under a public key $pk = (P, Y = sk \cdot P)$, and the public key pk, ReE re-randomizes the ciphertext c to c', where $c' = (U', V') = (U + r \cdot P, V + r \cdot Y)$, for $r \in_R \mathbb{F}_q$. ElGamal system preserves the semantic security property under re-encryption [7]. Let $O_{re-encrypt}$ be an oracle that, provided with two ciphertexts c_0, c_1, randomly chooses $b \in \{0,1\}$, re-encrypts c_b using ElGamal and public key pk, and returns the resulting ciphertext c_b. The semantic security of ElGamal under re-encryption implies that guessing the value of b is as difficult for $\mathcal{A}$ as the decisional Diffie-Hellman (DDH) problem [7].

4.2 Protocol

Assume that an RFID-enabled supply chain path authentication system consists of a set of n tags, an issuer I, a set of l managers $\mathcal{M}$, and a set of m normal readers $\mathcal{R}$. Our protocol has three steps: initialization, updating and verification. In the initialization step, the issuer and the managers setup the system together and initialize the tags. When the tags enter the supply chain, the corresponding reader updates the tags on each step. Finally, when a tag reaches a manager in $\mathcal{M}$, the manager reads out the content of the tag and checks the validity of the tag. Each tag stores an encrypted ID and an encrypted credential generated by the readers in its path.

Initialization: The managers $\mathcal{M}$ generate $(sk, pk) = (x, y = g^x)$ for ElGamal encryption and send pk to the issuer and the readers. The underlying elliptic curve of the ElGamal system is denoted as $\mathcal{E}$. The issuer $\mathcal{I}$ selects a secret-key $k_0 \in \{0,1\}^\kappa$, where κ is the system parameter. $\mathcal{I}$ sets for each reader R_j a secret key k_j, where $1 \leq j \leq m$. $\mathcal{I}$ distributes k_j to R_j. The issuer selects a pseudorandom function PRF, and sends PRF to all the normal readers.

Each tag T_i has an unique identity ID_i, where $ID_i \in \mathcal{E}$. For each T_i, the issuer $\mathcal{I}$ sets its initial state to be $\{c_i = E(ID_i), t_i = PRF_{k_0}(ID_i)\}$. We denote the path which T_i will go through as P_i. Suppose $P_i = (R_{i_0}, R_{i_1}, R_{i_2}, \cdots, R_{i_l})$, for any $0 \leq j \leq l$, where i_j denotes the reader ID in the position j of path P_i. Then for T_i, the issuer $\mathcal{I}$ computes $v_i = PRF_{k_{i_l}}(PRF_{k_{i_{l-1}}}(\cdots(PRF_{k_0}(ID_i)))$, and stores a copy of (ID_i, v_i) on the databases of the managers $\mathcal{M}$.

Interaction between Reader and Tag: When tag T_i reaches R_j, reader R_j reads out T_i's current state $S_{T_i} = \{c_i, t_i\}$. R_j computes the new state $\{c_i', t_i'\}$, where c_i' is re-randomization of c_i under the public key pk and $t_i' = PRF_{k_j}(t_i)$, and then writes $\{c_i', t_i'\}$ to the tag.

Check the Validity of Tag: Only the managers $\mathcal{M}$ can check the validity of tags. Upon the arrival of a tag at a check point, with state $\{c_i, t_i\}$, M decrypts c_i to get ID_i, and searches its database; if and only if it can find a tuple (ID_i, v_i) that satisfies $t_i = v_i$, then T_i is considered as a valid tag.

4.3 Security and Privacy Analysis

The security and privacy of the proposed protocol are based on the pseudorandomness of PRF and the semantic security of Elgamal Encryption scheme under re-encryption. In the following, we provide a formal security and privacy analysis.

Suppose PRF is a pseudorandom function that mapps $m(\kappa)$-bit-long strings to $l(\kappa)$-bit-long strings. We call the function $CPRF(m) = PRF_{k_l}(PRF_{k_{l-1}}(\cdots(PRF_{k_0}(m)))$ as "cascaded" pseudorandom function, where $k_0, ..., k_l$ are randomly chosen keys for the pseudorandom function PRF. If for all k_i, $0 \leq i \leq l$, PRF_{k_i} is a pseudorandom function, $CPRF(m)$ is a pseudorandom function (formal proof please refer to [4]).

Lemma 1. *Producing a new valid pair $\{c_i, t_i\}$ contradicts with the pseudorandomness property of CPRF. Here a new pair of $\{c_i, t_i\}$ means that c_i is a ciphertext of a new ID_i under the public key of the system, or c_i is a ciphertext of an existing ID_i in the system while t_i is a new value that has not appeared in the system.*

Proof (sketch). The security of our system is based on the pseudorandomness of $CPRF(m)$. Suppose there is an oracle $O_{CPRF}^{distinguish}$, given a message m, the oracle randomly returns the value of $CPRF(m)$ or $H(m)$, denoted as m', where $H()$ is an arbitrarily selected function among all functions mapping $m(\kappa)$-bit-long strings to $l(\kappa)$-bit-long strings. After getting m', the adversary outputs 1 if it guesses $m' = CPRF(m)$, else he outputs 0. $Pr[\mathcal{A}^{CPRF}(\kappa) = 1]$ denotes the probability that the adversary outputs 1 when the oracle $O_{CPRF}^{distinguish}$ returns value $CPRF(m)$. $Pr[\mathcal{A}^{H}(\kappa) = 1]$ denotes the probability that the adversary outputs 1 when the oracle $O_{CPRF}^{distinguish}$ returns value $H(m)$. Since $CPRF$ is a pseudorandom function, given $\mathcal{A}$ with limited access to the function $CPRF$, we have $|\Pr[\mathcal{A}^{CPRF}(\kappa) = 1] - \Pr[\mathcal{A}^{H}(\kappa) = 1]| \geq \varepsilon$, where ε is negligible. We will show that if an adversary $\mathcal{A}'$ can successfully forge a new pair $\{c_i, t_i\}$, then using $\mathcal{A}'$ as a subroutine, there exists an adversary $\mathcal{A}$ that breaks $CPRF(m)$'s pseudorandomness, namely the value of $|Pr[\mathcal{A}^{CPRF}(\kappa) = 1] - Pr[\mathcal{A}^{H}(\kappa) = 1]|$ will be non-negligible.

$\mathcal{A}$ sets up a supply chain system with public key pk, private key sk for Elgamal encryption system, and a valid path in which the readers have the keys $k_0, \cdots, k_l$, respectively, where l is the length of the path. $\mathcal{A}$ does not know the keys $k_0, \cdots, k_l$, while it is provided with $PRF_{k_0}, \cdots, PRF_{k_i}$ by the oracle $O_{CPRF}^{distinguish}$. $\mathcal{A}$ transfers the public system parameters to $\mathcal{A}'$ which runs two algorithms $\mathcal{A}'_1$ and $\mathcal{A}'_2$ in $\mathbf{Exp}_{\mathcal{A}'}^{\mathbf{Path-Privacy}}[\kappa]$, . In the learning phase, $\mathcal{A}'_1$ accesses the supply chain system without exceeding the constraints defined in $\mathbf{Exp}_{\mathcal{A}'}^{\mathbf{Path-Privacy}}[\kappa]$. In the challenge phase, $\mathcal{A}'_2$ outputs a new pair $\{c_i, t_i\}$. $\mathcal{A}$ decrypts c_i to get ID. Then $\mathcal{A}$ queries $O_{CPRF}^{distinguish}$ with ID. $O_{CPRF}^{distinguish}$ returns a message mes_{ID}. In case $\{c_i, t_i\}$ is valid, then by checking whether $mes_{ID} = t_i$, $\mathcal{A}$ knows whether $O_{CPRF}^{distinguish}$ has chosen the function $CPRF$ or a random function $H()$. As a result, if $\mathcal{A}'$ $(t, n_1, n_2, n_3, n_4, \epsilon)$-breaks the the security of path authentication, then $\mathcal{A}$ (t, ϵ)-breaks the pseudorandomness of function $CPRF$.

Theorem 3. *If PRF is pseudorandom, then our system has path privacy property under the semantic security of ElGamal re-encryption.*

Proof (sketch). Assume that our system is not path private, namely, there exists an adversary $\mathcal{A}$ that breaks the path privacy of our system. Then we can construct an adversary $\mathcal{B}$ to break the semantic security of ElGamal encryption system under re-encryption. $\mathcal{B}$ uses $\mathcal{A}$ as a subroutine and maintains a list L to answer $\mathcal{A}$'s queries as follows.

Suppose the public key of an ElGamal encryption cryptosystem is pk, and its corresponding private key is sk. Adversary $\mathcal{B}$ can break the semantic security of the system under re-encryption. $\mathcal{B}$ firstly simulates a path authentication system; it initializes the system the same as Initialization step defined in Section 4.2, except that the public and private keys of the manager are set to pk and sk, respectively. Note that $\mathcal{B}$ knows all the secret keys of the readers, but it does not know the value of sk. Then an adversary $\mathcal{A}$ starts the path-privacy experiment. In the learning phase of $\mathcal{A}_1$, when $\mathcal{A}_1$ queries the oracles, $\mathcal{B}$ answers the queries. $\mathcal{B}$ can answer the queries to O_1, O_2 and O_4 directly. However, $\mathcal{B}$ does not have the private key sk, hence in case $\mathcal{A}_1$ queries the O_3 with a state $\{c_i, t_i\}$, $\mathcal{B}$ cannot decrypt c_i to get ID_i and compare the value of t_i with v_i in the database. In order to answer the queries to $\mathcal{O}_3$, $\mathcal{B}$ maintains a list L that records the history of each oracle's operations. Firstly, $\mathcal{B}$ inserts the tuples (ID_i, c_i, t_i, v_i) for $i = 1$ to n into list L. Then, each time a tag's state is changed, $\mathcal{B}$ adds a link between the tag's new state and old state. With the list L, given a tag's state, even through $\mathcal{B}$ cannot decrypt the ciphertext, it can get the tag's ID through the records of the tag's state in list L. Thus $\mathcal{B}$ can answer the queries to $\mathcal{O}_3$ by searching the database and comparing t_i with v_i. At the end of the learning phase, $\mathcal{A}_1$ outputs two tags T_0 and T_1, a path P with no less than k steps, st. Suppose that the state of T_0 is $\{c_0, t_0\}$, the state of T_1 is $\{c_1, t_1\}$. $\mathcal{B}$ firstly submits the two messages $\{c_0, c_1\}$ to $\mathcal{O}_{re-encrypt}$. $\mathcal{O}_{re-encrypt}$ randomly chooses $b \in \{0, 1\}$, and re-encrypts c_b to c'_b under the public key pk. Then $\mathcal{B}$ sends $S = \{c'_b, r\}$ to $\mathcal{A}_2$, where r is a random string. Note that actually, $\mathcal{B}$ should provide $\{c'_b, t_b\}$ to $\mathcal{A}_2$, where t'_b is the new value of t_b after been processed by k readers in path P. We argue that $\{c'_b, r\}$ and $\{c'_b, t'_b\}$ contain same information that can be used by $\mathcal{A}_2$. $\mathcal{A}_2$ cannot get any information from t'_b since the function PRF is a pseudorandom function. $\mathcal{A}_2$ guesses the value of b by analyzing $\{c'_b, r\}$. $\mathcal{B}$ outputs whatever output by $\mathcal{A}$.

Assuming the pseudorandomness of PRF, the advantage of $\mathcal{B}$ to break the semantic security of ElGamal under re-encryption is the same as the advantage of $\mathcal{A}$ to break the path privacy of the system. Since the ElGamal encryption scheme under re-encryption is semantic secure, hence our system is path private. □

4.4 Performance

Computational requirement: Our scheme does not require the tags to perform any computation. All the computation will be performed at the reader side. To update a tag, each reader requires one re-encryption operation and one computation on PRF. For a manager to verify a tag's validity, it requires one decrypting operation and one comparison.

Storage requirement: Each tag T_i's state S_i consists of $\{c_i, t_i\}$. c_i is ElGamal ciphertext on ID_i which requires $2 \cdot 160$ bits. t_i is the path mark, generated by the PRF, thus 160 bits is sufficient. Therefore 480 bits storage is required for each tag. The protocol can thus be implemented with the standard EPC Class 1 Gen 2 tag with an extensible EPC memory bank (scalable between 16-480 bits), a scalable user memory bank (64-512 bits), which are available in the market [1].

On the reader side, the issuer stores a copy of system parameters including pk and k_j, for $0 \leq j \leq m$, m is the number of normal readers. So the storage requirement for the issuer is $O(1)$. Each normal reader R_j at step v_j needs to store the public key pk of the system and its own key k_j, the storage requirement for each normal reader is $O(1)$. Each manager stores a copy of sk. It also maintains a database DB, for each tag T_i, DB stores the verification information (ID_i, v_i). The storage requirement for a manager is $O(n)$, n is the number of the tags. As a tag's record takes 480 bits, a manager with 1GB storage can stores more than 17 million tags' records.

Compare to TRACKER [3,4], our system is more practical. Since the tags' paths are predetermined in the initial stage, there is no need to store the path information on tag. A manage can perform path verification by simple comparison. Consequently, the storage and computational requirements on updating tags are reduced. The comparisons of storage and computational requirements between our protocol and TRACKER are shown in Table 1. Note that in comparing the computational load, we omit the cheap operations such as hash operation, computing PRF, and point addition on elliptic curve etc. We only count the relative expensive operations such as point multiplication on elliptic curve.

Table 1. Comparisons of TRACKER and Our Protocol

	TRACKER [3]	Our protocol
storage requirement		
tag	960 bits	480 bits
issuer	$O(1)$	$O(1)$
normal reader	$O(1)$	$O(1)$
manager	$O(n+vp)$, vp is the number of valid paths n is the number of tags	$O(n)$, n is the number of tags
computational requirement of processing a tag (operation on elliptic curve)		
issuer	8 point multiplication	2 point multiplication
normal reader	10 point multiplication	2 point multiplication
manager	5 point multiplication	1 point multiplication

5 Conclusions

In this paper, we analyzed the existing security and privacy notions for RFID-enabled path authentication in [3]. We provided refined versions of the notions. We proposed the first single-game-based privacy notion for path authentication which implies the existing notions. We also proposed a path authentication protocol that satisfies the privacy notion. Our protocol can be implemented on standard EPC class 1 Generation 2 tags, and it outperforms the existing path authentication solutions [3,4] for RFID-enabled supply chains.

References

1. `http://www.alientechnology.com/tags/index.php`
2. Arbit, A., Oren, Y., Wool, A.: Toward Practical Public Key Anti-Counterfeiting for Low-Cost EPC Tags. In: IEEE RFID 2011, Orlando, FL, pp. 184–191 (April 2011)

3. Blass, E.O., Elkhiyaoui, K., Molva, R.: Tracker: Security and Privacy for RFID-Based Supply Chains. Cryptology ePrint Archive, Report 2010/219 (2010)
4. Blass, E.O., Elkhiyaoui, K., Molva, R.: Tracker: Security and Privacy for RFID-Based Supply Chains. In: NDSS 2011, San Diego, California, USA, pp. 455–472 (2011)
5. Cai, S., Li, Y., Li, T., Deng, R.H.: Attacks and Improvements to an RFID Mutual Authentication Protocol and its Extensions. In: WiSec 2009, Zurich, Switzerland, pp. 51–58 (2009)
6. Cai, S., Li, Y., Zhao, Y.: Distributed Path Authentication for Dynamic RFID-Enabled Supply Chains. In: IFIP SEC 2012, Crete, Greece (2012)
7. Golle, P., Jakobsson, M., Juels, A., Syverson, P.: Universal Re-encryption for Mixnets. In: Okamoto, T. (ed.) CT-RSA 2004. LNCS, vol. 2964, pp. 163–178. Springer, Heidelberg (2004)
8. Li, Y., Ding, X.: Protecting RFID Communications in Supply Chains. In: ASIACCS 2007, New York, NY, USA, pp. 234–241 (2007)
9. Molnar, D., Wagner, D.: Privacy and Security in Library RFID: Issues, Practices, and Architectures. In: CCS 2004, New York, NY, USA, pp. 210–219 (2004)
10. Peris-Lopez, P., Hernandez-Castro, J.C., Tapiador, J.M.E., Li, T., Li, Y.: Vulnerability Analysis of RFID Protocols for Tag Ownership Transfer. Computer Networks 54(9), 1502–1508 (2010)
11. Piramuthu, S.: RFID Mutual Authentication Protocols. Decision Support Systems (2010), `http://dx.doi.org/10.1016/j.dss.2010.09.005`
12. Rizomiliotis, P., Rekleitis, E., Gritzalis, S.: Security Analysis of the Song-Mitchell Authentication Protocol for Low-Cost RFID Tags. IEEE Communications Letters 13(4), 274–276 (2009)
13. Song, B., Mitchell, C.J.: RFID Authentication Protocol for Low-Cost Tags. In: WiSec 2008, Alexandria, Virginia, USA, pp. 140–147 (2008)
14. Wang, H., Li, Y., Zhang, Z., Cao, Z.: Two-Level Path Authentication in EPCglobal Network. In: IEEE RFID 2012, Orlando, Florida, pp. 24–31 (2012)
15. Yao, A.C., Yung, M., Zhao, Y.: Adaptive Concurrent Non-Malleability with Bare Public-Keys. CoRR, abs/0910.3282 (2009)

$GHB^{\#}$: A Provably Secure *HB*-Like Lightweight Authentication Protocol

Panagiotis Rizomiliotis and Stefanos Gritzalis

Dep. of Information and Communication Systems Engineering
University of the Aegean
Karlovassi, Samos, GR 83200, Greece
{prizomil,sgritz}@aegean.gr

Abstract. RFID technology constitutes a fundamental part of what is known as the Internet of Things; i.e. accessible and interconnected machines and everyday objects that form a dynamic and complex environment. In order to secure RFID tags in a cost-efficient manner, the last few years several lightweight cryptography-based tag management protocols have been proposed. One of the most promising proposals is the HB^+ protocol, a lightweight authentication protocol that is supported by an elegant security proof against all passive and a subclass of active attackers based on the hardness of the Learning Parity with Noise (LPN) problem. However, the HB^+ was shown to be weak against active man-in-the-middle (MIM) attacks and for that several variants have been proposed. Yet, the vast majority of them has been broken.

In this paper, we introduce a new variant of the HB^+ protocol that can provably resist MIM attacks. More precisely, we improve the security of another recently proposed variant, the $HB^{\#}$ protocol by taking advantage of the properties of the well studied Gold power functions. The new authentication protocol is called $GHB^{\#}$ and its security can be reduced to the LPN problem. Finally, we show that the $GHB^{\#}$ remains practical and lightweight.

1 Introduction

Radio Frequency Identification (RFID) technology constitutes a fundamental part and key enabler of what is known as the Internet of Things (IoT); i.e. accessible and interconnected machines and everyday objects that form a dynamic and complex environment. In the IoT vision, the Internet extends into our everyday lives through a wireless network of uniquely identifiable objects or 'things'. RFID tags are much "smarter" and more efficient than the classical barcode and can provide us with the data needed to manage 'things', unmanageable until today; thus rendering RFID the most pervasive technology in human history. Each physical object is accompanied by a rich, globally accessible virtual object that contains both current and historical information on that object's physical properties, origin, ownership. When available ubiquitously and in real time, this information can dramatically streamline how we manufacture, distribute, manage, and recycle our goods.

F. Bao, P. Samarati, and J. Zhou (Eds.): ACNS 2012, LNCS 7341, pp. 489–506, 2012.

Applications ranging from inventory monitoring, and payment systems to supply-chain management and smart home devices are already taking advantage of the RFID technology. However, this rapid proliferation of RFID tags raises several security and privacy concerns. Given also that, in order to sustain the pervasiveness, the cost of the tag must remain as low as possible; i.e. space, as well as, peak and average power consumption limitations must be instituted, it was identified early on that new lightweight cryptographic protocols have to be deployed for their management.

In this context, several new lightweight schemes have been proposed in the last few years ([1]), mainly for secure tag authentication. Amongst the proposed solutions, the most prominent ones are the authentication schemes that are based on the conjectured hardness of the *Learning Parity in the presence of Noise (LPN)* problem, which is closely related to the well-studied problem of decoding random linear codes.

Definition 1. *(LPN Problem) Let* $\boldsymbol{A}$ *be a random* $(q \times k)$*-binary matrix, let* $\boldsymbol{x}$ *be a random* k*-bit vector, let* $\eta \in (0, 1/2)$ *be a noise parameter, and let* $\boldsymbol{\nu}$ *be a random* q*-bit vector such that* $wt(\boldsymbol{\nu}) \leq \eta q$*. Given* $\boldsymbol{A}$*,* η*, and* $\boldsymbol{z} = \boldsymbol{A} \cdot \boldsymbol{x}^t + \boldsymbol{\nu}^t$*, find a* k*-bit vector* $\boldsymbol{y}^t$ *such that* $wt(A \cdot \boldsymbol{y}^t + \boldsymbol{z}) \leq \eta q$*.*

In [12], Juels and Weis proposed *HB*$^+$, a symmetric key authentication scheme, inspired by *HB* ([11]), the work of Hopper and Blum for the secure identification of human beings. The *HB*$^+$ has very simple circuit representation, as it performs only a few dot-product and bit exclusive-or computations. In more detail, the prover (the Tag) and the verifier (the Reader) exchange random binary vectors $\boldsymbol{a}$ and $\boldsymbol{b}$, and the prover based on this exchanged information and two secret vectors $\boldsymbol{x}$ and $\boldsymbol{y}$, produces and transmits to the verifier one bit $z = \boldsymbol{ax} \oplus \boldsymbol{by} + \nu$, where ν is one bit that follows a Bernoulli distribution with parameter $\eta \in (0, \frac{1}{2})$. The verifier accepts $z = \boldsymbol{ax} \oplus \boldsymbol{by}$. This basic interaction has soundness $\frac{1}{2}$ and completeness $1 - \eta$ and it is improved via sequential or parallel composition, i.e. the verifier accepts if after r repetitions of the basic round at most t times the condition is not satisfied.

Certainly, the most interesting feature of the protocol is the elegant proof that supports its security analysis. Specifically, in [12], a concrete reduction of the LPN problem to the security of the *HB*$^+$ protocol in two attack models was shown. In the first model the attacker is passive and can only eavesdrop the communication between the prover and the verifier, while in the second model she is active and she can also send queries to the prover. The original proof was further simplified and extended in [13], [14].

However, the above described attack models do not include more powerful adversaries, like the ones that can manipulate messages exchanged between the reader and the tag. Thus, it came as no surprise that soon after the introduction of the *HB*$^+$, it was shown ([7]) that there is a simple man-in-the-middle (MIM) attack that can easily reveal the secret vectors $\boldsymbol{x}$ and $\boldsymbol{y}$. Motivated by this MIM attack, several variants of *HB*$^+$ have been proposed ([6],[8], [2], [24], [22], [3], [20], [16], [25]). However, most of these schemes have been shown to be weak against a MIM attacker.

Recently, the first two HB^+ variants that can provably resist MIM attacks have been proposed and both are based on equivalent to LPN problems. In [15], Kiltz et al. built on the Subspace LWE problem to construct two secure Message Authentication Code (MAC) schemes. However, both these schemes require the application of Pairwise Independent Permutation, while the secret keys are very long. In [4], Bosley et al., introduced the Learning Subspace with Noise (LSN) problem and they showed the equivalence between the LPN and LSN. Based on the LSN problem, they introduced an authentication protocol and they proved its security against MIM attacks. This protocol is equivalent to the second MAC scheme introduced in [15].

1.1 Our Contribution

In this paper, we propose a new variant of the HB^+ protocol that can provably resist all known MIM attacks. More precisely, we improve the security of another recently proposed variant, the $HB^{\#}$ protocol ([8]), by taking advantage of the properties of the Gold power functions. The $HB^{\#}$ protocol was introduced by Gilbert et al. and it was provably secure against the attack that succeeded against HB^+([7]). However, Oaufi et al. [21] presented another MIM attack on $HB^{\#}$. The main objective of our work is to enhance the security of $HB^{\#}$ by adding some nonlinear components without increasing significantly its complexity.

The idea of using nonlinear functions to built secure LPN-based authentication protocols is not new. In [2], Bringer et al. proposed the HB^{++} protocol, a modified version of the HB^+ protocol that could resist the attack in [7] using a specific family of nonlinear multi-output Boolean functions, the Gold functions. Gold functions can be efficiently implemented in hardware, they have been extensively studied in the literature and they possess very good cryptographic properties, like high nonlinearity and good derivative behaviour, and for that they constitute an excellent choice. However, HB^{++} was shown to be weak [9]. A more recent attempt to introduce nonlinear HB-like protocols by Madhavan et al. [18] was also unsuccessful ([23]).

Our protocol, called $GHB^{\#}$, is the first nonlinear variant of HB^+ that it is provably resistant against MIM attacks. Our reduction is using rewinding, like in the case of the HB^+ and $HB^{\#}$ protocol and the security is based on the hardness of the LPN problem. Moreover, we show that, despite the use of nonlinear functions, the $GHB^{\#}$ protocol remains as practical and lightweight as its direct ancestor $HB^{\#}$.

1.2 Outline

The paper is organized as follows. In Section 2, we establish the necessary background on vectorial Boolean functions with emphasis in the family of Gold functions. In the same section, we describe the $HB^{\#}$ protocol. In Section 3, we present the new authentication protocol and in Section 4, we provide efficient implementation guidelines and we compute the overall complexity. In Section 5, we provide the security analysis and we prove that the new protocol is secure against active attackers that can interrogate a tag and/or modify all the exchanged messages between a tag and the reader. Finally, conclusions and topics for further research can be found in Section 6.

2 Background

2.1 Gold Functions

Vectorial Boolean functions constitute fundamental building blocks for many cryptographic algorithms and have been extensively studied in the literature. In this paper, we use a specific family of such functions, the so called *Gold functions* that possess very good cryptographic properties ([10], [5]). First we introduce some notation and then we present the necessary background.

Let $\mathbb{F}_2$ be the finite field with two elements and $\mathcal{B}_{n,m}$ the set of vectorial Boolean function with n inputs and m outputs; i.e. the set of multi-output Boolean from $\mathbb{F}_2^n$ to $\mathbb{F}_2^m$. We use normal, bold and capital bold letters, x, $\boldsymbol{x}$ and $\boldsymbol{M}$ to denote single elements, vectors and matrices, respectively. Also, normal and capital bold letters are used for single input (univariate) and multi input Boolean functions, respectively. The Hamming weight $\mathrm{wt}(\boldsymbol{x})$ of a vector $\boldsymbol{x} = [x(0), x(1), \cdots, x(n-1)]$ is the number of nonzero elements. Finally, $\mathbf{0}_m$ denotes the all zeros vector of length m and for real numbers $\eta, \psi \in \Re$, $]\eta, \psi[= \{x \in \Re | \eta < x < \psi\}$.

Definition 2. *([5]) A vectorial function $\boldsymbol{F} \in \mathcal{B}_{n,m}$ is balanced if it takes all values $\boldsymbol{y} \in \mathbb{F}_2^m$ the same number of times; i.e. 2^{n-m} times.*

Definition 3. *([5]) The derivative of a vectorial Boolean function $\boldsymbol{F} \in \mathcal{B}_{n,m}$ is defined as $D_{\boldsymbol{a}}\boldsymbol{F}(\boldsymbol{x}) = \boldsymbol{F}(\boldsymbol{x}) + \boldsymbol{F}(\boldsymbol{x}+\boldsymbol{a})$, $\boldsymbol{a} \in \mathbb{F}_2^n$ and $\boldsymbol{a} \neq \mathbf{0}_n$.*

Definition 4. *A vectorial Boolean function $\boldsymbol{F} \in \mathcal{B}_{n,m}$ is called almost perfect nonlinear (APN) if and only if for every $\boldsymbol{a} \in \mathbb{F}_2^n$, $\boldsymbol{a} \neq \mathbf{0}_n$ and $\boldsymbol{b} \in \mathbb{F}_2^m$ the equation $D_{\boldsymbol{a}}\boldsymbol{F}(\boldsymbol{x}) = \boldsymbol{b}$ has zero or 2 solutions.*

In [10], R. Gold introduced the so-called *Gold functions*, the power functions $x \to x^d$ on the field $\mathbb{F}_{2^n}$, where n odd and $d = 2^i + 1$, with $\gcd(i, n) = 1$ and $1 \leq i < \frac{n-1}{2}$. Gold proved that these univariate polynomials are APN functions, with very high nonlinearity, balanced and have quadratic algebraic degree. (Note that Gold functions have good cryptographic properties when n is an even integer.)

Gold functions have been defined and analysed as univariate functions, but it is well known, that they can be easily transformed to a vectorial Boolean function. Let $\{\alpha_0, \cdots, \alpha_{n-1}\}$ be a basis of $\mathbb{F}_{2^n}$ over $\mathbb{F}_2$, then any element $x \in \mathbb{F}_{2^n}$ can be written as $x = \sum_{i=0}^{n-1} x_i \alpha_i$, $x_i \in \mathbb{F}_2$. In this paper, we are going to use a normal basis $\{\gamma^{2^0}, \gamma^{2^1}, \cdots, \gamma^{2^{n-1}}\}$ of $\mathbb{F}_{2^n}$ over $\mathbb{F}_2$, where $\gamma \in \mathbb{F}_{2^n}$. It is well known that there is such basis for any $m > 1$ ([17]). Note that depending on the choice of the basis, the mapping from the univariate functions to the vectorial

Boolean functions differs. Thus, a Gold function $g = x^d$ can be written as a multi input Boolean function $\boldsymbol{G}$ as follows,

$$\boldsymbol{G}(x_0, x_1, \cdots, x_{n-1}) = g(\sum_{i=0}^{n-1} x_i \gamma^{2^i}).$$

Similarly, the output of the function can be written as a linear combination of the elements of the basis.

Definition 5. *([5]) Two functions are affine equivalent if one derives from the other with some left and right compositions with an affine permutation.*

We denote by $\mathcal{G}(n, d)$ the set of all multi-input vectorial Boolean functions that are affine equivalent to a Gold function with n inputs and outputs and with exponent d. That is, if $\boldsymbol{G}$ is the Gold multi-input Boolean function, then for every vectorial Boolean function $\boldsymbol{\Phi} \in \mathcal{G}(n, d)$, there are affine permutations $\boldsymbol{P}_1$ and $\boldsymbol{P}_2$ such that $\boldsymbol{\Phi} = \boldsymbol{P}_1 \circ \boldsymbol{G} \circ \boldsymbol{P}_2$. Every function $\boldsymbol{\Phi} \in \mathcal{G}(n, d)$ has all the aforementioned properties of Gold functions; i.e. $\boldsymbol{\Phi}$ is APN, balanced and quadratic ([5]).

Since every $\boldsymbol{\Phi} \in \mathcal{G}(n, d)$ is quadratic, it can be written as

$$\boldsymbol{\Phi}(x(0), \cdots, x(n-1)) = \boldsymbol{L}(x(0), \cdots, x(n-1)) \oplus \boldsymbol{Q}(x(0), \cdots, x(n-1)),$$

where $\boldsymbol{L}$ is a linear vectorial Boolean function and $\boldsymbol{Q}$ a purely quadratic vectorial Boolean function. We denote by $I_{\boldsymbol{\Phi}} \subset \{0, 1, \cdots, n-1\}$ the smallest subset of the input variable indexes of $\boldsymbol{\Phi}$, such that

$$\boldsymbol{Q}(x(0), x(1), \cdots, x(n-1)) = \boldsymbol{0}_m,$$

for all $\boldsymbol{x} \in K(\boldsymbol{\Phi})$, where

$$K(\boldsymbol{\Phi}) = \{\boldsymbol{x} \in \mathbb{F}_2^n |\ x(i) = 0,\ \forall\ i \in I_{\boldsymbol{\Phi}}\}.$$

Clearly, $K(\boldsymbol{\Phi})$ is the subspace of equations $x_i = 0$, $i \in I_{\boldsymbol{\Phi}}$, and the restriction of $\boldsymbol{\Phi}$ to this subspace is a linear vectorial function, i.e. $\boldsymbol{\Phi}(\boldsymbol{x}_1 \oplus \boldsymbol{x}_2) = \boldsymbol{\Phi}(\boldsymbol{x}_1) \oplus \boldsymbol{\Phi}(\boldsymbol{x}_2)$, for $\boldsymbol{x}_1, \boldsymbol{x}_2 \in K(\boldsymbol{\Phi})$.

2.2 The $HB^{\#}$ protocol

In this section, we briefly describe the $HB^{\#}$ protocol ([8]). We try to apply, as possible, the established notation. We use $x \overset{\$}{\leftarrow} X$ to denote the assignment to x of a value sampled from the uniform distribution on the finite set X. We use $Ber(\eta)$ to denote the Bernoulli distribution with parameter η, meaning that a bit $\nu \in Ber(\eta)$, then $Pr[\nu = 1] = \eta$ and $Pr[\nu = 0] = 1 - \eta$. A vector $\boldsymbol{\nu}$ randomly chosen among all the vectors of length m, such that $\boldsymbol{\nu}(i) \in Ber(\eta)$ and $\eta \in (0, 1/2)$, for $0 \le i \le m-1$, is denoted as $\boldsymbol{\nu} \overset{\$}{\leftarrow} Ber(m, \eta)$. Finally, we use $\boldsymbol{b} \overset{\$}{\leftarrow} \mathbb{F}_2^k$ to denote a random binary vector $\boldsymbol{b}$ of length k.

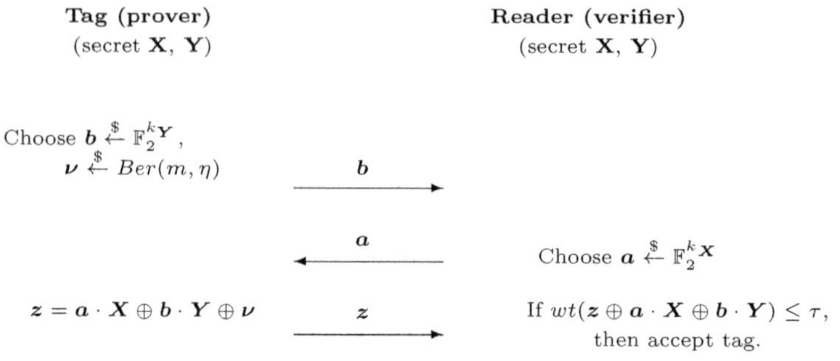

Fig. 1. The *HB*$^{\#}$ protocol

The *HB*$^{\#}$ protocol can be seen as a natural matrix extension of the *HB*$^{+}$ protocol, where the prover and the verifier, instead of vectors, they share two binary matrices $\boldsymbol{X}$ and $\boldsymbol{Y}$ with size $k_{\boldsymbol{X}} \times m$ and $k_{\boldsymbol{Y}} \times m$ respectively. The protocol is again a three pass one, but now the verifier and prover need only one round to interact (Fig. 1). Like the *HB*$^{+}$ protocol, the *HB*$^{\#}$ has low computational complexity $\mathcal{O}(k_{\boldsymbol{X}} \cdot m + k_{\boldsymbol{Y}} \cdot m)$, while it reduces the transmission costs to $(k_{\boldsymbol{Y}} + k_{\boldsymbol{X}} + m)$ bits in total and it provides more practical error rates. However, at the same time, it needs more memory bits for the secret keys, as the tag has to store the two secret matrices, i.e. $(k_{\boldsymbol{X}} \cdot m + k_{\boldsymbol{Y}} \cdot m)$ bits in total.

Security analysis. The *HB*$^{\#}$ protocol was designed to resist the attack introduced in [7] and for that it is supported by a proof of security against attackers that can modify only the messages sent by the reader to the tag during an execution of the protocol. To prove the security of the scheme, a natural matrix-based extension of the *HB* problem was introduced, the *MHB puzzle.*

Definition 6. *((k, m, η, q)-MHB puzzle, [8]) Let $\eta \in (0, 1/2)$ and m and q be polynomials in k. On input the security parameter 1^k, the puzzle generator G draws a random secret $(k \times m)$-binary matrix $\boldsymbol{X}$, q random vectors $(\boldsymbol{a}_1, \cdots, \boldsymbol{a}_q)$ of length k, computes for $1 \leq i \leq q$ the set of answers $\boldsymbol{z}_i = \boldsymbol{a}_i \cdot \boldsymbol{X} + \boldsymbol{\nu}_i$, where each bit of $\boldsymbol{\nu}_i$ is 1 with probability η, and draws a random vector a of length k constituting the challenge to the adversary. It outputs $\{(\boldsymbol{a}_i, \boldsymbol{z}_i)\}_{1 \leq i \leq q}$ and $\boldsymbol{a}$. The solver returns a vector $\boldsymbol{z}$. The secret is $\boldsymbol{X}$, and the verifier V accepts, if and only if, $\boldsymbol{z} = \boldsymbol{a} \cdot \boldsymbol{X}$.*

Using the theory of weakly verifiable puzzles the hardness of this extended problem was proved (Lemma 1) and a concrete reduction of the MHB puzzle to the security of the *HB*$^{\#}$ protocol was provided in [8]. We are going to use the MHB puzzle in our proposal as well.

Lemma 1. *([8]) Assume the hardness of the LPN problem. Then, the MHB puzzle is $(1 - \frac{1}{2^m})$-hard.*

Attack against *HB*$^{\#}$. In [21], it was shown that the protocol is not secure against a more general MIM attack. That is, when the attacker can manipulate all the messages exchanged between a legitimate tag and the reader, and not only the messages sent by the reader, there is a key recovery attack that she can mount. In a few words, the attack goes as follows. The attacker obtains a valid triplet $(\hat{\boldsymbol{b}}, \hat{\boldsymbol{a}}, \hat{\boldsymbol{z}})$; i.e. a triplet that satisfies $\text{wt}(\hat{\boldsymbol{z}} \oplus \hat{\boldsymbol{b}} \cdot \boldsymbol{X} \oplus \hat{\boldsymbol{a}} \cdot \boldsymbol{Y}) \leq \tau$. Using this triplet to modify several executions of the protocol between the reader and the same legitimate tag, the success of the i-th authentication depends on the condition $\text{wt}(\boldsymbol{\nu}_i \oplus \hat{\boldsymbol{\nu}}) \leq \tau$, where $\hat{\boldsymbol{\nu}} = \hat{\boldsymbol{z}} \oplus \hat{\boldsymbol{b}} \cdot \boldsymbol{X} \oplus \hat{\boldsymbol{a}} \cdot \boldsymbol{Y}$ and $\boldsymbol{\nu}_i \in Ber(m, \eta)$ is the masking vector used by the tag. The overall success probability leaks information on the Hamming weight of $\hat{\boldsymbol{\nu}}$. In the second phase of the attack, the attacker modifies one bit of $\hat{\boldsymbol{y}}$ and computes the Hamming weight of the new vector $\hat{\boldsymbol{\nu}}'$. Thus, one bit of the vector $\hat{\boldsymbol{\nu}}$ can be estimated from the difference between $\text{wt}(\hat{\boldsymbol{\nu}})$ and $\text{wt}(\hat{\boldsymbol{\nu}}')$. After repeating the same procedure several times a set of linear equations is constructed involving $\hat{\boldsymbol{b}} \cdot \boldsymbol{X} + \hat{\boldsymbol{a}} \cdot \boldsymbol{Y}$ and the solution of this linear system reveals the secret keys $\boldsymbol{X}$ and $\boldsymbol{Y}$.

During the application of this attack many unsuccessful executions of protocol occur and a mechanism that detects this abnormal behaviour could provide a sufficient countermeasure. However, such a mechanism is not built-in property of the protocol and a new protocol has to be proposed. In the following section, we will show that the $GHB^{\#}$ is such a proposal.

3 The $GHB^{\#}$ protocol

In this section, we introduce a nonlinear variant of the $HB^{\#}$ protocol, a one round symmetric key protocol called the $GHB^{\#}$. Following the notation introduced in Section 2.2, the tag and the reader share two secret binary matrices $\boldsymbol{X}$ and $\boldsymbol{Y}$ of size $k_{\boldsymbol{X}} \times m$ and $k_{\boldsymbol{Y}} \times m$ respectively. The single round of the protocol appears in Fig. 2.

The tag and the reader exchange the randomly selected vectors $\boldsymbol{b}$ and $\boldsymbol{a}$ of length $k_{\boldsymbol{Y}}$ and $k_{\boldsymbol{X}}$, respectively. Then, the tag computes and sends the vector $\boldsymbol{z} = \boldsymbol{\Phi}(\boldsymbol{a} \cdot \boldsymbol{X}) \oplus \boldsymbol{\Phi}(\boldsymbol{b} \cdot \boldsymbol{Y}) \oplus \boldsymbol{\nu}$ of length m, where $\boldsymbol{\nu} \in Ber(m, \eta)$. If $\text{wt}(\boldsymbol{z} \oplus \boldsymbol{\Phi}(\boldsymbol{a} \cdot \boldsymbol{X}) \oplus \boldsymbol{\Phi}(\boldsymbol{b} \cdot \boldsymbol{Y})) \leq \tau$, then the reader accepts the tag as authentic. Otherwise, the tag is rejected. The threshold $\tau = um$, where $u \in \left]\eta, \frac{1}{2}\right[$.

The function $\boldsymbol{\Phi}$ is publicly known and it can be any multi-input function that belongs to $\mathcal{G}(m, d)$; i.e. it is affine equivalent to a Gold multi-input vectorial Boolean function $\boldsymbol{G}$, as defined in Section 2.1. The choice of the specific univariate Gold function x^d used for the construction of $\boldsymbol{G}$ does not influence the security of the protocol or its complexity. In Section 4, we give design directives for the efficient hardware implementation of $\boldsymbol{\Phi}$ and we compute the hardware cost that brings to the protocol.

The error rates of the new protocol are computed similarly to the ones of $HB^{\#}$. In more detail, the false rejection rate P_{FR} of the protocol; i.e. the probability to reject a legitimate tag, equals the probability $\text{wt}(\boldsymbol{\nu}) > \tau$ and it is given by

$$P_{FR} = \sum_{i=\tau+1}^{m} \binom{m}{i} \eta^i (1-\eta)^{m-i}.$$

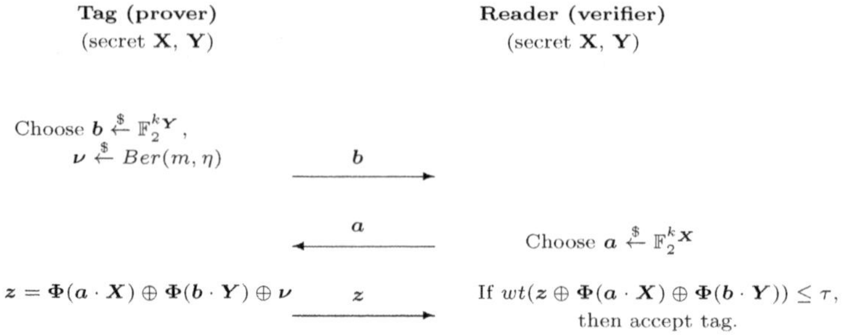

Fig. 2. The $GHB^{\#}$ protocol

It is common practice, in most *HB*-like protocols, to use an extra step in which $\boldsymbol{\nu}$ is used only when its Hamming weight is at most τ; i.e. the completeness error is $P_{FR} = 0$.

Finally, the false acceptance rate P_{FA}; i.e. the probability to accept a randomly selected response $\boldsymbol{z}$, equals the probability a binary vector of length m to have Hamming weight at most τ. That is that, the soundness error is given by:

$$P_{FA} = \sum_{i=0}^{\tau} \binom{m}{i} 2^{-m}.$$

4 Complexity Analysis and Implementation Issues

Next, we compute the overall storage, communication and computation complexity. The main challenge for the *GHB* protocol is to efficiently implement the function $\boldsymbol{\Phi}$ and for that we provide implementation directives.

Storage Complexity. The memory cost for the tag; i.e. the storage for the two secret matrices, is $(k_{\mathbf{X}} \cdot m + k_{\mathbf{Y}} \cdot m)$ bits.

Communication Complexity. The protocol requires $(k_{\mathbf{Y}} + k_{\mathbf{X}} + m)$ bits to be transfered in total.

Computational Complexity. We concentrate on the computationally weaker of the two entities; i.e. the tag. We distinguish two main operations, the multiplication of the random vectors with the secret matrices and the application of the function $\boldsymbol{\Phi}$ for the computation of $\boldsymbol{z}$. The two multiplications require in total approximately $\mathcal{O}(k_{\mathbf{X}} \cdot m + k_{\mathbf{Y}} \cdot m)$ basic binary operations. This is, also, the computational complexity of the $HB^{\#}$ protocol. For the implementation of $\boldsymbol{\Phi}$, we propose the following approach.

Let $\{\gamma, \gamma^2, \gamma^{2^2}, \cdots, \gamma^{2^{m-1}}\}$ be a normal basis of $\mathbb{F}_{2^m}$ over $\mathbb{F}_2$, where $\gamma \in \mathbb{F}_{2^m}$. It is well known that there is such basis for any $m > 1$ ([17]). The implementation of a Gold function x^{2^i+1} requires one exponentiation and one multiplication. By $\otimes$ we demote the multiplication of two field elements of the field. When a field element $x \in \mathbb{F}_{2^m}$ is represented in normal form; i.e. $x = \sum_{i=0}^{m-1} x(i)\gamma^{2^i}$, the

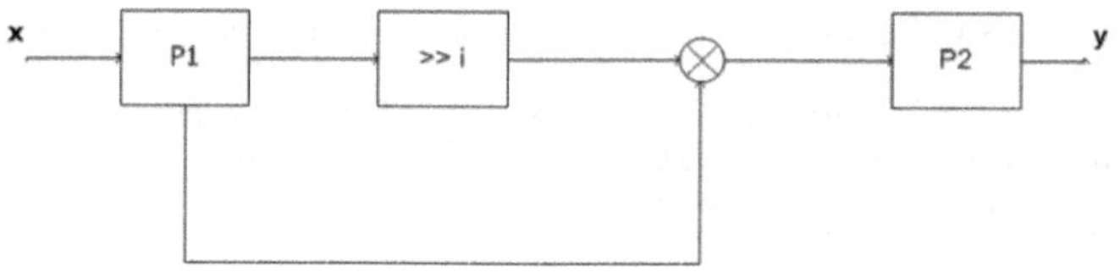

Fig. 3. The implementation of the function $\boldsymbol{\Phi}$

exponentiation can be performed by right cyclic shift of the binary representation $\boldsymbol{x}$. Thus, x^{2^i} is obtained by simply shifting $\boldsymbol{x}$ to the right by i steps.

Concerning the multiplication $x \cdot x^{2^i}$, one of the most straightforward ways to perform efficiently a normal basis multiplication is the one proposed by Massey and Omura [19]. More precisely, for each normal basis there is an $m \times m$ matrix M called the multiplication matrix of the normal basis and if $\boldsymbol{x}_1$ and $\boldsymbol{x}_2$ is, respectively, the binary vector representation of the elements $x_1, x_2 \in \mathbb{F}_{2^m}$ with respect to the basis, then the binary representation of the product $y = x_1 x_2$ is computed as $y(m-1-i) = h(x_1^{2^i}, x_2^{2^i})$, for $0 \leq i \leq m-1$, where $h(x_1, x_2) = \boldsymbol{x}_1 M \boldsymbol{x}_2^T$. The complexity of the operation is determined by the number C_N of ones of M. It is proved that the number of required AND and XOR gates is C_N and $C_N - 1$ respectively. When an optimal normal basis is used, then $C_N = 2m+1$, and we have the least possible complexity.

Finally, since the function $\boldsymbol{\Phi}$ belongs to $\mathcal{G}(m,d)$, any function affine equivalent to a Gold function can be used. This function can be implemented from the proposed construction for the Gold functions by multiplying the input and the output vectors by the $m \times m$ matrix that corresponds to the left and right affine permutations, respectively. If P_1 and P_2 are these two matrices, then the total computation is given in Fig. 3. The complexity for each one of the permutations can vary from constant to at most $\mathcal{O}(m^2)$. Thus, the computational complexity of the $\boldsymbol{\Phi}$ function varies from $\mathcal{O}(m)$ to $\mathcal{O}(m^2)$ depending on the choice of the permutations and the total computational complexity of the protocol is at most $\mathcal{O}(k_{\mathbf{X}} \cdot m + k_{\mathbf{Y}} \cdot m + m^2)$.

To summarize the $GHB^{\#}$ protocol has the same communication and storage complexity as the its predecessor $HB^{\#}$, while it requires at least $\mathcal{O}(m)$ (and at most $\mathcal{O}(m^2)$) more basic binary computations (Table 1). In Table 2, we use practical parameters that have been proposed for the $HB^{\#}$ protocol in order to compare the efficiency of the two protocols.

Table 1. Complexity Comparison between $GHB^{\#}$ and $HB^{\#}$

	Security	Stor. Compl.	Comm. Compl.	Comp. Compl.
$HB^{\#}$	Active	$\mathcal{O}(k_{\mathbf{X}} \cdot m + k_{\mathbf{Y}} \cdot m)$	$\mathcal{O}(k_{\mathbf{X}} + k_{\mathbf{Y}} + m)$	$\mathcal{O}(k_{\mathbf{X}} \cdot m + k_{\mathbf{Y}} \cdot m)$
$GHB^{\#}$	MIM	$\mathcal{O}(k_{\mathbf{X}} \cdot m + k_{\mathbf{Y}} \cdot m)$	$\mathcal{O}(k_{\mathbf{X}} + k_{\mathbf{Y}} + m)$	$\mathcal{O}(k_{\mathbf{X}} \cdot m + k_{\mathbf{Y}} \cdot m + m)$

Table 2. Comparison between the $GHB^{\#}$ and $HB^{\#}$ protocols for practical parameters

k_X	k_Y	m	η	τ	Stor. $GHB^{\#}$	Stor. $HB^{\#}$	Comm. $GHB^{\#}$	Comm. $HB^{\#}$	Comp. $GHB^{\#}$	Comp. $HB^{\#}$
80	512	1163	0.25	405	$688k$	$688k$	$1.7k$	$1.7k$	$689k$	$688k$
80	512	441	0.125	113	$261k$	$261k$	$1k$	$1k$	$261k$	$261k$

5 Security Analysis

5.1 Definition of Security Models

Following the notation used in [8], we use $\mathcal{R}_{\boldsymbol{X},\boldsymbol{Y},\tau}$ to denote the algorithm that it is run by the reader (verifier) and $\mathcal{T}_{\boldsymbol{X},\boldsymbol{Y},\eta}$ the one run by a legitimate tag (prover). We use $\boldsymbol{X} \xleftarrow{\$} \mathbb{F}_2^{(m_1,m_2)}$ to indicate the random selection of a $m_1 \times m_2$ binary matrix $\boldsymbol{X}$.

All the attacks against HB^+ and its variants are active ones; i.e. the attacker can interact with the reader and/or the tag and change some of the messages exchanged between the two legitimate entities. We distinguish two models of security, the $DET-model$ and the $MIM-model$. In each of the models the adversary runs in two stages. In the first stage she has some interaction with the prover and/or the verifier and in the second she interacts only with the verifier and wins if the verifier returns *accept*. We define the advantage of an attacker $\mathcal{A}$ against $GHB^{\#}$ in the models as the overhead success probability over P_{FA}; i.e. the best possible soundness error we can hope for is the success probability when the attacker does not perform any action during the first phase of the attack and just sends a randomly selected $\boldsymbol{z}$ in the second phase. Note that, P_{FA} is negligible for the chosen values of τ and for security $m = \Theta(k)$, where k is the security parameter. In the $DET-model$ the attacker interacts only with an honest prover for a polynomial number of times. More precisely,

Definition 7. *(DET-model). In the $DET-model$ the attack is carried in two phases:*

- **Phase 1.** *Adversary $\mathcal{A}$ interacts q times with the honest tag $\mathcal{T}_{\boldsymbol{X},\boldsymbol{Y},\eta}$. More precisely, on the i-th invocation, $\mathcal{T}_{\boldsymbol{X},\boldsymbol{Y},\eta}$ internally generates a random blinding vector $\boldsymbol{b}_i$, it takes a challenge $\boldsymbol{a}_i$ from $\mathcal{A}$ as input and outputs $\boldsymbol{z}_i = \boldsymbol{\Phi}(\boldsymbol{a}_i \cdot \boldsymbol{X}) \oplus \boldsymbol{\Phi}(\boldsymbol{b}_i \cdot \boldsymbol{Y}) \oplus \boldsymbol{\nu}_i$ and sends the message to $\mathcal{A}$.*
- **Phase 2.** *Adversary $\mathcal{A}$ interacts with the reader $\mathcal{R}_{\boldsymbol{X},\boldsymbol{Y},\tau}$ trying to impersonate the tag $\mathcal{T}_{\boldsymbol{X},\boldsymbol{Y},\eta}$ with advantage*

$$Adv_{\mathcal{A}}^{DET}(k_{\boldsymbol{X}}, k_{\boldsymbol{Y}}, m, \eta, \tau, q) =$$

$$Pr[\boldsymbol{X} \xleftarrow{\$} \mathbb{F}_2^{(k_{\boldsymbol{X}},m)}, \boldsymbol{Y} \xleftarrow{\$} \mathbb{F}_2^{(k_{\boldsymbol{Y}},m)}, \mathcal{A}^{\mathcal{T}_{\boldsymbol{X},\boldsymbol{Y},\eta}}(1^k) : \langle \mathcal{A}, \mathcal{R}_{\boldsymbol{X},\boldsymbol{Y},\tau} \rangle = ACC] - P_{FA}.$$

In the $MIM-model$ the attacker can interact with both the prover and the verifier and learn the verifier's decision, *accept* or *reject*.

Definition 8. *(MIM-model). In the $MIM-model$ the attack is carried in two phases and the adversary can manipulate all messages exchanged between the tag and the reader:*

- **Phase 1.** *On the i-th invocation,* $\mathcal{T}_{\boldsymbol{X},\boldsymbol{Y},\eta}$ *internally generates a random blinding vector* $\boldsymbol{b}_i$ *and sends it to the adversary* $\mathcal{A}$*. The reader* $\mathcal{R}_{\boldsymbol{X},\boldsymbol{Y},\tau}$ *receives a modified blinding vector* $\hat{\boldsymbol{b}}_i = \bar{\boldsymbol{b}} \oplus \boldsymbol{b}_i$ *from* $\mathcal{A}$*. Then, the reader generates a challenge vector* $\boldsymbol{a}_i$ *and sends it to the adversary* $\mathcal{A}$*. The tag receives a modified challenge vector* $\hat{\boldsymbol{a}}_i = \bar{\boldsymbol{a}} \oplus \boldsymbol{a}_i$ *from* $\mathcal{A}$ *and replies with* $\boldsymbol{z}_i = \boldsymbol{\Phi}(\hat{\boldsymbol{a}}_i \cdot \boldsymbol{X}) \oplus \boldsymbol{\Phi}(\boldsymbol{b}_i \cdot \boldsymbol{Y}) \oplus \boldsymbol{\nu}_i$, $\boldsymbol{\nu}_i \in Ber(m,\eta)$*. The reader receives a modified vector* $\hat{\boldsymbol{z}}_i = \bar{\boldsymbol{z}} \oplus \boldsymbol{z}_i$ *and if* $\mathrm{wt}(\hat{\boldsymbol{z}}_i \oplus \boldsymbol{\Phi}(\boldsymbol{a}_i \cdot \boldsymbol{X}) \oplus \boldsymbol{\Phi}(\hat{\boldsymbol{b}}_i \cdot \boldsymbol{Y})) \le \tau$*, then the reader outputs accept. Otherwise, it outputs reject. The adversary interferes for q executions of the of the protocol.*
- **Phase 2.** *Adversary* $\mathcal{A}$ *interacts with the reader* $\mathcal{R}_{\boldsymbol{X},\boldsymbol{Y},\tau}$ *trying to impersonate the tag* $\mathcal{T}_{\boldsymbol{X},\boldsymbol{Y},\eta}$ *with advantage*

$$Adv_{\mathcal{A}}^{MIM}(k_{\boldsymbol{X}}, k_{\boldsymbol{Y}}, m, \eta, \tau, q) = Pr[\boldsymbol{X} \xleftarrow{\$} \mathbb{F}_2^{(k_{\boldsymbol{X}},m)},$$

$$\boldsymbol{Y} \xleftarrow{\$} \mathbb{F}_2^{(k_{\boldsymbol{Y}},m)}, \mathcal{A}^{\mathcal{T}_{\boldsymbol{X},\boldsymbol{Y},\eta},\mathcal{R}_{\boldsymbol{X},\boldsymbol{Y},\tau}}(1^k) : \langle \mathcal{A}, \mathcal{R}_{\boldsymbol{X},\boldsymbol{Y},\tau}\rangle = ACC] - P_{FA}.$$

Note 1. As we have all ready described in Section 1, most variants of the HB^+ protocol are secure under the $DET-model$. However, the attack presented in [7], the *GRS attack*, against the HB^+ protocol was easily applied to most of these variants ([9]). This attack is included in the $MIM-model$, but the adversary in the first phase is limited to modify only the messages that the reader sends. The $HB^{\#}$ protocol was the first one provably secure against the GRS attack, but it was shown to be weak under the $MIM-model$ ([21]).

5.2 Security Proofs

Next we prove that the $GHB^{\#}$ protocol is secure under both the $DET-model$ and the $MIM-model$ given the hardness of the LPN problem. Our reduction is using rewinding, like in the case of the HB^+ and $HB^{\#}$ protocols and the security is based on the MHB puzzle. We say that a function in x is negligible if it vanishes faster then the inverse of any polynomial in x.

Lemma 2. *There is* $\hat{\boldsymbol{\Phi}} \in \mathcal{G}(m,d)$*, such that for each* $\boldsymbol{x}_1, \boldsymbol{x}_2 \in K(\hat{\boldsymbol{\Phi}})$*, it holds that* $\hat{\boldsymbol{\Phi}}(\boldsymbol{x}_1 \oplus \boldsymbol{x}_2) = \hat{\boldsymbol{\Phi}}(\boldsymbol{x}_1) \oplus \hat{\boldsymbol{\Phi}}(\boldsymbol{x}_2) = (\boldsymbol{y}||\boldsymbol{0}_{|I_{\hat{\boldsymbol{\Phi}}}|})$*, for some* $\boldsymbol{y} \in \mathbb{F}_2^{m-|I_{\hat{\boldsymbol{\Phi}}}|}$.

Proof. The linearity derives directly for the definition of the subspace $K(\hat{\boldsymbol{\Phi}})$. Next, we prove the existence of such $\hat{\boldsymbol{\Phi}}$.

By definition, every $\boldsymbol{x} \in K(\boldsymbol{\Phi})$ has $|I_{\boldsymbol{\Phi}}|$ entries fixed to 0 and every Boolean function $\boldsymbol{\Phi} \in \mathcal{G}(m,d)$ is a linear function in the subspace $\boldsymbol{x} \in K(\boldsymbol{\Phi})$. That is, for $\boldsymbol{x} \in K(\boldsymbol{\Phi})$, $\boldsymbol{\Phi}$ can be seen as a function with $m - |I_{\boldsymbol{\Phi}}|$ input variables and m outputs.

Since the number of inputs is less than the number of outputs, $|I_{\boldsymbol{\Phi}}|$ of the output bits can written as a linear combination of the other $m - |I_{\boldsymbol{\Phi}}|$; i.e. there is a linear transformation $\boldsymbol{M}$ that can be applied to the output of $\boldsymbol{\Phi}$ and results to $|I_{\hat{\boldsymbol{\Phi}}}|$ zero output bits, for $\boldsymbol{x} \in K(\boldsymbol{\Phi})$. Also, as any permutation $\boldsymbol{P}$ of the outputs is acceptable, these zero outputs can be put last. From the composition $\hat{\boldsymbol{\Phi}} = \boldsymbol{P} \circ \boldsymbol{M} \circ \boldsymbol{\Phi}$, of $\boldsymbol{\Phi}$ with the linear transformation and the permutation with $\hat{\boldsymbol{\Phi}}$, the result follows. □

Theorem 1. *(Security in the DET-model) If there is an adversary $\mathcal{A}^{\#}$ that can attack the GHB$^{\#}$ protocol, with parameters $(k_{\boldsymbol{X}}, k_{\boldsymbol{Y}}, m^{\#}, \eta, \tau)$, in the DET-model by interacting with an honest tag $q^{\#}$ times, running time $T^{\#}$ and achieving advantage at least $\delta^{\#}$, then, there is an adversary $\mathcal{A}$ that can solve the $(k_{\boldsymbol{Y}}, m^{\#} - |I_{\boldsymbol{\Phi}}|, \eta, q) -$ MHB puzzle with parameters in running time $T = 2qT^{\#}$ and success probability $\delta > (\frac{1}{2^m} + \frac{\delta^{\#}}{4})$, where $q = m^{\#} q^{\#} L(2 + \log_2 q^{\#})$, $L \geq \frac{2}{\epsilon'^2} \ln(\frac{1}{1-e^{-\frac{ln2}{m}}})$, $\epsilon' = \frac{\delta^{\#3}}{16}(\frac{1}{2} - \frac{\tau}{m})^3(\frac{1}{2} - \frac{1}{k_{\boldsymbol{X}}})$ and $\boldsymbol{\Phi}$ is the Gold linearly equivalent function used in GHB$^{\#}$.*

Proof. The adversary $\mathcal{A}$ has obtained q pairs $(\boldsymbol{b}_i, \boldsymbol{z}_i)$ from the *MHB* puzzle generator, where $\boldsymbol{z}_i = \boldsymbol{b}_i \cdot \boldsymbol{Y} \oplus \boldsymbol{\nu}_i$, $1 \leq i \leq q$, and $\boldsymbol{Y}$ is a randomly selected $k \times m$ binary matrix. Let $\boldsymbol{b}$ be the k-bit challenge vector of the puzzle; i.e. she has to compute $\boldsymbol{z} = \boldsymbol{b} \cdot \boldsymbol{Y}$. We will show how the adversary $\mathcal{A}$ can solve the *MHB* puzzle using the algorithm of the adversary $\mathcal{A}^{\#}$. The proof is a modified version of the proof introduced for the security reduction in the DET-model in [8].

During the attack, $\mathcal{A}^{\#}$ interrogates a legitimate tag and $\mathcal{A}$ simulates the behaviour of the tag algorithm. The function $\boldsymbol{\Phi}$ is chosen to be one that has the properties described in Lemma 2. Let $\boldsymbol{X}^{\#}$ and $\boldsymbol{Y}^{\#}$ be the two secret matrices shared between the tag and the reader. The $k_{\boldsymbol{X}} \times m^{\#}$ matrix $\boldsymbol{X}^{\#}$ has all the entries randomly selected except the i-th column $\boldsymbol{X}^{\#}(:, i)$, for all $i \in I_{\boldsymbol{\Phi}}$, and the s-th row $\boldsymbol{X}^{\#}(s, :)$, for a random row $1 \leq s \leq k_{\boldsymbol{X}}$, that are all zero. Similarly, the matrix $\boldsymbol{Y}^{\#}$ is a $k_{\boldsymbol{Y}} \times m^{\#}$ binary matrix that has also the i-th column $\boldsymbol{Y}^{\#}(:, i)$ all zero, for all $i \in I_{\boldsymbol{\Phi}}$, while all the other entries of the matrix are randomly selected.

$\mathcal{A}$ divides the q pairs, that she has obtained from the *MHB* puzzle, in m sets with $Lq^{\#}(1 + r)$ pairs each. L is the number of estimations for each bit of the vector $\boldsymbol{z}$ that the adversary gets from each set and r defines the size of a pool of extra pairs that she can use in each estimation. The vector $\boldsymbol{e}$ of length m stores intermediate values and it is initialised $\boldsymbol{e} = \boldsymbol{0}_m$. We use '||' to denote the concatenation of vectors.
For $j_0 = 0$ to $m - 1$ do:

1. For $j_1 = 0$ to $L - 1$ do:
   (a) Phase I: For $j_2 = 0$ to $q^{\#} - 1$ do:
      i. $\mathcal{A}$ selects random bit $c \in \mathbb{F}_2$ and sends $\boldsymbol{b}^{\#}_{j_0,j_1,j_2} = \boldsymbol{b}_{(j_0 L + j_1)q^{\#} + j_2} \oplus c \cdot \boldsymbol{b}$.
      ii. Let $\boldsymbol{a}^{\#}_{j_0,j_1,j_2}$ be the challenge vector send by the attacker $\mathcal{A}^{\#}$.
      iii. If $\boldsymbol{a}^{\#}_{j_0,j_1,j_2}(r) = c$, then $\mathcal{A}$ replies with

$$\boldsymbol{z}^{\#}_{j_0,j_1,j_2} = \boldsymbol{\Phi}(\boldsymbol{a}^{\#}_{j_0,j_1,j_2} \boldsymbol{X}^{\#}) \oplus \boldsymbol{z}^{\#}_{(j_0 L + j_1)q^{\#} + j_2}, \qquad (1)$$

where the second term is given by

$$\boldsymbol{z}^{\#}_{(j_0 L + j_1)q^{\#} + j_2} = \left(\boldsymbol{z}_{(j_0 L + j_1)q^{\#} + j_2} || \boldsymbol{\mu}^{\#}_{(j_0 L + j_1)q^{\#} + j_2}\right),$$

and $\boldsymbol{\mu}^{\#}_{(j_0 L + j_1)q^{\#} + j_2} \in Ber(|I_{\boldsymbol{\Phi}}|, \eta)$. Otherwise, she rewinds adversary $\mathcal{A}^{\#}$ to the beginning of the current query and uses a new pair. If the available pairs are exhausted, guess the message $\boldsymbol{z}^{\#}_{(j_0 L + j_1)q^{\#} + j_2}$.

(b) Phase II: $\mathcal{A}$ proceeds to the second, impersonation, phase of the DET-model attack.
   i. Adversary $\mathcal{A}^{\#}$ sends the commitment vector $\boldsymbol{b}'$.
   ii. Adversary $\mathcal{A}$ chooses two challenges $\boldsymbol{a}_0'$ and $\boldsymbol{a}_1'$ with complement values at the s-th bit; i.e. $\boldsymbol{a}_0'(s) \oplus \boldsymbol{a}_1'(s) = 1$.
   iii. Adversary $\mathcal{A}$ transmits $\boldsymbol{a}_0'$ and gets the reply $\boldsymbol{z}_0'$.
   iv. Adversary $\mathcal{A}$ rewinds the attacker $\mathcal{A}^{\#}$ just after the transmission of $\boldsymbol{b}'$ and sends $\boldsymbol{a}_1'$ to get the reply $\boldsymbol{z}_1'$.
   v. Adversary $\mathcal{A}$ computes the sum

$$\boldsymbol{z}' = \boldsymbol{z}_0' \oplus \boldsymbol{z}_1' \oplus \boldsymbol{\Phi}\left(\sum_{i=1, i\neq s}^{k_{\boldsymbol{X}}} \boldsymbol{a}_0'(i)\boldsymbol{X}^{\#}(:,i)\right) \oplus \boldsymbol{\Phi}\left(\sum_{i=1, i\neq s}^{k_{\boldsymbol{X}}} \boldsymbol{a}_1'(i)\boldsymbol{X}^{\#}(:,i)\right) \tag{2}$$

and adds the value of the j_0-th bit to $\boldsymbol{e}(j_0)$, i.e $\boldsymbol{e}(j_0) = \boldsymbol{e}(j_0) + \boldsymbol{z}'(j_0)$.

2. The estimation of the bit $\boldsymbol{z}(j_0)$ is given by majority decision; i.e.

$$\boldsymbol{z}(j_0) = \boldsymbol{e}(j_0)/L \mod 2.$$

We will show that the attacker $\mathcal{A}$ successfully simulates a tag algorithm that uses a $k_{\boldsymbol{Y}} \times m^{\#}$ binary matrix $\boldsymbol{Y}^{\#}$ that has the i-th column $\boldsymbol{Y}^{\#}(:,i)$ all zero, for all $i \in I_{\boldsymbol{\Phi}}$, i.e. from Lemma 2, $\boldsymbol{\Phi}(\boldsymbol{b}_1 \cdot \boldsymbol{Y}^{\#}) \oplus \boldsymbol{\Phi}(\boldsymbol{b}_2 \cdot \boldsymbol{Y}^{\#}) = \boldsymbol{\Phi}((\boldsymbol{b}_1 \oplus \boldsymbol{b}_2) \cdot \boldsymbol{Y}^{\#})$ and the last $|I_{\boldsymbol{\Phi}}|$ of the output of are all zero. Finally, the secret matrix $\boldsymbol{Y}^{\#}$ is such that $\boldsymbol{\Phi}(\boldsymbol{b}_i \cdot \boldsymbol{Y}^{\#}) = (\boldsymbol{b}_i \cdot \boldsymbol{Y} \mid\mid \boldsymbol{0}_{|I_{\boldsymbol{\Phi}}|})$.

When $\boldsymbol{a}^{\#}_{j_0,j_1,j_2}(r) = 0$, the reply (1) of $\mathcal{A}$ can be written as

$$\begin{aligned}
\boldsymbol{z}^{\#}_{j_0,j_1,j_2} &= \boldsymbol{\Phi}(\boldsymbol{a}^{\#}_{j_0,j_1,j_2}\boldsymbol{X}^{\#}) \oplus \left(\boldsymbol{z}_{(j_0L+j_1)q^{\#}+j_2} || \boldsymbol{\mu}^{\#}_{(j_0L+j_1)q^{\#}+j_2}\right) \\
&= \boldsymbol{\Phi}(\boldsymbol{a}^{\#}_{j_0,j_1,j_2}\boldsymbol{X}^{\#}) \oplus (\boldsymbol{b}_{(j_0L+j_1)q^{\#}+j_2} \cdot \boldsymbol{Y} || \boldsymbol{0}_{I_{\boldsymbol{\Phi}}}) \\
&\oplus \left(\boldsymbol{\nu}_{(j_0L+j_1)q^{\#}+j_2} || \boldsymbol{\mu}^{\#}_{(j_0L+j_1)q^{\#}+j_2}\right)
\end{aligned}$$

and since $\boldsymbol{\nu}_{(j_0L+j_1)q^{\#}+j_2} \in Ber(m - |I_{\boldsymbol{\Phi}}|, \eta)$ and $\boldsymbol{\mu}^{\#}_{(j_0L+j_1)q^{\#}+j_2} \in Ber(|I_{\boldsymbol{\Phi}}|, \eta)$, it holds that, $\left(\boldsymbol{\nu}_{(j_0L+j_1)q^{\#}+j_2} || \boldsymbol{\mu}^{\#}_{(j_0L+j_1)q^{\#}+j_2}\right) \in Ber(m, \eta)$.

Due to the specific choice of the matrix $\boldsymbol{X}^{\#}$, the function $\boldsymbol{\Phi}$ is restricted to $K(\boldsymbol{\Phi})$ and behaves like a linear function. Thus, for $\boldsymbol{a}^{\#}_{j_0,j_1,j_2}(r) = 1$, the reply (1) of $\mathcal{A}$ can be written as

$$\begin{aligned}
\boldsymbol{z}^{\#}_{j_0,j_1,j_2} = \boldsymbol{\Phi}\left(\sum_{i=1, i\neq s}^{k_{\boldsymbol{X}}} \boldsymbol{a}^{\#}_{j_0,j_1,j_2}(i)\boldsymbol{X}^{\#}(:,i)\right) \oplus \boldsymbol{\Phi}(\boldsymbol{X}^{\#}(s,:)) \\
\oplus (\boldsymbol{b} \cdot \boldsymbol{Y} || \boldsymbol{0}_{I_{\boldsymbol{\Phi}}}) \oplus (\boldsymbol{b}_{(j_0L+j_1)q^{\#}+j_2} \cdot \boldsymbol{Y} || \boldsymbol{0}_{I_{\boldsymbol{\Phi}}}) \oplus \left(\boldsymbol{\nu}_{(j_0L+j_1)q^{\#}+j_2} || \boldsymbol{\mu}^{\#}_{(j_0L+j_1)q^{\#}+j_2}\right)
\end{aligned}$$

and, again, it holds that, $\left(\boldsymbol{\nu}_{(j_0L+j_1)q^{\#}+j_2} || \boldsymbol{\mu}^{\#}_{(j_0L+j_1)q^{\#}+j_2}\right) \in Ber(m, \eta)$. From the above, we have that only the first $m - |I_{\boldsymbol{\Phi}}|$ entries of $\boldsymbol{z}'$ given in (2) provide an estimation of the puzzle's answer $\boldsymbol{z}$.

Next, we compute necessary amount of estimations L for the majority strategy to give the correct value of $\boldsymbol{z}$ with significant advantage δ. For the computation of r, L, δ, we follow mainly the approach presented in [8].

For each of the L estimations of a single bit of $\boldsymbol{z}$, the attacker has $r \cdot q^{\#}$ extra pairs, where $r = 1 + log_2 q^{\#}$ and these pairs will be sufficient with probability more than 1/2.

The guess of $\boldsymbol{z}$ is correct if either both $\boldsymbol{z}'_0$ and $\boldsymbol{z}'_1$ are correct or if both are false. From [12], the probability this to happen is greater than $p = \frac{1}{2} + \frac{\epsilon^3}{2} - \frac{\epsilon^3+1}{k_{\boldsymbol{X}}}$, where $\epsilon = \frac{\delta^{\#}}{2}(\frac{1}{2} - \frac{\tau}{m})$. Since, the probability of guessing the message $\boldsymbol{z}^{\#}_{(j_0L+j_1)q^{\#}+j_2}$ is less than 1/2, the probability of correct guessing one of the m bits of $\boldsymbol{z}$ is lower bounded by $1/4 + p/2 \geq 1/2 + \epsilon'$, where $\epsilon' = \frac{\epsilon^3}{4} - \frac{\epsilon^3+1}{2k_{\boldsymbol{X}}}$.

Finally, from Chernoff bound on the majority of the L experiments, the guess of all m bits is lower bounded by $p^{MHB} \geq \left(1 - e^{-\frac{L\epsilon'^2}{2}}\right)^m$. Thus, for the probability p^{MHB} to be greater than 1/2, the number of experiments must be at least

$$L \geq \frac{2}{\epsilon'^2} \ln\left(\frac{1}{1 - e^{-\frac{\ln(2)}{m}}}\right).$$

□

From Theorem 1, any efficient adversary achieving a noticeable advantage $\delta^{\#}$ against the *GHB*$^{\#}$ protocol in the DET-model can be turned into an efficient solver of the MHB puzzle with a success probability greater than $\frac{1}{2^m}$ by $\frac{\delta^{\#}}{4}$, and, from Lemma 1, this contradicts the hardness assumption of the LPN problem.

Lemma 3. *Let* $\boldsymbol{\Phi} \in \mathcal{G}(m,d)$ *and let* $\mathcal{X}$ *and* $\mathcal{Y}$ *be two sets of randomly selected binary vectors of length* m *with cardinality* $|\mathcal{X}| = 2^{k_{\boldsymbol{X}}}$ *and* $|\mathcal{Y}| = 2^{k_{\boldsymbol{Y}}}$*, respectively, and* $k_{\boldsymbol{X}} \leq k_{\boldsymbol{Y}}$*. Then, for given* $(\bar{\boldsymbol{b}}, \bar{\boldsymbol{a}}, \bar{\boldsymbol{z}}) \in \mathbb{F}_2^m \times \mathbb{F}_2^m \times \mathbb{F}_2^m$*, the probability*

$$p(d) = Pr\left[\mathrm{wt}(D_{\bar{\boldsymbol{a}}}\boldsymbol{\Phi}(\boldsymbol{x}) \oplus D_{\bar{\boldsymbol{b}}}\boldsymbol{\Phi}(\boldsymbol{y}) \oplus \bar{\boldsymbol{z}}) \leq d\right],\ 1 \leq d \leq n$$

where $\boldsymbol{x} \in \mathcal{X}$ *and* $\boldsymbol{y} \in \mathcal{Y}$*, is upper bounded by*

$$p(d) \leq 2^{-\min(k_{\boldsymbol{X}},m)+2+mH(\frac{d}{m})}.$$

$H(s) = s \cdot \log_2(\frac{1}{s}) - (1-s) \cdot \log_2(\frac{1}{1-s})$ *is the entropy function.*

Proof. From the Definition 4 of APN functions, for given $\bar{\boldsymbol{a}}$ there is a subset $\mathcal{S}_{\bar{\boldsymbol{a}}} \subseteq \mathbb{F}_2^m$, such that for every $\boldsymbol{c} \in \mathcal{S}_{\bar{\boldsymbol{a}}}$ there is $\boldsymbol{x} \in \mathcal{X}$ satisfying $D_{\bar{\boldsymbol{a}}}\boldsymbol{\Phi}(\boldsymbol{x}) = \boldsymbol{c}$. Since each $\boldsymbol{c} \in \mathcal{S}_{\bar{\boldsymbol{a}}}$ can appear at most twice, it holds that $\min(2^{k_{\boldsymbol{X}}-1}, 2^{m-1}) \leq |\mathcal{S}_{\bar{\boldsymbol{a}}}| \leq \min(2^{k_{\boldsymbol{X}}}, 2^{m-1})$. Similarly, we define $\mathcal{S}_{\bar{\boldsymbol{b}}} \subseteq \mathbb{F}_2^m$, such that for every $\boldsymbol{c} \in \mathcal{S}_{\bar{\boldsymbol{b}}}$ there is $\boldsymbol{y} \in \mathcal{Y}$ such that $D_{\bar{\boldsymbol{b}}}\boldsymbol{\Phi}(\boldsymbol{y}) = \boldsymbol{c}$, with $\boldsymbol{c} \in \mathcal{S}_{\bar{\boldsymbol{b}}}$ and $\min(2^{k_{\boldsymbol{Y}}-1}, 2^{m-1}) \leq |\mathcal{S}_{\bar{\boldsymbol{b}}}| \leq \min(2^{k_{\boldsymbol{Y}}}, 2^{m-1})$.

All the sums $\boldsymbol{c} = \boldsymbol{c}_1 \oplus \boldsymbol{c}_2$ of a given vector $\boldsymbol{c}_1 \in \mathcal{S}_{\bar{\boldsymbol{a}}}$ with any $\boldsymbol{c}_2 \in \mathcal{S}_{\bar{\boldsymbol{b}}}$, are different. Thus, the sum $D_{\bar{\boldsymbol{a}}}\boldsymbol{\Phi}(\boldsymbol{x}) \oplus D_{\bar{\boldsymbol{b}}}\boldsymbol{\Phi}(\boldsymbol{y}) = \boldsymbol{c}$, with $\boldsymbol{x} \in \mathcal{X}$ and $\boldsymbol{y} \in \mathcal{Y}$ can take the same value $\boldsymbol{c}$ with probability at most

$$\frac{2}{2^{\min(k_{\boldsymbol{X}},m)}} \frac{2}{2^{\min(k_{\boldsymbol{Y}},m)}} 2^{\max(|\mathcal{S}_{\bar{\boldsymbol{a}}}|,|\mathcal{S}_{\bar{\boldsymbol{b}}}|)} \leq \frac{1}{2^{\min(k_{\boldsymbol{X}},m)-2}}.$$

Given that the number of binary vectors of length m and Hamming weight less than d is $\sum_{i=0}^{d} \binom{m}{i} \leq 2^{mH(\frac{d}{m})}$, the probability $D_{\bar{\boldsymbol{a}}}\boldsymbol{\Phi}(\boldsymbol{x}) \oplus D_{\bar{\boldsymbol{b}}}\boldsymbol{\Phi}(\boldsymbol{y}) = \boldsymbol{c}$ to have Hamming weight less than $d \leq n$ is upper bounded by $2^{-\min(k_{\boldsymbol{X}},m)+2+mH(\frac{d}{m})}$. To conclude, clearly, the constant value $\bar{\boldsymbol{z}}$ does not influence this probability. □

Theorem 2. *(Security in the MIM-model) If there is an adversary $\mathcal{A}^{\#}$ that can attack the $GHB^{\#}$ protocol with parameters $(k_{\boldsymbol{X}}, k_{\boldsymbol{Y}}, m^{\#}, \eta, \tau)$ in the MIM-model by modifying $q^{\#}$ protocol executions between an honest tag and the reader, with running time $T^{\#}$ and achieving advantage at least $\delta^{\#}$, then, there is an adversary $\mathcal{A}$ that can attack the $GHB^{\#}$ protocol in the DET-model with the same parameters by interrogating an honest tag $q^{\#}$ times, with running time at most $T^{\#}$ and with advantage at least $\delta \geq \delta^{\#} - (P_{FA} + \delta^{\#})q^{\#}p_r$, where p_r is a negligible function.*

Proof. The attacker $\mathcal{A}$ has a legitimate tag at her disposal that she can interrogate. We will show how $\mathcal{A}$ can attack $GHB^{\#}$ protocol in the DET-model using the algorithm that the adversary $\mathcal{A}^{\#}$ executes.

During the MIM attack, $\mathcal{A}^{\#}$ is modifying all messages between the legitimate tag and reader. While, the adversary $\mathcal{A}$ has access to an honest tag, she has to simulate the behavior of the reader. More precisely, her strategy goes as follows.

1. $\mathcal{A}$ receives from the honest tag $\mathcal{T}_{\boldsymbol{X},\boldsymbol{Y},\eta}$ a blinding vector $\boldsymbol{b}$ and sends this vector to the $\mathcal{A}^{\#}$.
2. $\mathcal{A}^{\#}$ produces a new blinding vector $\hat{\boldsymbol{b}} = \boldsymbol{b} \oplus \bar{\boldsymbol{b}}$ and sends this vector to the simulated reader; i.e. to the adversary $\mathcal{A}$.
3. $\mathcal{A}$ produces a random challenge vector $\boldsymbol{a}$, on behalf of the reader and sends it to $\mathcal{A}^{\#}$.
4. $\mathcal{A}^{\#}$ produces a new challenge vector $\hat{\boldsymbol{a}} = \boldsymbol{a} \oplus \bar{\boldsymbol{a}}$ and sends this vector to the honest tag, via $\mathcal{A}$.
5. the tag responds with $\boldsymbol{z}$ and $\mathcal{A}$ sends the response to $\mathcal{A}^{\#}$.
6. $\mathcal{A}^{\#}$ produces a new response vector $\hat{\boldsymbol{z}} = \boldsymbol{z} \oplus \bar{\boldsymbol{z}}$ and sends this vector to the simulated reader; i.e. to adversary $\hat{\mathcal{A}}$.
7. If the triplet $(\bar{\boldsymbol{b}}, \bar{\boldsymbol{a}}, \bar{\boldsymbol{z}})$ is all-zero, the simulated reader; i.e. $\mathcal{A}$, notifies adversary $\mathcal{A}^{\#}$ that the tag has been accepted. Otherwise, it is rejected.

The previous steps are repeated $q^{\#}$ times. The adversary $\mathcal{A}$ impersonates the tag to an honest reader in the DET-attack, by using the second phase of $\mathcal{A}^{\#}$.

The probability of successfully simulating a reader's behavior depends on the ability of the adversary to simulate the last step; i.e. the acceptance or rejection of the tag. Let p_{auth} be this probability, then the overall probability of the attack is given by

$$p_{MIM} = p_{auth} \cdot (P_{FA} + \delta). \tag{3}$$

We will compute p_{MIM}. In order for the attack to be successful, the adversary $\mathcal{A}$ must be able to simulate the reader's behavior for $q^{\#}$ consecutive executions of the protocol. Let p_r be the probability to fail in one execution. Then,

$$p_{auth} = (1 - q^{\#} \cdot p_r). \tag{4}$$

The probability of false rejecting a tag when $(\bar{\boldsymbol{b}}, \bar{\boldsymbol{a}}, \bar{\boldsymbol{z}})$ is all zero is P_{FR}. That is, $p_r \geq P_{FR} = 0$, since we have assumed that the Hamming weight of ν is checked. The value p_r is also defined by the probability that the condition $\mathrm{wt}(\hat{\boldsymbol{z}} \oplus \boldsymbol{\Phi}(\boldsymbol{a} \cdot \boldsymbol{X}) \oplus \boldsymbol{\Phi}(\hat{\boldsymbol{b}} \cdot \boldsymbol{Y})) \leq \tau$ is satisfied when $(\bar{\boldsymbol{b}}, \bar{\boldsymbol{a}}, \bar{\boldsymbol{z}}) \neq (\mathbf{0}_{k_Y}, \mathbf{0}_{k_X}, \mathbf{0}_m)$. The sum can be written as

$$\begin{aligned} \hat{\boldsymbol{z}} \oplus \boldsymbol{\Phi}(\boldsymbol{a} \cdot \boldsymbol{X}) \oplus \boldsymbol{\Phi}(\hat{\boldsymbol{b}} \cdot \boldsymbol{Y}) &= \boldsymbol{\Phi}(\hat{\boldsymbol{a}} \cdot \boldsymbol{X}) \oplus \boldsymbol{\Phi}(\boldsymbol{b} \cdot \boldsymbol{Y}) \oplus \boldsymbol{\nu} \oplus \bar{\boldsymbol{z}} \oplus \boldsymbol{\Phi}(\boldsymbol{a} \cdot \boldsymbol{X}) \oplus \boldsymbol{\Phi}(\hat{\boldsymbol{b}} \cdot \boldsymbol{Y}) \\ &= D_{\bar{\mathbf{a}}}\boldsymbol{\Phi}(\boldsymbol{a} \cdot \boldsymbol{X}) \oplus D_{\bar{\boldsymbol{b}}}\boldsymbol{\Phi}(\boldsymbol{b} \cdot \boldsymbol{Y}) \oplus \boldsymbol{\nu} \oplus \bar{\boldsymbol{z}}. \end{aligned}$$

Let $\boldsymbol{y}_{\bar{\mathbf{a}},\bar{\boldsymbol{b}},\bar{\boldsymbol{z}}} = D_{\bar{\mathbf{a}}}\boldsymbol{\Phi}(\boldsymbol{a} \cdot \boldsymbol{X}) \oplus D_{\bar{\boldsymbol{b}}}\boldsymbol{\Phi}(\boldsymbol{b} \cdot \boldsymbol{Y}) \oplus \bar{\boldsymbol{z}}$ and let $\beta_{\bar{\mathbf{a}},\bar{\boldsymbol{b}},\bar{\boldsymbol{z}}}$ be the Hamming weight of $\boldsymbol{y}_{\bar{\mathbf{a}},\bar{\boldsymbol{b}},\bar{\boldsymbol{z}}}$. Then, $m - \beta_{\bar{\mathbf{a}},\bar{\boldsymbol{b}},\bar{\boldsymbol{z}}}$ bits of $\boldsymbol{y}_{\bar{\mathbf{a}},\bar{\boldsymbol{b}},\bar{\boldsymbol{z}}} \oplus \boldsymbol{\nu}$ follow a Bernoulli distribution of parameter η and the rest $\beta_{\bar{\mathbf{a}},\bar{\boldsymbol{b}},\bar{\boldsymbol{z}}}$ bits follow a Bernoulli distribution of parameter $1-\eta$. That is, the Hamming weight $\mathrm{wt}(\boldsymbol{y}_{\bar{\mathbf{a}},\bar{\boldsymbol{b}},\bar{\boldsymbol{z}}} \oplus \boldsymbol{\nu})$ follows a binomial distribution of expected value $\mu = (m - \beta_{\bar{\mathbf{a}},\bar{\boldsymbol{b}},\bar{\boldsymbol{z}}})\eta + (1 - \beta_{\bar{\mathbf{a}},\bar{\boldsymbol{b}},\bar{\boldsymbol{z}}}\eta)$ and variance $\sigma^2 = m\eta(1-\eta)$. Since, the expected value is a function of $\beta_{\bar{\mathbf{a}},\bar{\boldsymbol{b}},\bar{\boldsymbol{z}}}$ we can easily verify that for $\beta_{\bar{\mathbf{a}},\bar{\boldsymbol{b}},\bar{\boldsymbol{z}}} \geq 1 + \lfloor \frac{\tau - \eta m}{1-2\eta} \rfloor$, it holds that $\mu > \tau$.

When, $\mu > \tau$; i.e. $\beta_{\bar{\mathbf{a}},\bar{\boldsymbol{b}},\bar{\boldsymbol{z}}} \geq 1 + \lfloor \frac{\tau - \eta m}{1-2\eta} \rfloor$ from the Chernoff bound we have that $\mathrm{wt}(\boldsymbol{y}_{\bar{\mathbf{a}},\bar{\boldsymbol{b}},\bar{\boldsymbol{z}}} \oplus \boldsymbol{\nu}) < \tau$ with probability $p_1 < e^{-\frac{(\mu-\tau)^2}{2\mu}}$. When, $\mu \leq \tau$; i.e. $\beta_{\bar{\mathbf{a}},\bar{\boldsymbol{b}},\bar{\boldsymbol{z}}} < 1 + \lfloor \frac{\tau - \eta m}{1-2\eta} \rfloor$, trivially we have that $\mathrm{wt}(\boldsymbol{y}_{\bar{\mathbf{a}},\bar{\boldsymbol{b}},\bar{\boldsymbol{z}}} \oplus \boldsymbol{\nu}) < \tau$ with probability p_2. By combining the two cases, $\mathrm{wt}(\boldsymbol{y}_{\bar{\mathbf{a}},\bar{\boldsymbol{b}},\bar{\boldsymbol{z}}} \oplus \boldsymbol{\nu}) \leq \tau$ with probability

$$\hat{p}_r = p_1 \cdot Pr[\mu > \tau] + p_2 \cdot Pr[\mu \leq \tau] \leq e^{-\frac{(\mu-\tau)^2}{2\mu}} \cdot Pr[\mu > \tau] + P[\mu \leq \tau].$$

From Lemma 3, we have that $P\left[\beta_{\bar{\mathbf{a}},\bar{\boldsymbol{b}},\bar{\boldsymbol{z}}} \leq 1 + \lfloor \frac{\tau - \eta m}{1-2\eta} \rfloor\right]$ is upper bounded by

$$Pr\left[\beta_{\bar{\mathbf{a}},\bar{\boldsymbol{b}},\bar{\boldsymbol{z}}} \leq 1 + \lfloor \frac{\tau - \eta m}{1 - 2\eta} \rfloor\right] \leq 2^{mH(\frac{1+\lfloor \frac{\tau-\eta m}{1-2\eta} \rfloor}{m})+2-\min(k_{\boldsymbol{X}},m)}.$$

Thus,

$$\begin{aligned} p_r &\leq e^{-\frac{(\mu-\tau)^2}{2\mu}} \cdot (1 - 2^{mH(\frac{1+\lfloor \frac{\tau-\eta m}{1-2\eta} \rfloor}{m})+2-\min(k_{\boldsymbol{X}},m)}) + 2^{mH(\frac{1+\lfloor \frac{\tau-\eta m}{1-2\eta} \rfloor}{m})+2-\min(k_{\boldsymbol{X}},m)} \\ &\leq e^{-\frac{(\mu-\tau)^2}{2\mu}} + 2^{mH(\frac{1+\lfloor \frac{\tau-\eta m}{1-2\eta} \rfloor}{m})+2-\min(k_{\boldsymbol{X}},m)}. \end{aligned}$$

The exponent of the second term is negative for practical values of the parameters and a decreasing function of d. Also, similarly to [8], in order to ascertain that the first term is negligible, we define $\hat{d}$ the least integer such that $\mu((\hat{d}) > (1+c)\tau$ for some $c > 0$ and for all $d \geq \hat{d}$, $e^{-\frac{(\mu-\tau)^2}{2\mu}} \leq e^{-\frac{(c\tau)^2}{2(c+1)}}$. From (3) and (4), the overall probability of the attack is lower bounded by

$$(1 - q^{\#} \cdot (e^{-\frac{(\mu-\tau)^2}{2\mu}} + 2^{mH(\frac{1+\lfloor \frac{\tau-\eta m}{1-2\eta} \rfloor}{m})+2-\min(k_{\boldsymbol{X}},m)})) \cdot (P_{FA} + \delta) \leq p_{MIM}.$$

□

From Theorem 2, any efficient attacker achieving a noticeable advantage $\delta^{\#}$ against the *GHB*$^{\#}$ protocol in the MIM-model can be turned into an efficient attacker against the same protocol in the DET-model. However, from Theorem 1 this contradicts the conjectured hardness assumption of the LPN problem.

6 Conclusions

The design of lightweight protocols for RFID tag authentication is a challenging task. In this paper, we introduced a new secure authentication protocol, the *GHB*$^{\#}$, that it is supported by a security proof based on the conjectured hardness of the LPN problem. The new protocol belongs to the family of *HB*-like protocols that have been extensively analysed in the last few years. The *GHB*$^{\#}$ protocol is shown to be secure against all the attacks that have been proposed so far against LPN-based authentication protocols, including the MIM attacks in which the attacker is able to modify all messages exchanged between an honest tag and the reader. These MIM attacks has been the Achilles heel of almost all the HB-like protocols with only two very recent exceptions ([15], [4]).

As further research, it is interesting to investigate the relation between the *GHB*$^{\#}$ protocol and other recently proposed LPN based protocols ([15], [4]), that can resist MIM attacks.

Acknowledgement. The authors would like to thank the anonymous reviewers for their valuable comments and suggestions.

References

1. Avoine, G.: RFID Security and Privacy Lounge, The list of papers is available at `http://www.avoine.net/rfid/download/bib/bibliography-rfid.pdf`
2. Bringer, J., Chabanne, H., Dottax, E.: *HB*$^{++}$: a Lightweight Authentication Protocol Secure against Some Attacks. In: Proceedings of the IEEE Int. Conference on Pervasive Sevices, Workshop - SecPerU (2006)
3. Bringer, J., Chabanne, H.: *Trusted-HB*: A Low-Cost Version of HB Secure Against Man-in-the-Middle Attack*HB*$^{++}$. IEEE Transactions on Information Theory 54, 4339–4342 (2008)
4. Bosley, C., Haralambiev, K., Nicolosi, A.: HB^{N}: An HB-like protocol secure against man-in-the-middle attacks. Cryptology ePrint Archive, Report 2011/350 (2011), `http://eprint.iacr.org`
5. Carlet, C.: Vectorial Boolean Functions for Cryptography. In: Crama, Y., Hammer, P.L. (eds.) Boolean Models and Methods in Mathematics, Computer Science, and Engineering, pp. 398–469. Cambridge University Press (2010)
6. Duc, D.N., Kim, K.: Securing *HB*$^{+}$ against GRS Man-in-the-Middle Attack. In: Proceedings of the Symp. on Cryptography and Information Security (2007)
7. Gilbert, H., Robshaw, M., Silbert, H.: An Active Attack against *HB*$^{+}$-a Provable Secure Lightweighted Authentication Protocol. Cryptology ePrint Archive, Report 2005/237 (2005), `http://eprint.iacr.org`

8. Gilbert, H., Robshaw, M., Seurin, Y.: *$HB^{\#}$*: Increasing the Security and Efficiency of *HB^+*. In: Smart, N.P. (ed.) EUROCRYPT 2008. LNCS, vol. 4965, pp. 361–378. Springer, Heidelberg (2008)
9. Gilbert, H., Robshaw, M., Seurin, Y.: Good Variants of HB^+ Are Hard to Find. In: Tsudik, G. (ed.) FC 2008. LNCS, vol. 5143, pp. 156–170. Springer, Heidelberg (2008)
10. Gold, R.: Maximal recursive sequences with 3-valued recursive crosscorrelation functions. IEEE Transactions on Information Theory 14, 154–156 (1968)
11. Hopper, N.J., Blum, M.: Secure Human Identification Protocols. In: Boyd, C. (ed.) ASIACRYPT 2001. LNCS, vol. 2248, pp. 52–66. Springer, Heidelberg (2001)
12. Juels, A., Weis, S.A.: Authenticating Pervasive Devices with Human Protocols. In: Shoup, V. (ed.) CRYPTO 2005. LNCS, vol. 3621, pp. 293–308. Springer, Heidelberg (2005)
13. Katz, J., Shin, J.S.: Parallel and Concurrent Security of the HB and HB^+ Protocols. In: Vaudenay, S. (ed.) EUROCRYPT 2006. LNCS, vol. 4004, pp. 73–87. Springer, Heidelberg (2006)
14. Katz, J., Shin, J.: Analyzing the *HB* and *HB^+* Protocols in the Large Error Case. Cryptology ePrint Archive, Report 2006/326 (2006), `http://eprint.iacr.org/`
15. Kiltz, E., Pietrzak, K., Cash, D., Jain, A., Venturi, D.: Efficient Authentication from Hard Learning Problems. In: Paterson, K.G. (ed.) EUROCRYPT 2011. LNCS, vol. 6632, pp. 7–26. Springer, Heidelberg (2011)
16. Leng, X., Mayes, K., Markantonakis, K.: *HP-MP^+*: An Improvement on the *HB-MP* Protocol. In: Proceedings of the IEEE Int. Conference on RFID 2008, pp. 118–124. IEEE Press (2008)
17. Lidl, R., Niederreiter, H.: Introduction to Finite Fields and Their Applications. Cambridge University Press (1994)
18. Madhavan, M., Thangaraj, A., Sankarasubramaniam, Y., Viswanathan, K.: *NLHB*: A Non-Linear Hopper Blum Protocol. In: Proceedings of IEEE National Conference on Communications, NCC (2010), CoRR abs/1001.2140:2010.
19. Massey, J.L., Omura, J.K.: Computational Method and Apparatus for Finite Field Arithmetic. US Patent No. 4,587,627 (1986)
20. Munilla, J., Peinado, A.: *HP-MP*: A Further Step in the *HB*-family of Lightweight authentication protocols. Computer Networks 51, 2262–2267 (2007)
21. Ouafi, K., Overbeck, R., Vaudenay, S.: On the Security of $HB^{\#}$ against a Man-in-the-Middle Attack. In: Pieprzyk, J. (ed.) ASIACRYPT 2008. LNCS, vol. 5350, pp. 108–124. Springer, Heidelberg (2008)
22. Piramuthu, S.: *HB* and Related Lightweight Authentication Protocols for Secure RFID Tag/Reader Authentication. In: Proceedings of CollECTeR Europe Conference, Basel, Switzerland, June 9-10 (2006)
23. Reza, M., Abyaneh, S., On, S.: the Security of Non-Linear *HB* (*NLHB*) Protocol Against Passive Attack. Cryptology ePrint Archive, Report 2010/402 (2010), `http://eprint.iacr.org/`
24. Rizomiliotis, P.: *HB-MAC*: Improving the Random - *$HB^{\#}$* Authentication Protocol. In: Fischer-Hübner, S., Lambrinoudakis, C., Pernul, G. (eds.) TrustBus 2009. LNCS, vol. 5695, pp. 159–168. Springer, Heidelberg (2009)
25. Yoon, B., Sung, M.Y., Yeon, S.H., Oh, S., Kwon, Y.: Kim, Ch., Kim, K.-H.: *HB-MP^{++}* protocol: An ultra light-weight authentication protocol for RFID system. In: Proceedings of the IEEE Int. Conference on RFID, pp. 186–191 (2009)

Knox: Privacy-Preserving Auditing for Shared Data with Large Groups in the Cloud

Boyang Wang[1,2], Baochun Li[2], and Hui Li[1]

[1] State Key Laboratory of Integrated Services Networks, Xidian University, China
{bywang,lihui}@mail.xidian.edu.cn
[2] Department of Electrical and Computer Engineering,
University of Toronto, Canada
bli@eecg.toronto.edu

Abstract. With cloud computing and storage services, data is not only stored in the cloud, but routinely shared among a large number of users in a group. It remains elusive, however, to design an efficient mechanism to audit the integrity of such shared data, while still preserving identity privacy. In this paper, we propose Knox, a privacy-preserving auditing mechanism for data stored in the cloud and shared among a large number of users in a group. In particular, we utilize group signatures to construct homomorphic authenticators, so that a third party auditor (TPA) is able to verify the integrity of shared data for users without retrieving the entire data. Meanwhile, the identity of the signer on each block in shared data is kept private from the TPA. With Knox, the amount of information used for verification, as well as the time it takes to audit with it, are not affected by the number of users in the group. In addition, Knox exploits homomorphic MACs to reduce the space used to store such verification information. Our experimental results show that Knox is able to efficiently audit the correctness of data, shared among a large number of users.

Keywords: Privacy-Preserving, Auditing, Shared Data, Cloud Computing.

1 Introduction

With cloud computing and storage, users are able to access and to share resources offered by cloud service providers at a lower marginal cost. With Dropbox, for example, data is stored in the cloud (operated by Amazon), and shared among a group of users in a collaborative manner. It is natural for users to wonder whether their data remain intact over a prolonged period of time: due to hardware failures and human errors in an untrusted cloud environment [2], the integrity of data stored in the cloud can become compromised. To protect the integrity of data in the cloud and to offer "peace of mind" to users, it is best to introduce a third party auditor (TPA) to perform auditing tasks on behalf of users. Such a third party auditor enjoys amply computation/communication resources that users may not possess.

F. Bao, P. Samarati, and J. Zhou (Eds.): ACNS 2012, LNCS 7341, pp. 507–525, 2012.

Provable data possession (PDP) [3], first proposed by Ateniese *et al.*, allows a verifier to perform public auditing on the integrity of data stored in an untrusted server without retrieving the entire data. Subsequent work focused on how dynamic data [5, 11, 20, 24] and data privacy [19] can be supported during the public auditing process. However, most of previous work only focus on auditing the integrity of *personal data*. Recently, Wang *et al.* [16] first design a privacy-preserving public auditing mechanism (named Oruta) for *shared data* in an untrusted cloud, so that the identity of the signer on each block in shared data is not disclosed to the third party auditor (TPA) during an auditing task. Without knowing the identities of signers, the TPA cannot learn which user in the group or which block in shared data is a higher valuable target than others [16].

Unfortunately, Oruta [16] fails to scale well to a large number of users sharing data in a group. In Oruta, information used for verification are computed with ring signatures [8]; as a result, the size of verification information, as well as the time it takes to audit with it, are linearly increasing with the number of users in a group. To make matters worse, when adding new users to a group, all the existing verification information will need to be re-computed if ring signatures are used, introducing a significant computation burden to all users. In addition, the identities of signers are unconditional [8] protected by ring signatures, which prevent the group manager to trace the identity when someone in the group is misbehaved.

In this paper, we propose Knox, a new privacy-preserving mechanism to audit data stored in an untrusted cloud and shared among a large number of users in a group. In Knox, we take advantage of group signatures [6, 12] to construct homomorphic authenticators [3, 15], so that the third party auditor is able to verify the integrity of shared data without retrieving the entire data, but cannot reveal the identities of signers on all blocks in shared data. Meanwhile, the size of verification information, as well as the time it takes to audit with it, are not affected when the number of users sharing the data increases. The original user, who creates and shares the data in the cloud, is able to add new users into a group without re-computing any verification information. In addition, the original user (acts as the group manager) can trace group signatures on shared data, and reveal the identities of signers when it is necessary. We also utilize homomorphic MACs [1] to effectively reduce the amount of storage space needed to store verification information. As a necessary trade-off, we allow the third party auditor to share a secret key pair with users, which we refer to as *authorized auditing*. Although we allow an authorized TPA to possess the secret key pair, the TPA cannot compute valid group signatures as group users because this secret key pair is only a part of a group user's private key. To our best knowledge, we present the first mechanism designed with scalability in mind when it comes to support auditing data shared among a large number of users in a privacy-preserving fashion. A high-level comparison between Knox and previous work [16] is shown in Table 1.

The remainder of this paper is organized as follows. In Section 2, we briefly discuss related work. Then, we present the system model, threat model and

Table 1. Comparison between Previous Work [16] and Knox

	Previous work [16]	Knox
Public Auditing	Yes	No
Identity Privacy	Yes	Yes
Support for Large Groups	No	Yes
Traceability	No	Yes

design goals in Section 3. In Section 4, we introduce complexity assumptions and cryptographic primitives used in Knox. Detailed design and security analysis of Knox are presented in Section 5 and 6. Finally, we evaluate the performance of Knox in Section 7, and conclude this paper in Section 8.

2 Related Work

Ateniese *et al.* [3] first proposed provable data possession (PDP), which allows a client to verify the integrity of her data stored at an untrusted server without retrieving the entire file. However, this mechanism is only suitable for static data. To improve the efficiency of verification, Ateniese *et al.* [5] constructed scalable and efficient PDP using symmetric keys. Unfortunately, it cannot support public verifiability, and only offers each user a limited number of verification requests.

Juels and Kaliski [14] defined another similar model called proofs of retrievability (POR), which is also able to check the correctness of data on an untrusted server. The original file is added with a set of randomly-valued check blocks called sentinels. The verifier challenges the untrusted server by specifying the positions of a collection of sentinels, and by asking the untrusted server to return the associated sentinel values. Shacham and Waters [15] designed two improved POR mechanisms, which are built on BLS signatures and pseudo-random functions.

Wang *et al.* [20] leveraged the Merkle Hash Tree to construct a public auditing mechanism with fully dynamic data. Hao *et al.* [13] also designed a dynamic public auditing mechanism based on RSA. Erway *et al.* [11] presented a dynamic PDP based on the rank-based authenticated dictionary. Zhu *et al.* [24] exploited index hash tables to support fully dynamic data. To ensure the correctness of users' data stored on multiple servers, Wang *et al.* [18] utilized homomorphic tokens and erasure codes in the auditing process. An excellent survey of previous work about data auditing can be found in [21].

Wang *et al.* [19] considered data privacy with public auditing in the cloud. In their mechanism, the TPA is able to check the integrity of cloud data but cannot obtain any private data. Zhu *et al.* [23] also designed a mechanism to preserve data privacy from the TPA. Recent work [16], Oruta, represents the first privacy-preserving public auditing mechanism for shared data in the cloud. In this mechanism, the TPA can verify the integrity of shared data, but is not able to reveal the identity of the signer on each block. Unfortunately, it is not readily scalable to auditing the integrity of data shared among a large number of users in the group.

3 Problem Statement

3.1 System Model

In this paper, we consider data storage and sharing services in the cloud with three entities: the cloud, the third party auditor (TPA), and users who participate as a group (as shown in Fig. 1). Users in a group include one original user and a number of group users. The original user is the original owner of data, and shares data in the cloud with other users. Based on access control policies [22], other users in the group are able to access, download and modify shared data. The cloud provides data storage and sharing services for users, and has ample storage space. The third party auditor is able to verify the integrity of shared data based on requests from users, without downloading the entire data.

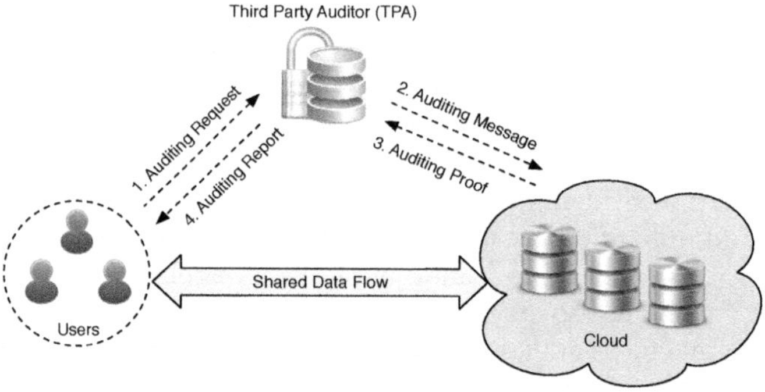

Fig. 1. The system model includes the cloud, the third party auditor and users

When a user (either the original user or a group user) wishes to check the integrity of shared data, she first sends an auditing request to the TPA. After receiving the auditing request, the TPA generates an auditing message to the cloud, and retrieves an auditing proof of shared data from the cloud. Then the TPA verifies the correctness of the auditing proof. Finally, the TPA sends an auditing report to the user based on the result of the verification.

3.2 Threat Model

Integrity Threats. In general, two kinds of threats related to the integrity of shared data are possible. First, an external adversary may try to pollute shared data in the cloud, and prevent users from using shared data correctly. Second, the cloud service provider may inadvertently corrupt or even remove shared data in the cloud due to hardware failures and human errors. To make matters worse, in order to avoid jeopardizing its reputation, the cloud service provider may be reluctant to inform users about such corruption of data.

Privacy Threats. During an auditing task, a semi-trusted TPA, who is only responsible for verifying the integrity of shared data, may try to reveal the identity of the signer on each block in shared data based on verification information (i.e. signatures). The identity of the signer on each block is private and sensitive information, which users do not wish to be revealed to any third party.

3.3 Design Goals

To make it efficient and secure for the TPA to verify shared data with a large number of users in a group, Knox should be designed to achieve the following properties: (1) **Correctness**: The TPA is able to correctly audit the integrity of shared data. (2) **Efficiency**: The TPA is able to verify the integrity of shared data without retrieving the entire data from the cloud. (3) **Identity privacy**: During an auditing task, the TPA cannot distinguish the identity of the signer on each block. (4) **Support for large groups**: The TPA is able to efficiently audit data that are shared among a large number of users. In particular, the size of verification information, as well as the time it takes to audit with it, are not affected by the number of users in the group; the original user can add new users to the group without re-computing existing verification information. (5) **Traceability**: The original user is able to trace a signature on a block and reveal the identity of the signer.

4 Preliminaries

4.1 Bilinear Maps

Let G_1, G_2 and G_T be three multiplicative cyclic groups of prime order p, g_1 be a generator of G_1, and g_2 be a generator of G_2. A bilinear map e is a map e: $G_1 \times G_2 \rightarrow G_T$ with the following properties: 1) **Computability**: there exists an efficiently computable algorithm for computing map e. 2) **Bilinearity**: for all $u \in G_1$, $v \in G_2$ and $a, b \in Z_p$, $e(u^a, v^b) = e(u, v)^{ab}$. 3) **Non-degeneracy**: $e(g_1, g_2) \neq 1$. For ease of exposition, we assume $G_1 = G_2$ in the rest of this paper.

4.2 Complexity Assumptions

Definition 1. ***Computational Diffie-Hellman (CDH) Problem.*** *For* $(a, b) \in Z_p^2$, *given* $(g_1, g_1^a, g_1^b) \in G_1^3$ *as input, output* $g_1^{ab} \in G_1$.

The CDH assumption holds in G_1 if no t-time algorithm has advantage at least ϵ in solving the CDH problem in G_1, which means it is *computational infeasible* to solve the CDH problem in G_1.

Definition 2. ***q-Strong Diffie-Hellman (q-SDH) Problem.*** *For* $\gamma \in Z_p$, *given a* $(q+2)$*-tuple* $(g_1, g_2, g_2^{\gamma}, g_2^{\gamma^2}, ..., g_2^{\gamma^q}) \in G_1 \times G_2^{q+1}$ *as input, output a pair* $(g_1^{1/(\gamma+x)}, x) \in G_1 \times Z_p$.

The q-SDH assumption holds in (G_1, G_2) if no t-time algorithm has advantage at least ϵ in solving the q-SDH problem in (G_1, G_2).

Definition 3. ***Decision Linear (DL) Problem.*** *For* $(a, b, c) \in Z_p^3$*, given* $(u, v, h, u^a, v^b, h^c) \in G_1^6$ *as input, output yes if* $a + b = c$ *and no otherwise.*

The DL assumption holds in G_1 if no t-time algorithm has advantage at least ϵ in solving the DL problem in G_1.

4.3 Group Signatures

Group signatures, first introduced by Chaum and van Heyst [10], aim to provide anonymity of signers, who are from a same group. A verifier is convinced that messages are correct and signed by one of the group members, but cannot reveal the identity of the signer. Meanwhile, only the group manager is able to trace these group signatures and reveal the identity of the signer. Boneh *et al.* [6] (denoted as BBS) proposed a short group signature scheme based on the q-SDH assumption. In their scheme, the length of each group signature is independent from the number of group members.

4.4 Homomorphic MACs

Homomorphic MACs, introduced by Agrawal and Boneh [1], provide data integrity for network coding. Using homomorphic MACs, an intermediate node can construct a valid MAC of an output block based on the MACs of input blocks without knowing the secret keys. More specifically, given a block $\boldsymbol{m}_j = (m_{j,1}, ..., m_{j,n}) \in Z_p^n$, the homomorphic MAC of this block can be computed as $t_j = \sum_{i=1}^{n} \delta_i m_{j,i} + b_j \in Z_p$, where $\boldsymbol{\delta} = (\delta_1, ..., \delta_n)$ is generated by a pseudo-random generator and a secret key k_{prg}, and b_j is calculated by a pseudo-random function and a secret key k_{prf}. Given t_1 and t_2, an intermediate node can compute a valid MAC of a new block $\boldsymbol{m}' = \boldsymbol{m}_1 + \boldsymbol{m}_2$ by calculating $t' = t_1 + t_2$ without knowing the secret key pair (k_{prg}, k_{prf}).

4.5 Homomorphic Authenticators

Homomorphic authenticators (also denoted as homomorphic verifiable tags) are basic tools to construct data auditing mechanisms [3,13,15,16,19,23,24]. Besides unforgeability (only a user with a private key can generate valid signatures), a homomorphic authenticable signature scheme, which denotes a homomorphic authenticator based on signatures, should also satisfy the following properties:

Let $(\mathbf{pk}, \mathbf{sk})$ denote the signer's public/private key pair, σ_1 denote the signature on message $m_1 \in Z_p$, σ_2 denote the signature on message $m_2 \in Z_p$.

- **Blockless verification:** Given σ_1 and σ_2, two random values α_1, $\alpha_2 \in Z_p$ and a message $m' = \alpha_1 m_1 + \alpha_2 m_2 \in Z_p$, a verifier is able to check the correctness of message m' without knowing message m_1 and m_2.

- **Non-malleability:** Given σ_1 and σ_2, two random values α_1, $\alpha_2 \in Z_p$ and a message $m' = \alpha_1 m_1 + \alpha_2 m_2 \in Z_p$, a user who does not have private key **sk**, is not able to generate a valid signature σ' for message m' by combining signature α_1 and α_2.

The first property allows a verifier to check the correctness of data in the cloud with a linear combination of all the blocks, while the entire data does not need to be downloaded to the verifier. The second property prevents an attacker from generating signatures for invalid messages by combining existing signatures. Other cryptographic techniques related to homomorphic authenticable signatures includes aggregate signatures [8], homomorphic signatures [7] and batch-verification signatures [12]. If a signature scheme is blockless verifiable and malleable, it is a homomorphic signature scheme. In the construction of data auditing mechanisms, we should use homomorphic authenticable signatures, not homomorphic signatures. Otherwise, based on malleability of homomorphic signatures, an adversary can successfully corrupt data in the cloud by linearly combining existing blocks and corresponding signatures.

5 Homomorphic Authenticable Group Signatures

5.1 Overview

As introduced at the beginning of this paper, we expect to utilize group signatures for computing verification information, so that the identity of the signer on each block can be kept private from the TPA. However, traditional group signatures [4,6,10] cannot be directly used in Knox, since they are not blockless verifiable. Without blockless verification, a verifier has to download the entire data to check the integrity of shared data, which consumes excessive bandwidth and takes long verification times. Therefore, we first build a homomorphic authenticable group signature (HAGS) scheme in this section. Then we will present the full construction of our privacy-preserving auditing mechanism for shared data among a large number of users based on HAGS in the next section.

In HAGS, we extend BBS group signatures [6] to achieve blockless verification. Meanwhile, to keep unforgeability (nobody outside the group can produce valid signatures) of HAGS, we leverage BLS signatures [9] as a part of our group signatures. BLS signatures, which are based on bilinear maps, are used in previous work [15,19] to audit data integrity of personal users. In addition, we exploit batch verification methods of group signatures in [12] to improve the efficiency of HAGS for verifying multiple group signatures. Note that if only using BLS signatures among a group of users, which means that all the users in the group generate signatures on messages only with a common private key, it is also possible to achieve identity privacy on messages. Unfortunately, traceability of the group manager on signatures generated by group members will be immediately lost.

5.2 Construction of HAGS

Following general constructions of group signatures in [4,6], HAGS contains five algorithms: **KeyGen**, **Join**, **Sign**, **Verify** and **Open**. In **KeyGen**, the group manager generates her private key and a group public key. In **Join**, the group manager is able to compute a private key for a new group user and add this user to the group user list. A group user signs messages using her private key and the group public key in **Sign**. In **Verify**, a verifier is able to check the correctness of a message using the group public key, but she cannot reveal the identity of the signer. The group manager can reveal the identity of the signer on a message in **Open**.

Scheme Details: Let G_1, G_2 and G_T be multiplicative cyclic groups of order p, g_1 and g_2 be generators of G_1 and G_2 respectively, $G_1 \times G_2 \rightarrow G_T$ be a bilinear map. There are two hash functions, $H_1 : \{0,1\}^* \rightarrow Z_p$ and $H_2 : \{0,1\}^* \rightarrow G_1$. The total number of group users is d.

KeyGen. The group manager randomly selects $h \in G_1 \backslash \{1_{G_1}\}$ and ξ_1, $\xi_2 \in Z_p^*$, and sets u, $v \in G_1$ such that $u^{\xi_1} = v^{\xi_2} = h$. Then, she randomly selects $\gamma, \pi, \eta \in Z_p^*$, and sets $w = g_2^{\gamma}, \rho = g_2^{\pi} \in G_2$.

The group public key is $\mathbf{gpk} = (g_1, g_2, h, u, v, w, \rho, \eta)$, the group manager's private key is $\mathbf{gmsk} = (\xi_1, \xi_2)$. The group manager keeps γ private. And π will be a part of a group user's private key, which is issued to group users later.

Join. For user i, $1 \leq i \leq d$, the group manager randomly selects $x_i \in Z_p^*$ with $x_i + \gamma \neq 0$, and sets $A_i = g_1^{1/(\gamma + x_i)} \in G_1$. The private key of user i is $\mathbf{gsk[i]} = (A_i, x_i, \pi)$. The group manager secretly issues $\mathbf{gsk[i]}$ to user i, and adds this user into the group user list.

Sign. Given a group public key $\mathbf{gpk} = (g_1, g_2, h, u, v, w, \rho, \eta)$, a private key $\mathbf{gsk[i]} = (A_i, x_i, \pi)$, a message $m \in Z_p$ and this message's identifier id, user i computes the signature σ as follows:

1. Randomly select $\alpha, \beta, r_\alpha, r_\beta, r_x, r_{\gamma_1}, r_{\gamma_2} \leftarrow Z_p$.
2. Compute T_1, T_2 and T_3 as $T_1 = u^\alpha, T_2 = v^\beta, T_3 = A_i \cdot h^{\alpha+\beta}$.
3. Compute $\gamma_1 = x_i \cdot \alpha$ and $\gamma_2 = x_i \cdot \beta$.
4. Compute R_1, R_2, R_3, R_4 and R_5 as

$$R_1 = u^{r_\alpha}, \quad R_2 = v^{r_\beta}, \quad R_4 = T_1^{r_x} \cdot u^{-r_{\gamma_1}}, \quad R_5 = T_2^{r_x} \cdot v^{-r_{\gamma_2}},$$

$$R_3 = e(T_3, g_2)^{r_x} \cdot e(h, w)^{-r_\alpha - r_\beta} \cdot e(h, g_2)^{-r_{\gamma_1} - r_{\gamma_2}}.$$

5. Compute a challenge $c \in Z_p$ as $c = \eta^m H_1(T_1, T_2, T_3, R_1, R_2, R_3, R_4, R_5)$. For ease of exposition, we use $H_1(T_1, ..., R_5)$ instead of $H_1(T_1, T_2, T_3, R_1, R_2, R_3, R_4, R_5)$ in the remainder of this paper.
6. Compute $s_\alpha, s_\beta, s_x, s_{\gamma_1}, s_{\gamma_2} \in Z_p$ as:

$$s_\alpha = r_\alpha + c \cdot \alpha, \quad s_\beta = r_\beta + c \cdot \beta, \quad s_x = r_x + c \cdot x_i,$$

$$s_{\gamma_1} = r_{\gamma_1} + c \cdot \gamma_1, \quad s_{\gamma_2} = r_{\gamma_2} + c \cdot \gamma_2.$$

7. Compute a tag θ as $\theta = [H_2(id)g_1^m]^\pi \in G_1$.
8. Output the signature of this message m as $\sigma = (T_1, T_2, T_3, \theta, R_3, c, s_\alpha, s_\beta, s_x, s_{\gamma_1}, s_{\gamma_2}) \in G_1^4 \times G_T \times Z_p^6$.

Verify. Given a group public key $\mathbf{gpk} = (g_1, g_2, h, u, v, w, \rho, \eta)$, a message m, an identifier id and a group signature $\sigma = (T_1, T_2, T_3, \theta, R_3, c, s_\alpha, s_\beta, s_x, s_{\gamma_1}, s_{\gamma_2})$, a verifier checks the integrity of this message as follows:

1. Re-compute values R_1, R_2, R_4 and R_5 from σ as:

$$\tilde{R_1} = u^{s_\alpha} \cdot T_1^{-c}, \quad \tilde{R_2} = v^{s_\beta} \cdot T_2^{-c}, \quad \tilde{R_4} = T_1^{s_x} \cdot u^{-s_{\gamma_1}}, \quad \tilde{R_5} = T_2^{s_x} \cdot v^{-s_{\gamma_2}}$$

2. Check the following equations as:

$$R_3 \stackrel{?}{=} e(T_3, g_2)^{s_x} e(h, w)^{-s_\alpha - s_\beta} e(h, g_2)^{-s_{\gamma_1} - s_{\gamma_2}} \cdot \left(e(T_3, w) \cdot e(g_1, g_2)^{-1}\right)^c, \quad (1)$$

$$c \stackrel{?}{=} \eta^m \cdot H_1(T_1, T_2, T_3, \tilde{R_1}, \tilde{R_2}, R_3, \tilde{R_4}, \tilde{R_5}), \quad (2)$$

$$e(\theta, g_2) \stackrel{?}{=} e(H_2(id) \cdot g_1^m, \rho). \quad (3)$$

If all the three equations hold, the verifier accepts message m. Otherwise, she rejects this message.

Open. Only the group manager can trace a group signature and reveal the identity of the signer. Given a group public key $\mathbf{gpk} = (g_1, g_2, h, u, v, w, \rho, \eta)$, the group manager's private key $\mathbf{gmsk} = (\xi_1, \xi_2)$, a message m and a signature σ, the group manager reveals the identity of the signer as follows:

1. Verify that the signature σ is a valid signature on message m.
2. Decrypt user i's A_i as $A_i = T_3/(T_1^{\xi_1} \cdot T_2^{\xi_2})$.
3. Given A_i, which is a part of user i's private key, the group manager is able to reveal the identity of the signer on message m.

5.3 Security Analysis of HAGS

Theorem 1. *Given a message m and its group signature σ, a verifier is able to correctly check the integrity of message m under HAGS.*

Proof. Equation (1) is correct because $e(T_3, g_2)^{r_x} \cdot e(h, w)^{-r_\alpha - r_\beta} \cdot e(h, g_2)^{-r_{\gamma_1} - r_{\gamma_2}}$ $= e(T_3, g_2)^{s_x} \cdot e(h, w)^{-s_\alpha - s_\beta} \cdot e(h, g_2)^{-s_{\gamma_1} - s_{\gamma_2}} \cdot \left(e(T_3, w) \cdot e(g_1, g_2)^{-1}\right)^c$. Because R_1, R_2, R_4, R_5 can be successfully recomputed [6], Equation (2) is correct. Equation (3) is correct since $e(\theta, g_2) = e([H_2(id)g_1^m]^\pi, g_2) = e(H_2(id)g_1^m, \rho)$. Further explanations about correctness can be found in [6,12].

Theorem 2. *Suppose $\mathcal{F}$ is a (t', ϵ')-algorithm that can generate a forgery of a group signature under HAGS. Then there exists a (t, ϵ)-algorithm $\mathcal{A}$ that can solve the CDH problem with $t \leq t' + (q_H + q_S + 1)c_{G_1}$ and $\epsilon \geq \epsilon'/(\mathbf{e} + q_S\mathbf{e})$, where $\mathcal{F}$ issues at most q_H hash queries and at most q_S signing queries, $\mathbf{e} = \lim_{q_S \to \infty}(1 + 1/q_S)^{q_S}$, exponentiation and inversion on G_1 take time c_{G_1}.*

Proof. Details of this proof can be found in our technical report [17].

Theorem 3. *HAGS is a homomorphic authenticable group signature scheme.*

Proof. We first prove that HAGS has the property of blockless verification. Then we show HAGS is also non-malleable.

Given a group public key $\mathbf{gpk} = (g_1, g_2, h, u, v, w, \rho, \eta)$, two signatures $\sigma_1 = (T_{1,1}, T_{1,2}, T_{1,3}, \theta_1, R_{1,3}, c_1, s_{1,\alpha}, s_{1,\beta}, s_{1,x}, s_{1,\gamma_1}, s_{1,\gamma_2})$, $\sigma_2 = (T_{2,1}, T_{2,2}, T_{2,3}, \theta_2, R_{2,3}, c_2, s_{2,\alpha}, s_{2,\beta}, s_{2,x}, s_{2,\gamma_1}, s_{2,\gamma_2})$, and a message $m' = \sum_{j=1}^{2} y_j m_j \in Z_p$, where $y_j \in Z_p^*$, a verifier is able to check the correctness of message m' without knowing message m_1 and m_2. More specifically, she first recomputes $R_{j,1}$, $R_{j,2}$, $R_{j,4}$ and $R_{j,5}$ as in **Verify**. Then she checks:

$$\prod_{j=1}^{2} R_{j,3}^{y_j} \stackrel{?}{=} e(\prod_{j=1}^{2} (T_{j,3}^{s_{j,x}} \cdot h^{-s_{j,\gamma_1}-s_{j,\gamma_2}} \cdot g_1^{-c_j})^{y_j}, g_2) e(\prod_{j=1}^{2} (h^{-s_{j,\alpha}-s_{j,\beta}} \cdot T_{j,3}^{c_j})^{y_j}, w), \tag{4}$$

$$\prod_{j=1}^{2} c_j^{y_j} \stackrel{?}{=} \eta^{m'} \cdot \prod_{j=1}^{2} H_1(T_{j,1}, ..., \tilde{R}_{j,5})^{y_j}, \tag{5}$$

$$e(\prod_{j=1}^{2} \theta_j^{y_j}, g_2) \stackrel{?}{=} e(\prod_{j=1}^{2} H_2(id_j)^{y_j} \cdot g_1^{m'}, \rho). \tag{6}$$

Only if all the three equations hold, then the verifier accepts message m'.

Note that only Equation (5) and (6) are related to message m', while Equation (4) is independent from the content of message m'. The correctness of Equation (4) can be proved using batch verification methods of group signatures [12]. Based on Theorem 1, the correctness of Equation (5) and (6) can be proved as:

$$\begin{aligned}\prod_{j=1}^{2} c_j^{y_j} &= \prod_{j=1}^{2} \left(\eta^{m_j} \cdot H_1(T_{j,1}, ..., \tilde{R}_{j,5})\right)^{y_j} \\ &= \eta^{y_1 m_1 + y_2 m_2} \cdot \prod_{j=1}^{2} H_1(T_{j,1}, ..., \tilde{R}_{j,5})^{y_j} = \eta^{m'} \cdot \prod_{j=1}^{2} H_1(T_{j,1}, ..., \tilde{R}_{j,5})^{y_j}.\end{aligned}$$

$$\begin{aligned}e(\prod_{j=1}^{2} \theta_j^{y_j}, g_2) &= e(\prod_{j=1}^{2} (H_2(id_j) \cdot g_1^{m_j})^{\pi \cdot y_j}, g_2) \\ &= e(\prod_{j=1}^{2} H_2(id_j)^{y_j} \cdot g_1^{y_1 m_1 + y_2 m_2}, g_2^{\pi}) = e(\prod_{j=1}^{2} H_2(id_j)^{y_j} \cdot g_1^{m'}, \rho).\end{aligned}$$

Because all the three equations are correct, HAGS is blockless verifiable.

Meanwhile, an attacker, who does not have a private key, cannot generate a valid signature σ' for message m' by combining existing signatures. More specifically, this user cannot construct a tag θ' by linearly combining θ_1 and θ_2 with y_1 and y_2. Because $\theta_1^{y_1}\theta_2^{y_2} = [H_2(id_1)^{y_j}H_2(id_2)^{y_2}g_1^{m'}]^{\pi}$, $\theta' = [H_2(id')g_1^{m'}]^{\pi}$ and $H_2(id') \neq H_2(id_1)^{y_j}H_2(id_2)^{y_2}$, then we have $\theta_1^{y_1} \cdot \theta_2^{y_2} \neq \theta'$. Therefore, HAGS is non-malleable.

Theorem 4. *Given a message m and its group signature σ, only the group manager can reveal the identity of the signer on this message. For a verifier, it is computational infeasible to reveal the identity of the signer on message m.*

Proof. For the group manager, she can always successfully recover the identity of the signer on message m using her manager private key **gmsk** $= (\xi_1, \xi_2)$. Because $T_3/(T_1^{\xi_1} \cdot T_2^{\xi_2}) = A_i \cdot h^{\alpha+\beta}/(u^{\alpha\cdot\xi_1} \cdot v^{\beta\cdot\xi_2}) = A_i \cdot h^{\alpha+\beta}/h^{\alpha+\beta} = A_i$. For a verifier, if she can successfully choose a value c with $c = \alpha+\beta$, then she is able to decrypt A_i and reveal the identity of the signer by computing $T_3/h^c = A_i \cdot h^{a+b}/h^c$. However, given $u, v, h, T_1 = u^{\alpha}, T_2 = v^{\beta}, h^c \in G_1$, deciding whether $c = \alpha + \beta$ is as hard as solving Decision Linear problem in G_1. Further proofs about anonymity and traceability of group signatures can be found in [6].

6 Privacy-Preserving Auditing for Shared Data

6.1 Overview

We now present Knox, a privacy-preserving auditing mechanism for shared data among a large number of users. Using HAGS, we can preserve the identity of the signer on each block from the TPA. Meanwhile, the original user, who is the group manager and shares data with other group users, can reveal an identity of a signer when it is necessary. Moreover, the length of each group signature is independent from the number of group users, which is a desirable property for large groups to share their data in the cloud. If users wish to protect the privacy of shared data during an auditing task, users can encrypt data using encryption techniques, such as the combination of symmetric encryption and attribute-based encryption [22], before outsourcing data to the cloud server. The main objective of designing Knox is to provide identity privacy for users.

To reduce the storage space of group signatures on shared data, we utilize homomorphic MACs [1] to compress each block into a small value, and then sign this small value instead of signing the entire block. As a necessary trade-off, Knox does not support public auditing, since the TPA in our mechanism needs to share a secret key pair with all group users, referred to as authorized auditing. This secret key pair is used to compute homomorphic MACs. Although we allow an authorized TPA to possess this secret key pair, the TPA cannot compute valid group signatures as group users because this secret key pair is only a part of a group user's private key.

Because the computation of a signature includes an identifier of a block (as we described in HAGS), conventional methods, which only use the index of a block as its identifier, are not suitable for dynamic data. The reason is that when a user modifies shared data by performing an insert or delete operation on a single block, the indices of blocks that after the modified block are all changed, and the change of these indices requires users to re-compute the signatures of these blocks, even though the content of these blocks are not modified [16]. To avoid this type of re-computation and support dynamic data for users, we take advantage of index hash tables [16, 24] as identifiers of blocks. Further explanations about index hash tables can be found in [16, 24].

In addition, we continue to use sampling strategies as previous work [3] to detect any corrupted block in shared data with a high probability, by only choosing a subset of all blocks in each auditing task. For example, if 1% of all the blocks are corrupted, the TPA can detect this misbehavior with a probability greater than 99% by choosing only 460 random selected blocks, where the number of selected blocks is independently with the total number of blocks in shared data if the percentage of corrupted blocks is fixed [3]. To improve the detection probability, the TPA can increase the number of selected blocks in each auditing task [3, 23]. In some emerging applications, the auditor may need to achieve a 100% detection probability if only one corrupted block exists, then all the blocks in shared data should be selected during an auditing task. As a trade-off, the computation and communication cost are significantly increased.

6.2 Construction of Knox

Knox includes six algorithms: **KeyGen**, **Join**, **Sign**, **ProofGen**, **ProofVerify** and **Open**. In **KeyGen**, the original user of shared data generates a group public key and a group manager private key. In **Join**, the original user, who acts as the group manager, is able to issue private keys to users. A user (either the original user or a group user) is able to sign blocks using her private key and the group public key in **Sign**. In **ProofGen**, the cloud generates a proof of possession of shared data to the TPA. **ProofVerify** is operated by the TPA to verify the correctness of the proof. The original user can reveal the identity of the signer on each block in **Open**.

Scheme Details: Let G_1, G_2 and G_T be multiplicative cyclic groups of order p, g_1 and g_2 be generators of G_1 and G_2 respectively, $G_1 \times G_2 \rightarrow G_T$ be a bilinear map. There are two hash functions, $H_1 : \{0,1\}^* \rightarrow Z_p$ and $H_2 : \{0,1\}^* \rightarrow G_1$. The total number of group users is d. Data M, which is going to be shared by users, is divided into n blocks. Each block is further divided into k elements of Z_p. Therefore, shared data M can be presented as:

$$M = \begin{pmatrix} \boldsymbol{m}_1 \\ \vdots \\ \boldsymbol{m}_n \end{pmatrix} = \begin{pmatrix} m_{1,1} & \dots & m_{1,k} \\ \vdots & \ddots & \vdots \\ m_{n,1} & \dots & m_{n,k} \end{pmatrix} \in Z_p^{n \times k}.$$

There are also a pseudo-random generator PRG: $\mathcal{K}_{prg} \rightarrow Z_p^k$ and a pseudo-random function PRF: $\mathcal{K}_{prf} \times \mathcal{I} \rightarrow Z_p$, where $\mathcal{K}_{prg}$ and $\mathcal{K}_{prf}$ denote the set of secret keys for PRG and PRF respectively, and $\mathcal{I}$ is the set of all identifiers in the index hash table of data M.

KeyGen. The original user, who acts as the group manager, first selects system parameters as in HAGS. Meanwhile, she also randomly computes a secret key pair $\mathbf{skp} = (sk_{prg}, sk_{prf})$, where $sk_{prg} \in \mathcal{K}_{prg}$ and $sk_{prf} \in \mathcal{K}_{prf}$. The group public key is $\mathbf{gpk} = (g_1, g_2, h, u, v, w, \rho, \eta)$, the group manager's private key is $\mathbf{gmsk} = (\xi_1, \xi_2)$. The original user keeps γ private. Both π and $\mathbf{skp}$ will be a part of a group user's private key, which is issued to group users later. The original user also privately shares $\mathbf{skp} = (sk_{prg}, sk_{prf})$ with an authorized TPA.

Join. For user i, $1 \leq i \leq d$, the original user randomly selects $x_i \in Z_p^*$ with $x_i + \gamma \neq 0$, and sets $A_i = g_1^{1/(\gamma+x_i)} \in G_1$. The private key of user i is $\mathbf{gsk[i]} = (A_i, x_i, \pi, \mathbf{skp})$. The original user secretly issues $\mathbf{gsk[i]}$ to user i, and adds this user into the group user list.

Sign. Given a group public key $\mathbf{gpk} = (g_1, g_2, h, u, v, w, \rho, \eta)$, a private key $\mathbf{gsk[i]} = (A_i, x_i, \pi, \mathbf{skp})$, a block $\boldsymbol{m}_j \in Z_p^k$ and this block's identifier $id_j \in \mathcal{I}$, user i computes the signature σ_j as follows:

1. Select $\alpha_j, \beta_j, r_{j,\alpha}, r_{j,\beta}, r_{j,x}, r_{j,\gamma_1}, r_{j,\gamma_2}$ as in HAGS.
2. Compute $T_{j,1}$, $T_{j,2}$, $T_{j,3}$, $\gamma_{j,1}$, $\gamma_{j,2}$, $R_{j,1}$, $R_{j,2}$, $R_{j,3}$, $R_{j,4}$ and $R_{j,5}$ as in HAGS.
3. Compute $\boldsymbol{\delta} = (\delta_1, ..., \delta_k) \leftarrow \mathtt{PRG}(sk_{prg}) \in Z_p^k$ and $b_j \leftarrow \mathtt{PRF}(sk_{prf}, id_j) \in Z_p$, then calculate the homomorphic MAC of block $\boldsymbol{m}_j = (m_{j,1}, ..., m_{j,k})$ as $t_j = \sum_{l=1}^{k} \delta_l \cdot m_{j,l} + b_j \in Z_p$.
4. Compute a challenge c_j for block $\boldsymbol{m}_j$ as $c_j = \eta^{t_j} \cdot H_1(T_{j,1}, ..., R_{j,5}) \in Z_p$.
5. Compute $s_{j,\alpha}$, $s_{j,\beta}$, $s_{j,x}$, s_{j,γ_1}, s_{j,γ_2} as in HAGS.
6. Compute a tag θ_j as $\theta_j = [H_2(id_j)g_1^{t_j}]^{\pi} \in G_1$.
7. Output a signature σ_j of this block $\boldsymbol{m}_j$ as $\sigma_j = (T_{j,1}, T_{j,2}, T_{j,3}, \theta_j, R_{j,3}, c_j, s_{j,\alpha}, s_{j,\beta}, s_{j,x}, s_{j,\gamma_1}, s_{j,\gamma_2})$.

ProofGen. To audit the integrity of shared data, a user first sends an auditing request to the TPA. After receiving an auditing request, the TPA generates an auditing message as follows:

1. Randomly pick a q-element subset $\mathcal{J}$ of set $[1, n]$ to locate the q selected blocks in this auditing task.
2. Generate a random $y_j \in Z_p$, for $j \in \mathcal{J}$.
3. Output an auditing message $\{(j, y_j)\}_{j \in \mathcal{J}}$, and send it to the cloud.

After receiving an auditing message, the cloud generates a proof of possession of selected blocks in shared data as follows:

1. Compute $\mu_l = \sum_{j \in \mathcal{J}} y_j m_{j,l} \in Z_p$, for $l \in [1, k]$, and aggregate the selected tags as $\Theta = \prod_{j \in \mathcal{J}} \theta_j^{y_j} \in G_1$.
2. Output Φ and $\phi_j = (T_{j,1}, T_{j,2}, T_{j,3}, R_{j,3}, c_j, s_{j,\alpha}, s_{j,\beta}, s_{j,x}, s_{j,\gamma_1}, s_{j,\gamma_2})$ based on σ_j, where $j \in \mathcal{J}$ and Φ is the set of all ϕ_j.
3. Generate an auditing proof $\{\boldsymbol{\mu}, \Theta, \Phi, \{id_j\}_{j \in \mathcal{J}}\}$, and send it to the TPA, where $\boldsymbol{\mu} = (\mu_1, ..., \mu_k)$.

ProofVerify. Given an auditing proof $\{\boldsymbol{\mu}, \Theta, \Phi, \{id_j\}_{j\in\mathcal{J}}\}$, an auditing message $\{(j, y_j)\}_{j\in\mathcal{J}}$, a group public key $\mathbf{gpk} = (g_1, g_2, h, u, v, w, \rho, \eta)$, a secret key pair $\mathbf{skp} = (sk_{prg}, sk_{prf})$, the TPA verifies this proof as follows:

1. Generate $\boldsymbol{\delta} = (\delta_1, ..., \delta_k) \leftarrow \texttt{PRG}(sk_{prg}) \in Z_p^k$ and $b_j \leftarrow \texttt{PRF}(sk_{prf}, id_j) \in Z_p$, where $j \in \mathcal{J}$.
2. Re-compute $R_{j,1}$, $R_{j,2}$, $R_{j,4}$, $R_{j,5}$ as in HAGS.
3. Compute $\lambda = \sum_{l=1}^{k} \delta_l \mu_l + \sum_{j\in\mathcal{J}} y_j b_j \in Z_p$.
4. Check the following equations

$$\prod_{j\in\mathcal{J}} R_{j,3}^{y_j} \stackrel{?}{=} e(\prod_{j\in\mathcal{J}} (T_{j,3}^{s_{j,x}} \cdot h^{-s_{j,\gamma_1} - s_{j,\gamma_2}} \cdot g_1^{-c_j})^{y_j}, g_2) \cdot e(\prod_{j\in\mathcal{J}} (h^{-s_{j,\alpha} - s_{j,\beta}} \cdot T_{j,3}^{c_j})^{y_j}, w), \quad (7)$$

$$\prod_{j\in\mathcal{J}} c_j^{y_j} \stackrel{?}{=} \eta^{\lambda} \cdot \prod_{j\in\mathcal{J}} H_1(T_{j,1}, ..., \tilde{R}_{j,5})^{y_j}, \quad (8)$$

$$e(\Theta, g_2) \stackrel{?}{=} e(\prod_{j\in\mathcal{J}} H_2(id_j)^{y_j} \cdot g_1^{\lambda}, \rho). \quad (9)$$

   If all three equations hold, the proof is valid. Otherwise, it is not.
5. If the proof is valid, the TPA sends a positive report to the user. Otherwise, she sends a negative report.

Open. Given a block $\boldsymbol{m}_j$ and a signature σ_j, the original user can reveal the identity of the signer on this block using her group manager private key $\mathbf{gmsk} = \{\xi_1, \xi_2\}$ as in HAGS.

Discussions. In Knox, the TPA is able to verify the integrity of shared data without retrieving the entire data. The original user can add new users to the group without re-computing any signature. Using the group manager's private key, the original user can reveal the identity of the signer on each block. While in previous work [16], the original user cannot disclose the identity of the signer because the identity is unconditional protected by ring signatures [8]. In addition, if the original user in previous work [16] wishes to add new users to the group, all signatures on shared data has to be recomputed, because the generation and verification of a ring signature require all the current group members' public keys.

User Revocation. Once a group user is misbehaved and her identity is revealed by the group manager, it is necessary to revoke this misbehaved user from the group. In our current mechanism, to revoke a group user from the group, the group manager needs to re-generate and re-distribute some parts of the private key for existing users, then all existing users need to re-sign their blocks in shared data with new private keys. The blocks previously signed by the revoked user should be re-signed by the group manager. Specifically, the group manager generates and distributes a new pair $(\pi', \mathbf{skp}')$ for existing users, then user i can compute group signatures with her new private key $\mathbf{gsk}'[\mathbf{i}] = (A_i, x_i, \pi', \mathbf{skp}')$; while the revoked user cannot compute valid group signatures anymore because she has no knowledge of $(\pi', \mathbf{skp}')$. The TPA will audit shared data with the new corresponding public key $\mathbf{gpk}' = (g_1, g_2, h, u, v, w, \rho', \eta)$, where $\rho' = g_2^{\pi'}$. In some special cases, the group manager herself may need to leave the group. Then the new group manager should compute new private keys for users and a new public key for the new group, and all the users in the new group need to re-sign blocks in shared data with their new private keys.

6.3 Security Analysis of Knox

Theorem 5. *Given shared data M and its group signatures, a verifier is able to correctly check the integrity of shared data M.*

Proof. To prove the correctness of Knox is equivalent of proving Equation (7), (8) and (9) are all correct. Because Equation (1) is correct, it is clear that Equation (7) is also correct. Equation (8) can be expanded as follows:

$$\begin{aligned}
\prod_{j\in\mathcal{J}} c_j^{y_j} &= \prod_{j\in\mathcal{J}} \left(\eta^{t_j}\cdot H_1(T_{j,1},...,\tilde{R}_{j,5})\right)^{y_j} \\
&= \prod_{j\in\mathcal{J}} \eta^{t_j y_j}\cdot \prod_{j\in\mathcal{J}} H_1(T_{j,1},...,\tilde{R}_{j,5})^{y_j} \\
&= \eta^{\sum_{j\in\mathcal{J}} y_j(\sum_{l=1}^{k}\delta_l m_{j,l}+b_j)}\cdot \prod_{j\in\mathcal{J}} H_1(T_{j,1},...,\tilde{R}_{j,5})^{y_j} \\
&= \eta^{\sum_{l=1}^{k}\delta_l\mu_l+\sum_{j\in\mathcal{J}} y_j b_j}\cdot \prod_{j\in\mathcal{J}} H_1(T_{j,1},...,\tilde{R}_{j,5})^{y_j} \\
&= \eta^{\lambda}\cdot \prod_{j\in\mathcal{J}} H_1(T_{j,1},...,\tilde{R}_{j,5})^{y_j}.
\end{aligned}$$

Similar to the proof of Equation (8), the correctness of Equation (9) can be presented as

$$\begin{aligned}
e(\Theta, g_2) &= e(\prod_{j\in\mathcal{J}} \left(H_2(id_j)\cdot g_1^{t_j}\right)^{y_j}, g_2^{\pi}) \\
&= e(\prod_{j\in\mathcal{J}} H_2(id_j)^{y_j}\cdot \prod_{j\in\mathcal{J}} g_1^{t_j y_j}, \rho) \\
&= e(\prod_{j\in\mathcal{J}} H_2(id_j)^{y_j}\cdot g_1^{\lambda}, \rho).
\end{aligned}$$

All three equations are correct, therefore, a verifier in Knox is able to correctly check the integrity of shared data M.

Theorem 6. *Given shared data M and its group signatures, it is computational infeasible for an untrusted cloud or adversary to generate an auditing proof based on corrupted data M', where this auditing proof can pass the verification under Knox.*

Proof. Details of this proof can be found in our technical report [17].

Theorem 7. *Given shared data M and its group signatures, only the original user (the group manager) can reveal the identity of the signer on each block. For the TPA, it is computational infeasible to reveal the identity of the signer on each block during the auditing process.*

Proof. According to Theorem 4, for the TPA, who does not possess group manager's private key $\mathbf{gmsk} = (\xi_1, \xi_2)$, revealing the identity of the signer on each block during the auditing process is as hard as solving Decision Linear problem in G_1. The original user, who acts as the group manager, is able to trace the identity of the signer on each block using her group manager's private key.

7 Experimental Results

We now compare the performance of Knox with previous work, Oruta [16]. Due to space limitations, we only provide some experimental results in this section. Detailed analysis of computation and communication cost of Knox can be found in [17]. In our experiments, we utilize GMP and PBC libraries to implement cryptographic operations in Knox. All our experiments are tested on a 2.26 GHz Linux system over 1,000 times. The security level is $|p| = 160$ bits. We also assume the total number of blocks in shared data is $n = 1,000,000$, each block contains $k = 100$ elements, the size of each block is 2KB and total size of shared data is 2GB. In the following experiments, we assume the number of selected blocks is $q = 300$, which allows the TPA to keep the detection probability greater than 95% if 1% of all the blocks are corrupted [3].

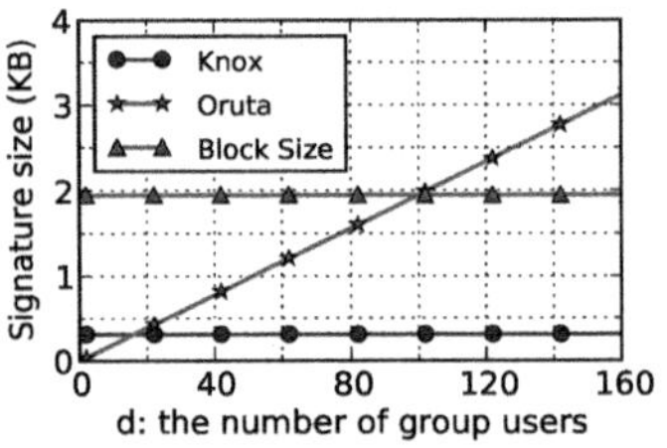

Fig. 2. Impact of group size d on signature size (KB)

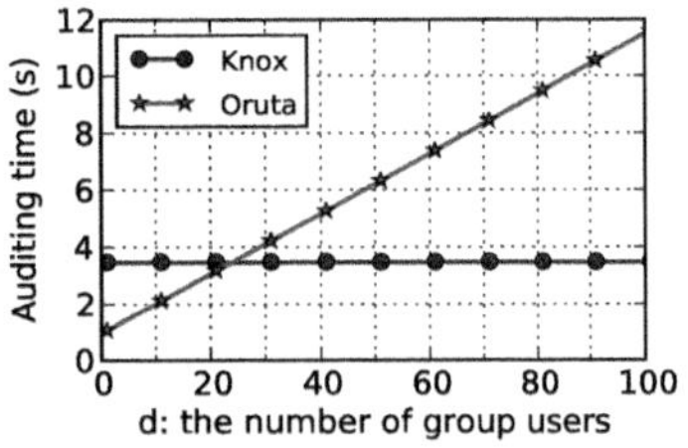

Fig. 3. Impact of group size d on auditing time (s)

As shown in Fig. 2, the signature size of Knox is independent from the number of users in a group. On the contrary, the signature size of Oruta is linearly increasing with the size of the group. Specifically, when $d = k$ in Oruta, the size of a signature is even the same as the size of a block.

In Fig. 3, we compare the auditing time of Knox and Oruta. In Knox, the auditing time is independent from the group size, while the auditing time in Oruta linearly increases with the size of the group. When the data in the cloud are shared by a large group, Knox requires less auditing time than Oruta. More specifically, when the group size is 100, Knox is able to finish an auditing task in less than 4 seconds while Oruta requires nearly 12 seconds to finish the same auditing task.

A detailed comparison of the auditing performance between Knox and Oruta is illustrated in Table 2, where $d = 100$ and $k = 100$. Although Knox requires less auditing time than Oruta, the communication cost of Knox is higher. However, it is still a small percentage of the entire size of shared data, which means the TPA can efficiently audit shared data without downloading the entire data. Our experimental results show that Knox has a better performance when auditing data shared among a large number of users.

Table 2. Comparison of Auditing Performance

	Oruta [16]	Knox
Data Storage Usage (GB)	2	
Signature Storage Usage (GB)	2	0.33
Communication Cost (KB)	18	106.4
Auditing Time (seconds)	11.49	3.44

8 Conclusion

In this paper, we propose Knox, a privacy-preserving auditing scheme for shared data with large groups in the cloud. We utilize group signatures to compute verification information on shared data, so that the TPA is able to audit the correctness of shared data, but cannot reveal the identity of the signer on each block. With the group manager's private key, the original user can efficiently add new users to the group and disclose the identities of signers on all blocks. The efficiency of Knox is not affected by the number of users in the group.

Acknowledgement. We are grateful to the anonymous reviewers for their helpful comments. This work is supported by the National Science and Technology Major Project (No. 2012ZX03002003), Fundamental Research Funds for the Central Universities (No. K50511010001), National 111 Program (No. B08038) and Doctoral Foundation of Ministry of Education of China (No. 20100203110002).

References

1. Agrawal, S., Boneh, D.: Homomorphic MACs: MAC-Based Integrity for Network Coding. In: Abdalla, M., Pointcheval, D., Fouque, P.-A., Vergnaud, D. (eds.) ACNS 2009. LNCS, vol. 5536, pp. 292–305. Springer, Heidelberg (2009)
2. Armbrust, M., Fox, A., Griffith, R., Joseph, A.D., Katz, R.H., Konwinski, A., Lee, G., Patterson, D.A., Rabkin, A., Stoica, I., Zaharia, M.: A View of Cloud Computing. Communications of the ACM 53(4), 50–58 (2010)

3. Ateniese, G., Burns, R., Curtmola, R., Herring, J., Kissner, L., Peterson, Z., Song, D.: Provable Data Possession at Untrusted Stores. In: Proc. ACM CCS, pp. 598–610 (2007)
4. Ateniese, G., Camenisch, J., Joye, M., Tsudik, G.: A Practical and Provably Secure Coalition-Resistant Group Signature Scheme. In: Bellare, M. (ed.) CRYPTO 2000. LNCS, vol. 1880, pp. 255–270. Springer, Heidelberg (2000)
5. Ateniese, G., Pietro, R.D., Mancini, L.V., Tsudik, G.: Scalable and Efficient Provable Data Possession. In: Proc. ICST SecureComm, pp. 1–10 (2008)
6. Boneh, D., Boyen, X., Shacham, H.: Short Group Signatures. In: Franklin, M. (ed.) CRYPTO 2004. LNCS, vol. 3152, pp. 41–55. Springer, Heidelberg (2004)
7. Boneh, D., Freeman, D.M.: Homomorphic Signatures for Polynomial Functions. In: Paterson, K.G. (ed.) EUROCRYPT 2011. LNCS, vol. 6632, pp. 149–168. Springer, Heidelberg (2011)
8. Boneh, D., Gentry, C., Lynn, B., Shacham, H.: Aggregate and Verifiably Encrypted Signatures from Bilinear Maps. In: Biham, E. (ed.) EUROCRYPT 2003. LNCS, vol. 2656, pp. 416–432. Springer, Heidelberg (2003)
9. Boneh, D., Lynn, B., Shacham, H.: Short Signatures from the Weil Pairing. In: Boyd, C. (ed.) ASIACRYPT 2001. LNCS, vol. 2248, pp. 514–532. Springer, Heidelberg (2001)
10. Chaum, D., van Heyst, E.: Group Signatures. In: Davies, D.W. (ed.) EUROCRYPT 1991. LNCS, vol. 547, pp. 257–265. Springer, Heidelberg (1991)
11. Erway, C., Kupcu, A., Papamanthou, C., Tamassia, R.: Dynamic Provable Data Possession. In: Proc. ACM CCS, pp. 213–222 (2009)
12. Ferrara, A.L., Green, M., Hohenberger, S., Pedersen, M.Ø.: Practical Short Signature Batch Verification. In: Fischlin, M. (ed.) CT-RSA 2009. LNCS, vol. 5473, pp. 309–324. Springer, Heidelberg (2009)
13. Hao, Z., Zhong, S., Yu, N.: A Privacy-Preserving Remote Data Integrity Checking Protocol with Data Dynamics and Public Verifiability. IEEE Transactions on Knowledge and Data Engineering 23(9), 1432–1437 (2011)
14. Juels, A., Kaliski, B.S.: PORs: Proofs pf Retrievability for Large Files. In: Proc. ACM CCS, pp. 584–597 (2007)
15. Shacham, H., Waters, B.: Compact Proofs of Retrievability. In: Pieprzyk, J. (ed.) ASIACRYPT 2008. LNCS, vol. 5350, pp. 90–107. Springer, Heidelberg (2008)
16. Wang, B., Li, B., Li, H.: Oruta: Privacy-Preserving Public Auditing for Shared Data in the Cloud. Tech. rep., University of Toronto (2011), `http://iqua.ece.toronto.edu/~bli/techreports/oruta.pdf`
17. Wang, B., Li, B., Li, H.: Knox: Privacy-Preserving Auditing for Shared Data with Large Groups in the Cloud. Tech. rep., University of Toronto (2012), `http://iqua.ece.toronto.edu/~bli/techreports/knox.pdf`
18. Wang, C., Wang, Q., Ren, K., Lou, W.: Ensuring Data Storage Security in Cloud Computing. In: Proc. IEEE IWQoS, pp. 1–9 (2009)
19. Wang, C., Wang, Q., Ren, K., Lou, W.: Privacy-Preserving Public Auditing for Data Storage Security in Cloud Computing. In: Proc. IEEE INFOCOM, pp. 525–533 (2010)
20. Wang, Q., Wang, C., Li, J., Ren, K., Lou, W.: Enabling Public Verifiability and Data Dynamics for Storage Security in Cloud Computing. In: Backes, M., Ning, P. (eds.) ESORICS 2009. LNCS, vol. 5789, pp. 355–370. Springer, Heidelberg (2009)

21. Yang, K., Jia, X.: Data storage auditing service in cloud computing: challenges, methods and opportunities. World Wide Web 15(4), 409–428 (2012)
22. Yu, S., Wang, C., Ren, K., Lou, W.: Achieving Secure, Scalable, and Fine-grained Data Access Control in Cloud Computing. In: Proc. IEEE INFOCOM, pp. 534–542 (2010)
23. Zhu, Y., Hu, H., Ahn, G.-J., Yau, S.S.: Efficient Audit Service Outsourcing for Data Integrity in Clouds. Journal of System and Software 85(5), 1083–1095 (2012)
24. Zhu, Y., Wang, H., Hu, Z., Ahn, G.-J., Hu, H., Yau, S.S.: Dynamic Audit Services for Integrity Verification of Outsourced Storage in Clouds. In: Proc. ACM Symposium On Applied Computing, pp. 1550–1557 (2011)

SPICE – Simple Privacy-Preserving Identity-Management for Cloud Environment*

Sherman S.M. Chow[1], Yi-Jun He[2,**], Lucas C.K. Hui[2], and Siu Ming Yiu[2]

[1] University of Waterloo, Ontario, Canada
smchow@math.uwaterloo.ca
[2] University of Hong Kong, Hong Kong
{yjhe,hui,smyiu}@cs.hku.hk

Abstract. Identity security and privacy have been regarded as one of the top seven cloud security threats. There are a few identity management solutions proposed recently trying to tackle these problems. However, none of these can satisfy all desirable properties. In particular, *unlinkability* ensures that none of the cloud service providers (CSPs), even if they collude, can link the transactions of the same user. On the other hand, *delegatable authentication* is unique to the cloud platform, in which several CSPs may join together to provide a packaged service, with one of them being the source provider which interacts with the clients and performs authentication while the others will be transparent to the clients. Note that CSPs may have different authentication mechanisms that rely on different attributes. Moreover, each CSP is limited to see only the attributes that it concerns.

This paper presents SPICE – the first digital identity management system that can satisfy these properties in addition to other desirable properties. The novelty of our scheme stems from combining and exploiting two group signatures so that we can randomize the signature to make the same signature look different for multiple uses of it and hide some parts of the messages which are not the concerns of the CSP. Our scheme is quite applicable to cloud systems due to its simplicity and efficiency.

Keywords: Cloud Computing, Digital Identity Management, Interoperability, Delegation, Privacy, Unlinkability.

1 Introduction

Cloud computing is a new computing platform where thousands of users around the world can have network access of a shared pool of computing resources

* The work described in this paper was partially supported by the General Research Fund from the Research Grants Council of the Hong Kong Special Administrative Region, China (Project No. RGC GRF HKU 713009E), the NSFC/RGC Joint Research Scheme (Project No. N_HKU 722/09), HKU Seed Fundings for Applied Research 201102160014, and HKU Seed Fundings for Basic Research 201011159162 and 200911159149.

** Corresponding author.

F. Bao, P. Samarati, and J. Zhou (Eds.): ACNS 2012, LNCS 7341, pp. 526–543, 2012.

(e.g., storage, applications, services, *etc.*) with minimal requirement for the users' computers, and minimal management effort from the service provider. Various popular cloud computing services have emerged including Amazon's Simple Storage Service (S3), Box.net, CloudSafe and Internap XIPCloud Storage. On the other hand, there are quite a number of security and privacy concerns in this computing platform that are yet to be resolved [20,21]. In particular, user identity security was regarded as one of the top seven cloud security threats by the Cloud Security Alliance in 2010 [3].

1.1 Extended DIM Framework in the Cloud

Existing solutions for tackling the identity security problem in the cloud are usually based on the framework of a digital identity management (DIM) system [4,6,19], which allows the cloud service providers to provide services only to authenticated users. A typical DIM system in a cloud environment consists of four components: *cloud service providers* (CSPs), *identity providers* (IdPs), *registrars*, and *users* (cloud clients). IdPs are responsible for assigning attributes to users. Registrars are able to verify the attributes given by an IdP to a user, and then issue a certificate to the user. Based on these certificates, users can authenticate themselves to CSPs and gain access to the authorized services.

In an *extended* DIM framework depicted in Figure 1, CSPs may not work alone as independent providers. Several CSPs may join together to provide a packaged service to the clients. One of them, called the *source CSP*, acts as the interface to the clients while the others, called the *receiving CSPs*, can be transparent to the clients and provide services in the back-end. A CSP can be both a source and a receiving CSP, and a receiving CSP can also interact with different source CSPs providing different services.

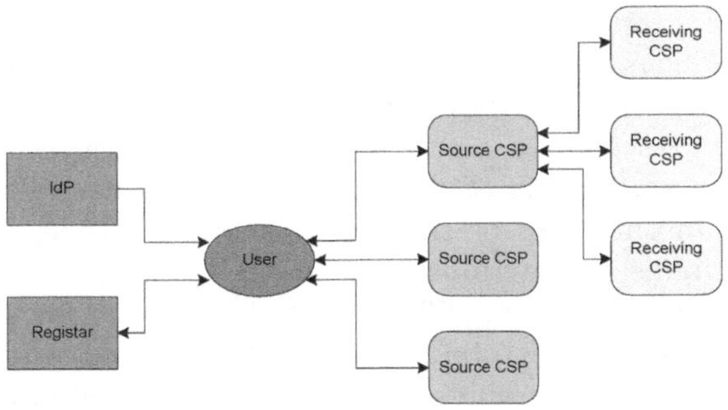

Fig. 1. Extended DIM Structure

1.2 Desirable Properties for DIM in the Cloud

A DIM system should have the following desirable security/privacy and functional properties for authentication.

- **Unlinkability.** In cloud computing, a user may purchase multiple services associated with the same or different CSPs. Unlinkability ensures that no CSPs, even if they collude, can link different transactions, whether they are of the same service or different services, of the same user.
- **Delegatable Authentication.** Each CSP would like to authenticate the user in its own way before providing their packaged services. The authentication should be *delegatable* such that the receiving CSP can authenticate a user without a direct communication with either the user[1] or the registrar, and without fully trusting the source CSP.[2]
- **Anonymity.** The users should be able to anonymously authenticate themselves, as authorized users to the CSP, without letting the CSP know about their real identity or exact attributes.
- **Accountability.** The users may abuse their anonymity. If needed, a trusted party can revoke the anonymity so the users can be held accountable for their malicious action.
- **User Centric Access Control.** Users should be able to control what information they want to reveal about themselves over the cloud or to a CSP, and to control who can access that information, and how this information would be used in order to minimize the risk of identity theft and fraud.
- **Single registration.** The users only need to register themselves once for getting the credentials without the need of contacting the registrar every time authentication is needed.

1.3 Existing Solution and Limitations

In Bertino *et al.* [6]'s DIM system, the registrar stores a set of signatures, each signing on a commitment of a user's attribute. To authenticate to a CSP, the user first retrieves from the registrar a set of signatures on the attributes that the CSP requires (matching of names may be needed to determine the naming of the attributes). Then, the user executes a zero-knowledge proof (ZKP) protocol to prove to the CSP that the signatures presented by the user signs on the committed attributes. Finally, a new credential will be generated to this user by the CSP upon its verification of the registrar's signatures. This credential can be presented to another CSP by the user to show that there was a CSP which verified the required set of attributes.

Different CSPs may require a different set of attributes for authentication, so many different credentials will be issued. In their scheme, the registrar should remain (practically always) online to return different signature sets to a user whenever the user subscribes to new services. The number of times that the

[1] It is more convenient to the users if all receiving CSPs are transparent. On the other hand, the source CSP may want to keep the operation private and may not be willing to let the users know the CSPs which it collaborated for providing indirect services.

[2] The receiving CSP cannot fully trust the source CSP since it is also regarded as a big client to the receiving CSP. E.g., PayPal (providing a payment gateway between the merchant and the customer) does not rely on eBay (providing shopping "service") for authenticating consumers although they are partners.

registrar responds to a user is linear to the total number of services that the user wants to invoke. Apart from the scalability issue, this also increases the risk of exposing the "master secret key" (used by the registrar to certify attributes) from a variety of network attacks.

The second non-negligible consequence is the linkability issue that the attacker may link registrar's signatures of the same user to compromise the privacy of that user. Even though their authentication mechanism uses zero-knowledge proof of knowledge (ZKPoK) of a "correct"[3] attribute (that is contained in the commitment signed by the registrar) to avoid showing the attributes in clear, the user simply shows a signature (which is an aggregation of the signatures given by the registrar) on the commitments of the attributes, which makes different signatures easily linkable. One may wonder if we can just also use ZKPoK to hide the signature. Looking ahead, our solution leverages the re-randomizability of the signature, which is a very lightweight approach when compared with ZKPoK.

The third consequence is that it is hard to delegate authentications (not to say making the delegatable authentications unlinkable) based on this framework. We want that a source CSP who just verified the credential of a user can convince a receiving CSP that the user in question indeed has (more than) what the receiving CSP expects, without requiring the receiving CSP to directly interact with the registrar or the user. This feature is termed as delegatability in Bertino *et al.*'s work and was remarked as a future research direction. Finally, we remark that simultaneously addressing all the desirable requirements is more involved as they may conflict with each other.

There is an open source reference implementation called Shibboleth [1], which can be regarded as using a similar (in a high-level way) authentication mechanism. Their single sign-on feature allows service providers (SPs) in a federation, which must be explicitly configured *a priori*, to provide a packaged service to a user, which seems to be the delegatable authentication feature. However, the user needs to authorize a source SP to access a receiving SP's service on behalf, so the process is *not transparent.* Most importantly, their IdP (which plays the role of registrar in the terminologies in [6] and in this paper) needs to be online during the access and two certificates will be generated for both SPs. Unlinkability is also not a concern in Shibboleth.

There are other related works which try to provide extensions of the basic notion such as [4,11,16,19]. However, up to now, it is fair to say there still does not exist any scheme that can satisfy both the property of *unlinkability* and *delegatable authentication*, in an efficient yet cryptographically secure manner.

1.4 Our Contributions

We present SPICE – a DIM system in the cloud environment that can satisfy all the aforementioned six desirable properties for identity management, which is the first such system to our knowledge. In our scheme, the registrar issues only

[3] It is not clear that how the CSP can verify if an attribute is correct if it is hidden by the commitment and the zero-knowledge proof.

one credential to each user no matter how many services that the user subscribes or will subscribe later. For authentication, the user generates from the credential many different versions of certificates for proving the possession of (different sets of) attributes required by different service providers, without asking registrar for issuing new certificates each time.

The novelty of our proposed system stems from extending two group signature schemes [8,9] to their "full" potential with the help of the nice properties of Waters' signature [22] and Groth-Sahai proof system [15], both are implicitly used as building blocks in [8,9]. Group signatures are privacy-oriented signatures where a group manager can issue signing keys to many group members who in turn can sign on behalf of the group. Anyone can be convinced that the signature is indeed from the group, but not exactly who.

To satisfy the unlinkability property, we apply randomization to the signatures (which acts as the certificates in our DIM system) to make them look different and hence unlinkable for multiple requests. Both [15] and [22] are known to be re-randomizable (see Section 4 for the technical details). On the other hand, to control which attributes can be revealed to which CSPs, we want the messages being signed (correspond to the attributes in our DIM system) to be partitioned into different blocks. We thus need to extend the message space from a single message to many blocks of messages. As a result, some of the attributes can be hidden from the CSPs if they are not related to CSPs' verification. Finally, for delegated authentication, some of the attributes certified by a group signature can be hidden by the source CSP by a sanitization process.

The idea of using sanitizable and re-randomizable group signature for a privacy-preserving DIM is conceptually simple, and the run-time performance of our prototype is practically efficient, hence we believe our system is applicable in cloud environment nowadays.

2 Related Work

2.1 Identity Management Systems

After Bertino *et al.*'s work, Celesti *et al.* [11] constructed an InterCloud Identity Management Infrastructure focusing on the InterCloud scenario where clouds cooperate with other federated ones with the purpose to enlarge their computing and storage capabilities. In InterCloud, the users' home cloud should be able to perform a single sign-on in order to gain access to the resources offered by another cloud that participates in an InterCloud formation; each home cloud should be able to authenticate itself with foreign clouds using its digital identity guaranteed by a third party.

The PhD thesis of Hussain [16] studied secure anonymous authentication with Personas. Personas are defined as sets of statements, where each statement asserts the status of an individual's set of attributes. The main advantage of [16] is that the Personas of an individual are distributed among a number of servers which mitigate the damage of a single server compromise. However, they did not pick up the challenge to support unlinkability and delegation.

Angin *et al.* proposed an entity-centric approach for privacy and identity management in cloud computing [4]. It aims at achieving three requirements: (i) authenticating users without disclosing personal identity information (PII); (ii) ability to provide identity management services on untrusted hosts, such as public hosts; and (iii) not using trusted third parties. Their approach is based on an entity-centric DIM which uses active bundles scheme to protect PII from untrusted hosts; and anonymous identification which involves using zero-knowledge proofs for authentication of an entity without disclosing its identifier. However, the proposed approach could have a large communication overhead since each anonymous identification would involve a good number of communications between a CSP and the user. Further, an active bundle (a token) which is a container with a payload of sensitive data, metadata, and a virtual machine (VM), needs to be passed between a CSP and a user during a communication session, but no details mention about how to pass the token effectively. The token could have a large size. So, we are not certain about the practicality of the proposed approach.

Ranchal *et al.* [19] proposed another scheme to address the same privacy problem [4] by using the active bundles scheme and computing predicate over encrypted data for giving answer about PII. Their approach uses identity data on untrusted hosts without TTP. However, when users asking for a service from the CSP, the overhead is high since they need to compute a token for a CSP using secure multiparty computation involving the cooperation with a number of parties. Also, they did not mention how to pass effectively a big active bundle between a user and a CSP.

2.2 Other Credential-Based Authentication Systems

A group signature is essentially a ZKPoK of the signatures. Indeed, that is exactly the mechanism behind most cryptographic anonymous credentials. But many such systems often have their own specific features (e.g., ensuring a credential can only be used for at most k times [12]) and do not perfectly fit with our needs.

One may consider asking the user to "delegate" a credential to a source CSP using "delegatable anonymous credential", then the source CSP can show the credential to a receiving CSP on the user's behalf. Not every anonymous credential systems come with this property. A recent system (e.g., [5]) supports delegation of the credential itself, but it is different from delegating the verification process. In more details, the credential system in [5] requires that each user to have their own private key for exculpability reason, which is not a concern in our scenario but results in a much more complicated signing mechanism (an interactive protocol for obtaining a signature on a message hidden in the commitment) and the corresponding proof (the private key corresponding to the message being signed at the i-th level is used to sign the message at the $(i+1)$-th level). Besides, their application does not require any anonymity revocation. We stressed that their scheme is more on delegating the credential (i.e. the signing process) than delegating the verification. To conclude, delegatable anonymous credentials are too heavyweight for our usage on one hand, but still lack the anonymity management mechanism required in our scenario.

Finally, we remark that there are extensions of the basic group signature idea in terms of anonymity management mechanism, such as tracing and signature claiming or denial considered in traceable signature [2,12].

3 Overview of SPICE

3.1 Framework of SPICE

We will describe the main entities in our system, and different phases (corresponding to different tasks) of a typical run-time scenario. We will list the actions performed by different entities, the objects created, and how these objects flow between them.

In our framework, the registrar will first create a user-specific credential, which is then used to generate a *source certificate* certifying a number of attributes. That is the *user enrollment* phase.

In the *authentication phase*, a user uses the source certificate obtained from the registrar to create an *authentication certificate*, or simply a *certificate* according to the set of attributes expected by a CSP. Upon successful *verification*, the CSP either provides the service to the user, or contacts other receiving CSP(s) for help in providing a packaged service.

In order to perform *delegatable authentication*, the CSP makes use of the certificate obtained to generate a *"new" certificate* depending on the set of attributes expected by a receiving CSP.

3.2 Framework of Basic Group Signatures

A group signature scheme, GS, consists of five algorithms. The first two and the last one are used by the group manager (GM), the third is used by group members (U), and the fourth one is used by any verifiers.

- $\mathsf{GS.Setup}$: $(\mathsf{gpk}, \mathsf{gsk}, \mathsf{gok}) \leftarrow \mathsf{GS.Setup}(1^\kappa)$ is an algorithm that generates a group public (verification) key gpk, certificate issuing key gsk and opening-key gok.
- $\mathsf{GS.UG}$: $\mathsf{sk} \leftarrow \mathsf{GS.UG}_{\mathsf{gpk}}(\mathsf{gsk}, \mathsf{id})$ is an algorithm that generates a user signing key sk to the user having identity id.
- $\mathsf{GS.Sign}$: $\sigma \leftarrow \mathsf{GS.Sign}_{\mathsf{gpk}}(\mathsf{sk}, M)$ is a probabilistic algorithm that generates a signature σ for message M by using signing-key sk.
- $\mathsf{GS.Ver}$: $1/0 \leftarrow \mathsf{GS.Ver}_{\mathsf{gpk}}(\sigma, M)$ is a deterministic algorithm that decides the correctness of signature σ on M. It outputs 1 for any input created through GS.Setup, GS.UG and GS.Sign, 0 otherwise.
- $\mathsf{GS.Open}$: $\mathsf{id}/\bot \leftarrow \mathsf{GS.Open}_{\mathsf{gpk}}(\mathsf{gok}, \sigma)$ is an algorithm that identifies the signer of a valid signature σ by using the opening-key gok.

3.3 The Key Ideas

The authentication mechanism in SPICE will be instantiated by a group signature scheme with special properties denoted by GS. The registrar will act as the group manager in GS. As revealed previously in Section 1.4, there are multiple attributes associated with a source certificate (which is a signature on $(\ell - 1)$ attributes given by member signing key in GS). These attributes (hosted in the message blocks associated with GS) will be accompanied by certificate created by the credential under the normal circumstance.

For accessing a service, the user creates a certificate by the signing algorithm of GS, signing on one more block of message, which is the session's information such as the time of access, the description of the service, *etc.* When needs arise, the opening mechanism of GS can be used by the registrar to revoke the anonymity of a certificate. From now on, we will use the terminologies in the context of GS.

One of the key concepts in our scheme is randomization of the group signatures. It is either applied on the signature so that multiple requests issued by using the same signature are unlinkable, or applied on the message-component of the signature, such that the attribute being signed can be hidden. In order to design such a group signature scheme, we combine and extend two group signature schemes from [8,9]. First, we extend [9] by adding two algorithms: randomization algorithm (GS.Rand), and hiding algorithm (GS.Hide) to make the signature re-randomizable and some blocks of the signatures can be hidden (sanitized). Second, we extend the idea of two-level hierarchical signature in [8] to make our group signature feasible to be signed on blocks of messages instead of a single message.

Authentication is done by presenting a randomized copy of the certificate in general, as depicted in Figure 2. Take a user A with only three attributes as a simple example. User A's certificate is a signature of attributes A_1, A_2 and A_3. Every time A randomizes the signature before authenticating himself to CSPs. We use dark color to represent the original signature, and light color for the randomized signature. The re-randomizable property prevents the certificates from the same user from being linked.

SPICE classifies attributes into the following types for possible signature randomization and sanitization.

I) Sensitive personal information. This class of attributes, depicted by A_1 in Figure 2 is relatively stable and has a common representation across different CSPs. There is no special treatment for this class of attributes in our scheme.[4]

II) Service-specific attributes. Some attributes may be of interest to CSPs who are providing a similar kind of services (e.g., whether HTML or attachment is allowed in out-going e-mail) or providing a collective / coupled service (e.g., geographic information for various social networks or other location-based services).

[4] We defer to Section 5.2 for a possible strengthening of privacy concern when getting a credential from the registrar.

This class is depicted by A_2 in Figure 2. The CSPs may share similar authentication policy, but they may employ heterogeneous naming when establishing their authentication policies.

A nice feature considered in Bertino *et al.*'s work is *naming heterogeneity*, i.e., the existence of variations between attributes associated with the user's credential and attributes specified in the policies required by different CSPs. They used a matching technique based on look-up tables, dictionaries (WordNet 2.1 English Lexical Database[5]), and ontology mapping (Falcon-AO[6]) to resolve syntactic (e.g., "ID" vs. "Identity"), terminological (e.g., "email address" vs. "email account") and semantic (e.g., "privacy level" vs. "sharing setting") variations. SPICE will also use the same tools.

This class of attributes will be certified by a group signature, comes from another instance of GS, i.e., instead of the group of cloud clients, now we consider a group of CSPs. Source CSPs and receiving CSPs are group members. Being another instance of GS, the group signature generated is supposed to be certificates on the class of attributes that may need naming resolution. In particular, they cannot be used as typical user certificates. Whenever there is a variation of attributes representing the same concept, the source CSP generates a group signature certifying the same attribute with a different naming. These signatures may be sent to the client or stored by the source CSP for future usage (and re-randomization can also be done to ensure the unlinkability). In this way, the only thing that the registrar needs to know is who are concerned with a selected subset of attributes, instead of knowing the ontology mapping between all these CSPs, which has been offloaded to the CSPs effectively. The operation is depicted in Figure 2(b).

We remark that we just take a simple approach of appending another group signature to the original one. Updating the attributes associated with a cryptographic key is a tricky problem (except a recent work [10] that works with a particular group signature scheme [7]), which hinders the mismatch resolution considered by Bertino *et al*, and especially difficult when we also want to satisfy unlinkability.

III) Irrelevant attributes. There are many other attributes, which are unrelated to a specific (class) of services in question, depicted by A_3 in Figure 2. On the other hand, even when the CSPs are sharing authentication information, it is likely that some of the attributes are irrelevant for some of these CSPs (e.g., when sharing geographic information of a user logged into a social network, it is inappropriate to share other information such as the posting privilege of the user in the social network to other location-based services).

The privacy of this class of attributes can be taken care of by the hiding algorithm of our group signature scheme. The operation depicted in Figure 2(c) is hiding attribute A_3 which is not required by the receiving CSP, and Figure 2(d) is re-randomizing a sanitized signature.

[5] `http://wordnet.princeton.edu`

[6] `http://iws.seu.edu.cn/projects/matching`

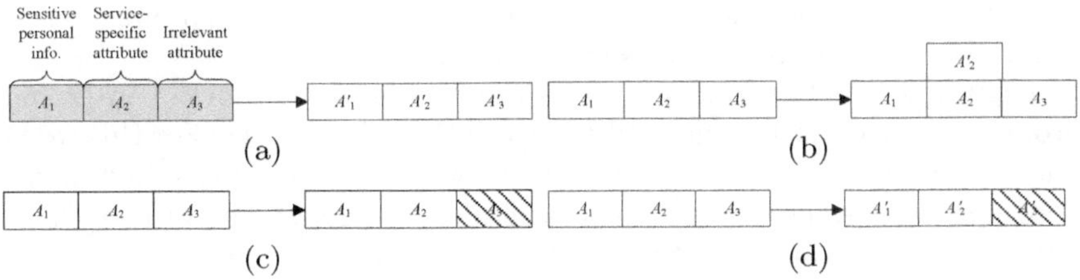

Fig. 2. (a) Re-randomization (b) Naming conversion (c) Attribute hiding (d) Hide-and-randomize

4 Randomizable Group Signatures

This section provides the details of instantiating the group signature scheme GS, which is the main building block in SPICE as described in previous section.

4.1 Design of Randomizable Group Signatures

Recall that hierarchical signatures with non-interactive zero knowledge (NIZK) proofs can be used to construct group signatures [8]. Two schemes given by Boyen and Waters [8,9] fall into this paradigm, with Groth-Sahai proofs [15] as the NIZK proofs. The basic idea is that, the message at the first level of the hierarchy will serve as a group member's identity, and the lower level corresponds to the message that a group member is going to sign. To hide the identity, an (extractable) NIZK proof is used to hide the message at the first level.

We use a group signature scheme which can be seen as a combination of these two schemes [8,9]. Indeed, one can just use the scheme from [8] which uses the hash functions of the same form (instantiated with different parameters) for both the group members' identities and the actual messages to be signed. As [9], the scheme to be presented here is a hybrid of two schemes, one at each level. For our application, we added three extra properties.

1. We want the group signatures can be re-randomizable by a public algorithm, so we add the algorithm GS.Rand. For this we require two properties from our building blocks. Recall that the group signature here is just an NIZK proof of a "regular" signature. First, the implicit regular signatures should be re-randomizable. This property can be satisfied by Waters' signatures – the signing mechanism used for signing the actual messages. Second, the NIZK proofs should be re-randomizable. The Groth-Sahai proof system [15] being used here is known to be re-randomizable (e.g., [5,14]), hence re-randomization of the group signatures can be done.
2. We want the group signatures to be able to sign blocks of messages instead of a single message. For this we rely on the fact that one may add a hierarchy of messages to be signed by extending the public key with a new set of parameters for the Waters-hashes [22] for other levels, as illustrated in the

two-level hierarchical signatures schemes in [8]. This technique has been utilized in other applications such as [13,18].
3. For multiple blocks of message, it is desirable if some blocks can be hidden from the verifiers when they only concern the other blocks. For this we introduce another algorithm GS.Hide which transforms a given group signature into one with some selected blocks hidden, in the same spirit as how the identity at the first level of the hierarchy is being hidden.

4.2 Concrete Construction

Below we supply the details of GS. In our DIM structure, registrar takes the role of the group manager (GM), users and the CSPs are group members (U).

Basic Algebraic Settings. Let $(\mathbb{G} = \langle g \rangle, \mathbb{G}_T)$ be a bilinear group pair of composite order $n = pq$ where p and q are random primes of bit size at least κ. Let $\mathbb{G}_q = \langle h \rangle$ be the cyclic subgroup of $\mathbb{G}$ of order q.

GS.Setup. This algorithm is used by the registrar. It generates the bilinear groups parameter listed above. The opening-key is $\mathsf{gok} = q$. Then, it picks two random exponents $\alpha, \omega \in \mathbb{Z}_n$, defines $\hat{A} = \hat{e}(g, g)^{\alpha}$ and $\Omega = g^{\omega}$. It also picks a random generator $u \in \mathbb{G}$. The certificate issuing key is $\mathsf{gsk} = \langle g^{\alpha}, \omega \rangle$.

Suppose the number of message blocks to be supported is ℓ, where each of them is m bit. It picks ℓ vectors, each vector $\boldsymbol{v}_j$ consists of $(m+1)$ random generators, $v_j, v_{j,1}, \cdots, v_{j,m} \in \mathbb{G}$. Define $F_j(M_j) = v_j \prod v_{j,i}{}^{\mu_{j,i}}$ where $M_j = \mu_{j,1} \cdots \mu_{j,m}$. The group public key is $\mathsf{gpk} = \langle \mathbb{G}, \mathbb{G}_T, n, g, u, \Omega, h, \hat{A}, \boldsymbol{v}_1, \cdots, \boldsymbol{v}_\ell \rangle$.

GS.UG. This algorithm is used by the registrar to generate signing keys for group members (users and CSPs). The system supports 2^k members and each of their identities are mapped to integer id where $0 \leq \mathsf{id} < 2^k < p$, a unique value $s_{\mathsf{id}} \in \mathbb{Z}_n$ is assigned to each member and will be stored for later opening purpose. This value must be chosen so that $(\omega + s_{\mathsf{id}})$ lies in $\mathbb{Z}_n^{\times}$, the multiplicative group modulo n. The member signing key $\mathsf{sk}_{\mathsf{id}}$ is in the form of $\langle K_1, K_2, K_3 \rangle = \langle (g^{\alpha})^{\frac{1}{\omega + s_{\mathsf{id}}}}, g^{s_{\mathsf{id}}}, u^{s_{\mathsf{id}}} \rangle$, which is of the same form as [9].

GS.Sign. This algorithm has three usages. When it is used by the registrar to generate a source certificate, it will sign on $(\ell - 1)$ attributes. When it is used by a user, it will sign on the information for the authentication session (by appending the source certificate with a last block of message, i.e. the ℓ-th block). If used by a CSP, it will generate a sanitized certificate for solving naming conflict.

I) For signing on blocks of message $\{M_j\}_{j=1}^{\ell-1}$ where $M_j = \mu_{j,1} \cdots \mu_{j,m}$, it first picks $r_1, \cdots, r_{\ell-1} \in \mathbb{Z}_n$ and creates an ordinary signature by

$$\begin{aligned} \theta &= \langle \theta_1, \theta_2, \theta_3, \{\theta_{j'}\}_{j'=4}^{\ell+2}, \{\theta'_{j'}\}_{j'=4}^{\ell+2} \rangle \\ &= \langle K_1, K_2, K_3 \prod_{j=1}^{\ell-1} F_j(M_j)^{r_j}, \{g^{-r_j}\}, \{h^{r_j}\} \rangle. \end{aligned}$$

This step extends [9] in two ways. First, instead of signing on a single message M_1, a product of $F_j(M_j)^{r_j}$'s is attached to sign on multiple messages, similar to the 2-level structure of [8]. Second, h^{r_j} is also attached for future hiding of the component $F_j(M_j)^{r_j}$, similar to how the "first-level" message is being hidden in [8].

Then, it turns θ into a group signature with s_{id} hidden by randomly picking $t_1, t_2, t_3, \{t_{j'}\}_{j'=4}^{\ell+2} \in \mathbb{Z}_n$, computing $\{\sigma_j = \theta_j \cdot h^{t_j}\}$ for $j = 1, \cdots, \ell+2$ and generating the corresponding proofs.

$$\pi_1 = h^{t_1 t_2} \cdot (\theta_1)^{t_2} \cdot (\theta_2 \Omega)^{t_1},$$
$$\pi_2 = u^{t_2} \cdot g^{-t_3} \cdot (\prod_{j=1}^{\ell-1} (F_j(M_j))^{-t_{j+3}})$$

This is the exact same step in [9], apart from the fact that a product of $F_j(M_j)^{r_j}$'s is attached.

The final signature is $\langle \{\sigma_j\}_{j=1}^{\ell+2}, \{\theta'_{j'}\}_{j'=4}^{\ell+2}, \pi_1, \pi_2 \rangle$.

II) When given a source certificate / signature in the above form signed on $(\ell - 1)$ message blocks, one can generate a certificate / signature on ℓ blocks by first appending the last message M_ℓ following the format of θ, then followed by a re-randomization using GS.Rand to be described. Due to the lack of space we omit an explicit description of this procedure, but that can be easily deduced from the above description and follows a similar signing procedure in the hierarchical signature schemes [22,8] based on Waters' signature.

GS.Ver. This algorithm is used to verify a source certificate. The verifier outputs 1 if and only if both of the following equations hold, 0 otherwise.

$$\hat{e}(h, \pi_1) = \hat{A}^{-1} \cdot \hat{e}(\sigma_1, \sigma_2 \Omega),$$
$$\hat{e}(h, \pi_2) = \hat{e}(\sigma_2, u) \hat{e}(\sigma_3, g)^{-1} \prod_{j=1}^{\ell} (\hat{e}(\sigma_{j+3}, F_j(M_j))^{-1})$$

(One may defer the checking of $\{\theta'_{j'}\}_{j'=4}^{\ell+2}$ from GS.Ver to GS.Hide since they are used when the $(j'-3)$-th message block $M_{j'-3}$ should be hidden.)

GS.Open. This algorithm is used by the registrar to ensure accountability. Given a valid signature, this algorithm takes out its σ_2 component, and tests whether $(\sigma_2)^q = (g^{s_{\mathsf{id}_i}})^q$ for each suspected id_i. We note that $(g^{s_{\mathsf{id}_i}})^q$ can be precomputed and stored when each user secret key is generated, as suggested in [9].

GS.Hide$(\sigma, M_j, j) \rightarrow \sigma'$. This algorithm is used by users and source CSPs to hide some attributes of the certificate. Given a valid signature σ that may have been re-randomized and may have some blocks already hidden, one may hide the j-th block of message $M_j = \mu_{j,1} \cdots \mu_{j,m}$ from being required in GS.Ver as follows. For simplicity the below description assumes none of the blocks have been hidden, but it can be easily seen that the hiding action of the j-th block is independent of whether any other block $j' \neq j$ was hidden or not.

1. Abort if $\hat{e}(\sigma_{j+3}, h) \cdot \hat{e}(g, \theta'_{j+3}) \neq 1$.
2. Randomly pick $\tau_1, \cdots, \tau_m \in \mathbb{Z}_n$.
3. Create the commitments $c_{j,k} = v_{j,k}^{\mu_{j,k}} \cdot h^{\tau_k}$ for $k = 1, \cdots, m$.
4. Construct the proofs $\pi_{j,k} = (v_{j,k}^{2\mu_{j,k}-1} \cdot h^{\tau_k})^{\tau_k}$ for $k = 1, \cdots, m$.
5. Define $\tau = \sum_{k=1}^{m} \tau_k$.
6. Set $c_j = v_j \prod_{k=1}^{m} c_{j,k} = F_j(M_j) \cdot h^{\tau}$.
7. Randomly pick $\tilde{r} \in \mathbb{Z}_n$.
8. Compute $\sigma'_3 = \sigma_3 \cdot (\theta'_j)^{\tau} \cdot (c_j)^{\tilde{r}}$.
9. Compute $\sigma'_{j+3} = \sigma_{j+3} \cdot g^{-\tilde{r}}$.
10. Output $\mathsf{GS.Rand}(\langle \{\sigma_1, \sigma_2, \sigma'_3, \sigma_4, \cdots, \sigma_{j+2}, \sigma'_{j+3}, \cdots, \sigma_{\ell+3}, \{\theta'_{j'}\}_{j'=4}^{\ell+2}, \pi_1, \pi_2, \{c_{j,k}\}, \{\pi_{j,k}\}\rangle)$.

In essence, that is the conceptual step (for hiding the group member's identity) in [8] which changes a regular signature to a proof of signature on a hidden message. Broadly speaking, that is the technique from Groth-Sahai system [15].

GS.Rand$(\sigma) \to \sigma'$. This algorithm is used by user to randomize the certificate in order to provide certificate unlinkability. Given a valid signature σ, anyone can output its randomized version σ' as follows. For simplicity, we first shown how to randomize when σ has no hidden message component.

1. First randomly pick $t'_1, t'_2, t'_3, \{t'_{j'}\}_{j'=4}^{\ell+2} \in \mathbb{Z}_n$.
2. Randomize the commitment parts of the signature by computing $\sigma'_j = \sigma_j \cdot h^{t'_j}$.
3. Then, for the proof component π_1, compute $\pi'_1 = \pi_1 \cdot \sigma_1^{t'_2} \cdot \sigma_2^{t'_1} \cdot h^{t'_1 t'_2} \cdot \Omega^{t'_1}$.
4. Update the corresponding proof $\pi'_2 = \pi_2 \cdot u^{t'_2} \cdot g^{-t'_3} \cdot (\prod_{j=1}^{\ell} (F_j(M_j))^{-t'_{j+3}})$.

Now the signature with message block hidden can also be randomized in a similar manner, by returning $\pi'_{j,k} = \pi_{j,k} \cdot c_{j,k}^{2\tau'_k} \cdot v_{j,k}^{-\tau'_k} \cdot h^{\tau'_k \tau'_k}$ for randomly picked $\tau'_k \in \mathbb{Z}_n$. The commitment can be updated correspondingly by $c'_{j,k} = c_{j,k} \cdot h^{\tau'_k}$. However, the proof π'_2 should be updated in a slightly different manner. For each hidden message block j, $F'_j(\cdot) = v_j \prod_{k=1}^{m} c_{j,k}$ (which is a constant function) should be used instead of $F_j(M_j)$.

Essentially, this algorithm just re-randomizes the commitments and the proofs using the re-randomization procedures [5,14] of the Groth-Sahai systems [15].

Finally, it is trivial to re-randomize $\{\theta'_{j'}\}_{j'=4}^{\ell+3}$ and hence they are omitted.

GS.Ver'. This algorithm is used to verify a randomized and sanitized certificate.

1. For all hidden message block j:
   (a) Check whether the proofs are correct, abort if $\hat{e}(c_{j,k}, c_{j,k} \cdot v_{j,k}^{-1}) = \hat{e}(h, \pi_{j,k})$.
   (b) Define $F'_j(\cdot) = v_j \prod_{k=1}^{m} c_{j,k}$ (which is a constant function).
2. Define $F'_j(M_j) = F_j(M_j)$ for $F'_j(\cdot)$ which has not been defined previously.

3. Output 1 if and only if both of the following equations hold, 0 otherwise.

$$\hat{e}(h, \pi_1) = \hat{A}^{-1} \cdot \hat{e}(\sigma_1, \sigma_2 \Omega)$$

$$\hat{e}(h, \pi_2) = \hat{e}(\sigma_2, u)\hat{e}(\sigma_3, g)^{-1} \prod_{j=1}^{\ell} (\hat{e}(\sigma_{j+3}, F'_j(M_j))^{-1})$$

The hybrid definition of $F'_j(\cdot)$ just corresponds to the hybrid signing methods in [8] – one on a known message and another on a hidden message.

The scheme described above is a natural combination and/or extension of existing techniques and hence its analysis will be largely based on the existing works. Correctness of verification follows from those in [8,9] and is trivial to see. Security analysis of our scheme will be deferred to the full version of this paper.

5 Privacy-Preserving Identity-Management

5.1 SPICE for Web Authentication

Here we show a typical run-time scenario of our system for a typical web authentication scenario. The whole system architecture is shown in Figure 3. The white color boxes represent the components from existing technologies (e.g., [6,17]), and the boxes with diagonals represent our new components. There are five components: *registrar*, *source CSP*, *receiving CSP*, *user* and *web browser*. The web browser interacts with the CSPs on behalf of the user to perform online authentication. The registrar is a trusted third party, which generates source certificate for each user. It has no interaction with the web browser, and will not involve in the authentication process. It can remain offline unless there may be new users joining the system.

The registrar manages user attributes, and uses the algorithm GS.UG to generate users' credentials. Then a source certificate is generated from the credentials for the user by registrar. Each user keeps the source certificate locally. When user accesses the browser to request for a service from a source CSP, the CSP will send back an authentication request, which includes the attributes names to be authenticated. These requests can be sent using an HTTP GET packet for example. When a user receives the authentication request, the functions GS.Rand and GS.Hide will be used to randomize and sanitize a certificate. The user uploads the randomized certificate to source CSP through web browser. The source CSP uses function GS.Ver' to authenticate the certificate. If successful, source CSP provides the service to user.

If a source CSP and a receiving CSP join together to provide a packaged service, the receiving CSP needs to authenticate user too. Firstly, the source CSP uses the *vocabulary conflicts handler module* [6] to check if there are attributes naming mismatch. If so, the CSP may generate another certificate using GS.Sign accompanying this certificate. Then, the source CSP generates another certificate by sanitizing (i.e., using GS.Hide to hide) the attributes that receiving CSP does not concern and re-randomizing (i.e., applying GS.Rand on) the certificate. The sanitized certificate is sent to receiving CSP. A receiving CSP authenticates the certificate using GS.Ver'.

Additionally, the *naming management service* stores the mappings with other service provider ontologies, and the sets of synonyms (Synsets). *Policy repository* stores attributes verification policies. The *request manager component* asks the users for the attributes necessary for verification. The vocabulary conflicts handler checks if there are receiving CSP required attributes names that match the Synsets of itself.

A receiving CSP should have same modules as a source CSP. However, in order to distinguish between their roles, we omit some of the components which are only useful when they take another side of role.

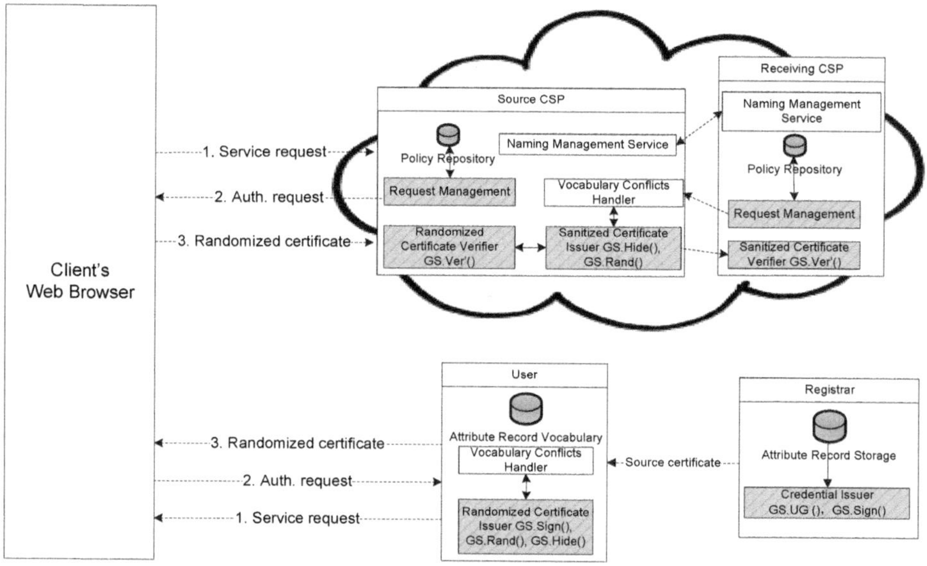

Fig. 3. System Architecture based on Randomizable Group Signatures

5.2 Security, Privacy and Functional Requirements

Unlinkability. The source certificate will be randomized by user for each request, so that the CSPs cannot obtain a certificate of a special form. If any CSP can link these randomized certificates to the same source certificate, this breaks the perfect unlinkability provided by GS.Rand which leads to a contradiction.

Delegatable Authentication. In order to allow a receiving CSP to authenticate a user without directly contacting with the user, firstly re-randomization is done, then sanitization is done by issuing a new group signature which is also re-randomizable and unlinkable.

Anonymity. CSPs are not able to see the clear values of the user attributes from the certificate whenever the user chose to hide them using GS.Hide, which is guaranteed by the hiding property of the commitment and the zero-knowledge property of the proof implicitly used by GS.

Registrar is the trusted party who controls which user has which set of attributes, so we do not consider anonymity against this party. In case the registrar is not allowed to see this sensitive personal information, one may actually use a group signature scheme which supports signing committed messages to realize a signing process similar to that of a blind signature scheme. The hiding algorithm presented in the previous section essentially changes a signature signing directly on a message to one signing on a commitment of that message, but for simplicity of our presentation, an explicit description of this variant is omitted.

Accountability. From the identifiability of GS, the registrar can trace which user (or which source CSP in the case of naming mismatch) issued the certificate, by the GS.Open algorithm using its group secret key.

User Centric Access Control. In our system, the user can choose which attributes to reveal when presenting a certificate, the disclosure of user identity or attributes is no longer arbitrarily or at the will of CSP.

Single Registration. The users only need to contact the registrar once in our system, instead of contacting the registrar every time authentication is about to occur for preserving the unlinkability in previous systems.

Efficiency. For unlinkable authentication, re-randomization is needed and its time complexity is in the total number of attributes a user possesses. However, it only involves a small number of exponentiations for each attribute. In particular, no pairing operation is involved on the user side. Also, we note that the randomization can be pre-computed. The only online operation for authentication is signing appending the certificate with the last block of the message, which takes constant time (in the number of attributes). We will report a performance evaluation of our prototype implementation for different settings in the full version of this paper.

6 Conclusion

Privacy and security have become a critical concern with the immense growth in the popularity of cloud computing, and digital identity management (DIM) being one of the critical components. We proposed a privacy-aware interoperable DIM system for the cloud based on the simple use of randomizable group signatures, which solved two open problems (unlinkability and delegatable authentication) left by Bertino *et al.* [6]. Our scheme relies on the conceptually simple use of extended group signatures. Most part of the operations in our system can be performed offline and we remove the need of contacting the registrar before every authentication or storing a large amount of certificates. We believe the overhead is quite minimal for the privacy concern and we leave the design of more efficient anonymous credential systems as our future work.

References

1. Shibboleth, http://shibboleth.internet2.edu
2. Abe, M., Chow, S.S.M., Haralambiev, K., Ohkubo, M.: Double-Trapdoor Anonymous Tags for Traceable Signatures. In: Lopez, J., Tsudik, G. (eds.) ACNS 2011. LNCS, vol. 6715, pp. 183–200. Springer, Heidelberg (2011)
3. Alliance, C.S.: Top Threats to Cloud Computing V1.0 (March 2010), https://cloudsecurityalliance.org/topthreats/csathreats.v1.0.pdf
4. Angin, P., Bhargava, B.K., Ranchal, R., Singh, N., Linderman, M., Othmane, L.B., Lilien, L.: An Entity-Centric Approach for Privacy and Identity Management in Cloud Computing. In: IEEE Symposium on Reliable Distributed Systems (SRDS), pp. 177–183 (2010)
5. Belenkiy, M., Camenisch, J., Chase, M., Kohlweiss, M., Lysyanskaya, A., Shacham, H.: Randomizable Proofs and Delegatable Anonymous Credentials. In: Halevi, S. (ed.) CRYPTO 2009. LNCS, vol. 5677, pp. 108–125. Springer, Heidelberg (2009)
6. Bertino, E., Paci, F., Ferrini, R., Shang, N.: Privacy-preserving Digital Identity Management for Cloud Computing. IEEE Data Eng. Bull. 32(1), 21–27 (2009)
7. Boneh, D., Boyen, X., Shacham, H.: Short Group Signatures. In: Franklin, M. (ed.) CRYPTO 2004. LNCS, vol. 3152, pp. 41–55. Springer, Heidelberg (2004)
8. Boyen, X., Waters, B.: Compact Group Signatures Without Random Oracles. In: Vaudenay, S. (ed.) EUROCRYPT 2006. LNCS, vol. 4004, pp. 427–444. Springer, Heidelberg (2006)
9. Boyen, X., Waters, B.: Full-Domain Subgroup Hiding and Constant-Size Group Signatures. In: Okamoto, T., Wang, X. (eds.) PKC 2007. LNCS, vol. 4450, pp. 1–15. Springer, Heidelberg (2007)
10. Camenisch, J., Kohlweiss, M., Soriente, C.: Solving Revocation with Efficient Update of Anonymous Credentials. In: Garay, J.A., De Prisco, R. (eds.) SCN 2010. LNCS, vol. 6280, pp. 454–471. Springer, Heidelberg (2010)
11. Celesti, A., Tusa, F., Villari, M., Puliafito, A.: Security and Cloud Computing: InterCloud Identity Management Infrastructure. In: WETICE, pp. 263–265. IEEE Computer Society (2010)
12. Chow, S.S.M.: Real Traceable Signatures. In: Jacobson Jr., M.J., Rijmen, V., Safavi-Naini, R. (eds.) SAC 2009. LNCS, vol. 5867, pp. 92–107. Springer, Heidelberg (2009)
13. Chow, S.S.M., Phan, R.C.-W.: Proxy Re-signatures in the Standard Model. In: Wu, T.-C., Lei, C.-L., Rijmen, V., Lee, D.-T. (eds.) ISC 2008. LNCS, vol. 5222, pp. 260–276. Springer, Heidelberg (2008)
14. Fuchsbauer, G., Pointcheval, D.: Proofs on Encrypted Values in Bilinear Groups and an Application to Anonymity of Signatures. In: Shacham, H., Waters, B. (eds.) Pairing 2009. LNCS, vol. 5671, pp. 132–149. Springer, Heidelberg (2009)
15. Groth, J., Sahai, A.: Efficient Non-interactive Proof Systems for Bilinear Groups. In: Smart, N.P. (ed.) EUROCRYPT 2008. LNCS, vol. 4965, pp. 415–432. Springer, Heidelberg (2008)

16. Hussain, M.: The Design and Applications of a Privacy-Preserving Identity and Trust-Management System. PhD thesis, School of Computing, Queen's University, Canada (2010), http://hdl.handle.net/1974/5520
17. Kalfoglou, Y., Schorlemmer, W.M.: If-map: An ontology-mapping method based on information-flow theory. J. Data Semantics 1, 98–127 (2003)
18. Paterson, K.G., Schuldt, J.C.N.: Efficient Identity-Based Signatures Secure in the Standard Model. In: Batten, L.M., Safavi-Naini, R. (eds.) ACISP 2006. LNCS, vol. 4058, pp. 207–222. Springer, Heidelberg (2006)
19. Ranchal, R., Bhargava, B.K., Othmane, L.B., Lilien, L., Kim, A., Kang, M.H., Linderman, M.: Protection of Identity Information in Cloud Computing without Trusted Third Party. In: IEEE Symposium on Reliable Distributed Systems (SRDS), pp. 368–372 (2010)
20. Rocha, F., Correia, M.: Lucy in the Sky without Diamonds: Stealing Confidential Data in the Cloud. In: DSN Workshop – Dependability of Clouds, Data Centers and Virtual Computing Env., pp. 129–134 (2011)
21. Takabi, H., Joshi, J.B.D., Ahn, G.-J.: Security and Privacy Challenges in Cloud Computing Environments. IEEE Security & Privacy 8(6), 24–31 (2010)
22. Waters, B.: Efficient Identity-Based Encryption Without Random Oracles. In: Cramer, R. (ed.) EUROCRYPT 2005. LNCS, vol. 3494, pp. 114–127. Springer, Heidelberg (2005)

A Practical Smart Metering System Supporting Privacy Preserving Billing and Load Monitoring

Hsiao-Ying Lin[1], Wen-Guey Tzeng[2], Shiuan-Tzuo Shen[2], and Bao-Shuh P. Lin[1]

[1] Intelligent Information and Communications Research Center, National Chiao Tung University, Taiwan
hsiaoying.lin@gmail.com, bplin@mail.nctu.edu.tw

[2] Department of Computer Science, National Chiao Tung University, Taiwan
{vink,wgtzeng}@cs.nctu.edu.tw

Abstract. Fine-grained meter readings enable applications in an advanced metering infrastructure. However, those meter readings threaten personal privacy by implying a sketch of daily activities of households. The privacy issue has been addressed in smart metering systems by either a trusted third party assumption or cryptographic primitives. We address the privacy issue by using a trusted platform module and lightweight cryptographic primitives. Our smart metering system simultaneously supports the billing and load monitoring applications in a privacy preserving manner. It allows an electricity service provider obtain sums of meter readings over a time period and a monitoring center obtain sums of meter readings from meters in an area at some recent time unit while keeping individual meter reading private. Moreover, we formally prove that our system is privacy preserving. Our system provides a simple yet very practical solution to a privacy preserving smart metering system.

Keywords: Trusted platform module, smart metering, privacy preserving technique, secure aggregation, pseudorandom number generator.

1 Introduction

The emergence of smart grids has established a trend towards building our next generation of power grid systems. As shown in Fig. 1, new features include two-way power flows and mutual communications between electricity entities. Smart grids integrate intelligence and automation into the conventional power grid system to increase energy efficiency and improve system reliability and quality. We can build advanced applications upon smart grids, such as load monitoring, automatic billing, dynamic pricing, and power generation planning.

One essential technology of smart grids is fine-grained meter reading within a very short period of time per household. However, meter readings of a household reveal detailed information about daily activities of the household and used appliances during a specific time period [7,14,11]. Fine-grained meter readings cause serious privacy issues. Actually, the granularity of meter readings often exceeds the need of some underlying applications. Current smart meters record

F. Bao, P. Samarati, and J. Zhou (Eds.): ACNS 2012, LNCS 7341, pp. 544–560, 2012.

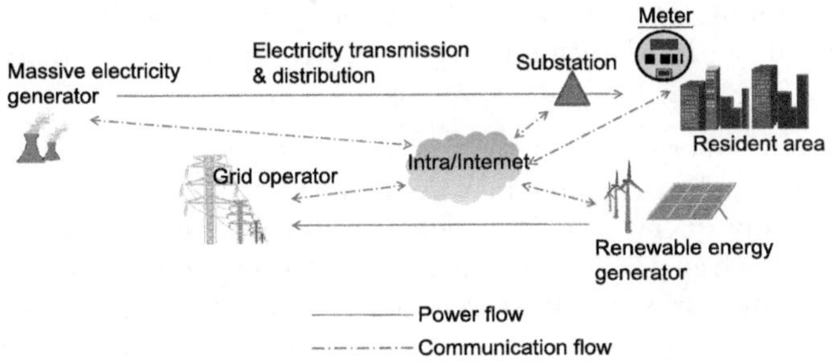

Fig. 1. Power is massively generated by a power station and transmitted by a grid operator from the generator to end-consumers. Local renewable energy can also be transmitted to other entities.

electricity usage every 5 to 60 minutes [8]. The next generation of smart meters will upgrade a time unit to seconds. In billing applications, the electricity service provider (ESP) only needs the amount of power consumption per hour to compute a bill. For example, in Ontario, Canada, the time-of-use price service during winters only needs the consumption data over two hours in an on-peak time period, six hours in a mid-peak time period, and 12 hours in an off-peak time period [1]. In load monitoring applications, the load monitoring center (LMC) collects the amount of electricity usage over a local area in order to monitor current activities of the power grid. LMC requires consumption data in much finer time granularity than ESP does. Nevertheless, LMC only needs the total power consumption over the area at recent time units.

To address the privacy issues against service providers, an approach of secure aggregation is proposed. By secure aggregation techniques, a service provider can only get an aggregated result of meter readings while individual meter reading remains private. For the billing application, previous works use public key homomorphic encryption schemes, commitment schemes, or a trusted third party to securely aggregate meter readings of a meter. For the load monitoring application, previous works use public key homomorphic encryption schemes, secret sharing techniques, or distributed random noise generation to securely aggregate meter readings of meters in an area.

On the other hand, many manufactures of smart meters use a hardware component to address various cyber-security issues. For example, Atmel provides electricity meters with a hardware security component for cryptographic authentication. Embedding a trusted platform module chip (TPM) into a smart meter is a general practice for securing metering services [12,13,9]. We shall assume that a TPM is embedded into a smart meter for providing securing functions.

It is a challenge to design a smart metering system that simultaneously supports multiple privacy preserving applications without using a trusted third party and public key cryptographic primitives. We focus on the billing and load monitoring applications and consider the privacy requirements for them. Our main contribution is to propose a practical privacy preserving smart metering system that supports billing and load monitoring applications with TPM technologies. Our system uses a pseudorandom number generator and hash functions supported by TPM technologies. Features of our smart metering system are as follows:

- ESP can only query a meter for a sum of meter readings over a time period. Each meter reading remains private against ESP.
- LMC can only query a sum of meter readings from meters in an area at a time unit. Each meter reading remains private against LMC.
- Meter readings are securely stored in a semi-trusted storage system.
- Meters can freely join or leave our smart metering system without overhead.

Moreover, we formally define a privacy model with respect to time-series meter readings to capture privacy requirements and prove that our smart metering system meets the requirements.

2 Related Work

We briefly introduce existing privacy preserving protocols of smart metering systems and TPM technologies.

Privacy preserving metering protocols. Anonymous technology is suggested by NIST to anonymizing traces of meter readings [2]. For the billing application, Petrlic proposed a solution by using pseudonym of households against ESP where the grid operator to be a fully trusted intermediate translator [15]. Jawurek et al. constructed a privacy preserving billing protocol by integrating a homomorphic commitment scheme, zero knowledge proofs and a tamper-evident meter [9]. Meter readings are committed and aggregated by using the homomorphic commitment scheme. Only the final bill is opened to ESP and the correctness of the computation is verified by using zero knowledge proofs. Rial and Danezis took a similar approach [16], where they replaced the tamper-evident meter by TPM.

For the load monitoring application, Garcia and Jacobs proposed a solution by using a trusted aggregator in a substation and an additively homomorphic encryption scheme [6]. Each meter encrypts meter readings by using LMC's public key. The aggregator aggregates encrypted meter readings and only sends the aggregated result to LMC. Shi et al. proposed a privacy model for aggregation of time-series data (such as meter readings) while individual datum remains private [17]. In their system, the number of meters is fixed after the system is setup. The system must to be reset when meters join or leave. Later, Shi et al. proposed a new solution by using the subset cover technique to tolerate leaving meters [5]. Kursawe et al. proposed a privacy friendly aggregation method [10]. An aggregator and meters secretly share 0 for multiple times in parallel such that no share of a meter is revealed. Ács and Castelluccia [3] proposed a solution by using random noise and secret sharing. Meters independently generate random noise and pairs of meters secretly share 0. Meter readings are masked by random noise and encrypted by secret shares. The sum of masked and encrypted meter readings gives a noisy sum of meter readings.

Bohi et al. proposed a privacy model and two approaches for the billing and load monitoring applications, respectively [4]. First, they used a trusted third party to compute the bill for the billing application. Second, they introduced random noises on meter readings, where the distribution of the noise has a known mean and variance. LMC gets only an approximate sum of meter readings while individual reading is private.

Our work is distinguished as it simultaneously addresses both applications but only requires a simple and lightweight use of TPM for generating pseudorandom numbers without a trusted third party and without mutual communications among meters.

TPM technologies. TPM is a microcontroller that offers facilities for secure generation of cryptographic keys, the ability to limit the use of keys, non-volatile storage and a hardware pseudorandom number generator. It enables platform attestation and cryptographic primitives, such as RSA and SHA-1. The TPM specification is defined by the trusted computing group and the latest version is TPM 1.2 revision 116[1].

A TPM chip itself is a solid component through platform attestation. It employs platform configuration registers to record configurations of platform and software, and prevents unauthorized modifications on these configurations. By verifying configurations, TPM assures that the platform is initialized from a secure and correct condition.

3 System Model

We describe our time notation, smart metering system, and the billing and load monitoring applications. We also brief privacy requirements. Detailed descriptions of privacy requirements are provided in Section 5.

3.1 Time Notations

Time is divided into basic *time units* $t_1, t_2, \cdots$. Let l be a fixed positive integer, where $l \geq 2$. We set l to be the minimum number of time units where ESP gets the sum of meter readings. Based on the parameter l, we define *time periods* and the *current time window*. A time period T consists of al continuous time units for any positive integer a. The current time window W is the latest continuous l time units t_{z-l+1},t_{z-l+2}, $\cdots$, t_z, where t_z is the current time unit.

3.2 Smart Metering System

Our smart metering system consists of meters, a storage system, ESP and LMC, as shown in Fig. 2. We assume that meters are purchased by households and deployed by the grid operator. Households trust the grid operator that it honestly deploys meters. We assume that meters are trusted, that is, meters honestly follow defined steps. We also assume that ESP and LMC are honest-but-curious, that is, they follow defined steps but try to dig out individual meter readings

[1] The specification is available as international standard ISO/IEC 11889.

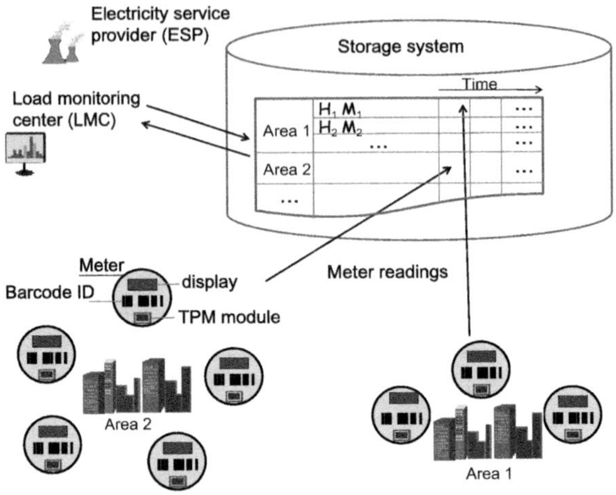

Fig. 2. Our system model consists of meters, a storage system, ESP, and LMC

from what they obtain from communications. Moreover, we assume that ESP and LMC do not collude.

A household $\mathcal{H}_i$ has a meter $\mathcal{M}_i$ that records power consumption $d_{i,j}$ of $\mathcal{H}_i$ at time unit t_j. Households may physically move in or out an area. A meter $\mathcal{M}_i$ has a serial number $\mathcal{SN}_i$ assigned by the meter manufacture. A meter $\mathcal{M}_i$ *encrypts* a meter reading $d_{i,j}$ as $c_{i,j}$ and stores $c_{i,j}$ into the storage system. The storage system stores the encrypted meter reading according to the meter and the time unit. We assume that ESP and LMC can freely access the storage system after being authenticated by the storage system.

Meter readings are conceptually arranged in a matrix in the storage system, where a row represents meter readings of a household over time and a column represents meter readings of households in an area at a time unit. An example is shown in Fig. 3. From the time unit t_1, Areas 1 and 2 have 3 and 5 households, respectively. Each household $\mathcal{H}_i$ has a meter $\mathcal{M}_i$ for $1 \leq i \leq 9$. At t_4, new household $\mathcal{H}_9$ moves in Area 3 and then a row of $\mathcal{M}_9$ is added in the matrix. When household $\mathcal{H}_7$ moves out Area 2 at t_9, the row of $\mathcal{H}_7$ in Area 2 is deleted from the matrix.

3.3 Supporting Billing Applications

ESP is allowed to query the meter for decryption information of a sum of meter readings over a time period T. ESP sums up encrypted meter readings over T. By the decryption information, ESP decrypts the encrypted sum to obtain the power consumption of the household over T. In the example in Fig. 3, l is set to 4. ESP queries the meter $\mathcal{M}_1$ for decryption information of the time period $T = (t_2, t_3, t_4, t_5, t_6, t_7, t_8, t_9)$ and decrypts the encrypted sum $c = \sum_{j=2}^{9} c_{1,j}$ to obtain the sum $\xi = \sum_{j=2}^{9} d_{1,j}$ of the meter readings between t_2 and t_9.

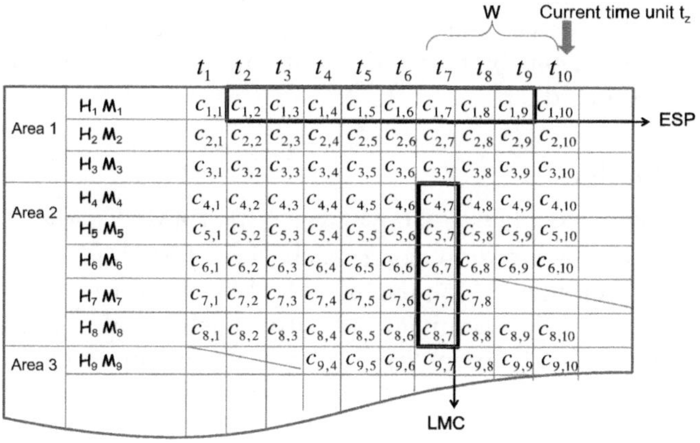

Fig. 3. Meter readings are conceptually arranged in a matrix

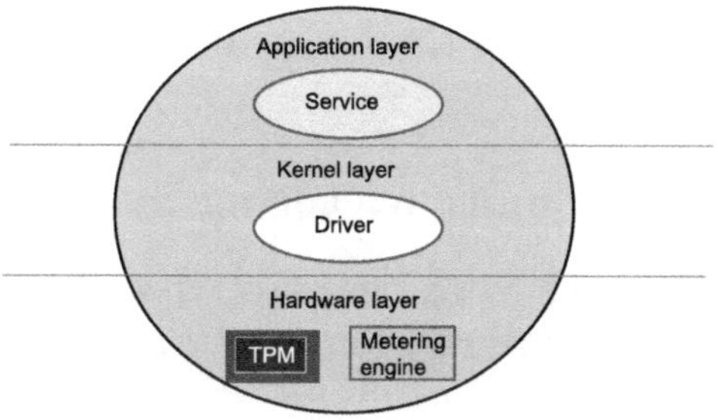

Fig. 4. We model a meter in three layers

The correctness for ESP is that ESP obtains the correct sum ξ of a household over T. The privacy requirement for ESP is that it cannot get individual meter readings of a household.

3.4 Supporting Load Monitoring Applications

LMC is allowed to query meters in an area for *approximate* decryption information of a time unit t_j in the current time window W. LMC sums up encrypted meter readings in the area at t_j to get the encrypted sum. By the approximate decryption information, LMC decrypts and gets an approximate overall power consumption of households in the area at t_j. In the same example, the current time unit is t_{10} and the current time window is (t_7, t_8, t_9, t_{10}). LMC can query meters in Area 2 for decryption information at t_7 and decrypts the encrypted sum $c = \sum_{i=4}^{8} c_{i,7}$ to obtain an approximate sum for $\xi = \sum_{i=4}^{8} d_{i,7}$.

The correctness for LMC is that LMC obtains a good approximate sum $\tilde{\xi}$ for ξ. We formulate the approximation by the error ratio $\omega = |\tilde{\xi} - \xi|/\xi$, a threshold value ϵ and a confidence probability δ as $\Pr[\omega \leq \epsilon] > 1-\delta$. With sufficiently small ϵ and δ, LMC obtains a good approximate sum $\tilde{\xi}$ for ξ with a higher probability. The privacy requirement for LMC is that it cannot get exact individual meter readings of a household.

3.5 Meter Model

As shown in Fig. 4, we have a three-layer model for a meter. The hardware layer consists of hardware components, such as a TPM chip, a metering engine, a processing and communication engine. The kernel layer consists of drivers of hardware components. We assume that a driver is in charge of meter readings of the metering system. The application layer is built upon the kernel layer to provide services, such as a web interface for observing current meter readings.

The power consumption is often measured in kWh. Since we consider a finer time granularity, the unit of measurement is changed to Wh so that integer representation is enough. Moreover, we assume that meter readings (in integers) at a time unit are much less than a defined number p. From the statistics of U.S. Energy Information Administration, in 2009, the average power consumption per household per month is 908 kWh. That is $105Wh$ per 5 minutes. Thus, we set p to be of length 64 bits.

4 Privacy Preserving Smart Metering System

We describe our smart metering system and two types of queries supporting billing and load monitoring applications.

4.1 Metering System Construction

We assume that a meter is deployed or reset by the grid operator when a household moves in an area. At the beginning, the metering system consists of a storage system, ESP and LMC. Later, meters join in. Choose a large number p, where $p \geq 2\sqrt{p}$. Let the initial time unit be t_1. Let the pseudorandom number generator be g, where $g : \{0,1\}^\tau \times \{0,1\}^\lambda \rightarrow \mathcal{Z}_p$. Let h and h' be cryptographic hash functions, where $h : \{0,1\}^* \rightarrow \{0,1\}^\lambda$ and $h' : \{0,1\}^* \rightarrow \{0,1\}^\tau$. A meter $\mathcal{M}_i$ runs as follows.

Meter initialization.

1. $\mathcal{M}_i$ takes a user input as a seed s_i. The TPM of $\mathcal{M}_i$ generates a master key k_i by using the seed s_i, the serial number $\mathcal{SN}_i$, and the hash function h', where $k_i = h'(s_i||\mathcal{SN}_i)$ and $||$ is the operator of concatenation. The master key k_i is then securely stored in non-volatile storage of the TPM of $\mathcal{M}_i$.
2. The driver of $\mathcal{M}_i$ creates and initializes l first-in first-out memory slots as 0.
3. The TPM of $\mathcal{M}_i$ generates l pseudorandom numbers $r_{i,1}, r_{i,2}, \cdots, r_{i,l-1}$ and $R_{i,1}$, where

$$r_{i,j} = g(k_i, t_j), 1 \leq j \leq l-1, \text{ and}$$
$$R_{i,1} = g(k_i, h(t_1||t_2||\cdots||t_l))$$

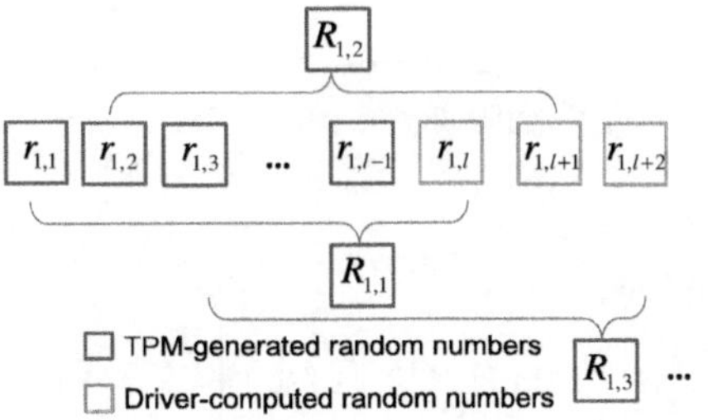

Fig. 5. The TPM generates random numbers $r_{1,1}, r_{1,2}, \cdots, r_{1,l-1}$ and $R_{1,1}$ in the initialization part and a random number $R_{1,i+1}$ at time unit t_i. Random numbers $r_{1,i}$ for $i \geq l$ are computed on the fly.

Then, the TPM of $\mathcal{M}_i$ passes all pseudorandom numbers to the driver of $\mathcal{M}_i$.

4. The driver of $\mathcal{M}_i$ computes $r_{i,l}$ as follows:

$$r_{i,l} = \left(R_{i,1} - \sum_{j=1}^{l-1} r_{1,j} \right) \mod p$$

Then, the driver stores l pseudorandom numbers $r_{i,1}, r_{i,2}, \cdots, r_{i,l}$ in memory slots.

Storage of meter readings at t_j, $j \geq 1$.

1. $\mathcal{M}_i$ measures the consumption $d_{i,j}$ and encrypts it as $c_{i,j}$, where

$$c_{i,j} = (d_{i,j} + r_{i,j}) \mod p$$

$c_{i,j}$ is sent and stored to the storage system.

2. The TPM of $\mathcal{M}_i$ generates a random number $R_{i,j+1}$ and passes it to the driver of $\mathcal{M}_i$, where

$$R_{i,j+1} = g(k_i, h(t_{j+1}||t_{j+2}|| \cdots ||t_{j+l})).$$

The driver of $\mathcal{M}_i$ computes

$$r_{i,j+l} = \left(R_{i,j+1} - \sum_{\alpha=j+1}^{j+l-1} r_{i,\alpha} \right) \mod p.$$

The driver then replaces $r_{i,j}$ with $r_{i,j+l}$ in the memory slot.

An example of $\mathcal{M}_1$ is shown in Fig. 5. The TPM of $\mathcal{M}_1$ generates l random numbers in the initialization part and generates a random number at each time unit on the fly. For any time period T with l continuous time units, a random number generated by the TPM of $\mathcal{M}_1$ helps decrypt the sum of meter readings over T. The snapshots at t_l of $\mathcal{M}_1$ is shown in Fig. 6. At any time unit, the driver of $\mathcal{M}_1$ maintains the random numbers used for time units in W.

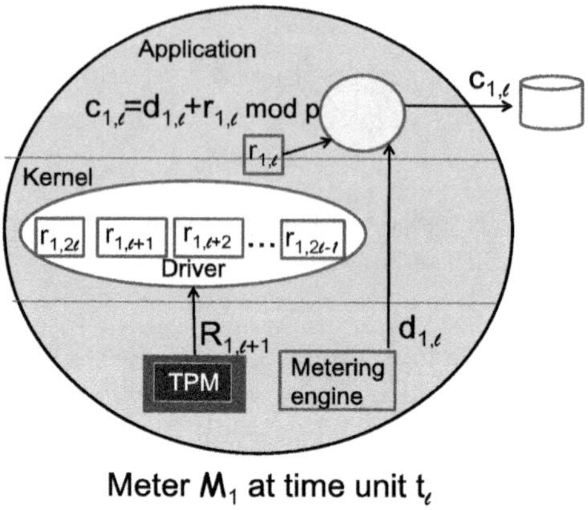

Fig. 6. Snapshot of the meter $\mathcal{M}_1$ at t_l

4.2 Supporting Billing Application

To compute a bill of a household $\mathcal{H}_i$, ESP queries the meter $\mathcal{M}_i$ of $\mathcal{H}_i$ for a time period T of al continuous time units, where $T = (t_\beta, t_{\beta+1}, \cdots, t_{\beta+al-1})$. The application layer of $\mathcal{M}_i$ divides the time period into a sub-periods and asks the TPM to regenerate the corresponding random numbers $R_{i,\beta}$, $R_{i,\beta+l}$, $R_{i,\beta+2l}$, $\cdots$, $R_{i,\beta+(a-1)l}$. The application layer of $\mathcal{M}_i$ computes the sum B of random numbers and sends B to ESP, where

$$B = \left(\sum_{j=0}^{a-1} R_{i,\beta+jl} \right) \mod p$$

ESP gets encrypted meter readings $(c_{i,\beta}, c_{i,\beta+1}, \cdots, c_{i,\beta+al-1})$ over T from the storage system and computes

$$\sum_{j=\beta}^{\beta+al-1} d_{i,j} = \left(\sum_{j=\beta}^{\beta+al-1} c_{i,j} - B \right) \mod p \qquad (1)$$

ESP correctness is that ESP can obtain the consumption of a household $\mathcal{H}_i$ over T. Note that for $0 \leq j \leq a-1$,

$$R_{i,\beta+jl} = (\sum_{k=0}^{l-1} r_{i,\beta+jl+k}) \mod p.$$

Thus,

$$\left(\sum_{j=\beta}^{\beta+al-1} c_{i,j} - B\right) \mod p = \left(\sum_{j=\beta}^{\beta+al-1} (d_{i,j} + r_{i,j}) - B\right) \mod p$$
$$= \left(\sum_{j=\beta}^{\beta+al-1} d_{i,j}\right) \mod p \tag{2}$$

By Equation (2), Equation (1) holds when

$$\sum_{j=\beta}^{\beta+al-1} d_{i,j} = (\sum_{j=\beta}^{\beta+al-1} d_{i,j}) \mod p$$

That is, the sum of meter readings must be less than p. By choosing a large p, ESP correctness is guaranteed. In practice, it is sufficient to set p to be of 64-bits when each meter reading is of 32-bits and al is up to 2^{32}.

4.3 Supporting Load Monitoring Application

To monitor the power consumption in an area, LMC queries meters in an area at t_j in current time widow W. A meter $\mathcal{M}_i$ in the area should reply. The driver of $\mathcal{M}_i$ chooses a random number n according to a normal distribution $N(0, \sigma^2)$ with the mean 0 and the variance σ^2 and computes the noise $n_{i,j}$ as the floor of the chosen random number n, i.e. $n_{i,j} = \lfloor n \rfloor$. The variance σ shall be defined later. The driver of $\mathcal{M}_i$ then passes a noised random number $\tilde{r}_{i,j}$ to LMC, where

$$\tilde{r}_{i,j} = (r_{i,j} + n_{i,j} - \lceil\sqrt{p}\rceil) \mod p$$

Recall that the stored meter reading $c_{i,j} = (d_{i,j} + r_{i,j}) \mod p$. By $\tilde{r}_{i,j}$, LMC computes a noised meter reading $\tilde{d}_{i,j}$ of the meter $\mathcal{M}_i$ as follows:

$$\tilde{d}_{i,j} = (c_{i,j} - \tilde{r}_{i,j} \mod p) - \lceil\sqrt{p}\rceil$$
$$= (d_{i,j} - n_{i,j} + \lceil\sqrt{p}\rceil \mod p) - \lceil\sqrt{p}\rceil$$

The number $\sqrt{p}$ is used to prevent an overflowing issue for correctness. Note that $\tilde{d}_{i,j}$ may be negative. To obtain $\tilde{d}_{i,j} = d_{i,j} - n_{i,j}$, we need

$$d_{i,j} + \lceil\sqrt{p}\rceil \geq n_{i,j} \geq d_{i,j} - p + \lceil\sqrt{p}\rceil$$

Since $p \geq 2\sqrt{p}$, we bound the probability by

$$\Pr[|n_{i,j}| \leq d_{i,j} + \lceil\sqrt{p}\rceil] \geq 1 - \frac{\sigma^2}{p}$$

Since p is very large and σ is sufficiently small, the error probability is negligible.

Let LMC obtain m noised meter readings $\tilde{d}_{i_\alpha,j}$ from m meters $\mathcal{M}_{i_\alpha}$, where $1 \le \alpha \le m$. LMC computes an approximate value $\tilde{\xi}$ for the overall consumption ξ at t_j in the area, where

$$\tilde{\xi} = \sum_{\alpha=1}^{m} \tilde{d}_{i_\alpha,j} \quad \text{and} \quad \xi = \sum_{\alpha=1}^{m} d_{i_\alpha,j}$$

Since some meters may fail to reply due to various reasons, LMC needs to set a maximal waiting time period T_{max}.

LMC correctness requires that LMC obtains an approximate value for the overall consumption. The error between the approximate sum $\tilde{\xi}$ and ξ depends on the number m and the variance σ^2. For m meter readings at t_j, let $x = \tilde{\xi} - \xi = \sum_{\alpha=1}^{m} n_{i_\alpha,j}$. We measure the error by using the error ratio $\omega = |x|/\xi$. Let $\hat{d}$ be the average value of meter readings per time unit. Thus, we assume $\xi = m\hat{d}$.

Since each noise is randomly chosen from a normal distribution $N(0, \sigma^2)$, the distribution of x is a normal distribution $N(0, m\sigma^2)$. By the Chebyshev inequality, we have

$$\Pr[\omega \le \epsilon] = \Pr[|x|/\xi \le \epsilon] = \Pr[|x - 0| \le \xi\epsilon] \ge 1 - \frac{m\sigma^2}{(\xi\epsilon)^2} = 1 - \frac{\sigma^2}{m\hat{d}^2\epsilon^2} \quad (3)$$

Let $\delta = \frac{\sigma^2}{m\hat{d}^2\epsilon^2}$. Equation (3) shows that when σ is sufficiently small and m is sufficiently large, LMC obtains a good approximate with high probability $1 - \delta$. We set $\hat{d}$ to be 105 (an average meter reading in Wh per 5 minutes) according to the statistics of U.S. Energy Information Administration. We fix $\delta = 1\%$ and present values of m and σ for achieving $\epsilon = 10\%$, $\epsilon = 7\%$, and $\epsilon = 5\%$ in Fig. 7. When $m = 600$, σ is about 25, 18 and 12, respectively. The parameter σ is a tradeoff between LMC correctness and LMC privacy requirement. Here we obtain that a better approximate needs a smaller σ. We will see that LMC privacy requirement needs a larger σ in the subsection 5.2.

5 Privacy Requirements and Analysis

We formally define ESP and LMC privacy requirements and show that our system meets the requirements. We also show that meter readings are securely stored.

5.1 ESP Privacy Requirement and Analysis

ESP privacy requirement is that ESP cannot get individual meter readings of a household, where ESP gets sums of meter readings in time periods and accesses encrypted meter readings. We capture the ESP privacy requirement in a security game $\mathcal{G}$, where the power of ESP is enlarged to adaptively decide meter readings for non-challenge time periods. Even having the ability of adaptively setting meter readings and observing resulting encrypted meter readings, ESP

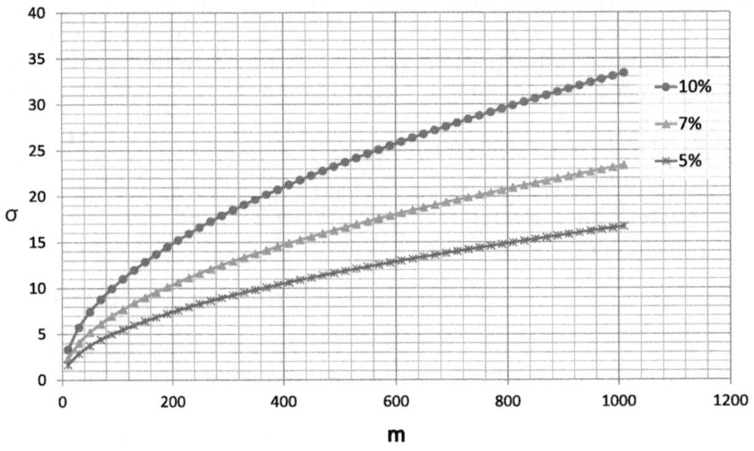

Fig. 7. Values of m and σ for achieving the error ratio ω less than 10%, 7%, and 5%, respectively, where $\hat{d} = 105$ and $\delta = 1\%$

still cannot distinguish meter readings from two possible sets of meter readings in the game $\mathcal{G}$. The security game $\mathcal{G}$ is described in the following.

The challenger $\mathcal{C}$ represents the metering system of a household $\mathcal{H}_i$ and the adversary $\mathcal{A}$ represents the honest-but-curious ESP. A query phase proceeds at beginning. $\mathcal{A}$ adaptively chooses meter readings $d_{i,j}$ at time units t_j from $j = 1$ and $\mathcal{C}$ returns the encrypted meter readings $c_{i,j}$ back to $\mathcal{A}$. $\mathcal{A}$ then decides to enter the challenge phase at t_{j_1}. This phase simulates that $\mathcal{A}$ reveals meter readings and their decryption information at time units earlier than t_{j_1}. In the challenge phase, $\mathcal{A}$ chooses a time period from the time unit t_{j_1} to a later time unit t_{j_2}, where $j_2 - j_1 = al$ for a positive integer a, and two challenge sets D_0, D_1 of meter readings for time units between t_{j_1} and t_{j_2}, where $D_v = \{d^v_{i,j_1}, d^v_{i,j_1+1}, \cdots, d^v_{i,j_2}\}$ for $v \in \{0, 1\}$ and $\sum_{s=j_1}^{j_2} d^0_{i,s} = \sum_{s=j_1}^{j_2} d^1_{i,s}$. $\mathcal{A}$ sends D_0 and D_1 to $\mathcal{C}$. $\mathcal{C}$ throws a random coin b and encrypts meter reading $d^b_{i,s}$ in D_b as $c_{i,s}$ for $s \in [j_1, j_2]$. Let $C = \{c_{i,j_1}, c_{i,j_1+1}, \cdots, c_{i,j_2}\}$. After getting encrypted meter readings C, $\mathcal{A}$ enters the second query phase. Again, $\mathcal{A}$ adaptively chooses a meter reading $d_{i,j'}$ for arbitrary time unit $t_{j'}$ where $j' > j_2$ and $\mathcal{C}$ adaptively returns the encrypted meter reading $c_{1,j'}$ back to $\mathcal{A}$. $\mathcal{A}$ then outputs a guessing b' for b.

If $b' = b$, $\mathcal{A}$ wins the game $\mathcal{G}$. That is, $\mathcal{A}$ successfully distinguishes which set D_b is encrypted. The advantage of an adversary is defined as $|\Pr[b' = b] - 1/2|$.

Definition 1. *A smart metering system satisfies ESP privacy requirement if for any probabilistic polynomial time algorithm $\mathcal{A}$ and a negligible function ε, $|\Pr_{\mathcal{A}}[b' = b] - 1/2| < \varepsilon$.*

A similar game is defined in [17], where the adversary needs to choose challenge sets D_0 and D_1 at the very beginning. It only captures a snapshot of meter readings at a time unit. In our security game, the queries from the adversary is adaptive. As a result, our security game models a stronger security requirement.

ESP Privacy Analysis. We first rephrase the description of the pseudorandom number generator in our smart metering system in Definition 2. Consider a set E_0 of a polynomial number $f(\tau)$ of elements randomly and uniformly chosen from $\mathcal{Z}_p$ and a set E_1 of the same number $f(\tau)$ of elements generated by g.

Definition 2. *Let $\tilde{b} \in \{0,1\}$ be a random coin. A function $g : \{0,1\}^\tau \times \{0,1\}^\lambda \to \mathcal{Z}_p$ is a pseudorandom number generator if given a set $E_{\tilde{b}}$ of elements, no probabilistic polynomial time algorithm guesses $\tilde{b}$ with an advantage more than ε', where ε' is a negligible function in τ.*

Theorem 1 states that our system satisfies ESP privacy requirement.

Theorem 1. *Let g be a pseudorandom number generator. Our smart metering system satisfies ESP privacy requirement, where $\varepsilon \leq 2\varepsilon'$.*

Proof. We prove by contradiction. Assume that an adversary $\mathcal{A}$ wins the game with an advantage at least $2\varepsilon'$. We construct an algorithm $\mathcal{S}$ such that given $E_{\tilde{b}}$, where $\tilde{b} = 0$ and $\tilde{b} = 1$ with equal probabilities, $\mathcal{S}$ guesses $\tilde{b}$ with an advantage more than ε' by using $\mathcal{A}$ as follows.

$\mathcal{S}$ acts as $\mathcal{C}$ and interacts with $\mathcal{A}$ in the security game. $\mathcal{S}$ embeds elements in $E_{\tilde{b}}$ as random numbers $\{r_{i,1}, r_{i,2}, \cdots, r_{i,l-1}, R_{i,1}, R_{i,2}, \cdots, R_{i,f(\tau)-l+1}\}$. For queries $d_{i,j}$ from $\mathcal{A}$, $\mathcal{S}$ returns $c_{i,j}$, where $c_{i,j} = (d_{i,j} + r_{i,j}) \mod p$. For $j \geq l-1$, $\mathcal{S}$ computes $r_{i,j}$ as $(R_{j-l+2} - \sum_{k=j-l+2}^{j-1} r_{i,k}) \mod p$. For challenges D_0 and D_1 from $\mathcal{A}$, $\mathcal{S}$ chooses D_b, which is either D_0 or D_1 with equal probabilities, and computes $C = \{c_{i,s} | c_{i,s} = (d_{i,s}^b + r_{i,s}) \mod p, s \in [j_1, j_2]\}$. After $\mathcal{S}$ sends C to $\mathcal{A}$, again, $\mathcal{S}$ answers queries from $\mathcal{A}$. Finally, if $\mathcal{A}$ successfully guesses b' for b, i.e., $b' = b$, $\mathcal{S}$ outputs 1. Otherwise, $\mathcal{S}$ outputs 0.

When $\tilde{b} = 1$ (E_1 contains pseudorandom numbers), the simulated environment is identical to our system. Thus, $\mathcal{A}$ outputs b' , where $b' = b$, with a probability at least $1/2 + 2\varepsilon'$ by our assumption. When $\tilde{b} = 0$ (E_0 contains truly random numbers), for each possible set D_b, there exists a unique set of values $r_{i,j_1}, r_{i,j_1+1}, \cdots, r_{i,j_2}$ satisfying that $c_{i,s} = (d_{i,s}^b + r_{i,s}) \mod p$ for $s \in [j_1, j_2]$. The distributions of C conditioned on D_0 and D_1 are identical. Thus, $\mathcal{A}$ has no advantage. Therefore, $\mathcal{A}$ correctly guesses b with probability 1/2. As a result, $\mathcal{S}$ outputs 1 with probability 1/2. $\mathcal{S}$ guesses $\tilde{b}$ with an advantage at least ε':

$$
\begin{aligned}
&\Pr[\mathcal{S} \text{ correctly guesses } \tilde{b}] \\
&= \Pr[\mathcal{S} \text{ outputs } 0 | \tilde{b} = 0] \Pr[\tilde{b} = 0] + \Pr[\mathcal{S} \text{ outputs } 1 | \tilde{b} = 1] \Pr[\tilde{b} = 1] \\
&= \Pr[\mathcal{A} \text{ outputs } b', b' \neq b | \tilde{b} = 0] \Pr[\tilde{b} = 0] \\
&\quad + \Pr[\mathcal{A} \text{ outputs } b', b' = b | \tilde{b} = 1] \Pr[\tilde{b} = 1] \\
&\geq (\frac{1}{2})\frac{1}{2} + (\frac{1}{2} + 2\varepsilon')\frac{1}{2} \\
&= \frac{1}{2} + \varepsilon'
\end{aligned}
$$

It contradicts with the assumption. □

5.2 LMC Privacy Requirement and Analysis

LMC Privacy Requirement. For LMC privacy requirement, we require that LMC only gets an approximate value $\tilde{d}_{i,j}$ for $d_{i,j}$ with a non-negligible probability.

Definition 3. *A metering smart system satisfies LMC privacy requirement if LMC guesses value $\tilde{d}_{i,j}$ for $d_{i,j}$ with $\Pr[\tilde{d}_{i,j} \neq d_{i,j}] \geq \eta$ for some significant probability η.*

LMC privacy requirement is slightly weak. Nevertheless, it is practical enough for smart grids. In smart grid deployments, the load monitoring system is often bundled with the grid operator. The grid operator can physically measure the power consumption at a power substation. By cooperating with the grid operator, LMC can get individual meter readings. Our LMC privacy requirement guarantees that when LMC does not get help from the grid operator, LMC cannot get exactly individual meter readings with a significant probability.

LMC Privacy Analysis.

Theorem 2. *Let a noise be the floor of a randomly chosen number from the normal distribution $N(0, \sigma^2)$ in our system. Our smart metering system satisfies LMC privacy requirement, where $\delta = 1/2 - 1/(4\pi\sigma^2)$*

Proof. We analyze $\Pr[\tilde{d}_{i,j} \neq d_{i,j}]$. The event of $\tilde{d}_{i,j} \neq d_{i,j}$ implies that the noise $n_{i,j}$ is not 0. Since the noise $n_{i,j}$ is the floor of a randomly chosen value n from $N(0, \sigma^2)$, the event $n_{i,j} = 0$ implies that $0 \leq n < 1$. Since $\Pr[n_{i,j} \neq 0]$ and $\Pr[n < 0] = (1 - \Pr[n = 0])/2$,

$$\Pr[n_{i,j} \neq 0] > \frac{1}{2} - \frac{1}{4\pi\sigma^2}$$

Thus, for a meter reading $d_{i,j}$ and a noised meter reading $\tilde{d}_{i,j}$, we have $\eta = \frac{1}{2} - \frac{1}{4\pi\sigma^2}$, which is significant for properly chosen σ. It concludes the proof for the LMC privacy requirement. □

When σ is larger, LMC has less probability to get a correct meter reading. Nevertheless, when σ is small, LMC has a better guess $\tilde{\xi}$ for ξ. Based on the previous chosen condition of $\epsilon = 10\%$ and $\delta = 10\%$ for LMC correctness, we consider $\sigma = 25$ and $m = 600$ for achieving that $\Pr[\omega \leq 10\%] \geq 99\%$. Under this setting, we have $\Pr[n_{i,j} \neq 0] > 0.4998$. Similarly, when $\epsilon = 7\%$ ($\sigma = 18$) and $\epsilon = 5\%$ ($\sigma = 12$), we have $\Pr[n_{i,j} \neq 0] > 0.4997$ and $\Pr[n_{i,j} \neq 0] > 0.4994$, respectively.

5.3 Storage Security

We show that meter readings are computationally securely stored in the storage system. Note that ESP has more information than the storage system does and our smart metering system satisfies the ESP privacy requirement.

Table 1. Summary of performance analysis

Actor	Storage system	Meter				
Subject	Storage	Computation			Communication	
		for storing	for ESP	for LMC	for ESP	for LMC
Result	9MB	$7(l+1)$ms	$14a$ms	14ms	64 bits	64 bits

We define the security requirement of storage in a security game $\mathcal{G}'$. The security game $\mathcal{G}'$ is the same as $\mathcal{G}$ except that the attacker $\mathcal{A}$ in $\mathcal{G}'$ represents the storage system. Thus, the security game $\mathcal{G}'$ captures that the storage system colludes with ESP. The storage security requirement is then defined:

Definition 4. *A metering system satisfies secure storage requirement if for any probabilistic polynomial time algorithm $\mathcal{A}$ and a negligible function ε, $|\Pr_{\mathcal{A}}[b' = b] - 1/2| < \varepsilon$.*

Theorem 3 states that our smart metering system satisfies secure storage requirement.

Theorem 3. *Let g be a pseudorandom number generator. Our smart metering system satisfies secure storage requirement, where $\varepsilon \leq 2\varepsilon'$.*

Proof. Since the proof is the same as the proof of Theorem 1, here we refer readers to the proof of Theorem 1. □

6 Performance Analysis

We use the previous setting of $\lceil \log_2 p \rceil = 64$ and set a time unit as 5 minutes. We evaluate the storage cost, computation cost, and communication cost in the following. Table 1 gives a summary.

Storage cost. Inside each meter $\mathcal{M}_i$, l pseudorandom numbers $r_{i,z-l+1}, r_{i,z-l+2}, \cdots, r_{i,z}$ are stored. The total storage size is $l\lceil \log_2 p \rceil$, i.e. $8l$ bytes.

For the storage system, each household uses $\lceil \log_2 p \rceil$ bits per time unit. Let a time unit be 5 minutes. The total storage size for meter readings of a household over 10 years is about 9 MB.

Computation cost. Computation operations of ESP and LMC are modular additions, which are efficient in modern computers. We focus on the computation cost of a meter. For a meter $\mathcal{M}_i$, to store a meter reading $d_{i,j}$, one pseudorandom number $R_{i,j}$ is generated by the TPM and l modular additions are performed by the driver of $\mathcal{M}_i$. To reply a query of al continuous time units $t_\beta, t_{\beta+1}, \cdots, t_{\beta+al-1}$ from ESP, a pseudorandom numbers $R_{i,\beta+kl}$ for $0 \leq k \leq a-1$ are generated and a modular additions are performed by the driver of $\mathcal{M}_i$. To reply a query of a recent time unit t_j from LMC, the driver of $\mathcal{M}_i$ generates a random noise $n_{i,j}$ and performs a modular addition to compute the noised meter reading $\tilde{d_{i,j}}$.

A recent commercial TPM chip[2] consists of a cryptographic accelerator capable of computing a 1024-bit RSA signature in 100 ms. Since generating a 1024-bit random number by using the pseudorandom number generator is no slower than the task of generating a 1024-bit RSA signature, each 64-bit random number can be generated in less than 7 ms. Similarly, we assume that a modular addition over $\mathcal{Z}_p$ can be done in less than 7 ms. Thus, the smart meter can store a meter reading in less than a time unit (5 minutes) when $l < 42856$ and reply a query in less than a time unit when $a < 21428$. The numerical results show that the computation of our system is well supported by current hardware technologies.

The *communication cost* between a meter and ESP or LMC for a query is $\lceil \log_2 p \rceil$ bits. That is, for a query from ESP or LMC, a meter transmits a 64-bit sum B of random numbers or a 64-bit noised random number $\tilde{r}_{i,j}$.

7 Conclusion and Future Works

We proposed a smart metering system that simultaneously supports the billing and load monitoring applications in a privacy preserving manner. ESP can only query for consumption of a household over a time period. LMC can only query an approximate consumption in an area at a recent time unit. Our construction is based on the layered meter model and uses the pseudorandom number generator in the TPM. According to our performance analysis, based on current TPM technologies, our construction is a practical and feasible solution to privacy preserving smart metering systems.

In addition to the billing and load monitoring applications, fine-grained consumption data contribute to other intelligent smart grid applications, such as demand prediction and power distribution planning. It is interesting to design a secure smart metering system that supports more applications.

Acknowledgement. We thank anonymous reviewers for their comments. The research was supported in part by projects ICTL-101-Q707, ATU-101-W958, NSC 101-2218-E-009-003-, NSC 98-2221-E-009-068-MY3, and NSC 100-2218-E-009-006.

References

1. Ontario energy board, smart meters and time of use (tou) prices, webpage available at `http://www.ontarioenergyboard.ca/OEB/Consumers/Electricity/Smart+Meters`
2. Smart grid cyber security strategy and requirements. National Institute of Standards and Technology (U.S.) Interagency Reports, DRAFT-NISTIR-7628 (2009)
3. Ács, G., Castelluccia, C.: I Have a DREAM (DiffeRentially privatE smArt Metering). In: Filler, T., Pevný, T., Craver, S., Ker, A. (eds.) IH 2011. LNCS, vol. 6958, pp. 118–132. Springer, Heidelberg (2011)
4. Bohli, J.M., Sorge, C., Ugus, O.: A privacy model for smart metering. In: Proceedings of the 1st IEEE International Workshop on Smart Grid Communications (2010)

[2] Atmel AT97SC3203S.

5. Chan, T.H.H., Shi, E., Song, D.: Privacy preserving stream aggregation with fault tolerance. In: Proceedings of the 16th International Conference on Financial Cryptography and Data Security - FC 2012. LNCS. Springer (2012)
6. Garcia, F.D., Jacobs, B.: Privacy-Friendly Energy-Metering via Homomorphic Encryption. In: Cuellar, J., Lopez, J., Barthe, G., Pretschner, A. (eds.) STM 2010. LNCS, vol. 6710, pp. 226–238. Springer, Heidelberg (2011)
7. Hart, G.W.: Nonintrusive appliance load monitoring. In: Proceedings of the IEEE, pp. 1870–1891. IEEE Press (1992)
8. Inc., A.P.L.S.: Aclara ami industry glossary (2008)
9. Jawurek, M., Johns, M., Kerschbaum, F.: Plug-in privacy for smart metering billing. Computing Research Repository - CoRR (2010)
10. Kursawe, K., Danezis, G., Kohlweiss, M.: Privacy-Friendly Aggregation for the Smart-Grid. In: Fischer-Hübner, S., Hopper, N. (eds.) PETS 2011. LNCS, vol. 6794, pp. 175–191. Springer, Heidelberg (2011)
11. Laughman, C., Lee, K., Cox, R., Shaw, S., Leeb, S., Norford, L., Armstrong, P.: Power signature analysis. IEEE Power and Energy Magazine 1, 56–63 (2003)
12. LeMay, M., Gross, G., Gunter, C.A., Garg, S.: Unified architecture for large-scale attested metering. In: Proceedings of the 40th Hawaii International International Conference on Systems Science - HICSS 2007, p. 115. IEEE Computer Society (2007)
13. Metke, A.R., Ekl, R.L.: Security technology for smart grid networks. IEEE Transactions on Smart Grid 1(1), 99–107 (2010)
14. Molina-Markham, A., Shenoy, P., Fu, K., Cecchet, E., Irwin, D.: Private memoirs of a smart meter. In: Proceedings of the 2nd ACM Workshop on Embedded Sensing Systems for Energy-Efficiency in Building - BuildSys 2010, pp. 61–66. ACM (2010)
15. Petrlic, R.: A privacy preserving concept for smart grids. In: Sicherheit in vernetzten Systemen: 18. DFN Workshop, pp. B1–B14. Books on Demand GmbH (2010)
16. Rial, A., Danezis, G.: Privacy-preserving smart metering. Microsoft Technical Report, MSR-TR-2010-150 (2010)
17. Shi, E., Chan, T.H.H., Rieffel, E., Chow, R., Song, D.: Privacy-preserving aggregation of time-series data. In: Proceedings of the 18th Annual Network and Distributed System Security - NDSS 2011. The Internet Society (2011)

Private Computation of Spatial and Temporal Power Consumption with Smart Meters

Zekeriya Erkin[1] and Gene Tsudik[2]

[1] Information Security and Privacy Lab, Delft University of Technology, 2628 CD, Delft, The Netherlands
[2] Department of Computer Science, University of California, Irvine, CA, USA
z.erkin@tudelft.nl, gene.tsudik@uci.edu

Abstract. Smart metering of utility consumption is rapidly becoming reality for multitudes of people and households. It promises real-time measurement and adjustment of power demand which is expected to result in lower overall energy use and better load balancing. On the other hand, finely granular measurements reported by smart meters can lead to starkly increased exposure of sensitive information, including all kinds of personal attributes and activities. Reconciling smart metering's benefits with privacy concerns is a major challenge.

In this paper we explore some simple and relatively efficient cryptographic privacy techniques that allow spatial (group-wide) aggregation of smart meter measurements. We also consider temporal aggregation of multiple measurements for a single smart meter. While our work is certainly not the first to tackle this topic, we believe that proposed techniques are appealing due to their simplicity, few assumptions and peer-based nature, i.e., no need for any on-line aggregators or trusted third parties.

1 Introduction

Growing energy needs motivate both governments and industry to look for alternative energy resources and, more importantly, provide better management of existing power grids. However, improving efficiency of existing power grids and smart load-balancing are challenging tasks. One approach to smart load-balancing currently pursued by many developed countries is the deployment of so-called "smart meters" that measure and report power consumption on a regular basis, thus allowing for real-time management of the grid.

While smart meters offer some clear benefits, accurate and fine-grained measurements of household energy consumption trigger serious privacy concerns [2]. A plethora of sensitive information can be gleaned or derived from such measurements, e.g., types of electrical devices being used as well as presence (and number of) inhabitants. For example, due to privacy considerations, deployment of smart meters in the Netherlands has been cancelled by the Parliament. However, it is well under way in other European countries, the USA and Canada. It is anticipated that 80% of EU consumers will be using smart meters by year

F. Bao, P. Samarati, and J. Zhou (Eds.): ACNS 2012, LNCS 7341, pp. 561–577, 2012.

2020. Since their usage is essential to better grid management, it is important to develop technologies that reconcile privacy with desired utility and functionality of smart meters.

In this paper, we consider three privacy smart meter scenarios:

- **Spatial aggregation:** a local grid corresponding to a group of households each equipped with a smart meter, where owners are interested in aggregate (total) consumption in order to either adjust their own consumption according to the average or check whether there is enough energy in the grid to power an extra electrical device. This scenario is especially very important for self-sufficient, remote places, particularly, in developing countries, where renewable resources (such as wind turbines and solar panels) have become more affordable for local energy production, as an alternative to traditional carbon-based fuels.
- **Temporal aggregation:** a single household equipped with a smart meter that reports its power consumption on a regular basis, for billing purposes. In this scenario, the energy supplier charges the households for a certain time period.
- **Spatio-temporal data aggregation:** a hybrid setting that combines both of the above scenarios. In it, each node disseminates a single value for its measurement and this value is used for computing spatial aggregate consumption for the neighborhood, in that interval. At the same time, a number of such values per household allows computation of temporal aggregate consumption for each smart meter, for billing purposes.

In all aforementioned scenarios, individual smart meter measurements represent sensitive information. Our goal is to keep them private without impacting either utility or functionality of smart meters. We plan to achieve it by blending cryptographic secret sharing coupled with additively homomorphic encryption. To this end, the main contribution of this paper is an encryption scheme, wherein each smart meter encrypts its fine-grained power consumption measurement. However, *no one* can decrypt this individual encryption. Decryption only becomes possible when a fixed, predefined number of encryptions is aggregated. This scheme allows us to compute spatial consumption in a local grid with a fixed number of households (for one period) and/or temporal consumption of a single household (for a fixed number of periods).

Although this paper is framed in terms of smart meters and power consumption, our proposed scheme is quite general. It can be used in any scenario where there is a need to additively aggregate plaintexts and keep individual plaintext secret. In particular, clustering and collaborative filtering algorithms, used e.g. in social networks and e-commerce applications, rely on privacy-sensitive data of users like preferences, profiles and ratings. While there is a potential privacy risk for users since the service provider can process the private data for other purposes, re-sell them to third parties or fail to provide adequate physical security, the provided services is still very appealing for many users. In such situations, the ideas in this paper can be used to re-design the algorithms in a privacy-preserving way.

Moreover, the cost of our scheme is quite low and its security is not based on any non-standard cryptographic or adversarial assumptions. As shown in the complexity analysis, the computation performed by each smart meter is minimal compared to the existing works in the literature.

The rest of the paper is organized as follows. We discuss related work in Section 2 and summarize notation and adverserial model in Section 3. We present our protocol for computing spatial consumption in a neighborhood in Section 4 and temporal consumption of a single household in Section 5. We explain the protocol for computing both spatial and temporal consumptions in Section 6. We provide an informal discussion on the security of the proposed protocols in Section 7. We discuss complexity and how to adopt our protocols for different types of measurements in Section 8. We finally conclude the paper in Section 9.

2 Related Work

A number of research results tackled privacy issues in smart meters, including privacy-preserving billing [19,14] and aggregation of private data. Examples of techniques that compute the sum of multiple private inputs include [7,5], where encryption is done by modular addition (each player simply adds its key to the plaintext) and aggregation is very efficient, also performed via addition. However, this approach assumes a semi-trusted aggregator who knows the sum of all keys (for each reporting interval) and can thus decrypt the aggregated value by subtraction. This operation is not easily extensible to settings without the aggregator or where the latter is simply not trusted with any secrets.

Peter *et al.* [18] consider three methods of aggregating data in a wireless networks based on *homomorphic encryption* [11]. The first protocol uses the Domingo-Ferrer (DF) encryption scheme [8] that is allegedly both additively and multiplicatively homomorphic. However, there is no evidence that the underlying DF cryptosystem is secure. The second protocol is a minor modification of [7] and the third protocol is based on Elliptic Curve ElGamal, which is quite inefficient because of expensive algebraic operations.

Kursawe *et al.* [15] present cryptographic protocols for computing aggregated consumptions using Diffie-Hellman key exchange protocol and bilinear mapping, which also requires expensive elliptic curve operations. Kohlweiss and Danezis [14] propose a mechanism for privacy-preserving billing in a smart grid by using homomorphic encryption, secure multi-party computation (MPC) techniques and cryptographic commitment schemes [13]. It requires the use of certificates to obtain accountability. Since it involves heavy-weight cryptographic tools – such as MPC – the cost of this scheme is very high.

In a recent result, Shi *et al.* [20] introduce an interesting technique for aggregating private data using distributed differential privacy. Similar to our work, it blends secret sharing with homomorphic encryption. However, it also requires the aggregator to solve an instance of the discrete log problem (albeit, with limited range) to obtain plaintext.

Garcia and Jacobs [12] propose a scheme to compute aggregate consumption without revealing individual measurements using homomorphic encryption and secret sharing. For this purpose, every smart meter splits its measurement into random shares and encrypt each of them using the public key of another smart meter but keeps one share for itself. A substation collects all the encryptions and multiplies the ones which are encrypted with the same public key. Later, the substation sends the encrypted sums to the smart meters. Upon receiving the encrypted sum, smart meters decrypt and add their shares in plaintext. Finally, the substation collects the plain text sums and aggregate them all to obtain the total consumption. While this apporach is privacy preserving, the number of homomorphic encryptions per user is linear and the amount of data transferred is quadratic in the number of smart meters, which is clearly inefficient.

Another approach offering differential privacy in the context of smart meters is given by Árc and Castelluccia [1]. In addition to smart meters, the authors introduce two other parties: a supplier and an aggregator. Individual measurements are protected by adding Laplacian noise. This scheme uses efficient symmetric encryption. To prevent the aggregator from learning individual measurements, each encryption is masked with a random number, composed of dummy keys collectively generated by a (fixed) group of smart meters. Similar to [6], each encryptor also uses another key – shared by the aggregator and each smart meter – such that only the aggregator (or the supplier) can obtain the noise-altered sum of all measurements.

3 Preliminaries

In this section, we provide some background information on the envisaged operating environment, cryptographic schemes, the adversarial model and other assumptions.

3.1 Anticipated Setting

We assume an environment (e.g., a residential neighborhood) composed of a fixed (static) group of N tamper-resistant smart meters, one per household. (We use the terms *household* and *smart meter* interchangeably from here on.) Every smart meter – denoted by sm_i, $0 < i \leq N$ – is programmed to report its current measurement (power consumption) with certain fixed periodicity common to all other smart meters. All smart meters are loosely time-synchronized, i.e., report their current measurements at more-or-less the same time. Furthermore, a smart meter is assumed capable of performing simple public key operations and of generating high-quality (cryptographically strong) random numbers.

We do not assume any other *active* entities, such as aggregators, suppliers or trusted third parties. One of our goals is for any smart meter to be able to act as an aggregator, for the purpose of computing total (group-wide) consumption. On the other hand, we do not preclude the presence of *passive* entities, e.g., an aggregator that learns total consumption by overhearing messages, while not taking part in any protocol.

Moreover, we assume that all underlying communication channels are secure: both integrity and authentication of all messages are obtained via standard means, e.g., IPSec or SSL/TLS [9].

3.2 Notation

Our notation is summarized in Table 1.

Table 1. Notation Summary

Symbol	Definition	Symbol	Definition
N	number of smart meters	M	number of measurement intervals
n	product of two large primes	sm_i	smart meter i
g	generator	p, q	prime numbers
$\mathbb{Z}_n$	set of integers from 0 to $n-1$	$\mathbb{Z}_n^*$	set of integers co-prime to n
p	time interval	K_i	shared key of sm_i
$\mathcal{E}_{pk}(\cdot)$	encryption function	$\mathcal{D}_{sk}(\cdot)$	decryption function
PRF$(\cdot)$	pseudo random function	$H(\cdot)$	cryptographic hash function, e.g, SHA-2
h_i	hash of the sm_i using K_i	$Pr(F)$, α	probability of a malfunction at time interval F
$c_{(i,p)}$	measurement of sm_i in time interval p	C_p	total consumption of N smart meters for time interval p
$R_{(i,p)}$	composite random number of sm_i for time interval p	F	time interval when a smart meter malfunctions
$r_{(i\rightarrow j,p)}$	random number sent from sm_i to sm_j in time interval p	$h_{(i,p)}$	hash of the sm_i using the pth period identifier (time stamp)
k	bit length of each measurement	T	Number of colluding smart meters

3.3 Adversarial Model

We assume the semi-honest (also known as "Honest-but-Curious") adversarial model. Consequently, all smart meters faithfully follow all prescribed protocol steps. However, they may attempt to learn as much as possible information beyond what they are entitled to have. We claim that this is realistic, since we also assume that smart meters are (somewhat) tamper-resistant and interfering with measurements is not trivial.

We also allow adversarial smart meters to collude as long as their number does not exceed some fixed threshold $T < N - 1$. (This ensures the existence of at least two honest smart meters for which only their combined consumption is learned by the coalition of dishonest peers).

Although participants are assumed to follow all protocol steps and provide real measurements, we do not rule out so-called data pollution (or other DoS) attacks that can result in meaningless or incorrect measurement results. Since smart meters are assumed to be tamper-resistant, we do not consider such attacks. However, we note that they are more relevant to security rather than privacy. Also, some pollution attacks can be addressed by incorporating zero-knowledge proofs to show that measurements are within a certain sensible range [4].

3.4 Homomorphic Encryption

The Paillier cryptosystem presented in [17] is *additively homomorphic*. This means that there exists an operation over the ciphertexts $\mathcal{E}_{pk}(m_1)$ and $\mathcal{E}_{pk}(m_2)$ such that the result of that operation corresponds to a new ciphertext whose decryption yields the sum of the plaintext messages m_1 and m_2:

$$\mathcal{D}_{sk}\left(\mathcal{E}_{pk}(m_1) \times \mathcal{E}_{pk}(m_2)\right) = m_1 + m_2 . \tag{1}$$

As a consequence of additive homomorphism, exponentiation of any ciphertext yields the encrypted product of the original plaintext and the exponent:

$$\mathcal{E}_{pk}(m)^e = \mathcal{E}_{pk}(m \cdot e) . \tag{2}$$

Given message $m \in \mathbb{Z}_n$, Paillier encryption is defined as:

$$\mathcal{E}_{pk}(m, r) = g^m \cdot r^n \bmod n^2 , \tag{3}$$

where n is a product of two large primes p and q, g is a generator of order n and r is a random number in $\mathbb{Z}_n^*$. The tuple (g, n) is the public key. For decryption, we refer readers to [17].

The Paillier cryptosystem is semantically secure. This is particularly important for encryption of plaintext within a small range.

4 Aggregating Spatial Consumption

In this section, we describe a peer-based scheme for privately computing (spatial) aggregate consumption.

Total consumption of sm_i is defined as: $C_p = \sum_{i=1}^{N} c_{(i,p)}$, where $c_{(i,p)}$ is the measurement of sm_i in time interval p. The measurement interval p can take any value – from seconds to days – depending on the specific application requirements. Each smart meter stores only *one* Paillier public key, common to all N smart meters in the group.

One of the distinguishing features of our scheme is that the (normally private) Paillier decryption key is actually **public**. In other words, it is assumed to be known at least by all smart meters in the group. In fact, it can be known by any other party that is authorized to learn the total consumption. This feature is clearly unusual. However, the justification is very simple: we use homomorphic (Paillier) scheme not because of encryption but only for its homomorphic property.

Note: Although both Paillier encryption and decryption keys are "public" in our protocol, a secure instance Paillier scheme still needs to be set up correctly and securely. For this reason, we assume the existence of a trusted party (e.g., a CA) that bootstraps an instance of Paillier scheme, i.e., generates appropriate parameters, including primes, modulii and keys. This third party is no longer required after the set up phase.

The proposed scheme works as follows:

1. For each measurement interval p, sm_i generates a set of random numbers, one for every other smart meter. It then sends these numbers to all its peers using the underlying (secure) communication channel(s).
2. Upon receiving these random values, each smart meter encrypts its measurement using the Pailler scheme. (Recall that the main idea is to prevent smart meters from decrypting individual measurements.) All encryptions are then diseminated to the entire group.
3. Next, each smart meter combines all encryptions, including its own, to obtain the encrypted sum (using the homomorphic property), and decrypts it using the common private key. The resulting plaintext represents the total consumption for the p-th measurement interval.

The protocol is shown in more detail in Figure 1.

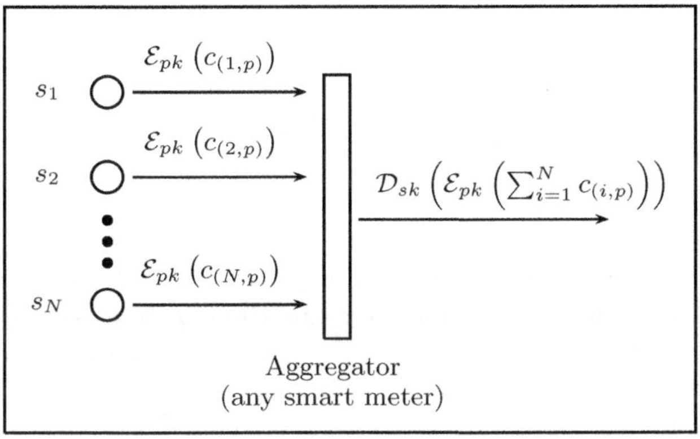

Fig. 1. Spatial Consumption

4.1 Generating and Exchanging Random Numbers

To compute total consumption for interval p, all smart meters initially exchange random values to be used for masking individual consumption measurements. For this purpose, each sm_i generates a random number $r_{(i \to j,p)}$ and sends it to a peer sm_j. We assume that all smart meters participate in the protocol by identifying themselves via valid certificates. At the end of this step, each sm_i receives $N - 1$ random values from its peers.

Note: Exchanging random numbers between smart meter pairs in each interval introduces unnecessary communication overhead. Instead, smart meters can exchange the seed of their pseudo-random number generators when they initially become active.

Next, each sm_i computes $R_{(i,p)}$ based on all collected randomness:

$$R_{(i,p)} = n + \sum_{j=1, i \neq j}^{N} r_{(i \to j,p)} - \sum_{j=1, i \neq j}^{N} r_{(j \to i,p)} , \tag{4}$$

where n is the Paillier modulus. $R_{(i,p)}$ is used later to encrypt sm_i's measurement for the pth interval.

4.2 Encrypting Measurements

Recall that we want to disseminate individual measurements such that, only when all of them are aggregated, the total can be retrieved. We achieve this by encrypting measurements, $c_{(i,p)}$ using a modified version of the Paillier cryptosystem. First, for each time interval p, each smart meter computes a hash: $h_{(i,p)} = H(p)$, where $H(\cdot)$ is a secure hash function such as SHA-2. It is required for $h_{(i,p)}$ to be in $\mathbb{Z}_n^*$, for the encryption scheme to work. This holds when $\gcd(h_{(i,p)}, n) = 1$.[1]

Next, sm_i encrypts its measurement, $c_{(i,p)}$, as follows

$$\mathcal{E}_{pk}\left(c_{(i,p)}\right) = g^{c_{(i,p)}} \cdot h_{(i,p)}^{R_{(i,p)}} , \tag{5}$$

using the common Paillier public key. Finally, each smart meter disseminates its encryption.

Encrypting measurements in this fashion has the following features. First, no one in the smart grid can decrypt individual encryptions due to $h_{(i,p)}^{R_{(i,p)}}$, even though everyone has the decryption key. Second, encryption remains semantically secure since $h_{(i,p)} \in \mathbb{Z}_n^*$ and $R_{(i,p)}$ is a random number in $\mathbb{Z}_n$, which is in accordance with the original scheme. Third, by using $h_{(i,p)}$, computation of total power consumption is bound to interval p.

4.3 Aggregation of Encrypted Measurements

To obtain total power consumption C_p, *any* sm_i multiplies all encrypted measurements, including its own:

$$\begin{aligned}\prod_{i=1}^{N} \mathcal{E}_{pk}\left(c_{(i,p)}\right) &= \prod_{i=1}^{N} g^{c_{(i,p)}} \cdot h_{(i,p)}^{R_{(i,p)}} \\ &= g^{\sum_{i=1}^{N} c_{(i,p)}} \cdot h_{(i,p)}^{\sum_{i=1}^{N} R_{(i,p)}} ,\end{aligned} \tag{6}$$

where,

$$\sum_{i=1}^{N} R_{(i,p)} = \sum_{i=1}^{N} n + \sum_{i=1}^{N} \sum_{j=1, i \neq j}^{N} r_{(i \to j,p)} - \sum_{i=1}^{N} \sum_{j=1, i \neq j}^{N} r_{(j \to i,p)} . \tag{7}$$

[1] The number of values in $\mathbb{Z}_{n^2}^*$ is $(\Phi(n))^2$, which is close to n^2 for large p and q.

Since $\sum_{i=1}^{N}\sum_{j=1,i\neq j}^{N} r_{(i\rightarrow j,p)}$ equals $\sum_{i=1}^{N}\sum_{j=1,i\neq j}^{N} r_{(j\rightarrow i,p)}$, the terms in (7) cancel each other out and the summation results in:

$$\sum_{i=1}^{N} R_{(i,p)} = \sum_{i=1}^{N} n = N \cdot n \,. \tag{8}$$

Replacing the above sum in Eq. (6), we obtain:

$$g^{\sum_{i=1}^{N} c_{(i,p)}} \cdot h_{(i,p)}^{\sum_{i=1}^{N} R_{(i,p)}} = g^{\sum_{i=1}^{N} c_{(i,p)}} \cdot h_{(i,p)}^{N\cdot n} \,, \tag{9}$$

which is the encryption of $\sum_{i=1}^{N} c_{(i,p)}$ with a random value $h_{(i,p)}^{N}$:

$$g^{\sum_{i=1}^{N} c_{(i,p)}} \cdot (h_{(i,p)}^{N})^{n} = \mathcal{E}_{pk}\left(\sum_{i=1}^{N} c_{(i,p)}\right) = \mathcal{E}_{pk}\left(C_p\right) \,. \tag{10}$$

This result decrypted to obtain the total power consumption.

5 Computing Temporal Consumption

We now consider privacy in the temporal dimension. In this setting, we envision a single smart meter that periodically reports its consumption totals, e.g., for the purpose of billing. However, as discussed earlier, such fine-grained reporting might be detrimental to privacy. We consider two scenarios.

1. The smart meter reports its measurements for billing purposes and the total is computed only when a pre-defined number of measurements is received by the supplier. In the case of a smart meter malfunction, the supplier asks for help from the manufacturer of that smart meter in order to obtain partial consumption, i.e., until the time malfunction occurred.
2. The smart meter reports its accumulated measurement, i.e., the total consumption: $\sum_{p=1} c_{(i,p)}$.
   Note that in this scenario, all incremental consumption measurements are encrypted using the public key of the manufacturer. In the last interval, the smart meter sends the total consumption to the supplier using the public key of the latter. In the event of a malfunction (i.e., the smart meter cannot report) the last consumption measurement encrypted with the public key of the manufacturer will be sent to the manufacturer for decryption.

Each scenario has its advantages. While, in the first, the manufacturer is not needed to encrypt any private data, the supplier has to store all encrypted messages sent by all smart meters. In the second scenario, however, the manufacturer decrypts a single ciphertext for the supplier. The supplier stores only the last message sent by each smart meter.

For these two scenarios, we define the following roles:

- **Manufacturer $\mathcal{M}$:** the entity that produces the smart meters. It is not involved in the billing process.

- **Supplier $\mathcal{S}$:** the authority that periodically bills the households for their consumption. In this setting, we assume that invoices are sent for every M intervals. The supplier also has a Paillier public key-pair. Its public key is available to all smart meters in the grid.
- **Smart meter:** Household with a smart meter as defined before, capable of reporting its consumption on a regular basis. Every smart meter has a bi-directional communication channel with the supplier that uses a secure and reliable transfer protocol.

We now present the first protocol. The second protocol is trivial to realize by following a similar approach.

5.1 Encrypting Measurements

We use a similar construction to that in Section 4 – a modified version of the Paillier cryptosystem: sm_i generates a random number, $r_{(i,p)}$, using a PRF that takes two inputs: (1) the secret key K_i unique key to each sm_i and shared with the manufacturer, and (2) the unique interval identifier – p. In other words: $R_{(i,p)} := \text{PRF}(K_i, p)$. (Note that p can be viewed as a corsely granular timestamp.) As in Section 4, sm_i also generates $h_i := H(K_i) \in \mathbb{Z}_n^*$ to be used *throughout* all M intervals. The consumption $c_{(i,p)}$ is then encrypted as: $\mathcal{E}_{pk}\left(c_{(i,p)}\right) = g^{c_{(i,p)}} \cdot h_i^{R_{(i,p)}}$.

With billing occurring every M measurement intervals, sm_i generates $R_{(i,p)}$ for the first $M-1$ intervals as described above. The value to be used in time interval M is computed as follows:

$$R_{(i,M)} := n - \sum_{p=1}^{M-1} R_{(i,p)} \,. \tag{11}$$

5.2 Obtaining Total Consumption

Upon receiving all encryptions for M time intervals from sm_i, the supplier aggregates them:

$$\begin{aligned}\prod_{p=1}^{M} \mathcal{E}_{pk}\left(c_{(i,p)}\right) &= \prod_{p=1}^{M} g^{c_{(i,p)}} \cdot h_i^{R_{(i,p)}} = g^{\sum_{p=1}^{M} c_{(i,p)}} \cdot h_i^{\sum_{i=p}^{M} R_{(i,p)}} \\ &= g^{\sum_{p=1}^{M} c_{(i,p)}} \cdot h_i^{n} = \mathcal{E}_{pk}\left(\sum_{p=1}^{M} c_{(i,p)}\right) \,. \end{aligned} \tag{12}$$

Since the sum of all $R_{(i,p)}$'s is n, the above encryption can be easily decrypted by the supplier.

5.3 Coping with Malfunctions

In the event of a malfunction, sm_i can not send its measurements after interval F. At the same time, with only encrypted measurements of the first F intervals, the supplier can not decrypt and determine total consumption. To remedy the situation, the supplier contacts the manufacturer, who has a unique secret key K_i pre-shared with sm_i. The manufacturer can re-generate all random numbers used for the first F intervals: $R_{(i,p)} := \mathrm{PRF}(K_i, p)$ and h_i. Having computed these values, the manufacturer then encrypts:

$$\mathcal{E}_{pk}\left(O\right) = g^0 \cdot h_i^{R_{(i,F+1)}} \ , \text{ where } R_{(i,F+1)} = n - \sum_{p=1}^{F} r_{(i,p)} \ . \tag{13}$$

Using the encryption sent by the manufacturer, the supplier can compute the total consumption for the first F intervals by multiplying the encryption received from the manufacturer and decrypting the result using its private key.

$$\mathcal{D}_{sk}\left(\mathcal{E}_{pk}\left(\sum_{p=1}^{F} c_{(i,p)}\right) \cdot \mathcal{E}_{pk}\left(O\right)\right) = \sum_{p=1}^{F} c_{(i,p)} \tag{14}$$

6 Computing Spatio-temporal Consumption

In prior sections, we focused on computing either spatial or temporal total consumption in a smart neighborhood grid. In this section, we turn to spatio-temporal total consumption.

The scheme involves three types of entities, as before: a manufacturer, a supplier and smart meters.

6.1 Encrypting Measurements

As in Section 4, each sm_i comes up with a secret value $R_{(i,p)}$ for interval p such that $\sum_{i=1}^{N} R_{(i,p)}$ is a multiple of n. Each such $R_{(i,p)}$ can be generated jointly by contributions from all smart meters, as described in Section 4. In cases where manufacturer's involvement is possible, $R_{(i,p)}$-s can be provided by the manufacturer, with the property of: $\sum_{i=1}^{N} R_{(i,p)} = 0$.

In interval p, sm_i encrypts its consumption, $c_{(i,p)}$ with the common Paillier public key:

$$\mathcal{E}_{pk}\left(c_{(i,p)}\right) = g^{c_{(i,p)}} \cdot h_p^{R_{(i,p)}} \ , \tag{15}$$

where $h_p \in \mathbb{Z}_n^*$ is the hash of the current interval, e.g., $h_p = H(p)$. Each ciphertext is then broadcasted to all peers.

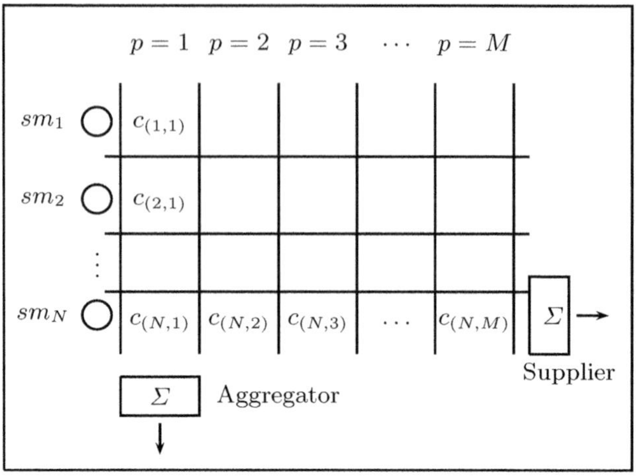

Fig. 2. Spatio-Temporal Consumption

6.2 Obtaining Spatial Consumption

Upon receiving N ciphertexts, sm_i computes total consumption as before, by multiplying all ciphertexts and decrypting the final value. Recall that individual encryptions cannot be decrypted by anyone.

$$\prod_{i=1}^{N} \mathcal{E}_{pk}\left(c_{(i,p)}\right) = \mathcal{E}_{pk}\left(\sum_{i=1}^{N} c_{(i,p)}\right) = g^{\sum_{i=1}^{N} c_{(i,p)}} \cdot h_p^{\sum_{i=1}^{N} R_{(i,p)}} . \tag{16}$$

Since $R_{(i,p)}$-s add up to a multiple of n (or sum up to 0 if the manufacturer is involved), the above multiplication results in proper encryption of total consumption, that can be decrypted using the common private key.

6.3 Obtaining Temporal Consumption

After each smart meter broadcasts the ciphertexts of its consumption for M intervals, temporal consumption can be computed. However, each sm_i uses a different hash , $h_{(i,p)}$, and $R_{(i,p)}$, for encryption in each interval p. Even after multiplying all M ciphertexts from the same sm_i, it is impossible to decrypt the resulting ciphertext:

$$\prod_{p=1}^{M} \mathcal{E}_{pk}\left(c_{(i,p)}\right) = g^{\sum_{p=1}^{M} c_{(i,p)}} \cdot \prod_{p=1}^{M} h_p^{R_{(i,p)}} , \tag{17}$$

since it does not represent a valid encryption. To decrypt it, an additional random value, $R_{(i,M+1)}$, must be provided by sm_i such that the following condition is satisfied:

$$R_{(i,M+1)} = \frac{r^n}{\prod_{p=1}^{M} h_p^{R_{(i,p)}}} , \tag{18}$$

where r is a random value in $\mathbb{Z}_n^*$. Note that, after multiplying the ciphertext in in Eq. (17) with $R_{(i,M+1)}$, we have:

$$\begin{aligned} \prod_{p=1}^{M} \mathcal{E}_{pk}\left(c_{(i,p)}\right) \cdot R_{(i,M+1)} =& g^{\sum_{p=1}^{M} c_{(i,p)}} \cdot \prod_{p=1}^{M} h_{(i,p)}^{R_{(i,p)}} \\ &\times \frac{r^n}{\prod_{p=1}^{M} h_p^{R_{(i,p)}}} \\ =& g^{\sum_{p=1}^{M} c_{(i,p)}} \cdot r^n , \end{aligned} \tag{19}$$

which can be decrypted properly.

6.4 Coping with Malfunctions

The scheme described above can be realized without any suppliers or manufacturers. However, in case of a malfunction, it becomes impossible to obtain the total consumption. To recover data, collaboration between the manufacturer and the supplier is necessary. In that case, the manufacturer should genrate and store the random values, $R_{(i,p)}$, and give them to the smart meters. When a malfunction occurs, supplier asks for the random value $R_{(i,M+1)}$ from the manufacturer, that could compute it to be used for decryption as in previous section.

7 Security Considerations

There are two basic flavours of security that we consider in this paper: semantic security of the modified Paillier cryptosystem and collisions. We give an informal discussion on these issues in this section.

The security of our schemes mainly based on the semantic security of the modified Paillier cryptosytem. Once a measurement is encrypted, ciphertext is disseminated, meaning that the encryption is accessibly by all of the smart meters in the grid. Assuming that the bit length of the measurements are small compared to the message space of the cryptosysten, semantic security is crucial.

The consumption measurement of sm_i, $c_{(i,p)}$, is encrypted by following the description of the Paillier scheme but randomized in a different way. Instead of using a random number $r \in \mathbb{Z}_n^*$ and raising it to the power of n, we generate a hash, by taking the hash of either the time interval $h_p = H(p)$ or the shared key of the smart meter $h_i = H(K_i)$, and raise this hash to the power of a random number, $R_{(i,p)}$. The way we generate the hash value guarantees that it is in $\mathbb{Z}_n^*$, matching the requirements of the original cryptosystem. Therefore, the encrypted message is uniformly distributed to the ciphertext space of the cryptosystem, satisfying the semantic security.

The security against the malicious coalition relies on the assumption that at least two out of N smart meters are acting accordingly to the protocol specifications. It is trivial to see that any smart meter can obtain the encrypted measurements of any other smart meter, assuming that these encryptions are disseminated in the network, and cannot decrypt the ciphertext even though every smart meter has the *public* decryption key. A coalition of $N-1$ malicious, or curious, smart meters can sum up the measurements of $N-1$ smart meters and obtain the measurement of the honest N^{th} smart meter by subtracting that sum from the total, which is computed by following the protocol steps. Only in the case of having two honest smart meters in the neighbourhood, the rest of the smart meters can not obtain the individual measurements of these two smart meters.

8 Complexity and Data Packing

In this section, we present complexity analysis and a way to compute different type of measurements using a single smart meter.

8.1 Complexity

We based our complexity analysis on the number of operations performed by a smart meter, that include: en/de-cryptions, generation of random numbers, PRF invocations and hashing. We denote the probability of malfunction for a smart meter (e.g., quoted at 0.08% in [10]) by Pr(F) $= \alpha$. The total number of operations performed by each party for different cases is summarized in Table 2.

Table 2. Numbers of cryptographic operations for: (1) smart meter ($\mathcal{SM}$), (2) aggregator ($\mathcal{A}$), (3) supplier ($\mathcal{S}$) and (4) manufacturer ($\mathcal{M}$)

	Spatial		Temporal			Spatio-Temporal			
	$\mathcal{SM}$	$\mathcal{A}$	$\mathcal{SM}$	$\mathcal{S}$	$\mathcal{M}$	$\mathcal{SM}$	$\mathcal{A}$	$\mathcal{S}$	$\mathcal{M}$
Encryption	1	-	M	-	$\alpha \cdot 1$	M	-	-	-
Decryption	-	1	-	1	-	-	1	1	-
Multiplication	-	$N-1$	-	$M-1$	-	-	$N-1$	$M-1$	$\alpha(M-1)$
Hash	1	-	M	-	$\alpha \cdot F$	M	-	-	$\alpha \cdot F$
PRF	$N-1$	-	M	-	$\alpha \cdot F$	M	-	-	$\alpha \cdot F$

As seen in Table 2, obtaining aggregated consumptions cost only 1 encryption and constant amount of hash and PRF functions per smart meter in each time interval. The computation of $R_{(i,M)}$, which is necessary for the decryption of total consumption, requires M multiplications over n and computing the inverse of that product. In practice, smart meters are supposed to report their consumptions as often as 5 minutes. Implementation results in [16] show that even more expensive cryptographic operations can be realized efficiently on smart meters. It is our conclusion that the proposed cryptographic protocols in this paper, which are only based on performing cryptographic primitives like encryption, hash functions and random number generation, present a highly efficient way of computing aggregated consumptions without disclosing individual measurements.

8.2 Multiple Utility Measurements

This paper focused on aggregating smart meter measurements, however, without specifying explicitly what kind of measurements are possible. In practice, for each type of basic utility – e.g., water, gas and electricity – there is a different metering device and (usually) a different supplier. However, if the same smart device is used for measuring multiple types of utilities, our approach can still be used.

Assume that for a given sm_i we have the following measurements: $c_{(ij,p)}$ for $j \in [1, L]$ each k bits, where $k \ll n$ and n is the Paillier modulus. Then, a number of such measurements can be *packed* into one plaintext: $\hat{c}_{(i,p)} := c_{(i1,p)}|c_{(i2,p)}|c_{(i3,p)}|\dots|c_{(iL,p)}$ as follows:

$$\hat{c}_{(i,p)} := \sum_{j=1}^{L} c_{(ij,p)} \cdot 2^{j \cdot (k + \lceil \log N \rceil)} . \tag{20}$$

This construction is similar to [21,3]. It assumes that each measurement from N smart meters is aggregated in subsequent steps. Therefore, each measurement type has a reserved "compartment" of $k + \lceil \log N \rceil$ bits. With $N > M$, compartments are sufficient for computing temporal measurements. However, the number of measurements that can fit into one plaintext is $\frac{n}{k + \lceil \log N \rceil}$. Therefore, more than one encryption might be needed in some cases where a vast number of measurements are needed to be packed.

9 Conclusion

Fine granular reporting in smart metering systems causes serious privacy considerations and thus creates resistance against wide-deployment of such systems. In this paper, we have addressed computing total consumption in a privacy-preserving way in three scenarios: spatial, temporal and spatio-temporal total consumption computations, in which individual measurements of the households are kept secret from any party but the total consumption in the neighbourhood and/or of a particular smart meter is obtained accurately. The methods we have presented rely only on the capability of performing public-key operations on the smart metering device. The complexity analysis shows that with the currently existing smart metering device configurations, deployment of the proposed methods is realistic.

References

1. Ács, G., Castelluccia, C.: I Have a DREAM (DiffeRentially privatE smArt Metering). In: Filler, T., Pevný, T., Craver, S., Ker, A. (eds.) IH 2011. LNCS, vol. 6958, pp. 118–132. Springer, Heidelberg (2011)
2. Anderson, R., Fuloria, S.: On the security economics of electricity metering. In: The 9th Workshop on the Economics of Information Security (2010)
3. Bianchi, T., Piva, A., Barni, M.: Composite signal representation for fast and storage-efficient processing of encrypted signals. IEEE Transactions on Information Forensics and Security 5(1), 180–187 (2010)
4. Boudot, F.: Efficient Proofs that a Committed Number Lies in an Interval. In: Preneel, B. (ed.) EUROCRYPT 2000. LNCS, vol. 1807, pp. 431–444. Springer, Heidelberg (2000)
5. Castelluccia, C., Chan, A.C.-F., Mykletun, E., Tsudik, G.: Efficient and provably secure aggregation of encrypted data in wireless sensor networks. ACM Trans. Sen. Netw. 5, 20:1–20:36 (2009)
6. Castelluccia, C., Chan, A.C.-F., Mykletun, E., Tsudik, G.: Efficient and provably secure aggregation of encrypted data in wireless sensor networks. TOSN 5(3) (2009)
7. Castelluccia, C., Mykletun, E., Tsudik, G.: Efficient aggregation of encrypted data in wireless sensor networks. In: Proceedings of the The Second Annual International Conference on Mobile and Ubiquitous Systems: Networking and Services, pp. 109–117. IEEE Computer Society, Washington, DC (2005)
8. Domingo-Ferrer, J.: A Provably Secure Additive and Multiplicative Privacy Homomorphism. In: Chan, A.H., Gligor, V.D. (eds.) ISC 2002. LNCS, vol. 2433, pp. 471–483. Springer, Heidelberg (2002)
9. Doraswamy, N., Harkins, D.: IPSec: The New Security Standard for the Internet, Intranets, and Virtual Private Networks. Prentice Hall PTR, Upper Saddle River (1999)
10. Fine, J.: Malfunctioning smart meters demonstrate their intelligence (May 16, 2011), `http://blogs.edf.org/energyexchange/2011/05/16/malfunctioning-smart-meters-demonstrate-their-intelligence/`
11. Fontaine, C., Galand, F.: A survey of homomorphic encryption for nonspecialists. EURASIP Journal on Information Security (2007)
12. Garcia, F.D., Jacobs, B.: Privacy-Friendly Energy-Metering via Homomorphic Encryption. In: Cuellar, J., Lopez, J., Barthe, G., Pretschner, A. (eds.) STM 2010. LNCS, vol. 6710, pp. 226–238. Springer, Heidelberg (2011)
13. Goldreich, O.: Foundations of Cryptography. Basic Applications, 1st edn., vol. 2. Cambridge University Press (May 2004) ISBN 0-521-83084-2
14. Danezis, G., Kohlweiss, M., Rial, A.: Differentially Private Billing with Rebates. In: Filler, T., Pevný, T., Craver, S., Ker, A. (eds.) IH 2011. LNCS, vol. 6958, pp. 148–162. Springer, Heidelberg (2011)
15. Kursawe, K., Danezis, G., Kohlweiss, M.: Privacy-Friendly Aggregation for the Smart-Grid. In: Fischer-Hübner, S., Hopper, N. (eds.) PETS 2011. LNCS, vol. 6794, pp. 175–191. Springer, Heidelberg (2011)
16. Molina-Markham, A., Danezis, G., Fu, K., Shenoy, P.J., Irwin, D.E.: Designing privacy-preserving smart meters with low-cost microcontrollers. IACR Cryptology ePrint Archive 2011, 544 (2011)
17. Paillier, P.: Public-Key Cryptosystems Based on Composite Degree Residuosity Classes. In: Stern, J. (ed.) EUROCRYPT 1999. LNCS, vol. 1592, pp. 223–238. Springer, Heidelberg (1999)

18. Peter, S., Piotrowski, K., Langendoerfer, P.: On concealed data aggregation for wireless sensor networks. In: 4th IEEE Consumer Communications and Networking Conferences (2007)
19. Rial, A., Danezis, G.: Privacy-preserving smart metering. In: Proceedings of the 10th Annual ACM Workshop on Privacy in the Electronic Society, WPES 2011, pp. 49–60. ACM, New York (2011)
20. Shi, E., Chan, T.-H.H., Rieffel, E.G., Chow, R., Song, D.: Privacy-preserving aggregation of time-series data. In: Proceedings of 18th Annual Network and Distributed System Security Symposium (NDSS 2011) (February 2011)
21. Troncoso-Pastoriza, J.R., Katzenbeisser, S., Celik, M.U., Lemma, A.N.: A secure multidimensional point inclusion protocol. In: ACM Workshop on Multimedia and Security, pp. 109–120 (2007)

Author Index

GPSR Compliance

The European Union's (EU) General Product Safety Regulation (GPSR) is a set of rules that requires consumer products to be safe and our obligations to ensure this.

If you have any concerns about our products, you can contact us on ProductSafety@springernature.com

In case Publisher is established outside the EU, the EU authorized representative is:

Springer Nature Customer Service Center GmbH
Europaplatz 3
69115 Heidelberg, Germany

Batch number: 10844619

Printed by Printforce, the Netherlands